APPLIED NUMERICAL ANALYSIS

FOURTH EDITION

Curtis F. Gerald
Patrick O. Wheatley

California Polytechnic State University
San Luis Obispo

▲▼ ADDISON-WESLEY PUBLISHING COMPANY

Reading, Massachusetts • Menlo Park, California • New York
Don Mills, Ontario • Wokingham, England • Amsterdam • Bonn
Sydney • Singapore • Tokyo • Madrid • San Juan

To my wife, Elsie, who has been
a sustaining influence from the beginning

C.F.G.

To my wife, Jo Ann,
and my son, Patrick

P.O.W.

Sponsoring Editor: Thomas N. Taylor

Editorial and Production Services: The Book Company

Text Design: The Book Company

Illustrator: Folium

Manufacturing Supervisor: Roy Logan

Cover Designer: Marshall Henrichs

Library of Congress Cataloging-in-Publication Data

Gerald, Curtis F., 1915–
    Applied numerical analysis / by Curtis Gerald and Patrick
Wheatley.—4th ed.
       p. cm.
    Bibliography: p.
    Includes index.
    ISBN 0-201-11583-2
    1. Numerical analysis. I. Wheatley, Patrick O. II. Title.
QA297.G47   1989
519.4—dc19

Reprinted with corrections, June 1989

CDEFGHIJ—HA—89

Preface

Applied Numerical Analysis is written for sophomores and juniors in engineering, science, mathematics, and computer science. It is also valuable as a sourcebook for practicing engineers and scientists who need to use numerical procedures to solve problems. We have been gratified to find that many who purchased this book as students keep it as a permanent part of their reference library because its readability and breadth allow them to expand their knowledge of the subject by self-study.

While it is assumed that the reader has a good knowledge of calculus, appropriate topics are reviewed in the context of their use, and an appendix gives a summary of the most important items that are used to develop and analyze numerical procedures. The mathematical notation is purposely kept simple for clarity.

PURPOSE

The purpose of this fourth edition is the same as in previous editions: to give a broad coverage of the field of numerical analysis, emphasizing its practical applications rather than theory. At the same time, methods are compared, errors are analyzed, and relationships to the fundamental mathematical basis for the procedures are presented so that a true understanding of the subject is attained. Clarity of exposition, development through examples, and logical arrangement of topics aid the student to become more and more adept at applying the methods.

CONTENT FEATURES

Applied Numerical Analysis has enjoyed significant success because of several outstanding features. These are retained and amplified in this fourth edition:

- The unusually large number of exercises allows the instructor to select those that are appropriate for the background and interests of the students. These are keyed to the corresponding section of the chapter to assist in this. When the reader is using the book for self-study, the many exercises are an important supplement to the text. In addition to the practice exercises, each chapter has "Applied Problems and Projects" that are more challenging and that illustrate many fields of application of the various numerical procedures.

- By solving the same problem with several different methods, the relative efficiency and effectiveness of the methods become clear (pp. 11, 12, 13, and 27; pp. 98 and 111; pp. 350, 353, 358, 362, 365, and 368).

- A short summary of its contents is given at the beginning of each chapter to provide a preview of what is coming (pp. 1, 86, 180, 264, 347, and so on).

- Most chapters are introduced with an easily understood example that applies the material of the chapter. This motivates the student and shows that the material is of real utility (pp. 2, 87, 181, 349, 472, and so forth).

- Each chapter ends with a summary that reminds the student of what has been covered and suggests that appropriate review be done to ensure that nothing has been overlooked or not learned (pp. 49, 147, 235, 321, and so on).

- The coverage of partial-differential equations in an easily understood manner is unusual in a book at this level.

- Postponing the treatment of computer arithmetic and errors until after the student has been exposed to some numerical methods puts this important topic into proper context and helps in the appreciation of its significance as one factor in the accuracy of the computed result (p. 38).

- Computer programs in FORTRAN are given at the conclusion of each chapter. These implement the more important algorithms and serve as easily understood examples of how the computer can be used to carry out the computations. They do not pretend to be at a professional level of programming, because their purpose is illustrative and clarity would otherwise be sacrificed. We recognize that many instructors prefer a more structured language, but the presence of many subroutines in FORTRAN dictates our choice. We do provide a Pascal version of the programs in the supplements to the text; both versions are provided on diskettes for ease of entering into the local computer system.

FEATURES OF THIS EDITION

We have benefited by suggestions from those who have used the book in its previous editions. Significant new material has been added, and some chapters have been improved through rearrangement or rewriting. The format of the book has been improved. The use of a second color highlights important items and adds interest and variety to the text.

Important revisions include the following:

- The section on Muller's method in Chapter 1 has been moved up, and the difference between this and methods based on a linear approximation to the function has been emphasized (p. 18).

- Sections on computer arithmetic and errors have been rewritten and expanded (pp. 38–46).

- The use of matrix algebra has been broadened throughout the book. This is especially apparent in the discussion of iterative methods in Chapters 2 and 6.

- The relation between Gauss, Gauss–Jordan, and other *LU* methods has been clarified (p. 106).

- The chapter on interpolation has been streamlined and the use of divided differences has been emphasized.

- Other spline-based methods have been added and their use in approximating surfaces has been discussed (pp. 217 and 233).

- The section on adaptive integration has been improved (p. 305).

- We have given more modern versions of higher-order Runge–Kutta methods (p. 358).

- A most significant addition has been an elementary treatment of the finite-element method. This is introduced by an explanation of variational methods in one-dimensional boundary-value problems (p. 426). Finite elements are then applied to elliptic equations (p. 512) and to parabolic equations (p. 573).

- A discussion of the *QR* method for finding eigenvalues is included in Chapter 6.

- The section on fast Fourier transforms has been rewritten and expanded (p. 652).

- More of the programs at the end of each chapter have sample output for the reader to examine. Moreover, we have included new programs that implement adaptive integration, divided differences, the Runge–Kutta–Fehlberg method, and the *QR* technique. We have replaced a program in Chapter 2 with one that implements the *LU* decomposition method for Gaussian elimination.

- The appendix that describes software packages has been updated to cover some newer software, especially that for personal computers.

PEDAGOGICAL FEATURES

We recognize that the student is the most important part of the teaching/learning system and have tried to facilitate his or her understanding by several items, some of which are new in this edition:

- Sections that preview the material lead off each chapter.

- A chapter summary at the end of each chapter reminds the student of what should have been learned.

- The second color makes it easier to recognize the more important equations and algorithms. Color in the illustrations adds interest and makes their message clearer.

- The sections that have been rewritten eliminate some items that may have caused confusion for some students.

- The bibliography has been updated, and selected references have been given at the end of each chapter. This makes it easy for the student to find other treatments of the material, some of which go into more detail than we are able to.

SUPPLEMENTS

A number of supplements are available to assist the instructor:

- A *Solutions Manual* gives the answers to nearly all the exercises and to some of the "Applied Problems and Projects." Hints for other problems or projects are also provided.

- Copies of the programs, both in the FORTRAN version that is printed in the text and in an equivalent Pascal version, are available to adopters. These are on diskettes in IBM PC–compatible form so that they can readily be entered into the local computer system if the instructor wishes to make them available to the student on-line. Alternatively, copies can be provided to students when personal computers are to be used.

- The *Solutions Manual* gives suggestions for how the instructor can select from the text when he or she does not have time to cover all of it. Since our coverage of topics in numerical analysis is unusually broad, such selection is frequently necessary.

ACKNOWLEDGMENTS

We especially want to thank the multitude of our own students whose feedback has helped us to improve over previous editions. Our wives have been supportive during this revision and have helped us with proofreading. Many instructors have given valuable suggestions and constructive criticism.

In addition, we specifically mention those whose thorough reviews have been of material help:

Neil E. Berger, University of Illinois at Chicago

George Davis, Georgia State University

S. K. Dey, Eastern Illinois University

George Fix, University of Texas, Arlington

Richard Franke, Naval Postgraduate School

Vincent P. Manno, Tufts University

William Margules, California State University

Michael Pilant, Texas A&M University

Paul Schnare, Eastern Kentucky University

H. Doyle Thompson, Purdue University

Peter Welcher, U.S. Naval Academy

San Luis Obispo, California

C.F.G.
P.O.W.

Contents

2 Solving Sets of Equations 86

3 Interpolating Polynomials 180

4 Numerical Differentiation and Numerical Integration 264

5   Numerical Solution of Ordinary Differential Equations   347

6   Boundary-Value Problems and Characteristic-Value Problems   411

7 Numerical Solution of Partial-Differential Equations 471

8 Parabolic Partial-Differential Equations 544

9   Hyperbolic Partial-Differential Equations   592

10   Curve-fitting and Approximation of Functions   624

Appendixes   A-1

References   A-22

Answers to Selected Exercises   A-25

Index   I-1

1

Solving Nonlinear Equations

1.0 CONTENTS OF THIS CHAPTER

Chapter 1 introduces you to numerical analysis by explaining how methods of successive approximations can find the solution of a single nonlinear equation. It explains how computer programs can be written to obtain such solutions and gives numerous examples that implement the methods of the chapter.

1.1 THE LADDER IN THE MINE

In this book we will begin most chapters with an example that illustrates the application of the numerical techniques covered in the chapter. We will frame these in the context of the real world but simplified. This first example is typical and defines the problem, describes how it can be solved, and ends by pointing out how numerical methods are useful in getting the solution.

It is not uncommon, in applied mathematics, to have to solve a nonlinear equation. If you worked for a mining company the following might be a typical problem.

EXAMPLE   There are two intersecting mine shafts that meet at an angle of 123°, as shown in Fig. 1.1. The straight shaft has a width of 7 ft, while the entrance shaft is 9 ft wide. What is the longest ladder that can negotiate the turn? You can neglect the thickness of the ladder members and assume it is not tipped as it is maneuvered around the corner. Your solution should provide for the general case in which the angle A is a variable, as well as for the widths of the shafts.

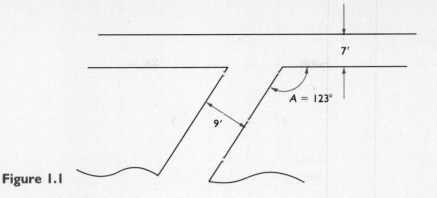

Figure 1.1

Whenever a scientific or engineering problem is solved, there are four general steps to be followed:

1. State the problem clearly, including any simplifying assumptions.

2. Develop a mathematical statement of the problem in a form that can be solved for a numerical answer. This may involve, as in the present case, the use of calculus. In other situations, other mathematical procedures may be employed.

3. Solve the equation(s) that result from step 2. Sometimes this is a method from algebra, but frequently more advanced methods will be needed. The subject matter of this text is numerical procedures that are quite powerful and of general applicability. The result of this step is a numerical answer or set of answers.

4. Interpret the numerical result to arrive at a decision. This will require experience and an understanding of the situation in which the problem was embedded. This interpretation is the hardest part of solving problems and must be learned on the job. This book will emphasize step 3 and will deal to some extent with steps 1 and 2, but step 4 cannot be meaningfully treated in the classroom.

The above description of the problem has taken care of step 1. Now for step 2:

Here is one way to analyze our ladder problem. Visualize the ladder in successive locations as we carry it around the corner; there will be a critical position in which the two ends of the ladder touch the walls while a point along the ladder touches the corner where the two shafts intersect. (See Fig. 1.2.) Let C be the angle between the ladder and the wall when in this critical position. It is usually preferable to solve problems in general terms, so we work with variables C, A, B, w_1, and w_2.

Consider a series of lines drawn in this critical position—their lengths vary with the angle C, and the following relations hold (angles are expressed in radian measure):

$$\ell_1 = \frac{w_2}{\sin B}; \quad \ell_2 = \frac{w_1}{\sin C};$$

$$B = \pi - A - C;$$

$$\ell = \ell_1 + \ell_2 = \frac{w_2}{\sin(\pi - A - C)} + \frac{w_1}{\sin C}.$$

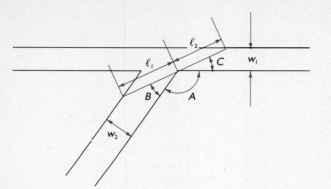

Figure 1.2

The maximum length of ladder that can negotiate the turn is the minimum of ℓ as a function of angle C. We hence set $d\ell/dC = 0$.

$$\frac{d\ell}{dC} = \frac{w_2 \cos(\pi - A - C)}{\sin^2(\pi - A - C)} - \frac{w_1 \cos C}{\sin^2 C} = 0.$$

We can solve the general problem if we can find the value of C that satisfies this equation. With the critical angle determined, the ladder length is given by

$$\ell = \frac{w_2}{\sin(\pi - A - C)} + \frac{w_1}{\sin C}.$$

As this analysis shows, to solve the specific problem we must solve a transcendental equation for the value of C:

$$\frac{9 \cos(\pi - 2.147 - C)}{\sin^2(\pi - 2.2147 - C)} - \frac{7 \cos C}{\sin C} = 0,$$

and then substitute C into

$$\ell = \frac{9}{\sin(\pi - 2.2147 - C)} + \frac{7}{\sin C},$$

where we have converted $123°$ into 2.147 radians.

Finding the solution to an algebraic or transcendental equation, as we must do here, is the topic of this first chapter.

In this chapter we study methods to find the roots of an equation, such as in our ladder-in-the-mine example. Much of algebra is devoted to the "solution of equations." In simple situations, this consists of a rearrangement to exhibit the value of the unknown variable as a simple arithmetic combination of the constants of the equation. For second-degree polynomials, this can be expressed by the familiar quadratic formula. For third- and fourth-degree polynomials, formulas exist but are so complex as to be rarely used; for higher-degree equations it has been proved that finding the solution through a formula is impossible. Most transcendental equations (involving trigonometric or exponential functions) are likewise intractable.

Even though it is difficult if not impossible to exhibit the solution of such equations in explicit form, numerical analysis provides a means where a solution may be found, or at least approximated as closely as desired. Many of these numerical procedures follow a scheme that may be thought of as providing a series of successive approximations, each more precise than the previous one, so that enough repetitions of the procedure eventually give an approximation that differs from the true value by less than some arbitrary error tolerance. Numerical procedures are thus seen to resemble the limit concept of mathematical analysis.

When a numerical solution that satisfies the transcendental equation above has been obtained, we have completed step 3 of the general procedure. The rest of this chapter treats several methods for doing this. We will not do step 4, but in this case it would consist of deciding if the maximum-length ladder that can be carried into the mine is long enough. If it is not, a decision must be made about the remedy. Perhaps this would be to use an extension ladder or to cut a notch in the corner of the wall.

This example shows how important calculus can be in solving practical problems. You will also find that calculus is a critical component in the analysis of numerical methods. Appendix A provides a summary of some of the most important elements of calculus. Look this over now to see if there are items that you should review.

1.2 METHOD OF HALVING THE INTERVAL

This first chapter describes methods for solving equations; that is, given an equation of the form $f(x) = 0$, what value(s) of x satisfy the equation? There are several obvious possibilities that we will not cover, such as trial and error (trying various values of x until we discover one that works) and drawing a graph of values of $f(x)$ versus x-values, seeing where the plot crosses the x-axis. Our methods will be more systematic than these, although a rough graph is frequently helpful in understanding the nature of the function and approximately where the function has roots.

The first numerical procedure that we will study is that of *interval halving*.* Consider the cubic

$$f(x) = x^3 + x^2 - 3x - 3 = 0.$$

At $x = 1$, f has the value -4. At $x = 2$, f has the value $+3$. Since the function is continuous, it is obvious that the change in sign of the function between $x = 1$ and $x = 2$ guarantees at least one root on the interval $(1, 2)$. (See Fig. 1.3.)

Suppose we now evaluate the function at $x = 1.5$ and compare the result to the function values at $x = 1$ and $x = 2$. Since the function changes sign between $x = 1.5$ and $x = 2$, a root lies between these values. We can obviously continue this interval halving to determine a smaller and smaller interval within which a root must lie. For this example, continuing the process leads eventually to an approximation to the root at $x = \sqrt{3} = 1.7320508075. \ldots$ The process is illustrated in Fig. 1.4.

---

*The method, also known as the *Bolzano method*, is of ancient origin. Some authors call it the *bisection method*.

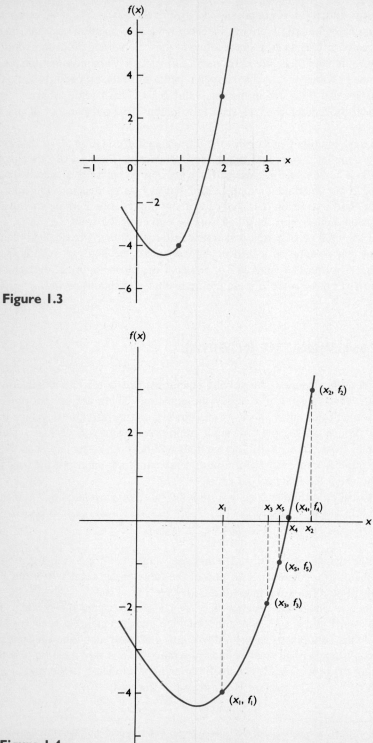

Figure 1.3

Figure 1.4

While a graphic method, as illustrated in Fig. 1.4, may be suitable if we want only an approximate answer, to obtain more accuracy we need to write a rule to do it mathematically. We should also express our algorithm (the technical name for a systematic procedure) in a way that makes it easy to implement the method with a computer program. We shall adopt a style of expressing algorithms that emphasizes the orderly structure.

Method of Halving the Interval (Bisection Method)

To determine a root of $f(x) = 0$ that is accurate within a specified tolerance value, given values of x_1 and x_2 such that $f(x_1) * f(x_2) < 0$,

        REPEAT
            Set $x_3 = (x_1 + x_2)/2$.
            IF $f(x_3) * f(x_1) < 0$:
                Set $x_2 = x_3$.
            ELSE   Set $x_1 = x_3$.
            ENDIF.
        UNTIL ($|x_1 - x_2| <$ tolerance value) OR $f(x_3) = 0$.

The final value of x_3 approximates the root; it is in error by not more than $\frac{1}{2}|x_1 - x_2|$.

Note. The method may give a false root if $f(x)$ is discontinuous on $[x_1, x_2]$.

Applying the method to $f(x) = x^3 + x^2 - 3x - 3 = 0$, we get the results of Table 1.1. The repetition of our algorithm is called *iteration*, and the successive approximations are termed the *iterates*.

Table 1.1   Method of halving the interval for $f(x) = x^3 + x^2 - 3x - 3 = 0$

Iteration number	x_1	x_2	x_3	$f(x_1)$	$f(x_2)$	$f(x_3)$	Maximum error in x_3
1	1	2	1.5	−4.0	3.0	−1.875	0.5
2	1.5	2	1.75	−1.875	3.0	0.17187	0.25
3	1.5	1.75	1.625	−1.875	0.17187	−0.94335	0.125
4	1.625	1.75	1.6875	−0.94335	0.17187	−0.40942	0.0625
5	1.6875	1.75	1.71875	−0.40942	0.17187	−0.12478	0.03125
6	1.71875	1.75	1.73437	−0.12478	0.17187	0.02198	0.015625*
7	1.71875	1.73437	1.72656				0.0078125
⋮	⋮	⋮	⋮			⋮	
∞			1.73205			−0.00000 ⋯	

*Actual error in x_3 after five iterations is 0.01330.

In this first chapter we expect you to do most of your calculations using a hand calculator.* Later in this chapter we will discuss computer programs that carry out the computations. Please verify some of the values in Table 1.1 now.

The entries in Table 1.1 indicate the necessity of representing values of the argument x as well as of the function $f(x)$ only approximately when we carry a limited number of decimal figures. In floating-point operations on digital computers, there is a similar inaccuracy in our work because computers retain only a limited number of significant digits. Accuracy of numbers in computers is discussed in Section 1.12. Note that this is true in all computations, not just in numerical methods. We will give attention to such "round-off errors" later. The distinction between numerical methods and numerical analysis is that the latter term implies the consideration of errors in the procedure used. Certainly the blind use of any calculation method without concern for its accuracy is foolish.

Whether one rounds to the nearest fractional value or chops off, the extra digits will make a difference in the effect of the round-off errors. In Table 1.1, the figures have been chopped after five places, which is similar to the action of many digital computers.

In addition to the limitation on accuracy because we retain only a limited number of figures in our work, there is an obvious limitation if we terminate the procedure itself too soon. One important advantage of the interval-halving method, beyond its simplicity, is our knowledge of the accuracy of the current approximation to the root. Since a root must lie between the x-values where the function changes a sign,[†] the error in the last approximation can be no more than one-half the last interval of which it is the midpoint. This interval is known exactly, since the original difference, $|x_1 - x_2|$, is halved at each iteration. For other methods, the accuracy determination is more difficult.

The accuracy of a computed value is usually expressed either as the absolute error (true value minus approximate value) or as the relative error (absolute error divided by true value). The relative error is often the better measure of accuracy for very large or very small values. Sometimes the accuracy is expressed as the number of digits that are correct; in other cases, the number of correct digits after the decimal point is used. When the true value is unknown, it is impossible to express the accuracy with exactness, and approximate accuracy must be specified.

The method of halving the interval applies equally well to transcendental equations, as do the other methods of this chapter. Table 1.2 shows the results when we apply the method to $f(x) = e^x - 3x = 0$, which has a root between $x = 1$ and $x = 2$.

The method of interval halving requires that starting values be obtained before the method can begin. This is true of most methods for root finding. Getting these starting values can be done by making a rough graph, by trial calculations, or by writing a search program on a computer or programmable calculator. Perhaps the best way is through interactive graphics, letting the computer draw curves at the direction of the user and varying the parameters at the console to find approximate values of roots.

---

*There are several good reasons for this. First, we want you to concentrate your attention on the algorithms, and if you were struggling to get a program to work, it might divert your attention. Second, you will get a better "feel" for what is happening if you are deeply involved in the successive steps of the computations. Finally, not every reader of the book is already an expert at programming. Postponing the use of programs can help you get up to speed in your programming at the same time that you start on what we think is a fascinating subject—numerical analysis.

[†]Observe that, if the function is discontinuous, $f(x)$ may change sign without having a root in the interval. Unknown functions should be examined for continuity before you attempt to evaluate their roots.

Table 1.2   Halving the interval for $f(x) = e^x - 3x = 0$

Iteration number	x_1	x_2	x_3	$f(x_1)$	$f(x_2)$	$f(x_3)$	Maximum error in x_3
1	1.0	2.0	1.5	−0.28172	1.38906	−0.01831	0.5
2	1.5	2.0	1.75	−0.01831	1.38906	0.50460	0.25
3	1.5	1.75	1.625	−0.01831	0.50460	0.20342	0.125
4	1.5	1.625	1.5625	−0.01831	0.20342	0.08323	0.0625
5	1.5	1.5625	1.53125	−0.01831	0.08323	0.03020	0.03125
6	1.5	1.53125	1.51562	−0.01831	0.03020	0.00539	0.015625*
⋮	⋮	⋮	⋮				
∞			1.51213				

*Actual error in x_3 after five iterations is −0.01912.

1.3   METHOD OF LINEAR INTERPOLATION

While the interval-halving method is easy and has simple error analysis, it is not very efficient. For most functions, we can improve the rate at which we converge to the root. One such method is the *method of linear interpolation*.* Suppose we assume that the function is linear over the interval (x_1, x_2), where $f(x_1)$ and $f(x_2)$ are of opposite sign. From the obvious similar triangles in Fig. 1.5 we can write[†]

$$\frac{x_2 - x_3}{x_2 - x_1} = \frac{f(x_2)}{f(x_2) - f(x_1)},$$

$$x_3 = x_2 - \frac{f(x_2)}{f(x_2) - f(x_1)}(x_2 - x_1).$$

We then compute $f(x_3)$ and again interpolate linearly between the values at which the function changes sign giving a new value for x_3. Repetition of this will give improving estimates of the root. Table 1.3 shows the results of this method for the same polynomial discussed in Section 1.2. The method appears to be somewhat faster than the method of halving the interval, giving about the same accuracy after three steps as was obtained

---

*This is also known as the *method of false position*, and by the Latinized version *regula falsi*. It is also a very old method.

[†]Note that, since $[f(x_2) - f(x_1)]/(x_2 - x_1)$ is the slope of the secant line, which approximates the slope of the function in the neighborhood of the root, the equation can be considered to be $x_3 = x_2 - f(x_2)/(\text{slope of function})$. Compare to Newton's method, in the next section.

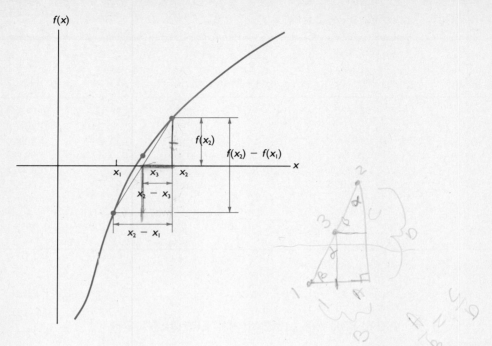

Figure 1.5

there in seven. It is intuitively obvious that the speed with which the successive approximations approach the zero of the function will depend on the degree to which the function departs from a straight line in the interval of consideration. In other words, the rate of convergence will be related to the rate of change of the slope of the curve, which is measured by the magnitude of the second derivative.

An algorithmic statement of this method is shown below.

Method of Linear Interpolation (*Regula Falsi* Method)

To determine a root of $f(x) = 0$, given values of x_1 and x_2 such that $f(x_1)$ and $f(x_2)$ are of opposite sign,

REPEAT

$$\text{Set } x_3 = x_2 - f(x_2) \frac{x_2 - x_1}{f(x_2) - f(x_1)}.$$

IF $\quad f(x_3)$ of opposite sign to $f(x_1)$:

Set $x_2 = x_3$.

ELSE Set $x_1 = x_3$.

ENDIF.

UNTIL $|f(x_3)| <$ tolerance value.

Note. The method may give a false root if $f(x)$ is discontinuous on $[x_1, x_2]$.

Table 1.3   Method of linear interpolation for $f(x) = x^3 + x^2 - 3x - 3 = 0$

Iteration number	x_1	x_2	x_3	$f(x_1)$	$f(x_2)$	$f(x_3)$
1	1.0	2.0	1.57142	−4.0	3.0	−1.36449
2	1.57142	2.0	1.70540	−1.36449	3.0	−0.24784
3	1.70540	2.0	1.72788	−0.24784	3.0	−0.03936
4	1.72788	2.0	1.73140	−0.03936	3.0	−0.00615
5	1.73140	2.0	1.73194*			

*Error in x_3 after five iterations is 0.00011.

Table 1.3 discloses a serious fault of the interpolation method: The approach to the root is one-sided. If $f(x)$ has significant curvature between x_1 and x_2, this can be most damaging to the speed with which we approach the root, as shown in Fig. 1.6.

A remedy for this is the *modified linear interpolation method*. We replace the value of $f(x)$ at the stagnant end position with $f(x)/2$.* This helps, as Fig. 1.7 shows.

An algorithm for this modification to the method of linear interpolation is shown below.

Modified Linear Interpolation Method

To determine a root of $f(x) = 0$, given values of x_1 and x_2 such that $f(x_1)$ and $f(x_2)$ are of opposite sign,

    Set SAVE = $f(x_1)$; set F1 = $f(x_1)$; set F2 = $f(x_2)$.
    REPEAT

$$\text{Set } x_3 = x_2 - F2 \frac{x_2 - x_1}{F2 - F1}.$$

        IF $f(x_3)$ of opposite sign to F1:
            Set $x_2 = x_3$.
            Set F2 = $f(x_3)$.
            IF $f(x_3)$ of same sign as SAVE:
                Set F1 = F1/2.
            ENDIF.
        ELSE Set $x_1 = x_3$.
            Set F1 = $f(x_3)$.
            IF $f(x_3)$ of same sign as SAVE:
                Set F2 = F2/2.
            ENDIF.
        ENDIF.
        Set SAVE = $f(x_3)$.
      UNTIL $\lfloor f(x_3) \rfloor <$ tolerance value.

*This halving of the ordinate at the other end of the interval is omitted when we step "beyond" the root, as occurs on the third iteration in Fig. 1.7.

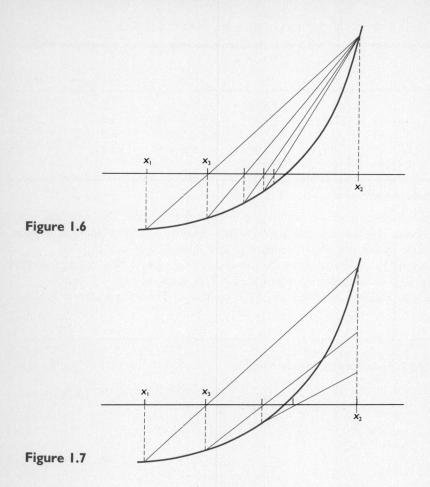

Figure 1.6

Figure 1.7

Table 1.4 shows the modified method applied to our example polynomial problem. While the modified method normally converges faster, in this example there is no difference. The changing values of x_1 and x_2 do better confine the root, however.

Table 1.4   Modified linear interpolation for $f(x) = x^3 + x^2 - 3x - 3 = 0$

Iteration number	x_1	x_2	x_3	F1	F2	$f(x)$	SAVE
1	1.0	2.0	1.57142	−4.0	3.0	−1.36449	−4.0
2	1.57142	2.0	1.77557	−1.36449	1.5*	0.42369	−1.36449
3	1.57142	1.77557	1.72720	−1.36449	0.42369	−0.04576	0.42369
4	1.72720	1.77557	1.73191	−0.04576	0.42369	-1.332×10^{-3}	−0.04576
5	1.73191	1.77557	1.732183†	-1.332×10^{-3}	0.21184*		

*These function values are old F2/2.
†Error in x_3 after five iterations is −0.00013.

Table 1.5 Secant method for $f(x) = x^3 - x^2 - 3x - 3 = 0$

Iteration number	x_1	x_2	x_3	$f(x_1)$	$f(x_2)$	$f(x_3)$
1	1.0	2.0	1.57142	−4.0	3.0	−1.36449
2	2.0	1.57142	1.70540	3.0	−1.36449	−0.24784
3	1.57142	1.70540	1.73513	−1.36449	−0.24784	0.02920
4	1.70540	1.73513	1.73199	−0.24784	0.02920	−0.0005755
5	1.73513	1.73199	1.73205*			

*Error in x_3 after five iterations is $<10^{-6}$.

There is one other way that we can improve the method of linear interpolation. Instead of requiring that the function have opposite signs at the two values used for interpolation, we can choose the two values nearest the root (as indicated by the magnitude of the function at the various points) and interpolate or extrapolate from these. Usually the nearest values to the root will be the last two values calculated. This makes the interval under consideration shorter and hence improves the assumption that the function can be represented by the line through the two points.

Table 1.5 shows the calculations according to this scheme, which is known as the *secant method*.* The example illustrates a more rapid convergence: x_6 is more accurate than was x_7 by linear interpolation.

We leave the development of the algorithm for the secant method as an exercise for the student.

It is important not to extrapolate to a root from two points whose functional values are of the same sign when knowledge is lacking that a real root is nearby. Figure 1.8 illustrates the futility of searching for a root that is not there. This is especially important in a computer program, since the successive calculated values are usually not apparent as soon as they are computed, as they are in a hand computation. In addition, it will be observed that the secant method can lead to a division by zero when $f(x_2) = f(x_1)$. It may also shoot off to find a root different from the expected one.

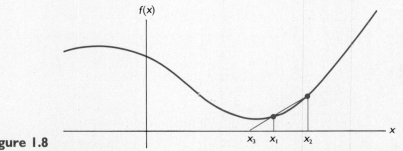

Figure 1.8

*So called because the line through two points on a curve is the secant line.

Table 1.6 Comparison of methods for finding the roots of $f(x) = 3x + \sin x - e^x = 0$

Iteration number	Interval halving		Linear interpolation		Modified linear interpolation		Secant method	
	x_3	$f(x_3)$	x_3	$f(x_3)$	x_3	$f(x_3)$	x_3	$f(x_3)$
1	0.5	0.330704	0.470990	0.265160	0.470990	0.265160	0.470990	0.265160
2	0.25	−0.286621	0.372277	0.029533	0.372277	0.029533	0.372277	0.029533
3	0.375	0.036281	0.361598	2.94×10^{-3}	0.351514	−0.022356	0.359904	-1.29×10^{-3}
4	0.3125	−0.121899	0.360538	2.90×10^{-4}	0.360727	7.64×10^{-4}	0.360424	5.53×10^{-6}
5	0.34375	−0.041956	0.3604334	2.93×10^{-5}	0.3604224	1.80×10^{-6}	0.3604218	2.13×10^{-7}
Error after five iterations	0.01667		-1.17×10^{-5}		-7×10^{-7}		-1×10^{-7}	

The methods based on linear interpolation are not limited to polynomials, of course. Table 1.6 compares the methods so far discussed when each is used to find the root of the equation

$$3x + \sin x - e^x = 0.$$

The trigonometric term is, of course, evaluated with the x-value in radians. Each method began with $x_1 = 0$ and $x_2 = 1$.

Note that all the methods we have been using require an initial estimate of the root we are computing. It often requires as much thought and effort to get a good starting value as it does to refine it to acceptable accuracy. Sometimes one's knowledge of the physical problem will suggest a starting value. When this is not available, one normally finds starting values by initial trial-and-error computations, or by making a rough graph of the function. We later discuss some methods that are self-starting for polynomials.

1.4 NEWTON'S METHOD

One of the most widely used methods of solving equations is Newton's method.* Like the previous ones, this method is also based on a linear approximation of the function, but does so using a tangent to the curve. Figure 1.9 gives a graphical description. Starting from an initial estimate that is not too far from a root x, we extrapolate along the tangent to its intersection with the x-axis, and take that as the next approximation. This is continued until either the successive x-values are sufficiently close, or the value of the function is sufficiently near zero.[†]

The calculation scheme follows immediately from the right triangle shown in Fig. 1.9, which has the angle of inclination of the tangent line to the curve at $x = x_1$ as one of its acute angles:

$$\tan \theta = f'(x_1) = \frac{f(x_1)}{x_1 - x_2}, \qquad x_2 = x_1 - \frac{f(x_1)}{f'(x_1)}.$$

We continue the calculation scheme by computing

$$x_3 = x_2 - \frac{f(x_2)}{f'(x_2)},$$

or, in more general terms,

$$x_{n+1} = x_a - \frac{f(x_n)}{f'(x_n)}, \qquad n = 1, 2, 3, \ldots$$

---

*Newton did not publish an extensive discussion of this method, but he solved a cubic polynomial in *Principia* (1687). The version given here is considerably improved over his original example.

[†]Which criterion should be used often depends on the particular physical problem to which the equation applies. Customarily, agreement of successive x-values to a specified tolerance is required.

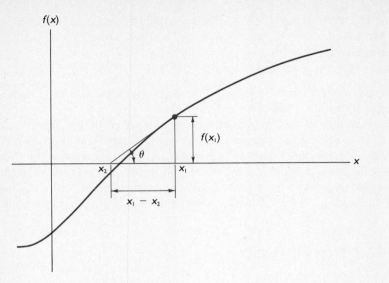

Figure 1.9

Newton's algorithm is widely used because, at least in the near neighborhood of a root, it is more rapidly convergent than any of the methods so far discussed. We show in a later section that the method is quadratically convergent, by which we mean that the error of each step approaches a constant K times the square of the error of the previous step. The net result of this is that the number of decimal places of accuracy nearly doubles at each iteration. However, offsetting this is the need for two function evaluations at each step, $f(x_n)$ and $f'(x_n)$.

When Newton's method is applied to $f(x) = 3x + \sin x - e^x = 0$, we have the following calculations:

$$f(x) = 3x + \sin x - e^x,$$
$$f'(x) = 3 + \cos x - e^x.$$

If we begin with $x_1 = 0.0$, we have

$$x_2 = x_1 - \frac{f(x_1)}{f'(x_1)} = 0.0 - \frac{-1.0}{3.0} = 0.33333;$$

$$x_3 = x_2 - \frac{f(x_2)}{f'(x_2)} = 0.33333 - \frac{-0.068418}{2.54934} = 0.36017;$$

$$x_4 = x_3 - \frac{f(x_3)}{f'(x_3)} = 0.36017 - \frac{-6.279 \times 10^{-4}}{2.50226} = 0.3604217.$$

After three iterations, the root is correct to seven significant digits. Comparing this with the results in Table 1.6, we see that Newton's method converges considerably more rapidly than the previous methods. In comparing numerical methods, however, one usually counts the number of times functions must be evaluated. Because Newton's method requires two function evaluations per step, the comparison is not as one-sided in favor of Newton's method as at first appears; the three iterations with Newton's method required six function evaluations. Five iterations with the previous methods required seven evaluations.

A more formal statement of the algorithm for Newton's method, suitable for implementation in a computer program, is shown here.

Newton's Method

To determine a root of $f(x) = 0$, given a value x_1 reasonably close to the root,

> Compute $f(x_1)$, $f'(x_1)$.
> Set $x_2 = x_1$.
> IF $(f(x_1) \neq 0)$ AND $(f'(x_1) \neq 0)$
>> REPEAT
>>> Set $x_1 = x_2$.
>>> Set $x_2 = x_1 - f(x_1)/f'(x_1)$.
>>
>> UNTIL $(|x_1 - x_2| <$ tolerance value 1) OR
>> $(|f(x_2)| <$ tolerance value 2).

Note: The method may converge to a root different from the expected one or diverge if the starting value is not close enough to the root.

Newton's method can be applied to polynomial functions, of course, and special techniques facilitate such application. We consider these in a later section of this chapter.

In some cases Newton's method will not converge. Figure 1.10 illustrates this situation. Starting with x_1, one never reaches the root r. We will develop the analytical condition for this in a later section and show that Newton's method is quadratically convergent in most cases

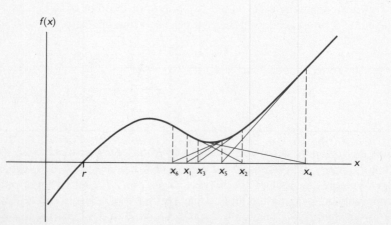

Figure 1.10

1.5   MULLER'S METHOD

Each of the root-finding methods that we have considered so far has approximated the function in the neighborhood of the root by a straight line. Obviously this is never true; if the function were linear, finding the root would take practically no effort. Muller's method is based on approximating the function in the neighborhood of the root by a quadratic polynomial. This gives a much closer match to the actual curve.

A second-degree polynomial is made to fit three points near a root, $[x_0, f(x_0)]$, $[x_1, f(x_1)]$, $[x_2, f(x_2)]$, and the proper zero of this quadratic, using the quadratic formula, is used as the improved estimate of the root. The process is then repeated using the set of three points nearest the root being evaluated.

The procedure for Muller's method is developed by writing a quadratic equation that fits through three points in the vicinity of a root, in the form $av^2 + bv + c$. (See Fig. 1.11.) The development is simplified if we transform axes to pass through the middle point, by letting $v = x - x_0$.

Let $h_1 = x_1 - x_0$ and $h_2 = x_0 - x_2$. We evaluate the coefficients by evaluating $p_2(v)$ at the three points:

$$v = 0: \quad a(0)^2 + b(0) + c = f_0;$$

$$v = h_1: \quad ah_1^2 + bh_1 + c = f_1;$$

$$v = -h_2: ah_2^2 - bh_2 + c = f_2.$$

From the first equation, $c = f_0$. Letting $h_2/h_1 = \gamma$, we can solve the other two equations for a and b:

$$a = \frac{\gamma f_1 - f_0(1 + \gamma) + f_2}{\gamma h_1^2(1 + \gamma)}, \quad b = \frac{f_1 - f_0 - ah_1^2}{h_1}.$$

After computing a, b, and c, we solve for the root of $av^2 + bv + c = 0$ by the quadratic formula, choosing the root nearest to the middle point x_0. This value is

$$\text{Root} = x_0 - \frac{2c}{b \pm \sqrt{b^2 - 4ac}},$$

with the sign in the denominator taken to give the largest absolute value of the denominator (that is, if $b > 0$, choose plus; if $b < 0$, choose minus; if $b = 0$, choose either).

We take the root of the polynomial as one of a set of three points for the next approximation, taking the three points that are most closely spaced (that is, if the root is to the right of x_0, take x_0, x_1, and the root; if to the left, take x_0, x_2, and the root). We always reset the subscripts to make x_0 be the middle of the three values.

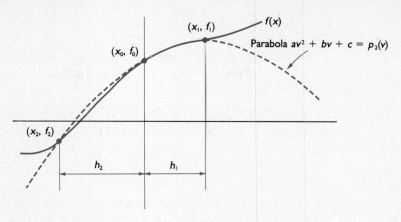

Figure 1.11

An algorithm for Muller's method is

Muller's Method

Given the points x_2, z_0, x_1 in increasing value,

1. Evaluate the corresponding function values f_2, f_0, f_1.
2. Find the coefficients of the parabola determined by the three points.
3. Compute the two roots of the parabolic equation.
4. Choose the root closest to x_0 and label it x_r.
5. IF $x_r > x_0$ THEN rearrange x_0, x_r, x_1 into x_2, x_0, x_1
    ELSE rearrange x_2, x_r, x_0 into x_2, x_0, x_1.
6. IF $|f(x_r)| <$ FTOL, THEN RETURN (x_r)
    ELSE go to 1.

EXAMPLE   Find a root of $f(x) = \sin z - x/2$ near $x = 2.0$.
Let

$$x_0 = 2.0, \quad f(x_0) = -0.09070, \quad h_1 = 0.2;$$
$$x_1 = 2.2, \quad f(x_1) = -0.29150, \quad h_2 = 0.2;$$
$$x_2 = 1.8, \quad f(x_2) = 0.07385, \quad \gamma = 1.0.$$

Then

$$a = \frac{(1.0)(-0.29150) - (-0.09070)(2.0) + 0.07385}{(1.0)(0.2)^2(2.0)} = -0.45312,$$

$$b = \frac{-0.29150 - (-0.09070) - (-0.45312)(0.2)^2}{0.2} = -0.91338,$$

$$c = -0.09070;$$

and

$$\text{Root} = 2.0 - \frac{2(-0.09070)}{-0.91338 - \sqrt{(0.91338)^2 - 4(-0.45312)(-0.09070)}} = 1.89526.$$

For the next iteration, we have

$$x_0 = 1.89526, \qquad f(x_0) = 1.9184 \times 10^{-4}, \qquad h_1 = 0.10474;$$
$$x_1 = 2.0, \qquad f(x_1) = -0.09070, \qquad h_2 = 0.09526;$$
$$x_2 = 1.8, \qquad f(x_2) = 0.07385, \qquad \gamma = 0.9095.$$

Then

$$a = \frac{(0.9095)(-0.09070) - (1.9184 \times 10^{-4})(1.9095) + 0.07385}{(0.9095)(0.10474)^2(1.9095)} = -0.47280,$$

$$b = \frac{-0.09070 - 1.9184 \times 10^{-4} - (-0.47280)(0.10474)^2}{0.10474} = -0.81826,$$

$$c = 1.9184 \times 10^{-4};$$

and

$$\text{Root} = 1.89526 - \frac{(2)(1.9184 \times 10^{-4})}{-0.81826 - \sqrt{(0.81826)^2 - 4(-0.47280)(1.9184 \times 10^{-4})}}$$

$$= 1.895494. \quad \blacksquare$$

After this second iteration, the root is accurate to seven significant digits. Experience shows that Muller's method converges at a rate that is similar to that for Newton's method.* It does not require the evaluation of derivatives, however, and (after we have obtained the starting values) needs only one function evaluation per iteration. There is an initial penalty in that one must evaluate the function three times, but this is frequently overcome by the time the required precision is attained.

1.6 USE OF $x = g(x)$ METHOD†

We now discuss another method that is of general applicability and that also lets us develop some necessary theory. We begin with the equation $f(x) = 0$, and rearrange it into an equivalent expression of the form

$$x = g(x), \text{ such that if } f(r) = 0, r = g(r).$$

Under suitable conditions, which we will develop, the algorithm

$$\boxed{x_{n+1} = g(x_n), \qquad n = 1, 2, 3, \ldots}$$

---

*Atkinson (1978) shows that each error is about proportional to the previous error to the 1.85th power.
†Some authors simply call the method the *iteration method*.

will converge to a zero of $f(x)$. Consider a simple example:

$$f(x) = x^2 - 2x - 3 = 0,$$

which has obvious roots at $x = 3$, $x = -1$.

Rearranging yields

$$x = \sqrt{2x + 3},$$

so $g(x) = \sqrt{2x + 3}$. Starting with $x = 4$, we get

$$x_2 = \sqrt{11} = 3.316,$$
$$x_3 = \sqrt{9.632} = 3.104,$$
$$x_4 = \sqrt{9.208} = 3.034,$$
$$x_5 = \sqrt{9.068} = 3.011,$$
$$x_6 = \sqrt{9.022} = 3.004.$$

The various iterates appear to converge to $x_\infty = 3$.

The equation $f(x) = x^2 - 2x - 3 = 0$ can be rearranged in other ways also. For example, $x = 3/(x - 2)$ is an alternative rearrangement of form $x = g(x)$. If $x_1 = 4$,

$$x_2 = 1.5,$$
$$x_3 = -6,$$
$$x_4 = -0.375,$$
$$x_5 = -1.263,$$
$$x_6 = -0.919,$$
$$x_7 = -1.028,$$
$$x_8 = -0.991,$$
$$x_9 = -1.003.$$

Note that this converges, but to the root at $x = -1$, and that the iterates oscillate rather than converging monotonically.

Consider a third rearrangement:

$$x = \frac{x^2 - 3}{2}.$$

For $x_1 = 4$ we get

$$x_2 = 6.5,$$
$$x_3 = 19.635,$$
$$x_4 = 191.0,$$

which obviously is diverging.

Figure 1.12 illustrates the various cases; (a) shows monotonic convergence, (b) shows oscillatory convergence, and (c) shows divergence. For a function $x = g(x)$, the solution

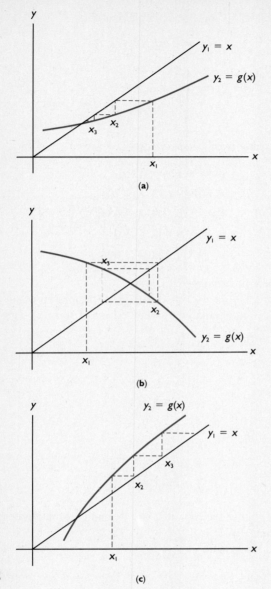

Figure 1.12

is at the intersection of the line $y_1 = x$ with the curve $y_2 = g(x)$. In every case, we move vertically to the curve and then horizontally to the line, and repeat. The point r where $g(r) = r$ is often called a *fixed point* of the function g. An algorithm for this method is shown on the facing page.

We now study the conditions that are needed for convergence. We iterate with

$$x_{n+1} = g(x_n).$$

Let $x = r$ be a solution to $f(x) = 0$, so $f(r) = 0$ and $r = g(r)$. Subtracting, and then multiplying and dividing by $(x_n - r)$, we have

Iteration with the Form $x = g(x)$

To determine a root of $f(x) = 0$, given a value x_1 reasonably close to the root,

> Rearrange the equation to an equivalent form $x = g(x)$.
> Set $x_2 = x_1$.
> REPEAT
> > Set $x_1 = x_2$.
> > Set $x_2 = g(x_1)$.
> UNTIL $|x_1 - x_2| <$ tolerance value.

Note: The method may converge to a root different from the expected one, or it may diverge. Different rearrangements will converge at different rates.

$$x_{n+1} - r = g(x_n) - g(r) = \frac{g(x_n) - g(r)}{(x_n - r)}(x_n - r).$$

If $g(x)$ and $g'(x)$ are continuous on the interval from r to x_n, the mean-value theorem* lets us write

$$x_{n+1} - r = g'(\xi_n)(x_n - r),$$

where ξ_n lies between x_n and r.

If we define the error of the ith iterate as $e_i = x_i - r$, and $e_{i+1} = x_{i+1} - r$, we can write

$$e_{i+1} = g'(\xi_i)e_i.$$

Taking absolute values, we get

$$|e_{i+1}| = |g'(\xi_i)| * |e_i|.$$

Now suppose that $|g'(x)| \le K < 1$ for all values of x in an interval of radius h about r. If x_1 is chosen in this interval, x_2 will also be in the interval and the algorithm will converge, since

$$|e_{n+1}| \le K|e_n| \le K^2|e_{n-1}| \le \cdots \le K^n|e_1|.$$

In summary, if $g(x)$ and $g'(x)$ are continuous on an interval about a root r of the equation $x = g(x)$, and if $|g'(x)| < 1$ for all x in the interval, then $x_{n+1} = g(x_n)$, $n = 1, 2, 3, \ldots$, will converge to the root $x = r$, provided that x_1 is chosen in the interval. Note that this is a sufficient condition only, since for some equations convergence is secured even though not all the conditions hold.[†]

---

*Appendix A reviews certain calculus principles, including this theorem.
[†]The analytical test that $|g'(x)| < 1$ is often awkward to apply. A constructive test is merely to observe whether the successive x_i values converge. In a computer program it is worthwhile to determine whether $|x_3 - x_2| < |x_2 - x_1|$.

Since the preceding demonstration shows that $e_{i+1} = g'(\xi_i) * e_i$, it is obvious that the rate of convergence is rapid if $|g'(x)|$ is small in the interval. If the derivative is negative, the errors alternate in sign, giving oscillatory convergence. The curves of Figure 1.12 give visual confirmation of this. As the iterates get closer to the root r, the values of $g'(\xi_i)$ approach the constant value $g'(r)$ because the ξ's are squeezed into smaller intervals about r. In the limit, each error becomes proportional to the previous error. For this reason, the method is sometimes called *linear iteration*.

Even though the proportionality between successive errors is true only in the limiting situation, if we assume that each of the errors is proportional to the previous one, we can develop an acceleration technique called *Aitken acceleration* that is often useful:

Assume that

$$e_n = x_n - r = K^{n-1}e_1$$

or

$$x_n = r + K^{n-1}e_1.$$

Similarly, if

$$e_{n+1} = x_{n+1} - r = K^n e_1,$$

and

$$e_{n+2} = x_{n+2} - r = K^{n+1}e_1,$$

then

$$x_{n+1} = r + K^n e_1,$$

and

$$x_{n+2} = r + K^{n+1}e_1.$$

Substitute these expressions into

$$\frac{x_n x_{n+2} - x_{n+1}^2}{x_{n+2} - 2x_{n+1} + x_n}.$$

It is found that

$$\frac{x_n x_{n+2} - x_{n+1}^2}{x_{n+2} - 2x_{n+1} + x_n} = \frac{(r + K^{n-1}e_1)(r + K^{n+1}e_1) - (r + K^n e_1)^2}{(r + K^{n+1}e_1) - 2(r + Ke_1) + (r + K^{n-1}e_1)}$$

$$= \frac{r(K^{n+1} - 2K^n + K^{n-1})e_1}{(K^{n+1} - 2K^n + K^{n-1})e_1} = r.$$

From three successive estimates of the root, x_1, x_2, and x_3, we extrapolate to an improved estimate. Since the assumption of constant ratio between successive errors is not normally true, our extrapolated value is not exact, but it is usually improved. One uses this technique by calculating two new values, extrapolating again, and so on.

A different form is useful to avoid the round-off problem that occurs in subtracting large numbers of nearly the same magnitude. Define

$$\Delta x_i = x_{i+1} - x_i,$$

$$\Delta^2 x_i = \Delta(\Delta x_i) = \Delta(x_{i+1} - x_i) = x_{i+2} - 2x_{i+1} + x_i.$$

Our acceleration scheme becomes

$$r = x_n - \frac{(\Delta x_n)^2}{\Delta^2 x_n} = \frac{x_n x_{n+2} - x_{n+1}^2}{x_{n+2} - 2x_{n+1} + x_n}.$$

The differences are most readily computed in a table. We illustrate with the iterates from the first example in this section:

$$f(x) = x^2 - 2x - 3 = 0,$$
$$x_{n+1} = \sqrt{2x_n + 3}, \qquad x_1 = 4,$$

x	Δx	$\Delta^2 x$
$x_1 = 4.000$		
	0.684	
$x_2 = 3.316$		0.472
	0.212	
$x_3 = 3.104$		

The accelerated estimate is

$$r = 4.000 - \frac{(0.684)^2}{0.472} = 3.009.$$

We have jumped ahead about two iterations. If $g(x)$ is expensive to compute, we have gained. Often Aitken acceleration gives bigger jumps than this. In fact, there is an excellent measure to determine when to use Aitken acceleration.

Suppose for some n we have $x_n, x_{n+1}, x_{n+2}, x_{n+3}$. Then evaluate

$$C = \frac{\sum x_i x_{i+1} - \frac{1}{3}\sum x_i \sum x_{i+1}}{\sqrt{\left(\sum x_i^2 - \frac{1}{3}\left(\sum x_i\right)^2\right)\left(\sum x_{i+1}^2 - \frac{1}{3}\left(\sum x_{i+1}\right)^2\right)}},$$

where the sums are from $i = n$ to $i = n + 2$. If C is close to ± 1, then Aitken acceleration is most effective. In the present example we have the values

$$x_0 = 4.000,$$
$$x_1 = 3.316,$$
$$x_2 = 3.104,$$
$$x_3 = 3.034.$$

We find that with $n = 0$ in the formula,

$$C = \frac{32.974 - \frac{1}{3}(10.42)(9.454)}{\sqrt{(36.631 - 36.1921)(29.8358 - 29.7927)}}$$
$$= 0.99992.$$

See Jones (1982) for this and further extensions for improving the acceleration method.

1.7   CONVERGENCE OF NEWTON'S METHOD

We now use the result of the previous section to show a criterion for convergence of Newton's method. The algorithm

$$x_{n+1} = x_n - \frac{f(x_n)}{f'(x_n)}, \qquad n = 1, 2, 3, \ldots$$

is of the form $x_{n+1} = g(x_n)$. Successive iterations converge if $|g'(x)| < 1$. Since

$$g(x) = x - \frac{f(x)}{f'(x)},$$

$$g'(x) = 1 - \frac{f'(x)f'(x) - f(x)f''(x)}{[f'(x)]^2} = \frac{f(x)f''(x)}{[f'(x)]^2}.$$

Hence if

$$\left| \frac{f(x)f''(x)}{[f'(x)]^2} \right| < 1$$

on an interval about the root r, the method will converge for any initial value x_1 in the interval. The condition is sufficient only, and requires the usual continuity and existence of $f(x)$ and its derivatives. Note that $f'(x)$ must not be zero.

Now we show that Newton's method is quadratically convergent. Since r is a root of $f(x) = 0$, $r = g(r)$. Since $x_{n+1} = g(x_n)$, we can write

$$x_{n+1} - r = g(x_n) - g(r).$$

Let us expand $g(x_n)$ as a Taylor series* in terms of $(x_n - r)$, with the second-derivative term as the remainder:

$$g(x_n) = g(r) + g'(r)(x_n - r) + \frac{g''(\xi)}{2}(x_n - r)^2,$$

where ξ lies in the interval from x_n to r.

Since

$$g'(r) = \frac{f(r)f''(r)}{[f'(r)]^2} = 0$$

because $f(r) = 0$ (r is a root), we have

$$g(x_n) = g(r) + \frac{g''(\xi)}{2}(x_n - r)^2.$$

---

*See Appendix A for a review of Taylor series.

Letting $x_n - r = e_n$, we have

$$e_{n+1} = x_{n+1} - r = g(x_n) - g(r) = \frac{g''(\xi)}{2} e_n^2.$$

Each error is (in the limit) proportional to the square of the previous error; that is, Newton's method is quadratically convergent.*

1.8   METHODS FOR POLYNOMIALS

Polynomial functions are of special importance. We will see throughout the remainder of this book that many valuable numerical procedures are based on polynomials. This important role of polynomial functions is due to their "nice" behavior: They are everywhere continuous, they are smooth, their derivatives are also continuous and smooth, and they are readily evaluated. Descartes' rule of signs (see Appendix A) lets us predict the number of positive roots. Polynomials are particularly well adapted to computers because the only mathematical operations they require for evaluation are addition, subtraction, and multiplication, all of which are speedy operations on computers.

Because of this special importance of polynomials, we now consider how our root-finding methods can be applied to them. For most of the methods previously discussed there is nothing new to say, but for Newton's method there are significant new ideas to consider. We begin on a historical note, with a procedure that saves time in hand computations. However, we will see that this same procedure is also the basis for computer calculations as well.

In applying Newton's method to polynomials, it is most efficient to evaluate $f(x_n)$ and $f'(x_n)$ by use of synthetic division.[†] We illustrate this by the same cubic polynomial that we used before, $x^3 + x^2 - 3x - 3 = 0$, which has a root at $x = \sqrt{3}$. We begin with the value $x = 2$. We utilize the remainder theorem to evaluate $f(2)$, and evaluate $f'(2)$ as the remainder when the reduced polynomial (of degree 2 here) is divided by $(x - 2)$:

$$
\begin{array}{r|rrrl}
x_1 = 2 & 1 & 1 & -3 & -3 \\
        &   & 2 & 6 & 6 \\
\hline
        & 1 & 3 & 3 & 3 \leftarrow \text{remainder} = f(2) \\
        &   & 2 & 10 & \\
\hline
        & 1 & 5 & 13 \longleftarrow \text{second remainder} = f'(2)
\end{array}
$$

---

*If $f'(x) = 0$ at $x = r$ (hence a multiple root), the rate of convergence will not be quadratic. For a multiple root, it can be shown that convergence is linear.

[†]The mechanics of synthetic division, whereby we divide a polynomial by the factor $x - x_i$, are explained in most algebra books. In the example, when $x^3 + x^2 - 3x - 3$ is divided by $(x - 2)$, the result is $x^2 + 3x + 3$, with a remainder of 3.

$$x_2 = 2 - \frac{3}{13} = 1.76923\ldots$$

$$
\begin{array}{r|rrrr}
x_2 = 1.76923 & 1 & 1 & -3 & -3 \\
 & & 1.76923 & 4.89940 & 3.36048 \\
\hline
 & 1 & 2.76923 & 1.89940 & 0.36048 \\
 & & 1.76923 & 8.02957 & \\
\hline
 & 1 & 4.53846 & 9.92897 & \\
\end{array}
$$

$$x_3 = 1.76923 - \frac{0.36048}{9.92897} = 1.73292.$$

Similarly,

$$x_4 = 1.73292 - \frac{0.00823}{9.47487} = 1.73205.$$

The value of x_4 is correct to five decimals. To observe the improvement in accuracy, consider the successive errors:

	Error	**Number of Correct Figures**
$x_1 = 2$	0.26895	1
$x_2 = 1.76923$	0.03718	2
$x_3 = 1.73292$	0.00087	4
$x_4 = 1.73205$	0.00000	6+

In order to compute with five decimal places, as in this example, we used a desk calculator. (If you have access to a calculator with storage for two or more values, you will find it especially well adapted to this method.)

The initial value at which Newton's method is begun can make a considerable difference. For example, if this problem is started with $x = 1$, the following values result:

x	$f(x)$	$f'(x)$
$x_1 = 1$	-4	2
$x_2 = 3$	24	30
$x_3 = 2.2$	5.888	15.92
$x_4 = 1.83015$		

From here on, the convergence is rapid, for we are using iterates just about as near the root as in the previous example.

After a first root is found (as shown by a remainder that is very small), one normally proceeds to determine additional roots from the reduced polynomial (whose coefficients are in the third row of the synthetic-division tableau). This makes the computations somewhat shorter. In the example, the reduced equation is a quadratic, so the quadratic formula would be used, but if a higher-degree polynomial were being solved, Newton's method employing synthetic division would be employed to improve an initial estimate of a second root. The process is then repeated until the reduced equation is of second degree.

This technique of working with the reduced function can be used even if the function is not a polynomial. After a root r of $f(x) = 0$ has been found, the new function $F(x) = f(x)/(x - r)$ will have all the roots of $f(x)$ except the root r. This procedure is called *deflating the function*. One must remember that a discontinuity has been introduced at $x = r$, however. We suggest that you explore how deflation works on nonpolynomial functions by graphing $f(x) = (x - 1)(e^x - \cos x)$, then $g(x) = f(x)/x$, $h(x) = f(x)/(x - 1)$, and comparing the graphs. You may also wish to compare to $h(x)/x$ and to functions derived by deflating $f(x)$ with other roots than $x = 0$ and $x = 1$.

It should be observed that using deflated functions can result in unexpected errors. If the first root is determined only approximately, the coefficients of the reduced equation are themselves not exact and the succeeding roots are subject not only to round-off errors and the errors that occur when iterations are terminated too soon, but also to inherited errors residing in the nonexact coefficients. Some functions are extremely sensitive in that small changes in the value of the coefficients cause large differences in the roots. Removing roots in order of increasing magnitude is said to minimize the difficulty, and the use of double-precision arithmetic will further help preserve accuracy.

It is of interest to develop the synthetic-division algorithm and to establish the remainder theorems. The scheme is also the most efficient way to evaluate polynomials and their derivatives in a computer program.

Write the nth-degree polynomial as

$$P_n(x) = a_1 x^n + a_2 x^{n-1} + \cdots + a_n x + a_{n+1}.$$

We wish to divide this by the factor $(x - x_1)$, giving a reduced polynomial $Q_{n-1}(x)$ of degree $n - 1$, and a remainder, b_{n+1}, which is a constant:

$$\frac{P_n(x)}{x - x_1} = Q_{n-1}(x) + \frac{b_{n+1}}{x - x_1}.$$

Rearranging yields

$$P_n(x) = (x - x_1)Q_{n-1}(x) + b_{n+1}.$$

Note that at $x = x_1$,

$$P_n(x_1) = (0)[Q_{n-1}(x_1)] + b_{n+1},$$

which is the remainder theorem: The remainder on division by $(x - x_1)$ is the value of the polynomial at $x = x_1$, $P_n(x_1)$.

If we differentiate the last equation, we get

$$P_n'(x) = (x - x_1)Q_{n-1}'(x) + (1)Q_{n-1}(x) + 0.$$

Letting $x = x_1$, we have

$$P_n'(x_1) = Q_{n-1}(x_1).$$

We evaluate the Q-polynomial at x_1 by a second division whose remainder equals $Q_{n-1}(x_1)$. This verifies that the second remainder from synthetic division yields the value for the derivative of the polynomial.

We now develop the synthetic-division algorithm, writing $Q_{n-1}(x)$ in form similar to $P_n(x)$:

$$P_n(x) = a_1 x^n + a_2 x^{n-1} + \cdots + a_n x + a_{n+1}$$
$$= (x - x_1)Q_{n-1}(x) + b_{n+1}$$
$$= (x - x_1)(b_1 x^{n-1} + b_2 x^{n-2} + \cdots + b_{n-1}x + b_n) + b_{n+1}.$$

Multiplying out and equating coefficients of like terms in x, we get

$$
\begin{array}{llcl}
\text{Coef. of } x^n: & a_1 = b_1 & & b_1 = a_1 \\
x^{n-1}: & a_2 = b_2 - b_1 x_1 & & b_2 = a_2 + b_1 x_1 \\
x^{n-2}: & a_3 = b_3 - b_2 x_1 & \text{or} & b_3 = a_3 + b_2 x_1 \\
x: & a_n = b_n - b_{n-1} x_1 & & b_n = a_n + b_{n-1} x_1 \\
\text{Const.}: & a_{n+1} = b_{n+1} - b_n x_1 & & b_{n+1} = a_{n+1} + b_n x_1.
\end{array}
$$

The general form is $b_i = a_i + b_{i-1}x_1$, by which all the b's except b_1 may be calculated. If this is compared to the synthetic divisions above, it is seen to be identical, except that we now have a vertical array. The horizontal layout is easier for hand computation. For evaluation of the derivative, a set of c-values is computed from the b's in the same way in which the b's are computed from the a's.

Synthetic division is also known as the *nested multiplication method* of evaluating polynomials. Consider the fifth-degree polynomial, evaluated at $x = x_1$:

$$a_1 x_1^5 + a_2 x_1^4 + a_3 x_1^3 + a_4 x_1^2 + a_5 x_1 + a_6.$$

We can rewrite this as

$$((((a_1 x_1 + a_2)x_1 + a_3)x_1 + a_4)\, x_1 + a_5)\, x_1 + a_6.$$

In the original form, $5 + 4 + 3 + 2 + 1 = 15$ multiplications are required, plus five additions. In the nested form, only five multiplications are required, plus five additions; it is obviously the more efficient method.

Comparing this with the equations $b_2 = a_2 + b_1 x_1$ and $b_i = a_i + b_{i-1}x_1$ for synthetic division, we see that the successive terms are formed in exactly the same way, so that synthetic division and nested multiplication are two names for the same thing.

1.9   BAIRSTOW'S METHOD FOR QUADRATIC FACTORS

The methods considered so far are difficult to use to find a complex root of a polynomial. It is true that Newton's and Muller's methods work satisfactorily, provided that one begins with initial estimates that are complex-valued; however, in a hand computation, performing the multiplications and divisions of complex numbers is awkward. There is no problem in a computer program if complex arithmetic capabilities exist, but the execution is slower.

For polynomials, the complex roots occur in conjugate pairs if the coefficients are all real-valued. For this case, if we extract the quadratic factors that are the products of the pairs of complex roots, we can avoid complex arithmetic because such quadratic factors have real coefficients. We first develop the algorithm for synthetic division by a trial quadratic, $x^2 - rx - s$, which is hopefully near to the desired factor of the polynomial:

$$P_n(x) = a_1 x^n + a_2 x^{n-1} + \cdots + a_{n-1}$$
$$= (x^2 - rx - s)Q_{n-2}(x) + \text{remainder}$$
$$= (x^2 - rx - s)(b_1 x^{n-2} + b_2 x^{n-3} + \cdots + b_{n-2}x + b_{n-1})$$
$$- b_n(x - r) + b_{n+1}.$$

(The remainder is the linear term $b_n(x - r) + b_{n+1}$, written in this form to provide later simplicity. If $x^2 - rx - s$ is an exact divisor of $P_n(x)$, then b_n and b_{n+1} will both be zero.) The negative signs in the factor are also for later simplification.

On multiplying out and equating coefficients of like powers of x, we get

$$
\begin{array}{lll}
a_1 = b_1 & b_1 = a_1 & \\
a_2 = b_2 - rb_1 & b_2 = a_2 + rb_1 & \\
a_3 = b_3 - rb_2 - sb_1 & b_3 = a_3 + rb_2 + sb_1 & \\
a_4 = b_4 - rb_3 - sb_2 \quad \text{or} & b_4 = a_4 + rb_3 + sb_2 & \quad \textbf{(1.1)} \\
\vdots & \vdots & \\
a_n = b_n - rb_{n-1} - sb_{n-2} & b_n = a_n + rb_{n-1} + sb_{n-2} & \\
a_{n+1} = b_{n+1} - rb_n - sb_{n-1} & b_{n+1} = a_{n+1} + rb_n + sb_{n-1}. &
\end{array}
$$

We would like both b_n and b_{n+1} to be zero, for that would show $x^2 - rx - s$ to be a quadratic factor of the polynomial. This will normally not be so; if we properly change the values of r and s, we can make the remainder zero, or at least make its coefficients smaller. Obviously b_n and b_{n+1} are both functions of the two parameters r and s. Expanding these as a Taylor series for a function of two variables* in terms of $(r^* - r)$ and

---

*Appendix A reviews this.

$(s^* - s)$ where $(r^* - r)$ and $(s^* - s)$ are presumed small so that terms of higher order than the first are negligible, we obtain

$$b_n(r^*, s^*) = b_n(r, s) + \frac{\partial b_n}{\partial r}(r^* - r) + \frac{\partial b_n}{\partial s}(s^* - s) + \cdots,$$

$$b_{n+1}(r^*, s^*) = b_{n+1}(r, s) + \frac{\partial b_{n+1}}{\partial r}(r^* - r) + \frac{\partial b_{n+1}}{\partial s}(s^* - s) + \cdots.$$

Let us take (r^*, s^*) as the point at which the remainder is zero, and

$$r^* - r = \Delta r, \qquad s^* - s = \Delta s.$$

(Δr and Δs are increments to add to the original r and s to get the new values r^* and s^* for which the remainder is zero.) Then

$$b_n(r^*, s^*) = 0 \doteq b_n + \frac{\partial b_n}{\partial r}\Delta r + \frac{\partial b_n}{\partial s}\Delta s,$$

$$b_{n+1}(r^*, s^*) = 0 \doteq b_{n+1} + \frac{\partial b_{n+1}}{\partial r}\Delta r + \frac{\partial b_{n+1}}{\partial s}\Delta s.$$

All the terms on the right are to be evaluated at (r, s). We wish to solve these two equations simultaneously for the unknown Δr and Δs, so we need to evaluate the partial derivatives.

Bairstow showed that the required partial derivatives can be obtained from the b's by a second synthetic division by the factor $x^2 - rs - s$ in just the same way that the b's are obtained from the a's. Define a set of c's by the relations shown below at the left, and compare these to the partial derivatives in the right columns:

$c_1 = b_1$

$$\frac{\partial b_1}{\partial r} = \frac{\partial a_1}{\partial r} = 0 \qquad\qquad \frac{\partial b_1}{\partial s} = \frac{\partial a_1}{\partial s} = 0$$

$c_2 = b_2 + rc_1$

$$\frac{\partial b_2}{\partial r} = r\frac{\partial b_1}{\partial r} + b_1 = b_1 = c_1 \qquad\qquad \frac{\partial b_2}{\partial s} = \frac{\partial a_2}{\partial s} + r\frac{\partial b_1}{\partial s} = 0$$

$c_3 = b_3 + rc_2 + sc_1$

$$\frac{\partial b_3}{\partial r} = r\frac{\partial b_2}{\partial r} + b_2 = c_2 \qquad\qquad \frac{\partial b_3}{\partial s} = r\frac{\partial b_2}{\partial s} + s\frac{\partial b_1}{\partial s} + b_1$$
$$= b_1 = c_1$$

$c_4 = b_4 + rc_3 + sc_2$

$$\frac{\partial b_4}{\partial r} = r\frac{\partial b_3}{\partial r} + b_3 + s\frac{\partial b_2}{\partial r} \qquad\qquad \frac{\partial b_4}{\partial s} = r\frac{\partial b_3}{\partial s} + s\frac{\partial b_2}{\partial s} + b_2$$
$$= b_3 + rc_2 + sc_1 = c_3 \qquad\qquad = b_2 + rc_1 = c_2$$

$$\vdots \qquad\qquad\qquad \vdots \qquad\qquad\qquad \vdots$$

$c_n = b_n + rc_{n-1} + sc_{n-2}$

$$\frac{\partial b_n}{\partial r} = r\frac{\partial b_{n-1}}{\partial r} + b_{n-1} + s\frac{\partial b_{n-2}}{\partial r} \qquad\qquad \frac{\partial b_n}{\partial s} = r\frac{\partial b_{n-1}}{\partial s} + s\frac{\partial b_{n-2}}{\partial s} + b_{n-2}$$
$$= b_{n-1} + rc_{n-2} + sc_{n-3} \qquad\qquad = b_{n-2} + rc_{n-3} + sc_{n-4}$$
$$= c_{n-1} \qquad\qquad\qquad\qquad = c_{n-2}$$

Hence the partial derivatives that we need are equal to the properly corresponding c's. Our simultaneous equations become, where Δr and Δs are unknowns to be solved for,

$$-b_n = c_{n-1}\Delta r + c_{n-2}\Delta s,$$
$$-b_{n+1} = c_n\Delta r + c_{n-1}\Delta s.$$

We express the solution as ratios of determinants:

$$\Delta r = \frac{\begin{vmatrix} -b_n & c_{n-2} \\ -b_{n+1} & c_{n-1} \end{vmatrix}}{\begin{vmatrix} c_{n-1} & c_{n-2} \\ c_n & c_{n-1} \end{vmatrix}},$$

$$\Delta s = \frac{\begin{vmatrix} c_{n-1} & -b_n \\ c_n & -b_{n-1} \end{vmatrix}}{\begin{vmatrix} c_{n-1} & c_{n-2} \\ c_n & c_{n-1} \end{vmatrix}}.$$

A statement of the algorithm for Bairstow's method is given in Section 1.15.

E X A M P L E   Find the quadratic factors of

$$x^4 - 1.1x^3 + 2.3x^2 + 0.5x + 3.3 = 0.$$

Use $x^2 + x + 1$ as starting factor ($r = -1$, $s = -1$). (Frequently $r = s = 0$ are used as starting values if no information as to an approximate factor is known.) Equations (1.1) lead to a double synthetic-division scheme as follows:

	a_1	a_2	a_3	a_4	a_5
	1	−1.1	2.3	0.5	3.3
$r = -1$		−1.0	2.1	−3.4	0.8
$s = -1$		—	−1.0	2.1	−3.4
	1	−2.1	3.4	−0.8	0.7
		−1.0	3.1	−5.5	
		—	−1.0	3.1	
	1	−3.1	5.5	−3.2	

$$c_{n-2} \qquad c_{n-1} \qquad c_n$$

b_{n+1}

b_n

Note that the equations for b_2 and c_2 have no term involving s. The dashes in the preceding tableau represent these missing factors. Then

$$\Delta r = \frac{\begin{vmatrix} 0.8 & -3.1 \\ -0.7 & 5.5 \end{vmatrix}}{\begin{vmatrix} 5.5 & -3.1 \\ -3.2 & 5.5 \end{vmatrix}} = \frac{2.23}{20.33} = 0.11, \quad r^* = -1 + 0.11 = -0.89,$$

$$\Delta s = \frac{\begin{vmatrix} 5.5 & 0.8 \\ -3.2 & -0.7 \end{vmatrix}}{20.33} = \frac{-1.29}{20.33} = -0.06, \quad s^* = -1 - 0.06 = -1.06.$$

The second trial yields

	1	−1.1	2.3	0.5	3.3
−0.89		−0.89	1.77	−2.68	0.06
−1.06		—	−1.06	2.11	−3.17
	1	−1.99	3.01	−0.07	0.17
		−0.89	2.56	−4.01	
		—	−1.06	3.05	
		−2.88	4.51	−1.03	

$$\Delta r = \frac{\begin{vmatrix} 0.07 & -2.88 \\ -0.17 & 4.51 \end{vmatrix}}{\begin{vmatrix} 4.51 & -2.88 \\ -1.03 & 4.51 \end{vmatrix}} = \frac{-0.175}{17.374} = -0.010, \quad r^* = -0.89 - 0.010 = -0.900,$$

$$\Delta s = \frac{\begin{vmatrix} 4.51 & 0.07 \\ -1.03 & -0.17 \end{vmatrix}}{17.374} = \frac{-0.694}{17.374} = -0.040, \quad s^* = -1.06 - 0.040 = -1.100.$$

The exact factors are $(x^2 + 0.9x + 1.1)(x^2 - 2x + 3)$. ∎

1.10 OTHER METHODS FOR POLYNOMIALS

In this section we discuss two methods that do not seem to be widely used but that have the special advantage of not requiring a reasonably good starting value. We first discuss the QD algorithm, then mention Graeffe's method.

A relatively efficient method to determine all the roots of a polynomial without starting values is the QD or quotient-difference algorithm. We present the method without elaboration.*

---

*Henrici (1964) discusses the method in some detail.

For the nth-degree polynomial

$$P_n(x) = a_1 x^n + a_2 x^{n-1} + \cdots + a_n x + a_{n+1}$$

we form an array of q and ϵ terms, starting the tableau by calculating a first row of q's and a second row of e's:

$$q^{(1)} = -a_2/a_1, \quad \text{all other } q\text{'s are zero,}$$
$$e^{(i)} = a_{i+2}/a_{i+1}, \quad i = 1, 2, \ldots, n - 1,$$
$$e^{(0)} = e^{(n)} = 0.$$

The start of the array is

$e^{(0)}$	$q^{(1)}$	$e^{(1)}$	$q^{(2)}$	$e^{(2)}$	$q^{(3)}$	$\cdots$	$e^{(n-1)}$	$q^{(n)}$	$e^{(n)}$
	$\dfrac{-a_2}{a_1}$		0		0	$\cdots$		0	
0		$\dfrac{a_3}{a_2}$		$\dfrac{a_4}{a_3}$		$\cdots$	$\dfrac{a_{n+1}}{a_n}$		0

A new row of q's is computed by the equation

$$\text{New } q^{(i)} = e^{(i)} - e^{(i-1)} + q^{(i)},$$

using terms from the e and q rows just above. Note that this algorithm is "e to right minus e to left plus q above."

A new row of e's is now computed by the equation

$$\text{New } e^{(i)} = \left(\frac{q^{(i+1)}}{q^{(i)}} \right) e^{(i)};$$

"q to right over q to left times e above." The example in Table 1.7 isolates the roots of the quartic

$$P_4(x) = 128x^4 - 256x^3 + 160x^2 - 32x + 1$$

by continuing to compute rows of q's and then e's until all the e-values approach zero. When this occurs, the q-values assume the values of the roots. Since the method is slow to converge, it is generally used only to get approximate values, which are then improved by Newton's method.

If the polynomial has a pair of conjugate complex roots, one of the e's will not approach zero but will fluctuate in value. The sum of the two q-values on either side of this e will approach r and the product of the q above and to the left times the q below and to the right approaches $-s$ in the factor $x^2 - rx - s$. Two equal roots behave similarly.

Table 1.8 shows the result of the method for the polynomial

$$(x - 1)(x - 4)(x^2 - x + 3) = x^4 - 6x^3 + 12x^2 - 19x + 12.$$

Table 1.7   Example of QD method for $P(x) = 128x^4 - 256x^3 + 160x^2 - 32x + 1$

$e^{(0)}$	$q^{(1)}$	$e^{(1)}$	$q^{(2)}$	$e^{(2)}$	$q^{(3)}$	$e^{(3)}$	$q^{(4)}$	$e^{(4)}$
	2.000		0		0		0	
0		−0.625		−0.200		−0.031		0
	1.375		0.425		0.169		0.031	
0		−0.193		−0.079		−0.006		0
	1.182		0.539		0.242		0.037	
0		−0.088		−0.036		−0.001		0
	1.094		0.591		0.277		0.038	
0		−0.048		−0.017		−0.000		0
	1.046		0.622		0.294		0.038	
0		−0.028		−0.008		−0.000		0
	1.018		0.642		0.302		0.038	
0		−0.018		−0.004		−0.000		0
	1.000		0.656		0.304		0.038	
0		−0.012		−0.002		−0.000		0
	0.988		0.666		0.306		0.038	
0		−0.008		−0.001		−0.000		0
	0.980		0.673		0.307		0.038	
0		−0.005		−0.001		−0.000		0
	0.975		0.677		0.308		0.038*	

*The true values of the roots are 0.96194, 0.69134, 0.30866, and 0.03806.

Factors are $(x - 4)(x - 1)(x^2 - x + 3)$.

$$q^{(1)} \text{ converging to } 4.$$
$$q^{(4)} \text{ converging to } 1.$$

Since $e^{(2)}$ does not approach zero, $q^{(2)}$ and $q^{(3)}$ represent a quadratic factor:

$$r \doteq q^{(2)} + q^{(3)} = 1.456 - 0.466 = 0.990;$$
$$s \doteq -(-6.426)(-0.466) = -2.995.$$

This quadratic factor is $x^2 - rx - s = x^2 - 0.990x - (-2.995)$.

Note that one cannot compute the first q and e rows if one of the coefficients in the polynomial is zero, for division by zero is undefined. In such a case, we change the variable to $y = x - 1$. (Subtracting 1 from the roots of the equation is an arbitrary choice, but this facilitates the reverse change of variable to get the roots of the original equation after the roots of the new equation in y have been found.)

Table 1.8 QD method with complex roots, for $P(x) = x^4 - 6x^3 + 12x^2 - 19x + 12$

$e^{(0)}$	$q^{(1)}$	$e^{(1)}$	$q^{(2)}$	$e^{(2)}$	$q^{(3)}$	$e^{(3)}$	$q^{(4)}$	$e^{(4)}$
	6.000		0		0		0	
0		−2.000		−1.583		−0.632		0
	4.000		0.417		0.951		0.632	
0		−0.208		−3.610		−0.420		0
	3.792		−2.985		4.141		1.052	
0		0.164		5.008		−0.107		0
	3.956		1.859		−0.974		1.159	
0		0.077		−2.624		0.127		0
	4.033		−0.842		1.777		1.032	
0		−0.016		5.538		0.074		0
	4.017		4.712		−3.687		0.958	
0		−0.019		−4.333		−0.019		0
	3.998		0.398		0.627		0.977	
0		−0.002		−6.826		−0.030		0
	4.000		−6.426		7.423		1.007	
0		0.003		7.885		−0.004		0
	4.003		1.456		−0.466		1.010	

For example, if $f(x) = x^4 - 2x^2 + x - 1 = 0$, we let $y = x - 1$ and use repeated synthetic division to determine the coefficients of $f(y) = 0$. The successive remainders on dividing by $x - 1$ are the coefficients of $f(y)$:

$$
\begin{array}{rrrrrr}
1 & 0 & -2 & 1 & -1 & \lfloor 1 \\
  & 1 & 1 & -1 & 0 & \\
\hline
1 & 1 & -1 & 0 & -\textcircled{1} & \\
  & 1 & 2 & 1 & & \\
\hline
1 & 2 & 1 & \textcircled{1} & & \\
  & 1 & 3 & & & \\
\hline
1 & 3 & \textcircled{4} & & & \\
  & 1 & & & & \\
\hline
1 & \textcircled{4} & & & & \\
\hline
\textcircled{1} & & & & & \\
\end{array}
$$

Therefore,

$$f(y) = y^4 + 4y^3 + 4y^2 + y - 1.$$

We proceed to find the roots of $f(y) = 0$, and then get the roots of $f(x) = 0$ by adding 1.

Graeffe's method finds values for all the roots of a polynomial directly from its coefficients without requiring starting values. It is based on the fact that if the roots are all different and widely separated, then for the polynomial

$$P_n(x) = a_1 x^n + a_2 x^{n-1} + \cdots + a_n x + a_{n+1},$$

the roots are given by

$$r_1 \doteq -\frac{a_2}{a_1}, \qquad r_2 \doteq -\frac{a_3}{a_2}, \qquad \cdots, \qquad r_n \doteq -\frac{a_{n+1}}{a_n}. \qquad (1.2)*$$

In order to separate the roots of the given polynomial, it is converted to another polynomial whose roots are the negative squares of the original roots.[†] After enough repetitions of the root-squaring operation the relations of Eq. (1.2) give the values of the roots, provided no multiple roots occur.

1.11 COMPUTER ARITHMETIC AND ERRORS

Up to now we have not anticipated that you would use computer programs to solve the exercises. Many of the methods presented in this and the following chapters can readily be done with a calculator. The reason we have avoided a discussion of computers so far was to let you concentrate on the algorithms. However, in your professional careers you may need to solve large problems quickly, handle a large amount of data, or use certain methods frequently. To carry this out, a computer is essential. To use one, you will either write your own programs to implement the methods or make use of some of the excellent software already available.[‡] In this section and the next we examine the limitations of the computer as well as discussing programming languages you could use. A person with a solid knowledge of mathematics, of a programming language, and of the computer will find numerical analysis a powerful asset in solving real-world problems.

Many programs are presented in this book. These are meant mainly to show the implementation of the algorithms and are not truly professional codes. They should be useful to you as easily understood programming examples as well as in solving the exercises if you do not have access to libraries of numerical analysis subroutines.

---

*We use the notation $r_i \doteq -a_2/a_1$ to indicate approximate equality of the two quantities.
[†]See Scarborough (1950) for more information.
[‡]Appendix C discusses some software that is generally available, both for personal computers and larger multiuser systems.

We have previously observed that it is essential not to neglect the study of the errors of our numerical techniques. For several of our methods, we have analyzed how the errors of successive iterations decrease (or increase if the method diverges). It is time that we look more deeply into the various sources of error that interact to affect the accuracy of our result. We can list these sources as:

TRUNCATION ERROR

This term is given to the errors caused by the method itself (the term originates from the fact that numerical methods can usually be compared to a truncated Taylor series) and is the error to which we have so far paid most attention. For instance we may approximate e^x by the cubic

$$p_3(x) = 1 + \frac{x}{1!} + \frac{x^2}{2!} + \frac{x^3}{3!}.$$

However, we know that to compute e^x really requires an infinitely long series:

$$e^x = p_3(x) + \sum_{n=4}^{\infty} \frac{x^n}{n!}.$$

Note that an approximation of e^x with the cubic gives an inexact answer. The error is due to truncating the series and has nothing to do with a computer or calculator. For iterative methods, this error usually can be reduced by repeated iterations, but since life is finite and computer time is expensive, we must be satisfied with approximations to the exact analytical answer.

ROUND-OFF ERROR

All computing devices represent numbers with some imprecision. Digital computers, which are the normal devices for implementing numerical methods, will nearly always use floating-point numbers with a fixed word length. The true values are not exactly expressed by such representations. We call this error *round-off*, whether the decimal fraction is rounded or chopped after the final digit. We discuss this in more detail below.

ERROR IN ORIGINAL DATA

Real-world problems, in which an existing or proposed physical situation is modeled by a mathematical equation, frequently have coefficients that are imperfectly known. The model itself may not perfectly reflect the behavior of the situation either. The numerical analyst can do nothing to overcome such errors by any choice of method, but he or she needs to be aware of such uncertainties; in particular, one may need to perform tests to see how sensitive the results are to changes in the input information. Since the reason for performing the computation is to permit some decision with validity in the real world, sensitivity analysis is of extreme importance. As Hamming says, "the purpose of computing is insight, not numbers."

BLUNDERS

It is anticipated that you will use a digital computer (or at least a programmable calculator) in your professional use of numerical analysis. You will probably use such computing tools extensively while you are learning the topics covered in this text. Such machines make mistakes only very infrequently, but since humans are involved in programming, operation, preparing the input, and interpreting the output, blunders or gross errors do occur more frequently than we like to admit. The solution here is care, coupled with a careful examination of the results for reasonableness. Sometimes a test run with known results is worthwhile, but this is no guarantee of freedom from foolish error. When hand computation was more common, check sums were frequently computed—these would ordinarily reveal the mistake and permit its correction.

PROPAGATED ERROR

This is more subtle than the other errors. By propagated error we mean the error in the succeeding steps of the process due to the occurrence of an earlier error. This is in addition to the local error made at that step; it is somewhat analogous to errors in the initial conditions. The methods discussed in this chapter do not reflect this type of error, except in the case of finding additional zeros of a function using the reduced or deflated equation. Here the reduced equations reflect errors in the previous stages. In other examples of numerical methods treated in later chapters, propagated error is of critical importance. If errors are magnified continuously as the method continues, eventually they will completely overshadow the true value, destroying its validity; we call such a method *unstable*. (For a *stable* method—the desirable kind—errors made at early points die out as the method continues. This will be covered more thoroughly in later chapters.)

Each of these types of error, while interacting to a degree, may occur even in the absence of the other kinds. For example, round-off error will occur even if truncation error is absent, as in an analytical method. Likewise, truncation errors would cause inaccuracies even if one could attain perfect precision in the calculations. The usual error analysis of a numerical method treats the truncation error as if such perfect precision did exist.

1.12 FLOATING-POINT ARITHMETIC AND ERROR ESTIMATES

To examine round-off error in detail, we need to understand how numeric quantities are represented in computers. In nearly all cases, they are stored as floating-point quantities, which are very much like scientific notation.* Different computers use slightly different techniques, but the general procedure is similar.

---

*Another name often used for floating-point number is *real number*, but we here reserve the term *real* for the continuous (and infinite) set of numbers on the "number line." When printed as a number with a decimal point, it is called *fixed-point*. The essential concept is that these are in contrast to integers.

We can represent a floating-point number in the general form

$$\pm .d_1 d_2 \ldots d_p * B^e, \tag{1.3}$$

where d_i's are digits or bits with $0 \le d_i \le B - 1$; and

B: the number base being used (usually either 2, 16, or 10)

p: the number of significant digits (bits)—that is, precision

e: the integer exponent, with a range of values defined on the interval [Emin, Emax] (ordinarily the range will include negative as well as positive values)

f: $d_1 d_2 \ldots d_p$, the fractional part of the number

For hand calculators the base B is usually 10. In computers this base is often 2, but other bases, such as 16, are also used.

EXAMPLE   Suppose $B = 10$ and $p = 4$. Then these numbers would be represented as

$$
\begin{array}{rcl}
27.39 & \to & +.2739 \times 10^2; \\
-0.00124 & \to & -.1240 \times 10^{-2}; \\
37000 & \to & +.3700 \times 10^5. \quad \blacksquare
\end{array}
$$

In this example we have required that the first element in f, $d_1 \neq 0$. When this is true for a floating-point number system, we refer to them as *normalized* floating-point numbers. Moreover, although the number of reals on any finite interval is infinite, this is not true for floating-point numbers. It turns out that the actual number of normalized floating-point numbers represented in (1.3) is

$$2 * (B - 1) * B^{p-1} * (\text{Emax} - \text{Emin} + 1) + 1, \tag{1.4}$$

where the last term, $+1$, is for 0.0. In computers, a floating-point number system is the IEEE standard. Two levels of precision are defined:

Single-precision format (32 bits long):

1 bit	8 bits	23 bits
$\pm$	e	$d_1 d_2 \ldots d_p$

$$B = 2, e \text{ in } [-128, 127], p = 23$$

Double-precision format (64 bits long):

1 bit	11 bits	52 bits
$\pm$	e	$d_1 d_2 \ldots d_p$

$$B = 2, e \text{ in } [-1024, 1023], p = 52$$

The first bit represents the sign of the number. The sign of the exponent is handled by biasing the exponent value—that is, adding 128 (or 1024) to it so that unsigned values from 0 to 255 (2047) are actually stored. From (1.4), the number of normalized floating-point numbers in these formats is

Single format: $2 * 1 * 2^{22} * (2^7 - (-2^7)) + 1 \doteq 2.14 \times 10^9$

Double format: $2 * 1 * 2^{51} * (2^{10} - (-2^{10})) + 1 \doteq 9.11 \times 10^{18}$

There are other methods for storing floating-point numbers. In large IBM computers, for example, $B = 16$, with 24 bits (6 hex digits) for the fraction part and 7 bits (biased 64) for the binary exponent. Control Data CYBER computers have 60-bit words, 48 bits for the fraction part and 11 for exponent with $B = 2$. Some computers with $B = 2$ gain an extra bit of precision by not storing d_1 (which is always a 1 for normalized fractions), so it can be omitted in the stored numbers. It is then referred to as the *hidden bit*.

We now examine the question of the distribution of normalized floating-point numbers. Here we consider the case where $B = 2$, $p = 2$, and $-2 \le e \le 3$. In this example, all the normalized numbers would be of the form

$$\pm.10 * 2^e \text{ or } \pm.11 * 2^e, -2 \le e \le 3.$$

Since for binary fractions, $.10 = \frac{1}{2}$ and $.11 = \frac{1}{2} + \frac{1}{4} = \frac{3}{4}$, these numbers range from -6 to $+6$ ($.11 * 2^3 = 6$ and $-.11 * 2^3 = -6$). A list of all the positive numbers in this system is

$$\begin{array}{ll} .10 * 2^{-2} = \frac{1}{8}; & .11 * 2^{-2} = \frac{3}{16}; \\ .10 * 2^{-1} = \frac{1}{4}; & .11 * 2^{-1} = \frac{3}{8}; \\ .10 * 2^{0} = \frac{1}{2}; & .11 * 2^{0} = \frac{3}{4}; \\ .10 * 2^{1} = 1; & .11 * 2^{1} = \frac{3}{2}; \\ .10 * 2^{2} = 2; & .11 * 2^{2} = 3; \\ .10 * 2^{3} = 4; & .11 * 2^{3} = 6. \end{array}$$

In the diagram, we see the distribution of all of the nonnegative floating-point numbers on the interval $[0, 6]$.

There are 25 total numbers in this tiny floating-point system, exactly as given by the formula

$$2 * (2 - 1) * 2^{2-1} * (3 - (-2) + 1) + 1 = 2 * 1 * 2 * 6 + 1 = 25.$$

It is hard for us to think in number bases other than 10, so we will discuss the arithmetic accuracy of floating-point operations using normalized base-10 numbers. The other number bases that are actually used in computers behave analogously. To simplify our treatment, assume only three digits in the fraction part, and one decimal digit for the exponent. We supply signs for the fraction and the exponent as separate symbols. For these examples, $B = 10$, $p = 3$, e in $[-9, 9]$. We compare rounding and chopping the results.

When two floating-point numbers are added or subtracted, the digits in the fraction of the number with the smaller exponent must be shifted to align the decimal points; the sum may need to be shifted and the exponent adjusted to normalize the result. For example,

1. $.137 \times 10^1 + .269 \times 10^{-1}$:

$$
\begin{array}{ll}
.137 \ \ \times 10^1 & \\
+.00269 \times 10^1 & \text{Align decimal points} \\
\hline
.13969 \times 10^1 \rightarrow & \text{Chopped} = .139 \times 10^1 \\
+.0005 & \\
\hline
.14019 \times 10^1 \rightarrow & \text{Rounded*} = .140 \times 10^1
\end{array}
$$

2. $.485 \times 10^4 - .482 \times 10^4$:

$$
\begin{array}{ll}
.485 \ \times 10^4 & \\
-.482 \ \times 10^4 & \\
\hline
.003 \ \times 10^4 & \\
.300 \ \ \times 10^2 & \text{Normalized} \rightarrow \text{Chopped} = .300 \times 10^2 \\
+.0005 & \\
\hline
.3005 \times 10^2 \rightarrow & \text{Rounded} = .300 \times 10^2
\end{array}
$$

3. $.378 \times 10^4 + .727 \times 10^4$:

$$
\begin{array}{ll}
.378 \ \ \times 10^4 & \\
+.727 \ \ \times 10^4 & \\
\hline
1.105 \ \ \times 10^4 & \\
.1105 \times 10^5 & \text{Normalized} \rightarrow \text{Chopped} = .110 \times 10^5 \\
+.0005 & \\
\hline
.1110 \times 10^5 \rightarrow & \text{Rounded} = .111 \times 10^5
\end{array}
$$

Observe that rounding requires an extra operation, which can be done with either hardware or software. Either way adds cost, so most computers chop rather than round. Pay particular attention to the loss of precision in the second example. There is only one digit of accuracy in the result, although the difference is represented as if the trailing zeros were significant. This loss of significance when two nearly equal numbers are subtracted is the major reason for the loss of accuracy in floating-point operations.[†]

---

*To round to three digits after the decimal point, we add 0.0005 and then chop the digits beyond three.
[†]One way to detect such insignificant zeros is to provide for changing the entering digit on a left shift of the fraction during normalizing. Two runs with different fill digits will then give different results. This is pretty costly. Another way (also costly) is to repeat the computer run using double precision, and compare the results.

Multiplication of two n-digit numbers gives a result of $2n$ digits. The floating-point registers of a computer* normally allow for this, converting the $2n$-digit product to an n-digit result. Such double-length registers are also used in division. Some examples:

4. $.403 \times 10^6 * .197 \times 10^{-1}$:

Multiply fractions	Add exponents
.403	6
*.197	-1
.079391	5

$.079391 \times 10^5$
$.79391 \ \times 10^4$ Normalized → Chopped $= .793 \times 10^4$
$+.0005$

$.79441 \ \times 10^4 \rightarrow$ Rounded $= .794 \times 10^4$

5. $.356 \times 10^{-2} \div .156 \times 10^4$:

Divide fractions	Subtract exponents
.356	-2
$\div$ .156	-4
2.28205	-6

2.28205×10^{-6} Normalized → Chopped $= .228 \times 10^{-7}$
$.22820 \times 10^{-7}$
$+ \ .0005$

$.22870 \times 10^{-7} \rightarrow$ Rounded $= .228 \times 10^{-7}$

In multiplication and division, the initial shifting to align decimal points can be omitted. The multiplication step is usually considerably slower than addition; the overall time for floating-point multiplication is typically from 2.5 to 10 times that for floating-point addition or subtraction. Floating-point division is usually the slowest of all (4 to 25 times that for addition). These timing differences are even greater when software routines are required for multiplication and division.

Many computers provide for two (or even three) levels of precision by using a double-length word for double precision. Usually the number of bits used for the exponent remains the same, so the number of digits in the fraction more than doubles.[†]

Since decimal numbers are usually converted to a different number base when stored as a floating-point value, the number of significant decimal digits equivalent to the accuracy of the fraction is not an integer. Different computer manufacturers have approached the problem of floating-point representation with wide variations of word lengths, so the accuracy provided is considerably different (from 6 to 16 equivalent decimal digits in single precision, for example).

---

*Not all computers have hardware to multiply or divide floating-point numbers. This is particularly true for microprocessors. In such cases, software routines are used.
†Double precision therefore requires twice the memory space to store the numbers. The execution time is also increased, from 25% to several fold.

Converting numbers to the computer's internal number base will often introduce some error. Terminating decimal fractions may be nonterminating in binary [$(0.6)_{10}$ = $(0.100110011001 \ldots)_2$, for example]. In addition, representing floating-point numbers in a fixed finite word length has many "gaps" in its number set, as we have seen. In effect, we have to map the infinite set of mathematical reals into a finite set of computer numbers. For our simplified example of three-digit fractions, there are only 900 different fraction values—all the mathematical numbers between 0.1 and 1.0 must be translated to one of these 900 values. In each decade, as represented by a constant value of the exponent, there also are only 900 different values. The spacing between values in the different decades is therefore different.

Zero is a special situation among the floating-point numbers. It cannot be normalized, of course, so special conventions are adopted. In most systems, the fraction digits are all zeros. The exponent must also be set to the most negative value; if this is not done, alignment of decimal points when adding will shift out significant digits from the addend. Zero is relatively isolated from the other values. In our simplified example system, the nearest neighbors to zero are $\pm 0.1 * 10^{-9}$. Trying to represent any magnitudes smaller than this causes a program error called *exponent underflow*. In some FORTRAN systems, on underflow the number is replaced by floating-point zero and execution continues. This may produce other errors if subsequent division is performed. Similarly, an attempt to represent numbers of magnitude larger than $0.999 * 10^9$ in our system gives rise to exponent overflow. Normally, this terminates execution, but some systems replace the number with the largest possible floating-point quantity and then continue.

Peculiar things happen in floating-point arithmetic. For example, adding 0.001 one thousand times usually does not equal 1.0 exactly. In some instances, multiplying a number by unity does not reproduce the number. In many calculations, changing the order in which operations are performed will produce different results.

ABSOLUTE VERSUS RELATIVE ERROR, SIGNIFICANT DIGITS

We introduced the terms *absolute* and *relative error* informally in Section 1.2. More formally, the absolute error of a given result is usually defined as

$$\text{Absolute error} = \text{true value} - \text{approximate value}$$

so we get the true value by adding the absolute error to the approximation. However, a given error is usually much more serious when the magnitude of the true value is small. For example, 1036.52 ± 0.010 is accurate to five significant digits and is frequently of more than adequate precision, while 0.005 ± 0.010 is a clear disaster. The relative error,

$$\text{Relative error} = \frac{\text{true value} - \text{approximate value}}{\text{true value}}$$

is often a better indicator of the accuracy. It is more nearly scale-independent, a most desirable property. When the true value is zero, the relative error is undefined. It follows that the round-off error due to finite fraction length in floating-point numbers is more nearly constant when expressed as relative error than when expressed as absolute error. Observe that the loss of significant digits, when small, nearly equal floating-point numbers are subtracted, produces a particularly severe relative error. (Note that others define both absolute error and relative error in terms of absolute values, so that these errors can be only positive, or zero.)

In addition to the concept of relative and absolute error, we have used the term *significant digits*. Suppose we write

1. True value = $d_1 d_2 \ldots d_n d_{n+1} \ldots d_p$, and
2. Approximate value = $d_1 d_2 \ldots d_n e_{n+1} \ldots e_p$

where $d_1 \neq 0$ and the first difference of the digits is at the $(n + 1)$st digit. Then we say that (1) and (2) agree to n significant digits if $|d_{n+1} - e_{n+1}| < 5$. Otherwise, we say they agree to $n - 1$ significant digits.

EXAMPLE   Let the true value = 10/3 and the approximate value = 3.333.

Then the absolute error = 0.000333. . . = 1/3000;

the relative error = (1/3000)/(10/3) = 1/10000;

the number of significant digits = 4.   ∎

1.13 PROGRAMMING FOR NUMERICAL SOLUTIONS

In this and in all of the following chapters, we will present computer programs that implement some of the more important methods of the chapter by a fully written computer program. A well-written program can clarify an algorithm as well as provide a ready-to-use tool for solving problems. We encourage you to improve, modify, or create your own version of the program.

The programs in this book are almost all in the FORTRAN language, even though some people prefer to use a language that is more modern in its constructs. The reason for choosing FORTRAN is that most practitioners of numerical analysis are familiar with it and, more important, that there are lots of prewritten subroutines that are of professional quality, thoroughly tested, and optimized. In an appendix, you will find more information about subroutine libraries.

When one writes a program, it is not enough just to know some programming language. Before implementing the algorithm, you must also think about the limitations of the computer and also about efficiency. Consider the following example:

$$f(x) = e^x = 1 + x + x^2/2! + x^3/3! + \cdots .$$

An algorithm for computing this series is

1. Set SUM = 1, N = 0, NEXT_TERM = 1.
2. Save SUM value in OLDSUM.
   a) Increment N by 1.
   b) Compute NEXT_TERM = NEXT_TERM * X/N.
   c) Compute SUM = SUM + NEXT_TERM.
   d) Compare OLDSUM with SUM. If they are different, go back to 2, ELSE go to 3.
3. Set F = SUM.

(We know that FORTRAN and most other languages have a built-in exp(x) function that can get the result. Just bear with us—the example makes some important points.) The following is a correct FORTRAN subprogram that implements the algorithm.

```
      REAL FUNCTION F(X)
      REAL X
C
      REAL SUM, OLDSUM, NEXTRM
      INTEGER N
C
      SUM = 1.0
      NEXTRM = 1.0
C
      N = 0
10    OLDSUM = SUM
         N = N+1
         NEXTRM = NEXTRM*X/N
         SUM = SUM + NEXTRM
         IF (SUM .NE. OLDSUM) GO to 10
C
      F = SUM
      END
```

The program is easy to understand, but there are problems when it is run with values of x that are of large magnitude. Suppose we use it to compute $f(30.4)$ or, worse yet, $f(-30.4)$. Computing these values of the function requires an inordinate number of repetitions of the loop. With $x = -30.4$, the answer is also entirely incorrect! While the correct value for $f(-30.4)$, to five significant digits, is 0.00000, the printout of F was

```
C  ------------------------------------------------------------
C
C          SECOND VERSION OF THE CODE FOR
C          COMPUTING F(X) = EXP(X).
C
C  ------------------------------------------------------------
       REAL FUNCTION F(X)
C
       REAL X
C
       REAL SUM, OLDSUM, NEXTRM, FRACT, EXP1, N
       INTEGER M
       PARAMETER (EXP1 = 2.718281828459)
       SAVE EXP1
C
C  HERE WE EXTRACT THE FRACTIONAL PART OF X
C
       FRACT = ABS(X-AINT(X))
       SUM = 1.0;
       NEXTRM = 1.0
       N = 0.0
C  ------------------------------------------------------------
C
C  WE COMPUTE THE SERIES AS FAR OUT AS WE NEED TO; THAT IS,
C  WE SEE NO CHANGE IN THE SERIES VALUE WHEN
C  ADDING AN ADDITIONAL TERM.
C
C  ------------------------------------------------------------
   10  OLDSUM = SUM
       N = N+1.0
       NEXTRM = NEXTRM*FRACT/N
       SUM = SUM + NEXTRM
       IF (SUM .NE. OLDSUM) GO TO 10
C
C  WE STORE THE INTEGER PART OF X IN M
C
       M = INT( ABS(X) )
       IF (X .GE. 0.0) THEN
           F = SUM * (EXP1**M)
         ELSE
           F = 1.0/(SUM * (EXP1**M) )
         END IF
       RETURN
       END
```

Figure 1.13   Improved program for e^x in FORTRAN.

2.34872! The result varies depending on the computer that is used. This was from an HP-150 computer and 104 iterations were required before the program terminated. (Another version, with exactly the same program steps but written in BASIC, gave 91887.2 after 85 iterations!!!)

Perhaps you have already recognized why this happened. The successive terms of the series alternate in sign and are large. As we have previously described, subtraction of large numbers can generate very large errors. That is precisely what happens here.

We can avoid the alternating-sign problem, if we compute $f(-30.4)$ as $1/f(+30.4)$. Doing so, we get $1,594,234,671 * 10^4$ for $f(+30.4)$ (exact to 10 digits), and its reciprocal is the correct value of $f(-30.4)$. However, 77 terms of the Taylor series were needed. We should consider modification to improve the efficiency. One way to do this is the following.

Since $f(x) = e^x$, it is true that $f(30.4) = f(30) * f(0.4)$. With $x = 0.4$, the series converges rapidly. When x is an integer, we can just raise e to that power and not use a series at all. This is the strategy that we will use. Figure 1.13 shows code that does it. There is one other improvement that you should note. We declare N as real rather than integer to avoid the (hidden) conversion from integer to real that occurred during each execution of the loop in the previous version.*

When $f(30.4)$ and $f(-30.4)$ were computed with the improved version, only 11 terms of the series were needed to converge to the result, and it was accurate to about 10 digits.

While FORTRAN remains the most widely used language for numerical analysis (due largely to the huge libraries of useful subroutines in that language), one may code the algorithms in almost any computer language. As an illustration of this, a Pascal implementation is given in Figure 1.14.

The Pascal version is written in Borland's Turbo Pascal—if you have that compiler on your personal computer, you can try it out. (Incidentally, there are Pascal versions of all the programs in this book available on a disk. If you are interested in these, see your instructor.) Since there is no exponentiation operator in Pascal, we have computed the integer power of e by repeated multiplications. A little trick is used to cut down the number of multiplies by about half.

1.14 CHAPTER SUMMARY

If you have understood this chapter on solving nonlinear equations, you are now able to

1. Explain the four steps in problem solving with reference to a typical problem situation.

2. Use these methods to find solutions to $f(x) = 0$:

   halving the interval
   linear interpolation
   modified linear interpolation
   Newton's method
   Muller's method

---

*Some FORTRAN compilers are smart enough to do this for us.

```
      (*************************************************************
      *                                                           *
      *              Pascal program for computing                 *
      *              f(x) = exp(x). This is based                 *
      *              on the improved algorithm for                *
      *              computing the series.                        *
      *                                                           *
      *************************************************************)

      CONST
         expl = 2.718281828459;
      VAR
         x, f,
         fractional_part_of_X,
         save                      : REAL;

         number_of_iterations,
         integer_part_of_x         : INTEGER;

      (*
         This functions computes x**n where n is an integer.
      *)
         FUNCTION x_to_nth_power(x : REAL; n : INTEGER) : REAL;
         VAR
            i          : INTEGER;
            product , square_x : REAL;
         BEGIN
            product := 1.0; i := n; square_x := x*x;

            WHILE i > 0 DO
               IF ODD(i) THEN          BEGIN
                  product := product*x;
                  i := i-1              END
               ELSE                     BEGIN
                  product := product*square*_x;
                  i := i-2              END;

            x_to_nth_power := product
         END; (* function *)


      (*************************************************************
      *  This function computes the series solution for e^x       *
      *  0 <= x <= 1.                                             *
      *************************************************************)


         FUNCTION series_value( x : REAL) : REAL;
         VAR
            sum, oldsum,
            nexterm, s : real;
            n : integer;
```

Figure 1.14   Pascal version of improved program.

Figure 1.14   *(continued)*

```
            BEGIN
               n := 0; sum := 1.0; nexterm := 1.0;

               REPEAT
                  n := n+1;   oldsum := sum;
                  nexterm := nexterm*x/n;
                  sum := sum + nexterm
                                          UNTIL sum = oldsum;
               number_of_iterations := n;
               series_value := sum
         END;
         BEGIN (* MAIN *)

            WRITELN; WRITELN;
            WRITE('    ENTER A GIVEN VALUE: '); READ(X);
            WRITELN; WRITELN; WRITELN;


         (****************************************************************
          *      Break x into its integer and fractional parts,       *
          *      then invoke functions for computing powers of eˣ      *
          *      for each part.                                        *
          ****************************************************************)


            fractional_part_of_X := ABS(x − TRUNC(x) );
            integer_part_of_X := ABS( TRUNC(x) );

            IF x <> 0.0 THEN
                save := x_to_nth_power(expl, integer_part_of_X) *
                        series_value(fractional_part_of_x)
                ELSE
                   f := 1.0;

            IF x > 0.0 THEN f := save;
            IF x < 0.0 THEN f := 1.0/save;

            WRITELN('X':3,'SERIES VALUE':24,'EXACT VALUE':20,
                    'ITERATIONS':16);   WRITELN;

            WRITELN( x:5:2,' ',f:19:5,' ',
                    EXF(x):19:5,number_of_iterations:13)
         END.
```

3. Compare the methods of (2) for efficiency, rate of convergence, certainty of finding a root, and reliability of estimates of accuracy of the solution.

4. Rearrange a function into a $x = g(x)$ form, test to show that the rearrangement will converge, and use it to find a root of $f(x) = 0$.

5. Explain how the Aitken acceleration can be used to minimize the number of iterations required to attain a desired degree of accuracy.

6. Outline the demonstration that shows when Newton's method is usually quadratically convergent and when it is only linearly convergent; tell what this means in terms of increase of precision per step.

7. Use Newton's and Bairstow's methods to find all the roots of a polynomial.

8. Show that nested multiplication is equivalent to synthetic division and tell why this method of evaluating a polynomial is advantageous.

9. Describe in general terms the QD method and Graeffe's method.

10. Distinguish between five types of errors and explain how each can be minimized.

11. Explain how floating-point numbers are stored in computers and tell what factors affect their accuracy and range.

12. Discuss the relative advantages of using absolute error versus relative error as measures of accuracy.

13. Use the programs and subroutines of Section 1.15 to solve the chapter exercises, writing driver programs as required.

14. Demonstrate that you can write a successful computer program that implements one or more of the algorithms of this chapter without referring to any of the example programs.

15. Critique the programs of Section 1.15 and point out how they might be improved.

SELECTED READINGS FOR CHAPTER 1

Atkinson (1978); Conte and de Boor (1980); Forsythe, Malcolm, and Moler (1977); Waser and Flynn (1982).

1.15   COMPUTER PROGRAMS

Numerical procedures for obtaining the roots of an equation are often extremely slow and tedious when done by hand, although calculators ease the task. In some cases, both of these techniques are so tedious as to make the task impossible. In contrast, computers are a useful and powerful tool used by mathematicians, scientists, and engineers and are ideally adapted to implementing the algorithms presented. The following programs will implement the methods presented in Chapter 1. As an alternative, you may wish to use canned programs such as those in IMSL (Appendix C); there is a large variety to choose from. Among them are ZBRENT, which uses linear interpolation and bisection; ZRPOLY, which uses Muller's method to find the roots of a polynomial with real coefficients; and ZSYSTM, which uses Newton's method.

We illustrate the program-development process with two examples, the first one applied to the ladder-in-the-mine problem that served to introduce this chapter. Recall that that problem required us to solve a transcendental equation for C:

$$\frac{w_2\cos(\pi - A - C)}{\sin^2(\pi - A - C)} - \frac{w_1\cos C}{\sin^2 C} = 0.$$

In this equation w_1, w_2, and A are supplied as constants. After finding the angle C, we get the maximum length from

$$\ell_{max} = \frac{w_2}{\sin(\pi - A - C)} + \frac{w_1}{\sin C}.$$

For our first example, we will find a root, $0° < C < (180° - A)$, using the modified linear interpolation algorithm. Our problem has $w_1 = 7$, $w_2 = 9$, and $A = 123°$. The algorithm was given in Section 1.3; we repeat it here but add a second termination criterion.

Modified Linear Interpolation Method

To determine a root of $f(x) = 0$, given values of x_1 and x_2 such that $f(x_1)$ and $f(x_2)$ are of opposite sign,

Set SAVE $= f(x_1)$; set F1 $= f(x_1)$; set F2 $= f(x_2)$.
REPEAT

Set $x_3 = x_2 - F2\dfrac{x_2 - x_1}{F2 - F1}.$

IF $f(x_3)$ of opposite sign to F1:
Set $x_2 = x_3$.
Set F2 $= f(x_3)$.
IF $f(x_3)$ of same sign as SAVE:
Set F1 $=$ F1/2.
ENDIF.
ELSE Set $x_1 = x_3$.
Set F1 $= f(x_3)$.
IF $f(x_3)$ of same sign as SAVE:
Set F2 $=$ F2/2.
ENDIF.
ENDIF.
Set SAVE $= f(x_3)$.
UNTIL $|x_1 - x_2| <$ XTOL
OR $|f(x_2)| <$ FTOL.

Before we can write our program, we must establish the design of our program modules. We must also refine the algorithm so that decisions are unambiguous.

1. A main program will read in the starting values of x_1 and x_2 and call the subroutine MDLNIN to execute the algorithm. Reflection reveals that the function is continuous in C and has only one zero. Suitable values for x_1 and x_2 are 5° and 118°; these certainly bracket the value that makes the function zero.

2. A function subprogram will compute values of $f(x)$. It then is easy to find roots of a different function by replacing the function subprogram. To provide flexibility, we will pass the name of the function subprogram as a parameter to the MDLNIN subroutine. This requires us to declare the name of the function subprogram as EXTERNAL. We pass the constants of the problem, w_1, w_2, and A, through COMMON. The values of x_1 and x_2 are passed as parameters; the value of the root is returned in XR.

3. In the subroutine, it will be appropriate to test initially whether $f(x_1)$ and $f(x_2)$ are indeed of opposite sign, printing a message if they are not.

4. The termination on $|x_1 - x_2|$ must be quantified. We will compare this value to an input parameter, XTOL, to provide for terminating the calculation. In some situations, it may be preferable to terminate when the relative error $|x_1 - x_2|/x_1$ is less than some predefined tolerance.

5. A second termination criterion is whether the function value is extremely small; it is possible that this indicates an x-value very near a root. The subroutine compares values of $f(x_3)$ to the parameter FTOL, to stop when the function is "sufficiently small."

6. It is always good practice to limit the maximum number of iterations. The parameter NLIM does this.

7. The utility of the subroutine is increased if the calling program can easily determine on which condition the subroutine terminated. The parameter I signals which of the various conditions was satisfied on termination. It is also desirable to sometimes print out the successive iterates, but at other times this is not needed. The parameter I is used (at the time the subroutine is called) to indicate whether or not to print out each iterate.

In our programs, we try to provide adequate documentation by comments. Some documentation systems also require a flowchart;* we provide one in Fig. 1.15.

Program 1 (see Fig. 1.16) was run with input data as follows:

$$
\begin{aligned}
X1 &= 5.0 \\
X2 &= 118.0 \\
XTOL &= 0.01 \\
FTOL &= 0.0001 \\
NLIM &= 20
\end{aligned}
$$

---

*The relative advantages of flowcharts as compared to the statement of the algorithm in a structured form are well illustrated by this example. The flowchart is much more specific as regards details of the program but is also much more dependent on the implementation language. The structured statement is much more general and presents the essentials of the procedure more clearly.

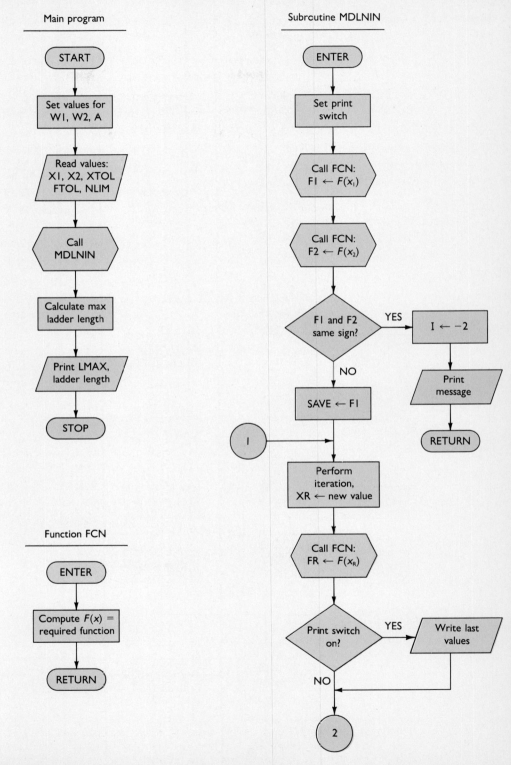

Figure 1.15

Figure 1.15 (*continued*)

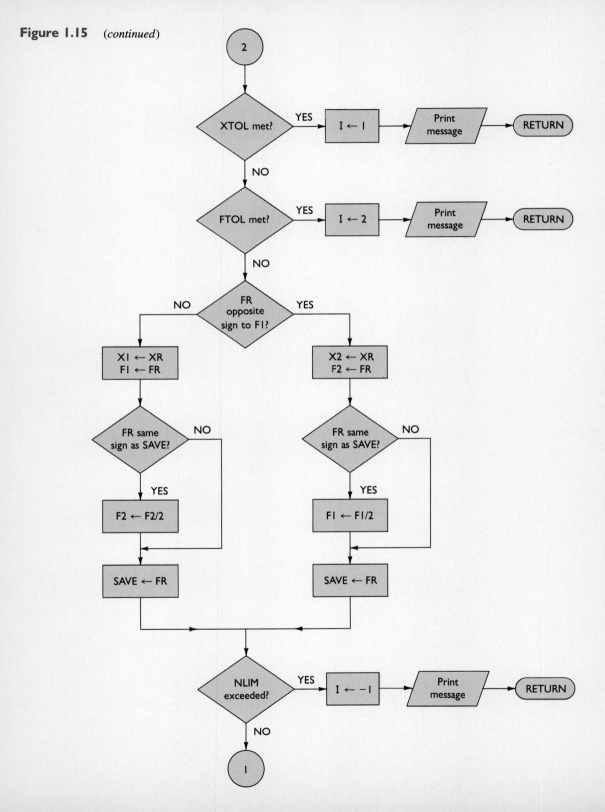

```
        PROGRAM LADDER(INPUT,OUTPUT)
C
C    ------------------------------------------------------------------
C
        REAL LMAX,W1,W2,A,PI,X1,X2,XR,XTOL,FTOL,FCN
        INTEGER I,NLIM
        EXTERNAL FCN
        COMMON W1,W2,A,PI
        DATA I/0/
        W1 = 7.0
        W2 = 9.0
        A = 123.0
        PI = 4.0 * ATAN(1.0)
C
C    ------------------------------------------------------------------
C
C  READ IN INITIAL GUESSES AND TOLERANCES
C
        READ*, X1,X2,XTOL,FTOL,NLIM
        CALL MDLNIN(FCN,X1,X2,XR,XTOL,FTOL,NLIM,I)
C
        IF ( I .GT. 0 ) THEN
C
C  THE ROOT IS RETURNED IN XR AND USED TO CALCULATE LMAX
C
        LMAX = W2 / SIN((180.0-A-XR)/180.0*PI) + W1 / SIN(XR/180.0*PI)
C
C    ------------------------------------------------------------------
C
C  PRINT MAXIMUM LENGTH OF THE LADDER
C
        PRINT 200, LMAX
        END IF
 200    FORMAT(/'  MAXIMUM LENGTH OF THE LADDER IN FEET IS ',F6.2)
        STOP
        END
C
C    ------------------------------------------------------------------
C
        REAL FUNCTION FCN(X)
        REAL X,W1,W2,A,PI
C
        REAL ANG1,ANG2
        COMMON W1,W2,A,PI
C
C  COMPUTE FUNCTION VALUE
C
        ANG1 = (180.0-A-X)/180.0*PI
        ANG2 = X/180.0*PI
        FCN = W2 * COS(ANG1)/SIN(ANG1)**2 - W1 * COS(ANG2)/SIN(ANG2)**2
        RETURN
        END
C
C    ------------------------------------------------------------------
C
        SUBROUTINE MDLNIN(FCN,X1,X2,XR,XTOL,FTOL,NLIM,I)
C
C    ------------------------------------------------------------------
C
C
C  SUBROUTINE MDLNIN :
C                        THIS SUBROUTINE FINDS A ROOT BY MODIFIED
C  LINEAR INTERPOLATION.
```

Figure 1.16   Program 1.

Figure I.16 (*continued*)

```
C
C       --------------------------------------------------------------------
C
C       PARAMETERS ARE :
C
C       FCN        -FUNCTION THAT COMPUTES VALUES FOR F(X). MUST BE DECLARED
C                   EXTERNAL IN CALLING PROGRAM. IT HAS ONE ARGUMENT, X.
C       X1,X2      -INITIAL VALUE OF X. F(X) MUST CHANGE SIGNS AT THESE PTS.
C       XR         -RETURNS THE ROOT TO THE MAIN PROGRAM.
C       XTOL,FTOL  -TOLERANCE VALUES FOR X, F(X) TO TERMINATE ITERATIONS.
C       NLIM       -LIMIT TO NUMBER OF ITERATIONS.
C       I          -A SIGNAL FOR HOW ROUTINE TERMINATED.
C        I=1         MEETS TOLERANCE FOR X VALUES.
C        I=2         MEETS TOLERANCE FOR F(X).
C        I=-1        NLIM EXCEEDED.
C        I=-2        F(X1) NOT OPPOSITE IN SIGN TO F(X2).
C
C       WHEN THE SUBROUTINE IS CALLED, THE VALUE OF I INDICATES WHETHER
C       TO PRINT EACH VALUE OR NOT. I=0 MEANS PRINT THEM, .NE. 0 MEANS
C       DO NOT PRINT THEM.
C
C       --------------------------------------------------------------------
C
        REAL XERR,FSAVE,F1,F2,FR
        INTEGER J
        LOGICAL OUTPUT
C
        OUTPUT = .TRUE.
        IF ( I .NE. 0 ) OUTPUT = .FALSE.
        F1 = FCN(X1)
        F2 = FCN(X2)
C
C       --------------------------------------------------------------------
C
C   INITIAL F(X1) & F(X2) MUST BE OPPOSITE IN SIGN
C
        IF ( F1*F2 .GT. 0.0 ) THEN
          I = -2
          PRINT 201
          RETURN
        END IF
C
        FSAVE = F1
        DO 20 J=1,NLIM
        XR = X2 - F2*(X2-X1)/(F2-F1)
        FR = FCN(XR)
        XERR = ABS(X1-X2)/2.0
        IF (OUTPUT) THEN
          PRINT 199, J,XR,FR
        END IF
C
C       --------------------------------------------------------------------
C
C   CHECK STOPPING CRITERIA
C
        IF ( XERR .LE. XTOL ) THEN
          I = 1
          PRINT 202, J,XR,FR
          RETURN
        END IF
C
        IF ( ABS(FR) .LE. FTOL ) THEN
```

Figure 1.16   (*continued*)

```
        I = 2
        PRINT 203, J,XR,FR
        RETURN
     END IF
C
C    ----------------------------------------------------------------
C
C FIND NEW POINT
C
     IF ( FR*F1 .GE. 0.0 ) THEN
        X1 = XR
        F1 = FR
        IF ( FR*FSAVE .GT. 0.0 ) F2 = F2/2.0
        FSAVE = FR
     ELSE
        X2 = XR
        F2 = FR
        IF ( FR*FSAVE .GT. 0.0 ) F1 = F1/2.0
        FSAVE = FR
     END IF
  20 CONTINUE
C
C    ----------------------------------------------------------------
C
C WHEN LOOP IS NORMALLY COMPLETED, NLIM IS EXCEEDED.
C
     I = -1
     PRINT 200, NLIM,XR,FR
     RETURN
C
C    ----------------------------------------------------------------
C
 199 FORMAT(' AT ITERATION',I3,3X,' X = ',E10.5,3X,' F(X) = ',E12.5)
 200 FORMAT(/' TOLERANCE NOT MET AFTER ',I4,' ITERATIONS. X = ',E12.5,
    +        ' AND F(X) = ',E12.5)
 201 FORMAT(/' FUNCTION HAS SAME SIGN AT INITIAL X1 & X2')
 202 FORMAT(/' X TOLERANCE MET IN ',I4,' ITERATIONS. X = ',E12.5,
    +        ' AND F(X) = ',E12.5)
 203 FORMAT(/' F TOLERANCE MET IN ',I4,' ITERATIONS. X = ',E12.5,
    +        ' AND F(X) = ',E12.5)
     END



                    OUTPUT FOR PROGRAM 1

AT ITERATION  1     X = .11678E+03     F(X) =    .10024E+02
AT ITERATION  2     X = .11556E+03     F(X) =    .10160E+02
AT ITERATION  3     X = .11314E+03     F(X) =    .10524E+02
AT ITERATION  4     X = .10836E+03     F(X) =    .11660E+02
AT ITERATION  5     X = .98739E+02     F(X) =    .16241E+02
AT ITERATION  6     X = .77902E+02     F(X) =    .64523E+02
AT ITERATION  7     X = .27286E+02     F(X) =    .22130E+01
AT ITERATION  8     X = .24282E+02     F(X) =   -.11813E+02
AT ITERATION  9     X = .26812E+02     F(X) =    .61146E-01
AT ITERATION 10     X = .26799E+02     F(X) =    .18591E-02
AT ITERATION 11     X = .26798E+02     F(X) =   -.17452E-02
AT ITERATION 12     X = .26799E+02     F(X) =    .42591E-08

F TOLERANCE MET IN   12 ITERATIONS. X =    .26799E+02 AND F(X) =    .42591E-08

  MAXIMUM LENGTH OF THE LADDER IN FEET IS  33.42
```

The output is shown with the program. We conclude that quite long ladders can negotiate this turn in the mine shaft; the calculated length is over 33 ft.

For our second example of program development, we exhibit a program to solve for the roots of a polynomial up to the 10th degree, using Bairstow's method to get the quadratic factors. The algorithm, which is an adaptation of the equations of Section 1.9, can be written as shown below. Subscripts are advanced by two, so the equations for all the b's and c's become identical when $b_1 = b_2 = c_1 = c_2 = 0$. (See Program 2, Fig. 1.17.)

Bairstow's Method

To determine a quadratic factor, $x^2 - rx - s$, of the nth-degree polynomial $a_3x^n + a_4x^{n-1} + \cdots + a_{n+2}x + a_{n+3}$, choose initial coefficients R and S of the quadratic factor, then

Set B(1) = 0, B(2) = 0, C(1) = 0, C(2) = 0, DELR = 1000, DELS = 1000.
DO WHILE DELR $\geq$ tolerance value, or
    DELS $\geq$ tolerance value,
 DO FOR J = 3 to n + 3 step 1,
  Set B(J) = A(J) + R $*$ B(J $-$ 1) + S $*$ B(J $-$ 2).
  Set C(J) = B(J) + R $*$ C(J $-$ 1) + S $*$ C(J $-$ 2).
  Set DENOM = C(N + 1) $*$ C(N + 1) $-$ C(N + 2) $*$ C(N).
  IF DENOM = 0:
   Set R = R + 1.
   Set S = S + 1.
   Repeat from beginning.
  ENDIF.
  Set DELR = [$-$B(N + 2) $*$ C(N + 1) + B(N + 3)
               $*$ C(N)]/DENOM.
  Set DELS = [$-$C(N + 1) $*$ B(N + 3) + C(N + 2)
               $*$ B(N + 2)]/DENOM.
  Set R = R + DELR.
  Set S = S + DELS.
 ENDDO.
ENDWHILE

Because the program handles only one specific type of equation, there is no advantage to subdividing it into subroutines. The number of iterations needed to determine the quadratic factor within a given tolerance is arbitrarily limited to 20. In case the denominator during the step where we solve for new R and S values is zero, the trial values of R and S are arbitrarily increased by unity and the search for a quadratic factor is begun again—this will ordinarily avoid the attempted division by zero. When a quadratic factor is found successfully, the deflated polynomial (the b's are its coefficients) is used to find additional factors. (The algorithm applies only to obtaining each quadratic factor.) The program could be used for polynomials of degree higher than 10 by changing the DIMENSION statement, but this is not advisable because accumulated errors will often be too large.

```
      PROGRAM BRSTOW(INPUT,OUTPUT)
C
C THIS PROGRAM USES BAIRSTOW'S METHOD OF EXTRACTING THE
C QUADRATIC FACTORS OF A POLYNOMIAL.
C
C -----------------------------------------------------------------
C
C READ IN THE DEGREE OF THE ORIGINAL POLYNOMIAL AND THE
C ESTIMATES R & S FOR THE QUADRATIC FACTOR. (IF ESTIMATES NOT AVAILABLE,
C R & S ARE TAKEN AS ZERO.) ALSO READ IN ERROR TOLERANCE.
C
      REAL  A(13),B(13),C(13),R,S,DELR,DELS,TOL,DENOM
      INTEGER I,J,NP3,K
C
      READ *, N,R,S,TOL
C
C -----------------------------------------------------------------
C
C READ IN COEFFICIENTS OF THE POLYNOMIAL
C
      NP3 = N + 3
      READ *, ( A(I), I=3,NP3 )
C
C -----------------------------------------------------------------
C
C PRINT HEADING AND ECHO PRINT THE COEFFICIENTS
C
      PRINT 102
      DO 5 I=3,NP3
          J = NP3 - I
          PRINT 103, J,A(I)
    5 CONTINUE
      PRINT 104, TOL
C
C -----------------------------------------------------------------
C
C COMPUTE B AND C ARRAYS. LIMIT NUMBER OF ITERATIONS TO 20.
C
      DATA B,C,K / 26*0.0,1 /
    8 IF ( K .LE. 20 ) THEN
          DO 10 J=3,NP3
              B(J) = A(J) + R*B(J-1) + S*B(J-2)
              C(J) = B(J) + R*C(J-1) + S*C(J-2)
   10     CONTINUE
C
C -----------------------------------------------------------------
C
C COMPUTE DENOMINATOR AND CHECK IF ZERO. IF NOT ZERO, COMPUTE NEW R
C AND S VALUES. CHECK IF ACCURACY IS OK.
C
          DENOM = C(N+1) * C(N+1)  -  C(N+2) * C(N)
          IF ( DENOM .NE. 0.0 ) THEN
              DELR = ( -B(N+2) * C(N+1) + B(N+3) * C(N) ) / DENOM
              DELS = ( -C(N+1) * B(N+3) + C(N+2) * B(N+2) ) / DENOM
              R = R + DELR
              S = S + DELS
              IF ( ABS(DELR) + ABS(DELS) .LE. TOL ) THEN
                 PRINT 106, R,S
                 N = N - 2
                 IF ( N - 2 ) 22,23,24
```

Figure 1.17   Program 2.

Figure 1.17 (*continued*)

```
C
C     --------------------------------------------------------------------
C
C     REDUCED EQUATION IS OF DEGREE ONE - PRINT IT
C
   22          PRINT 107, B(N+2),B(N+3)
               STOP
C
C     REDUCED EQUATION IS OF DEGREE TWO - PRINT IT
C
   23          PRINT 108, B(N+1),B(N+2),B(N+3)
               STOP
C
C     --------------------------------------------------------------------
C
C     DEGREE OF REDUCED EQUATION MORE THAN TWO. SET COEFFICIENTS INTO AN
C     ARRAY AND GET NEXT FACTOR.
C
   24          NP3 = N + 3
               DO 30 I=3,NP3
               A(I) = B(I)
   30          CONTINUE
               K = 1
               GO TO 8
            ELSE
               K = K + 1
               GO TO 8
            END IF
         ELSE
C
C     --------------------------------------------------------------------
C
C     IF DENOM IS ZERO, PERTURB R AND S VALUES. START OVER.
C
            R = R + 1.0
            S = S + 1.0
            K = 1
            GO TO 8
         END IF
      END IF
C
C     --------------------------------------------------------------------
C
C     TOLERANCE NOT MET. PRINT FACTOR AND QUIT.
C
      PRINT 105, R,S
      STOP
C
C     --------------------------------------------------------------------
C
  102 FORMAT(//,' QUADRATIC FACTORS BY BAIRSTOW METHOD.'/
     +        ' ORIGINAL POLYNOMIAL IS - ',//' POWER OF X ',
     +        5X,' COEFFICIENT'/)
  103 FORMAT(I6,10X,F10.3)
  104 FORMAT(/' FACTORS ARE, WITH TOLERANCE OF ',E14.6)
  105 FORMAT(' TOLERANCE NOT MET IN 20 ITERATIONS. LAST FACTOR FOUND ',
     +        'WAS ',//'   X**2 - ',F10.5,' X - ',F10.5)
  106 FORMAT('   X**2 - ',F8.5,' X - ',F8.5)
  107 FORMAT(1X,F10.5,' X +  ',F10.5)
  108 FORMAT(1X,F10.5,' X**2 +  ',F10.5,' X + ',F10.5)
      END
```

Figure 1.17   (*continued*)

```
        OUTPUT FOR PROGRAM 2

QUADRATIC FACTORS BY BAIRSTOW METHOD.
ORIGINAL POLYNOMIAL IS -

POWER OF X          COEFFICIENT
    5                   1.000
    4                 -17.800
    3                  99.410
    2                -261.218
    1                 352.611
    0                -134.106


FACTORS ARE, WITH TOLERANCE OF     .100000E-03
  X**2 -   4.20000 X - -2.10000
  X**2 -   3.30002 X - -6.20007
    1.00000 X +    -10.30000
```

When the program was tested with the polynomial $x^5 - 17.8x^4 + 99.41x^3 - 261.218x^2 + 352.611x - 134.106$, the results were as shown after the program. The exact factors are

$$x^2 - 4.2x + 2.1, \ x^2 - 3.3x + 6.2, \text{ and } x - 10.3,$$

and the program finds them with excellent precision.
The input data were

$$N = 5 \text{ (degree)}$$
$$R = 0.0$$
$$S = 0.0$$
$$\text{TOL} = 0.0001$$
$$\text{Coefficients} = 1.0, -17.8, 99.41, -261.218,$$
$$352.611, -134.106$$

Programs 3 through 7 (Figs. 1.18 through 1.22) are FORTRAN implementations of other algorithms of this chapter:

Program	Method	Subroutine Name
3	Interval halving (bisection)	INTHV
4	Linear interpolation (*regula falsi*)	LNINTP
5	Newton's method	NEWTN
6	Iteration with $x = g(x)$	XGXIT
7	Muller's method	MULLR

Each of the subroutines is tested with the function $f(x) = 3x + \sin(x) - e^x = 0$.

These subroutines might appropriately be placed in the FORTRAN library at your installation so that they may be called automatically by your programs.

```
      PROGRAM PINTHV(INPUT,OUTPUT)
C
C DRIVER  FOR INTERVAL HALVING SUBROUTINE, INTHV
C
C -----------------------------------------------------------------------
C
      REAL FCN,X1,X2,XTOL,FTOL
      INTEGER I,NLIM
      EXTERNAL FCN
C
C -----------------------------------------------------------------------
C
C INITIALIZE VARIABLES FOR SUBROUTINE INTHV
C
      DATA X1,X2,XTOL,FTOL,I,NLIM/0.0,1.0,0.0001,0.00001,0,50/
C
C -----------------------------------------------------------------------
C
      CALL INTHV(FCN,X1,X2,XR,FTOL,NLIM,I)
C
C THE ROOT HAS BEEN FOUND AT XR IF I EQUALS 1 OR 2
C
      STOP
      END
C
C -----------------------------------------------------------------------
C
      REAL FUNCTION FCN(X)
      REAL X
      FCN = 3.0*X + SIN(X) - EXP(X)
      RETURN
      END
C
C -----------------------------------------------------------------------
C
      SUBROUTINE INTHV(FCN,X1,X2,XR,FTOL,NLIM,I)
C
C -----------------------------------------------------------------------
C
C SUBROUTINE INTHV :
C                   THIS SUBROUTINE FINDS THE ROOT OF A FUNCTION,
C     F(X) = 0, BY INTERVAL HALVING.
C
C -----------------------------------------------------------------------
C
C PARAMETERS ARE :
C
C FCN        -FUNCTION THAT COMPUTES VALUES FOR F(X). MUST BE DECLARED
C              EXTERNAL IN CALLING PROGRAM. IT HAS ONE ARGUMENT, X.
C X1,X2      -INITIAL VALUES OF X. F(X) MUST CHANGE SIGN AT THESE PTS.
C XR         -RETURNS THE ROOT TO THE MAIN PROGRAM.
C XTOL,FTOL -TOLERANCE VALUES FOR X, F(X) TO TERMINATE ITERATIONS.
C NLIM       -LIMIT TO NUMBER OF ITERATIONS.
C I          -A SIGNAL FOR HOW ROUTINE TERMINATED ON OUTPUT.
C  I = 1      MEETS TOLERANCE FOR F(X) VALUES.
C  I = -1     NLIM EXCEEDED.
C  I = -2     F(X1) NOT OPPOSITE IN SIGN TO F(X2)
C            -PRINT CONTROL ON INPUT
C  I = 0      PRINT RESULTS AT EACH ITERATION
C
C WHEN THE SUBROUTINE IS CALLED, THE VALUE OF I INDICATES WHETHER TO
C PRINT EACH VALUE OR NOT. I=0 MEANS PRINT THEM, I.NE.0 MEANS DON'T.
```

Figure I.18   Program 3.

Figure 1.18   (*continued*)

```
C
C     -----------------------------------------------------------------
C
      REAL FCN,X1,X2,XR,FTOL
      INTEGER NLIM,I,J
      REAL F1,F2,FR
C
C     -----------------------------------------------------------------
C
C  CHECK THAT F(X1) & F(X2) DIFFER IN SIGN
C
      F1 = FCN(X1)
      F2 = FCN(X2)
      IF ( F1*F2 .GT. 0.0) THEN
        I = -2
        PRINT 201
        RETURN
      END IF
C
C     -----------------------------------------------------------------
C
C  COMPUTE SEQUENCE OF POINTS CONVERGING TO THE ROOT
C
      DO 20 J=1,NLIM
      XR = (X1 + X2) / 2.0
      FR = FCN(XR)
      IF ( I .EQ. 0 ) THEN
         PRINT 199, J,XR,FR
      END IF
C
C     -----------------------------------------------------------------
C
C  CHECK ON STOPPING CRITERIA
C
      IF ( ABS(FR) .LE. FTOL ) THEN
        I = 1
        PRINT 203, J,XR,FR
        RETURN
      END IF
C
C     -----------------------------------------------------------------
C
C  COMPUTE NEXT X
C
      IF ( FR*F1 .GT. 0.0 ) THEN
        X1 = XR
        F1 = FR
      ELSE
        X2 = XR
        F2 = FR
      END IF
   20 CONTINUE
C
C     -----------------------------------------------------------------
C
C  WHEN LOOP IS NORMALLY COMPLETED, NLIM IS EXCEEDED
C
      I = -1
      PRINT 200, NLIM,XR,FR
      RETURN
C
C     -----------------------------------------------------------------
```

Figure 1.18 (*continued*)

```
C
  199  FORMAT(' AT ITERATION',I3,3X,' X =',E12.5,4X,' F(X) =',E12.5)
  200  FORMAT(//' TOLERANCE NOT MET AFTER ',I4,' ITERATIONS  X = ',
       +        E12.5,' F(X) = ',E12.5)
  201  FORMAT(//' FUNCTION HAS SAME SIGN AT X1 & X2 ')
  202  FORMAT(//' TOLERANCE MET IN ',I4,' ITERATIONS   X = ',E12.5,
       +         ' F(X) = ',E12.5)
  203  FORMAT(//' F TOLERANCE MET IN ',I4,' ITERATIONS   X = ',E12.5,
       +         ' F(X) = ',E12.5)
C
       END

                    OUTPUT FOR PROGRAM 3

AT ITERATION  1    X =  .50000E+00      F(X) =  .33070E+00
AT ITERATION  2    X =  .25000E+00      F(X) = -.28662E+00
AT ITERATION  3    X =  .37500E+00      F(X) =  .36281E-01
AT ITERATION  4    X =  .31250E+00      F(X) = -.12190E+00
AT ITERATION  5    X =  .34375E+00      F(X) = -.41956E-01
AT ITERATION  6    X =  .35938E+00      F(X) = -.26196E-02
AT ITERATION  7    X =  .36719E+00      F(X) =  .16886E-01
AT ITERATION  8    X =  .36328E+00      F(X) =  .71467E-02
AT ITERATION  9    X =  .36133E+00      F(X) =  .22670E-02
AT ITERATION 10    X =  .36035E+00      F(X) = -.17548E-03
AT ITERATION 11    X =  .36084E+00      F(X) =  .10460E-02
AT ITERATION 12    X =  .36060E+00      F(X) =  .43529E-03
AT ITERATION 13    X =  .36047E+00      F(X) =  .12992E-03
AT ITERATION 14    X =  .36041E+00      F(X) = -.22780E-04

TOLERANCE MET IN   14 ITERATIONS   X =   .36041E+00 F(X) =  -.22780E-04
```

```
       PROGRAM PLNINT(INPUT,OUTPUT)
C
C DRIVER  FOR LINEAR INTERPOLATION (I.E. REGULA FALSI) SUBROUTINE, LNINTP
C
C -------------------------------------------------------------------------
C
       REAL FCN,X1,X2,XTOL,FTOL
       INTEGER I,NLIM
       EXTERNAL FCN
C
C -------------------------------------------------------------------------
C
C INITIALIZE VARIABLES FOR SUBROUTINE LNINTP
C
       DATA X1,X2,XTOL,FTOL,I,NLIM/0.0,1.0,0.0001,0.00001,0,50/
C
C -------------------------------------------------------------------------
C
       CALL LNINTP(FCN,X1,X2,XR,XTOL,FTOL,NLIM,I)
C
C THE ROOT HAS BEEN FOUND AT XR IF I EQUALS 1 OR 2
C
       STOP
       END
C
C -------------------------------------------------------------------------
```

Figure 1.19   Program 4.

Figure 1.19   (*continued*)

```
      REAL FUNCTION FCN(X)
      REAL X
      FCN = 3.0*X  + SIN(X) - EXP(X)
      RETURN
      END
C
C     -----------------------------------------------------------------
C
      SUBROUTINE LNINTP(FCN,X1,X2,XR,XTOL,FTOL,NLIM,I)
C
C     -----------------------------------------------------------------
C
C
C     SUBROUTINE LNINTP :
C                     THIS SUBROUTINE FINDS THE ROOT OF A FUNCTION,
C     F(X) = 0, BY LINEAR INTERPOLATION (I.E. REGULA FALSI).
C
C     -----------------------------------------------------------------
C
C     PARAMETERS ARE :
C
C     FCN        -FUNCTION THAT COMPUTES VALUES FOR F(X). MUST BE DECLARED
C                 EXTERNAL IN CALLING PROGRAM. IT HAS ONE ARGUMENT, X.
C     X1,X2      -INITIAL VALUES OF X. F(X) MUST CHANGE SIGN AT THESE PTS.
C     XR         -RETURNS THE ROOT TO THE MAIN PROGRAM.
C     XTOL,FTOL  -TOLERANCE VALUES FOR X, F(X) TO TERMINATE ITERATIONS.
C     NLIM       -LIMIT TO NUMBER OF ITERATIONS.
C     I          -A SIGNAL FOR HOW ROUTINE TERMINATED.
C      I = 1       MEETS TOLERANCE FOR X VALUES.
C      I = 2       MEETS TOLERANCE FOR F(X)
C      I = -1      NLIM EXCEEDED.
C      I = -2      F(X1) NOT OPPOSITE IN SIGN TO F(X2)
C
C  WHEN THE SUBROUTINE IS CALLED, THE VALUE OF I INDICATES WHETHER TO
C  PRINT EACH VALUE OR NOT. I=0 MEANS PRINT THEM, .NE. 0 MEANS DON'T.
C
C     -----------------------------------------------------------------
C
      REAL FCN,X1,X2,XR,XTOL,FTOL
      INTEGER NLIM,I,J
      REAL F1,F2,FR,XERR
C
C     -----------------------------------------------------------------
C
C CHECK THAT F(X1) & F(X2) DIFFER IN SIGN
C
      F1 = FCN(X1)
      F2 = FCN(X2)
      IF ( F1*F2 .GT. 0.0) THEN
        I = -2
        PRINT 201
        RETURN
      END IF
C
C     -----------------------------------------------------------------
C
C COMPUTE SEQUENCE OF POINTS CONVERGING TO THE ROOT
C
      DO 20 J=1,NLIM
      XR = X2 - F2*(X2-X1)/(F2-F1)
      FR = FCN(XR)
      XERR = ABS(X1 - X2) / 2.0
      IF ( I .EQ. 0 ) THEN
        PRINT 199, J,XR,FR
```

Figure 1.19   (*continued*)

```
        END IF
C
C     --------------------------------------------------------------------------
C
C
C  CHECK ON STOPPING CRITERIA
C
        IF ( XERR .LE. XTOL ) THEN
          I = 1
          PRINT 202, J,XR,FR
          RETURN
        END IF
C
        IF ( ABS(FR) .LE. FTOL ) THEN
          I = 2
          PRINT 203, J,XR,FR
          RETURN
        END IF
C
C     --------------------------------------------------------------------------
C
C  COMPUTE NEXT X
C
        IF ( FR*F1 .GT. 0.0 ) THEN
          X1 = XR
          F1 = FR
        ELSE
          X2 = XR
          F2 = FR
        END IF
   20 CONTINUE
C
C     --------------------------------------------------------------------------
C
C  WHEN LOOP IS NORMALLY COMPLETED, NLIM IS EXCEEDED
C
        I = -1
        PRINT 200, NLIM,XR,FR
        RETURN
C
C     --------------------------------------------------------------------------
C
  199 FORMAT(' AT ITERATION',I3,3X,'  X = ',E10.5,4X,' F(X) = ',E12.5)
  200 FORMAT(/' TOLERANCE NOT MET AFTER ',I4,' ITERATIONS  X = ',
     +        E12.5,' F(X) = ',E12.5)
  201 FORMAT(/' FUNCTION HAS SAME SIGN AT X1 & X2 ')
  202 FORMAT(/' TOLERANCE MET IN ',I4,' ITERATIONS   X = ',E12.5,
     +        ' F(X) = ',E12.5)
  203 FORMAT(/' F TOLERANCE MET IN ',I4,' ITERATIONS   X = ',E12.5,
     +        ' F(X) = ',E12.5)
C
        END
```

```
                OUTPUT FOR PROGRAM 4

AT ITERATION  1     X = .47099E+00     F(X) =   .26516E+00
AT ITERATION  2     X = .37228E+00     F(X) =   .29534E-01
AT ITERATION  3     X = .36160E+00     F(X) =   .29410E-02
AT ITERATION  4     X = .36054E+00     F(X) =   .28945E-03
AT ITERATION  5     X = .36043E+00     F(X) =   .28454E-04
AT ITERATION  6     X = .36042E+00     F(X) =   .27968E-05

F TOLERANCE MET IN   6 ITERATIONS   X =    .36042E+00 F(X) =    .27968E-05
```

```
      PROGRAM PNEWTN(INPUT,OUTPUT)
C
C  DRIVER  FOR NEWTON'S METHOD SUBROUTINE, NEWTN
C
C  ----------------------------------------------------------------------
C
      REAL FCN,FDER,X,XTOL,FTOL
      INTEGER I,NLIM
      EXTERNAL FCN,FDER
C
C  ----------------------------------------------------------------------
C
C  INITIALIZE VARIABLES FOR SUBROUTINE NEWTN
C
      DATA X,XTOL,FTOL,I,NLIM/0.5,0.0001,0.00001,0,50/
C
C  ----------------------------------------------------------------------
C
      CALL NEWTN(FCN,FDER,X,XTOL,FTOL,NLIM,I)
C
C  THE ROOT HAS BEEN FOUND AT X IF I EQUALS 1 OR 2
C
      STOP
      END
C
C  ----------------------------------------------------------------------
C
      REAL FUNCTION FCN(X)
      REAL X, SIN, EXP
      FCN = 3.0*X + SIN(X) - EXP(X)
      RETURN
      END
C
C  ----------------------------------------------------------------------
C
      REAL FUNCTION FDER(X)
      REAL X, COS, EXP
      FDER = 3.0 + COS(X) - EXP(X)
      RETURN
      END
C
C  ----------------------------------------------------------------------
C
      SUBROUTINE NEWTN(FCN,FDER,X,XTOL,FTOL,NLIM,I)
C
C  ----------------------------------------------------------------------
C
C
C  SUBROUTINE NEWTN :
C                    THIS SUBROUTINE FINDS THE ROOT OF A FUNCTION,
C  F(X) = 0, BY NEWTON'S METHOD.
C
C  ----------------------------------------------------------------------
C
C  PARAMETERS ARE :
C
C  FCN       -FUNCTION THAT COMPUTES VALUES FOR F(X). MUST BE DECLARED
C             EXTERNAL IN CALLING PROGRAM. IT HAS ONE ARGUMENT, X.
C  FDER      -FUNCTION THAT COMPUTES THE DERIVATIVE OF F, ALSO
C             DECLARED EXTERNAL.
C  X         -INITIAL VALUE OF X, SHOULD BE NEAR ROOT. ALSO RETURNS VALUE
C             OF THE ROOT TO THE CALLER.
C  XTOL,FTOL -TOLERANCE VALUES FOR X, F(X) TO TERMINATE ITERATIONS.
```

Figure 1.20   Program 5.

Figure I.20   (*continued*)

```
C      NLIM       -LIMIT TO NUMBER OF ITERATIONS.
C      I          -A SIGNAL FOR HOW ROUTINE TERMINATED.
C       I = 1      MEETS TOLERANCE FOR X VALUES.
C       I = 2      MEETS TOLERANCE FOR F(X).
C       I = -1     NLIM EXCEEDED.
C
C  WHEN THE SUBROUTINE IS CALLED, THE VALUE OF I INDICATES WHETHER TO
C  PRINT EACH VALUE OR NOT. I=0 MEANS PRINT THEM, .NE. 0 MEANS DON'T.
C
C
C  -------------------------------------------------------------------
C
       REAL FCN,FDER,X,XTOL,FTOL
       INTEGER NLIM,I,J
       REAL FX,DELX
C
C  -------------------------------------------------------------------
C
C  COMPUTE SEQUENCE OF POINTS CONVERGING TO THE ROOT
C
       FX = FCN(X)
       DO 20 J=1,NLIM
       DELX = FX / FDER(X)
       X = X - DELX
       FX = FCN(X)
       IF ( I .EQ. 0 ) THEN
         PRINT 199, J,X,FX
       END IF
C
C  -------------------------------------------------------------------
C
C  CHECK ON STOPPING CRITERIA
C
       IF ( ABS(DELX) .LE. XTOL ) THEN
         I = 1
         PRINT 202, J,X,FX
         RETURN
       END IF
C
       IF ( ABS(FX) .LE. FTOL ) THEN
         I = 2
         PRINT 203, J,X,FX
         RETURN
       END IF
  20   CONTINUE
C
C  -------------------------------------------------------------------
C
C  WHEN LOOP IS NORMALLY COMPLETED, NLIM IS EXCEEDED
C
       I = -1
       PRINT 200, NLIM,X,FX
       RETURN
C
C  -------------------------------------------------------------------
C
 199   FORMAT(' AT ITERATION',I3,3X,' X =',E12.5,3X,' F(X) =',E12.5)
 200   FORMAT(/' TOLERANCE NOT MET AFTER ',I4,' ITERATIONS  X = ',
      +        E12.5,' F(X) = ',E12.5)
 202   FORMAT(/' TOLERANCE MET IN',I3,' ITERATIONS:   X =',E12.5,
      +        3X,' F(X) =',E12.5)
 203   FORMAT(/' F TOLERANCE MET IN',I3,' ITERATIONS    X =',E12.5,
      +        3X,' F(X) =',E12.5)
```

Figure 1.20   (*continued*)

```
C       END
      AT ITERATION   1    X =  .35163E+00    F(X) = -.22073E-01
      AT ITERATION   2    X =  .36039E+00    F(X) = -.68143E-04
      AT ITERATION   3    X =  .36042E+00    F(X) = -.66267E-09

      TOLERANCE MET IN  3 ITERATIONS:    X =  .36042E+00    F(X) = -.66267E-09
```

```
        PROGRAM PXGXIT(INPUT,OUTPUT)
C
C  THIS DRIVER CALLS SUBROUTINE XGXIT
C
C  ---------------------------------------------------------------------
C
      REAL GFCN,X
      INTEGER I
      EXTERNAL GFCN
      DATA X,I / 0.6, 0 /
C
      CALL XGXIT(GFCN,X,0.0001,50,I)
C
C  IF I = 1, THE ROOT IS EQUAL TO X.
C
      STOP
      END
C
C  ---------------------------------------------------------------------
C
      REAL FUNCTION GFCN(X)
C
      REAL X
      GFCN = ( EXP(X) - SIN(X) ) / 3.0
      RETURN
      END
C
C  ---------------------------------------------------------------------
C
      SUBROUTINE XGXIT(GFCN,X,XTCL,NLIM,I)
C
C  ---------------------------------------------------------------------
C
C
C  SUBROUTINE XGXIT :
C                    THIS SUBROUTINE FINDS THE ROOT OF F(X) = 0
C  BY ITERATION WITH X = G(X).
C
C
C  ---------------------------------------------------------------------
C
C  PARAMETERS ARE :
C
C  GFCN       -FUNCTION TO COMPUTE G(X), MUST BE DECLARED EXTERNAL IN
C              CALLING PROGRAM. IT HAS ONE ARGUMENT, X.
C  X          -INITIAL VALUE TO BEGIN ITERATIONS. X ALSO RETURNS THE
C              VALUE OF THE ROOT TO THE CALLER.
C  XTOL       -TOLERANCE VALUE FOR CHANGE IN X TO TERMINATE ITERATIONS.
C  NLIM       -LIMIT TO NUMBER OF ITERATIONS.
C  I          -A SIGNAL FOR HOW TERMINATED
C    I = 1    MEETS TOLERANCE FOR X VALUES.
C    I =-1    NLIM EXCEEDED.
C    I =-2    ITERATIONS APPEAR TO DIVERGE IN INITIAL CALCULATIONS.
```

Figure 1.21   Program 6.

Figure I.21   (*continued*)

```
C
C        WHEN THE SUBROUTINE IS CALLED, THE VALUE OF I INDICATES WHETHER TO
C        PRINT EACH VALUE OR NOT. I=0 MEANS PRINT, .NE. 0 MEANS DON'T.
C
C   ---------------------------------------------------------------------
C
        REAL GFCN,X,XTOL
        INTEGER NLIM,I,J
        REAL SAVEX,DEL1,DEL2
C
C   ---------------------------------------------------------------------
C
C   CHECK INITAL VALUES
C
        J = 1
        SAVEX = X
        X = GFCN(X)
        DEL1 = ABS(X - SAVEX)
        IF ( DEL1 .LE. XTOL ) THEN
          I = 1
          PRINT 202, J,SAVEX,X
          RETURN
        END IF
        IF ( I .EQ. 0 ) THEN
          PRINT 199, J,SAVEX,X
        END IF
C
C   ---------------------------------------------------------------------
C
C   GENERATE SEQUENCE OF POINTS
C
        DO 20 J=2,NLIM
        SAVEX = X
        X = GFCN(X)
        DEL2 = ABS(X - SAVEX)
C
C   ---------------------------------------------------------------------
C
C   CHECK FOR XTOL CONDITION
C
        IF ( DEL2 .LE. XTOL ) THEN
          I = 1
          PRINT 202, J,SAVEX,X
          RETURN
        END IF
C
C   ---------------------------------------------------------------------
C
C   CHECK FOR DIVERGENCE AT START
C
        IF ( J .EQ. 2 ) THEN
          IF ( DEL1 .LE. DEL2 ) THEN
            I = -2
            PRINT 201
            RETURN
          END IF
        END IF
        IF ( I .EQ. 0 ) PRINT 199, J,SAVEX,X
  20 CONTINUE
C
C   ---------------------------------------------------------------------
C
C   WHEN LOOP IS NORMALLY TERMINATED, NLIM IS EXCEEDED.
```

Figure 1.21   (*continued*)

```
C
      I = -1
      PRINT 200, NLIM,SAVEX,X
      RETURN
C
C   --------------------------------------------------------------------
C
  199 FORMAT(' AT ITERATION',I3,4X,' X =',F9.4,4X,'G(X) =',F9.4)
  200 FORMAT(/' METHOD DID NOT CONVERGE IN ',I4,' ITERATIONS. FINAL ',
     +       'VALUES ARE X = ',F12.5,' G(X) = ',F12.5)
  201 FORMAT(/' FIRST THREE VALUES INDICATE DIVERGENCE.')
  202 FORMAT(/' XTOL MET IN ',I4,' ITERATIONS. X = ',F12.5,
     +       ' G(X) = ',F12.5)
      END

              OUTPUT FOR PROGRAM 6

AT ITERATION  1    X =     .5000    G(X) =     .4192
AT ITERATION  2    X =     .4192    G(X) =     .3712
AT ITERATION  3    X =     .3712    G(X) =     .3623
AT ITERATION  4    X =     .3623    G(X) =     .3607
AT ITERATION  5    X =     .3607    G(X) =     .3605

XTOL MET IN   6 ITERATIONS.  X =   .36047   G(X) =   .36043
```

```
      PROGRAM PMULLR(INPUT,OUTPUT)
C
C  THIS DRIVER CALLS SUBROUTINE MULLR
C
C   --------------------------------------------------------------------
C
      REAL FCN,XR,H,XTOL,FTOL
      INTEGER I,NLIM
      EXTERNAL FCN
      DATA XR,H,XTOL,FTOL,NLIM,I/0.5,0.2,0.0001,0.00001,50,0/
C
      CALL MULLR(FCN,XR,H,XTOL,FTOL,NLIM,I)
C
C  IF I = 1 OR 2, THE ROOT IS EQUAL TO  XR
C
      STOP
      END
C
C   --------------------------------------------------------------------
C
      REAL FUNCTION FCN(X)
C
      REAL X
      FCN = 3*X + SIN(X) - EXP(X)
      RETURN
      END
C
C   --------------------------------------------------------------------
C
      SUBROUTINE MULLR(FCN,XR,H,XTOL,FTOL,NLIM,I)
C
```

Figure 1.22   Program 7.

Figure 1.22   (*continued*)

```
C  -----------------------------------------------------------------------
C
C     SUBROUTINE MULLR :
C                        THIS SUBROUTINE FINDS THE ROOT OF F(X) = 0 BY
C     QUADRATIC INTERPOLATION ON THREE POINTS - MULLER'S METHOD.
C
C  -----------------------------------------------------------------------
C
C     PARAMETERS ARE :
C
C     FCN       -FUNCTION THAT COMPUTES VALUES FOR F(X). MUST BE DECLARED
C                EXTERNAL IN CALLING PROGRAM. IT HAS ONE ARGUMENT, X.
C     XR        -INITIAL APPROXIMATION TO THE ROOT. USED TO BEGIN
C                ITERATIONS. ALSO RETURNS THE VALUE OF THE ROOT.
C     H         -DISPLACEMENT FROM X USED TO BEGIN CALULATIONS. THE FIRST
C                QUADRATIC IS FITTED AT F(X), F(X+H), F(X-H).
C     XTOL,FTOL -TOLERANCE VALUES FOR X, F(X) TO TERMINATE ITERATIONS.
C     I         -A SIGNAL FOR HOW ROUTINE TERMINATED.
C      I = 1      MEETS TOLERANCE FOR X VALUES.
C      I = 2      MEETS TOLERANCE FOR F(X).
C      I = -1     NLIM EXCEEDED.
C
C  WHEN THE SUBROUTINE IS CALLED, THE VALUE OF I INDICATES WHETHER TO
C  PRINT EACH VALUE OR NOT. I=0 MEANS PRINT THEM, I.NE.0 MEANS DON'T.
C
C  -----------------------------------------------------------------------
C
      REAL FCN,XR,H,XTOL,FTOL
      INTEGER NLIM,I,J
      REAL Y(3),F1,F2,F3,H1,H2,G,A,B,C,DISC,FR,DELX
C
C  -----------------------------------------------------------------------
C
C  SET INITIAL VALUES INTO Y ARRAY AND EVALUATE  F  AT THESE POINTS
C
      Y(1) = XR - H
      Y(2) = XR
      Y(3) = XR + H
      F1   = FCN( Y(1) )
      F2   = FCN( Y(2) )
      F3   = FCN( Y(3) )
C
C  -----------------------------------------------------------------------
C
C  BEGIN ITERATIONS
C
      DO 20 J=1,NLIM
      H1 = Y(2) - Y(1)
      H2 = Y(3) - Y(2)
      G = H1 / H2
      A = ( F3*G - F2*(1.0 + G) + F1 ) / ( H1*(H1 + H2))
      B = ( F3 - F2 - A*H2*H2 ) / H2
      C = F2
      DISC = SQRT( B*B - 4.0*A*C )
      IF ( B .LT. 0.0 ) DISC = -DISC
C
C  -----------------------------------------------------------------------
C
C  FIND ROOT OF QUADRATIC :  A * V**2  +  B * V  +  C  =  0
C
      DELX = -2.0 * C / ( B + DISC )
C
```

Figure 1.22   (*continued*)

```
C  UPDATE XR
C
      XR = Y(2) + DELX
      FR = FCN(XR)
      IF ( I .EQ. 0 ) PRINT 199, J,XR,FR
C
C  -------------------------------------------------------------------
C
C
C  CHECK STOPPING CRITERIA
C
      IF ( ABS(DELX) .LE. XTCL ) THEN
        I = 1
        PRINT 202, J,XR,FR
        RETURN
      END IF
C
      IF ( ABS(FR) .LE. FTOL ) THEN
        I = 2
        PRINT 203, J,XR,FR
        RETURN
      END IF
C
C  -------------------------------------------------------------------
C
C  SELECT THE THREE POINTS FOR THE NEXT ITERATION. WHEN DELX .GT. 0, CHOOSE
C  Y(2), Y(3), & XR, BUT WHEN DELX .LT. 0 CHOOSE Y(1), Y(2), & XR.
C
C  ENTER THE PROPER SET INTO Y ARRAY SO THEY ARE IN ASCENDING ORDER.
C
      IF ( DELX .GE. 0 ) THEN
        Y(1) = Y(2)
        F1   = F2
        IF ( DELX .GT. H2 ) THEN
          Y(2) = Y(3)
          F2   = F3
          Y(3) = XR
          F3   = FR
        ELSE
          Y(2) = XR
          F2   = FR
        END IF
      ELSE
        Y(3) = Y(2)
        F3   = F2
        IF ( ABS(DELX) .GT. H1 ) THEN
          Y(2) = Y(1)
          F2   = F1
          Y(1) = XR
          F1   = FR
        ELSE
          Y(2) = XR
          F2   = FR
        END IF
C
      END IF
C
  20  CONTINUE
C
C  -------------------------------------------------------------------
C
C  WHEN LOOP IS NORMALLY TERMINATED, NLIM IS EXCEEDED
C
```

Figure I.22 (*continued*)

```
        I = -1
        PRINT 200, NLIM,XR,FR
        RETURN
C
C  -----------------------------------------------------------------
C
 199  FORMAT(' AT ITERATION',I3,3X,' X =',E12.5,4X,'F(X) = ',E12.5)
 200  FORMAT(/' TOLERANCE NOT MET AFTER ',I4,' ITERATIONS  X = ',
      +         E12.5,' F(X) = ',E12.5)
 201  FORMAT(/' FUNCTION HAS SAME SIGN AT X1 & X2 ')
 202  FORMAT(/' TOLERANCE MET IN ',I2,' ITERATIONS   X = ',E10.5,
      +         '   F(X) = ',E10.5)
 203  FORMAT(/' F TOLERANCE MET IN ',I4,' ITERATIONS   X = ',E10.5,
      +         '   F(X) = ',E10.5)
C
        END



                    OUTPUT FOR PROGRAM 7


AT ITERATION  1    X =  .35995E+00    F(X) =  -.11837E-02
AT ITERATION  2    X =  .36042E+00    F(X) =   .15922E-05

F TOLERANCE MET IN  2 ITERATIONS   X = .36042E+00   F(X) = .15922E-05
```

EXERCISES

It is impossible to learn numerical analysis without lots of practice. We provide you with good opportunities to do this because this book is richer than most textbooks in examples for you to work on. There are two types of these. We call a drill problem (one that makes you repeat the kind of calculations that are used in our examples) an "exercise." These test your ability to solve simple problems without offering much in the way of difficulty. Some of these have answers given at the back of the book (these exercises are marked).

However, the real world is full of problems that are not so simple. We provide a second kind of problem, in a section labeled "Applied Problems and Projects," that should be more challenging. While we make no claim that truly realistic problems will be found there, many are taken from various fields of science and engineering. You should find these more exciting, especially because some have no "answer" in the normal sense. These are provided to make you think. Some are extensions of things that we only touch on in the text.

Some remarks on computation techniques are in order. We think you will get a better feel for the various algorithms if you do some or all of the exercises by hand, using a calculator and writing down the intermediate results. We also think you should write some programs of your own to solve the exercises; doing so will assure you that you really understand the algorithm. A third method is to use a prewritten program or subroutine. In addition to those given (and copies of these are available to your instructor on disks in both FORTRAN and Pascal), you should gain some experience in using routines from one or more of the standard libraries such as IMSL. Which of these methods you use probably depends on your interests and what your instructor feels is most important. Hopefully you will do more than use this book as a cookbook of numerical recipes.

EXERCISES*

Section 1.2

▶1. The equation $e^x - 3x$ has a root at $r = 0.61906129$. Beginning with the interval $[0, 1]$, use six iterations of the method of halving the interval to find this root. How many iterations would you need to evaluate the root correct to four significant places—that is, $|x - r| < 0.00005$? How many for eight places?

2. The quadratic $(x - 0.4)(x - 0.6) = x^2 - x + 0.24$ has zeros at $x = 0.4$ and $x = 0.6$, of course. Observe that the endpoints of the interval $[0, 1]$ are not satisfactory to begin the interval-halving method. Graph the function, and from this deduce the boundaries of intervals that will converge to each of the zeros. If the endpoints of the interval $[0.5, 1.0]$ are used to begin the search, what is a bound to the error after five iterations? What is the actual error after five repetitions of interval halving?

3. Interval halving applies to any continuous function, not just to polynomials. Find where the graphs of $y = x - 2$ and $y = \ln x$ intersect by finding the root of $\ln x - x + 2 = 0$ correct to four decimals.

4. Use interval halving to find the smallest positive root of these equations. In each case first determine a suitable interval, then compute the root with relative accuracy of 0.5%.

   a) $e^x - x - 2 = 0$ ▶b) $x^3 - x^2 - 2x + 1 = 0$

   ▶c) $2e^{-x} - \sin x = 0$ d) $3x^3 + 4x^2 - 8x - 1 = 0$

Section 1.3

5. The polynomial $x^3 + x^2 - 3x - 3 = 0$, used as an example in Sections 1.2 and 1.3 where the root at $x = \sqrt{3}$ was approximated, has its other roots at $x = -1$ and $x = -\sqrt{3}$. Beginning with two suitable values that bracket the value $-\sqrt{3}$, show that the method of linear interpolation converges to that root.

6. In Exercise 5, if one tried as starting values $x = -1.5$ and $x = -1.7$, the function would not change sign, and, hence, they do not qualify for beginning the method of linear interpolation. However, the secant method can begin with these values. Use them to begin the secant method. How many iterations are needed to estimate the root correct to four decimals? Suppose the starting values are -1.5 and -1.1; which root is obtained by the secant method? What root if we begin with -1.5 and -1.25?

▶7. Find where the cubic $y = x^3 - x + 1$ intersects the parabola $y = 2x^2$. Make a sketch of the two curves to locate the intersections, and then use linear interpolation and/or the secant method to evaluate the x-values of the points of intersection.

8. a) Use the method of linear interpolation to solve the equations in Exercise 4.
   b) Use modified linear interpolation to solve these equations and compare the rates of convergence with those obtained in part (a).

9. Write the algorithm for the secant method, following the model of the other algorithms of this chapter.

Section 1.4

▶10. Find a root near $x = -0.5$ of $e^x - 3x^2 = 0$ by Newton's method, to six-digit accuracy.

---

*Answers are given at the end of the text for exercises marked by ▶.

11. The equation $e^x - 3x^2 = 0$ has a root not only near $x = -0.5$, but also near $x = 4.0$. Find the positive root by Newton's method.

12. Use Newton's method to solve the equations in Exercise 4. How many iterations are required to attain the specified accuracy?

13. a) Use Newton's method on the equation $x^2 = N$ to derive the algorithm for the square root of N:

$$x_{i+1} = \frac{1}{2}\left(x_i + \frac{N}{x_i}\right),$$

where x_0 is an initial approximation to $\sqrt{N}$.
b) Derive similar formulas for the third and fourth roots of N.

14. a) If the algorithm of Exercise 13 is applied twice, show that

$$\sqrt{N} \doteq \frac{A + B}{4} + \frac{N}{A + B}, \qquad \text{where } N = AB.$$

b) Show also that the relative error (error/true value) in (a) is approximately

$$\frac{1}{8}\left(\frac{A - B}{A + B}\right)^4.$$

15. Expand $f(x)$ about the point $x = a$ in a Taylor series. (See Appendix A if you have forgotten this.) Using appropriate terms from this, derive the formula for Newton's method.

16. $(x - 1)^3(x - 2) = x^4 - 5x^3 + 9x^2 - 7x + 2 = 0$ obviously has a root at $x = 2$, and a triple root at $x = 1$. Beginning with $x = 2.1$, use Newton's method once, and observe the degree of improvement. Then start with $x = 0.9$, and note the much slower convergence to the triple root even though the initial error is only 0.1 in each case. Use the secant method beginning with $f(0.9)$ and $f(1.1)$, and observe that just one application brings one quite close to the root in contrast to Newton's method. Explain.

Section 1.5

17. Use Muller's method to solve the problems below. Apply four iterations and determine how the successive errors are related to each other. Use starting values that differ by 0.2 from each other.

a) $2x^3 + 4x^2 - 2x - 5 = 0$, root near 1.0 (exact value is 1.07816259);
b) $e^x - 3x^2 = 0$, root near 4.0 (exact value is 3.73307903);
c) $e^x - 3x^2 = 0$, root near 1.0;
d) $\tan x - x - 1 = 0$, root near 1.1.

18. Muller's method is sometimes used in a "self-starting" form. Instead of specifying starting values that are near a root, the algorithm arbitrarily begins with $x_0 = 0$, $x_1 = 0.5$, $x_2 = -0.5$. Use these starting values on the equations of Exercise 17. One is supposed to obtain the root nearest the origin by this technique. Will this always be true?

19. An extension of the self-starting principle in Exercise 18 is to deflate the function after a first root is found. (*Deflating* means forming a new function that has the same roots as the original function except the root $x = r$, by dividing $f(x)$ by $(x - r)$.) Test this by comparing the graphs of $f(x) = x(x - 3)(x - 1)$, $g(x) = x(x - 3)$, $h(x) = x(x - 1)$, and $j(x) = (x - 1)(x - 3)$. Use this deflating technique to obtain all the roots of the equations in Exercise 17, beginning as in Exercise 18.

20. Since Muller's method subtracts function values that may be very nearly the same, there is a possibility of large relative errors. Investigate this by using starting values that are very

close to each other (but not near a root) in some of the equations in Exercise 17. The difficulty should be exaggerated when $f(x)$ changes very slowly near a root. See if this exaggerated effect is exhibited by a polynomial function that has a triple root.

▶21. Muller's method works on complex roots if complex arithmetic is used. Find a complex root of $x^4 + 4x^3 + 21x^2 + 4x + 20 = 0$. (Newton's method can also find complex roots; you may wish to make the comparison with Newton's method also. Complex arithmetic is laborious by hand; you will probably want to employ computer programs in this exercise.)

Section 1.6

22. $f(x) = e^x - 3x^2 = 0$ has three roots. An obvious rearrangement is

$$x = \pm\sqrt{e^x/3}.$$

Show, beginning with $x_0 = 0$, that this will converge to a root near -0.5 if the negative value is used, and that it converges to a root near 1.0 if the positive value is used. Show, however, that this form does not converge to the third root near 4.0 even when a nearly exact starting value is used. Find another form that will converge to the root near 4.0.

▶23. One root of the quadratic $x^2 + x - 1 = 0 = x(x + 1) - 1$ is at $x = 0.6180$. The equivalent form $x = 1/(x + 1)$ converges to this root beginning at $x_0 = 1$. Carrying four or five decimals, how many steps are required to reach the root (correct to four decimals) by linear iteration? If Aitken acceleration is used after three approximations are available, how many iterations are required?

24. The form $x = 1/(x + 1)$ of Exercise 23 will converge to a root of the quadratic for many starting values in addition to $x_0 = 1$. For what starting values will it not converge to a root? (For this problem do not stop with a division by zero; that is, for $x_0 = -1$, $x_1 = 1/0$. Use $x_3 = \lim_{x_2 \to \infty} 1/(x_2 + 1) = 0$.)

▶25. The cubic $2x^3 + 4x^2 - 2x - 5 = 0$ has a root near $x = 1$. Find at least three rearrangements that will converge to this root beginning with $x_0 = 1.0$.

26. Show that for the points x_0, x_1, x_2, x_3 in Exercises 22 and 23 the value

$$C = \frac{\sum x_i x_{i+1} - \frac{1}{3}\sum x_i \sum x_{i+1}}{\sqrt{\left(\sum x_i^2 - \frac{1}{3}\left(\sum x_i\right)^2\right)\left(\sum x_{i+1}^2 - \frac{1}{3}\left(\sum x_{i+1}\right)^2\right)}}$$

is close to 1 in absolute value where the sums are from $i = 0$ to $i = 2$.

Section 1.7

27. a) Show that if $f(x)$ has a double root at $x = r$, then $f'(r) = 0$.
     b) Show that if $f(x)$ has a root of multiplicity m at $x = r$, then

$$f^{(i)}(r) = 0, \qquad i = 1, 2, \ldots, m - 1.$$

28. If $f(x) = 0$ has a double root at $x = r$, so that $f'(r) = 0$, then the condition

$$\left|\frac{f(x)f''(x)}{[f'(x)]^2}\right| < 1$$

may not hold for any interval including r. According to Section 1.7, we therefore have, for Newton's method, no assurance of convergence to the root at r. The simple equation $(x - 2)^2 = 0 = x^2 - 4x + 4$ has such a double root at $x = 2$, and $f'(x) = 2x - 4$ is zero at $x = 2$. Still, beginning at any finite value of x_0, Newton's method will converge! Reconcile this fact with the convergence criterion of Section 1.7.

▶29. For the quadratic $x^2 - 4x + 4 = 0$, which has a double root at $x = 2$, begin with $x_0 = 1$ and compute successive approximations to the root by Newton's method. Tabulate the errors at each step and compare each with the next. Is Newton's method quadratically or only linearly convergent in this case? How could one accelerate the convergence?

30. If $f(x)$, $f'(x)$, $f''(x)$ are continuous and bounded on a certain interval containing $x = r$, and if both $f(r) = 0$ and $f'(r) = 0$, but $f''(r) \neq 0$, show that the form

$$x_{n+1} = x_n - 2\frac{f(x_n)}{f'(x_n)}$$

will converge quadratically if x_n is in the interval. (*Hint:* The algorithm is of the form $x_{n+1} = g(x_n)$. Show that $g'(r) = 0$, using L'Hospital's rule.)

31. The method suggested by Exercise 30 extends to a root of multiplicity m:

$$x_{n+1} = x_n - m\frac{f(x_n)}{f'(x_n)}.$$

Show, with suitable restrictions on $f(x)$, that this is quadratically convergent.

Section 1.8

▶32. Two of the four roots of the quartic $x^4 + 2x^3 - 7x^2 + 3 = 0$ are positive. Find these by Newton's method correct to seven decimal places using a calculator, and then determine the other roots from the reduced equation. Use synthetic division.

33. a) If $p(x)$ has a root of multiplicity m at $x = r$, show that the function $q(x) = p(x)/p'(x)$ has the same roots as $p(x)$ but just a simple root at $x = r$.
b) Show that in fact all the roots of $q(x)$ are simple.

▶34. Let $P(x) = x^4 + 4.6x^3 + 6.6x^2 - 11x - 14$. Use synthetic division to evaluate $P(-1)$. Write $P(x)$ as a product of two polynomials, one of degree 3. Find the one positive real root of this cubic polynomial and then use the quadratic formula to find the last two roots. This process is called *deflation*.

35. Let $P(x) = a_1x^n + a_2x^{n-1} + \cdots + a_nx + a_{n+1}$. Write a computer program (any language) that evaluates $P(x)$ and $P'(x)$ by synthetic division.

Section 1.9

▶36. Beginning with the trial factor $x^2 - 4x + 5$, improve by successive applications of Bairstow's method to find the quadratic factors of

$$x^4 - 3.1x^3 + 2.1x^2 + 1.1x + 5.2.$$

What are the four zeros of this polynomial?

37. Solve Exercise 32 by Bairstow's method to get quadratic factors.

38. None of the roots of this fourth-degree polynomial is real. Find these complex roots by resolving into quadratic factors:

$$x^4 + 4x^3 + 21x^2 + 4x + 20 = 0.$$

39. When the modulus of one pair of complex roots is the same as for another pair of complex roots, the Bairstow method is slow to converge. Try your patience on

$$x^4 - x^3 + x^2 - x + 1 = 0,$$

which has as factors $(x^2 + 0.618034x + 1)(x^2 - 1.618034x + 1)$. Start with $x^2 + 0.6x + 1$. What is the modulus of the roots?

Section 1.10

40. Use the QD algorithm to approximate the roots of

    ▶a) $x^3 - x^2 - 2x + 1 = 0$ b) $3x^3 + 3x^2 - 8x - 1 = 0$
    c) $x^4 - 5x^3 + 9x^2 - 7x + 2 = 0$ ▶d) $x^4 - 3.1x^3 + 2.1x^2 + 1.1x + 5.2 = 0$

41. The QD algorithm is also not very efficient on Exercise 39. Solve that problem by QD.

Section 1.11

42. In Section 1.11, an example of a truncated Taylor series for e^x is given. Use that cubic equation to approximate e^x for values of x from 0 to 0.5, using at least eight-decimal precision. Make a graph of the errors versus x. Also graph the upper bounds on the errors versus x. (See Appendix A if you need to refresh your memory of the error term for a Taylor series.)

43. Repeat Exercise 42 but now truncate at each step in the computations after five decimals. Why are the error curves different?

44. In Exercise 34 you solved for the zeros of a fourth-degree polynomial. Suppose the coefficients as given there are inexact. What influence does this have on the values of the roots? Answer this question by determining the effect on each of the roots due to a 1% change in each of the five coefficients; study the effect of the change in each coefficient separately from the changes in the others. Which coefficient makes the greatest difference in the roots when varied by 1%?

45. Exercise 32 has a quartic polynomial for which you found all four roots. Repeat that exercise, but now terminate your computations of the first two roots after attaining three decimal places of accuracy. This means that the reduced equations are different. What is the effect on the succeeding roots?

Section 1.12

46. a) For a simplified normalized floating-point system, $B = 10$, $p = 3$, $-9 \le e \le +9$. How many distinct numbers are there?
    b) How many if the numbers are not normalized?
    c) What if $p = 4$?
    d) What if $-3 \le e \le +3$?

In Exercises 47 through 50, use the floating-point system of Exercise 46(a).

47. Find the absolute error of each result, both with "chopping" and with rounding to three digits.

    a) $3.26 \times 10^{-3} + 2.07 \times 10^{-4}$
    ▶b) $1.96 \times 10^5 - 1.94 \times 10^4$
    c) $(3.26 \times 10^{-3} + 2.07 \times 10^{-4}) - 2.01 \times 10^{-4}$
    d) $3.26 \times 10^{-3} + (2.07 \times 10^{-4} - 2.01 \times 10^{-4})$

48. Find the relative errors, both with "chopping" and with rounding to three digits.

    a) $3.28 \times 10^{-2} * 6.98 \times 10^3$
    b) $3.28 \times 10^{-8} * 6.98 \times 10^{-7}$
    c) $(3.28 \times 10^{-2} * 6.98 \times 10^2) \div 4.82 \times 10^{-8}$
    d) $3.28 \times 10^{-2} * (6.98 \times 10^3 \div 4.82 \times 10^{-8})$

49. Find the absolute and relative errors with "chopping" to three digits.

    a) $4.82 \times 10^2 \div 8.81 \times 10^8$
    b) $1.06 \times 10^{-9} \div 4.06 \times 10^2$
    ▶c) $4.82 \times 10^2 \div (3.81 \times 10^8 * 4.06 \times 10^{-2})$
    d) $(4.82 \times 10^2 \div 8.81 \times 10^8) \div 4.06 \times 10^{-2}$

50. Find examples where:

    a) $(a + b) + c \neq a + (b + c)$
    ▶b) $(a * b) * c \neq a * (b * c)$
    c) $a * (b + c) \neq a * b + a * c$

51. Evaluate the polynomial $2.75x^3 - 2.95x^2 + 3.16x - 4.67$ for $x = 1.07$, using both chopping after three digits and rounding to three digits. What are the absolute and relative errors? Does nested multiplication differ from evaluation in "standard form"? What is the error of $2.75x^3 - (2.95x^2 - 3.16x) - 4.67$?

52. Determine how floating-point numbers are stored in the computer systems available to you. How many decimal digits of accuracy for single-precision numbers? for double-precision? What are the execution times for each of the arithmetic operations for single-precision? for double-precision?

▶53. Write a computer program to determine the effect of the sums in (a), (b), and (c):

    a) 0.01 added 100 times;
    b) 0.001 added 1000 times;
    c) 0.0001 added 10,000 times.

    Print out values for sums that should equal 0.1, 0.2, . . . , 1.0.

    d) Evaluate the divergent infinite series, $\Sigma\ 1 + \frac{1}{2} + \frac{1}{3} + \frac{1}{4} + \ldots$, with a computer program. Why is it not divergent in a computer?

    e) Compare the sum in part (d) when evaluated from right to left with the sum evaluated from left to right. Explain the differences.

APPLIED PROBLEMS AND PROJECTS

54. Given

$$x'' + x + 2y' + y = f(t),$$
$$x'' - x + y = g(t), \qquad x(0) = x'(0) = y(0) = 0.$$

In solving this pair of simultaneous second-order differential equations by the Laplace transform method, it becomes necessary to factor the expression

$$(S^2 + 1)(S) - (2S + 1)(S^2 - 1) = -S^3 - S^2 + 3S + 1,$$

so that partial fractions can be used in getting the inverse transform. What are the factors?

55. DeSantis (1976) has derived a relationship for the compressibility factor of real gases of the form

$$z = \frac{1 + y + y^2 - y^3}{(1 - y)^3},$$

where $y = b/(4v)$, b being the van der Waals correction and v the molar volume. If $z = 0.892$, what is the value of y?

56. In studies of solar-energy collection by focusing a field of plane mirrors on a central collector, Vant-Hull (1976) derives an equation for the geometrical concentration factor C:

$$C = \frac{\pi(h/\cos A)^2 F}{0.5\pi D^2 (1 + \sin A - 0.5 \cos A)}$$

where A is the rim angle of the field, F is the fractional coverage of the field with mirrors, D is the diameter of the collector, and h is the height of the collector. Find A if $h = 300$, $C = 1200$, $F = 0.8$, and $D = 14$.

57. Lee and Duffy (1976) relate the friction factor for flow of a suspension of fibrous particles to the Reynolds number by this empirical equation:

$$\frac{1}{\sqrt{f}} = \left(\frac{1}{k}\right)\ln(RE\sqrt{f}) + \left(14 - \frac{5.6}{k}\right).$$

In their relation, f is the friction factor, RE is the Reynolds number, and k is a constant determined by the concentration of the suspension. For a suspension with 0.08% concentration, $k = 0.28$. What is the value of f if RE $= 3750$?

58. The Redlich–Kwong equation is

$$P = \frac{RT}{v - b} - \frac{A(T)}{v(v + b)}.$$

Measurements of $P = 87.3$, $T = 486.9$, and $v = 12.005$ have been made. It is known that $A(T) = 0.0837$ under these conditions. $R = 1.98$, a constant. Find the value of b that satisfies the equation.

59. Based on the work of Frank–Kamenetski in 1955, temperatures in the interior of a material with embedded heat sources can be determined if we solve this equation:

$$e^{-(1/2)t}\cosh^{-1}(e^{(1/2)t}) = \sqrt{\tfrac{1}{2}L_{cr}}.$$

Given that $L_{er} = 0.088$, find t.

60. Suppose we have the 555 Timer Circuit

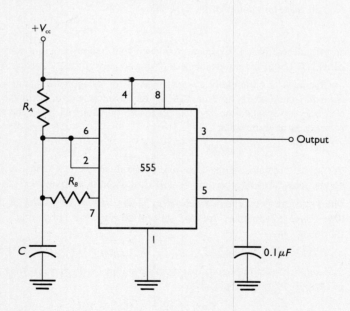

whose output waveform is

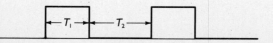

where

$$T_1 + T_2 = \frac{1}{f}$$

$$f = \text{frequency}$$

$$\text{Duty cycle} = \frac{T_1}{T_1 + T_2} \times 100\%.$$

It can be shown that

$$T_1 = R_A C \ln(2)$$

$$T_2 = \frac{R_A R_B C}{R_A + R_B} * \ln\left(\left|\frac{R_A - 2R_B}{2R_A - R_B}\right|\right).$$

Given that $R_A = 8670$, $C = 0.01 \times 10^{-6}$, $T_2 = 1.4 \times 10^{-4}$, find

a) T_1, f, and the duty cycle,
b) R_B using SUBROUTINE PLNINT.
c) Select an f and duty cycle, find T_1 and T_2.

61. The solution of boundary-value problems by an analytical (Fourier series) method often involves finding the roots of transcendental equations to evaluate the coefficients. For example,

$$y'' + \lambda y = 0, \qquad y(0) = 0, \qquad y(1) = y'(1),$$

involves solving $\tan z = z$. Find three values of z other than $z = 0$.

▶62. Find all max/min points of the function

$$f(x) = (\sin(x))^6 * e^{20x} * \tan(1 - x)$$

on the interval $[0, 1]$. Compare your own root-finding program with the IMSL subroutine ZBRENT. (Note the disadvantage in trying to solve $f'(x) = 0$ using Newton's method.)

63. In Chapter 4, a particularly efficient method for numerical integration of a function, called *Gaussian quadrature*, is discussed. In the development of formulas for this method it is necessary to evaluate the zeros of Legendre polynomials. Find the zeros of the Legendre polynomial of sixth order:

$$P_6(x) = \tfrac{1}{48}(693x^6 - 945x^4 + 315x^2 - 15).$$

(*Note:* All the zeros of the Legendre polynomials are less than one in magnitude and, for polynomials of even order, are symmetrical about the origin.)

64. The Legendre polynomials of Problem 63 are one set of a class of polynomials known as *orthogonal* polynomials. Another set are the *Laguerre* polynomials. Find the zeros of the following:

a) $L_3(x) = x^3 - 9x^2 + 18x - 6$ b) $L_4(x) = x^4 - 16x^3 + 72x^2 - 96x + 24$

65. Still another set of orthogonal polynomials are the *Chebyshev* polynomials. (We will use these in Chapter 10.) Find the roots of

$$T_6(x) = 32x^6 - 48x^4 + 18x^2 - 1 = 0.$$

(Note the symmetry of this function. All the roots of Chebyshev polynomials are also less than one in magnitude.)

66. A sphere of density d and radius r weighs $\frac{4}{3}\pi r^3 d$. The volume of a spherical segment is $\frac{1}{3}\pi(3rh^2 - h^3)$. Find the depth to which a sphere of density 0.6 sinks in water as a fraction of its radius. (See Fig. 1.23.)

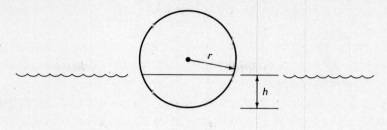

Figure 1.23

67. Write a subroutine that performs the QD algorithm of Section 1.10. Include an algorithm in the style of this chapter and a flowchart. Your routine should be able to handle both real and complex roots. Test it with several well-chosen examples.

68. a) Rewrite the subroutine NEWTN to handle complex roots and test it.
    b) Do the same for subroutine MULLR.

69. Rewrite program BRSTOW so it will handle polynomials with complex coefficients. Test it with

$$x^5 - (1.5 - 2.2i)x = 0.$$

70. Write a program that uses linear interpolation to get a first approximation to a root of $f(x) = 0$ and then calls subroutine NEWTN to refine it.

2

Solving Sets of Equations

2.0 CONTENTS OF THIS CHAPTER

Chapter 2 covers the most important topic of how to solve large systems of linear equations with efficiency and accuracy. A linear system is perhaps the most widely applied mathematical technique when real-world situations are simulated.

2.1 VOLTAGES AND CURRENTS IN A NETWORK
Is a typical application of simulating a real-world problem by a system of equations

2.2 MATRIX NOTATION
Describes how matrices can represent a system of equations in a compact form that facilitates their manipulation

2.3 ELIMINATION METHOD
Is a review of the classical procedure for solving equations simultaneously

2.4 GAUSS AND GAUSS–JORDAN METHODS
Are adaptations of the familiar elimination method that give improved accuracy and efficiency, and lead to using a computer to do the computations

2.5 OTHER *LU* METHODS
Shows that the computations used to reduce a system of equations can be saved within the coefficient matrix, allowing one to resolve the system of equations with new right-hand-side values with a minimum of work

2.6 PATHOLOGY IN LINEAR SYSTEMS—SINGULAR MATRICES
Points out that there are times when an accurate solution to a system of equations is extremely difficult to obtain and gives techniques to minimize the problem; also introduces you to some important properties of matrices

2.1   VOLTAGES AND CURRENTS IN A NETWORK

Electrical engineers often must find the currents flowing and voltages existing in a complex resistor network. Here is a typical problem.

Seven resistors are connected as shown, and voltage is applied to the circuit at points 1 and 6 (see Figure 2.1). You may recognize the network as a variation on a Wheatstone bridge.

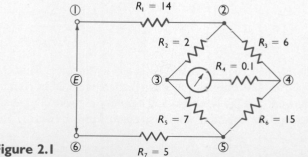

Figure 2.1

Suppose we want to know the current that would flow between points 3 and 4. One method would be to construct the network, apply the voltage, and measure the current that flows with an ammeter. A preferred technique is to compute the current by applying the laws of physics.

While we are especially interested only in finding the current that flows through the ammeter, the computational method can give the voltages at each numbered point (these are called *nodes*) and the current through each of the branches of the circuit. Two laws are involved:

Kirchhoff's law: the sum of all currents flowing into a node is zero.

Ohm's law: the current through a resistor equals the voltage across it divided by its resistance.

We can set up eleven equations using these laws and from these solve for eleven unknown quantities (the four voltages and seven currents):

Currents flowing into the four nodes:

$$i_{12} - i_{23} - i_{24} = 0 \text{ (node 2)}$$
$$i_{23} - i_{34} - i_{35} = 0 \text{ (node 3)}$$
$$i_{24} - i_{34} - i_{45} = 0 \text{ (node 4)}$$
$$i_{35} - i_{45} - i_{56} = 0 \text{ (node 5)}$$

Currents through the resistors:

$$i_{12} = \frac{5 - V_2}{14}$$

$$i_{23} = \frac{V_2 - V_3}{2}$$

$$i_{24} = \frac{V_2 - V_4}{6}$$

$$i_{34} = \frac{V_3 - V_4}{0.1}$$

$$i_{35} = \frac{V_3 - V_5}{7}$$

$$i_{45} = \frac{V_4 - V_5}{15}$$

$$i_{56} = \frac{V_5 - 0}{5}$$

These equations provide 11 linear equations in 11 unknown variables. This chapter describes methods for solving for these variables in ways that are adapted to computers. Without a computer to do the work, the job is tedious and it is easy to make computational errors. We will usually rearrange the equations into a standard form, however. Here is how they should be rearranged.

$$
\begin{aligned}
i_{12} - i_{23} - i_{24} && = 0 \\
i_{23} \quad - i_{34} - i_{35} && = 0 \\
i_{24} + i_{34} \quad - i_{45} && = 0 \\
i_{35} + i_{45} - i_{56} && = 0 \\
14i_{12} \qquad\qquad\qquad + V_2 && = 5 \\
2i_{23} \qquad\qquad - V_2 + V_3 && = 0 \\
6i_{24} \qquad\qquad - V_2 \quad + V_4 && = 0 \\
0.1i_{34} \qquad\qquad - V_3 + V_4 && = 0 \\
7i_{35} \qquad\qquad - V_3 \quad + V_5 && = 0 \\
15i_{45} \qquad\qquad - V_4 + V_5 && = 0 \\
5i_{56} \qquad\qquad - V_5 && = 0
\end{aligned}
$$

2.2 MATRIX NOTATION

Our discussion of methods to solve sets of linear equations will be facilitated by some of the concepts and notation of matrix algebra. Only the more elementary ideas will be needed.

A *matrix* is a rectangular array of numbers in which not only the value of the number is important but also its position in the array. The size of the matrix is described by the number of its rows and columns. A matrix of n rows and m columns is said to be $n \times m$. The elements of the matrix are generally enclosed in brackets, and double-subscripting is the common way of indexing the elements. The first subscript always denotes the row and the second denotes the column in which the element occurs. Capital letters are used to refer to matrices. For example,

$$
A = \begin{bmatrix}
a_{11} & a_{12} & \ldots & a_{1m} \\
a_{21} & a_{22} & \ldots & a_{2m} \\
\vdots & & & \\
a_{n1} & a_{n2} & \ldots & a_{nm}
\end{bmatrix}
= [a_{ij}], \quad i = 1, 2, \ldots, n, \quad j = 1, 2, \ldots, m.
$$

Enclosing the general element a_{ij} in brackets is another way of representing matrix A, as shown above.

Two matrices of the same size may be added or subtracted. The sum of

$$
A = [a_{ij}] \quad \text{and} \quad B = [b_{ij}]
$$

is the matrix whose elements are the sum of the corresponding elements of A and B,

$$C = A + B = [a_{ij} + b_{ij}] = [c_{ij}].$$

Similarly we get the *difference* of two equal-sized matrices by subtracting corresponding elements. If two matrices are not equal in size, they cannot be added or subtracted. Two matrices are equal if and only if each element of one is the same as the corresponding element of the other. Obviously, equal matrices must be of the same size. Some examples will help make this clear.

If

$$A = \begin{bmatrix} 3 & 2 & -1 \\ -4 & 0 & 2 \end{bmatrix} \quad \text{and} \quad B = \begin{bmatrix} 0 & 3 & 2 \\ 4 & -2 & 1 \end{bmatrix},$$

we say that A is 2×3 because it has two rows and three columns. B is also 2×3. Their sum C is also 2×3:

$$C = A + B = \begin{bmatrix} 3 & 5 & 1 \\ 0 & -2 & 3 \end{bmatrix}.$$

The difference D of A and B is

$$D = A - B = \begin{bmatrix} 3 & -1 & -3 \\ -8 & 2 & 1 \end{bmatrix}.$$

Multiplication of two matrices is defined as follows, when A is $n \times m$ and B is $m \times r$:

$$[a_{ij}][b_{ij}] = [c_{ij}] = \begin{bmatrix} (a_{11}b_{11} + a_{12}b_{21} + \cdots + a_{1m}b_{m1}) \ldots (a_{11}b_{1r} + \cdots + a_{1m}b_{mr}) \\ (a_{21}b_{11} + a_{22}b_{21} + \cdots + a_{2m}b_{m1}) \ldots (a_{21}b_{1r} + \cdots + a_{2m}b_{mr}) \\ \vdots \\ (a_{n1}b_{11} + a_{n2}b_{21} + \cdots + a_{nm}b_{m1}) \ldots (a_{n1}b_{1r} + \cdots + a_{nm}b_{mr}) \end{bmatrix},$$

$$c_{ij} = \sum_{k=1}^{m} a_{ik}b_{kj}, \quad i = 1, 2, \ldots, n, \quad j = 1, 2, \ldots, r.$$

It is simplest to select the proper elements if one counts across the rows of A with the left hand while counting down the columns of B with the right. Unless the number of columns of A equals the number of rows of B (so the counting comes out even), the matrices cannot be multiplied. Hence if A is $n \times m$, B must have m rows or else they are said to be "nonconformable for multiplication" and their product is undefined. In general $AB \neq BA$, so the order of factors must be preserved in matrix multiplication.

If a matrix is multiplied by a scalar (a pure number), the product is a matrix, each element of which is the scalar times the original element. We can write

$$\text{If } kA = C, \qquad c_{ij} = ka_{ij}.$$

A matrix with only one column, $n \times 1$ in size, is termed a *column vector*, and one of only one row, $1 \times m$ in size, is called a *row vector*. When the unqualified term *vector* is used, it nearly always means a *column* vector. Frequently the elements of vectors are only singly subscripted.

Some examples of matrix multiplication are

$$\text{Suppose } A = \begin{bmatrix} 2 & 4 & 0 \\ -1 & 2 & 3 \end{bmatrix}, \qquad B = \begin{bmatrix} -1 & 2 \\ 0 & 1 \\ 4 & -1 \end{bmatrix}, \qquad x = \begin{bmatrix} 4 \\ -2 \\ 1 \end{bmatrix}, \qquad y = \begin{bmatrix} y_1 \\ y_2 \\ y_3 \end{bmatrix}.$$

$$A * B = \begin{bmatrix} -2 & 8 \\ 13 & -3 \end{bmatrix}; \qquad B * A = \begin{bmatrix} -4 & 0 & 6 \\ -1 & 2 & 3 \\ 9 & 14 & -3 \end{bmatrix};$$

$$Ax = \begin{bmatrix} 0 \\ -5 \end{bmatrix}; \qquad Ay = \begin{bmatrix} 2y_1 + 4y_2 \\ -y_1 + 2y_2 + 3y_3 \end{bmatrix}.$$

Since A is 2×3 and B is 3×2, they are conformable for multiplication and their product is 2×2. When we form the product of $B * A$, it is 3×3. Observe that not only is $AB \neq BA$; AB and BA are not even the same size. The product of A and the vector x (a 3×1 matrix) is another vector, one with two components. Similarly, Ay has two components. We cannot multiply B times x or B times y; they are nonconformable.

The product of the scalar number 2 and A is

$$2A = \begin{bmatrix} 4 & 8 & 0 \\ -2 & 4 & 6 \end{bmatrix}.$$

Since a vector is just a special case of a matrix, a column vector can be multiplied by a matrix, so long as they are conformable in that the number of columns of the matrix equals the number of elements (rows) in the vector. The product in this case will be another column vector. The size of a product of two matrices, the first $m \times n$ and the second $n \times r$, is $m \times r$. An $m \times n$ matrix times an $n \times 1$ vector gives an $m \times 1$ product.

The general relation for $Ax = b$ is

$$b_i = \sum_{k=1}^{\text{No. of cols.}} a_{ik}x_k, \qquad i = 1, 2, \ldots, \text{No. of rows.}$$

Two vectors, each with the same number of components, may be added or subtracted. Two vectors are equal if each component of one equals the corresponding component of the other.

This definition of matrix multiplication permits us to write the set of linear equations

$$a_{11}x_1 + a_{12}x_2 + \cdots + a_{1n}x_n = b_1,$$
$$a_{21}x_1 + a_{22}x_2 + \cdots + a_{2n}x_n = b_2,$$
$$\vdots$$
$$a_{n1}x_1 + a_{n2}x_2 + \cdots + a_{nn}x_n = b_n,$$

much more simply in matrix notation, as $\boxed{Ax = b,}$ where

$$A = \begin{bmatrix} a_{11} & a_{12} \cdots a_{1n} \\ a_{21} & a_{22} \cdots a_{2n} \\ \vdots & \\ a_{n1} & a_{n2} \cdots a_{nn} \end{bmatrix}, \quad x = \begin{bmatrix} x_1 \\ x_2 \\ \vdots \\ x_n \end{bmatrix}, \quad b = \begin{bmatrix} b_1 \\ b_2 \\ \vdots \\ b_n \end{bmatrix}.$$

For example,

$$\begin{bmatrix} 3 & 2 & 4 \\ 1 & -2 & 0 \\ -1 & 3 & 2 \end{bmatrix} * x = \begin{bmatrix} 14 \\ -7 \\ 2 \end{bmatrix}$$

is the same as the set of equations

$$3x_1 + 2x_2 + 4x_3 = 14,$$
$$x_1 - 2x_2 \quad\quad = -7,$$
$$-x_1 + 3x_2 + 2x_3 = 2.$$

A very important special case is the multiplication of two vectors. The first must be a row vector if the second is a column vector, and each must have the same number of components. For example,

$$[1 \quad 3 \quad -2] * \begin{bmatrix} 4 \\ -1 \\ 3 \end{bmatrix} = [-5]$$

gives a "matrix" of one row and one column. The result is a pure number, a scalar. This product is called the *scalar product* of the vectors, also called the *inner product*.

Certain square matrices have special properties. The diagonal elements are the line of elements a_{ii} from upper left to lower right of the matrix. If only the diagonal terms are nonzero, the matrix is called a *diagonal matrix*. When the diagonal elements are each equal to unity while all off-diagonal elements are zero, the matrix is said to be the *identity matrix of order n.* The usual symbol for such a matrix is I_n, and it has properties similar to unity. For example, the order-4 identity matrix is

$$\begin{bmatrix} 1 & 0 & 0 & 0 \\ 0 & 1 & 0 & 0 \\ 0 & 0 & 1 & 0 \\ 0 & 0 & 0 & 1 \end{bmatrix} = I_4.$$

The subscript is omitted when the order is clear from the context.

A vector that has all its elements equal to zero except one element, which has a value of unity, is called a *unit vector*. There are three distinct unit vectors for order-3 vectors; they are

$$\begin{bmatrix} 1 \\ 0 \\ 0 \end{bmatrix}, \qquad \begin{bmatrix} 0 \\ 1 \\ 0 \end{bmatrix}, \qquad \text{and} \qquad \begin{bmatrix} 0 \\ 0 \\ 1 \end{bmatrix}.$$

If all the elements above the diagonal are zero, a matrix is called *lower-triangular*; it is called *upper-triangular* when all the elements below the diagonal are zero. For example, these order-3 matrices are lower- and upper-triangular:

$$L = \begin{bmatrix} 1 & 0 & 0 \\ 4 & 6 & 0 \\ -2 & 1 & -4 \end{bmatrix}, \qquad U = \begin{bmatrix} 1 & -3 & 3 \\ 0 & -1 & 0 \\ 0 & 0 & 1 \end{bmatrix}.$$

Tridiagonal matrices are those that have nonzero elements only on the diagonal and in the positions adjacent to the diagonal; they will be of special importance in certain partial-differential equations. An example of a tridiagonal matrix is

$$\begin{bmatrix} -4 & 2 & 0 & 0 & 0 \\ 1 & -4 & 1 & 0 & 0 \\ 0 & 1 & -4 & 1 & 0 \\ 0 & 0 & 1 & -4 & 1 \\ 0 & 0 & 0 & 2 & -4 \end{bmatrix}.$$

The *transpose* of a matrix is the matrix that results when the rows are written as columns (or, alternatively, when the columns are written as rows). The symbol A^T is used for the *transpose of A*.

E X A M P L E

$$A = \begin{bmatrix} 3 & -1 & 4 \\ 0 & 2 & -3 \\ 1 & 1 & 2 \end{bmatrix}; \qquad A^T = \begin{bmatrix} 3 & 0 & 1 \\ -1 & 2 & 1 \\ 4 & -3 & 2 \end{bmatrix}.$$

It should be obvious that the transpose of A^T is just A itself.

When a matrix is square, a quantity called its *trace* is defined. The trace of a square matrix is the sum of the elements on its main diagonal. For example, the traces of the above matrices are

$$\text{tr}(A) = 3 + 2 + 2 = 7; \qquad \text{tr}(A^T) = 3 + 2 + 2 = 7.$$

It should be obvious that the trace remains the same if a square matrix is transposed. ■

We present here some additional examples of arithmetic operations with matrices.

$$(3) * \begin{bmatrix} 1 & 2 \\ 3 & 4 \end{bmatrix} = \begin{bmatrix} 3 & 6 \\ 9 & 12 \end{bmatrix}.$$

$$\begin{bmatrix} 1 & 3 & 2 \\ -1 & 0 & 4 \end{bmatrix} + \begin{bmatrix} -1 & 0 & 2 \\ 4 & 1 & -3 \end{bmatrix} = \begin{bmatrix} 0 & 3 & 4 \\ 3 & 1 & 1 \end{bmatrix};$$

$$\begin{bmatrix} 2 & 1 \\ 0 & -4 \\ 7 & 2 \end{bmatrix} - \begin{bmatrix} 3 & -2 \\ 4 & 1 \\ 0 & -2 \end{bmatrix} = \begin{bmatrix} -1 & 3 \\ -4 & -5 \\ 7 & 4 \end{bmatrix}.$$

$$\begin{bmatrix} 2 & 0 & -1 \\ 3 & 2 & 6 \end{bmatrix} * \begin{bmatrix} -1 \\ 2 \\ 1 \end{bmatrix} = \begin{bmatrix} -3 \\ 7 \end{bmatrix} \quad \text{but} \quad \begin{bmatrix} 6 & 1 \\ 3 & -2 \end{bmatrix} * \begin{bmatrix} -1 \\ 2 \\ 1 \end{bmatrix} = \text{not defined.}$$

$$\begin{bmatrix} 1 & 3 \\ 2 & -1 \end{bmatrix} * \begin{bmatrix} 0 & 3 \\ -1 & 1 \end{bmatrix} = \begin{bmatrix} -3 & 6 \\ 1 & 5 \end{bmatrix} \quad \text{but} \quad \begin{bmatrix} 0 & 3 \\ -1 & 1 \end{bmatrix} * \begin{bmatrix} 1 & 3 \\ 2 & -1 \end{bmatrix} = \begin{bmatrix} 6 & -3 \\ 1 & -4 \end{bmatrix}.$$

Division of a matrix by another matrix is not defined, but we will discuss the *inverse* of a matrix later in this chapter.

The *determinant* of a square matrix is a number. For a 2×2 matrix, the determinant is computed by subtracting the product of the elements on the minor diagonal (from upper right to lower left) from the product of terms on the major diagonal. For example

$$A = \begin{bmatrix} 3 & -1 \\ 2 & 4 \end{bmatrix}, \quad \det(A) = (3)(4) - (-1)(2) = 14;$$

$\det(A)$ is the usual notation for the determinant of A. Sometimes the determinant is symbolized by writing the elements of the matrix between vertical lines (similar to representing the absolute value of a number).

For a 3×3 matrix, you may have learned a crisscross way of forming products of terms (we call it the "spaghetti rule") that probably should be forgotten, for it applies only to the special case of a 3×3 matrix; it won't work for larger systems. The general rule that applies in all cases is to expand in terms of the minors of some row or column. The minor of any term is the matrix of lower order formed by striking out the row and column in which the term is found. The determinant is found by adding the product of each term in any row or column by the determinant of its minor, with signs alternating $+$ and $-$. We expand each of the determinants of the minor until we reach 2×2 matrices. For example,

$$\text{Given } A = \begin{bmatrix} 3 & 0 & -1 & 2 \\ 4 & 1 & 3 & -2 \\ 0 & 2 & -1 & 3 \\ 1 & 0 & 1 & 4 \end{bmatrix},$$

$$\det(A) = 3\begin{vmatrix} 1 & 3 & -2 \\ 2 & -1 & 3 \\ 0 & 1 & 4 \end{vmatrix} - 0\begin{vmatrix} 4 & 3 & -2 \\ 0 & -1 & 3 \\ 1 & 1 & 4 \end{vmatrix} + (-1)\begin{vmatrix} 4 & 1 & -2 \\ 0 & 2 & 3 \\ 1 & 0 & 4 \end{vmatrix} - 2\begin{vmatrix} 4 & 1 & 3 \\ 0 & 2 & -1 \\ 1 & 0 & 1 \end{vmatrix}$$

$$= 3\left\{(1)\begin{vmatrix} -1 & 3 \\ 1 & 4 \end{vmatrix} - (3)\begin{vmatrix} 2 & 3 \\ 0 & 4 \end{vmatrix} + (-2)\begin{vmatrix} 2 & -1 \\ 0 & 1 \end{vmatrix}\right\}$$

$$+ (-1)\left\{(4)\begin{vmatrix} 2 & 3 \\ 0 & 4 \end{vmatrix} - (1)\begin{vmatrix} 0 & 3 \\ 1 & 4 \end{vmatrix} + (-2)\begin{vmatrix} 0 & 2 \\ 1 & 0 \end{vmatrix}\right\}$$

$$- 2\left\{(4)\begin{vmatrix} 2 & -1 \\ 0 & 1 \end{vmatrix} - (1)\begin{vmatrix} 0 & -1 \\ 1 & 1 \end{vmatrix} + (3)\begin{vmatrix} 0 & 2 \\ 1 & 0 \end{vmatrix}\right\}$$

$$= 3\{(1)(-7) - (3)(8) + (-2)(2)\} + (-1)\{(4)(8) - (1)(-3) + (-2)(-2)\}$$
$$- 2\{(4)(2) - (1)(1) + (3)(-2)\}$$
$$= 3(-7 - 24 - 4) + (-1)(32 + 3 + 4) - 2(8 - 1 - 6)$$
$$= 3(-35) + (-1)(39) - 2(1) = -146.$$

In computing the determinant, the expansion can be about the elements of any row or column. To get the signs, give the first term a plus sign if the sum of its column number and row number is even; give it a minus if the sum is odd, with alternating signs thereafter. (For example, in expanding about the elements of the third row we begin with a plus; the first element a_{31} has $3 + 1 = 4$, an even number.) Judicious selection of rows and columns with many zeros can hasten the process, but this method of calculating determinants is a lot of work if the matrix is of large size. Methods that triangularize a matrix, as described below, are good methods for getting determinants.

2.3 ELIMINATION METHOD

The first method we will study for the solution of a set of equations is just an enlargement of the familiar method of eliminating one unknown between a pair of simultaneous equations. It is generally called *Gaussian elimination* and is the basic pattern of a large number of methods that can be classed as *direct methods*. (This is to distinguish them from indirect, or iterative, methods, which we will discuss later.)

Consider the simple example of three equations

$$3x_1 - x_2 + 2x_3 = 12,$$
$$x_1 + 2x_2 + 3x_3 = 11,$$
$$2x_1 - 2x_2 - x_3 = 2.$$

Multiplying the first equation by -1 and the second by 3 and adding will eliminate x_1. Similarly, multiplying the first by -2 and the third by 3 and adding also eliminates x_1. (We prefer, in hand calculations, to multiply by the negative values and add, to avoid making mistakes when subtracting quantities of unlike sign.) The result is

$$3x_1 - x_2 + 2x_3 = 12,$$
$$7x_2 + 7x_3 = 21,$$
$$-4x_2 - 7x_3 = -18.$$

We eliminate x_2 between the second and third equations by multiplying the second by 4 and the third by 7 and adding. (Of course, just adding them as they stand would eliminate x_3, which is equally satisfactory, but we wish to keep our method systematic to lead up to an algorithm that can be readily programmed.) After this operation we have the upper-triangular system

$$
\begin{aligned}
3x_1 - x_2 + 2x_3 &= 12, \\
7x_2 + 7x_3 &= 21, \\
-21x_3 &= -42.
\end{aligned}
$$

Obviously $x_3 = 2$ from the third equation, and back-substitution gives $x_2 = 1$ (from the second equation), and $x_1 = 3$ (from the first equation).

We now present the same problem, solved in exactly the same way, in matrix notation:

$$
\begin{bmatrix} 3 & -1 & 2 \\ 1 & 2 & 3 \\ 2 & -2 & -1 \end{bmatrix} \begin{bmatrix} x_1 \\ x_2 \\ x_3 \end{bmatrix} = \begin{bmatrix} 12 \\ 11 \\ 2 \end{bmatrix}.
$$

The arithmetic operations we have performed affect only the coefficients and the constant terms, of course, so we work with the matrix of coefficients "augmented" with the b-vector:

$$
A \vdots b = \begin{bmatrix} 3 & -1 & 2 & \vdots & 12 \\ 1 & 2 & 3 & \vdots & 11 \\ 2 & -2 & -1 & \vdots & 2 \end{bmatrix}.
$$

(The dotted line is usually omitted.)

We perform elementary row transformations* to convert A to upper-triangular:

$$
\begin{bmatrix} 3 & -1 & 2 & 12 \\ 1 & 2 & 3 & 11 \\ 2 & -2 & -1 & 2 \end{bmatrix} \begin{matrix} \\ 3R_2 + (-1)R_1 \to \\ 3R_3 + (-2)R_1 \to \end{matrix} \begin{bmatrix} 3 & -1 & 2 & 12 \\ 0 & 7 & 7 & 21 \\ 0 & -4 & -7 & -18 \end{bmatrix}
$$

$$
7R_3 + 4R_2 \to \begin{bmatrix} 3 & -1 & 2 & 12 \\ 0 & 7 & 7 & 21 \\ 0 & 0 & -21 & -42 \end{bmatrix}. \tag{2.1}
$$

The steps are to add 3 times the second row to -1 times the first row and 3 times the third row to -2 times the first row. The next phase adds 7 times the third row to 4 times the second row.

We are now ready for back-substitution. Note that, except for notation and terminology, there is nothing new here. We depend on our memory to know which numbers in the converted augmented matrix are coefficients and which are the right-hand sides (the constant terms).

---

*Elementary row operations are arithmetic operations that are obviously valid rearrangements of a set of equations: (1) any equation can be multiplied by a constant; (2) the order of the equations can be changed; (3) any equation can be replaced by its sum with another of the equations.

The back-substitution step can be performed quite mechanically by eliminating the coefficients above the diagonal in (2.1). Adding the third row of (2.1) to 3 times the second row, and adding twice the third row to 21 times the first row gives

$$\begin{bmatrix} 63 & -21 & 0 & 168 \\ 0 & 21 & 0 & 21 \\ 0 & 0 & -21 & -42 \end{bmatrix}.$$

We finish the elimination of off-diagonal elements by adding the second row to the first:

$$\begin{bmatrix} 63 & 0 & 0 & 189 \\ 0 & 21 & 0 & 21 \\ 0 & 0 & -21 & -42 \end{bmatrix}.$$

If we divide each row by the diagonal element, we get a form in which the elements of x, the vector whose components are the unknowns x_i, $i = 1, 2, \ldots, n$, are equal to the components of the transformed b-vector:

$$\begin{bmatrix} 1 & 0 & 0 & 3 \\ 0 & 1 & 0 & 1 \\ 0 & 0 & 1 & 2 \end{bmatrix}, \quad x = \begin{bmatrix} 3 \\ 1 \\ 2 \end{bmatrix}, \quad x_1 = 3, \quad x_2 = 1, \quad x_3 = 2.$$

Thinking of this procedure in terms of matrix operations, we transform the augmented coefficient matrix by elementary row operations until the identity matrix is created on the left. The x-vector then stands as the rightmost column.

Note that there exists the possibility that the set of equations has no solution, or that the above procedure will fail to find it. During the triangularization step, if a zero is encountered on the diagonal, we cannot use that row to eliminate coefficients below that zero element. However, in that case, we can continue by interchanging rows and eventually achieve an upper-triangular matrix of coefficients. The real stumbling block is finding a zero on the diagonal after we have triangularized. If that occurs, the back-substitution fails for we cannot divide by zero.

It is worthwhile to explain in more detail what we mean by the *elementary row operations* that we have used above, and to see why they can be used in solving a linear system. There are three of these operations:

1. We may multiply any row of the augmented coefficient matrix by a constant.
2. We can add a multiple of one row to a multiple of any other row.
3. We can interchange the order of any two rows (this was not used above).

The validity of these row operations is intuitively obvious if we think of them applied to a set of linear equations. Certainly, multiplying one equation through by a constant does not change the truth of the equality. Adding equal quantities to both sides of an equality results in an equality, and this is the equivalent of the second transformation. Obviously the order of the set is arbitrary, so rule 3 is valid.

These operations, which do not change the relationships represented by a set of equations, can be applied to an augmented matrix, because this is only a different notation for the equations. (We need to add one proviso to the above. Since round-off error is related to the magnitude of the values when we express them in fixed-word-length computer representations, some of the above operations may have an effect on the accuracy of the computed solution.)

We should also observe that the "back-substitution" phase, when it is done by making the coefficients above the diagonal zero and then reducing the coefficient matrix to the identity matrix, is exactly the same as back-substitution in the more explicit sense. The order of operations is changed but each of the steps is identical.

2.4   GAUSS AND GAUSS–JORDAN METHODS

While the procedure of the previous section is satisfactory for hand calculations on small systems, there are several objections that we should eliminate before we write a computer program to perform Gaussian elimination. In a large set of equations, and that is the situation we must prepare for, the multiplications will give very large and unwieldy numbers that may overflow the computer's registers. We will therefore eliminate the first coefficient in the ith row by subtracting a_{i1}/a_{11} times the first equation from the ith equation. (This is equivalent to making the leading coefficient 1 in the equation that retains that leading term.) We use similar ratios of coefficients in eliminating coefficients in the other columns.

We must also guard against dividing by zero. Observe that zeros may be created in the diagonal positions even if they are not present in the original matrix of coefficients. A useful strategy to avoid (if possible) such zero divisors is to rearrange the equations so as to put the coefficient of largest magnitude on the diagonal at each step. This is called *pivoting*. Complete pivoting may require both row and column interchanges. This is not frequently done. Partial pivoting, which places a coefficient of larger magnitude on the diagonal by row interchanges only, will guarantee a nonzero divisor if there is a solution to the set of equations, and will have the added advantage of giving improved arithmetic precision. The diagonal elements that result are called *pivot elements*. (When there are large differences in magnitude of coefficients in one equation compared to the other equations, we may need to *scale* the values; we consider this later.)

We repeat the example of the previous section, incorporating these ideas and carrying four significant digits in our work. We begin with the augmented matrix.

$$\begin{bmatrix} 3 & -1 & 2 & 12 \\ 1 & 2 & 3 & 11 \\ 2 & -2 & -1 & 2 \end{bmatrix}$$

$$\begin{matrix} \\ \text{Row } 2 - (\tfrac{1}{3})\text{Row } 1 \rightarrow \\ \text{Row } 3 - (\tfrac{2}{3})\text{Row } 1 \rightarrow \end{matrix} \begin{bmatrix} 3 & -1 & 2 & 12 \\ 0 & 2.333 & 2.334 & 7.004 \\ 0 & -1.334 & -2.332 & -5.992 \end{bmatrix}$$

$$\text{Row } 3 - \left(\frac{-1.334}{2.333}\right)\text{Row } 2 \to \begin{bmatrix} 3 & -1 & 2 & 12 \\ 0 & 2.333 & 2.334 & 7.004 \\ 0 & 0 & -1.000 & -1.993 \end{bmatrix}$$

The method we have just illustrated is called *Gaussian elimination*. (In this example, no pivoting was required to make the largest coefficients be on the diagonal.) Back-substitution, beginning with the third equation and then moving "backward" to the second and first equations, gives $x_3 = 1.993$, $x_2 = 1.008$, $x_1 = 3.007$. The differences of these values from 2, 1, and 3 are due to the effects of round-off error. In this example, we have truncated after the fourth digit rather than rounding to the nearest fourth digit, which is the same as the arithmetic performed in many computers. In either case, the errors similarly affect the accuracy of the results. When there are many equations, the effects of round-off (the term is applied to the error due to chopping as well as when rounding is used) may cause large effects. In certain cases, the coefficients are such that the results are particularly sensitive to round-off; such systems are called *ill-conditioned*.

In the example just presented, the zeros below the main diagonal show that we have reduced the problem to solving an upper-triangular system of equations as in Section 2.3. However, at each stage, if we had stored the ratio of coefficients in place of zero, our final form would have been

$$\begin{bmatrix} 3 & -1 & 2 & 12 \\ (0.3333) & 2.333 & 2.334 & 7.004 \\ (0.6667) & (-0.5711) & -1.000 & -1.993 \end{bmatrix}.$$

Then in addition to solving the problem as we have done, one finds that the original matrix

$$A = \begin{bmatrix} 3 & -1 & 2 \\ 1 & 2 & 3 \\ 2 & -2 & -1 \end{bmatrix}$$

can be written as the product

$$\begin{bmatrix} 1 & 0 & 0 \\ 0.3333 & 1 & 0 \\ 0.6667 & -0.5711 & 1 \end{bmatrix} * \begin{bmatrix} 3 & -1 & 2 \\ 0 & 2.333 & 2.334 \\ 0 & 0 & -1.000 \end{bmatrix},$$

to four decimal places. This is called an *LU* decomposition of *A*. In Section 2.5, other methods for finding the *LU* decomposition of a matrix will be presented. If a matrix can be written in the *LU* form such as that presented here, the number of computations needed in solving the system of equations is reduced significantly.

Let us summarize the operations of Gaussian elimination in a form that will facilitate writing a computer program. We use a less formal method of stating this algorithm.

Gaussian Elimination

To solve a system of linear equations,

1. Augment the $n \times n$ coefficient matrix with the vector of right-hand sides to form a $n \times (n + 1)$ matrix.

2. Interchange rows if necessary to make the value of a_{11} the largest magnitude of any coefficient in the first column.

3. Create zeros in the second through nth rows in the first column by subtracting a_{i1}/a_{11} times the first row from the ith row. Store the a_{i1}/a_{11} in a_{i1}, $\quad i = 2, \ldots, n$.

4. Repeat steps 2 and 3 for the second through the $(n - 1)$st rows, putting the largest-magnitude coefficient on the diagonal by interchanging rows (considering only rows j to n), and then subtracting a_{ij}/a_{jj} times the jth row from the ith row so as to create zeros in all positions of the jth column below the diagonal. Store the a_{ij}/a_{jj} in a_{ij}, $i = j + 1, \ldots, n$. At the conclusion of this step, the system is upper-triangular.

5. Solve for x_n from the nth equation by

$$x_n = a_{n,n+1}/a_{nn}.$$

6. Solve for $x_{n-1}, x_{n-2}, \ldots, x_1$ from the $(n - 1)$st through the first equation in turn, by

$$x_i = \frac{a_{i,n+1} - \sum\limits_{j=i+1}^{n} a_{ij}x_j}{a_{ii}}.$$

Some computer programs do not actually interchange all the elements of the rows when pivoting. In these programs, one keeps track of the order in which the rows are to be used in a vector whose elements represent row order. When an interchange is indicated, only the elements of this ordering vector are changed. These numbers are then used to locate the positions of the elements in the matrix of coefficients that are to be operated on, both during the reduction step and during the back-substitution. This can reduce the computer time for large systems, but adds to the complexity of the program.

The algorithm for Gaussian elimination will be clarified by an additional numerical example.

Solve

$$2x_2 \qquad + \ x_4 = \quad 0,$$
$$2x_1 + 2x_2 + 3x_3 + 2x_4 = -2,$$
$$4x_1 - 3x_2 \qquad + \ x_4 = -7,$$
$$6x_1 + \ x_2 - 6x_3 - 5x_4 = \quad 6.$$

The augmented coefficient matrix is

$$\begin{bmatrix} 0 & 2 & 0 & 1 & 0 \\ 2 & 2 & 3 & 2 & -2 \\ 4 & -3 & 0 & 1 & -7 \\ 6 & 1 & -6 & -5 & 6 \end{bmatrix}.$$

We cannot permit a zero in the a_{11} position because that element is the pivot in reducing the first column. We could interchange the first row with any of the other rows to avoid a zero divisor, but interchanging the first and fourth rows is our best choice. This gives

$$\begin{bmatrix} 6 & 1 & -6 & -5 & 6 \\ 2 & 2 & 3 & 2 & -2 \\ 4 & -3 & 0 & 1 & -7 \\ 0 & 2 & 0 & 1 & 0 \end{bmatrix}.$$

We make all the elements in the first column zero by subtracting the appropriate multiple of row one:

$$\begin{bmatrix} 6 & 1 & -6 & -5 & 6 \\ 0 & 1.6667 & 5 & 3.6667 & -4 \\ 0 & -3.6667 & 4 & 4.3333 & -11 \\ 0 & 2 & 0 & 1 & 0 \end{bmatrix}.$$

We again interchange before reducing the second column, not because we have a zero divisor, but because we want to preserve accuracy.* Interchanging the second and third rows puts the element of largest magnitude on the diagonal. (We could also interchange the fourth column with the second, giving an even larger diagonal element, but we do not do this.) After the interchange, we have

$$\begin{bmatrix} 6 & 1 & -6 & -5 & 6 \\ 0 & -3.6667 & 4 & 4.3333 & -11 \\ 0 & 1.6667 & 5 & 3.6667 & -4 \\ 0 & 2 & 0 & 1 & 0 \end{bmatrix}.$$

Now we reduce in the second column

$$\begin{bmatrix} 6 & 1 & -6 & -5 & 6 \\ 0 & -3.6667 & 4 & 4.333 & -11 \\ 0 & 0 & 6.8182 & 5.6364 & -9.0001 \\ 0 & 0 & 2.1818 & 3.3636 & -5.9999 \end{bmatrix}.$$

---

*A numerical example that demonstrates the improved accuracy when partial pivoting is used will be found in Section 2.9.

No interchange is indicated in the third column. Reducing, we get

$$\begin{bmatrix} 6 & 1 & -6 & -5 & 6 \\ 0 & -3.6667 & 4 & 4.3333 & -11 \\ 0 & 0 & 6.8182 & 5.6364 & -9.0001 \\ 0 & 0 & 0 & 1.5600 & -3.1199 \end{bmatrix}.$$

Back-substitution gives

$$x_4 = \frac{-3.1199}{1.5600} = -1.9999,$$

$$x_3 = \frac{-9.0001 - 5.6364(-1.9999)}{6.8182} = 0.33325,$$

$$x_2 = \frac{-11 - 4.3333(-1.9999) - 4(0.33325)}{-3.6667} = 1.0000,$$

$$x_1 = \frac{6 - (-5)(-1.9999) - (-6)(0.33325) - (1)(1.0000)}{6} = -0.50000.$$

The correct answers are -2, $\frac{1}{3}$, 1, and $-\frac{1}{2}$ for x_4, x_3, x_2, and x_1. In this calculation we have carried five significant figures and rounded each calculation. Even so, we do not have five-digit accuracy in the answers. The discrepancy is due to round-off. The question of the accuracy of the computed solution to a set of equations is a most important one, and at several points in the following discussion we will discuss how to minimize the effects of round-off and avoid conditions that can cause round-off errors to be magnified.

In this example, if one had replaced the zeros below the main diagonal with the ratio of coefficients at each step, the resulting augmented matrix would be

$$\begin{bmatrix} 6 & 1 & -6 & -5 & 6 \\ (0.66667) & -3.6667 & 4 & 4.3333 & -11 \\ (0.33333) & (-0.45454) & 6.8182 & 5.6364 & -9.0001 \\ (0.0) & (-0.54545) & (0.32) & 1.5600 & -3.1199 \end{bmatrix}.$$

This gives the LU decomposition as

$$\begin{bmatrix} 1 & 0 & 0 & 0 \\ 0.66667 & 1 & 0 & 0 \\ 0.33333 & -0.45454 & 1 & 0 \\ 0.0 & -0.54545 & 0.32 & 1 \end{bmatrix} * \begin{bmatrix} 6 & 1 & -6 & -5 \\ 0 & -3.6667 & 4 & 4.3333 \\ 0 & 0 & 6.8182 & 5.6364 \\ 0 & 0 & 0 & 1.5600 \end{bmatrix}.$$

You should check that the product of these matrices is indeed the original matrix except that the rows are interchanged owing to pivoting. The next section explores the usefulness of *LU* formulations more fully.

There are many variants to the Gaussian elimination scheme. The back-substitution step can be performed by eliminating the above-diagonal elements, using elementary row operations and proceeding upward from the last row, after the triangularization has been finished. This is similar to an example presented in the previous section. The diagonal elements may all be made ones as a first step before creating zeros in their column; this does the divisions of the back-substitution phase at an earlier time.

One variant that is sometimes used is the *Gauss–Jordan* scheme. In it, the elements above the diagonal are made zero at the *same time* that zeros are created below the diagonal. Usually the diagonal elements are made ones at the same time that the reduction is performed; this transforms the coefficient matrix into the identity matrix. When this has been accomplished, the column of right-hand sides has been transformed into the solution vector. Pivoting is normally employed to preserve arithmetic accuracy.

The previous example, solved by the Gauss–Jordan method, gives this succession of calculations:

The original augmented matrix is

$$
\begin{bmatrix}
0 & 2 & 0 & 1 & 0 \\
2 & 2 & 3 & 2 & -2 \\
4 & -3 & 0 & 1 & -7 \\
6 & 1 & -6 & -5 & 6
\end{bmatrix}.
$$

Interchanging rows one and four, dividing the first row by 6, and reducing the first column gives

$$
\begin{bmatrix}
1 & 0.16667 & -1 & -0.83335 & 1 \\
0 & 1.6667 & 5 & 3.6667 & -4 \\
0 & -3.6667 & 4 & 4.3334 & -11 \\
0 & 2 & 0 & 1 & 0
\end{bmatrix}.
$$

Interchanging rows two and three, dividing the second row by −3.6667, and reducing the second column (operating above the diagonal as well as below) gives

$$
\begin{bmatrix}
1 & 0 & -1.5000 & -1.2000 & 1.4000 \\
0 & 1 & 2.9999 & 2.2000 & -2.4000 \\
0 & 0 & 15.000 & 12.400 & -19.800 \\
0 & 0 & -5.9998 & -3.4000 & 4.8000
\end{bmatrix}.
$$

No interchanges are required for the next step. We divide the third row by 15.000 and make the other elements in the third column into zeros:

$$
\begin{bmatrix}
1 & 0 & 0 & 0.04000 & -0.58000 \\
0 & 1 & 0 & -0.27993 & 1.5599 \\
0 & 0 & 1 & 0.82667 & -1.3200 \\
0 & 0 & 0 & 1.5599 & -3.1197
\end{bmatrix}.
$$

We now divide the fourth row by 1.5599 and create zeros above the diagonal in the fourth column:

$$
\begin{bmatrix}
1 & 0 & 0 & 0 & -0.49999 \\
0 & 1 & 0 & 0 & 1.0001 \\
0 & 0 & 1 & 0 & 0.33326 \\
0 & 0 & 0 & 1 & -1.9999
\end{bmatrix}.
$$

The solution is essentially the same as with the usual Gaussian method; round-off errors have created inaccuracies in a slightly different way than did the previous computation.

While the above Gauss–Jordan scheme appears to be a duplicate of the work done in the standard procedure, a count of the arithmetic operations shows that the Gauss–Jordan method requires almost 50% more operations. We therefore do not recommend its use.

A frequently occurring situation is to have to solve a set of equations with the same coefficient matrix but with a number of different right-hand sides. For example, in the case of interconnected resistances at the beginning of this chapter, we may desire to study the effects of different voltages applied at points 1 and 6. In the design of a truss (see Problems 62 and 64), one usually wishes to determine the stresses under a variety of external loads—this causes only the right-hand-side terms to vary.

If all of the different right-hand-side vectors are known in advance, the multiple solutions of the system can be obtained simultaneously using our Gaussian elimination method. One augments the coefficient matrix with all of the right-hand-side vectors, and treats each augmentation column in the same way as a single added column. In the back-substitution phase, each of the columns is employed to give a solution vector. At the end of this chapter, a subroutine is exhibited that does this.

EXAMPLE Solve the system $Ax = b$, with multiple values of b, by Gaussian elimination:

$$
A = \begin{bmatrix}
3 & 2 & -1 & 2 \\
1 & 4 & 0 & 2 \\
2 & 1 & 2 & -1 \\
1 & 1 & -1 & 3
\end{bmatrix}, \quad
b^{(1)} = \begin{bmatrix} 0 \\ 0 \\ 1 \\ 0 \end{bmatrix}, \quad
b^{(2)} = \begin{bmatrix} -2 \\ 1 \\ 3 \\ 4 \end{bmatrix}, \quad
b^{(3)} = \begin{bmatrix} 2 \\ 2 \\ 0 \\ 0 \end{bmatrix}.
$$

We augment A with all of the b's, then triangularize:

$$
\begin{bmatrix}
3 & 2 & -1 & 2 & 0 & -2 & 2 \\
1 & 4 & 0 & 2 & 0 & 1 & 2 \\
2 & 1 & 2 & -1 & 1 & 3 & 0 \\
1 & 1 & -1 & 3 & 0 & 4 & 0
\end{bmatrix} \rightarrow
$$

$$
\begin{bmatrix}
3 & 2 & -1 & 2 & 0 & -2 & 2 \\
(3.333) & 3.333 & 0.333 & 1.333 & 0 & 1.667 & 1.333 \\
(0.667) & -0.333 & 2.667 & -2.333 & 1 & 4.333 & -1.333 \\
(0.333) & 0.333 & -0.667 & 2.333 & 0 & 4.667 & -0.667
\end{bmatrix} \rightarrow
$$

$$\begin{bmatrix} 3 & 2 & -1 & 2 & 0 & -2 & 2 \\ (3.333) & 3.333 & 0.333 & 1.333 & 0 & 1.667 & 1.333 \\ (0.667) & (-0.100) & 2.700 & -2.200 & 1 & 4.500 & -1.200 \\ (0.333) & (0.100) & -0.700 & 2.200 & 0 & 4.500 & -0.800 \end{bmatrix} \rightarrow$$

$$\begin{bmatrix} 3 & 2 & -1 & 2 & 0 & -2 & 2 \\ (3.333) & 3.333 & 0.333 & 1.333 & 0 & 1.667 & 1.333 \\ (0.667) & (-0.100) & 2.700 & -2.200 & 1 & 4.500 & -1.200 \\ (0.333) & (0.100) & (0.259) & 1.630 & 0.259 & 5.667 & -1.111 \end{bmatrix} \cdot$$

$$\underset{c^{(1)}}{\uparrow} \qquad \underset{c^{(2)}}{\uparrow} \qquad \underset{c^{(3)}}{\uparrow}$$

We obtain the three solution vectors by back-substitution, employing the proper b' vector. (These are indicated above as $c^{(i)}$.)

$$x^{(1)} = \begin{bmatrix} 0.137 \\ -0.114 \\ 0.500 \\ 0.159 \end{bmatrix}, \qquad x^{(2)} = \begin{bmatrix} -0.591 \\ -1.340 \\ 4.500 \\ 3.477 \end{bmatrix}, \qquad x^{(3)} = \begin{bmatrix} 0.273 \\ 0.773 \\ -1.000 \\ -0.682 \end{bmatrix}. \qquad ∎$$

SCALING

We have mentioned that the rows of the augmented coefficient matrix may need to be scaled before a proper choice of pivot element can be made. *Scaling* is the operation of adjusting the coefficients of a set of equations so that they are all of the same order of magnitude. In some instances, a set of equations may involve relationships between quantities measured in widely different units (microvolts versus kilovolts, for example, or nanoseconds versus years). This may result in some of the equations having very large numbers and others very small. If we select the pivot elements without scaling, pivoting may put numbers on the diagonal that are not large in comparison to others in their row; this can actually create the round-off errors that pivoting was supposed to avoid. An example will clarify the concept.

$$\text{Given} \quad \begin{bmatrix} 3 & 2 & 100 \\ -1 & 3 & 100 \\ 1 & 2 & -1 \end{bmatrix} x = \begin{bmatrix} 105 \\ 102 \\ 2 \end{bmatrix}.$$

Carrying only three digits to emphasize round-off, and using partial pivoting, we find that the triangularized system is

$$\begin{bmatrix} 3 & 2 & 100 & 105 \\ 0 & 3.66 & 133 & 137 \\ 0 & 0 & -82.6 & -82.7 \end{bmatrix},$$

from which $x_3 = 1.00$, $x_2 = 1.09$, $x_1 = 0.94$; the exact solution vector should be $[1.00, 1.00, 1.00]$.

If we scale the values before reduction by dividing each row by the magnitude of the largest coefficient, so that the system is

$$\begin{bmatrix} 0.03 & 0.02 & 1.00 \\ -0.01 & 0.03 & 1.00 \\ 0.50 & 1.00 & -0.50 \end{bmatrix} x = \begin{bmatrix} 1.05 \\ 1.02 \\ 1.00 \end{bmatrix},$$

we get, with the same arithmetic precision, the triangularized system

$$\begin{bmatrix} 0.50 & 1.00 & -0.50 & 1.00 \\ 0 & 0.05 & 0.99 & 1.04 \\ 0 & 0 & 1.82 & 1.82 \end{bmatrix},$$

from which $x_3 = 1.00$, $x_2 = 1.00$, $x_1 = 1.00$. The reason for the improvement is that rows are interchanged after scaling has been done. No interchanges are indicated in the unscaled equations.

Whenever the coefficients in one column are widely different from those in another column, scaling is beneficial. When all values are about the same order of magnitude, scaling should be avoided, for the additional round-off error incurred during the scaling operation itself may adversely affect the accuracy. The usual way to scale is as we have done here, by dividing each row by the magnitude of the largest term. Some authorities recommend scaling so that the sum of the magnitudes of the coefficients in each row is the same. This is probably very slightly more economical to do in a computer program.

2.5   OTHER *LU* METHODS

With a modification of the elimination method in Section 2.4, we have other *LU* decomposition methods. The best known is named *Crout reduction*, or, after another discoverer, *Cholesky reduction*. In this method the matrix of coefficients *A* is transformed into the product of two matrices *L* and *U*, where *L* is a lower-triangular and *U* is an upper-triangular matrix with ones on its main diagonal. An equivalent method transforms *A* into an *LU* pair in which *L* has ones on its diagonal. This is called *Doolittle's method*. In this section we will concentrate on the Crout reduction method and refer to it simply as the *LU* decomposition method.

We have previously seen that a matrix that has been triangularized combined with the lower-triangular matrix formed from the ratios used in its reduction form an *LU* pair. But *LU* pairs take many other forms. In fact, any matrix that has all diagonal elements nonzero can be written as a product of a lower-triangular and an upper-triangular matrix in an infinity of ways. For example,

$$\underset{A}{\begin{bmatrix} 2 & -1 & -1 \\ 0 & -4 & 2 \\ 6 & -3 & 1 \end{bmatrix}} = \underset{L_1}{\begin{bmatrix} 2 & 0 & 0 \\ 0 & -4 & 0 \\ 6 & 0 & 4 \end{bmatrix}} \underset{U_1}{\begin{bmatrix} 1 & -\frac{1}{2} & -\frac{1}{2} \\ 0 & 1 & -\frac{1}{2} \\ 0 & 0 & 1 \end{bmatrix}}$$

$$= \begin{bmatrix} 1 & 0 & 0 \\ 0 & 1 & 0 \\ 3 & 0 & 1 \end{bmatrix} \begin{bmatrix} 2 & -1 & -1 \\ 0 & -4 & 2 \\ 0 & 0 & 4 \end{bmatrix}$$

$$\qquad\qquad L_2 \qquad\qquad\quad U_2$$

$$= \begin{bmatrix} 1 & 0 & 0 \\ 0 & 2 & 0 \\ 3 & 0 & 1 \end{bmatrix} \begin{bmatrix} 2 & -1 & -1 \\ 0 & -2 & 1 \\ 0 & 0 & 4 \end{bmatrix}$$

$$\qquad\qquad L_3 \qquad\qquad\quad U_3$$

= and so on.

Of the entire set of $L\bar{U}$s whose product equals matrix A, we choose the pair in which U has only ones on its diagonal, as in the first pair above. We get the rules for such an LU decomposition from the relationship that $LU = A$. In the case of a 4×4 matrix:

$$\begin{bmatrix} \ell_{11} & 0 & 0 & 0 \\ \ell_{21} & \ell_{22} & 0 & 0 \\ \ell_{31} & \ell_{32} & \ell_{33} & 0 \\ \ell_{41} & \ell_{42} & \ell_{43} & \ell_{44} \end{bmatrix} \begin{bmatrix} 1 & u_{12} & u_{13} & u_{14} \\ 0 & 1 & u_{23} & u_{24} \\ 0 & 0 & 1 & u_{34} \\ 0 & 0 & 0 & 1 \end{bmatrix} = \begin{bmatrix} a_{11} & a_{12} & a_{13} & a_{14} \\ a_{21} & a_{22} & a_{23} & a_{24} \\ a_{31} & a_{32} & a_{33} & a_{34} \\ a_{41} & a_{42} & a_{43} & a_{44} \end{bmatrix}.$$

Multiplying the rows of L by the first column of U, we get $\ell_{11} = a_{11}$, $\ell_{21} = a_{21}$, $\ell_{31} = a_{31}$, $\ell_{41} = a_{41}$; the first column of L is the same as the first column of A.

We now multiply the first row of L by the columns of U:

$$\ell_{11}u_{12} = a_{12}, \qquad \ell_{11}u_{13} = a_{13}, \qquad \ell_{11}u_{14} = a_{14}, \tag{2.2}$$

from which

$$u_{12} = \frac{a_{12}}{\ell_{11}}, \qquad u_{13} = \frac{a_{13}}{\ell_{11}}, \qquad u_{14} = \frac{a_{14}}{\ell_{11}}. \tag{2.3}$$

Thus the first row of U is determined.

In this method we alternate between getting a column of L and a row of U, so we next get the equations for the second column of L by multiplying the rows of L by the second column of U:

$$\begin{aligned} \ell_{21}u_{12} + \ell_{22} &= a_{22}, \\ \ell_{31}u_{12} + \ell_{32} &= a_{32}, \\ \ell_{41}u_{12} + \ell_{42} &= a_{42}, \end{aligned} \tag{2.4}$$

which gives

$$\begin{aligned} \ell_{22} &= a_{22} - \ell_{21}u_{12}, \\ \ell_{32} &= a_{32} - \ell_{31}u_{12}, \\ \ell_{42} &= a_{42} - \ell_{41}u_{12}. \end{aligned} \tag{2.5}$$

Proceeding in the same fashion, the equations we need are

$$u_{23} = \frac{a_{23} - \ell_{21}u_{13}}{\ell_{22}}, \qquad u_{24} = \frac{a_{24} - \ell_{21}u_{14}}{\ell_{22}},$$

$$\ell_{23} = a_{33} - \ell_{31}u_{13} - \ell_{32}u_{23}, \qquad \ell_{43} = a_{43} - \ell_{41}u_{13} - \ell_{42}u_{23},$$

$$u_{34} = \frac{a_{34} - \ell_{31}u_{14} - \ell_{32}u_{24}}{\ell_{33}},$$

$$\ell_{44} = a_{44} - \ell_{41}u_{14} - \ell_{42}u_{24} - \ell_{43}u_{34}.$$

The general formula for getting elements of L and U corresponding to the coefficient matrix for n simultaneous equations can be written

$$\ell_{ij} = a_{ij} - \sum_{k=1}^{j-1} \ell_{ik}u_{kj}, \qquad j \le i, \qquad i = 1, 2, \ldots, n, \qquad (2.6)$$

$$u_{ij} = \frac{a_{ij} - \sum_{k=1}^{i-1} \ell_{ik}u_{kj}}{\ell_{ii}}, \qquad i \le j, \qquad j = 2, 3, \ldots, n. \qquad (2.7)$$

$\left(\text{For } j = 1, \text{ the rule for } \ell \text{ reduces to}\right.$

$$\ell_{i1} = a_{i1}.$$

For $i = 1$, the rule for u reduces to

$$u_{1j} = \frac{a_{1j}}{\ell_{11}} = \frac{a_{1j}}{a_{11}}.\Bigg)$$

The reason this method is popular in programs is that storage space may be economized. There is no need to store the zeros in either L or U, and the ones on the diagonal of U can also be omitted. (Since these values are always the same and are always known, it is redundant to record them.) One can then store the essential elements of U where the zeros appear in the L array. Examination of Eqs. (2.2) through (2.7) shows that, after any element of A, a_{ij}, is once used, it never again appears in the equations. Hence its place in the original $n \times n$ array A can be used to store an element of either L or U. In other words, the A array can be transformed by the above equations and becomes

$$\begin{bmatrix} a_{11} & a_{12} & a_{13} & a_{14} \\ a_{21} & a_{22} & a_{23} & a_{24} \\ a_{31} & a_{32} & a_{33} & a_{34} \\ a_{41} & a_{42} & a_{43} & a_{44} \end{bmatrix} \rightarrow \begin{bmatrix} \ell_{11} & u_{12} & u_{13} & u_{14} \\ \ell_{21} & \ell_{22} & u_{23} & u_{24} \\ \ell_{31} & \ell_{32} & \ell_{33} & u_{34} \\ \ell_{41} & \ell_{42} & \ell_{43} & \ell_{44} \end{bmatrix}.$$

Because we can condense the *L* and *U* matrices into one array and store their elements in the space of *A*, this method is often called a *compact scheme*.

E X A M P L E Consider the matrix *A*

$$A = \begin{bmatrix} 3 & -1 & 2 \\ 1 & 2 & 3 \\ 2 & -2 & -1 \end{bmatrix}.$$

Applying the equations for the ℓ's and u's, we obtain

$$\ell_{11} = 3, \quad \ell_{21} = 1, \quad \ell_{31} = 2; \quad u_{12} = -\tfrac{1}{3}, \quad u_{13} = \tfrac{2}{3}.$$

$$\ell_{22} = 2 - (1)\left(-\tfrac{1}{3}\right) = \tfrac{7}{3}, \quad \ell_{32} = -2 - (2)\left(-\tfrac{1}{3}\right) = -\tfrac{4}{3}.$$

$$u_{23} = \frac{3 - (1)\left(\tfrac{2}{3}\right)}{\tfrac{7}{3}} = 1, \quad \ell_{33} = -1 - (2)\left(\tfrac{2}{3}\right) - \left(-\tfrac{4}{3}\right)(1) = -1.$$

$$L = \begin{bmatrix} 3 & 0 & 0 \\ 1 & \tfrac{7}{3} & 0 \\ 2 & -\tfrac{4}{3} & -1 \end{bmatrix}, \quad U = \begin{bmatrix} 1 & -\tfrac{1}{3} & \tfrac{2}{3} \\ 0 & 1 & 1 \\ 0 & 0 & 1 \end{bmatrix}.$$

If the quantities are written in the compact form as they are computed, we have

$$LU = \begin{bmatrix} 3 & -\tfrac{1}{3} & \tfrac{2}{3} \\ 1 & \tfrac{7}{3} & 1 \\ 2 & -\tfrac{4}{3} & -1 \end{bmatrix} \begin{matrix} \leftarrow ② \\ \leftarrow ④ \\ \\ \end{matrix}$$
$$\quad\quad\quad \underset{①}{\uparrow} \quad \underset{③}{\uparrow} \quad \underset{⑤}{\uparrow}$$

The circled numbers show the order in which columns and rows of the new matrix are obtained. ∎

Here is an algorithm for *LU* decomposition. It does not compute the *L* and *U* matrices in place, but sets them up as separate matrices.

LU Decomposition

To transform an $n \times n$ matrix *A* into the equivalent product of *L* and *U* matrices,

DO FOR I = 1 to *n* step 1,
    L(I,1) = A(I,1).
ENDDO.
DO FOR J = 1 to *n* step 1,
    U(1,J) = A(1,J)/L(1,1).
ENDDO.

```
DO FOR J = 2 to n step 1,
    DO FOR I = J to n step 1,
        DO FOR K = 1 to J − 1 step 1,
            Accumulate sum of L(I,K) * U(K,J), in double precision.
        ENDDO.
        L(I,J) = A(I,J) − SUM.
    ENDDO.
    U(J,J) = 1.
    DO FOR I = J + 1 to n step 1,
        DO FOR K = 1 to J − 1 step 1,
            Accumulate sum of L(J,K) * U(K,I), in double precision.
        ENDDO.
        U(J,I) = [A(J,I) − SUM]/L(J,J).
    ENDDO.
ENDDO.
```

The solution of the set of equations $Ax = b$ is readily obtained with the L and U matrices. Once the coefficient matrix has been converted to its LU equivalent, we are prepared to find the solution to the set of equations that corresponds to any given right-hand-side vector b. The L matrix is really a record of the operations required to make the coefficient matrix A into the upper-triangular matrix U. We apply these same transformations to the RHS vector b, converting it to a new vector b'. If we augment b' to U and back-substitute, the solution appears.

The general equation for the reduction of b (it is exactly the same as the rule for forming the elements of U) is

$$b'_1 = b_1/\ell_{11},$$

$$b'_i = \frac{b_i - \sum_{k=1}^{i-1} \ell_{ik}b'_k}{\ell_{ii}}, \qquad i = 2, 3, \ldots, n.$$

The equations for the back-substitution are

$$x_n = b'_n,$$

$$x_j = b'_j - \sum_{k=j+1}^{n} u_{jk}x_k, \qquad j = n - 1, n - 2, \ldots, 1.$$

For example, if

$$A = \begin{bmatrix} 3 & -1 & 2 \\ 1 & 2 & 3 \\ 2 & -2 & -1 \end{bmatrix},$$

we get

$$L = \begin{bmatrix} 3 & 0 & 0 \\ 1 & \frac{7}{3} & 0 \\ 2 & -\frac{4}{3} & -1 \end{bmatrix} \quad \text{and} \quad U = \begin{bmatrix} 1 & -\frac{1}{3} & \frac{2}{3} \\ 0 & 1 & 1 \\ 0 & 0 & 1 \end{bmatrix}.$$

For $b = \begin{bmatrix} 12 \\ 11 \\ 2 \end{bmatrix}$,

$$b' = \begin{bmatrix} 4 \\ 3 \\ 2 \end{bmatrix},$$

because

$$b_1' = \frac{12}{3} = 4,$$

$$b_2' = \frac{11 - (1)(4)}{\frac{7}{3}} = 3,$$

$$b_3' = \frac{2 - (2)(4) - \left(-\frac{4}{3}\right)(3)}{-1} = 2.$$

Augmenting b' to U and back-substituting,

$$\begin{bmatrix} 1 & -\frac{1}{3} & \frac{2}{3} & \vdots & 4 \\ 0 & 1 & 1 & \vdots & 3 \\ 0 & 0 & 1 & \vdots & 2 \end{bmatrix}$$

we get

$$x_3 = 2,$$

$$x_2 = 3 - 1(2) = 1,$$

$$x_1 = 4 - \frac{2}{3}(2) - \left(-\frac{1}{3}\right)(1) = 3.$$

Examination of the above operations reveals that we can get b' by augmenting L with b and solving that triangular system (a kind of "forward" substitution):

$$L \vdots b = \begin{bmatrix} 3 & 0 & 0 & \vdots & 12 \\ 1 & \frac{7}{3} & 0 & \vdots & 11 \\ 2 & -\frac{4}{3} & -1 & \vdots & 2 \end{bmatrix}$$

and

$$b_1' = \frac{12}{3} = 4,$$

$$b_2' = [11 - (1)(4)]/\left(\frac{7}{3}\right) = 3,$$

$$b_3' = [2 - (2)(4) - \left(-\frac{4}{3}\right)(3)]/(-1) = 2.$$

We do not write the algorithm for reducing the b vector and back-substitution. This is left as an exercise for the student.

A special advantage of these LU methods is that we can accumulate the sums in double precision. This gives us greater accuracy by just using one or two double-precision variables. We have indicated this in the algorithm. This is not easily done with the Gaussian elimination method of Section 2.4. Moreover, the LU method can be easily adapted to solve a system of new right-hand-side vectors with great economy of effort.* The number of arithmetic operations to get the solution corresponding to each b turns out to be exactly the same as to multiply an $n \times n$ matrix by an n-component vector.

Pivoting with the LU method is somewhat more complicated than with Gaussian elimination because we do not usually handle the right-hand-side vector simultaneously with our reduction of the A matrix. This means we must keep a record of any row interchanges made during the formation of L and U so that the elements of the right-hand-side vector can be similarly interchanged. We do the interchanges immediately after computing each column of L, choosing the value to appear on the diagonal so as to have the one of largest magnitude. We illustrate with an example:

$$\text{Given } A = \begin{bmatrix} 0 & 2 & 1 \\ 1 & 0 & 0 \\ 3 & 0 & 1 \end{bmatrix}.$$

We will keep a record of row order in a vector: $O = [1, 2, 3]$, representing the original ordering.

The first column of L is

$$\begin{bmatrix} 0 \\ 1 \\ 3 \end{bmatrix};$$

we need to interchange rows 3 and 1. To keep track of this, we interchange the first and third elements of O, so O becomes $[3, 2, 1]$.

Interchange the rows of A and compute the first row of the U matrix. (We use the compact scheme.)

$$\begin{bmatrix} 3 & 0 & \frac{1}{3} \\ 1 & 0 & 0 \\ 0 & 2 & 1 \end{bmatrix}, \quad \text{with } O = [3, 2, 1].$$

---

*A numerical example that illustrates using the LU method with multiple right-hand sides will be found in Section 2.7.

Now compute the second column of L; it is $\begin{bmatrix} 0 \\ 0 \\ 2 \end{bmatrix}$.

We must interchange again, the second row with the third, and O becomes [3, 1, 2]. Making the interchange of rows and computing the second row of U gives

$$\begin{bmatrix} 3 & 0 & \frac{1}{3} \\ 0 & 2 & \frac{1}{2} \\ 1 & 0 & 0 \end{bmatrix}, \quad \text{with } O = [3, 1, 2].$$

Completing the reduction, we find $\ell_{33} = 0 - (1)\left(\frac{1}{3}\right) = -\frac{1}{3}$, giving

$$LU = \begin{bmatrix} 3 & 0 & \frac{1}{3} \\ 0 & 2 & \frac{1}{2} \\ 1 & 0 & -\frac{1}{3} \end{bmatrix}, \quad \text{with } O = [3, 1, 2].$$

To solve the problem $Ax = b$, with $b^T = [5, -1, -2]$, we rearrange the elements of b in the order given by O and compute b':

$$L \mid b = \begin{bmatrix} 3 & 0 & 0 & \mid & -2 \\ 0 & 2 & 0 & \mid & 5 \\ 1 & 0 & -\frac{1}{3} & \mid & -1 \end{bmatrix}, \quad b' = \begin{bmatrix} -\frac{2}{3} \\ \frac{5}{2} \\ 1 \end{bmatrix},$$

so

$$U \mid b' = \begin{bmatrix} 1 & 0 & \frac{1}{3} & \mid & -\frac{2}{3} \\ 0 & 1 & \frac{1}{2} & \mid & \frac{5}{2} \\ 0 & 0 & 1 & \mid & 1 \end{bmatrix}, \quad \text{giving } x = \begin{bmatrix} -1 \\ 2 \\ 1 \end{bmatrix}.$$

2.6 PATHOLOGY IN LINEAR SYSTEMS—SINGULAR MATRICES

When a real physical situation is modeled by a set of linear equations, one can anticipate that the set of equations will have a solution that matches the values of the quantities in the physical problem, at least so far as the equations truly do represent it.* Because of round-off errors, the solution vector that is calculated may imperfectly predict the physical quantity, but there is assurance that a solution exists, at least in principle. Consequently, it must always be theoretically possible to avoid divisions by zero when the set of equations has a solution.

---

*There are certain problems for which values of interest are determined from a set of equations that do not have a unique solution; these are called *eigenvalue* problems and are discussed in another chapter.

An arbitrary set of equations may not have such a guaranteed solution, however. There are several such possible situations, which we term "pathological." In each case, there is *no unique solution* to the set of equations.

First, if the number of equations relating the variables is less than the number of *unknowns*, we certainly cannot solve for unique values of the unknown variables. It turns out, in this case, that there is an infinite set of solutions, for we may arrange the n equations with all but n of the variables on the right-hand sides, grouped with the constant terms. We may assign almost any desired values to these segregated variables (combining them with the constant terms) and then solve for the n remaining variables.* Assigning new values to the variables on the right-hand sides gives another set of values for the unknowns, and so on. For example,

$$\begin{cases} x_1 - 2x_2 + x_3 = 4, \\ x_1 - x_2 + x_3 = 5. \end{cases}$$

Rewrite as

$$x_1 - 2x_2 = 4 - x_3,$$
$$x_1 - x_2 = 5 - x_3.$$

If $x_3 = 0$,

$$x_1 = 6, \quad x_2 = 1.$$

If $x_3 = 1$,

$$x_1 = 5, \quad x_2 = 1.$$

If $x_3 = -1$,

$$x_1 = 7, \quad x_2 = 1, \quad \text{and so on.}$$

A second situation where we might not expect a set of equations to have a solution is that in which the number of equations is greater than the number of unknowns. If there are n unknowns, we can normally find a subset of the equations that can be solved for the unknowns. There are two subcases to consider:

If the remaining equations are satisfied by the values of the unknowns we have just determined (we would say these equations are *consistent* with the others), there exists a unique solution to the set of equations. Really, of course, there are not truly more equations than unknowns in this case; the extra equations are *redundant*.

The other subcase is a pathological one. If the solution to the first n equations does not satisfy the remaining ones, the set is clearly *inconsistent,* and *no* solution exists that satisfies the system.

---

*An important instance of this situation is the solution of a linear programming problem by the simplex method. In this method, the segregated variables are all assigned the value of zero.

Realizing that an equation may be redundant in the above situation makes us re-examine our more standard case of n equations in n unknowns. What if there is redundancy there? How can we recognize redundancy when it is present? In this example it is obvious:

$$x + y = 3, \qquad 2x + 2y = 6.$$

The second equation is clearly redundant and contains no information not already given by the first. This system will then have an *infinity* of values for x and y; it is an example of fewer equations than unknowns.

Inconsistancy may also be present:

$$x + y = 3, \qquad 2x + 2y = 7.$$

In this case there is *no solution*.

If $n \times n$ systems do not have a unique solution, they have a (square) coefficient matrix that is called *singular*. If the coefficient matrix can be triangularized without having zeros on the diagonal (hence the set of equations has a solution), the matrix is said to be *nonsingular*.

Larger systems may have redundancy or inconsistency even though it is not obvious at a glance. Even in a 3×3 system, it is not easy to tell

$$\begin{aligned}
x_1 - 2x_2 + 3x_3 &= 5, \\
2x_1 + 4x_2 - x_3 &= 7, \\
-x_1 - 14x_2 + 11x_3 &= 2.
\end{aligned}$$

Are these inconsistent or redundant? (In other words, is the coefficient matrix singular?) Or do they have a unique solution? (That is, is the matrix nonsingular?) Is there a rule that we can apply, especially one that works for large systems? The answer is *yes*, there is a rule, or rather, there are several tests we can apply. The standard response from mathematics is to determine the *rank* of the coefficient matrix. If this value is less than n, the number of equations, then no unique solution exists; the equations are either inconsistent or one or more is redundant, depending on the right-hand-side values.

But how does one determine the rank? One practical method is to triangularize by Gaussian elimination: If no zeros show up on the diagonal of the final triangularized coefficient matrix, the rank is equal to n (the matrix is said to be of *full rank*) and a unique solution exists. If, in spite of pivoting, one or more zeros occurs on the final diagonal, there is no unique solution. The set will be consistent (and have redundancy) if back-substitution gives $(0/0)$ indeterminate forms. When one would need to divide a nonzero term by zero in back-substituting, inconsistency occurs. Let us apply this to our earlier 3×3 example:

$$\begin{bmatrix} 1 & -2 & 3 & \vdots & 5 \\ 2 & 4 & -1 & \vdots & 7 \\ -1 & -14 & 11 & \vdots & 2 \end{bmatrix}.$$

On reduction we get

$$\begin{bmatrix} 1 & -2 & 3 & \vdots & 5 \\ 0 & 8 & -7 & \vdots & -3 \\ 0 & 0 & 0 & \vdots & 1 \end{bmatrix}.$$

We see that there is no solution and the equations are inconsistent. If the constant term in the third equation were 1 rather than 2, reduction gives

$$\begin{bmatrix} 1 & -2 & 3 & \vdots & 5 \\ 0 & 8 & -7 & \vdots & -3 \\ 0 & 0 & 0 & \vdots & 0 \end{bmatrix}.$$

This system is redundant.

Another way to find whether a set of equations has a unique solution is to test the rows or columns of the coefficient matrix for *linear dependency*. Vectors are called linearly dependent if a linear combination of them can be found that equals the zero vector (one with all components equal to zero). (Of course the linear combination $a\bar{x} + b\bar{y} + c\bar{z}$ always equals zero if all the coefficients are zero; we rule out this possibility in our test for linear dependency. When $a = b = c = 0$, we say we have the *trivial case*.)

If the vectors are linearly independent, the only way a weighted sum of them can equal the zero vector is to weight each of them with a zero coefficient.

Our singular 3×3 system has columns that form vectors that are linearly dependent:

$$(-10)\begin{bmatrix} 1 \\ 2 \\ -1 \end{bmatrix} + (7)\begin{bmatrix} -2 \\ 4 \\ -14 \end{bmatrix} + (8)\begin{bmatrix} 3 \\ -1 \\ 11 \end{bmatrix} = \begin{bmatrix} 0 \\ 0 \\ 0 \end{bmatrix}.$$

Similarly, the rows form linearly dependent vectors:

$$(-3)[1 \quad -2 \quad 3] + (2)[2 \quad 4 \quad -1] + (1)[-1 \quad -14 \quad 11] = [0 \quad 0 \quad 0].$$

In the general case, we say that vectors $\bar{x}_1, \bar{x}_2, \bar{x}_3, \ldots, \bar{x}_n$ are *linearly dependent* if we can find scalar coefficients, $a_1, a_2, \ldots, a_n$ (with not all the a_i simultaneously zero), for which

$$\sum_{i=1}^{n} a_i \bar{x}_i = \bar{0}. \tag{2.8}$$

If the only linear combination of the $\bar{x}_i$ that equals the zero vector requires that all the a_i be zero, the set of vectors is called linearly independent, It follows that, if a set of vectors is linearly dependent, at least one of the vectors can be written as a linear combination of the others. If the set is linearly independent, none of the vectors can be written as a linear combination of the others. As a practical matter, we do not usually test the columns (or rows) for linear dependency to determine whether a matrix is singular.

If we are interested in determining the coefficients $a_1, a_2, \ldots, a_n$ that appear in the linear combination of Eq. (2.8), it turns out that we have to solve a set of linear equations to obtain them.

It is worthwhile to summarize the concepts and terminology of this section. The following lists of terms are all equivalent expressions. If a square matrix can be shown to have one property, it has all the others.

Equivalent Properties of Singular or Nonsingular Matrices

The matrix is singular.	The matrix is nonsingular.
A set of equations with these coefficients has no unique solution.	A set of equations with these coefficients has a unique solution.
Gaussian elimination cannot avoid a zero on the diagonal.	Gaussian elimination proceeds without a zero on the diagonal.
The rank of the matrix is less than n.	The rank of the matrix equals n.
The rows form linearly dependent vectors.	The rows form linearly independent vectors.
The columns form linearly dependent vectors.	The columns form linearly independent vectors.

In the next section we will consider two other properties of the matrix: its determinant and its inverse. This adds two more attributes to our lists: A singular matrix has a zero determinant and a nonsingular matrix has a nonzero determinant. A singular matrix has no inverse and a nonsingular matrix does have an inverse.

2.7 DETERMINANTS AND MATRIX INVERSION

You have perhaps wondered why there has been no reference so far in this chapter to the solution of linear equations by determinants (Cramer's rule). The reason for this is that, except for systems of only two or three equations, the determinant method is too inefficient. For example, for a set of 10 simultaneous equations, about 70,000,000 multiplications and divisions are required if the usual method of expansion in terms of minors is used. A more efficient method of evaluating the determinants can reduce this to about 3000 multiplications, but even this is inefficient compared to Gaussian elimination, which would require about 380.

In fact, the evaluation of a determinant can perhaps best be done by adapting the Gaussian elimination procedure. Its utility derives from the fact that the determinant of a triangular matrix (either upper- or lower-triangular) is just the product of its diagonal elements. This is easily seen, in the case of an upper-triangular matrix, by expansion in terms of minors of the first column at each step. For example,

$$\begin{vmatrix} a_{11} & a_{12} & a_{13} & a_{14} \\ 0 & a_{22} & a_{23} & a_{24} \\ 0 & 0 & a_{33} & a_{34} \\ 0 & 0 & 0 & a_{44} \end{vmatrix} = a_{11} \begin{vmatrix} a_{22} & a_{23} & a_{24} \\ 0 & a_{33} & a_{34} \\ 0 & 0 & a_{44} \end{vmatrix} - 0 + 0 - 0$$

$$= a_{11}(a_{22} \begin{vmatrix} a_{33} & a_{34} \\ 0 & a_{44} \end{vmatrix} - 0 + 0)$$

$$= a_{11}a_{22}(a_{33}a_{44} - 0) = a_{11}a_{22}a_{33}a_{44}.$$

Adding a multiple of one row to another row of a matrix does not change the value of its determinant. The other row transformations change the value in predictable ways: interchanging two rows changes its sign and multiplying a row by a constant multiplies the value of the determinant by the same constant. If these changes are allowed for, using the procedure of Gaussian elimination to convert to upper-triangular is a simple way to evaluate the determinant.

EXAMPLE Find the value of the determinant by using elementary row transformations to make it upper-triangular.

$$\begin{vmatrix} 1 & 4 & -2 & 3 \\ 2 & 2 & 0 & 4 \\ 3 & 0 & -1 & 2 \\ 1 & 2 & 2 & -3 \end{vmatrix} = \begin{vmatrix} 1 & 4 & -2 & 3 \\ 0 & -6 & 4 & -2 \\ 0 & -12 & 5 & -7 \\ 0 & -2 & 4 & -6 \end{vmatrix} = \begin{vmatrix} 1 & 4 & -2 & 3 \\ 0 & -6 & 4 & -2 \\ 0 & 0 & -3 & -3 \\ 0 & 0 & \frac{8}{3} & -\frac{16}{3} \end{vmatrix}$$

$$= \begin{vmatrix} 1 & 4 & -2 & 3 \\ 0 & -6 & 4 & -2 \\ 0 & 0 & -3 & -3 \\ 0 & 0 & 0 & -8 \end{vmatrix} = (1)(-6)(-3)(-8) = -144.$$

For programming, an easy and very efficient method for computing the determinant is to use steps 2 through 4 of the algorithm in Section 2.4. Then the determinant of the matrix is just the product of the diagonal elements, with a reversed sign if there were an odd number of row interchanges: $\pm a_{11} * a_{22} * \ldots * a_{nn}$, where $+$ is used if there were 0 or an even number of row interchanges in steps 2 through 4 (otherwise we use $-$).

Applying this algorithm to the example, but using row interchanges, we see

$$\begin{bmatrix} 1 & 4 & -2 & 3 \\ 2 & 2 & 0 & 4 \\ 3 & 0 & -1 & 2 \\ 1 & 2 & 2 & -3 \end{bmatrix} \rightarrow \begin{bmatrix} 3 & 0 & -1 & 2 \\ (0.667) & 4 & -1.677 & 2.333 \\ (0.333) & (0.5) & 3.167 & -4.833 \\ (0.333) & (0.5) & (0.474) & 3.789 \end{bmatrix}.$$

Since $3 * 4 * 3.167 * 3.789 = 144$ and there were 3 row interchanges in the process, we have the determinant $= -144$. ∎

While division of matrices is not defined, the matrix inverse gives the equivalent result. If the product of two square matrices is the identity matrix, the matrices are said to be inverses. If $AB = I$, we write $B = A^{-1}$; also $A = B^{-1}$. Inverses commute on multiplication, which is not true for matrices in general: $AB = BA = I$. Not all square matrices have an inverse. Singular matrices do not have an inverse, and these are of extreme importance in connection with the coefficient matrix of a set of equations, as discussed earlier.

The inverse of a matrix can be defined in terms of the matrix of the minors of its determinant, but this is not a useful way to find an inverse. The Gauss–Jordan technique can be adapted to provide a practical way to invert a matrix. The procedure is to augment the given matrix with the identity matrix of the same order. One then reduces the original matrix to the identity matrix by elementary row transformations, performing the same operations on the augmentation columns. When the identity matrix stands as the left half of the augmented matrix, the inverse of the original stands as the right half. It should be apparent that this is equivalent to solving a set of equations with n different right-hand sides; each of the right-hand sides is a *unit vector*, in which the position of the element whose value is unity changes from row 1 to row 2 to row 3 . . . to row n.

EXAMPLE Find the inverse of

$$A = \begin{bmatrix} 1 & -1 & 2 \\ 3 & 0 & 1 \\ 1 & 0 & 2 \end{bmatrix}.$$

Augment A with the identity matrix and then reduce:

$$\begin{bmatrix} 1 & -1 & 2 & 1 & 0 & 0 \\ 3 & 0 & 1 & 0 & 1 & 0 \\ 1 & 0 & 2 & 0 & 0 & 1 \end{bmatrix} \rightarrow \begin{bmatrix} 1 & -1 & 2 & 1 & 0 & 0 \\ 0 & 3 & -5 & -3 & 1 & 0 \\ 0 & 1 & 0 & -1 & 0 & 1 \end{bmatrix}$$

$$\overset{(1)}{\rightarrow} \begin{bmatrix} 1 & -1 & 2 & 1 & 0 & 0 \\ 0 & 1 & 0 & -1 & 0 & 1 \\ 0 & 0 & -5 & 0 & 1 & -3 \end{bmatrix} \overset{(2)}{\rightarrow} \begin{bmatrix} 1 & -1 & 0 & 1 & \frac{2}{5} & -\frac{6}{5} \\ 0 & 1 & 0 & -1 & 0 & 1 \\ 0 & 0 & 1 & 0 & -\frac{1}{5} & \frac{3}{5} \end{bmatrix}$$

$$\rightarrow \begin{bmatrix} 1 & 0 & 0 & 0 & \frac{2}{5} & -\frac{1}{5} \\ 0 & 1 & 0 & -1 & 0 & 1 \\ 0 & 0 & 1 & 0 & -\frac{1}{5} & \frac{3}{5} \end{bmatrix}.$$

[1] Interchange the third and second rows before eliminating from the third row.
[2] Divide the third row by -5 before eliminating from the first row.

We confirm the fact that we have found the inverse by multiplication:

$$\begin{bmatrix} 1 & -1 & 2 \\ 3 & 0 & 1 \\ 1 & 0 & 2 \end{bmatrix} \begin{bmatrix} 0 & \frac{2}{5} & -\frac{1}{5} \\ -1 & 0 & 1 \\ 0 & -\frac{1}{5} & \frac{3}{5} \end{bmatrix} = \begin{bmatrix} 1 & 0 & 0 \\ 0 & 1 & 0 \\ 0 & 0 & 1 \end{bmatrix}. \quad \blacksquare$$

However, it is more efficient to use the Gaussian elimination algorithm of Section 2.4 by adding additional unit vectors to the augmented matrix.

Doing steps 2 through 4 gives us

$$
\begin{bmatrix}
1 & -1 & 2 & 1 & 0 & 0 \\
3 & 0 & 1 & 0 & 1 & 0 \\
1 & 0 & 2 & 0 & 0 & 1
\end{bmatrix}
\rightarrow
\begin{bmatrix}
3 & 0 & 1 & 0 & 1 & 0 \\
(0.333) & -1 & 1.667 & 1 & -0.333 & 0 \\
(0.333) & (0) & 1.667 & 0 & -0.333 & 1
\end{bmatrix}.
$$

Now applying back-substitution on the last three columns, we get

$$
\begin{bmatrix}
3 & 0 & 1 & 0 & 0.4 & -0.2 \\
(0.333) & -1 & 1.667 & -1 & 0 & 1 \\
(0.333) & (0) & 1.667 & 0 & -0.2 & 0.6
\end{bmatrix},
$$

where the last three columns store the inverse matrix. This method is actually more efficient than the Gauss–Jordan method, which takes about $3n^3/2$ versus $4n^3/3$ multiplications and/or divisions to compute the inverse of a nonsingular matrix. Moreover, in the Gaussian elimination method, we have also found the LU matrix, which we can use to solve the system for other right-hand sides.

The inverse of the coefficient matrix provides a way of solving the set of equations $Ax = b$ because, when we multiply both sides of the relation by A^{-1}, we get

$$
A^{-1}Ax = A^{-1}b,
$$
$$
x = A^{-1}b.
$$

The second equation follows because the product $A^{-1}A = I$, the identity matrix, and $Ix = x$. If we know the inverse of A, we can solve the system for any right-hand side b simply by multiplying the b vector by A^{-1}. This would seem like a good way to solve systems of equations, and one finds frequent references to it.

If we care about the efficiency of our method of solving the equations, however, this is not the preferred method, because solving the system with the LU decomposition of A, and doing the equivalent of two back-substitutions, requires exactly the same effort as multiplying b by the matrix. We compare the efficiency of the two schemes, then, by comparing the work needed to get the inverse and that to get the LU equivalent. Getting the inverse is more work, because it is the equivalent of solving the system with n right-hand sides, while getting the LU is the equivalent of doing only the reduction to triangular form.

Even though the inverse is not the most efficient way to solve a set of simultaneous equations, the inverse is very important for theoretical reasons and is essential to the understanding of many situations in applied mathematics. The use of the inverse concept and notation often simplifies the development of some fundamental relationships. We illustrate this by again considering the LU decomposition.

Find a pair of matrices such that $LU = A$.

Then $Ax = b$ can be written as

$$LUx = b.$$

Multiply both sides by L^{-1},

$$(L^{-1}L)Ux = L^{-1}b,$$

so $Ux = L^{-1}b$ because $L^{-1}L = I$.

We see that Ux (which is a vector because the product of a matrix times a vector always yields a vector) is equal to the vector formed by $L^{-1}b$. Call this vector b'. Then

$$Ux = b', \quad b' = L^{-1}b, \quad \text{or} \quad Lb' = b.$$

We can get the vector b' by solving the system $Lb' = b$. This is particularly easy to do because L is triangular; all we need to do is the back-substitution phase (actually a forward-substitution because L is lower-triangular).

Once we have b', we can solve for x from the system $Ux = b'$. This is also easy because U is triangular.

Observe how using the concept and notation of inverses helps to clarify and prove the validity of the LU method.

2.8 NORMS

When we discuss multicomponent entities like matrices and vectors, we frequently need a way to express their magnitude—some measure of "bigness" or "smallness." For ordinary numbers, the absolute value tells us how large the number is, but for a matrix there are many components, each of which may be large or small in magnitude. (We are not talking about the *size* of a matrix, meaning the number of elements it contains.)

Any good measure of the magnitude of a matrix (the technical term is *norm*) must have four properties that are intuitively essential:

1. The norm must always have a value greater than or equal to zero, and must be zero only when the matrix is the zero matrix (one with all elements equal to zero).

2. The norm must be multiplied by k if the matrix is multiplied by the scalar k.

3. The norm of the sum of two matrices must not exceed the sum of the norms.

4. The norm of the product of two matrices must not exceed the product of the norms.

More formally, we can state these conditions, using $\|A\|$ to represent the *norm of matrix* A:

$$
\begin{aligned}
&1. \quad \|A\| \geq 0 \text{ and } \|A\| = 0 \text{ if and only if } A = 0. \\
&2. \quad \|kA\| = |k| \, \|A\| \\
&3. \quad \|A + B\| \leq \|A\| + \|B\|. \\
&4. \quad \|AB\| \leq \|A\| \, \|B\|.
\end{aligned}
\qquad \textbf{(2.9)}
$$

The third relationship is called the *triangle inequality*. The fourth is important when we deal with the product of matrices.

For the special kind of matrices that we call vectors, our past experience can help us. For vectors in two- or three-space, the length satisfies all four requirements and is a good value to use for the norm of a vector. This norm is called the *Euclidean norm*, and is computed by $\sqrt{x_1^2 + x_2^2 + x_3^2}$.

We compute the Euclidean norm of vectors with more than three components by generalizing:

$$
\|x\|_e = \sqrt{x_1^2 + x_2^2 + \cdots + x_n^2} = \left(\sum_{i=1}^{n} x_i^2 \right)^{1/2}.
$$

This is not the only way to compute a vector norm, however. The sum of the absolute values of the x_i can be used as a norm; the maximum value of the magnitudes of the x_i will also serve. These three norms can be interrelated by defining the *p*-norm as

$$
\|x\|_p = \left(\sum_{i=1}^{n} |x_i|^p \right)^{1/p}.
$$

From this it is readily seen that

$$
\|x\|_1 = \sum_{i=1}^{n} |x_i| \qquad = \text{Sum of magnitudes;}
$$

$$\|x\|_2 = \left(\sum_{i=1}^{n} x_i^2 \right)^{1/2} = \text{Euclidean norm;}$$

$$\|x\|_\infty = \max_{1 \le i \le n} |x_i| = \text{Maximum-magnitude norm.}$$

Which of these vector norms is best to use may depend on the problem. In most cases, satisfactory results are obtained with any of these measures of the "size" of a vector.

EXAMPLE Compute the 1-, 2-, and ∞-norms of the vector x, if $x = (1.25, 0.02, -5.15, 0)$.

$$\|x\|_1 = |1.25| + |0.02| + |-5.15| + |0| = 6.42;$$
$$\|x\|_2 = [(1.25)^2 + (0.02)^2 + (-5.15)^2 + (0)^2]^{1/2} = 5.2996;$$
$$\|x\|_\infty = |-5.15| = 5.15. \quad \blacksquare$$

The norms of a matrix are developed by a correspondence to vector norms. Matrix norms that correspond to the above, for matrix A, can be shown to be

$$\|A\|_1 = \max_{1 \le j \le n} \sum_{i=1}^{n} |a_{ij}| = \text{Maximum column sum;}$$

$$\|A\|_\infty = \max_{1 \le i \le n} \sum_{j=1}^{n} |a_{ij}| = \text{Maximum row sum.}$$

The matrix norm $|A\|_2$ that corresponds to the 2-norm of a vector is not readily computed. It is related to the eigenvalues of the matrix (which we discuss in a later chapter). It sometimes has special utility because no other norm is smaller than this norm. It therefore provides the "tightest" measure of the magnitude of a matrix, but is also the most difficult to compute. This norm is also called the *spectral norm*.

For an $m \times n$ matrix, we can paraphrase the Euclidean (also called Frobenius) norm* and write

$$\|A\|_e = \left(\sum_{i=1}^{m} \sum_{j=1}^{n} a_{ij}^2 \right)^{1/2}$$

---

*Be alert to a situation that may be confusing: For a vector, the 2-norm is the same as the Euclidean norm, but for a matrix, the 2-norm is not the same as the Euclidean norm.

EXAMPLE Compute the Euclidean norms of A, B, and C, and the ∞-norms, given that

$$A = \begin{bmatrix} 5 & 9 \\ -2 & 1 \end{bmatrix}; \quad B = \begin{bmatrix} 0.1 & 0 \\ 0.2 & -0.1 \end{bmatrix}; \quad \text{and} \quad C = \begin{bmatrix} 0.2 & 0.1 \\ 0.1 & 0 \end{bmatrix}.$$

$$\|A\|_e = \sqrt{25 + 81 + 4 + 1} = \sqrt{111} = 10.53; \qquad \|A\|_\infty = 14.$$

$$\|B\|_e = \sqrt{0.01 + 0 + 0.04 + 0.01} = \sqrt{0.06} = 0.2449; \qquad \|B\|_\infty = 0.3.$$

$$\|C\|_e = \sqrt{0.04 + 0.01 + 0.01 + 0} = \sqrt{0.06} = 0.2449; \qquad \|C\|_\infty = 0.3.$$

The results of our examples look quite reasonable; certainly A is "larger" than B or C. While $B \neq C$, both are equally "small." The Euclidean norm is a good measure of the magnitude of a matrix. ∎

We see then that there are a number of ways that the norm of a matrix can be expresssed. Which way is preferred? There are certainly differences in their cost; for example, some will require more extensive arithmetic than others. The spectral norm is usually the most "expensive." Which norm is best? The answer to this question depends in part on the use for the norm. In most instances, we want the norm that puts the smallest upper bound on the magnitude of the matrix. In this sense, the spectral norm is "best." We observe, in the next example, that not all the norms give the same value for the magnitude of a matrix.

EXAMPLE

$$A = \begin{bmatrix} 5 & -5 & -7 \\ -4 & 2 & -4 \\ -7 & -4 & 5 \end{bmatrix}.$$

$$\|A\|_e = \text{Euclidean norm} = 15;$$

$$\|A\|_\infty = 17;$$

$$\|A\|_1 = 16;$$

$$\|A\|_2 = \text{Spectral norm} = 12.$$

If a matrix is a diagonal matrix, all p-norms have the same value, however. ∎

Why are norms important? For one thing, they let us express the accuracy of the solution to a set of equations in quantitative terms by stating the norm of the error vector (the true solution minus the approximate solution vector). Norms are also used to study quantitatively the convergence of iterative methods of solving linear systems (which we will cover in a later section).

2.9 CONDITION NUMBERS AND ERRORS IN SOLUTIONS

When we solve a set of linear equations, $Ax = b$, we hope that the calculated vector $\bar{x}$ is a close representation of the true solution vector x. In the previous examples we have seen how round-off can make the computed solution differ from the exact solution, and several times the use of pivoting has been recommended as a way to minimize the effect.

The next example will demonstrate how much the accuracy can be improved by pivoting. We will exaggerate the effect by using only four-digit arithmetic (rounding to four digits after each operation). The same effects are observed with more precise arithmetic in large systems.

Given $Ax = b$ with

$$A = \begin{bmatrix} -0.002 & 4.000 & 4.000 \\ -2.000 & 2.906 & -5.387 \\ 3.000 & -4.031 & -3.112 \end{bmatrix}$$

and

$$b = (7.998, -4.481, -4.143)^T.$$

After Gaussian elimination without pivoting, the triangularized matrix (augmented with the b vector) is

$$\begin{bmatrix} -0.002 & 4.000 & 4.000 & 7.998 \\ 0 & -3.997 & -4.005 & -8.002 \\ 0 & 0 & -10.00 & 0.000 \end{bmatrix},$$

which gives the computed solution

$$\bar{x} = (-1496, 2.000, 0.000)^T.$$

The exact solution is $x = (1.000, 1.000, 1.000)^T$, and we see that round-off errors (particularly obvious in the last row of the triangularized system) have given a completely incorrect result. In this example the very small value of a_{11} has been the source of difficulty.

If the equations are reordered to give

$$A = \begin{bmatrix} 3.000 & -4.031 & -3.112 \\ -0.002 & 4.000 & 4.000 \\ -2.000 & 2.906 & -5.387 \end{bmatrix}, \quad \text{and} \quad b = \begin{bmatrix} -4.413 \\ 7.998 \\ -4.481 \end{bmatrix},$$

and we repeat the four-digit computations, the triangularized system becomes

$$\begin{bmatrix} 3.000 & -4.031 & -3.112 & -4.143 \\ 0 & 3.997 & 3.998 & 7.995 \\ 0 & 0 & -7.681 & -7.681 \end{bmatrix},$$

which gives $\bar{x} = (1.000, 1.000, 1.000)^T$, whose error is nil.

We conclude from the above that the algorithm that we employ can have a very significant effect on the accuracy.

Unfortunately, the error due to round-off is sometimes large even with the best available algorithm, because the problem itself may be very sensitive to the effects of small errors. Consider this example:

Given $Ax = b$, with

$$A = \begin{bmatrix} 3.02 & -1.05 & 2.53 \\ 4.33 & 0.56 & -1.78 \\ -0.83 & -0.54 & 1.47 \end{bmatrix}, \quad \text{and} \quad b = \begin{bmatrix} -1.61 \\ 7.23 \\ -3.38 \end{bmatrix}.$$

If we solve the system by Gaussian elimination, using pivoting and carrying three digits rounded, the triangularized system is

$$\begin{bmatrix} 4.33 & 0.56 & -1.78 & 7.23 \\ 0 & -1.44 & 3.77 & -6.65 \\ 0 & -0 & -0.00362 & 0.00962 \end{bmatrix}.$$

The computed solution is $\bar{x} = (0.880, -2.35, -2.66)^T$ while the true solution vector is $x = (1, 2, -1)^T$. The very small number on the diagonal in the third row is a sign of such inherent sensitivity to round-off. Such a system is called ill-conditioned. One strategy to use with ill-conditioned systems is greater precision in the arithmetic operations. For the above example, when six-digit computations are used, we get marked improvement:

$$x = (0.9998, 1.9995, -1.002)^T;$$

but note that there is still an appreciable error. A large ill-conditioned system would require even more digits of accuracy than six if we wished to compute a solution anywhere near the exact answer.

The occurrence of an unavoidable small value on the diagonal in an ill-conditioned system indicates that the determinant of the original coefficient matrix tends to be a small number. (We can make this statement because we can evaluate the determinant of a matrix by triangularizing it and then multiplying the elements on the diagonal.) Since a zero value for the determinant occurs only for a singular matrix, we can say that an ill-conditioned matrix is "nearly singular." This is hardly a quantitative measure of the property, however, so we will look for a different way to express the degree of ill-conditioning.

There are some important problems involving linear systems in which the coefficient matrix is always nearly singular, and hence ill-conditioned. One example is fitting a polynomial of relatively high degree to a set of points by the least-squares method. This is covered in another chapter.

Another way to look at ill-conditioned systems is to examine the effect of small changes in the coefficients. In the above system, changing the a_{11} coefficient from 3.02 to 3.00 gives a solution (exact) of

$$x_3 = 0.05856, \qquad x_2 = 4.75637, \qquad x_1 = 1.07968.$$

This property of ill-conditioned systems—that their solution is extremely sensitive to small changes in the coefficients—explains why they are also so sensitive to round-off error. The small inaccuracies in values as the computation proceeds, caused by round-off errors, are equivalent to numbers we would encounter in using exact arithmetic on a problem with slightly altered coefficients.

A more vivid conception of ill-conditioning is obtained by examining a system of two equations with two unknowns.

Consider the system

$$\begin{bmatrix} 1.01 & 0.99 \\ 0.99 & 1.01 \end{bmatrix} \begin{bmatrix} x \\ y \end{bmatrix} = \begin{bmatrix} 2 \\ 2 \end{bmatrix},$$

which has the obvious solution $x = 1$, $y = 1$. The set of equations can be represented by two straight lines that intersect at the point $(1, 1)$. The lines are very nearly parallel, as Fig. 2.2(a) shows. Suppose there is some uncertainty in the coefficients: The position of the lines might be anywhere in the fuzzy region as indicated by Fig. 2.2(b), and a corresponding amplified uncertainty in the point of intersection results. On the other hand, as indicated by Fig. 2.2(c), lines that are not almost parallel do not show an amplified uncertainty in their point of intersection when the position of the lines is not known precisely. If two lines are perpendicular, the uncertainty in the point of intersection is no greater than the uncertainty in the lines themselves.

One can consider linearly dependent vectors of order 2 as representing parallel lines. The analogy of a consistent set of right-hand sides, with a corresponding redundancy in the set of equations, is two coincident straight lines. There is an infinite set of solutions because the lines "intersect" everywhere.

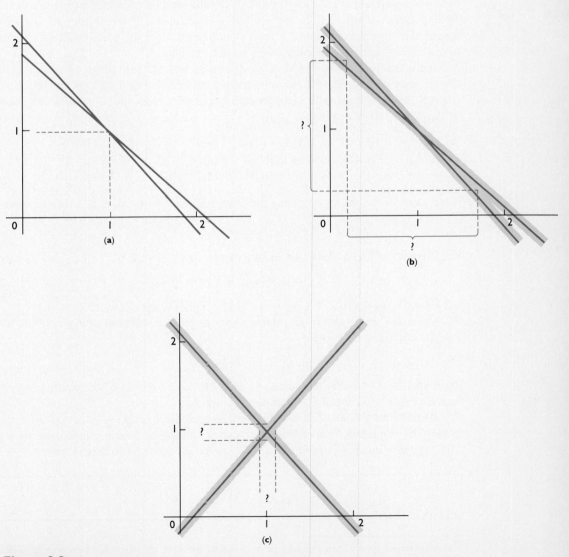

Figure 2.2

Two noncoincident parallel lines are analogous to an inconsistent set of equations with a singular coefficient matrix; there is no solution and no intersection of the lines. The visual picture of linear dependence in three-dimensional space would involve parallel planes. We need to think of parallel hyperplanes in higher-dimensional spaces.

A simple way to detect ill-conditioning would be to make a deliberate change in some of the coefficients and determine the degree of change in the solution. An even simpler way is to examine the values on the diagonal of the triangularized system, but a computer subroutine may not always exhibit these values. When ill-conditioning is suspected, one might compare the solution obtained using single precision with that when double precision is used. Then both the presence of ill-conditioning is observed and a better approximation of the solution is obtained.

In some situations, one can combat ill-conditioning by transforming the problem into an equivalent set of equations that are not ill-conditioned. The efficiency of this scheme is related to the relative amount of computation required for the transformation, compared to the cost of doing the calculations in higher precision.*

An interesting phenomenon of an ill-conditioned system is that we cannot test for the accuracy of the computed solution merely by substituting it into the equations to see whether the right-hand sides are reproduced. Consider again the ill-conditioned example we have previously examined:

$$A = \begin{bmatrix} 3.02 & -1.05 & 2.53 \\ 4.33 & 0.56 & -1.78 \\ -0.83 & -0.54 & 1.47 \end{bmatrix}, \qquad b = \begin{bmatrix} -1.61 \\ 7.23 \\ -3.38 \end{bmatrix}.$$

If we compute the vector Ax, using the exact solution $x = (1, 2, -1)^T$, we of course get

$$Ax = (-1.61, 7.23, -3.38)^T = b.$$

But if we substitute a clearly erroneous vector

$$\bar{x} = (0.880, -2.35, -2.66)^T,$$

we get $A\bar{x} = (-1.6047, 7.2292, -3.3716)^T$, which is very close to b.

We define the residual of a solution vector as the difference between b and $A\bar{x}$, where $\bar{x}$ is the computed solution:

$$r = b - A\bar{x}.$$

Our example shows that the norm of r is not a good measure of the norm of the error vector ($e = x - \bar{x}$) for an ill-conditioned system.

Because the degree of ill-condition of the coefficient matrix is so important in determining the magnitude of round-off effects, it is valuable to have a quantitative measure. The *condition number* is normally defined as the product of two matrix norms:

$$\text{Condition}(A) = \|A\| \, \|A^{-1}\|.$$

---

*Double precision is not required throughout the computations. When the system is solved through LU decomposition, the accumulation of inner products in double precision is sufficient.

Unfortunately this is not an inexpensive quantity to compute, for it requires us to invert A. Because inverting a matrix amounts to solving a linear system (solving it with n different right-hand sides, actually), and the computed solution for an ill-conditioned system may be inexact, we will not compute A^{-1} very accurately. This suggests that the condition number will not be computed very exactly either. Ordinarily this causes no great difficulty; if the conditon number is large, we know we are in serious trouble. Observe that condition numbers will always be at least unity, which corresponds to the condition number of the identity matrix.

For our previous example, we have

$$A = \begin{bmatrix} 3.02 & -1.05 & 2.53 \\ 4.33 & 0.56 & -1.78 \\ -0.83 & -0.54 & 1.47 \end{bmatrix}, \quad A^{-1} = \begin{bmatrix} 5.661 & -7.273 & -18.55 \\ 200.5 & -268.3 & -669.9 \\ 76.85 & -102.6 & -255.9 \end{bmatrix}.$$

Using matrix ∞-norms, we find that the condition number is

$$\|A\|\,\|A^{-1}\| = (6.67)(1138.7) = 7595.$$

The elements of A^{-1} will be large relative to the elements of A when A is ill-conditioned. However, this can also be true when the elements of A are small, even in the absence of ill-conditioning. Multiplying the two norms has a normalizing effect, so the condition number is large only for an ill-conditioned system.

The condition number lets us relate the magnitude of the error in the computed solution to the magnitude of the residual. We use norms to express the magnitude of the vectors.

Let $e = x - \bar{x}$, where x is the exact solution to $Ax = b$ and $\bar{x}$ is an approximate solution. Let $r = b - A\bar{x}$, the residual. Since $Ax = b$, we have

$$r = b - A\bar{x} = Ax - A\bar{x} = A(x - \bar{x}) = Ae. \tag{2.10}$$

Hence,

$$e = A^{-1}r.$$

Taking norms and recalling Eq. (2.9), line 4, for a product, we write

$$\|e\| \le \|A^{-1}\|\,\|r\|. \tag{2.11}$$

From $r = Ae$, we also have $\|r\| \le \|A\|\,\|e\|$, which combines with Eq. (2.11) to give

$$\frac{\|r\|}{\|A\|} \le \|e\| \le \|A^{-1}\|\,\|r\|. \tag{2.12}$$

Applying the same reasoning to $Ax = b$ and $x = A^{-1}b$, we get

$$\frac{\|b\|}{\|A\|} \le \|x\| \le \|A^{-1}\|\,\|b\|. \tag{2.13}$$

Taking Eqs. (2.12) and (2.13) together, we reach a most important relationship:

$$\frac{1}{\|A\| \, \|A^{-1}\|} \frac{\|r\|}{\|b\|} \leq \frac{\|e\|}{\|x\|} \leq \|A\| \, \|A^{-1}\| \frac{\|r\|}{\|b\|}, \qquad (2.14)$$

or

$$\frac{1}{(\text{Condition no.})} \frac{\|r\|}{\|b\|} \leq \frac{\|e\|}{\|x\|} \leq (\text{Condition no.}) \frac{\|r\|}{\|b\|}.$$

Equation (2.14) shows that the relative error in the computed solution vector $\bar{x}$ can be as great as the relative residual multiplied by the condition number. Of course it can also be as small as the relative residual divided by the condition number. Therefore, when the condition number is large, the residual gives little information about the accuracy of $\bar{x}$. Conversely, when the condition number is near unity, the relative residual is a good measure of the relative error of $\bar{x}$.

When we solve a linear system, we are normally doing so to determine values for a physical system for which the set of equations is a model. We use the measured values of the parameters of the physical system to evaluate the coefficients of the equations, so we expect these coefficients to be known only as precisely as the measurements. When these are in error, the solution of the equations will reflect these errors. We have already seen that an ill-conditioned system is extremely sensitive to small changes in the coefficients. The condition number lets us relate the change in the solution vector to such errors in the coefficients of the set of equations $Ax = b$.

Assume that the errors in measuring the parameters cause errors in the coefficients of A so that the actual set of equations being solved is $(A + E)\bar{x} = b$, where $\bar{x}$ represents the solution of the perturbed system and A represents the true (but unknown) coefficients. We let $\bar{A} = A + E$ represent the perturbed coefficient matrix. We desire to know how large $x - \bar{x}$ is.

Using $Ax = b$ and $\bar{A}\bar{x} = b$, we can write

$$\begin{aligned} x = A^{-1}b = A^{-1}(\bar{A}\bar{x}) &= A^{-1}(A + \bar{A} - A)\bar{x} \\ &= [I + A^{-1}(\bar{A} - A)]\bar{x} \\ &= \bar{x} + A^{-1}(\bar{A} - A)\bar{x}. \end{aligned}$$

Since $\bar{A} - A = E$, we have

$$x - \bar{x} = A^{-1}E\bar{x}.$$

Taking norms, we get

$$\|x - \bar{x}\| \leq \|A^{-1}\| \, \|E\| \, \|\bar{x}\| = \|A^{-1}\| \, \|A\| \, \frac{\|E\|}{\|A\|} \|\bar{x}\|,$$

so that

$$\frac{\|x - \bar{x}\|}{\|\bar{x}\|} \leq (\text{Condition no.}) \, \frac{\|E\|}{\|A\|}. \qquad (2.15)$$

This says that the error of the solution relative to the norm of the computed solution can be as large as the relative error in the coefficients of A multiplied by the condition number. The net effect is that, if the coefficients of A are known to only four-digit precision and the condition number is 1000, the computed vector x may have only one digit of accuracy.

When the solution to the system $Ax = b$ has been computed, and, because of round-off error, we obtain the approximate solution vector $\bar{x}$, it is possible to apply iterative improvement to correct $\bar{x}$ so that it more closely agrees with x. Define $e = x - \bar{x}$. Define $r = b - A\bar{x}$. As shown above [Eq. (2.10)],

$$Ae = r. \qquad (2.16)$$

If we could solve this equation for e, we could apply this as a correction to $\bar{x}$. Furthermore, if $\|e\|/\|\bar{x}\|$ is small, it means that $\bar{x}$ should be close to x. In fact, if the value of $\|e\|/\|\bar{x}\|$ is 10^{-p}, we know that $\bar{x}$ is probably correct to p digits.

The process of iterative improvement is based on solving Eq. (2.16). Of course this is also subject to the same round-off error as the original solution of the system for $\bar{x}$, so we actually get $\bar{e}$, an approximation to the true error vector. Even so, unless the system is so ill-conditioned that $\bar{e}$ is not a reasonable approximation to e, we will get an improved estimate of x from $\bar{x} + \bar{e}$. One special caution is important to observe: The computation of the residual vector r must be as precise as possible. One always uses double-precision arithmetic; otherwise iterative improvement will not be successful. An example will make this clear.

Given

$$A = \begin{bmatrix} 4.23 & -1.06 & 2.11 \\ -2.53 & 6.77 & 0.98 \\ 1.85 & -2.11 & -2.32 \end{bmatrix}, \qquad b = \begin{bmatrix} 5.28 \\ 5.22 \\ -2.58 \end{bmatrix},$$

whose true solution is

$$x = \begin{bmatrix} 1.000 \\ 1.000 \\ 1.000 \end{bmatrix}.$$

If three-digit chopped arithmetic is used, the approximate solution vector is $\bar{x} = (0.991, 0.997, 1.000)^T$. Using double precision, we compute $A\bar{x}$ and the residual as

$$\bar{A}x = \begin{bmatrix} 5.24511 \\ 5.22246 \\ -2.59032 \end{bmatrix}, \qquad r = \begin{bmatrix} 0.0348 \\ -0.00246 \\ 0.0103 \end{bmatrix}.$$

We now solve $A\bar{e} = r$, again using three-digit precision, and get

$$\bar{e} = \begin{bmatrix} 0.00822 \\ 0.00300 \\ -0.00000757 \end{bmatrix}.$$

Finally, correcting $\bar{x}$ with $\bar{x} + \bar{e}$ gives almost exactly the correct solution:

$$\bar{x} + \bar{e} = \begin{bmatrix} 0.999 \\ 1.000 \\ 1.000 \end{bmatrix}.$$

In the general case, the iterations are repeated until the corrections are negligible. Since we want to make the solution of Eq. (2.16) as economical as possible, we should use an *LU* method to solve the original system and apply the *LU* to Eq. (2.16).

2.10   ITERATIVE METHODS

As opposed to the direct method of solving a set of linear equations by elimination, we now discuss iterative methods. In certain cases, these methods are preferred over the direct methods—when the coefficient matrix is sparse (has many zeros) they may be more rapid.* They may be more economical in memory requirements of a computer. For hand computation they have the distinct advantage that they are self-correcting if an error is made; they may sometimes be used to reduce round-off error in the solutions computed by direct methods, as discussed above. They can also be applied to sets of nonlinear equations.

We illustrate the method by a simple example:

$$\begin{aligned} 8x_1 + x_2 - x_3 &= 8, \\ 2x_1 + x_2 + 9x_3 &= 12, \\ x_1 - 7x_2 + 2x_3 &= -4. \end{aligned} \qquad (2.17)$$

---

*When the occurrence of zeros follows some easy pattern, elimination methods can take advantage of this. See the program for a tridiagonal system at the end of the chapter.

The solution is $x_1 = 1$, $x_2 = 1$, $x_3 = 1$. We begin our iterative scheme by solving each equation for one of the variables, choosing, when possible, to solve for the variable with largest coefficient:

$$\begin{aligned} x_1 &= 1 & - 0.125x_2 + 0.125x_3 & \quad \text{(from first equation)}, \\ x_2 &= 0.571 + 0.143x_1 & + 0.286x_3 & \quad \text{(from third equation)}, \\ x_3 &= 1.333 - 0.222x_1 - 0.111x_2 & & \quad \text{(from second equation)}. \end{aligned} \qquad (2.18)$$

We begin with some initial approximation to the value of the variables. (Each component might be taken equal to zero if no better initial estimates are at hand.) Substituting these into the right-hand sides of the set of equations generates new approximations that are closer to the true value. The new values are substituted into the right-hand sides to generate a second approximation, and the process is repeated until successive values of each of the variables are sufficiently alike. For the set of equations given above we get

Successive Estimates of Solution (Jacobi Method)

	First	Second	Third	Fourth	Fifth	Sixth	Seventh	Eighth
x_1	0	1.000	1.095	0.995	0.993	1.002	1.001	1.000
x_2	0	0.571	1.095	1.026	0.990	0.998	1.001	1.000
x_3	0	1.333	1.048	0.969	1.000	1.004	1.001	1.000

Note that this method is exactly the same as the method of iteration for a single equation that was discussed in Chapter 1, but applied to a *set* of equations, for we may write Eq. (2.18) in the form

$$x^{(n+1)} = Gx^{(n)} = b' - Bx^{(n)}, \qquad (2.19)$$

which is identical in form to $x_{n+1} = g(x_n)$ as used in Chapter 1.

In the present context, of course, $x^{(n)}$ and $x^{(n+1)}$ refer to the nth and $(n + 1)$st iterates of a vector rather than a simple variable, and G is a linear transformation rather than a nonlinear function. For the example above, we restate Eq. (2.17) in matrix form after interchanging equations 2 and 3:

$$\begin{bmatrix} 8 & 1 & -1 \\ 1 & -7 & 2 \\ 2 & 1 & 9 \end{bmatrix} \begin{bmatrix} x \\ y \\ z \end{bmatrix} = \begin{bmatrix} 8 \\ -4 \\ 12 \end{bmatrix}, \qquad Ax = b. \qquad (2.20)$$

Now, let $A = L + D + U$, where

$$L = \begin{bmatrix} 0 & 0 & 0 \\ 1 & 0 & 0 \\ 2 & 1 & 0 \end{bmatrix}, \qquad D = \begin{bmatrix} 8 & 0 & 0 \\ 0 & -7 & 0 \\ 0 & 0 & 9 \end{bmatrix}, \qquad U = \begin{bmatrix} 0 & 1 & -1 \\ 0 & 0 & 2 \\ 0 & 0 & 0 \end{bmatrix}.$$

Then Eq. (2.20) above can be rewritten as

$$Ax = (L + D + U) x = b, \text{ or}$$
$$Dx = -(L + U) x + b, \text{ which gives}$$
$$x = -D^{-1}(L + U) x + D^{-1}b.$$

From this we have, identifying x on the left as the new iterate,

$$x^{(n+1)} = -D^{-1} (L + U) x^{(n)} + D^{-1}b. \tag{2.21}$$

In Eq. (2.19) we see that

$$b' = D^{-1}b = \begin{bmatrix} 1.000 \\ 0.571 \\ 1.333 \end{bmatrix},$$

$$B = D^{-1}(L + U) = \begin{bmatrix} 0 & 0.124 & -0.125 \\ -0.143 & 0 & -0.286 \\ 0.222 & 0.111 & 0 \end{bmatrix}.$$

The procedure we have just described is known as the *Jacobi method*, also called "the method of simultaneous displacements," because each of the equations is simultaneously changed by using the most recent set of x-values.

We can write the algorithm for the Jacobi iterative method as follows:

Algorithm for Jacobi Iteration

To solve a system of N linear equations, rearrange the rows so that the diagonal elements have magnitudes as large as possible relative to the magnitudes of other coefficients in the same row. Define the rearranged system as $Ax = b$. Beginning with an initial approximation to the solution vector, $x^{(1)}$, compute each component of $x^{(n+1)}$ for $i = 1, 2, \ldots, N$, by

$$x_i^{(n+1)} = \frac{b_i}{a_{ii}} - \sum_{\substack{j=1, \\ j \neq i}}^{N} \frac{a_{ij}}{a_{ii}} x_j^{(n)}, \quad n = 1, 2, \ldots$$

A sufficient condition for convergence is that

$$|a_{ii}| > \sum_{\substack{j=1, \\ j \neq i}}^{N} |a_{ij}|, \quad i = 1, 2, \ldots, N.$$

When this is true, $x^{(n)}$ will converge to the solution no matter what initial vector is used.

Actually, the x-values of the next trial are not all calculated "simultaneously" when we perform the Jacobi method. In the above, we calculated the second estimate of x_1 before we did the x_2, and new values of both x_1 and x_2 were available before we improved the value of x_3. In nearly all cases the new values are better than the old, and should be used in preference to the poorer values.* When this is done, the method is known by the name *Gauss–Seidel*. In this method our first step is to rearrange the set of equations by solving each equation for one of the variables in terms of the others, exactly as we have done above in the Jacobi method. One then proceeds to improve each x-value in turn, using always the most recent approximations to the values of the other variables. The rate of convergence is more rapid, as shown by reworking the same example as earlier (Eqs. (2.18), rearranged form of (2.17)):

Successive Estimates of Solution (Gauss–Seidel Method)

	First	Second	Third	Fourth	Fifth	Sixth
x_1	0	1.000	1.041	0.997	1.001	1.000
x_2	0	0.714	1.014	0.996	1.000	1.000
x_3	0	1.032	0.990	1.002	1.000	1.000

These values were computed by using Eq. (2.18) in the following way:

$$x_1^{(n+1)} = 1 \qquad\qquad -0.125x_2^{(n)} \qquad + 0.125x_3^{(n)},$$
$$x_2^{(n+1)} = 0.571 + 0.143x_1^{(n+1)} \qquad\qquad + 0.286x_3^{(n)},$$
$$x_3^{(n+1)} = 1.333 - 0.222x_1^{(n+1)} - 0.111x_2^{(n+1)},$$

beginning with $x^{(1)} = (0, 0, 0)^T$.

An algorithm for Gauss–Seidel iteration is as follows:

Algorithm for Gauss–Seidel Iteration

To solve a system of N linear equations, rearrange the rows so that the diagonal elements have magnitudes as large as possible relative to the magnitudes of other coefficients in the same row. Define the rearranged system as $Ax = b$. Beginning with an initial approximation to the solution vector, $x^{(1)}$, compute each component of $x^{(n+1)}$, for $i = 1, 2, \ldots, N$, by

$$x_i^{(n+1)} = \frac{b_i}{a_{ii}} - \sum_{j=1}^{i-1} \frac{a_{ij}}{a_{ii}} x_j^{(n+1)} - \sum_{j=i+1}^{N} \frac{a_{ij}}{a_{ii}} x_j^{(n)}, \qquad n = 1, 2, \ldots$$

A sufficient condition for convergence is that

$$|a_{ii}| > \sum_{\substack{j=1, \\ j\neq i}}^{N} |a_{ij}|, \qquad i = 1, 2, \ldots, N.$$

When this is true, $x^{(r)}$ will converge to the solution no matter what initial vector is used.

---

*There may be times when this obvious move is not beneficial, due to cancellation of errors, but in general this is good strategy in numerical analysis.

The Gauss–Seidel method will generally converge if the Jacobi method converges, and will do so more rapidly. However, the Jacobi method might still be the preferred method if we were running our program on parallel processors, since all n equations could be solved simultaneously at each iteration. In fact, other routines once thought "obsolete" may have new life due to parallel processing.

The matrix formulation for the Gauss–Seidel method is almost like the one given in Eq. (2.21). Here $Ax = b$ is rewritten as

$$(L + D)x = -Ux + b. \tag{2.22}$$

From this we have

$$x^{(n+1)} = -D^{-1}Lx^{(n+1)} - D^{-1}Ux^{(n)} + D^{-1}b. \tag{2.23}$$

(Compare this with Eq. (2.21).)

$$D^{-1}L = \begin{bmatrix} 0 & 0 & 0 \\ -0.143 & 0 & 0 \\ 0.222 & 0.111 & 0 \end{bmatrix}, \text{ and}$$

$$D^{-1}U = \begin{bmatrix} 0 & 0.125 & 0.125 \\ 0 & 0 & -0.286 \\ 0 & 0 & 0 \end{bmatrix}.$$

The usefulness of the matrix notation will be more apparent in Chapter 6, when we study the eigenvalues of the matrices $\boxed{D^{-1}(L + U)}$ in Eq. (2.21), and $\boxed{(L + D)^{-1} U}$ in Eq. (2.22). The eigenvalues of these matrices explain the rates of convergence of both methods.

These iteration methods will not converge for all sets of equations, nor for all possible rearrangements of the equations. When the equations can be ordered so that each diagonal entry is larger in magnitude than the sum of the magnitudes of the other coefficients in that row (such a system is called *diagonally dominant*), the iteration will converge for any starting values. This is easy to visualize, because all the equations can be put in the form

$$x_i = \frac{b_i}{a_{ii}} - \frac{a_1}{a_{ii}} x_1 - \frac{a_2}{a_{ii}} x_2 - \ldots \tag{2.24}$$

The error in the next value of x_i will be the sum of the errors in all other x's multiplied by the coefficients of Eq. (2.24), and if the sum of the magnitudes of the coefficients is less than unity, the error will decrease as the iteration proceeds. The preceding convergence condition is a sufficient condition only; that is, if the condition holds, the system always converges, but sometimes the system converges even if the condition is violated.

The speed with which the iterations converge is obviously related to the degree of dominance of the diagonal terms, for the coefficients in Eq. (2.24) are then smaller and x_i is less affected by the errors in the other components. When the initial approximation is close to the solution vector, relatively few iterations are needed to get an acceptable solution.

2.11   RELAXATION METHOD

There is an iteration method that is more rapidly convergent than Gauss–Seidel and that can be used to advantage for hand calculations. It is unfortunately not well adapted to computer application. The method is due to a British engineer, Richard Southwell, and has been applied to a wide variety of problems. (Allen (1954) is an excellent reference.) We discuss the method because of its historical importance and because it leads to an important acceleration technique called *overrelaxation*.

If we consider the Gauss–Seidel scheme, we realize that the order in which the equations are used is important. We should improve that x that is most in error, since, in the rearranged form, that variable does not appear on the right, and hence its own error will not affect the next iterate. By using that equation, then, we introduce lesser errors into the computation of the next iterate. The *method of relaxation* is a scheme that permits one to select the best equation to be used for maximum rate of convergence.

We illustrate the method by the same example we solved in Section 2.10. The original equations are

$$8x_1 + x_2 - x_3 = 8,$$
$$2x_1 + x_2 + 9x_3 = 12, \qquad (2.25)$$
$$x_1 - 7x_2 + 2x_3 = -4.$$

We again begin by a rearrangement of the equations, but different from that for the Gauss–Seidel or Jacobi methods. We transpose all the terms to one side, and then divide by the negative of the largest coefficient. Equations (2.25) become

$$-x_1 - 0.125x_2 + 0.125x_3 + 1 \quad = 0,$$
$$-0.222x_1 - 0.111x_2 - \quad x_3 + 1.333 = 0, \qquad (2.26)$$
$$0.143x_1 - \quad x_2 + 0.286x_3 + 0.571 = 0.$$

If we begin with some initial set of values, and substitute in Eqs. (2.26), the equations will not be satisfied (unless, by chance, we have stumbled onto the solution); the left sides will not be zero, but some other value that we call the *residual* and denote by R_i. It is also convenient to reorder the equation so the -1 coefficients are on the diagonal. Equations (2.26) become, with these rearrangements,

$$-x_1 - 0.125x_2 + 0.125x_3 + 1 \quad = R_1,$$
$$0.143x_1 - \quad x_2 + 0.286x_3 + 0.571 = R_2,$$
$$-0.222x_1 - 0.111x_2 - \quad x_3 + 1.333 = R_3.$$

For example, with $x_1 = 0$, $x_2 = 0$, $x_3 = 0$, we have

$$R_1 = 1, \quad R_2 = 0.571, \quad R_3 = 1.333.$$

The largest residual in magnitude, R_3, tells us that the third equation is most in error and should be improved first. The method gets its name "relaxation" from the fact that we make a change in x_3 to relax R_3 (the greatest residual) so as to make it zero. Observing the coefficients of the various equations, we see that increasing the value of x_3 by one, say, will decrease R_3 by one, will increase R_1 by 0.125, and increase R_2 by 0.286. To change R_3 from its initial value of 1.333 to zero, we should increase x_3 by that same amount.

We then select the new residual of greatest magnitude, and relax it to zero. We continue until all residuals are zero, and when this is true, the values of the x's will be at the exact solution. In implementing this method, there are some modifications that make the work easier. We illustrate in Fig. 2.3.

We make three double columns, one for each variable and for the residual of the equation in which that variable appears with -1 coefficient. The initial x values and the

Eq. No.	x_1	x_2	x_3
I	-1	-0.125	0.125
II	0.143	-1	0.286
III	-0.222	-0.111	-1

x_1	R_{I}	x_2	R_{II}	x_3	R_{III}
0	~~1000~~	0	~~571~~	0	~~1333~~
	+167		+381		
	~~1167~~		~~952~~	+1333	∅
			+167		−259
+1167	∅		~~1119~~		−259
	−140				−124
	~~−140~~	+1119	∅		~~−383~~
	− 48		−109		
	~~−189~~		~~−109~~	−383	∅
			− 27		+ 42
−189	∅		~~−136~~		~~42~~
	+ 17				+ 15
	~~17~~	−136	∅		~~57~~
	+ 7		+ 16		
	~~24~~		~~16~~	+57	∅
			+ 3		− 5
+24	∅		~~19~~		~~−5~~
	− 2				− 2
	~~−2~~	+19	∅		~~−7~~
	− 1		− 2		
	~~−3~~		~~−2~~	−7	∅
			0		+ 1
−3	∅		−2		~~1~~
	0				0
	∅	−2	∅		~~1~~
	0		0	+1	0
999		1000		1001	
Check residuals: 1		−1		0	

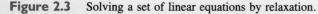

Figure 2.3 Solving a set of linear equations by relaxation.

initial residuals are entered as the first row of the table. It is convenient to work entirely with integers by multiplying the initial x-values and residuals by 1000, and then to scale down the solution by dividing by 1000 at the end of the computations. We avoid fractions; if a fractional change in a variable is needed to relax to zero we only relax to near zero.

In Fig. 2.3, we set down the increments to the x's but record the cumulative effect on the residuals. (The old values of the residuals are crossed out when replaced by a new value.) When the residuals are zero, we add the various increments to the initial value to get the final value. In this example, round-off errors cause an error of one in the third decimal.

It is important to make a final check by recomputing residuals at the end of the calculation to check for mistakes in arithmetic. The method is not usually programmed because searching on the computer for the largest residual is slow, and adds enough execution time that the acceleration gives no net benefit. The search can be done rapidly by scanning the residuals in a hand calculation, however.

Southwell and his coworkers observed, for many situations, that relaxing the residuals to zero was less efficient than relaxing beyond zero (*overrelaxing*) or relaxing short of zero (*underrelaxing*). The reason this is an improved strategy is that a zero residual doesn't stay zero; relaxing the residual of another equation affects the first residual, so it is appropriate to anticipate and allow for this by an appropriate under- or overrelaxation.

Table 2.1 shows that a significant improvement in the speed of convergence is obtained if R_1 is underrelaxed by 10% and R_3 is underrelaxed by 25%. Unfortunately the optimum

Table 2.1   Accelerated solution of linear equations by relaxation

Eq. No.	x_1	x_2	x_3
I	−1	−0.125	0.125
II	0.143	−1	0.286
III	−0.222	−0.111	−1

	R_{I}	x_2	R_{II}	x_3	R_{III}
0	1000	0	571	0	1333
	+125		+286		
	1125		857	+1000	333
			+144		−225
+1013	112		1001		108
	−125				−111
	−13	+1001	0		−3
			−2		+3
−12	−1		−2		0
	+0				+0
	−1	−2	0		0
			+0		+0
−1	0		0		0
	1000	999		1000	

degree of under- or overrelaxation is not easily determined. In many problems, acceleration is obtained by overrelaxing rather than underrelaxing.

Even though Southwell's relaxation method is not often used today, there is one aspect of it that has influence on the iterative solution of linear equations by computer. In using the Gauss–Seidel method, we can speed up the convergence by "overrelaxation," that is, by making the residuals go to the other side of zero instead of just relaxing to zero as in the first example. We can apply this to Gauss–Seidel iteration by modifying the algorithm.

The standard relationship for Gauss–Seidel iteration for the set of equations $Ax = b$, for variable x_i, can be written

$$x_i^{(k+1)} = \frac{1}{a_{ii}}\left(b_i - \sum_{j=1}^{i-1} a_{ij}x_j^{(k+1)} - \sum_{j=i+1}^{n} a_{ij}x_j^{(k)}\right), \qquad (2.27)$$

where the superscript $(k + 1)$ indicates that this is the $(k + 1)$st iterate. On the right side we use the most recent estimates of the x_j, which will be either $x_j^{(k)}$ or $x_j^{(k+1)}$.

An algebraically equivalent form for Eq. (2.27) is

$$x_i^{(k+1)} = x_i^{(k)} + \frac{1}{a_{ii}}\left(b_i - \sum_{j=1}^{i-1} a_{ij}x_j^{(k+1)} - \sum_{j=i}^{n} a_{ij}x_j^{(k)}\right),$$

because $x_i^{(k)}$ is both added to and subtracted from the right side. In this form, we see that Gauss–Seidel and Southwell's relaxation can have identical arithmetic: The term we add to $x_i^{(k)}$ to get $x_i^{(k+1)}$ is exactly the increment that relaxes the residual to zero. (Of course, we apply the relaxation to the x_i's in a different sequence in the two methods.) Overrelaxation can be applied to Gauss–Seidel if we will add to $x_i^{(k)}$ some multiple of the second term. It can be shown that this multiple should never be more than 2 in magnitude (to avoid divergence), and the optimum overrelaxation factor lies between 1.0 and 2.0. Our iteration equations take this form, where w is the *overrelaxation factor*:

$$x_i^{(k+1)} = x_i^{(k)} + \frac{w}{a_{ii}}\left(b_i - \sum_{j=1}^{i-1} a_{ij}x_j^{(k+1)} - \sum_{j=i}^{n} a_{ij}x_j^{(k)}\right). \qquad (2.28)$$

Table 2.2 shows how the convergence rate is influenced by the value of w for the system

$$\begin{bmatrix} -4 & 1 & 1 & 1 \\ 1 & -4 & 1 & 1 \\ 1 & 1 & -4 & 1 \\ 1 & 1 & 1 & -4 \end{bmatrix} x = \begin{bmatrix} 1 \\ 1 \\ 1 \\ 1 \end{bmatrix},$$

starting with an initial estimate of $x = 0$. The exact solution is

$$x_1 = -1, \qquad x_2 = -1, \qquad x_3 = -1, \qquad x_4 = -1.$$

Table 2.2   Acceleration of convergence of Gauss–Seidel iteration

w, the overrelaxation factor	Number of iterations to reach error $< 1 \times 10^{-5}$
1.0	24
1.1	18
1.2	13
1.3	11 ← Minimum
1.4	14 of iterations
1.5	18
1.6	24
1.7	35
1.8	55
1.9	100+

We see that the optimum value for the overrelaxation factor is about $w = 1.3$ for this example. The optimum value will vary between between 1.0 and 2.0 depending on the size of the coefficient matrix and the values of the coefficients. Overrelaxation is considered further in Chapter 7 in connection with methods to solve partial-differential equations.

2.12   SYSTEMS OF NONLINEAR EQUATIONS

As mentioned previously, the problem of finding the solution of a set of nonlinear equations is much more difficult than for linear equations. (In fact, some sets have no real solutions.) Consider the example of a pair of nonlinear equations:

$$\begin{aligned} x^2 + y^2 &= 4, \\ e^x + y &= 1. \end{aligned} \tag{2.29}$$

Graphically, the solution to this system is represented by the intersections of the circle $x^2 + y^2 = 4$ with the curve $y = 1 - e^x$. Figure 2.4 shows that these are near $(-1.8, 0.8)$ and $(1, -1.7)$. We can use the method of iteration to improve these approximations. Just as in Section 1.6, we rearrange both equations to a form of the pattern $x = f(x, y)$, $y = g(x, y)$, and use the method of iteration on each equation in turn. Under proper conditions, these will converge. For example, if we rearrange Eqs. (2.29) in the form

$$\begin{aligned} x &= \pm\sqrt{4 - y^2}, \qquad (- \text{ sign for leftmost root}) \\ y &= 1 - e^x, \end{aligned}$$

$x' = \pm 2y(4-y^2)^{-\frac{1}{2}} \qquad y' = -e^x$

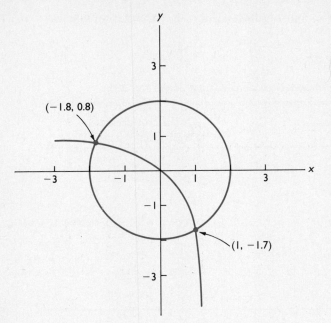

Figure 2.4

we get the following successive values, beginning with $y_1 = 0.8$ in the first equation:

x-values: -1.83 -1.815 -1.8163 -1.8162

y-values: 0.8   0.84   0.8372   0.8374   0.8374.

When we begin at $y = -1.7$ to find the root to the right of the origin, we get

x: 1.05   0.743   1.669   Imaginary value.

y: -1.7 -1.857 -1.102 -4.307

The equations diverge! (Beginning with $x = 1.0$ in the second equation is no help. This also diverges.) However, with a different rearrangement of the original equations, such as

$$x = \ln(1 - y),$$
$$y = \pm\sqrt{4 - x^2}, \quad (-\text{ sign for rightmost root}) \tag{2.30}$$

we get

x: 0.993   1.006   1.0038   1.0042   1.0042.

y: -1.7 -1.736 -1.7286 -1.7299 -1.7296

The pair of rearranged equations in (2.30) converges.

Some of the difficulties with sets of nonlinear equations are apparent from this simple example. If there are more than two equations in the system, finding a convergent form of the equations is increasingly difficult. A criterion for convergence (sufficiency condition only) is as follows:

The set of equations

$$x = f(x, y, z, \ldots), \qquad y = g(x, y, z, \ldots), \qquad z = h(x, y, z, \ldots), \qquad \ldots$$

will converge if, in an interval about the root,

$$|f_x| + |f_y| + |f_z| + \ldots < 1,$$
$$|g_x| + |g_y| + |g_z| + \ldots < 1,$$
$$|h_x| + |h_y| + |h_z| + \ldots < 1,$$
$$\cdots \cdots \cdots < 1.$$

In the above inequalities, the subscript notation designates partial derivatives. Computing all the partial derivatives and knowing where the root is are major problems. Getting starting values for the multidimensioned system is also correspondingly difficult.

Newton's method can also be applied to systems. We begin with the forms

$$f(x, y) = 0,$$
$$g(x, y) = 0.$$

Let $x = r$, $y = s$ be a root, and expand both functions as a Taylor series about the point (x_i, y_i) in terms of $(r - x_i)$, $(s - y_i)$, where (x_i, y_i) is a point near the root:

$$
\begin{aligned}
f(r, s) = 0 &= f(x_i, y_i) + f_x(x_i, y_i)(r - x_i) \\
&\quad + f_y(x_i, y_i)(s - y_i) + \cdots \\
g(r, s) = 0 &= g(x_i, y_i) + g_x(x_i, y_i)(r - x_i) \\
&\quad + g_y(x_i, y_i)(s - y_i) + \cdots
\end{aligned}
\tag{2.31}
$$

Truncating the series gives

$$
\begin{bmatrix} 0 \\ 0 \end{bmatrix} = \begin{bmatrix} f(x_i, y_i) \\ g(x_i, y_i) \end{bmatrix} + \begin{bmatrix} f_x(x_i, y_i) & f_y(x_i, y_i) \\ g_x(x_i, y_i) & g_y(x_i, y_i) \end{bmatrix} \begin{bmatrix} r - x_i \\ s - y_i \end{bmatrix}.
\tag{2.32}
$$

We rewrite this to solve as the system of equations

$$
\begin{bmatrix} f_x(x_i, y_i) & f_y(x_i, y_i) \\ g_x(x_i, y_i) & g_y(x_i, y_i) \end{bmatrix} \begin{bmatrix} \Delta x_i \\ \Delta y_i \end{bmatrix} = - \begin{bmatrix} f(x_i, y_i) \\ g(x_i, y_i) \end{bmatrix},
\tag{2.33}
$$

where $\Delta x_i = r - x_i$, and $\Delta y_i = s - y_i$.

We solve (2.33) by Gaussian elimination and then, if we set

$$\begin{bmatrix} x_{i+1} \\ y_{i+1} \end{bmatrix} = \begin{bmatrix} x_i \\ y_i \end{bmatrix} + \begin{bmatrix} \Delta x_i \\ \Delta y_i \end{bmatrix}, \tag{2.34}$$

we get an improved estimate of the root, (r, s). We repeat the above process with i replaced by $i + 1$ until f and g are close to 0. The extension to more than two simultaneous equations is straightforward. Program NLSYST is an implementation.

We illustrate by repeating the previous example:

$$f(x, y) = 4 - x^2 - y^2 = 0,$$
$$g(x, y) = 1 - e^x - y = 0.$$

The partials are

$$f_x = -2x, \qquad f_y = -2y,$$
$$g_x = -e^x, \qquad g_y = -1.$$

Beginning at $x_0 = 1$, $y_0 = -1.7$, we solve the equations

$$\begin{bmatrix} -2 & 3.4 \\ -2.7183 & -1.0 \end{bmatrix} \begin{bmatrix} \Delta x_0 \\ \Delta y_0 \end{bmatrix} = - \begin{bmatrix} 0.1100 \\ -0.0183 \end{bmatrix}, \tag{2.35}$$

to get $\Delta x_0 = 0.0043$, $\Delta y_0 = -0.0298$. This then gives us $x_1 = 1.0043$, $y_1 = -1.7298$. The results already agree with the true value of the root within two in the fourth decimal place. Repeating the process once more produces $x_2 = 1.004169$, $y_2 = -1.729637$. The function values at (x_2, y_2) are approximately -0.0000001 and -0.00000001 respectively.

We can write Newton's method for a system of n equations by expanding on Eq. (2.31):

$$
\begin{aligned}
0 &= (f_1) + (f_1)_x(r - x_0) + (f_1)_y(s - y_0) + (f_1)_z(t - z_0) + \ldots, \\
0 &= (f_2) + (f_2)_x(r - x_0) + (f_2)_y(s - y_0) + (f_2)_z(t - z_0) + \ldots, \\
0 &= (f_3) + (f_3)_x(r - x_0) + (f_3)_y(s - y_0) + (f_3)_z(t - z_0) + \ldots, \\
&\;\;\vdots \\
0 &= (f_n) + (f_n)_x(r - x_0) + (f_n)_y(s - y_0) + (f_n)_z(t - z_0) + \ldots
\end{aligned}
\tag{2.36}
$$

In these equations, each function is evaluated at the approximate root $(x_0, y_0, z_0, \ldots)$. Note that we have reduced the problem from solving a set of n nonlinear equations to solving a set of n linear equations. The unknowns are the improvements in each estimated variable $(r - x_0)$, $(s - y_0)$, $(t - z_0)$, $\ldots$.

Then Eq. (2.33) becomes

$$
\begin{bmatrix}
f_{1x} & f_{1y} & f_{1z} & \cdots \\
f_{2x} & f_{2y} & f_{2z} & \cdots \\
f_{3x} & f_{3y} & f_{3z} & \cdots \\
\vdots & \vdots & \vdots & \\
f_{nx} & f_{ny} & f_{nz} & \cdots
\end{bmatrix}
\begin{bmatrix}
\Delta x_i \\
\Delta y_i \\
\Delta z_i \\
\vdots \\
\;
\end{bmatrix}
= -
\begin{bmatrix}
f_1 \\
f_2 \\
f_3 \\
\vdots \\
f_n
\end{bmatrix}
\qquad (2.37)
$$

evaluated at $(x_i, y_i, z_i, \ldots)$. Solving this, we compute

$$
x_{i+1} = x_i + \Delta x_i, \; y_{i-1} = y_i + \Delta y_i, \; z_{i+1} = z_i + \Delta z_i, \ldots \qquad (2.38)
$$

In a computer program, it is awkward to introduce each of the partial-derivative functions (which must often be developed by hand, unless one has access to a program like MACSYMA) to use in Eqs. (2.37). An alternative technique is to approximate these partials by recalculating the function with a small perturbation to each of the variables in turn:

$$
(f_1)_x \doteq \frac{f_1(x + \delta, y, z, \ldots) - f_1(x, y, z, \ldots)}{\delta},
$$

$$
(f_1)_y \doteq \frac{f_1(x, y + \delta, z, \ldots) - f_1(x, y, z, \ldots)}{\delta},
$$

$$
\vdots
$$

$$
(f_i)_{x_j} \doteq \frac{f_i(x, y, z, \ldots, x_j + \delta, \ldots) - f_i(x, y, z, \ldots, x_j, \ldots)}{\delta}.
$$

Similar relations are used for each variable in each function.* A computer program in the next section exploits this idea.

It is interesting to observe that Newton's method, as applied to a set of nonlinear equations, reduces the problem to solving a set of *linear equations* in order to determine the values that improve the accuracy of the estimates. This points out quite dramatically how linear and nonlinear problems vary in difficulty.

Newton's method has the advantage of converging quadratically, at least when we are near a root, but it is expensive in terms of function evaluations. For the 2×2 system above, there are six function evaluations at each step, while for a 3×3 system, there are twelve. For n simultaneous equations, the number of function evaluations is $n^2 + n$. One can see why this is rarely applied to large systems. It is good strategy, in all cases of simultaneous nonlinear equations, to reduce the number of equations as much as possible by solving for one variable in terms of the others and eliminating that one by

---

*Approximation of derivatives by such difference quotients is discussed in Chapter 4. If the limiting value (as $\delta \to 0$) of this ratio were used, we would have exactly the definition of a derivative. Since limits are not used, we have approximate values.

substituting for it in the other equations. For example, we should attack the previous examples as follows:

$$\begin{cases} 4 - x^2 - y^2 = 0 \\ 1 - e^x - y = 0 \end{cases} \quad \text{Solve for } y: y = 1 - e^x.$$

Substituting in the first equation, we get

$$4 - x^2 - (1 - e^x)^2 = 0,$$
$$3 - x^2 + 2e^x - e^{2x} = 0.$$

We then use the methods of Chapter 1.

When we must solve a larger system of nonlinear equations, a modification of Newton's method is often used. It converges less than quadratically but usually faster than linearly. Unfortunately, it may *diverge* unless we start fairly near to a root. In this method we do not recompute the matrix of partials at each step. Rather we use the same matrix for several steps before recomputing it again. For a system of n equations we would recompute the matrix after every n steps. In this way we need only n function evaluations at each step, except when we occasionally have to update the matrix of partials, which then adds an additional n^2 function evaluations.

We first illustrate the method on the same 2×2 system above. Reworking Eq. (2.35) produces

$$\begin{bmatrix} \Delta x_0 \\ \Delta y_0 \end{bmatrix} = - \begin{bmatrix} -2 & 3.4 \\ -2.7183 & -1.0 \end{bmatrix}^{-1} \begin{bmatrix} 0.110 \\ -0.0183 \end{bmatrix} = \begin{bmatrix} 0.0043 \\ -0.0298 \end{bmatrix};$$

$$\begin{bmatrix} x_1 \\ y_1 \end{bmatrix} = \begin{bmatrix} 1.0 \\ -1.7 \end{bmatrix} + \begin{bmatrix} 0.0043 \\ -0.0298 \end{bmatrix} = \begin{bmatrix} 1.0043 \\ -1.7298 \end{bmatrix};$$

$$\begin{bmatrix} \Delta x_1 \\ \Delta y_1 \end{bmatrix} = - \begin{bmatrix} -2 & 3.4 \\ -2.7183 & -1.0 \end{bmatrix}^{-1} \begin{bmatrix} 0.000827 \\ 0.000196 \end{bmatrix} = \begin{bmatrix} -0.000133 \\ 0.000165 \end{bmatrix};$$

$$x_2 = 1.004167, \quad y_2 = -1.729635;$$
$$f(x_2, y_2) = -0.000011, \quad g(x_2, y_2) = -0.000002.$$

In addition we consider a different example. Here we have

$$e^x - y = 0,$$
$$xy - e^x = 0.$$

Eq. (2.33) for this system becomes

$$\begin{bmatrix} e^{x_i} & -1 \\ y_i - e^{x_i} & -x_i \end{bmatrix} \begin{bmatrix} \Delta x_i \\ \Delta y_i \end{bmatrix} = - \begin{bmatrix} f(x_i, y_i) \\ g(x_i, y_i) \end{bmatrix}.$$

Let $x_0 = 0.95$, $y_0 = 2.7$. Then solving the preceding with the inverse matrix

$$\begin{bmatrix} \Delta x_0 \\ \Delta y_0 \end{bmatrix} = - \begin{bmatrix} 2.5857 & -1 \\ 0.1143 & 0.95 \end{bmatrix}^{-1} \begin{bmatrix} -0.1143 \\ -0.0207 \end{bmatrix},$$

we get
$$x_1 = 1.00029, \qquad y_1 = 2.71575;$$

$$\begin{bmatrix} \Delta x_1 \\ \Delta y_1 \end{bmatrix} = - \begin{bmatrix} 2.5857 & -1 \\ 0.1143 & 0.95 \end{bmatrix}^{-1} \begin{bmatrix} 0.003325 \\ -0.002533 \end{bmatrix}, \text{ we get}$$

$$x_2 = 1.000048, \qquad y_2 = 2.718445.$$

Since $n = 2$, we update our matrix of partials so that

$$\begin{bmatrix} \Delta x_2 \\ \Delta y_2 \end{bmatrix} = - \begin{bmatrix} 2.7184 & -1 \\ 0.00003 & 1.0000 \end{bmatrix}^{-1} \begin{bmatrix} -0.000033 \\ -0.000163 \end{bmatrix}, \text{ we get}$$

$$x_3 = 1.000000, \qquad y_3 = 2.718282.$$

This agrees with the exact solution: $(1, e)$.

2.13   CHAPTER SUMMARY

To test your understanding of this chapter, ask yourself if you can do the following. If you cannot, restudy is indicated.

1. Manipulate matrices and express a set of linear equations in matrix form. You should be able to expand the matrix representation into the standard form for a set of simultaneous equations. You should know the definitions of *vector*, *scalar*, *identity matrix*, *triangular matrix*, *transpose*, *determinant*, *singular*, and *inverse* as they relate to matrices.

2. Solve a system of linear equations by row transformations, utilizing pivoting, and explain why pivoting is sometimes required and always helpful.

3. Tell how Gaussian elimination differs from the Gauss–Jordan method and explain why the former is generally preferred.

4. Rewrite a matrix as the product of *LU* pairs in several ways and create an *LU* matrix pair that implements Gauss elimination. You can apply this last to solve a system with multiple right-hand sides.

5. Explain what is meant by scaling during the solution process and why inaccurate solutions may result if no scaling is done.

6. Create examples of systems that have no solution for two different reasons and show that the matrix is singular in one case. You should be able to determine if a given matrix is singular or nearly so.

7. Find the determinant of a matrix efficiently.

8. Find the inverse of a matrix and use it to solve a system of equations. You should be able to explain why inverses are important even though they are not usually the preferred method of solving a system.

9. List the properties that all norms must have and compute the various norms of matrices and vectors. You should know which norm gives the tightest bound to the magnitude of a matrix.

10. Explain what the condition number of a matrix is and how it is involved in estimating the accuracy of the solution. You should be able to use iterative improvement to reduce the error of a solution.

11. Solve a system of equations by two iterative methods and explain why one usually converges faster. You should be able to explain when iteration is preferred over the direct methods.

12. Describe the method of relaxation and use overrelaxation to speed up the convergence of iteration.

13. Solve systems of nonlinear equations by iteration and by Newton's method; in the latter, explain how the number of function evaluations can be reduced and what sacrifice this entails.

14. Utilize subroutines for the solution of sets of equations, writing driver programs to interface with them.

15. Read, understand, and use subroutines for linear systems from one or more standard libraries.

SELECTED READINGS FOR CHAPTER 2

Forsythe and Moler (1967); Hageman and Young (1981); Stewart (1973).

2.14 COMPUTER PROGRAMS

Before we consider programs that apply to this chapter, it should be noted that there are several subroutines in IMSL (Appendix C) for solving a system of equations. For a system of nonlinear equations there is ZSPOW, which uses Newton's method. For a system of linear equations, LEQT2F is very easy to use. However, if one wishes to study a well-written program in this area, there is DECOMP and SOLVE in Forsythe, Malcolm, and Moler (1977). This does an LU decomposition and gives an approximation of the condition number of the matrix of coefficients. For iterative methods in solving $AX = b$ there is a package of subroutines called ITPACK developed at the University of Texas.

A number of examples of computer programs utilizing the methods of this chapter are exhibited. (These are grouped near the end of the chapter, as Figs. 2.6 through 2.13.) The first two employ the Gaussian elimination method—Program 1 for a general $n \times n$ array of coefficients and Program 2 for a tridiagonal system. The latter embodies an ingenious scheme to compress the matrix so that less storage is used by it for 500 equations than is required by the former for only 40. Programs 3, 4, and 5 are a combination of routines that form the LU equivalent of the coefficient matrix and then use it to solve the set of equations with one or several right-hand sides. Program 6 is a subroutine that can be used to scale the coefficient matrix when values that differ widely in magnitude are involved. Program 7 uses iteration to solve a set of equations; they must have the diagonal coefficients dominant. The last program solves n simultaneous nonlinear equations by Newton's method.

The first Gaussian elimination program proceeds quite conventionally, utilizing the straightforward method that we discussed earlier in the chapter (see Section 2.4). To use subroutine LUD (Program 1, Fig. 2.6) the coefficient matrix is augmented with the constant vector and passed to the subprogram. The set of equations can be solved with multiple right-hand sides. The coefficient matrix is made upper-triangular (although the elements below the diagonal are not converted to zeros because we know that the row reduction would create a zero; hence it need not actually be done). In order to minimize round-off error as well as to avoid division by zero, we scan each column to find the element of largest magnitude on or below the diagonal, and interchange rows to put this largest element on the diagonal. In other words, partial pivoting is performed. If the largest element is very small (less than 10^{-5}), the program terminates with a message, concluding that the matrix is either singular or nearly so. In the case of a small divisor, the accuracy will be poor. (It is, of course, true that if all the coefficients are very small in magnitude, this is not the correct conclusion. It is assumed that the values have been scaled so that they are not all small.)

(The pair of subprograms below, LUDCMP and SOLVLU, in Programs 3 and 5, can also be used to solve equations with multiple right-hand sides. These subroutines also can solve with a new right-hand-side vector at any time after LUDCMP has reduced the coefficient matrix, without repeating the reduction of the coefficient matrix.)

Subroutine TRIDG (Program 2) in Fig. 2.7 solves a system with a tridiagonal coefficient matrix. To conserve storage in the computer, the tridiagonal matrix of coefficients is compressed, together with the constant terms, into an $N \times 4$ matrix. Column one holds the coefficients to the left of the diagonal, column two holds the diagonal terms, column three holds the coefficient to the right of the diagonal, and column four holds the constant terms. The ith and $(i - 1)$st rows of the compressed matrix correspond to elements in the uncompressed matrix, as shown:

$$
\begin{bmatrix}
\cdot & & \cdot & & \cdot & \\
a_{i-1,1} & a_{i-1,2} & a_{i-1,3} & \\
& a_{i,1} & a_{i,2} & a_{i,3} \\
& & \cdot & & \cdot & & \cdot
\end{bmatrix}
\begin{bmatrix}
\vdots \\
a_{i-1,4} \\
a_{i,4} \\
\vdots
\end{bmatrix}
$$

We can eliminate $a_{i,1}$ by subtracting $a_{i,1}/a_{i-1,2}$ times the $(i - 1)$st row from the ith row. Since we know a zero will replace $a_{i,1}$ we don't need actually to perform the arithmetic. The values of $a_{i,2}$ and $a_{i,4}$ change as follows:

$$
a_{i,2} = a_{i,2} - \frac{a_{i,1}}{a_{i-1,2}} * a_{i-1,3},
$$

$$
a_{i,4} = a_{i,4} - \frac{a_{i,1}}{a_{i-1,2}} * a_{i-1,4}, \qquad \text{for } i = 2, 3, 4, \ldots, n.
$$

When the ith row is reduced, $a_{i,3}$ is unaffected because there is a zero above it. Elements a_{11} and $a_{n,3}$ are never referred to and their values can be anything, or they can be left undefined.

After reduction, a back-substitution is performed. The elements of the solution vector replace the constant vector in the fourth column of the matrix.

The equations for back-substitution are

$$a_{n,4} = \frac{a_{n,4}}{a_{n,2}},$$

$$a_{i,4} = \frac{a_{i,4} - a_{i,3} * a_{i+1,4}}{a_{i,2}}, \qquad i = n - 1, n - 2, \ldots, 1.$$

The third, fourth, and fifth subroutines (Figs. 2.8 through 2.10) will ordinarily be used together to solve a linear system. The first, LUDCMP, accepts the square matrix of coefficients and forms the LU decomposition that is equivalent to the coefficient matrix. This is returned in place of the coefficient matrix by compacting the L and U matrices as described in Section 2.5. The algorithm is a straightforward implementation of the formulas of Section 2.5. In this routine, partial pivoting is employed by a call to Program 4. This APVT subroutine does partial pivoting after a column of L's has been obtained. The order of the rows, as affected by pivoting, is recorded in a vector named ORDER, so that the elements of the b vector can be rearranged into the same order as the coefficient matrix. A very small diagonal element is signaled by printing a message.

After the LU decomposition has been obtained from LUDCMP, the solution to the system with a given right-hand-side vector is obtained by calling the SOLVLU subroutine, Program 5. In this program, the L-values are used to reduce the elements of the constant terms in the same way that the coefficient matrix was triangularized. The subroutine then does a back-substitution, using the reduced coefficients in the U matrix, and returns the solution vector to the caller.

Additional calls to SOLVLU will give the solutions corresponding to new right-hand sides with economy of computational effort, since the LU decomposition is done only once. This scheme is recommended when a given set of equations must be solved with a series of constant terms.

Program 6 (Fig. 2.11) is a utility subroutine that scales the values of the coefficient matrix so that the largest magnitude in each row is unity. This subroutine should be called before any of the previous routines are used if the coefficient matrix has values that are widely different in some rows than in other rows, as discussed in Section 2.3.

Program 7 (Fig. 2.12) is a subroutine named GSITRN, which finds the solution of a linear system by iteration. To make the convergence as rapid as possible, the coefficients must be arranged so that the terms of largest magnitude are on the diagonal. In some systems this cannot be totally achieved; in that case one should approach the condition as closely as possible. An initial approximation to the solution is passed to the subroutine as a parameter.

The first step in the subroutine is to divide each row by the diagonal term, including the corresponding element in the right-hand-side vector; this avoids later divisions. Iteration continues until each component of the approximate solution vector changes less than the input parameter TOL. If this does not occur within the number of iterations specified by NITER, the program prints a message and returns the last approximation.

Iteration is best adapted to sparse matrices.* The subroutine GSITRN does not illustrate this case. In fact, writing a general-purpose subroutine for iterative solutions to a sparse matrix is impossible, for this would require a repetitive pattern in the coefficients; such patterns cannot be anticipated. Many texts illustrate iteration with a tridiagonal system, but for these, iteration is not the best choice. A program like TRIDG, just discussed, has a minimum storage requirement and runs much faster than an iterative scheme.

Program 8 (Fig. 2.13), named NLSYST, is a subroutine that solves a nonlinear system. Newton's method (see Eqs. 2.37) is employed. The partial derivatives can be computed in either of two ways: numerically or analytically (for the former, METHOD = 0; for the latter, METHOD = 1). In the former case we check the ratio of a change in the function values for small changes in the independent variable. This estimation of the derivative by a difference quotient will be studied in greater detail in Chapter 4. The computation of the partials by either method is done in subroutine FCNJ. After all the partials have been approximated, the set of equations equivalent to Eq. (2.37) is solved by a call to subroutines LUD and SOLVE (Program 1), which give values for increments to the initial guesses for the variables that should make them closer to a solution of the nonlinear system. The subroutine repeats the procedure until convergence is obtained.

The calling program passes initial guesses for the variables, a value for DELTA (the change in variable value used to estimate the partial derivative), and tolerance values (FTOL and XTOL) to terminate iterations when the function values are all sufficiently small or the changes in the x-values are all sufficiently small. The maximum number of iterations is limited by MAXIT. Function values are supplied through an external subroutine.

Some judgment is needed to use NLSYST. First, the initial guesses for values of the variables must be near enough to a solution to give convergence. Second, the value for DELTA should be small enough to give a reasonable approximation to the partial derivative but not so small as to lead to excessive round-off. DELTA = 0.01 seems to give good results.

Calling the subroutines of Programs 1 to 7 is straightforward but a driver for Program 8 is more complex.

As an illustration of how the subroutine NLSYST is used, the program in Fig. 2.5, together with the function subroutine, will determine a root of the pair of equations

$$f(x, y) = x^2 + y^2 - 5 = 0,$$
$$g(x, y) = y - e^x - 1 = 0.$$

The starting value $(2, 4)$ is supplied, along with the values for DELTA, FTOL, XTOL, and MAXIT. After six iterations, the root at $(0.20437, 2.22672)$ was found.

---

*A major area of application is in solving certain partial-differential equations. This is discussed in later chapters.

```
      PROGRAM NLSYS(INPUT,OUTPUT)
C
C     THIS PROGRAM SOLVES THE EXAMPLE IN SECTION 2.12
C     USING EITHER NUMERICAL PARIALS FOR THE  JACOBIAN
C     MATRIX (METHOD = 0) OR ANALYTIC PARTIALS
C     (METHOD = 1). IN THE LATTER CASE THE MATRIX HAS
C     BEEN PROVIDED IN SUBROUTINE FCNJ.
C
      REAL X(2),F(2)
      INTEGER N
      EXTERNAL FCN
      DATA DELTA,XTOL,FTOL,I/0.125E-4,2*0.1E-8,0/
      DATA MAXIT / 50/
      DATA N,METHOD/2,0/
      DATA X/2.0,4.0/
C
      CALL NLSYST(FCN,N,MAXIT,X,F,DELTA,XTOL,FTOL,METHOD,I)
      PRINT 100
100   FORMAT(///'THE X-VALUES ARE:'/)
      PRINT 200, (X(I), I = 1,N)
200   FORMAT(T6,F8.4/)
      STOP
      END
```

Figure 2.5

```
      PROGRAM PDEC(INPUT,OUTPUT)
C
C     -------------------------------------------------------------
C
C     THIS PROGRAM SOLVES A SYSTEM OF EQUATIONS USING GAUSSIAN ELIM-
C     INATION WITH PARTIAL PIVOTING AND BACK SUBSTITUTION ACCORDING
C     TO THE ALGORITHM GIVEN IN SECTION 2.4.
C     THE ALGORITHM IS IMPLEMENTED IN TWO PROCEDURES:
C         DECOMP: RETURNS THE LU DECOMPOSITION OF THE COEFFICIENT MATRIX
C         SOLVE : SOLVE THE RIGHT HAND SIDES USING BOTH FORWARD AND
C                 BACK SUBSTITUTION
C
C     -------------------------------------------------------------
C
      REAL A(10,10), B(10), DET
      INTEGER IPVT(10), N,NDIM, I,J
C
      DATA N,NDIM/4,10/
C
C     -------------------------------------------------------------
C
C     THE MATRIX A IS INPUT VIA A DATA STATEMENT COLUMN BY COLUMN, SINCE
C     FORTRAN IS COLMUN-MAJOR.
C
C     -------------------------------------------------------------
C
```

Figure 2.6   Program 1.

Figure 2.6   (*continued*)

```
          DATA A/6,2,4,7*0,1,2,-3,2,6*0,-6,3,8*0,
        +      -5,2,1,1,6*0,60*0/
          DATA B/6,-2,-7,7*0/
C
C         PRINT OUT ORIGINAL MATRIX
C
          PRINT '(//)'
          PRINT 100
100       FORMAT(17X,'THE MATRIX A ', 13X,'    RIGHT HAND SIDE'/)
          DO 10 I = 1,N
              PRINT 200, (A(I,J), J = 1,N ), B(I)
10            CONTINUE
200       FORMAT(4F10.4, F15.4)
C
C         WE FIND THE LU DECOMPOSITION OF THE MATRIX A
C
          CALL LUD(A,N,IPVT,NDIM,DET)
C
C         PROCEDURE SOLVE NOW GETS SOLUTION BY FORWARD/BACK SUBSTITUTION
C
          CALL SOLVE(A,N,IPVT,B,NDIM)
C
C         PRINT LU MATRIX AND SOLUTION VECTOR
C
          PRINT '(//)'
          PRINT 300
300       FORMAT(10X,'THE LU DECOMPOSITION OF A ',8X,' SOLUTION VECTOR '/)
          DO 20 I = 1,N
              PRINT 200, (A(I,J), J = 1,N), B(I)
20            CONTINUE
C
          STOP
          END
C
C               SUBROUTINE LUD
C
          SUBROUTINE LUD(A,N,IPVT,NDIM,DET)
C
C    ------------------------------------------------------------------
C
C         SUBROUTINE LUD: THIS SUBROUTINE SOLVES A SET OF LINEAR EQUATIONS
C    RETURNS THE LU DECOMPOSITION OF THE COEFFICIENT MATRIX. THE METHOD
C    IS BASED ON THE ALGORITHM PRESENTED IN SECTION 4.
C
C    INPUT:  A  -   THE COEFFICIENT MATRIX
C            N  -   THE NUMBER OF EQUATIONS
C            NDIM  -   THE MAXIMUM ROW DIMENSION OF A
C
C    OUTPUT: A  -   THE LU DECOMPOSITION OF THE MATRIX A
C                   THE ORIGINAL MATRIX A IS LOST
C            IPVT  - A VECTOR CONTAINING THE ORDER OF THE ROWS OF THE
C                    REARRANGED MATRIX DUE TO PIVOTING
C        DET  -   THE DETERMINANT OF THE MATRIX. IT IS SET TO
C                 0 IF ANY PIVOT ELEMENT IS LESS THAN 0.00001.
C
C    ------------------------------------------------------------------
C
          REAL A(NDIM,N),  SAVE,RATIO,VALUE,DET
          INTEGER IPVT(N),N,NDIM,  I,IPVTMT,NLESS1,IPLUS1,J,L
          INTEGER KCOL,JCOL,JROW,TMPVT
C
          DET = 1.0
```

Figure 2.6 (*continued*)

```
        NLESS1 = N - 1
        DO 10 I = 1,N
            IPVT(I) = I
10          CONTINUE
C
        DO 20 I = 1,NLESS1
            IPLUS1 = I+1
            IPVTMT = I
C
C           FIND PIVOT ROW
C
            DO 30 J = IPLUS1,N
                IF (ABS(A(IPVTMT,I)) .LT. ABS(A(J,I))) IPVTMT = J
30              CONTINUE
C
C       ----------------------------------------------------------------
C
C       CHECK FOR SMALL PIVOT ELEMENT
C
C       ----------------------------------------------------------------
C
            IF (ABS(A(IPVTMT,I)). LT. 1.0E-05) THEN
                DET = 0.0
                PRINT '(//)'
                PRINT *, '   MATRIX IS SINGULAR OR NEAR SINGULAR
                PRINT '(//)'
                RETURN
            ENDIF
C
C           INTERCHANGE ROWS IF NECESSARY
C
            IF (IPVTMT .NE. I) THEN
                TMPVT = IPVT(I)
                IPVT(I) = IPVT(IPVTMT)
                DO 40 JCOL = 1,N
                    SAVE = A(I,JCOL)
                    A(I,JCOL) = A(IPVTMT,JCOL)
                    A(IPVTMT,JCOL) = SAVE
40                  CONTINUE
                IPVT(IPVTMT) = TMPVT
                DET = -DET
            END IF
C
C               REDUCE ALL ELEMENTS BELOW THE I'TH ROW
C
            DO 50 JROW = IPLUS1,N
                IF (A(JROW,I) .NE. 0.0) THEN
                    A(JROW,I) = A(JROW,I)/A(I,I)
                    DO 60 KCOL = IPLUS1,N
                        A(JROW,KCOL) = A(JROW,KCOL) - A(JROW,I)*A(I,KCOL)
60                      CONTINUE
                END IF
50              CONTINUE
C
20          CONTINUE
C
        IF (ABS(A(N,N)) .LT. 1.0E-5 ) THEN
            PRINT '(//)'
            PRINT *, '   MATRIX IS SINGULAR OR NEAR SINGULAR
            PRINT '(//)'
            DET = 0.0
            RETURN
```

Figure 2.6   (*continued*)

```
            END IF
C
C     ------------------------------------------------------------------
C
C     COMPUTE THE DETERMINANT OF THE MATRIX
C
C     ------------------------------------------------------------------
C
      DO 70 I = 1,N
          DET = DET * A(I,I)
70        CONTINUE
C
      RETURN
      END
C
C     ------------------------------------------------------------
C
C                   SUBROUTINE SOLVE
C
C     ------------------------------------------------------------
C
      SUBROUTINE SOLVE(A,N,IPVT,B,NDIM)
C
C     ------------------------------------------------------------
C
C     MAKING USE OF THE LU DECOMPOSITION OF THE MATRIX A, THIS SUB-
C     ROUTINE SOLVES THE SYSTEM BY FORWARD AND BACK SUBSTITUTION.
C
C     INPUT:    A - LU MATRIX FROM SUBROUTINE LUD
C               N - NUMBER OF EQUATIONS
C               IPVT - A RECORD OF THE REARRANGEMENT OF THE ROWS
C                      OR A FROM SUBROUTINE LUD
C               B - RIGHT HAND SIDE OF THE SYSTEM OF EQUATIONS
C
C     OUTPUT:   B - THE SOLUTION VECTOR
C
C     ------------------------------------------------------------
C
      REAL A(NDIM,N), B(N), X(10), SUM
      INTEGER IPVT(N),N,NDIM,   IROW,JCOL,I
C
C     ------------------------------------------------------------
C
C     REARRANGE THE ELEMENTS OF THE B VECTOR. STORE THEM IN THE
C     X VECTOR.
C
      DO 10 I = 1,N
          X(I) = B(IPVT(I))
10        CONTINUE
C
C     ------------------------------------------------------------
C
C     SOLVE USING FORWARD SUBSTITUTION--LY = B
C
C     ------------------------------------------------------------
C
      DO 20 IROW = 2,N
          SUM = X(IROW)
          DO 30 JCOL = 1,(IROW-1)
              SUM = SUM - A(IROW,JCOL)*X(JCOL)
30            CONTINUE
          X(IROW) = SUM
```

Figure 2.6   (*continued*)

```
20          CONTINUE
C
C    ----------------------------------------------------------------
C
C       SOLVE BY BACK SUBSTITUTION--UX = Y
C
C    ----------------------------------------------------------------
C
        B(N) = X(N)/A(N,N)
        DO 40 IROW = (N-1),1,-1
            SUM = X(IROW)
            DO 50 JCOL = (IROW+1),N
                SUM = SUM - A(IROW,JCOL)*B(JCOL)
50              CONTINUE
            B(IROW) = SUM/A(IROW,IROW)
40          CONTINUE

        RETURN
        END
```

```
                    OUTPUT FOR PROGRAM 1

            THE MATRIX A                    RIGHT HAND SIDE

    6.0000    1.0000   -6.0000   -5.0000        6.0000
    2.0000    2.0000    3.0000    2.0000       -2.0000
    4.0000   -3.0000     .0000    1.0000       -7.0000
     .0000    2.0000     .0000    1.0000         .0000

            THE LU DECOMPOSITION OF A       SOLUTION VECTOR

    6.0000    1.0000   -6.0000   -5.0000        -.5000
     .6667   -3.6667    4.0000    4.3333        1.0000
     .3333    -.4545    6.8182    5.6364         .3333
     .0000    -.5455     .3200    1.5600       -2.0000
```

```
        PROGRAM PDIAG(INPUT,OUTPUT)
        REAL A(4,4)
        INTEGER N,I
        DATA N,A/4,4*1.0,4*-4.0,1.0,2.0,1.0,0.0,-5,11,-11,-5/
        PRINT 300
        DO 10 I = 1,N
            PRINT 100, (A(I,J), J = 1,4)
10          CONTINUE
        NDIM = N
        CALL TRIDG(A,N,NDIM)
        PRINT 400
        DO 20 I = 1,N
            PRINT 100, (A(I,J), J = 1,4)
20          CONTINUE
100     FORMAT(T5,4(F10.4,3X))
300     FORMAT(//T10,10('*'),' ON INPUT THE MATRIX A IS ',10('*')/)
400     FORMAT(//T10,10('*'), ' ON OUTPUT THE MATRIX IS ',10('*')/)
        STOP
        END
C
C    ----------------------------------------------------------------
C
```

Figure 2.7   Program 2.

Figure 2.7   (*continued*)

```
C       SUBROUTINE TRIDG :
C                       THIS SUBROUTINE PERFORMS GAUSSIAN ELIMINATION
C       ON A TRIDIAGONAL SYSTEM. THE COEFFICIENTS OF N EQUATIONS ARE
C       STORED IN THE N X 4 ARRAY, A. THE FIRST COLUMN OF A HOLDS THE
C       ELEMENTS TO THE LEFT OF THE DIAGONAL, THE SECOND HOLDS THE DIA-
C       GONAL ELEMENTS, AND THE THIRD HOLDS THE ELEMENTS TO THE RIGHT.
C       THE FOURTH COLUMN HOLDS THE RIGHT HAND SIDE TERMS.
C
C       ------------------------------------------------------------------
C
C       PARAMETERS ARE :
C
C       A      - MATRIX OF COEFFICIENTS AND R.H.S. AS DESCRIBED
C       N      - NUMBER OF EQUATIONS
C       NDIM   - FIRST DIMENSION OF A IN THE CALLING PROGRAM
C
C       THE SOLUTION VECTOR IS RETURNED IN THE FOURTH COLUMN OF A.
C       THE FIRST THREE COLUMNS OF A ON OUTPUT WILL CONTAIN THE
C       LU DECOMPOSITION OF THE ORIGINAL TRIDIAGONAL MATRIX.
C
C       ------------------------------------------------------------------
C
C       THIS SUBROUTINE ASSUMES THAT NO PIVOTING IS NECESSARY
C
C       ------------------------------------------------------------------
C
        SUBROUTINE TRIDG(A,N,NDIM)
        REAL A(NDIM,4)
        INTEGER N,NDIM,I,NM1,M
C
C       ------------------------------------------------------------------
C
C FIRST WE ELIMINATE ALL BELOW DIAGONAL TERMS
C
        DO 10 I = 2,N
          A(I,1) = A(I,1)/A(I-1,2)
          A(I,2) = A(I,2) - A(I,1)*A(I-1,3)
          A(I,4) = A(I,4) - A(I,1)*A(I-1,4)
   10 CONTINUE
C
C       ------------------------------------------------------------------
C
C NOW WE DO THE BACK SUBSTITUTING
C
        NM1 = N - 1
        A(N,4) = A(N,4) / A(N,2)
        DO 20 I = NM1,1,-1
C
C THE INDEX M WILL COUNT UP THE ROWS
C
          A(I,4) = ( A(I,4) - A(I,3)*A(I+1,4) ) / A(I,2)
   20 CONTINUE
        RETURN
        END

                        OUTPUT FOR PROGRAM 2

        ********** ON INPUT THE MATRIX A IS **********

          1.0000      -4.0000       1.0000       -5.0000
          1.0000      -4.0000       2.0000       11.0000
          1.0000      -4.0000       1.0000      -11.0000
          1.0000      -4.0000       0.0000       -5.0000
```

Figure 2.7   (*continued*)

```
          ********** ON OUTPUT THE MATRIX IS **********
          1.0000      -4.0000      1.0000      1.0000
          -.2500      -3.7500      2.0000     -1.0000
          -.2667      -3.4667      1.0000      3.0000
          -.2885      -3.7115      0.0000      2.0000
```

```
          SUBROUTINE LUDCMP(A,N,NDIM,ORDER)
C
C     -----------------------------------------------------------------------
C
C     SUBROUTINE LUDCMP :
C                      THIS SUBROUTINE COMPUTES THE L AND U TRIANGULAR
C     MATRICES EQUIVALENT TO THE A MATRIX, SUCH THAT LU = A. THESE
C     MATRICES ARE RETURNED IN THE SPACE OF A, IN COMPACT FORM. THE U
C     MATRIX HAS ONES ON ITS DIAGONAL. PARTIAL PIVOTING IS USED TO GIVE
C     MAXIMUM VALUED ELEMENTS ON THE DIAGONAL OF L. THE ORDER OF THE ROWS
C     AFTER PIVOTING IS RETURNED IN VECTORS IN THE INTEGER VECTOR ORDER.
C     THIS SHOULD BE USED TO REORDER R.H.S. VECTORS BEFORE SOLVING THE
C     SYSTEM AX = B.
C
C     -----------------------------------------------------------------------
C
C     PARAMETERS ARE :
C
C     A       - THE N X N MATRIX OF COEFFICIENTS
C     N       - THE NUMBER OF EQUATIONS
C     NDIM    - THE FIRST DIMENSION OF A IN THE CALLING PROGRAM
C     ORDER   - INTEGER VECTOR HOLDING ROW ORDER AFTER PIVOTING
C
C     THIS ROUTINE CALLS A SUBROUTINE APVT TO LOCATE THE PIVOT ROW AND
C     MAKE INTERCHANGES.
C
C     -----------------------------------------------------------------------
C
          REAL A(NDIM,N)
          INTEGER N,NDIM,ORDER(N)
          REAL SUM
          INTEGER I,KCOL,NM1,JM1,JCOL,JP1,IROW
C
C     -----------------------------------------------------------------------
C
C     ESTABLISH INITIAL ORDERING IN ORDER VECTOR
C
          DO 10 I = 1,N
             ORDER(I) = I
   10 CONTINUE
C
C     DO PIVOTING FOR FIRST COLUMN BY CALL TO SUBROUTINE APVT
C
          CALL APVT(A,N,NDIM,ORDER,1)
C
C     -----------------------------------------------------------------------
C
```

Figure 2.8   Program 3.

Figure 2.8   (*continued*)

```
C  IF PIVOT ELEMENT VERY SMALL, PRINT ERROR MESSAGE AND RETURN
C
      IF ( ABS(A(1,1)) .LT. 1.0E-5 ) THEN
        PRINT 100
        RETURN
      END IF
C
C  NOW COMPUTE ELEMENTS FOR FIRST ROW OF U
C
      DO 20 KCOL = 2,N
        A(1,KCOL) = A(1,KCOL) / A(1,1)
   20 CONTINUE
C
C  ------------------------------------------------------------------------
C
C  COMPLETE THE COMPUTING OF L AND U ELEMENTS. THE GENERAL PLAN IS TO
C  COMPUTE A COLUMN OF L'S, THEN CALL APVT TO INTERCHANGE ROWS, AND THEN
C  GET A ROW OF U'S.
C
      NM1 = N - 1
      DO 80 JCOL = 2,NM1
C
C  FIRST COMPUTE A COLUMN OF L'S
C
      JM1 = JCOL - 1
      DO 50 IROW = JCOL,N
        SUM = 0
        DO 40 KCOL = 1,JM1
          SUM = SUM + A(IROW,KCOL)*A(KCOL,JCOL)
   40     CONTINUE
        A(IROW,JCOL) = A(IROW,JCOL) - SUM
   50   CONTINUE
C
C  ------------------------------------------------------------------------
C
C  NOW INTERCHANGE ROWS IF NEED BE, THEN TEST FOR TOO SMALL PIVOT
C
      CALL APVT(A,N,NDIM,ORDER,JCOL)
      IF ( ABS(A(JCOL,JCOL)) .LT. 1.0E-5 ) THEN
        PRINT 100
        RETURN
      END IF
C
C  NOW WE GET A ROW OF U'S
C
      JP1 = JCOL + 1
      DO 70 KCOL = JP1,N
        SUM = 0
        DO 60 IROW = 1,JM1
          SUM = SUM + A(JCOL,IROW)*A(IROW,KCOL)
   60     CONTINUE
        A(JCOL,KCOL) = ( A(JCOL,KCOL) - SUM ) / A(JCOL,JCOL)
   70   CONTINUE
   80 CONTINUE
C
C  ------------------------------------------------------------------------
C
C  STILL NEED TO GET LAST ELEMENT IN L MATRIX
C
      SUM = 0
      DO 90 KCOL = 1,NM1
        SUM = SUM + A(N,KCOL)*A(KCOL,N)
```

Figure 2.8 (*continued*)

```
   90 CONTINUE
      A(N,N) = A(N,N) - SUM
      RETURN
C
  100 FORMAT(/,' VERY SMALL PIVOT ELEMENT INDICATES A NEARLY',
     +        ' SINGULAR MATRIX. ')
      END
```

```
      SUBROUTINE APVT(A,N,NDIM,ORDER,JCOL)
C
C ------------------------------------------------------------------------------
C
C     SUBROUTINE APVT :
C                      THIS SUBROUTINE FINDS THE LARGEST ELEMENT FOR
C     PIVOT IN JCOL OF MATRIX A, PERFORMS INTERCHANGES OF ELEMENTS IN A
C     AND ALSO INTERCHANGES THE ELEMENTS IN THE ORDER VECTOR.
C
C ------------------------------------------------------------------------------
C
C     PARAMETERS ARE :
C
C     A      - MATRIX OF COEFFICIENTS WHOSE ROWS ARE TO BE INTERCHANGED
C     N      - NUMBER OF EQUATIONS
C     NDIM   - FIRST DIMENSION OF A IN THE MAIN PROGRAM
C     ORDER  - INTEGER VECTOR TO HOLD ROW ORDERING
C     JCOL   - COLUMN OF A BEING SEARCHED FOR PIVOT ELEMENT
C
C ------------------------------------------------------------------------------
C
      REAL A(NDIM,N)
      INTEGER ORDER(N),N,NDIM,JCOL
      REAL SAVE,ANEXT,BIG
      INTEGER IPVT,JP1,IROW,KCOL,ISAVE
C
C ------------------------------------------------------------------------------
C
C FIND PIVOT ROW, CONSIDERING ONLY THE ELEMENTS ON AND BELOW DIAGONAL
C
      IPVT = JCOL
      BIG = ABS(A(JCOL,JCOL))
      JP1 = JCOL + 1
      DO 10 IROW = JP1,N
         ANEXT = ABS(A(IROW,JCOL))
         IF ( ANEXT .GT. BIG ) IPVT = IROW
   10 CONTINUE
C
C ------------------------------------------------------------------------------
C
C NOW INTERCHANGE ROW ELEMENTS IN THE ROW WHOSE NUMBER EQUALS JCOL WITH
C THE PIVOT ROW UNLESS PIVOT ROW IS JCOL.
C
      IF ( IPVT .EQ. JCOL ) THEN
         RETURN
      END IF
      DO 20 KCOL = 1,N
```

Figure 2.9 Program 4.

Figure 2.9   (*continued*)

```
            SAVE = A(JCOL,KCOL)
            A(JCOL,KCOL) = A(IPVT,KCOL)
            A(IPVT,KCOL) = SAVE
      20 CONTINUE
C
C   ------------------------------------------------------------------------
C
C   NOW SWITCH ELEMENTS IN THE ORDER VECTOR
C
         ISAVE = ORDER(JCOL)
         ORDER(JCOL) = ORDER(IPVT)
         ORDER(IPVT) = ISAVE
         RETURN
         END
```

```
         SUBROUTINE SOLVLU(LU,B,X,N,NDIM,ORDER)
C
C   ------------------------------------------------------------
C
C   SUBROUTINE SOLVLU :
C                    THIS SUBROUTINE IS USED TO FIND THE SOLUTION
C   TO A SYSTEM OF EQUATIONS, AX = B, AFTER THE LU EQUIVALENT OF A HAS
C   BEEN FOUND.   BEFORE USING THIS ROUTINE, THE VECTOR B SHOULD BE
C   SCALED IF MATRIX A WAS SCALED, USING THE SAME SCALE FACTORS.   WITHIN
C   THIS ROUTINE, THE ELEMENTS OF B ARE REARRANGED IN THE SAME WAY THAT
C   THE ROWS OF A WERE INTERCHANGED, USING THE ORDER VECTOR WHICH HOLDS
C   THE ROW ORDERINGS.   THE SOLUTION IS RETURNED IN X.
C
C   ------------------------------------------------------------
C
C
C   PARAMETERS ARE :
C
C   LU    - THE LU EQUIVALENT OF THE COEFFICIENT MATRIX
C   B     - THE VECTOR OF RIGHT HAND SIDES
C   X     - SOLUTION VECTOR
C   N     - NUMBER OF EQUATIONS
C   NDIM  - FIRST DIMENSION OF A IN THE MAIN PROGRAM
C   ORDER - INTEGER ARRAY OF ROW ORDER AS ARRANGED DURING PIVOTING
C
C   ------------------------------------------------------------
C
         REAL LU(NDIM,N),B(N),X(N)
         INTEGER ORDER(N),N,NDIM
         REAL SUM
         INTEGER I,J,JCOL,IROW,NVBL,IM1,NP1
C
C   ------------------------------------------------------------
C
C   REARRANGE THE ELEMENTS OF THE B VECTOR.   X IS USED TO HOLD THEM.
C
         DO 10 I = 1,N
            J = ORDER(I)
            X(I) = B(J)
      10    CONTINUE
```

Figure 2.10   Program 5.

Figure 2.10 (*continued*)

```
C
C     ------------------------------------------------------------
C
C     COMPUTE THE B' VECTOR, STORING BACK IN X
C
      X(1) = X(1) / LU(1,1)
      DO 50 IROW = 2,N
        IM1 = IROW - 1
        SUM = 0.0
        DO 40 JCOL = 1,IM1
          SUM = SUM + LU(IROW,JCOL)*X(JCOL)
40      CONTINUE
        X(IROW) = (X(IROW) - SUM) / LU(IROW,IROW)
50    CONTINUE
C
C     ------------------------------------------------------------
C     NOW GET THE SOLUTION VECTOR,  X(N) = X(N) ALREADY.
C
      DO 70 IROW = 2,N
        NVBL = N - IROW + 1
        SUM = 0.0
        NP1 = NVBL + 1
        DO 60 JCOL = NP1,N
          SUM = SUM + LU(NVBL,JCOL)*X(JCOL)
60      CONTINUE
        X(NVBL) = X(NVBL) - SUM
70    CONTINUE
C
      RETURN
      END
```

```
      SUBROUTINE SCALES(A,N,NDIM,SCFAC)
C
C     ----------------------------------------------------------------------
C
C     SUBROUTINE SCALES :
C                       THIS SUBROUTINE SCALES THE VALUES IN AN N X N
C     COEFFICIENT MATRIX SO THAT THE LARGEST ELEMENT IN EACH ROW IS
C     UNITY. THE SCALED VALUES OF A ARE RETURNED IN THE A MATRIX AND THE
C     SCALE FACTORS FOR EACH ROW ARE RETURNED IN SCFAC VECTOR. USE SCFAC
C     TO SCALE THE ELEMENTS IN THE B VECTOR ( R.H. SIDES ) BEFORE
C     SOLVING THE SET OF EQUATIONS AX = B.
C
C     ----------------------------------------------------------------------
C
C     PARAMETERS ARE :
C
C     A      - MATRIX OF COEFFICIENTS
C     N      - NUMBER OF EQUATIONS
C     NDIM   - FIRST DIMENSION OF A IN THE CALLING PROGRAM
C     SCFAC  - ARRAY TO HOLD THE SCALE FACTORS
C
C     ----------------------------------------------------------------------
C
```

Figure 2.11 Program 6.

Figure 2.11   (*continued*)

```
        REAL A(NDIM,NDIM),SCFAC(N)
        INTEGER N,NDIM,I,J
        REAL BIG,ANEXT
C
C   ----------------------------------------------------------------
C
C   FIND THE LARGEST VALUE IN EACH ROW. IF ANY ROW HAS ONLY ZEROES,
C   PRINT MESSAGE AND RETURN.
C
        DO 20 I = 1,N
          BIG = ABS(A(I,1))
          DO 10 J = 2,N
            ANEXT = ABS(A(I,J))
            IF ( ANEXT .GT. BIG ) BIG = ANEXT
     10     CONTINUE
          IF ( BIG .EQ. 0 ) THEN
            PRINT 200, I
          END IF
          SCFAC(I) = 1.0 / BIG
     20 CONTINUE
C
C   ----------------------------------------------------------------
C
C   NOW SCALE THE A VALUES
C
        DO 40 I = 1,N
          DO 50 J = 1,N
            A(I,J) = A(I,J)*SCFAC(I)
     50     CONTINUE
     40 CONTINUE
C
    200 FORMAT(/' ALL ELEMENTS IN ROW ',I3,' ARE ZERO.')
        RETURN
        END
```

```
        SUBROUTINE GSITRN(A,E,X,N,NDIM,NITER,TOL)
C
C   ----------------------------------------------------------------
C
C   SUBROUTINE GSITRN :
C                        THIS SUBROUTINE OBTAINS THE SOLUTION TO N LINEAR
C   EQUATIONS BY GAUSS-SEIDEL ITERATION. AN INITIAL APPROXIMATION IS SENT
C   TO THE SUBROUTINE IN THE VECTOR X. THE SOLUTION, AS APPROXIMATED BY
C   THE SUBROUTINE, IS RETURNED IN X. THE ITERATIONS ARE CONTINUED UNTIL
C   THE MAXIMUM CHANGE IN ANY X COMPONENT IS LESS THAN TOL. IF THIS
C   CANNOT BE ACCOMPLISHED IN NITER ITERATIONS, A MESSAGE IS PRINTED
C   AND THE CURRENT APPROXIMATION IS RETURNED. THE COEFFICIENTS ARE
C   TO BE ARRANGED IN A SO AS TO HAVE THE LARGEST VALUES ON THE
C   DIAGONAL.
C
C   ----------------------------------------------------------------
C
C   PARAMETERS ARE :
C
```

Figure 2.12   Program 7.

Figure 2.12 *(continued)*

```
C     A       - COEFFICIENT MATRIX WITH LARGEST VALUES ON THE DIAGONAL
C     B       - RIGHT HAND SIDE VECTOR
C     X       - INITIAL APPROXIMATION TO SOLUTION, ALSO RETURNS RESULT
C     N       - THE NUMBER OF EQUATIONS
C     NDIM    - FIRST DIMENSION OF A IN THE CALLING PROGRAM
C     NITER   - LIMIT TO THE NUMBER OF ITERATIONS
C     TOL     - TEST VALUE TO STOP ITERATING THE SOLUTION
C
C     -------------------------------------------------------------------
C
      REAL A(NDIM,NDIM),B(N),X(N),TOL
      INTEGER N,NDIM,NITER,I,J
      REAL SAVE,XMAX
C
C     -------------------------------------------------------------------
C
C WE CAN SAVE SOME DIVISIONS BY MAKING THE DIAGONAL ELEMENTS EQUAL TO UNITY
C
      DO 10 I = 1,N
        SAVE = A(I,I)
        B(I) = B(I) / SAVE
        DO 5 J = 1,N
          A(I,J) = A(I,J) / SAVE
    5   CONTINUE
   10 CONTINUE
C
C     -------------------------------------------------------------------
C
C NOW WE PERFORM THE ITERATIONS. STORE MAX CHANGE IN X VALUES FOR TESTING
C AGAINST TOL. OUTER LOOP LIMITS ITERATIONS TO NITER.
C
      DO 40 ITER = 1,NITER
        XMAX = 0
        DO 30 I = 1,N
          SAVE = X(I)
          X(I) = B(I)
          DO 20 J = 1,N
            IF ( J .NE. I ) THEN
              X(I) = X(I) - A(I,J)*X(J)
            END IF
   20     CONTINUE
          IF ( ABS(X(I) - SAVE) .GT. XMAX ) THEN
            XMAX = ABS( X(I) - SAVE )
          END IF
   30   CONTINUE
        IF ( XMAX .LE. TOL ) RETURN
   40 CONTINUE
C
C     -------------------------------------------------------------------
C
C NORMAL EXIT FROM THE LOOP MEANS NON-CONVERGENT IN NITER ITERATIONS.
C PRINT MESSAGE AND RETURN.
C
      PRINT 200, TOL,NITER
  200 FORMAT(/' DID NOT MEET TOLERANCE OF ',E14.6,' IN ',I4,
     +          ' ITERATIONS'/' LAST VALUE OF X WAS RETURNED TO CALLER')
      RETURN
      END
```

```
C
        SUBROUTINE NLSYST(FCN,N,MAXIT,X,F,DELTA,XTOL,FTOL,METHOD,I)
C     -------------------------------------------------------------------
C
C     SUBROUTINE NLSYST :
C                          THIS SUBROUTINE SOLVES A SYSTEM OF N NON-
C     LINEAR EQUATIONS BY NEWTON'S METHOD. THE PARTIAL DERIVATIVES OF
C     THE FUNCTIONS ARE ESTIMATED BY DIFFERENCE QUOTIENTS WHEN A
C     VARIABLE IS PERTURBED BY AN AMOUNT EQUAL TO DELTA ( DELTA IS
C     ADDED ). THIS IS DONE FOR EACH VARIABLE IN EACH FUNCTION.
C     INCREMENTS TO IMPROVE THE ESTIMATES FOR THE X-VALUES ARE
C     COMPUTED FROM A SYSTEM OF EQUATIONS USING SUBROUTINES
C     LUD AND SOLVE
C
C     -------------------------------------------------------------------
C
C     PARAMETERS ARE :
C
C     FCN     - SUBROUTINE THAT COMPUTES VALUES OF THE FUNCTIONS. MUST
C               BE DECLARED EXTERNAL IN THE CALLING PROGRAM.
C     N       - THE NUMBER OF EQUATIONS
C     MAXIT   - LIMIT TO THE NUMBER OF ITERATIONS THAT WILL BE USED
C     X       - ARRAY TO HOLD THE X VALUES. INITIALLY THIS ARRAY HOLDS
C               THE INITIAL GUESSES. IT RETURNS THE FINAL VALUES.
C     F       - AN ARRAY THAT HOLDS VALUES OF THE FUNCTIONS
C     DELTA   - A SMALL VALUE USED TO PERTURB THE X VALUES SO PARTIAL
C               DERIVATIVES CAN BE COMPUTED BY DIFFERENCE QUOTIENT.
C     XTOL    - TOLERANCE VALUE FOR CHANGE IN X VALUES TO STOP ITERA-
C               TIONS. WHEN THE LARGEST CHANGE IN ANY X MEETS XTOL,
C               THE SUBROUTINE TERMINATES.
C     FTOL    - TOLERANCE VALUE ON F TO TERMINATE. WHEN THE LARGEST F
C               VALUE IS LESS THAN FTOL, SUBROUTINE TERMINATES.
C     I       - RETURNS VALUES TO INDICATE HOW THE ROUTINE TERMINATED
C
C       I=1      XTOL WAS MET
C       I=2      FTOL WAS MET
C       I=-1     MAXIT EXCEEDED BUT TOLERANCES NOT MET
C       I=-2     VERY SMALL PIVOT ENCOUNTERED IN GAUSSIAN ELIMINATION
C                STEP - NO RESULTS OBTAINED
C       I=-3     INCORRECT VALUE OF N WAS SUPPLIED - N MUST BE BETWEEN
C                2 AND 10
C     METHOD - ALLOWS ONE TO COMPUTE THE JACOBIAN MATRIX IN SUBROUTINE
C              FCNJ BY NUMERICAL OR ANALYTIC PARTIAL DERIVATIVES.
C       0        IMPLIES NUMERICAL PARTIALS ARE TO BE COMPUTED.
C       1        IMPLIES ANALYTIC PARTIALS. THIS REQUIRES THE USER TO
C                INPUT THESE PARTIAL DERIVATIVES IN FCNJ.
C
C     -------------------------------------------------------------------
C
        REAL X(N),F(N),DELTA,XTOL,FTOL
        INTEGER N,MAXIT,I
        REAL A(10,10), XSAVE(10),FSAVE(10),B(10)
        COMMON A,XSAVE,FSAVE
        INTEGER IPVT(10),IT,IVBL,ITEST,IFCN,IROW,JCOL
C
C     -------------------------------------------------------------------
C
C     CHECK VALIDITY OF VALUE OF N
C
        IF ( N .LT. 2 .OR. N .GT. 10 ) THEN
          I = -3
```

Figure 2.13   Program 8.

Figure 2.13 (*continued*)

```
            PRINT 1004, N
            RETURN
         END IF
C    ----------------------------------------------------------------------
C
C    BEGIN ITERATIONS - SAVE X VALUES, THEN GET F VALUES
C
         DO 100 IT = 1,MAXIT
            CALL FCN(X,F)
C
C    ----------------------------------------------------------------------
C
C    TEST F VALUES AND SAVE THEM
C
            ITEST = 0
            DO 20 IFCN = 1,N
               IF ( ABS(F(IFCN)) .GT. FTOL ) ITEST = ITEST + 1
   20       CONTINUE
            IF ( I .EQ. 0 ) THEN
               PRINT 1000, IT,X
               PRINT 1001, F
            END IF
C
C    ----------------------------------------------------------------------
C
C    SEE IF FTOL IS MET. IF NOT, CONTINUE. IF SO, SET I = 2 AND RETURN.
C
            IF ( ITEST .EQ. 0 ) THEN
               I = 2
               RETURN
            END IF
C
C    ----------------------------------------------------------------------
C
C    ----------------------------------------------------------------------
C
            CALL FCNJ(FCN,METHOD, B,N,X,F,DELTA)
C
C
          CALL LUD(A,N,IPVT,10)
         CALL SOLVE(A,N,IPVT,B,10)
C
C
C    BE SURE THAT THE COEFFICIENT MATRIX IS NOT TOO ILL-CONDITIONED
C
            DO 70 IROW = 1,N
               IF ( ABS(A(IROW,IROW)) .LE. 1.0E-6 ) THEN
                  I = -2
                  PRINT 1003
                  RETURN
               END IF
   70       CONTINUE
C
C    ----------------------------------------------------------------------
C
C    APPLY THE CORRECTIONS TO THE X VALUES, ALSO SEE IF XTOL IS MET.
C
            ITEST = 0
            DO 80 IVBL = 1,N
               X(IVBL) = XSAVE(IVBL) + B(IVBL)
               IF ( ABS(B(IVBL)) .GT. XTOL ) ITEST = ITEST + 1
   80       CONTINUE
```

Figure 2.13 *(continued)*

```
C
C      ----------------------------------------------------------------
C
C  IF XTOL IS MET, PRINT LAST VALUES AND RETURN, ELSE DO ANOTHER
C  ITERATION.
C
         IF ( ITEST .EQ. 0 ) THEN
           I = 1
           IF ( I .EQ. 0 ) PRINT 1002, IT,X
           RETURN
         END IF
  100 CONTINUE
C
C      ----------------------------------------------------------------
C
C  WHEN WE HAVE DONE MAXIT ITERATIONS, SET I = -1 AND RETURN
C
      I = -1
      RETURN
C
 1000 FORMAT(/' AFTER ITERATION NUMBER',I3,' X AND F VALUES ARE'
     +        //10F13.5)
 1001 FORMAT(/10F13.5)
 1002 FORMAT(/' AFTER ITERATION NUMBER',I3,' X VALUES (MEETING',
     +        ' XTOL) ARE '//10F13.5)
 1003 FORMAT(/' CANNOT SOLVE SYSTEM. MATRIX NEARLY SINGULAR.')
 1004 FORMAT(/' NUMBER OF EQUATIONS PASSED TO NLSYST IS INVALID.',
     +        ' MUST BE 1 < N < 11. VALUE WAS ',I3)
C
      END
C
C      ----------------------------------------------------------------
C
C      SUBROUTINE FCN:
C         THE TWO NONLINEAR FUNCTION ARE DEFINED
C
C      ----------------------------------------------------------------
C
      SUBROUTINE FCN(X,F)
      REAL X(*),F(*)

      F(1) =  X(1)*X(1) + X(2)*X(2) - 5.0
      F(2) = X(2) - EXP(X(1)) - 1.0
      RETURN
      END
C
C      ----------------------------------------------------------------
C
C      SUBROUTINE FCNJ:
C         HERE THE MATRIX OF PARTIALS ARE COMPUTED:
C         METHOD:  0 - IMPLIES NUMERICAL PARTIALS
C                  1 - IMPLIES ANALYTIC PARTIALS
C                      WHICH  MUST BE SUPPLIED.
C
C      ----------------------------------------------------------------
C
      SUBROUTINE FCNJ(FCN,METHOD,B, N,X,F,DELTA)
      INTEGER N
      REAL  X(N), F(N), DELTA
      INTEGER IROW, JCOL, NF
      REAL A(10,10), FSAVE(10), XSAVE(10), B(10)
      COMMON A,XSAVE,FSAVE
C
```

Figure 2.13 (*continued*)

```
        NP = N + 1
        DO 10 IROW = 1,N
            FSAVE(IROW) = F(IROW)
            XSAVE(IROW) = X(IROW)
10          CONTINUE
C
        IF (METHOD .EQ. 0) THEN
C
C                   COMPUTE NUMERICAL PARTIALS
C
C THIS DOUBLE LOOP COMPUTES THE PARTIAL DERIVATIVES OF EACH FUNCTION
C FOR EACH VARIABLE AND STORES THEM IN A COEFFICIENT ARRAY.
C
        DO 50 JCOL = 1,N
            X(JCOL) = XSAVE(JCOL) + DELTA
            CALL FCN(X,F)
            DO 40 IROW = 1,N
                A(IROW,JCOL) = (F(IROW) - FSAVE(IROW)) / DELTA
40          CONTINUE
C
C RESET X VALUES FOR NEXT COLUMN OF PARTIALS
C
            X(JCOL) = XSAVE(JCOL)
50      CONTINUE
        ELSE
C
C                   COMPUTE ANALYTIC PARTIALS
C
        A(1,1) = 2.0*X(1)
        A(2,1) = -EXP(X(1))
        A(1,2) = 2.0*X(2)
        A(2,2) = 1.0
C
        ENDIF
C
C
C -----------------------------------------------------------------------
C
C
C NOW WE PUT NEGATIVE OF F VALUES AS RIGHT HAND SIDES
C
        DO 60 IROW = 1 , N
            B(IROW) = - FSAVE(IROW)
60      CONTINUE
        RETURN
        END
C
C           SUBROUTINES LUD AND SOLVE
C           FROM PROGRAM 1 MUST BE
C           INCLUDED HERE.
C
```

Figure 2.13   (*continued*)

```
          OUTPUT FOR PROGRAM 8

 AFTER ITERATION NUMBER  1 X AND F VALUES ARE
        2.00000        4.00000
       15.00000       -4.38906
 AFTER ITERATION NUMBER  2 X AND F VALUES ARE
        1.20599        2.52201
        2.81494       -1.81804
 AFTER ITERATION NUMBER  3 X AND F VALUES ARE
         .58368        2.26151
         .45513        -.53112
 AFTER ITERATION NUMBER  4 X AND F VALUES ARE
         .27563        2.24040
         .09535        -.07696
 AFTER ITERATION NUMBER  5 X AND F VALUES ARE
         .20742        2.22751
         .00482        -.00300
 AFTER ITERATION NUMBER  6 X AND F VALUES ARE
         .20434        2.22671
         .00001        -.00001
 AFTER ITERATION NUMBER  7 X AND F VALUES ARE
         .20434        2.22671
         .00000         .00000


 THE X-VALUES ARE:
         .2043
        2.2267
```

EXERCISES

Section 2.2

1. Given the matrices A, B and vectors x, y where

$$A = \begin{bmatrix} 1 & -3 & 2 & 1 \\ 4 & 5 & 6 & 3 \\ -1 & 9 & 7 & 0 \end{bmatrix}, \quad x = \begin{bmatrix} -1 \\ 2 \\ 3 \\ -2 \end{bmatrix},$$

$$B = \begin{bmatrix} -3 & 2 & 1 & 4 \\ 6 & 1 & 9 & -2 \\ 13 & 4 & 1 & 2 \end{bmatrix}, \quad y = \begin{bmatrix} 4 \\ 1 \\ 5 \\ -3 \end{bmatrix}.$$

   a) Find $A + B$, $2A - 3B$, $5x - 3y$.
   ▶ b) Find Ax, By, $x^T y$.
   c) Find A^T, B^T.

2. Given the matrices

$$A = \begin{bmatrix} -2 & 3 & 1 & 5 \\ 2 & 0 & -4 & -1 \\ 1 & 6 & -3 & 2 \end{bmatrix}, \quad B = \begin{bmatrix} 4 & -2 & 1 \\ 1 & -3 & -4 \\ 2 & 5 & 7 \end{bmatrix}.$$

   a) Find BA, B^2, AA^T.
   b) Find det (B).
   c) Find tr (B).
   d) A square matrix can always be expressed as a sum of an upper-triangular and a lower-triangular matrix. Find a U and an L such that $U + L = B$.
   e) A square matrix can. also be expressed as a sum of a lower- and upper-triangular and a diagonal matrix. Find $L + D + U = B$.

3. Let

$$A = \begin{bmatrix} 1 & -2 & 2 \\ 3 & 1 & 1 \\ 2 & 0 & 1 \end{bmatrix}, \quad B = \begin{bmatrix} -1 & -2 & 4 \\ 1 & 3 & -5 \\ 2 & 4 & -7 \end{bmatrix}, \quad C = \begin{bmatrix} -1 & 1 & 2 \\ 0 & 1 & 3 \\ -1 & 0 & 1 \end{bmatrix}.$$

   a) Show that $AB = BA = I$ where I is the 3×3 identity matrix,

$$I = \begin{bmatrix} 1 & 0 & 0 \\ 0 & 1 & 0 \\ 0 & 0 & 1 \end{bmatrix}.$$

   We shall later define such a matrix B as the inverse of A.
   b) Show that $AI = IA = A$.
   ▶ c) Show that $AC \neq CA$ and also $BC \neq CB$. In general, matrices do not commute under multiplication.
   d) A square matrix can also be expressed as a sum of a lower- and upper-triangular and a diagonal matrix. Find $L + D + U = A$.

4. Write as a set of equations:

$$\begin{bmatrix} 1 & 3 & 1 & -1 \\ 2 & 0 & 1 & 1 \\ 0 & -1 & 4 & 1 \\ 0 & 1 & 1 & -5 \end{bmatrix} \begin{bmatrix} x_1 \\ x_2 \\ x_3 \\ x_4 \end{bmatrix} = \begin{bmatrix} 3 \\ 1 \\ 6 \\ 16 \end{bmatrix}.$$

Section 2.3

5. Solve by elimination:

$$\begin{bmatrix} 5 & -1 & 0 \\ -1 & 5 & -1 \\ 0 & -1 & 5 \end{bmatrix} \begin{bmatrix} x_1 \\ x_2 \\ x_3 \end{bmatrix} = \begin{bmatrix} 9 \\ 4 \\ -6 \end{bmatrix}.$$

6. a) Solve by back substitution:

$$\begin{aligned} 4x_1 - 2x_2 + x_3 &= 8, \\ -3x_2 + 5x_3 &= 3, \\ -2x_3 &= 6. \end{aligned}$$

▶ b) Solve by forward substitution (that is, find x_1, x_2, x_3 in that order).

$$\begin{aligned} 5x_1 &= -10, \\ 2x_1 - 3x_2 &= -13, \\ 4x_1 + 3x_2 - 6x_3 &= 7. \end{aligned}$$

7. Solve the set of equations in Exercise 4.

8. Show that the following set of equations does not have a solution:

$$\begin{aligned} 3x_1 + 2x_2 - x_3 - 4x_4 &= 10, \\ x_1 - x_2 + 3x_3 - x_4 &= -4, \\ 2x_1 + x_2 - 3x_3 &= 16, \\ -x_2 + 8x_3 - 5x_4 &= 3. \end{aligned}$$

9. If the constant vector in Exercise 8 has 2, 3, 1, 3 as components, show that the equations have an infinite set of solutions.

Section 2.4

10. Solve Exercise 4 by Gaussian elimination. Carry four decimals in your work.

11. Reorder the equations in Exercise 4 by putting the first equation last, and then solve using the Gauss–Jordan method. Are your answers the same as in Exercise 10?

▶ 12. a) Solve the system

$$\begin{aligned} 2.51x_1 + 1.48x_2 + 4.53x_3 &= 0.05, \\ 1.48x_1 + 0.93x_2 - 1.30x_3 &= 1.03, \\ 2.68x_1 + 3.04x_2 - 1.48x_3 &= -0.53, \end{aligned}$$

by Gaussian elimination, carrying just three significant digits and chopping. Do not interchange rows. Note that one divides by a small coefficient in reducing the third equation.
b) Solve the system again, using elimination with pivoting. Note now that the small coefficient is no longer a divisor.
c) Substitute each set of answers in the original equations and observe that the left- and right-hand sides match better with the (b) answers. The solution, when seven places are carried to eliminate round-off, is

$$x_1 = 1.45310, \qquad x_2 = -1.58919, \qquad x_3 -0.27489.$$

13. Solve the systems in Exercises 4, 5, and 12 by the Gauss–Jordan method.

▶14. By augmenting the coefficient matrix with each of the b-vectors simultaneously, solve $Ax = b$ with multiple right-hand sides, given that

$$A = \begin{bmatrix} 4 & 0 & -1 & 3 \\ 2 & 1 & -2 & 0 \\ 0 & 3 & 2 & -2 \\ 1 & 1 & 0 & 5 \end{bmatrix}, \quad b_1 = \begin{bmatrix} 0 \\ 1 \\ 4 \\ -2 \end{bmatrix}, \quad b_2 = \begin{bmatrix} 0 \\ 0 \\ -2 \\ 4 \end{bmatrix}, \quad b_3 = \begin{bmatrix} 7 \\ -1 \\ 4 \\ -2 \end{bmatrix}.$$

15. a) Show that for a general $n \times (n + 1)$ matrix the Gaussian elimination algorithm, steps 1–4, would take at most $n(n - 1)(2n - 1)/6 + n(n - 1)$ multiplications and/or divisions. You need to know that

$$1 + 2 + \cdots + n = n(n + 1)/2$$

and that

$$1^2 + 2^2 + \cdots + n^2 = n(2n + 1)(n + 1)/6.$$

   b) Also show that for the back-substitution part of this same algorithm, steps 4 and 5, the number of multiplications and/or divisions is $n(n + 1)/2$.

   c) Verify the statement in Section 2.4 that Gauss–Jordan takes about 50% more operations than Gaussian elimination for the specific case of a system of three equations. Add up the number of adds, multiplies, and divides (count a subtraction as an add).

16. Suppose we need to solve $Az = b$ where the a_{ij}, z_i, b_i are complex numbers. If the programming language being used does not allow for complex arithmetic,

   ▶ a) Show how one may solve this problem using only real arithmetic.

   ▶ b) Compare the amount of storage required using complex arithmetic versus solving this problem in real arithmetic. (*Hint:* The matrix A can be written as $A = B + Ci$ where the matrices B, C have only real entries; similarly, we can write $z = x + yi$ and $b = p + qi$.)

Section 2.5

17. Use LU decomposition to solve the equations in Exercise 5.

18. Use LU decomposition to solve the set of equations in Exercise 4.

19. Solve the system in Exercise 12(a) by the LU method. Note that the small divisor again appears and causes round-off error to be severe. Pivoting can be used with this scheme; one must interchange corresponding rows in both the working matrix (LU) and the original matrix (A).

20. a) Solve the tridiagonal system

$$\begin{bmatrix} 4 & -1 & 0 & 0 & 0 \\ -1 & 4 & -1 & 0 & 0 \\ 0 & -1 & 4 & -1 & 0 \\ 0 & 0 & -1 & 4 & -1 \\ 0 & 0 & 0 & -1 & 4 \end{bmatrix} x = \begin{bmatrix} 100 \\ 200 \\ 200 \\ 200 \\ 100 \end{bmatrix}$$

by finding the LU equivalent to the coefficient matrix. Note how readily the operations proceed for this particular case. The coefficient matrix occurs in the solution of partial-differential equations as discussed in Chapter 8.

   ▶ b) Solve the equations in part (a) with a new right-hand-side vector,

$$b = (100, 0, 0, 0, 200)^T.$$

Section 2.6

21. Which of these matrices are singular?

a)
$$A = \begin{bmatrix} 3 & 2 & -1 \\ 0 & -1 & 4 \\ 6 & 3 & 2 \end{bmatrix}.$$

b)
$$B = \begin{bmatrix} 1 & 0 & -2 & 3 \\ 3 & 1 & 1 & 4 \\ -1 & 0 & 2 & -1 \\ 4 & 2 & 6 & 0 \end{bmatrix}.$$

c)
$$C = \begin{bmatrix} 1 & 0 & -2 & 3 \\ 3 & 1 & 1 & 4 \\ -1 & 0 & 2 & -1 \\ 4 & 3 & 6 & 0 \end{bmatrix}.$$

22. a) Find values of x and y that make A singular:

$$A = \begin{bmatrix} 4 & 1 & 0 \\ 2 & -1 & 2 \\ x & y & -1 \end{bmatrix}.$$

b) Find values for x and y that make A nonsingular.

23. a) Matrix A in Exercise 21 is singular. Do its rows form linearly independent vectors? Find the values for the a_i in Eq. (2.8).
  ▶ b) Repeat part (a) for the elements of A considered as column vectors.

24. Do these sets of equations have a solution? Find a solution if it exists.

a) $\begin{cases} 3x - 2y + z = 2, \\ x - 3y + z = 5, \\ x + y - z = -5, \\ 3x \quad\quad + z = 0. \end{cases}$

b) $\begin{bmatrix} 1 & 1 & 0 \\ 0 & 1 & 1 \\ 1 & 0 & 1 \\ 1 & 1 & 1 \end{bmatrix} x = \begin{bmatrix} 1 \\ -2 \\ 0 \\ 4 \end{bmatrix}.$

c) $\begin{bmatrix} 2 & 1 & 3 \\ -1 & 0 & 2 \\ 6 & 2 & 2 \end{bmatrix} x = \begin{bmatrix} 1 \\ 0 \\ 0 \end{bmatrix}.$

d) $\begin{bmatrix} 2 & 1 & 3 \\ -1 & 0 & 2 \\ 6 & 2 & 2 \end{bmatrix} x = \begin{bmatrix} 1 \\ 0 \\ 2 \end{bmatrix}.$

▶25. The Hilbert matrix is a classical case of the pathological situation called "ill-conditioning." The 4×4 Hilbert matrix is

$$H = \begin{bmatrix} 1 & \frac{1}{2} & \frac{1}{3} & \frac{1}{4} \\ \frac{1}{2} & \frac{1}{3} & \frac{1}{4} & \frac{1}{5} \\ \frac{1}{3} & \frac{1}{4} & \frac{1}{5} & \frac{1}{6} \\ \frac{1}{4} & \frac{1}{5} & \frac{1}{6} & \frac{1}{7} \end{bmatrix}.$$

For the system $Hx = b$, with $b^T = [25/12, 77/60, 57/60, 319/420]$, the exact solution is $x^T = [1, 1, 1, 1]$.

a) Show that the matrix is ill-conditioned by showing that it is nearly singular.

b) Using only three significant digits (chopped) in your arithmetic, find the solution to $Hx = b$. Explain why the answers are so poor.

c) Using only three significant digits, but rounding, again find the solution and compare it to that obtained in part (b).

d) Using five significant digits in your arithmetic, again find the solution and compare it to those found in parts (b) and (c).

Section 2.7

26. Find the determinant of the matrix

$$\begin{bmatrix} 0 & 1 & -1 \\ 3 & 1 & -4 \\ 2 & 1 & 1 \end{bmatrix}$$

by row operations to make it (a) upper-triangular, (b) lower-triangular.

▶27. Find the determinant of the matrix

$$\begin{bmatrix} 1 & 4 & -2 & 1 \\ -1 & 2 & -1 & 1 \\ 3 & 3 & 0 & 4 \\ 4 & -4 & 2 & 3 \end{bmatrix}.$$

28. Invert the coefficient matrix in Exercise 5, and then use the inverse to generate the solution. Note that a symmetric matrix has a symmetric inverse.

29. If the constant vector in Exercise 5 is changed to one with components $(13, 11, -22)$, what is now the solution? Observe that the inverse obtained in Exercise 28 gives the answer readily.

30. Attempt to find the inverse of the coefficient matrix in Exercise 8. Note that a singular matrix has no inverse.

31. a) Find the determinant of the Hilbert matrix in Exercise 25. A small value of the determinant (when the matrix has elements of the order of unity) indicates ill-conditioning.

   ▶b) Find the inverse of the Hilbert matrix in Exercise 25. The inverse of an ill-conditioned matrix has some very large elements in comparison to the elements of the original matrix.

32. Both Gaussian elimination and the Gauss–Jordan method can be adapted to invert a matrix. In Section 2.7, we say "it is more efficient to use the Gaussian elimination algorithm." Verify this for the specific case of a 3×3 matrix by counting arithmetic operations for each method.

Section 2.8

33. Find the Euclidean norm of these vectors and matrices.

   a) $x = [1.06, -2.15, 14.05, 0.0]$.

   b) $y = [1, 0, 0, 0]$.

   c) $z^T = [1, 1, 1]$.

   d)
   $$A = \begin{bmatrix} 14 & 0 & 0 \\ 0 & 3 & 0 \\ 0 & 0 & -2 \end{bmatrix}.$$

   e)
   $$B = \begin{bmatrix} 1 & 0 & 1 \\ 1 & 1 & 1 \\ 0 & 1 & 1 \end{bmatrix}.$$

   f) Find the Euclidean norm of B^2, of B^3, of B^4.

g) Find the norms of AB; of Az.

h) Does the triangle inequality of Eq. (2.9) hold for $A + B$? for $B + B$? for $x + y$?

34. Find the Euclidean norm of the Hilbert matrix of Exercise 25.

▶35. Find the Euclidean norm of the inverse of the Hilbert matrix of order 4 (Exercise 31).

Section 2.9*

36. Consider the system $Ax = b$, where

$$A = \begin{bmatrix} 3.01 & 6.03 & 1.99 \\ 1.27 & 4.16 & -1.23 \\ 0.987 & -4.81 & 9.34 \end{bmatrix}, \qquad b = \begin{bmatrix} 1 \\ 1 \\ 1 \end{bmatrix}.$$

▶ a) Using double precision (or a calculator with 10 or more digits of accuracy), solve for x.

b) Solve the system using three-digit (chopped) arithmetic for each arithmetic operation; call this solution $\bar{x}$.

c) Compare x and $\bar{x}$ and compute $e = x - \bar{x}$. What is $\|e\|_2$?

d) Is the system ill-conditioned? What evidence is there to support your conclusion?

37. Repeat Exercise 36, except change the element a_{33} to -9.34.

▶38. Suppose, in Exercise 36, that uncertainties of measurement give slight changes in some of the elements of A. Specifically, suppose a_{11} is 3.00 instead of 3.01 and a_{31} is 0.99 instead of 0.987. What change does this cause in the solution vector (using precise arithmetic)?

39. Compute the residuals for the imperfect solutions in 36(b) and 37(b). Use double precision in this computation.

40. What are the condition numbers of the coefficient matrices in Exercises 36 and 37? Use the 1-norms.

▶41. Verify Eq. (2.14), using the results in Exercises 36, 39, and 40.

42. Verify Eq. (2.14), using the results in Exercises 37, 39, and 40.

43. Verify Eq. (2.15) for the results of Exercise 38.

44. Apply iterative improvement to the imperfect solution of Exercise 36.

▶45. Apply iterative improvement to the imperfect solution of Exercise 37.

Section 2.10

46. Solve Exercise 5 by the Jacobi method, beginning with the initial vector $(0, 0, 0)$. Compare the rate of convergence when Gauss–Seidel is used with the same starting vector.

47. Solve Exercise 5 by Gauss–Seidel iteration, beginning with approximate solution $(2, 2, -1)$.

▶48. The pair of equations

$$x_1 + 2x_2 = 3,$$
$$3x_1 + x_2 = 4,$$

can be rearranged to give $x_1 = 3 - 2x_2$, $x_2 = 4 - 3x_1$. Apply the Jacobi method to this rearrangement, beginning with a vector very close to the solution: $x^{(1)} = (1.01, 1.01)^T$ and observe divergence. Now apply Gauss–Seidel. Which method diverges more rapidly?

49. Solve Exercise 20(a) by Gauss–Seidel iteration.

---

*In certain exercises (36, 37, 38, 44, 45), imperfect solutions will result because low-precision arithmetic is used when the condition number is large. This exaggerates the condition number problem.

Section 2.11

▶50.   Beginning with $(0, 0, 0)$, use relaxation to solve the system

$$6x_1 - 3x_2 + x_3 = 11,$$
$$2x_1 + x_2 - 8x_3 = -15,$$
$$x_1 - 7x_2 + x_3 = 10.$$

51.   Solve the system in Exercise 4 by relaxation.

52.   Relaxation is especially well adapted to problems like Exercise 20. Solve by the relaxation method, starting with the vector $(45, 80, 90, 80, 45)$ which one obtains by inspection.

53.   Solve the system of equations in Exercise 20 by Gauss–Seidel iteration with overrelaxation (Eq. (2.13)). Vary the overrelaxation factor w to find its optimum value. (You will probably want to write a computer program for this.)

Section 2.12

▶54.   Find the two intersections nearest the origin of the two curves: $x^2 + x - y^2 = 1$ and $y - \sin x^2 = 0$. Use the method of iteration.

55.   Solve the system

$$x^2 + y^2 + z^2 = 9,$$
$$xyz = 1,$$
$$x + y - z^2 = 0,$$

by iteration to obtain the solution near $(2.5, 0.2, 1.6)$.

56.   Solve Exercise 54 by Newton's method.

▶57.   Solve by using Newton's method:

$$x^3 + 3y^2 = 21,$$
$$x^2 + 2y + 2 = 0.$$

Make sketches of the graphs to locate approximate values of the intersections.

58.   Apply Eq. (2.39) to compute partials and solve this system by Newton's method:

$$xyz - x^2 + y^2 = 1.34,$$
$$xy - z^2 = 0.09,$$
$$e^x - e^y + z = 0.41.$$

There should be a solution near $(1, 1, 1)$.

59.   At the end of Section 2.12, it is suggested that it would be more efficient to avoid recomputing the partials at each step of Newton's method for a nonlinear system, doing it only after each nth step when there are n equations. Redo Exercises 54 and 57 using this modification. Compare the rate of convergence with that when the partials are recomputed at each step.

APPLIED PROBLEMS AND PROJECTS

60.   In considering the movement of space vehicles, it is frequently necessary to transform coordinate systems. The standard inertial coordinate system has the N-axis pointed north, the E-axis pointed east, and the D-axis pointed toward the center of the earth. A second system is the vehicle's local coordinate system (with the i-axis straight ahead of the vehicle, the j-axis

to the right, and the k-axis downward). We can transform the vector whose local coordinates are (i, j, k) to the inertial system by multiplying by transformation matrices:

$$\begin{bmatrix} n \\ e \\ d \end{bmatrix} = \begin{bmatrix} \cos a & -\sin a & 0 \\ \sin a & \cos a & 0 \\ 0 & 0 & 1 \end{bmatrix} \begin{bmatrix} \cos b & 0 & \sin b \\ 0 & 1 & 0 \\ -\sin b & 0 & \cos b \end{bmatrix} \begin{bmatrix} 1 & 0 & 0 \\ 0 & \cos c & -\sin c \\ 0 & \sin c & \cos c \end{bmatrix} \begin{bmatrix} i \\ j \\ k \end{bmatrix}.$$

Transform the vector $(2.06, -2.44, -0.47)^T$ to the inertial system if $a = 27°$, $b = 5°$, $c = 72°$.

61. a) Exercise 25 shows the pattern for Hilbert matrices. Find the condition number of the 9×9 Hilbert matrix.

b) Suppose we have a system of nine linear equations whose coefficients are the 9×9 Hilbert matrix. Find the right-hand side (the b-vector) if the solution vector has ones for all components. Now increase the value of the first component of the b-vector by 1% and find the solution to the perturbed system. Which component of the solution vector is most changed?

62. In this statically determinate truss with pin joints, the tension F_i in each member can be obtained from the following matrix equation (the equations result from setting the sum of all forces acting horizontally or vertically at each pin equal to zero). (See Fig. 2.14.)

$$\begin{bmatrix} 0.7071 & 0 & 0 & -1 & -0.8660 & 0 & 0 & 0 & 0 \\ 0.7071 & 0 & 1 & 0 & 0.5 & 0 & 0 & 0 & 0 \\ 0 & 1 & 0 & 0 & 0 & -1 & 0 & 0 & 0 \\ 0 & 0 & -1 & 0 & 0 & 0 & 0 & 0 & 0 \\ 0 & 0 & 0 & 0 & 0 & 0 & 1 & 0 & 0.7071 \\ 0 & 0 & 0 & 1 & 0 & 0 & 0 & 0 & -0.7071 \\ 0 & 0 & 0 & 0 & 0.8660 & 1 & 0 & -1 & 0 \\ 0 & 0 & 0 & 0 & -0.5 & 0 & -1 & 0 & 0 \\ 0 & 0 & 0 & 0 & 0 & 0 & 0 & 1 & 0.7071 \end{bmatrix} F = \begin{bmatrix} 0 \\ -1000 \\ 0 \\ 0 \\ 500 \\ 0 \\ 0 \\ -500 \\ 0 \end{bmatrix}$$

a) Solve the system by Gauss elimination.

b) The matrix is quite sparse, so it is a candidate for Gauss–Seidel iteration. Can it be arranged into a diagonally dominant form? Is the system convergent for a starting vector with all elements 0?

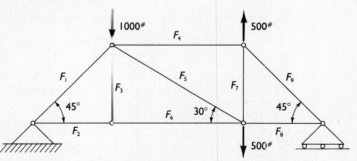

Figure 2.14

63. Mass spectrometry analysis gives a series of peak height readings for various ion masses. For each peak, the height h_j is contributed to by the various constituents. These make different contributions c_{ij} per unit concentration p_i so that the relation

$$h_j = \sum_{i=1}^{n} c_{ij} p_i$$

holds, with n being the number of components present. Carnahan (1964) gives the values shown in Table 2.3 for c_{ij}:

Table 2.3

Peak number	Component				
	CH$_4$	C$_2$H$_4$	C$_2$H$_6$	C$_3$H$_6$	C$_3$H$_8$
1	0.165	0.202	0.317	0.234	0.182
2	27.7	0.862	0.062	0.073	0.131
3		22.35	13.05	4.420	6.001
4			11.28	0	1.110
5				9.850	1.684
6					15.94

If a sample had measured peak heights of $h_1 = 5.20$, $h_2 = 61.7$, $h_3 = 149.2$, $h_4 = 79.4$, $h_5 = 89.3$, and $h_6 = 69.3$, calculate the values of p_i for each component. The total of all the p_i values was 21.53.

64. The truss in Problem 62 is called *statically determinate* because nine linearly independent equations can be established to relate the nine unknown values of the tensions in the members. If an additional cross brace is added, as sketched in Fig. 2.15, we have ten unknowns but still only nine equations can be written; we now have a statically *indeterminate* system. Consideration of the stretching or compression of the members permits a solution, however. We need to solve a set of equations that gives the displacements x of each pin, which is of the form $ASA^Tx = P$. We then get the tensions f by matrix multiplication: $SA^Tx = f$. The necessary matrices and vectors are

$$A = \begin{bmatrix} 0.7071 & 0 & 0 & -1 & -0.8660 & 0 & 0 & 0 & 0 & 0 \\ 0.7071 & 0 & 1 & 0 & 0.5 & 0 & 0 & 0 & 0 & 0 \\ 0 & 1 & 0 & 0 & 0 & -1 & 0 & 0 & 0 & -0.8660 \\ 0 & 0 & -1 & 0 & 0 & 0 & 0 & 0 & 0 & -0.5 \\ 0 & 0 & 0 & 0 & 0 & 0 & 1 & 0 & 0.7071 & 0.5 \\ 0 & 0 & 0 & 1 & 0 & 0 & 0 & 0 & -0.7071 & 0.8660 \\ 0 & 0 & 0 & 0 & 0.8660 & 1 & 0 & -1 & 0 & 0 \\ 0 & 0 & 0 & 0 & -0.5 & 0 & -1 & 0 & 0 & 0 \\ 0 & 0 & 0 & 0 & 0 & 0 & 0 & 1 & 0.7071 & 0 \end{bmatrix}.$$

S is a diagonal matrix with values (from upper left to lower right) of

$$4255, \quad 6000, \quad 6000, \quad 3670, \quad 3000,$$
$$3670, \quad 6000, \quad 6000, \quad 4255, \quad 3000.$$

(These quantities are the values of aE/L, where a is the cross-sectional area of a member, E is the Young's modulus for the material, and L is the length.)

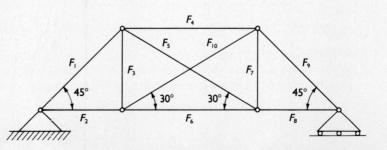

Figure 2.15

Solve the system of equations to determine the values of f for each of three loading vectors:

$$P_1 = [0, -1000, 0, 0, 500, 0, 0, -500, 0]^T;$$
$$P_2 = [1000, 0, 0, -500, 0, 1000, 0, -500, 0]^T;$$
$$P_3 = [0, 0, 0, -500, 0, 0, 0, -500, 0]^T.$$

65. For turbulent flow of fluids in an interconnected network (see Fig. 2.16), the flow rate V from one node to another is about proportional to the square root of the difference in pressures at the nodes. (Thus fluid flow differs from flow of electrical current in a network in that nonlinear equations result.) For the conduits in Fig. 2.16, we are to find the pressure at each node. The values of b represent conductance factors in the relation $v_{ij} = b_{ij}(p_i - p_j)^{1/2}$. These equations can be set up for the pressures at each node:

$$\text{At node 1: } 0.3\sqrt{500 - p_1} = 0.2\sqrt{p_1 - p_2} + 0.1\sqrt{p_1 - p_3};$$
$$\text{node 2: } 0.2\sqrt{p_1 - p_2} = 0.1\sqrt{p_2 - p_4} + 0.2\sqrt{p_2 - p_3};$$
$$\text{node 3: } 0.1\sqrt{p_1 - p_3} = 0.2\sqrt{p_2 - p_3} + 0.1\sqrt{p_3 - p_4};$$
$$\text{node 4: } 0.1\sqrt{p_2 - p_4} + 0.1\sqrt{p_3 - p_4} = 0.2\sqrt{p_4 - 0}.$$

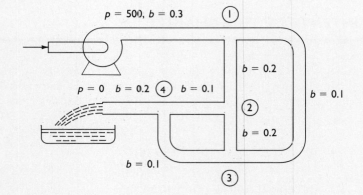

Figure 2.16

66. For a tridiagonal matrix of coefficients, the subroutine TRIDG is much more economical of memory space than the subroutine LUD. It should also be much faster in execution. Analyze each program to determine how the numbers of arithmetic operations vary with size of the matrix. Perform tests to find how the execution times with your computer system compare for systems of various sizes.

67. Make similar comparison between subroutine LUD and LUDCMP as described in Problem 66. Be sure to consider both sparse and dense matrices.

68. Compare the execution times to obtain the solution to some large sparse systems when using subroutine GSITRN in comparison to LUD or LUDCMP. Vary the TOL criterion that is input to find the effect of this parameter.

3

Interpolating Polynomials

3.0 CONTENTS OF THIS CHAPTER

Chapter 3 tells you how to estimate values for a function at points intermediate to those where its values are known. The techniques that are covered also have application in numerical integration and differentiation and in solving differential equations by numerical methods.

3.1 AN INTERPOLATION PROBLEM
Describes a typical situation where interpolation is necessary

3.2 INTERPOLATING POLYNOMIALS
Is the most widely used way to approximate unknown functions so that interpolation can be done; the Lagrangian polynomial, one way to construct the polynomial, is explained

3.3 DIVIDED DIFFERENCES
Are more efficient for constructing interpolating polynomials and allow you to readily change the degree of the polynomial

3.4 EVENLY SPACED DATA
Allow you to use simpler procedures for constructing the polynomials

3.5 OTHER INTERPOLATING POLYNOMIALS
Gives formulas for the many different forms of interpolating polynomials; they have the names of famous mathematicians: Newton, Gauss, Bessel, and Stirling

3.6 ERROR TERMS AND ERROR OF INTERPOLATION
Points out that these methods, like all numerical analysis procedures, are subject to error; you are shown how to estimate the magnitude of the error and how to minimize it

3.1 AN INTERPOLATION PROBLEM

Rita Laski and Ed Baker were at lunch in the cafeteria of Ruscon Engineering. Both had just started summer jobs and were excited at the prospect of finding out how their college training really applied in industry. Rita was explaining her first assignment.

"They have huge amounts of data on the performance of that new rocket. Telemetry signals are received every 10 sec, giving the position of the rocket as well as other information. My boss has asked me to look into how we can determine the position at intermediate times. In essence, it's a kind of interpolation problem."

"I see," said Ed, "something like what we did when we studied log tables in algebra. There we calculated the logs for intermediate values by assuming that a straight line went between the two values from the table."

"Exactly," Rita replied, "except that it isn't appropriate to assume that the positions are linear with time. Typically, the points look something like this when they are plotted." She drew on a paper napkin to illustrate her point. "As you can see, sometimes a signal is missed, like on the third point."

"You can see that a straight line is just impossible if we want to fit the data. And we can't just draw curves like I've done here because we want to get the intermediate values in the computer, and a computer can't look at a graph to find the values. What really bothers me is that Mr. Johnson, my boss, told me that we also must be able to get

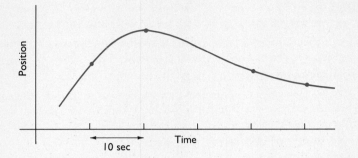

the intermediate values even when there is a maximum or minimum to the curve. Besides all this, we need a method of good efficiency because there are so many cases to handle."

In this chapter we will explore efficient techniques to interpolate, particularly for those situations where the data are far from linear. The principle that will be used is to fit a polynomial curve to the points. The problem is one of interest to many applications. Much of the development comes from work done by Newton and Kepler as they analyzed data on the positions of stars and planets.

In Chapters 1 and 2, we examined the question: "Given an explicit function of the independent variable x, what is the value of x corresponding to a certain value of the function?" In Chapter 1, x was a simple variable. In Chapter 2, x was a vector. We now want to consider a question somewhat the reverse of this: "Given values of an unknown function corresponding to certain values of x, what is the behavior of the function?" We would like to answer the question "What is the function?" but this is always impossible to determine with a limited amount of data.

Our purpose in determining the behavior of the function, as evidenced by the sample of data pairs $[x, f(x)]$, is severalfold. We will wish to approximate other values of the function at values of x not tabulated (interpolation or extrapolation) and to estimate the integral of $f(x)$ and its derivative. The latter objectives will lead us into ways of solving ordinary- and partial-differential equations.

The strategy we will use in approximating unknown values of the function is straightforward. We will find a polynomial that fits a selected set of points $(x_i, f(x_i))$ and assume that the polynomial and the function behave nearly the same over the interval in question. Values of the polynomial then should be reasonable estimates of the values of the unknown function. When the polynomial is of the first degree, this leads to the familiar linear interpolation. We will be interested in polynomials of degree higher than the first, so we can approximate functions that are far from linear, or so we can get good values from a table with wider spacing.

Interpolating in tables of data was once a very important topic. Today, with computers and calculators that recalculate values for most functions at electronic speeds, we look up values from published tables less frequently. However, if computer memory becomes inexpensive enough, it may be preferable to interpolate from tabulated values in computers rather than compute. For instance, on most hand calculators the trigonometric functions are evaluated with stored values through a method called CORDIC. Similarly, the logarithmic functions, like $\ln(x)$, are evaluated using stored values rather than a Taylor series. In both these cases the trade-off is to use cheap memory (ROM) to save computing time. Finally, a major reason for discussing this subject is to lay the groundwork for numerical integration and differentiation.

3.2   LAGRANGIAN POLYNOMIALS

If we desire to find a polynomial that passes through the same points as our unknown function, we could set up a system of equations involving the coefficients of the polynomial. For example, suppose we want to fit a cubic to these data:

x	$f(x)$
3.2	22.0
2.7	17.8
1.0	14.2
4.8	38.3
5.6	51.7

First, we need to select four points to determine our polynomial. (The maximum degree of the polynomial is always one less than the number of points.) Suppose we choose the first four points. If the cubic is $ax^3 + bx^2 + cx + d$, we can write four equations involving the unknown coefficients a, b, c, and d:

$$\text{when } x = 3.2\colon a(3.2)^3 + b(3.2)^2 + c(3.2) + d = 22.0,$$
$$\text{if } x = 2.7\colon a(2.7)^3 + b(2.7)^2 + c(2.7) + d = 17.8,$$
$$\text{if } x = 1.0\colon a(1.0)^3 + b(1.0)^2 + c(1.0) + d = 14.2,$$
$$\text{if } x = 4.8\colon a(4.8)^3 + b(4.8)^2 + c(4.8) + d = 38.3.$$

Solving these equations by the methods of the previous chapter gives us the polynomial. We can then estimate the values of the function at some value of x, say $x = 3.0$, by substituting 3.0 for x in the polynomial.

For this example, the set of equations gives

$$a = -0.5275$$
$$b = 6.4952$$
$$c = -16.117$$
$$d = 24.3499$$

and our polynomial is

$$-0.5275x^3 + 6.4952x^2 - 16.1177x + 24.3499.$$

At $x = 3.0$, the estimated value is 20.21.

We seek a better and simpler way of finding such interpolating polynomials. The above procedure is awkward, especially if we want a new polynomial that is also made to fit at the point (5.6, 51.7), or if we want to see what difference it would make to use a quadratic instead of a cubic. Furthermore, this technique leads to an ill-conditioned system of equations.*

---

*For this example, the system of four equations has a condition number of approximately 555. Adding the last point to solve for the coefficients of a quartic polynomial causes the condition number to jump to near 100,000!

We will first look at another very straightforward approach—the Lagrangian polynomial. The Lagrangian polynomial is perhaps the simplest way to exhibit the existence of a polynomial for interpolation with unevenly spaced data. Data where the x-values are not equispaced often occur as the result of experimental observations or when historical data are examined.

Suppose we have a table of data, of x- and $f(x)$-values:

x	$f(x)$
x_0	f_0
x_1	f_1
x_2	f_2
x_3	f_3

Here we do not assume uniform spacing between the x-values, nor do we need the x-values arranged in a particular order. The x-values must all be distinct, however. Through these four data pairs we can pass a cubic. The Lagrangian form for this is

$$P_3(x) = \frac{(x - x_1)(x - x_2)(x - x_3)}{(x_0 - x_1)(x_0 - x_2)(x_0 - x_3)} f_0 + \frac{(x - x_0)(x - x_2)(x - x_3)}{(x_1 - x_0)(x_1 - x_2)(x_1 - x_3)} f_1$$
$$+ \frac{(x - x_0)(x - x_1)(x - x_3)}{(x_2 - x_0)(x_2 - x_1)(x_2 - x_3)} f_2 + \frac{(x - x_0)(x - x_1)(x - x_2)}{(x_3 - x_0)(x_3 - x_1)(x_3 - x_2)} f_3$$

Note that it is made up of four terms, each of which is a cubic in x; hence the sum is a cubic. The pattern of each term is to form the numerator as a product of linear factors of the form $(x - x_i)$, omitting one x_i in each term, the omitted value being used to form the denominator by replacing x in each of the numerator factors. In each term, we multiply by the f_i corresponding to the x_i omitted in the numerator factors. The Lagrangian polynomial for other degrees of interpolating polynomials employs this same pattern of forming a sum of polynomials all of the desired degree.

It is easy to see that the Lagrangian polynomial does in fact pass through each of the points used in its construction. For example, in the above equation for $P_3(x)$, let $x = x_2$. All terms but the third vanish because of a zero numerator, while the third term becomes just $(1) f_2$. Hence $P_3(x_2) = f_2$. Similarly, $P_3(x_i) = f_i$ for $i = 0, 1, 3$.

E X A M P L E   Fit a cubic through the first four points of the preceding table and use it to find the interpolated value for $x = 3.0$.

$$P_3(3.0) = \frac{(3.0 - 2.7)(3.0 - 1.0)(3.0 - 4.8)}{(3.2 - 2.7)(3.2 - 1.0)(3.2 - 4.8)}(22.0) + \frac{(3.0 - 3.2)(3.0 - 1.0)(3.0 - 4.8)}{(2.7 - 3.2)(2.7 - 1.0)(2.7 - 4.8)}(17.8)$$
$$+ \frac{(3.0 - 3.2)(3.0 - 2.7)(3.0 - 4.8)}{(1.0 - 3.2)(1.0 - 2.7)(1.0 - 4.8)}(14.2) + \frac{(3.0 - 3.2)(3.0 - 2.7)(3.0 - 1.0)}{(4.8 - 3.2)(4.8 - 2.7)(4.8 - 1.0)}(38.3).$$

Carrying out the arithmetic, $P_3(3.0) = 20.21$. ■

Observe that we get the same result as before. The arithmetic in this method is tedious, although hand calculators are convenient for this type of computation. A computer program at the end of this chapter implements the method.

An interpolating polynomial, while passing through the points used in its construction, does not, in general, give exactly correct values when used for interpolation. The reason for this is that the underlying relationship is often not a polynomial of the same degree. We are therefore interested in the error of interpolation.

We will develop an expression for the error of $P_n(x)$, an nth-degree interpolating polynomial. We write the error function in a form that has the known property that it is zero at the $n + 1$ points, from x_0 through x_n, where $P_n(x)$ and $f(x)$ are the same. We call this function $E(x)$:

$$E(x) = f(x) - P_n(x) = (x - x_0)(x - x_1) \ldots (x - x_n)g(x).$$

The $n + 1$ linear factors give $E(x)$ the zeros we know it must have and $g(x)$ accounts for its behavior at values other than at $x_0, x_1, \ldots, x_n$. Obviously, $f(x) - P_n(x) - E(x) = 0$, so

$$f(x) - P_n(x) - (x - x_0)(x - x_1) \ldots (x - x_n)g(x) = 0. \tag{3.1}$$

In order to determine $g(x)$, we now use the interesting mathematical device of constructing an auxiliary function (the reason for its special form becomes apparent as the development proceeds). We call this auxiliary function $W(t)$, and define it as

$$W(t) = f(t) - P_n(t) - (t - x_0)(t - x_1) \ldots (t - x_n)g(x).$$

Note in particular that x has *not* been replaced by t in the $g(x)$ portion. (W is really a function of both t and x, but we are only interested in variations of t.) We now examine the zeros of $W(t)$.

Certainly at $t = x_0, x_1, \ldots, x_n$, the W function is zero ($n + 1$ times), but it is also zero if $t = x$ by virtue of Eq. (3.1). There are then a total of $n + 2$ values of t that make $W(t) = 0$. We now impose the necessary requirements on $W(t)$ for the *law of mean value* to hold. $W(t)$ must be continuous and differentiable. If this is so, there is a zero to its derivative $W'(t)$ between each of the $n + 2$ zeros of $W(t)$, a total of $n + 1$ zeros. If $W''(t)$ exists, and we suppose it does, there will be n zeros of $W''(t)$, and likewise $n - 1$ zeros of $W'''(t)$, and so on, until we reach $W^{(n+1)}(t)$, which must have at least one zero in the interval that has x_0, x_n, or x as endpoints. Call this value of $t = \xi$. We then have

$$W^{(n+1)}(\xi) = 0 = \frac{d^{n+1}}{dt^{n+1}}[f(t) - P_n(t) - (t - x_0) \ldots (t - x_n)g(x)]_{t=\xi}$$

$$= f^{(n+1)}(\xi) - 0 - (n + 1)!g(x). \tag{3.2}$$

The right-hand side of Eq. (3.2) occurs because of the following arguments. The $(n + 1)$st derivative of $f(t)$, evaluated at $t = \xi$, is obvious. The $(n + 1)$st derivative of $P_n(t)$ is zero because every time any polynomial is differentiated its degree is reduced by one, so that the nth derivative is of degree zero (a constant) and its $(n + 1)$st derivative is zero. We apply the same argument to the $(n + 1)$st degree polynomial in t that occurs in the last term—its $(n + 1)$st derivative is a constant and this constant results from the

t^{n+1} term and is $(n + 1)!$. Of course $g(x)$ is independent of t and goes through the differentiations unchanged. The form of $g(x)$ is now apparent:

$$g(x) = \frac{f^{(n+1)}(\xi)}{(n + 1)!}, \qquad \xi \text{ between } (x_0, x_n, x).$$

The conditions on $W(t)$ that are required for this development (continuous and differentiable $n + 1$ times) will be met *if* $f(x)$ has these same properties, because $P_n(x)$ is continuous and differentiable. We now have our error term:

$$E(x) = (x - x_0)(x - x_1) \ldots (x - x_n)\frac{f^{(n+1)}(\xi)}{(n + 1)!}, \tag{3.3}$$

with ξ on the smallest interval that contains $\{x, x_0, x_1, \ldots, x_n\}$.

The expression for error given in Eq. (3.3) is interesting but is not always extremely useful. This is because the actual function that generates the x_i, f_i values is often unknown; we obviously then do not know its $(n + 1)$st derivative. We can conclude, however, that if the function is "smooth," a low-degree polynomial should work satisfactorily. (A smooth function has small higher derivatives.) On the other hand, a "rough" function can be expected to have larger errors when interpolated. We can also conclude that extrapolation (applying the interpolating polynomial outside the range of x-values employed to construct it) will have larger errors than for interpolation.

3.3   DIVIDED DIFFERENCES

There are two problems when the Lagrangian polynomial method is used for interpolation. First, there are more arithmetic operations than for the divided-difference method we now discuss. More importantly, if we desire to add or subtract a point from the set used to construct the polynomial, we essentially have to start over in the computations. The divided-difference method permits one to reuse the previous computations.

Our treatment of divided-difference tables assumes that a function $f(x)$ is known at several distinct values for x:

$$
\begin{array}{cc}
x_0 & f_0 \\
x_1 & f_1 \\
x_2 & f_2 \\
x_3 & f_3
\end{array}
$$

We do not assume that the x's are equispaced nor even that the values are arranged in any particular order.

Consider the nth-degree polynomial:

$$P_n(x) = a_0 + (x - x_0)a_1 + (x - x_0)(x - x_1)a_2 + \cdots$$
$$+ (x - x_0)(x - x_1) \ldots (x - x_{n-1})a_n. \tag{3.4}$$

If we chose the a_i so that $P_n(x)$ equals $f(x)$ at the $n + 1$ known points, $x_0, x_1, \ldots, x_n$, then $P_n(x)$ is an interpolating polynomial. We will show that the a_i are readily determined by using what are called the divided differences of the tabulated values.

A special notion is used for divided differences:

$$f[x_0, x_1] = \frac{f_1 - f_0}{x_1 - x_0}$$

is called the first divided difference between x_0 and x_1.

$$f[x_1, x_2] = \frac{f_2 - f_1}{x_2 - x_1}$$

is the first divided difference between x_1 and x_2.

In general,

$$f[x_s, x_t] = \frac{f_t - f_s}{x_t - x_s}$$

is the first divided difference between x_s and x_t. (Observe that the ordering of the points is immaterial:

$$f[x_s, x_t] = \frac{f_t - f_s}{x_t - x_s} = \frac{f_s - f_t}{x_s - x_t} = f[x_t, x_s].)$$

Second- and higher-order divided differences are defined in terms of lower-order differences. For example:

$$f[x_0, x_1, x_2] = \frac{f[x_1, x_2] - f[x_0, x_1]}{x_2 - x_0},$$

$$f[x_0, x_1, \ldots, x_i] = \frac{f[x_1, x_2, \ldots, x_i] - f[x_0, x_1, \ldots, x_{i-1}]}{x_i - x_0}.$$

The concept is even extended to a zero-order difference:

$$f[x_s] = f_s.$$

Using this notation, a divided difference table, in symbolic form, is Table 3.1(a).

Table 3.1(a)

x_i	f_i	$f[x_i, x_{i+1}]$	$f[x_i, x_{i+1}, x_{i-2}]$	$f[x_i, x_{i+1}, x_{i+2}, x_{i+3}]$
x_0	f_0			
x_1	f_1	$f[x_0, x_1]$		
x_2	f_2	$f[x_1, x_2]$	$f[x_0, x_1, x_2]$	
x_3	f_3	$f[x_2, x_3]$	$f[x_1, x_2, x_3]$	$f[x_0, x_1, x_2, x_3]$
x_4	f_4	$f[x_3, x_4]$	$f[x_2, x_3, x_4]$	$f[x_1, x_2, x_3, x_4]$

A table with specific numerical values might be Table 3.1(b) (with the differences rounded to three decimal places). These are the same data as for the table in Section 3.2.

Table 3.1(b)

x_i	f_i	$f[x_i, x_{i+1}]$	$f[x_i, \ldots, x_{i+2}]$	$f[x_i, \ldots, x_{i+3}]$	$f[x_i, \ldots, x_{i+4}]$
3.2	22.0				
2.7	17.8	8.400			
1.0	14.2	2.118	2.856	−0.528	
4.8	38.3	6.342	2.012	0.0865	0.256
5.6	51.7	16.750	2.263		

We are now ready to establish that the a_i of Eq. (3.4) are given by these divided differences. We write Eq. (3.4) with $x = x_0, x = x_1, \ldots, x = x_n$, giving

with

$$x = x_0: \quad P_n(x_0) = a_0,$$
$$x = x_1: \quad P_n(x_1) = a_0 + (x_1 - x_0)a_1,$$
$$x = x_2: \quad P_n(x_2) = a_0 + (x_2 - x_0)a_1 + (x_2 - x_0)(x_2 - x_1)a_2,$$
$$\vdots$$
$$x = x_n: \quad P_n(x_n) = a_0 + (x_n - x_0)a_1 + (x_n - x_0)(x_n - x_1)a_2 + \cdots$$
$$+ (x_n - x_0) \ldots (x_n - x_{n-1})a_n.$$

If $P_n(x)$ is to be an interpolating polynomial, it must match the table for all $n + 1$ entries:

$$P_n(x_i) = f_i \quad \text{for} \quad i = 0, 1, 2, \ldots, n.$$

If the $P_n(x_i)$ in each equation is replaced by f_i, we get a triangular system, and each a_i can be computed in turn.

From the first equation,

$$a_0 = f_0 = f[x_0] \quad \text{makes} \quad P_n(x_0) = f_0.$$

If $a_1 = f[x_0, x_1]$, then

$$P_n(x_1) = f_0 + (x_1 - x_0)\frac{f_1 - f_0}{x_1 - x_0} = f_1.$$

If $a_2 = f[x_0, x_1, x_2]$, then

$$P_n(x_2) = f_0 + (x_2 - x_0)\frac{f_1 - f_0}{x_1 - x_0}$$
$$+ (x_2 - x_0)(x_2 - x_1)\frac{(f_2 - f_1)/(x_2 - x_1) - (f_1 - f_0)/(x_1 - x_0)}{x_2 - x_0}$$
$$= f_2.$$

One can show in similar fashion that each $P_n(x_i)$ will equal f_i if $a_i = f[x_0, x_1, \ldots, x_i]$.

EXAMPLE   Write the interpolating polynomial of degree 3 that fits the above table at all points from $x_0 = 3.2$ to $x_3 = 4.8$.

$$P_3(x) = 22.0 + 8.400(x - 3.2) + 2.856(x - 3.2)(x - 2.7)$$
$$- 0.528(x - 3.2)(x - 2.7)(x - 1.0).$$

What is the fourth-degree polynomial that fits at all five points? We only have to add one more term to $P_3(x)$:

$$P_4(x) = P_3(x) + 0.256(x - 3.2)(x - 2.7)(x - 1.0)(x - 4.8).$$

When this method is used for interpolation, one should observe that nested multiplication can be used to cut down on the number of arithmetic operations, for example, for $x = 3$:

$$P_3(3) = \{[-0.528(3 - 1.0) + 2.586](3 - 2.7) + 8.400\}(3 - 3.2) + 22.0. \quad \blacksquare$$

If we compute the interpolated value at $x = 3.0$ for each of the third-degree polynomials in Sections 3.2 and 3.3, we get the same result, $P_3(3.0) = 20.21$. This is not surprising, because all third-degree polynomials that pass through the same four points are identical. They may be of different form, but they can all be reduced to the same form.

This seems intuitively true, since there are $n + 1$ constants in the polynomial, and the $n + 1$ data pairs are exactly enough to determine them. More formally we can reason thus, the proof being by contradiction:

Suppose there are two different polynomials of degree n that are alike at the $n + 1$ points. Call these $P_n(x)$ and $Q_n(x)$ and write their difference:

$$D(x) = P_n(x) - Q_n(x),$$

where $D(x)$ is a polynomial of at most degree n. But since P and Q match at the $n + 1$ pairs of points, their difference $D(x)$ is equal to zero for all $n + 1$ of these x-values; that is, it is a polynomial of degree n at most but has $n + 1$ distinct zeros. This is impossible unless $D(x)$ is identically zero. Hence $P_n(x)$ and $Q_n(x)$ are not different—they must be the same polynomial.

One important consequence of this uniqueness property of interpolating polynomials is that their error terms are also identical (though we may want to express them in different forms). We then already know the error term for an interpolating polynomial derived from a table of divided differences. It is precisely that expression we derived in Section 3.2, Eq. (3.3).

3.4 EVENLY SPACED DATA

The problem of interpolation from tabulated data is considerably simplified if the values of the function are given at evenly spaced intervals of the independent variable. It is necessary here to arrange the data in a table with x-values in ascending order. In addition to columns for x and $f(x)$, we will tabulate differences of the functional values. Table 3.2 is a typical difference table.

Table 3.2 A difference table

x	$f(x)$	$\Delta f(x)$	$\Delta^2 f(x)$	$\Delta^3 f(x)$	$\Delta^4 f(x)$
0.0	0.000				
		0.203			
0.2	0.203		0.017		
		0.220		0.024	
0.4	0.423		0.041		0.020
		0.261		0.044	
0.6	0.684		0.085		0.052
		0.346		0.096	
0.8	1.030		0.181		0.211
		0.527		0.307	
1.0	1.557		0.488		
		1.015			
1.2	2.572				

Each of the columns to the right of the $f(x)$ column is computed by calculating the difference between two values in the column to its left.

Symbols that represent the entries in a difference table will be helpful in using the table of differences to determine coefficients in interpolating polynomials. It is conventional to let the letter h stand for the uniform difference in the x-values, $h = \Delta x$. Using subscripts to represent the order of the x and $f(x)$ values, we define the first differences of the function as

$$\Delta f_0 = f_1 - f_0, \quad \Delta f_1 = f_2 - f_1, \quad \Delta f_2 = f_3 - f_2, \quad \ldots, \quad \Delta f_i = f_{i+1} - f_i.$$

The second- and higher-order differences are similarly defined:*

$$\Delta^2 f_1 = \Delta(\Delta f_1) = \Delta(f_2 - f_1) = \Delta f_2 - \Delta f_1 = (f_3 - f_2) - (f_2 - f_1)$$
$$= f_3 - 2f_2 + f_1,$$
$$\Delta^2 f_i = f_{i+2} - 2f_{i+1} + f_i, \tag{3.5}$$
$$\Delta^3 f_1 = \Delta(\Delta^2 f_1) = f_4 - 3f_3 + 3f_2 - f_1,$$
$$\Delta^3 f_i = f_{i+3} - 3f_{i+2} + 3f_{i+1} - f_i,$$
$$\vdots$$
$$\Delta^n f_i = f_{i+n} - nf_{i+n-1} + \frac{n(n-1)}{2!} f_{i+n-2} - \frac{n(n-1)(n-2)}{3!} f_{i+n-3} + \cdots.$$

In Eqs. (3.5) and throughout this chapter, a subscript on f indicates the x-value at which it is evaluated: namely, $f_3 = f(x_3)$. The pattern of the coefficients in Eqs. (3.5) is the familiar array of coefficients in the binomial expansion. This fact we can prove most readily by symbolic methods, which we postpone until a later section. The second- and higher-order differences are generally obtained by differencing the previous differences,

---

*The differences that we define here are called *forward differences*. Some texts define differences that are called *backward differences* (written as ∇f_i, defined as $\nabla f_i = f_i - f_{i-1}$) and *central differences* (written δf_i). Our treatment uses only the forward difference.

but Eq. (3.5) shows how any difference can be calculated directly from the functional values.

Table 3.3 illustrates the formation of a difference table using symbolic representation of the entries. While it might be natural always to call the first x in the table x_0, we will frequently wish to refer to only a portion of the data pairs of the table, so the "first" x loses its significance. We then arbitrarily choose the origin for the subscripted variable, because by the use of negative values we can refer to x-entries that precede x_0. We will set the variable s equal to the subscript on x; it then serves to index the x-values.

In hand computation of a difference table, great care should be exercised to avoid arithmetic errors in the subtractions—the fact that we subtract the upper entry from the lower adds a real source of confusion. One of the best ways to check for mistakes is to add the sum of the numbers in each column to the top entry in the column to its left. This sum should equal the bottom entry in the column to the left.

Since each entry in the difference table is the difference of a pair of numbers in the column to its left, one could recompute one of this pair if it should be erased. As a consequence of this, the entire table could be reproduced, given only one value in each column, if the table is extended to the highest possible order of differences.

When $f(x)$ behaves like a polynomial for the set of data given, the difference table has special properties. In Table 3.4 a function is tabulated over the domain $x = 1$ to $x = 6$, and $f(x)$ obviously behaves the same as x^3. (Note carefully that this does *not* imply that $f(x) \equiv x^3$; the value of $f(x)$ at $x = 7$ might well be 17 instead of $7^3 = 343$. We *only* know the values of $f(x)$ as given in the table.)

Table 3.3   Difference table, using symbols

s	x	$f(x)$	Δf	$\Delta^2 f$	$\Delta^3 f$	$\Delta^4 f$
-2	x_{-2}	f_{-2}				
-1	x_{-1}	f_{-1}	Δf_{-2}			
0	x_0	f_0	Δf_{-1}	$\Delta^2 f_{-2}$		
			Δf_0	$\Delta^2 f_{-1}$	$\Delta^3 f_{-2}$	
1	x_1	f_1	Δf_1	$\Delta^2 f_0$	$\Delta^3 f_{-1}$	$\Delta^4 f_{-2}$
2	x_2	f_2		$\Delta^2 f_1$	$\Delta^3 f_0$	$\Delta^4 f_{-1}$
			Δf_2	$\Delta^2 f_2$	$\Delta^3 f_1$	$\Delta^4 f_0$
3	x_3	f_3	Δf_3			
4	x_4	f_4				

Table 3.4   Difference table for a function behaving like x^3

x	$f(x)$	Δf	$\Delta^2 f$	$\Delta^3 f$	$\Delta^4 f$
0	0				
1	1	1			
2	8	7	6		
3	27	19	12	6	0
4	64	37	18	6	0
5	125	61	24	6	0
6	216	91	30	6	

We observe that the third differences are constant. Consequently, the fourth and all higher differences will be zero. The fact that the nth-order differences of any nth-degree polynomial are constant is readily shown. To prove that these nth-order differences are constant, we first examine the differences of ax^n:

$$\Delta(ax^n) = a(x + h)^n - ax^n$$
$$= ax^n + anx^{n-1}h + \cdots + ah^n - ax^n$$
$$= (anh)x^{n-1} + \text{terms of lower degree in } x,$$
$$\Delta(anh\ x^{n-1}) = an(n - 1)h^2\ x^{n-2} + \text{terms of lower degree.}$$

Noting that every time a difference is taken the leading term has a power of x one less than originally and that the difference of a constant term will be zero, we have for a polynomial

$$\Delta P_n(x) = \Delta(a_1 x^n + a_2 x^{n-1} + \cdots + a_n x + a_{n+1})$$
$$= a_1 nh x^{n-1} + \text{terms of lower degree,}$$
$$\Delta^2 P_n(x) = a_1 n(n - 1)h^2\ x^{n-2} + \text{terms of lower degree,}$$
$$\vdots$$
$$\Delta^n P_n(x) = a_1 n(n - 1)(n - 2)\ldots(1)h^n x^{n-n} = a_1 n! h^n. \qquad (3.6)$$

This shows not only that the nth difference is a constant, but that its value is $a_1 n! h^n$. For $P_3(x) = x^3$,

$$\Delta^3 P_3(x) = (1)(3!)(1)^3 = 6 \qquad \text{when } h = 1.$$

This is exactly what was found in Table 3.4. Note a similarity between the nth difference and the nth derivative of $P_n(x)$:

$$P_n^{(n)}(x) = a_1 n!.$$

Note also that there are similarities between these differences in an evenly spaced table and the divided differences we computed in Section 3.3. In fact the first divided differences computed for Table 3.1(a) when multiplied by Δx are exactly the same as the column of Δf in Table 3.4. Similarly, the higher-order divided differences are related to the higher-order differences by simple multipliers. In an exercise, you are asked to show this relationship.

When the function that is tabulated behaves like a polynomial (and this we can tell by observing that its nth-order differences are constant or nearly so), we can approximate it by the polynomial it resembles. Our problem is to find the simplest means of writing the nth-degree polynomial that passes through $n + 1$ pairs of points, (x_i, f_i), $i = 0$, $1, \ldots, n$. Note that such a polynomial is unique—there is only one polynomial of degree n passing through $n + 1$ points, as we have previously observed.

Perhaps the easiest way to write a polynomial that passes through a group of equispaced points is the Newton–Gregory forward polynomial:

$$P_n(x_s) = f_0 + s\Delta f_0 + \frac{s(s-1)}{2!}\Delta^2 f_0 + \frac{s(s-1)(s-2)}{3!}\Delta^3 f_0 + \cdots$$

$$= f_0 + \binom{s}{1}\Delta f_0 + \binom{s}{2}\Delta^2 f_0 + \binom{s}{3}\Delta^3 f_0 + \binom{s}{4}\Delta^4 f_0 + \cdots \quad (3.7)$$

In this equation we have used the notation $\binom{s}{n}$, the number of combinations of s things taken n at a time, which is the same as the factorial ratios also shown. Referring to Table 3.3, we now observe that $P_n(x)$ does match the table at all the data pairs (x_i, f_i), $i = 0, 1, 2, \ldots, n$. When $s = 0$, $P_n(x_0) = f_0$. If $s = 1$,

$$P_n(x_1) = f_0 + \Delta f_0 = f_0 + f_1 - f_0 = f_1.$$

If $s = 2$,

$$P_n(x_2) = f_0 + 2\Delta f_0 + \Delta^2 f_0 = f_2.$$

Similarly we can demonstrate that $P_n(x)$ formed according to Eq. (3.7) matches at all $n + 1$ points.*

In Eq. (3.7), note that differences all fall on a descending diagonal line beginning at f_0, in Table 3.3.

We have previously observed that if, over the domain from x_0 to x_n, $P_n(x)$ and $f(x)$ have the same values at tabulated values of x, it is reasonable to assume that they will be nearly the same at intermediate points. This assumption is the basis for the use of $P_n(x)$ as an interpolating polynomial. We again emphasize that $f(x)$ and $P_n(x)$ will, in general, not be the same function. Hence there is some error to be expected in the estimate from such interpolation. We use the polynomial in Eq. (3.7) as an interpolating polynomial by letting s take on nonintegral values. This extends the definition of s so that, for any value of x,

$$s = \frac{x - x_0}{h}.$$

EXAMPLE Write a Newton–Gregory forward polynomial of degree 3 that fits Table 3.2 for the four points at $x = 0.4$ to $x = 1.0$. Use it to interpolate for $f(0.73)$.

To make the polynomial fit as specified, we must index the x's so that $x_0 = 0.4$. It follows then that $f_0 = 0.423$, $\Delta f_0 = 0.261$, $\Delta^2 f_0 = 0.085$, and $\Delta^3 f_0 = 0.096$. We compute s:

$$s = \frac{x - x_0}{h} = \frac{0.73 - 0.4}{0.2} = 1.65.$$

---

*This demonstration is not a proof, of course. The section on symbolic methods gives perhaps the neatest proof.

Applying these to Eq. (3.7) with terms through $\Delta^3 f_0$ to give a cubic, we have

$$f(0.73) = 0.423 + (1.65)(0.261) + \frac{(1.65)(0.65)}{2}(0.085)$$

$$+ \frac{(1.65)(0.65)(-0.35)}{6}(0.096)$$

$$= 0.423 + 0.4306 + 0.0456 - 0.0060$$

$$= 0.893. \quad \blacksquare$$

(3.8)

The function tabulated in Table 3.2 is $\tan x$, for which the true value is 0.895 at $x = 0.73$. We hence see that there is an error in the third decimal place. We should anticipate some error because, since the third differences are far from constant, our cubic polynomial is not a perfect representation of the function. Even so, our interpolating polynomial gave a fair estimate, certainly better than linear interpolation, which gives 0.911.

Even though the fourth differences are also not constant, we would hope for some improvement if we approximated $f(x)$ by a fourth-degree polynomial. We can just add one more term to Eq. (3.8) to do this:

$$\binom{s}{4}\Delta^4 f_0 = \frac{(1.65)(0.65)(-0.35)(-1.35)}{4!}(0.211) = 0.0044,$$

$$f(0.73) = 0.893 + 0.0044 = 0.898.$$

Normally the higher-degree polynomial is better, but in this instance, adding another term does not improve our estimate; in fact it worsens it slightly. This is due to round-off errors in the original values in this instance.

The domain over which an interpolating polynomial agrees with the function is most readily found by working backward from the last difference that is included in Eq. (3.7), drawing imaginary diagonals to the left between the entries. The x-values included between this fan of diagonals is the domain of the interpolating polynomial. It will always be found to include one more x-entry than the degree of the polynomial.

3.5   OTHER INTERPOLATING POLYNOMIALS

It is sometimes convenient to write the interpolating polynomial in other forms. The Newton–Gregory backward polynomial is

$$P_n(x) = f_0 + \binom{s}{1}\Delta f_{-1} + \binom{s+1}{2}\Delta^2 f_{-2} + \binom{s+2}{3}\Delta^3 f_{-3}$$

$$+ \binom{s+3}{4}\Delta^4 f_{-4} + \cdots.$$

(3.9)

Table 3.5   Difference table, using symbols. (This is similar to Table 3.4.)

s	x	$f(x)$	Δf	$\Delta^2 f$	$\Delta^3 f$	$\Delta^4 f$
-4	x_{-4}	f_{-4}				
-3	x_{-3}	f_{-3}	Δf_{-4}	$\Delta^2 f_{-4}$		
-2	x_{-2}	f_{-2}	Δf_{-3}	$\Delta^2 f_{-3}$	$\Delta^3 f_{-4}$	
-1	x_{-1}	f_{-1}	Δf_{-2}	$\Delta^2 f_{-2}$	$\Delta^3 f_{-3}$	$\Delta^4 f_{-4}$
0	x_0	f_0	Δf_{-1}	$\Delta^2 f_{-1}$	$\Delta^3 f_{-2}$	$\Delta^4 f_{-3}$
1	x_1	f_1	Δf_0	$\Delta^2 f_0$	$\Delta^3 f_{-1}$	$\Delta^4 f_{-2}$
2	x_2	f_2	Δf_1	$\Delta^2 f_1$	$\Delta^3 f_0$	$\Delta^4 f_{-1}$
3	x_3	f_3	Δf_2	$\Delta^2 f_2$	$\Delta^3 f_1$	$\Delta^4 f_0$
4	x_4	f_4	Δf_3			

In Table 3.5, it is seen that the differences used here form a diagonal row going upward and to the right,* in contrast to the downward sloping diagonal row of differences used in the Newton–Gregory forward formula. Trial with various negative integer values of s demonstrates that Eq (3.9) also matches with data pairs in the table from $x = x_0$ to $x = x_{-n}$.

If the subscripts are suitably chosen, the points where $P_n(x)$ matches the table will be the same as for the Newton–Gregory forward formula, however. When this is done, the two polynomials are really identical though of a different form. We illustrate by reworking the same problem as in Section 3.4.

Choosing $x_0 = 1.0$, so that

$$s = \frac{0.73 - 1.0}{0.2} = -1.35,$$

gives the cubic that fits Table 3.2 between $x = 0.4$ to $x = 1.0$; hence

$$f(0.73) = 1.557 + (-1.35)(0.527) + \frac{(-0.35)(-1.35)}{2}(0.181)$$

$$+ \frac{(0.65)(-0.35)(-1.35)}{6}(0.096)$$

$$= 1.557 - 0.7114 + 0.0428 + 0.0049$$

$$= 0.893.$$

One observes that the identical result is obtained as before.

If we again add one more term to make our interpolation correspond to a fourth-degree polynomial, we have

$$f(0.73) = 0.893 + \frac{(1.65)(0.65)(-0.35)(-1.35)}{24}(0.052) = 0.894.$$

---

*This ascending diagonal row of differences starting at f_i is equal to the backward differences of f_i; those in the downward sloping row are forward differences.

In this instance we do improve the estimate, coming closer to the true value of 0.895. Why did this fourth-degree polynomial not match the fourth-degree one of Section 3.4? In the present case, the domain is from $x = 0.2$ to $x = 1.0$, as is found by going back diagonally from the last difference, 0.052, in contrast to the domain of $x = 0.4$ to $x = 1.2$. The two fourth-degree polynomials are not identical.

There is much nonsense in many books about the application of the Newton–Gregory forward polynomial only at the beginning of the table, and the backward formula only at the end. As our examples clearly show, we may use either formula anywhere in the table by suitably subscripting the x's. Furthermore, the identical results are given by *any* interpolating polynomial that ends on the *same* difference entry. The reason that books tell us to use the forward polynomial at the beginning of a table is so we can increase the degree merely by adding a term. The backward polynomial can similarly be increased in degree very easily when applied to points near the end of the table.

There is a rich variety of interpolation formulas beyond the two we have so far discussed. They differ in the paths taken through the difference table. For example, the Gauss forward goes through the table in a zigzag path, the first step being a forward one. Stirling's and Bessel's formulas proceed horizontally, using averages of differences, one starting with f_0 and the other one starting halfway between f_0 and f_1.

Formulas for interpolation of equispaced data

Newton–Gregory Forward (fits at x_0 to x_n):

$$P_n(x) = f_0 + \binom{s}{1}\Delta f_0 + \binom{s}{2}\Delta^2 f_0 + \binom{s}{3}\Delta^3 f_0 + \binom{s}{4}\Delta^4 f_0 + \cdots + \binom{s}{n}\Delta^n f_0.$$

Newton–Gregory Backward (fits at x_{-n} to x_0):

$$P_n(x) = f_0 + \binom{s}{1}\Delta f_{-1} + \binom{s+1}{2}\Delta^2 f_{-2} + \binom{s+2}{3}\Delta^3 f_{-3} + \binom{s+3}{4}\Delta^4 f_{-4}$$

$$+ \cdots + \binom{s+n-1}{n}\Delta^n f_{-n}.$$

Gauss Forward (fits at $x_{-[n/2]}$ to $x_{[(n+1)/2]}$):

$$P_n(x) = f_0 + \binom{s}{1}\Delta f_0 + \binom{s}{2}\Delta^2 f_{-1} + \binom{s+1}{3}\Delta^3 f_{-1} + \binom{s+1}{4}\Delta^4 f_{-2}$$

$$+ \cdots + \binom{s+[(n-1)/2]}{n}\Delta^n f_{-[n/2]}.$$

Gauss Backward (fits at $x_{-[(n+1)/2]}$ to $x_{[n/2]}$):

$$P_n(x) = f_0 + \binom{s}{1}\Delta f_{-1} + \binom{s+1}{2}\Delta^2 f_{-1} + \binom{s+1}{3}\Delta^3 f_{-2} + \binom{s+2}{4}\Delta^4 f_{-2}$$

$$+ \cdots + \binom{s+[n/2]}{n}\Delta^n f_{-[(n+1)/2]}.$$

Stirling (fits at $x_{-[n/2]}$ to $x_{[n/2]}$ if n is even):

$$P_n(x) = f_0 + \binom{s}{1}\frac{\Delta f_{-1} + \Delta f_0}{2} + \frac{\binom{s+1}{2} + \binom{s}{2}}{2}\Delta^2 f_{-1}$$

$$+ \binom{s+1}{3}\frac{\Delta^3 f_{-2} + \Delta^3 f_{-1}}{2}$$

$$+ \frac{\binom{s-2}{4} + \binom{s+1}{4}}{2}\Delta^4 f_{-2} + \cdots$$

$$+ \begin{cases} \dfrac{\binom{s+[n/2]}{n} + \binom{s+[(n-1)/2]}{n}}{2}\Delta^n f_{-[n/2]} & \text{if } n \text{ is even;} \\[2ex] \binom{s+[(n-1)/2]}{n}\dfrac{\Delta^n f_{-[(n+1)/2]} + \Delta^n f_{-[n/2]}}{2} & \text{if } n \text{ is odd.} \end{cases}$$

Bessel (fits at $x_{-[n/2]}$ to $x_{[(n+1)/2]}$ if n is odd):

$$P_n = \frac{f_0 + f_1}{2} + \frac{\binom{s-1}{1} + \binom{s}{1}}{2}\Delta f_0$$

$$+ \binom{s}{2}\frac{\Delta^2 f_0 + \Delta^2 f_{-1}}{2} + \frac{\binom{s}{3} + \binom{s+1}{3}}{2}\Delta^3 f_{-1}$$

$$+ \binom{s+1}{4}\frac{\Delta^4 f_{-1} + \Delta^4 f_{-2}}{2} + \cdots$$

$$+ \begin{cases} \binom{s-[(n-2)/2]}{n}\dfrac{\Delta^n f_{-[(n-2)/2]} + \Delta^n f_{-[n/2]}}{2} & \text{if } n \text{ is even;} \\[2ex] \dfrac{\binom{s-[(n-2)]}{n} + \binom{s+[n/2]}{n}}{2}\Delta^n f_{-[n/2]} & \text{if } n \text{ is odd.} \end{cases}$$

In the above,

$$\binom{s+j}{k} = \frac{(s+j)(s+j-1)(s+j-2)\ldots(s+j-k+1)}{k!},$$

$\left[\dfrac{m}{2}\right]$ is greatest integer in $\dfrac{m}{2}$.

To use the formulas, choose x_0, compute $s = (x - x_0)/h$, and substitute values.

We illustrate the use of these formulas by writing several interpolating polynomials for the values in Table 3.6.

198   CHAPTER 3: INTERPOLATING POLYNOMIALS

Table 3.6

x	$f(x)$	Δf	$\Delta^2 f$	$\Delta^3 f$	$\Delta^4 f$
0.2	1.06894				
		0.11242			
0.5	1.18136		0.01183		
		0.12425		0.00123	
0.8	1.30561		0.01306		0.00015
		0.13731		0.00138	
1.1	1.44292		0.01444		0.00014
		0.15175		0.00152	
1.4	1.59467		0.01596		
		0.16771			
1.7	1.76238				

Newton–Gregory forward fitting   at $x = 0.5$ through $x = 1.4$ ($x_0 = 0.5$):

$$P_3(x) = 1.18136 + s(0.12425) + \binom{s}{2}(0.01306) + \binom{s}{3}(0.00138).$$

Newton–Gregory backward fitting   at $x = 1.1$ through $x = 0.2$ ($x_0 = 1.1$):

$$P_3(x) = 1.44292 + s(0.13731) + \binom{s+1}{2}(0.01306) + \binom{s+2}{3}(0.00123).$$

We could, however, also define $x_0 = 0.2$ and write:

$$P_3(x) = 1.44292 + \binom{s-3}{1}(0.13731) + \binom{s-2}{2}(0.01306) + \binom{s-1}{3}(0.00123).$$

These last two are really identical because the values of s are computed with different values of x_0 so that the coefficients turn out to be the same. For example, at $x = 0.35$, $s = -2.5$ for the first, but $s = 0.5$ for the second.

Gauss forward fitting   at $x = 0.5$ through $x = 1.1$ ($x_0 = 0.8$):

$$P_2(x) = 1.30561 + s(0.13731) + \binom{s}{2}0.01306.$$

Gauss backward fitting   at $x = 0.8$ through $x = 1.7$ ($x_0 = 1.4$):

$$P_3(x) = 1.59467 + s(0.15175) + \binom{s+1}{2}0.01596 + \binom{s+1}{3}0.00152.$$

Stirling fitting   at $x = 0.2$ through $x = 1.4$ ($x_0 = 0.8$):

$$P_4(x) = 1.30561 + \binom{s}{1}\frac{0.12425 + 0.13731}{2} + \frac{\binom{s}{2} + \binom{s+1}{2}}{2}(0.01306)$$

$$+ \binom{s+1}{3} \frac{0.00123 + 0.00138}{2}.$$

Bessel fitting   at $x = 0.8$ through $x = 1.1$ ($x_0 = 0.8$):

$$P_1(x) = \frac{1.30561 + 1.44292}{2} + \frac{\binom{s-1}{1} + \binom{s}{1}}{2}(0.13731).$$

(This is a pretty elaborate way to write a linear polynomial! It reduces to $1.30561 + s(0.13731)$, as given directly by the Newton–Gregory forward formula.)

We conclude this section with the observation that, since all polynomials through the same points are identical, we can use the familiar Newton–Gregory forward polynomial whenever the points to fit have been preselected from an equispaced table. There is no need to memorize any formula besides this one.

3.6   ERROR TERMS AND ERROR OF INTERPOLATION

Since the interpolating polynomial is not, in general, identical with the unknown function $f(x)$ even though they match at certain points, predicting the values of the function at nontabulated points is subject to error.

For instance, as the number of equispaced points increases, the interpolating polynomials for a given function may actually become less reliable over certain subintervals of the domain. An example that illustrates this is the function

$$f(x) = 1/(1 + 25x^2) \quad \text{on } [1, -1].$$

As the number of equispaced points $n = 1, 3, 5, 7, \ldots$ increases, so does the max error E_n, defined as

$$E_n = \max |f(x) - p_{n-1}(x)| \quad \text{for } x \text{ in } [-1, 1].$$

Here E_n is defined for the n (odd) equispaced points, and p is the $(n - 1)$th-degree interpolating polynomial defined by the n points. Values were computed by using Program 1 at the end of the chapter (see Fig. 3.11); 11, 13, 15, 17, 19 equispaced points on $[-1, 1]$ were used. The estimate of the max error for each n is

n	E_n (approximately)
11	1.9132554
13	3.4710393
15	7.0075584
17	14.3699041
19	29.0384012

From these estimates it is clear that the E_n's are actually increasing.

We leave it as an exercise for you to explain why this magnification of error occurs. You will find it instructive to compare the graph of $f(x)$ with the graphs of some of these higher-degree polynomials.

In Section 3.2 we derived the expression for the error of an interpolating polynomial and have earlier observed that all polynomials of degree n that pass through the same $n + 1$ points are identical. We can therefore just copy Eq. (3.3) to give our error expression:

$$E(x) = (x - x_0)(x - x_1) \ldots (x - x_n)\frac{f^{(n+1)}(\xi)}{(n + 1)!},$$

ξ in the interval (x_0, x_n).

We wish to modify this, expressing it in terms of $s = (x - x_0)/h$, to make it more compatible with our interpolating polynomials. Remembering that

$$x_1 = x_0 + h, \qquad x_2 = x_0 + 2h, \qquad \ldots,$$

so that

$$(x - x_0) = sh, \qquad (x - x_1) = sh - h = (s - 1)h,$$
$$(x - x_2) = sh - 2h = (s - 2)h, \qquad \ldots,$$

we find that Eq. (3.8) becomes

$$E(x_s) = \frac{(s)(s - 1)(s - 2) \ldots (s - n)}{(n + 1)!}h^{n+1}f^{(n+1)}(\xi)$$

$$= \binom{s}{n + 1}h^{n+1}f^{(n+1)}(\xi), \qquad \xi \text{ between } (x_0, x_n, x_s). \tag{3.10}$$

Referring again to Eq. (3.7), we observe that the next term after the last one included in an nth-degree Newton–Gregory forward interpolating polynomial is $\binom{s}{n+1}\Delta^{n+1}f_0$. One can get the error term of Eq. (3.10) by substituting $h^{n+1}f^{(n+1)}(\xi)$ for the $(n + 1)$st difference. This is true for all interpolating polynomials, not just the Newton–Gregory forward one.

E X A M P L E   The data below are for sin x. Interpolate to estimate sin (0.8), using a quadratic through the first three points; also estimate the error using Eq. (3.10).

x	$f(x)$	Δf	$\Delta^2 f$	$\Delta^3 f$
0.1	0.09983			
		0.37960		
0.5	0.47943		−0.07570	
		0.30390		−0.04797
0.9	0.78333		−0.12367	
		0.18023		−0.02846
1.3	0.96356		−0.15213	
		0.02810		
1.7	0.99166			

We take $x_0 = 0.1$, and $s = (0.8 - 0.1)/0.4 = 1.75$. Then

$$f(0.8) = 0.09983 + 1.75(0.37960) + \frac{1.75(0.75)}{2}(-0.07570)$$

$$= 0.71445.$$

$$\text{Error} = \frac{1.75(0.75)(-0.25)}{6}(0.4)^3(-\cos \xi); \; \xi = ?$$

We don't know what value to use for ξ. However, ξ lies in the interval bounded by $x_0 = 0.1$ and $x_2 = 0.9$. Because the cosine is monotonic in this interval, we can easily find maximum and minimum values for $\cos \xi$. (They occur at the endpoints.) This should bound the error:

$$\text{Error} \leq \frac{1.75(0.75)(-0.25)}{6}(0.4)^3[-\cos(0.1)] = 3.48 \times 10^{-3};$$

$$\text{Error} \geq \frac{1.75(0.75)(-0.25)}{6}(0.4)^3[-\cos(0.9)] = 2.18 \times 10^{-3}. \quad \blacksquare$$

The actual error is 2.90×10^{-3}, falling between the two calculated values, as expected. Note carefully that the error estimate we have made is for the truncation error only (the error that arises because we do not use a polynomial of infinite degree). The round-off error, which is negligible in this example, acts independently.

We now can determine which of several interpolating polynomials, all of the same degree, give the best estimates of $f(x)$. It makes no difference which form we use—any convenient one is satisfactory, but we will choose the domain where it fits the table in such a way as to minimize the error term. Changing the points where $P_n(x)$ and $f(x)$ match will cause two changes in the error term; the coefficient involving s will vary, and the value of $f^{(n+1)}(\xi)$ may vary because the interval in which ξ lies changes. Since the function $f(x)$ is, in general, unknown, we certainly do not know its derivatives, and further, the value of ξ is not known except for the intervals that contain it. We then choose the polynomial for which the coefficient involving s has the smallest value. This occurs if the value of x_s at which the polynomial is to be evaluated (the point we are interpolating) lies nearest the midpoint of the interval from x_0 to x_n. Note that this implies that extrapolation will normally be less accurate than interpolation, in accordance with our intuition.

In the preceding example where $f(x)$ is known, we reduce the error if we use the central three points to fit the polynomial.

The computations, with $x_0 = 0.5$, $s = 0.75$, give an estimate of $f(0.8) = 0.71895$. The error estimates are -0.7×10^{-3} and -2.19×10^{-3}, which bracket the actual error of -1.59×10^{-3}. We see that properly choosing the domain where the polynomial fits can reduce the error by nearly one-half.

Choosing $x_0 = 0.9$, so that the last three points are the domain, will be a bad choice, with the greatest error of all.

What if, as is usual, we do not know $f(x)$? In that situation, we cannot find bounds on $f^{(n+1)}(\xi)$. We will later see (Chapter 4) that the nth derivative and the nth differences are related. Anticipating some results from that chapter, we have

$$f^{(n+1)}(x) \doteq \frac{\Delta^{n+1} f(x)}{h^{n+1}}.$$

In the absence of knowledge of the function, one may use this relation in the error-term computation, provided that the $(n + 1)$st differences do not vary too greatly. This accounts for the rough rule of thumb that the error of interpolation is about the magnitude of the next term beyond that included in the formula, for with this approximation to $h^{n+1} f^{(n+1)}(\xi)$, the error term is this next term of the polynomial.

In the above examples, this rule gives the following results:

For the quadratic that fits at $x_0 = 0.1$ to $x_2 = 0.9$, the next term gives

$$\text{Error} \doteq \frac{1.75(0.75)(-0.25)}{6}(-0.04797)$$

$$= 2.62 \times 10^{-3}.$$

For the quadratic that fits at $x_0 = 0.5$ to $x_2 = 1.3$, we get

$$\text{Error} \doteq \frac{0.75(-0.25)(-1.25)}{6}(-0.02846)$$

$$= -1.11 \times 10^{-3}.$$

While these are not the same as the exact errors, they are certainly close enough to give a good working value. We recommend this technique when the derivative needed in Eq. (3.3) or Eq. (3.10) is unknown.

To keep the error of our polynomial interpolation as small as possible, we have seen that we must choose the points to fit our polynomial so that the x-value at which we do the interpolation is as well centered as possible. The Gauss forward and backward polynomials make it simple to keep the x-value centered in the range of fit, even though the degree is increased. Stirling's and Bessel's formulas act similarly.

For even-degree polynomials, Newton and Gauss polynomials are poorer in this respect than are Stirling's. For odd degrees, Bessel's formulas are preferred.

In addition to keeping the range symmetrical about the value for interpolation, a second most important decision must be made—the degree of the polynomial. The truncation error will decrease with the degree, but the round-off error that perturbs the differences in the table will increase. This argues for an intermediate degree as being most accurate.

Another problem occurs with high-degree interpolation in certain cases. A local irregularity in the tabulated function, due either to a local "bump" in the function or to a relatively large experimental error at one point, can cause amplified distortions in the interpolating polynomial at points remote from the point of disturbance. Such amplification increases with the degree; our only recourse is to keep the degree small, or else we will have to approximate the function by a different technique. This phenomenon of amplified distortions and alternative techniques for fitting functions to data is discussed later in this chapter.

To summarize, then, the error of polynomial interpolation is reduced by making its range as symmetrical as possible about the point of interpolation, and by choosing a higher degree of polynomial, up to the point where round-off or the effect of local irregularities cause offsetting errors. But another factor has the greatest importance of all—the step size h. With a given set of tabulated data, we may not be able to do much about the step size. However, if we are designing a new set of tables, the step size may be open to our selection; it is then advantageous to make it small. (This makes our tables more bulky and adds proportionately to the computational effort, of course.)

This effect of h is so important that a special notation is often used to focus attention on it. We write

$$\text{Error} = O(h^n)$$

if there is a constant K such that if h is "small enough," and

$$|\text{Error}| \leq Kh^n$$

where $h > 0$ and where K is some constant not equal to zero. The expression $O(h^n)$ is read "order of h to the nth power." For example, the error of a quadratic interpolating polynomial whose range is (x_0, x_2) would be of order h *cubed* because

$$\text{Error} = \frac{(s)(s - 1)(s - 2)}{6} h^3 f'''(\xi), \quad \text{or} \quad \text{Error} = O(h^3).$$

As h gets small, $f'''(\xi) \to f'''(x_0)$ because ξ is squeezed between x_0 and x_2, and hence approaches a constant value.

3.7 DERIVATION OF FORMULAS BY SYMBOLIC METHODS

In several instances we have presented formulas without proof, although we have demonstrated their validity in specific cases. Symbolic operator methods are a convenient way to establish these relations. We supplement the forward-differencing operator Δ with a backward-differencing operator ∇ and a *stepping* operator E. These are defined by their actions on a function:

$$\Delta f(x_0) = f(x_0 + h) - f(x_0),$$
$$\Delta^2 f(x_0) = \Delta[\Delta f(x_0)] = \Delta f(x_0 + h) - \Delta f(x_0),$$
$$\nabla f(x_0) = f(x_0) - f(x_0 - h),$$
$$\nabla^2 f(x_0) = \nabla[\nabla f(x_0)] = \nabla f(x_0) - \nabla f(x_0 - h),$$
$$E f(x_0) = f(x_0 + h),$$
$$E^2 f(x_0) = E[E f(x_0)] = E f(x_0 + h) = f(x_0 + 2h),$$
$$E^n f(x_0) = f(x_0 + nh).$$

$$\tag{3.11}$$

These obvious relationships exist:

$$\Delta f(x_0) = E f(x_0) - f(x_0) = (E - 1)f(x_0),$$
$$\nabla f(x_0) = f(x_0) - E^{-1}f(x_0) = (1 - E^{-1})f(x_0). \qquad (3.12)$$

We abstract from Eq. (3.12) the symbolic operator relations:

$$\Delta = E - 1,$$
$$\nabla = 1 - E^{-1}. \qquad (3.13)$$

Equations (3.13) are really meaningless, for neither side is defined as it stands. What they signify is that the effect of Δ when operating on a function is the same as the effect of operating with $(E - 1)$, and that ∇ and $(1 - E^{-1})$ have the same effect on a function, though all these quantities are without significance standing alone. We must apply them to a function to interpret them. This is no different, really, from other operational symbols that we use, such as $\sqrt{}$ for square root and d/dx for differentiation.

Since all the operators represented by Δ, ∇, and E are linear operators (the effect on a linear combination of functions is the same as the linear combination of the operator acting on the functions), the laws of algebra are obeyed in relationships between them. This means we can manipulate the relations of Eq. (3.13) by algebraic transformations, and then interpret the results by letting them operate on a function. For example, by raising $\Delta = E - 1$ to the nth power, we have

$$\Delta^n = (E - 1)^n = E^n - nE^{n-1} + \binom{n}{2}E^{n-2} - \binom{n}{3}E^{n-3} + \cdots,$$

so

$$\Delta^n f(x_0) = \left[E^n - nE^{n-1} + \binom{n}{2}E^{n-2} - \cdots \right]f(x_0)$$
$$= f(x_0 + nh) - nf[x_0 + (n - 1)h] + \binom{n}{2}f[x_0 + (n - 2)h] - \cdots,$$

or

$$\Delta^n f_0 = f_n - nf_{n-1} + \binom{n}{2}f_{n-2} - \cdots. \qquad (3.14)$$

Equation (3.14) is a proof of the alternating-sign, binomial-coefficient formula given in Eq. (3.5).

We can develop interesting relations between Δ and ∇, such as

$$E\nabla = E(1 - E^{-1}) = E - 1 = \Delta,$$
$$E^n\nabla^n = \nabla^n E^n = \Delta^n,$$
$$\Delta^n f_0 = \nabla^n E^n f_0 = \nabla^n f_n.$$

This illustrates the fact that a given difference entry in a table can be interpreted as either a forward or backward difference of the appropriate f-values. (This was already obvious from inspection of a difference table.) We can very simply derive the Newton–Gregory forward formula:

$$E = 1 + \Delta, \qquad E^s = (1 + \Delta)^s,$$

$$f_s = E^s f_0 = (1 + \Delta)^s f_0 = \left[1 + s\Delta + \binom{s}{2}\Delta^2 + \binom{s}{3}\Delta^3 + \cdots \right] f_0$$

$$= f_0 + s\Delta f_0 + \binom{s}{2}\Delta^2 f_0 + \binom{s}{3}\Delta^3 f_0 + \cdots.$$

The Newton–Gregory backward formula is similarly easy:

$$E^{-1} = 1 - \nabla, \qquad E^s = (1 - \nabla)^{-s},$$

$$f_s = E^s f_0 = (1 - \nabla)^{-s} f_0 = \left[1 + s\nabla + \binom{s+1}{2}\nabla^2 + \binom{s+2}{3}\nabla^3 + \cdots \right] f_0$$

$$= f_0 + s\nabla f_0 + \binom{s+1}{2}\nabla^2 f_0 + \binom{s+2}{3}\nabla^3 f_0 + \cdots$$

$$= f_0 + s\Delta f_{-1} + \binom{s+1}{2}\Delta^2 f_{-2} + \binom{s+2}{3}\Delta^3 f_{-3} + \cdots. \tag{3.15}$$

We will use symbolic methods to derive derivative formulas in the next chapter.

3.8 INVERSE INTERPOLATION

Suppose we have a table of data such as Table 3.7 and we are required to find the x-value corresponding to a certain value of the function, say at $y = 5.0$. There are two approaches we could use. We can consider the y's to be the independent variable (unevenly spaced) and interpolate for x with a divided-difference polynomial. Doing so gives $x = 2.312$. This approach is straightforward but in some instances gives poor results, the reason being that x considered as a function of y may not be well approximated by a polynomial. This may be true even though y itself behaves very much like a polynomial.

Table 3.7

x_i	f_i	$f[x_i, x_{i+1}]$	$f[x_i, \ldots, x_{i+2}]$	$f[x_i, \ldots, x_{i+3}]$	$f[x_i, \ldots, x_{i+4}]$
1.6	2.3756				
		2.9753			
1.9	3.2682		1.5863		
		3.7685		0.6317	
2.1	4.0219		2.0917		0.1823
		4.8143		0.8504	
2.4	5.4662		2.8570		
		6.8142			
2.8	8.1919				

(Try inverse interpolation among three or four points of the function $y = x^2$, especially for y-values outside the given range.)

The second method is to write y as a polynomial in x and then use the methods of Chapter 1. This will generally make it necessary to multiply out the interpolating polynomial so as to express the function in the usual polynomial form. If we use the divided-difference polynomial, with $x_0 = 1.6$, we have

$$\begin{aligned} y = P_4(x) = \ &2.3756 + 2.9753(x - 1.6) \\ &+ 1.5863(x - 1.6)(x - 1.9) \\ &+ 0.6317(x - 1.6)(x - 1.9)(x - 2.1) \\ &+ 0.1823(x - 1.6)(x - 1.9)(x - 2.1)(x - 2.4) \end{aligned} \tag{3.16}$$

To complete the problem, we must find the root near 2.3 of this fourth-degree polynomial:

$$f(x) = 5.0 - P_4(x) = 0.$$

The equivalent of this second technique for inverse interpolation can be accomplished more readily by a method of successive approximation. Our divided-difference polynomial is, in its general form,

$$\begin{aligned} y_1 = \ &f[x_i] + f[x_i, x_{i+1}](x - x_i) + f[x_i, x_{i+1}, x_{i+2}](x - x_i)(x - x_{i+1}) \\ &+ f[x_i, x_{i+1}, x_{i+2}, x_{i+3}](x - x_i)(x - x_{i+1})(x - x_{i+2}) \\ &+ \cdots \end{aligned} \tag{3.17a}$$

Rearrange to solve for x in the second term:

$$x = \frac{y - f[x_i] - f[x_i, x_{i+1}, x_{i+2}](x - x_i)(x - x_{i+1}) - \cdots}{f[x_i, x_{i+1}]} + x_i \tag{3.17b}$$

The method of successive approximation finds x_1 by first neglecting all the terms in x on the right. Taking $x_i = 2.1$, we get

$$x_1 = \frac{5.0 - 4.0219}{4.8143} + 2.1 = 2.3032.$$

The second approximation is obtained using x_1 on the right side of Eq. (3.17b), including now one more term. Taking $x_i = 1.9$, we get

$$x_2 = \frac{5.0 - 3.2682 - 2.0917(x_1 - 1.9)(x_1 - 2.1)}{3.7685} + 1.9$$
$$= 2.3141.$$

To get x_3, we use x_2 on the right-hand side and pick up another term:

$$x_3 = \frac{5.0 - 3.2682 - 2.0917(x_2 - 1.9)(x_2 - 2.1) - 0.8504(x_2 - 1.9)(x_2 - 2.1)(x_2 - 2.4)}{3.7686} + 1.9$$
$$= 2.3121.$$

In the same fashion, taking $x_i = 1.6$, we get

$$x_4 = \frac{\text{Numerator}}{3.7686} + 1.6 = 2.3127$$

where Numerator $= 5.0 - 2.3756 - 1.5863(x_3 - 1.6)(x_3 - 1.9)$
$- 0.6317(x_3 - 1.6)(x_3 - 1.9)(x_3 - 2.1) - 0.1823(x_3 - 1.6)(x_3 - 1.9)(x_3 - 2.1)(x_3 - 2.4)$.

Do you see why the values for x_i were changed? It was to use differences that are best centered on $y = 5.0$.

The data in Table 3.7 actually are for $y = \sinh x$. Substitution in the hyperbolic sine function gives $\sinh (2.3127) = 5.0013$. Our computation is off by 0.0013.

3.9   INTERPOLATING WITH A CUBIC SPLINE

The fitting of a polynomial curve to a set of data may be considered from another point of view, that of the draftsman. We wish to fit a "smooth curve" to the points. One could use a French curve on the drafting table, but this is a very subjective operation. Fitting a polynomial of high degree to a set of six or eight points, say, does not appeal to us, since we do not expect that the functional relationship is that complicated.

Not only do high-degree polynomials seem undesirable on the basis of our intuition; in some cases they show unexpectedly large deviations from a smooth curve through the data. As an extreme case, in which this effect is exaggerated, consider this function:

$$f(x) = 0, \qquad -1.0 \le x \le -0.2;$$
$$f(x) = 1 - |5x|, \qquad -0.2 < x < 0.2;$$
$$f(x) = 0, \qquad 0.2 \le x \le 1.0.$$

Suppose we fit exact polynomials of degree 2, 4, 6, and 8, all agreeing with $f(x)$ over the interval $-1 \le x \le 1$. Figure 3.1 shows the results. The wide swings when the degree is high, which are caused by a local "bump" in the function at $x = 0$ but which occur at points far removed from $x = 0$, are strong arguments against using high-degree polynomials in cases such as this, when a function is generally smooth but has some local roughness. (The roughness might be only apparent, being caused by an abnormally large error in one of the points.)

One solution to this problem is fitting subregions of the region $-1 \le x \le 1$ with low-degree polynomials. Figure 3.2 illustrates how quadratics would accomplish this. The difficulty with this approach is that the slope is discontinuous at the points where the quadratics join. If we have a generally smooth function, this is undesirable. We seek a method that retains smoothness where the function is smooth, but still can fit local irregularities without the violent misbehavior exhibited in our example above.

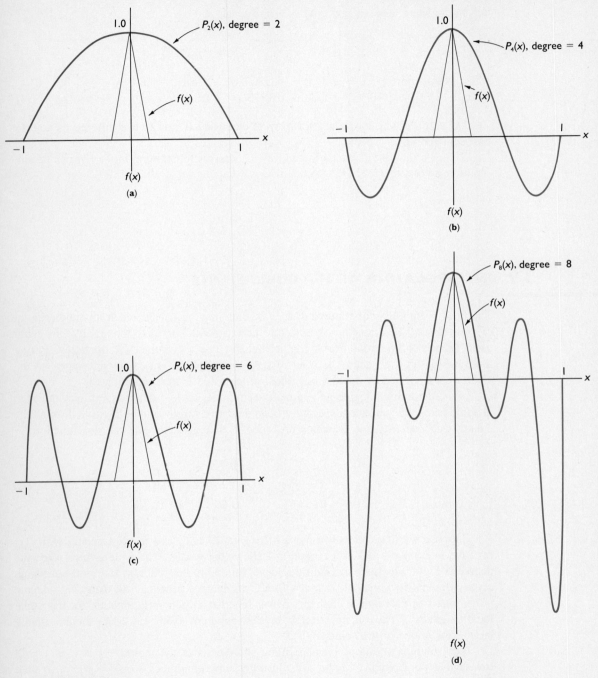

Figure 3.1

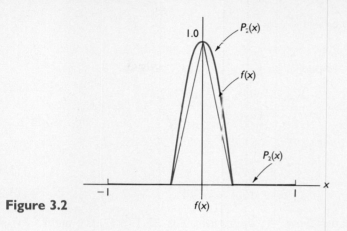

Figure 3.2

One technique that is becoming increasingly important is the so-called *spline fitting* of a curve. The name derives from another draftsman's device. A spline is a flexible strip that can be held by weights so that it passes through each of the given points but goes smoothly from each interval to the next according to the laws of beam flexure. The present mathematical procedure is an adaptation of this idea. It is particularly advantageous when we want to find derivatives of the data.

The conditions for a cubic spline fit are that we pass a set of cubics through the points, using a new cubic in each interval. To correspond to the idea of the draftsman's spline, we require that both the slope and the curvature be the same for the pair of cubics that join at each point. We now develop the equations subject to these conditions.

Write the cubic for the ith interval, which lies between the points (x_i, y_i) and (x_{i+1}, y_{i+1}) in the form

$$y = a_i(x - x_i)^3 + b_i(x - x_i)^2 + c_i(x - x_i) + d_i. \tag{3.18}$$

Since it fits at the two endpoints of the interval,

$$y_i = a_i(x_i - x_i)^3 + b_i(x_i - x_i)^2 + c_i(x_i - x_i) + d_i = d_i; \tag{3.19}$$

$$y_{i+1} = a_i(x_{i+1} - x_i)^3 + b_i(x_{i+1} - x_i)^2 + c_i(x_{i+1} - x_i) + d_i$$

$$= a_i h_i^2 + b_i h_i^2 + c_i h_i + d_i. \tag{3.20}$$

In the last equation, we use h_i for Δx in the ith interval. We need the first and second derivatives to relate the slopes and curvatures of the joining polynomials, so we differentiate Eq. (3.18):

$$y' = 3a_i(x - x_i)^2 + 2b_i(x - x_i) + c_i, \tag{3.21}$$

$$y'' = 6a_i(x - x_i) + 2b_i. \tag{3.22}$$

The mathematical procedure is simplified if we write the equations in terms of the second derivatives of the interpolating cubics. Let S_i represent the second derivative at the point (x_i, y_i) and S_{i+1} at the point (x_{i+1}, y_{i+1}).

From Eq. (3.22) we have

$$S_i = 6a_i(x_i - x_i) + 2b_i$$
$$= 2b_i;$$
$$S_{i+1} = 6a_i(x_{i+1} - x_i) + 2b_i$$
$$= 6a_ih_i + 2b_i.$$

Hence we can write

$$b_i = S_i/2, \tag{3.23}$$
$$a_i = (S_{i+1} - S_i)/6h_i. \tag{3.24}$$

We substitute the relations for a_i, b_i, d_i given by Eqs. (3.19), (3.23), and (3.24) into Eq. (3.20) and then solve for c_i:

$$y_{i+1} = \left(\frac{S_{i+1} - S_i}{6h_i}\right)h_i^3 + \frac{S_i}{2}h_i^2 + c_ih_i + y_i;$$

$$c_i = \frac{y_{i+1} - y_i}{h_i} - \frac{2h_iS_i + h_iS_{i+1}}{6}.$$

We now invoke the condition that the slopes of the two cubics that join at (x_i, y_i) are the same. For the equation in the ith interval, Eq. (3.21) becomes, with $x = x_i$,

$$y_i' = 3a_i(x_i - x_i)^2 + 2b_i(x_i - x_i) + c_i = c_i.$$

In the previous interval, from x_{i-1} to x_i, the slope at its right end will be

$$y_i' = 3a_{i-1}(x_i - x_{i-1})^2 + 2b_{i-1}(x_i - x_{i-1}) + c_{i-1}$$
$$= 3a_{i-1}h_{i-1}^2 + 2b_{i-1}h_{i-1} + c_{i-1}.$$

Equating these, and substituting for a, b, c, d their relationships in terms of S and y, we get

$$y_i' = \frac{y_{i+1} - y_i}{h_i} - \frac{2h_iS_i + h_iS_{i+1}}{6}$$
$$= 3\left(\frac{S_i - S_{i-1}}{6h_{i-1}}\right)h_{i-1}^2 + 2\left(\frac{S_{i-1}}{2}\right)h_{i-1} + \frac{y_i - y_{i-1}}{h_{i-1}} - \frac{2h_{i-1}S_{i-1} + h_{i-1}S_i}{6}.$$

On simplifying this equation we get

$$h_{i-1}S_{i-1} + (2h_{i-1} + 2h_i)S_i + h_iS_{i+1} = 6\left(\frac{y_{i+1} - y_i}{h_i} - \frac{y_i - y_{i-1}}{h_{i-1}}\right) \tag{3.25}$$
$$= 6(f[x_i, x_{i+1}] - f[x_{i-1}, x_i]).$$

Equation (3.25) applies at each internal point, from $i = 1$ to $i = n - 1$, there being $n + 1$ total points. This gives $n - 1$ equations relating the $n + 1$ values of S_i. We get two additional equations involving S_0 and S_n when we specify conditions pertaining to the end intervals of the whole curve. To some extent, these end conditions are arbitrary. Four* alternative choices are often used:

---

1. Take $S_0 = 0$, $S_n = 0$. This is equivalent to assuming that the end cubics approach linearity at their extremities.

2. Take $S_0 = S_1$, $S_n = S_{n-1}$. This is equivalent to assuming that the end cubics approach parabolas at their extremities.

3. Take S_0 as a linear extrapolation from S_1 and S_2, and S_n as a linear extrapolation from S_{n-1} and S_{n-2}. With this assumption, for a set of data that are fit by a single cubic equation, their cubic splines will all be this same cubic. The other conditions do not have this property.

The relations for end condition 3 are as follows:

At the left end:

$$\frac{S_1 - S_0}{h_0} = \frac{S_2 - S_1}{h_1}, \qquad S_0 = \frac{(h_0 + h_1)S_1 - h_0 S_2}{h_1}.$$

At the right end:

$$\frac{S_n - S_{n-1}}{h_{n-1}} = \frac{S_{n-1} - S_{n-2}}{h_{n-2}}, \qquad S_n = \frac{(h_{n-2} + h_{n-1})S_{n-1} - h_{n-1}S_{n-2}}{h_{n-2}}.$$

4. Force the slopes at each end to assume certain values. We may have to make estimates of these slopes.

The relations for end condition 4 are (with $f'(x_0) = A$ and $f'(x_n) = B$):

At the left end:

$$2h_0 S_0 + h_1 S_1 = 6(f[x_0, x_1] - A).$$

At the right end:

$$h_{n-1}S_{n-1} + 2h_n S_n = 6(B - f[x_{n-1}, x_n]).$$

Observe that we use divided differences here.

Relation 1, where $S_0 = 0$ and $S_n = 0$, is called a *natural spline*. It is often felt that this flattens the curve too much at the ends; in spite of this, it is frequently used. Relation 3 frequently suffers from the other extreme, giving too much curvature in the end intervals.

---

*A fifth condition is sometimes encountered—a function is periodic and the data cover a full period. In this case, $S_0 = S_n$ and the slopes are also the same at the first and last points.

Probably the best end condition to use is condition 4, provided reasonable estimates of the derivative are available.

If we write the equation of $S_1, S_2, \ldots, S_{n-1}$ (Eq. (3.25)) in matrix form, we get

$$
\begin{bmatrix}
h_0 & 2(h_0 + h_1) & h_1 & & & & \\
 & h_1 & 2(h_1 + h_2) & h_2 & & & \\
 & & h_2 & 2(h_2 + h_3) & h_3 & & \\
 & & & & \ddots & & \\
 & & & & h_{n-2} & 2(h_{n-2} + h_{n-1}) & h_{n-1}
\end{bmatrix}
\begin{bmatrix}
S_0 \\ S_1 \\ S_2 \\ S_3 \\ \vdots \\ S_{n-1} \\ S_n
\end{bmatrix}
$$

$$
= 6
\begin{bmatrix}
f[x_1, x_2] & - f[x_0, x_1] \\
f[x_2, x_3] & - f[x_1, x_2] \\
f[x_3, x_4] & - f[x_2, x_3] \\
\vdots \\
f[x_{n-1}, x_n] - f[x_{n-2}, x_{n-1}]
\end{bmatrix}
$$

In the matrix array above, there are only $n - 1$ equations, but $n + 1$ unknowns. We can eliminate two unknowns (S_0 and S_n) using the above relations that correspond to the end-condition assumptions. In the first three cases, this reduces the S vector to $n - 1$ elements, and the coefficient matrix becomes square, of size ($n - 1 \times n - 1$). Furthermore, the matrix is always tridiagonal (even in case 4), and hence is solved speedily and can be stored economically. A program is given at the end of this chapter that creates this tridiagonal matrix and augments it with the proper right-hand-side vector, then solves for the values of S in each interval. You will remember that S is the second derivative of the cubic in each interval, so there is more work to do to get the actual cubic that makes the interpolating curve.

For each end condition, the coefficient matrices become

Condition 1 $S_0 = 0, S_n = 0$:

$$
\begin{bmatrix}
2(h_0 + h_1) & h_1 & & & \\
h_1 & 2(h_1 + h_2) & h_2 & & \\
 & h_2 & 2(h_2 + h_3) & h_3 & \\
 & & & \ddots & \\
 & & & h_{n-2} & 2(h_{n-2} + h_{n-1})
\end{bmatrix}
$$

Condition 2 $S_0 = S_1, S_n = S_{n-1}$:

$$
\begin{bmatrix}
(3h_0 + 2h_1) & h_1 & & & \\
h_1 & 2(h_1 + h_2) & h_2 & & \\
 & h_2 & 2(h_2 + h_3) & h_3 & \\
 & & & \ddots & \\
 & & & h_{n-2} & (2h_{n-2} + 3h_{n-1})
\end{bmatrix}
$$

Condition 3 S_0 and S_n are linear extrapolations:

$$
\begin{bmatrix}
\dfrac{(h_0 + h_1)(h_0 + 2h_1)}{h_1} & \dfrac{h_1^2 - h_0^2}{h_1} & & & \\
h_1 & 2(h_1 + h_2) & h_2 & & \\
 & h_2 & 2(h_2 + h_3) & & h_3 \\
 & & & \ddots & \ddots \\
 & & h_{n-2}^2 - h_{n-1}^2 \over h_{n-2} & \dfrac{(h_{n-1} + h_{n-2})(h_{n-1} + 2h_{n-2})}{h_{n-2}}
\end{bmatrix}
$$

Condition 4 $f'(x_0) = A$ and $f'(x_n) = B$:

$$
\begin{bmatrix}
2h_0 & h_1 & & & \\
h_0 & 2(h_0 + h_1) & h_1 & & \\
 & h_1 & 2(h_1 + h_2) & h_2 & \\
 & & & \ddots & \\
 & & & h_{n-2} & 2h_{n-1}
\end{bmatrix}
$$

With condition 3, after solving the set of equations, we must compute S_0 and S_n using Eq. (3.26). For conditions 1, 2, and 4, no computations are needed. For each of the first three cases, the right-hand-side vector is the same; it is given in Eq. (3.27). If the data are evenly spaced, the matrices reduce to a simple form.

After the S_i values are obtained, we get the coefficients a_i, b_i, c_i, and d_i for the cubics in each interval. From these we can compute points on the interpolating curve.

$$a_i = \frac{S_{i+1} - S_i}{6h_i};$$

$$b_i = \frac{S_i}{2};$$

$$c_i = \frac{Y_{i+1} - Y_i}{h_i} - \frac{2h_i S_i + h_i S_{i+1}}{6};$$

$$d_i = Y_i.$$

EXAMPLE I   Fit the data of Table 3.8 by a cubic spline curve. (The true relation is just $y = x^3 - 8$.) Use all four end conditions.

Table 3.8

x	y
0	-8
1	-7
2	0
3	19
4	56

For end condition 1, we solve

$$\begin{bmatrix} 4 & 1 & \\ 1 & 4 & 1 \\ & 1 & 4 \end{bmatrix} \begin{bmatrix} S_1 \\ S_2 \\ S_3 \end{bmatrix} = \begin{bmatrix} 36 \\ 72 \\ 108 \end{bmatrix},$$

so that

$$S_0 = 0, \quad S_1 = 6.4285, \quad S_2 = 10.2857, \quad S_3 = 24.4285, \quad S_4 = 0.$$

For end condition 2, we solve

$$\begin{bmatrix} 5 & 1 & \\ 1 & 4 & 1 \\ & 1 & 5 \end{bmatrix} \begin{bmatrix} S_1 \\ S_2 \\ S_3 \end{bmatrix} = \begin{bmatrix} 36 \\ 72 \\ 108 \end{bmatrix},$$

so that

$$S_0 = 4.80, \quad S_1 = 4.80, \quad S_2 = 12.00, \quad S_3 = 19.20, \quad S_4 = 19.20.$$

For end condition 3, we solve

$$\begin{bmatrix} 6 & 0 & \\ 1 & 4 & 1 \\ & 0 & 6 \end{bmatrix} \begin{bmatrix} S_1 \\ S_2 \\ S_3 \end{bmatrix} = \begin{bmatrix} 36 \\ 72 \\ 108 \end{bmatrix},$$

so that

$$S_0 = 0.00, \quad S_1 = 6.00, \quad S_2 = 12.00, \quad S_3, = 18.00, \quad S_4 = 24.00.$$

For end condition 4, we solve

$$\begin{bmatrix} 2 & 1 & 0 & 0 & 0 \\ 1 & 4 & 1 & 0 & 0 \\ 0 & 1 & 4 & 1 & 0 \\ 0 & 0 & 1 & 4 & 1 \\ 0 & 0 & 0 & 1 & 2 \end{bmatrix} \begin{bmatrix} S_0 \\ S_1 \\ S_2 \\ S_3 \\ S_4 \end{bmatrix} = \begin{bmatrix} 6 \\ 36 \\ 72 \\ 108 \\ 66 \end{bmatrix},$$

so that

$$S_0 = 0.00, \quad S_1 = 6.00, \quad S_2 = 12.00, \quad S_3 = 18.00, \quad S_4 = 24.00.$$

The equation of the data, $y = x^3 - 8$, gives second-derivative values that match those obtained by using conditions 3 and 4. ∎

The next example is a more realistic situation in which a cubic spline might be used.

EXAMPLE 2   The data in the following table are from astronomical observations of a type of variable star called a *Cepheid variable,* and represent variations in its apparent magnitude with time:

Time	0.0	0.2	0.3	0.4	0.5	0.6	0.7	0.8	1.0
Apparent magnitude	0.302	0.185	0.106	0.093	0.240	0.579	0.561	0.468	0.302

Use each of the four end conditions to compute cubic splines, and compare the values interpolated from each spline function at intervals of time of 0.05.

The augmented matrices whose solutions give values for $S_1, S_2, \ldots, S_7$ are shown in Table 3.9(a). The computer program at the end of the chapter was used, giving the results shown in Table 3.9(b).

Table 3.9(a)

Condition 1

Matrix coefficients are

0.20	0.60	0.10	−1.23
0.10	0.40	0.10	3.96
0.10	0.40	0.10	9.60
0.10	0.40	0.10	11.52
0.10	0.40	0.10	−21.42
0.10	0.40	0.10	−4.50
0.10	0.60	0.20	0.60

Condition 2

Matrix coefficients are

0.20	0.80	0.10	−1.23
0.10	0.40	0.10	3.96
0.10	0.40	0.10	9.60
0.10	0.40	0.10	11.52
0.10	0.40	0.10	−21.42
0.10	0.40	0.10	−4.50
0.10	0.80	0.20	0.60

Condition 3

Matrix coefficients are

0.20	1.20	−0.30	−1.23
0.10	0.40	0.10	3.96
0.10	0.40	0.10	9.60
0.10	0.40	0.10	11.52
0.10	0.40	0.10	−21.42
0.10	0.40	0.10	−4.50
−0.30	1.20	0.20	0.60

Table 3.9(a) (*continued*)

	Condition 4		
Matrix coefficients are			
1.00	0.40	0.10	0.00
0.20	0.60	0.10	−1.23
0.10	0.40	0.10	3.96
0.10	0.40	0.10	9.60
0.10	0.40	0.10	11.52
0.10	0.40	0.10	−21.42
0.10	0.40	0.10	−4.50
0.10	0.60	0.20	0.60
0.10	0.40	0.00	0.00

Table 3.9(b)

t	Values, condition 1	Values, condition 2	Values, condition 3	Values, condition 4*
0.00	0.302	0.302	0.302	0.302
0.05	0.278	0.282	0.297	0.276
0.10	0.252	0.256	0.271	0.250
0.15	0.222	0.224	0.231	0.221
0.20	0.185	0.185	0.185	0.185
0.25	0.143	0.142	0.141	0.143
0.30	0.106	0.106	0.106	0.106
0.35	0.087	0.088	0.088	0.087
0.40	0.093	0.093	0.093	0.093
0.45	0.133	0.133	0.133	0.133
0.50	0.240	0.240	0.240	0.240
0.55	0.424	0.424	0.424	0.424
0.60	0.579	0.579	0.579	0.579
0.65	0.608	0.608	0.608	0.608
0.70	0.561	0.561	0.561	0.561
0.75	0.511	0.511	0.511	0.511
0.80	0.468	0.468	0.468	0.468
0.85	0.426	0.426	0.430	0.426
0.90	0.385	0.384	0.392	0.385
0.95	0.343	0.343	0.350	0.343
1.00	0.302	0.302	0.302	0.302

*Note that in the values for condition 4, we used forward and backward differences to approximate the slope at either end of the curve; that is, $V'(0.0) = -0.585$ and $V'(1.0) = -0.830$.

A graph of the four solutions is shown in Figure 3.3. The points all are so close to each other that we must magnify the portions near the ends to see the differences. In the central part of the curve, between $t = 0.2$ and 0.8, none differ by more than 0.001.

Figure 3.3

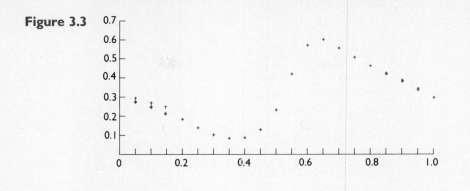

3.10   OTHER SPLINES

In addition to the splines we have studied in the previous section, there are others that are important. In particular, Bezier curves and B-splines are widely used in computer graphics and computer-aided design. These two types of curves are not really interpolating splines, since the curves do not normally pass through all of the points. In this respect they show some similarity to least-squares curves that are discussed in a later chapter. However, both Bezier curves and B-splines have the important property of staying within the polygon determined by the given points. We will be more explicit about this property later. In addition, these two new spline curves have a nice geometric property in that in changing one of the points we change only one portion of the curve, a "local" effect. For the cubic spline curve of the previous section, by changing just one point we would have a "global" effect, in that the curve from the first to the last point would be affected. Finally, for the cubic splines just studied the points were given data points. For the two curves we study in this section the points in question are more likely "control" points we select to determine the shape of the curve we are working on.

For simplicity, we consider mainly the cubic version of these two curves. In what follows, we will express $y = f(x)$ in parametric form. The parametric form represents a relation between x and y by two other equations, $x = F_1(u)$, $y = F_2(u)$. The independent variable u is called the *parameter*. For example, the equation for a circle can be written, with θ as the parameter, as

$$x = r \cos(\theta),$$

$$y = r \sin(\theta).$$

When y and x are expressed in terms of a parameter u, $(x(u), y(u))$, $0 \leq u \leq 1$, defines a set of points (x, y), associated with the values of u. We discuss Bezier curves before B-splines. Bezier curves are named after the French engineer, P. Bezier of the Renault Automobile Company. He developed them in the early 1960s to fill a need for curves whose shape can be readily controlled by changing a few parameters. Bezier's application was to construct pleasing surfaces for car bodies.

Suppose we are given a set of control points, $p_i = (x_i, y_i)$, $i = 0, 1, \ldots, n$. (These points are also referred to as *Bezier points*.) Figure 3.4 is an example.

Figure 3.4

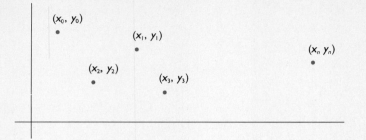

These points could be chosen on a computer screen either by a mouse or a light pen. The points do not necessarily progress from left to right. We treat the coordinates of each point as a two-component vector,

$$p_i = \begin{pmatrix} x_i \\ y_i \end{pmatrix}.$$

The set of points, in parametric form, is

$$P(u) = \begin{pmatrix} x(u) \\ y(u) \end{pmatrix}, \qquad 0 \le u \le 1.$$

The nth-degree Bezier polynomial determined by $n + 1$ points is given by

$$P(u) = \sum_{i=0}^{n} \binom{n}{i}(1 - u)^{n-i}u^i p^i, \text{ where}$$

$$\binom{n}{i} = \frac{n!}{i!(n - i)!}.$$

(The above really represents two other scalar equations, one for x_i and the other for y_i.) For $n = 2$, this would give the quadratic equation defined by three points, p_0, p_1, p_2:

$$P(u) = (1)(1 - u)^2 p_0 + 2(1 - u)(u)p_1 + (1)u^2 p_2,$$

since, for $n = 2$ and $i = 0, 1, 2$, we have $\binom{2}{0} = 1$, $\binom{2}{1} = 2$, $\binom{2}{2} = 1$. The above equation represents the pair of equations

$$x(u) = (1 - u)^2 x_0 + 2(1 - u)(u)x_1 + u^2 x_2,$$

$$y(u) = (1 - u)^2 y_0 + 2(1 - u)(u)y_1 + u^2 y_2.$$

Observe that, if $u = 0$, $x(0)$ is identical to x_0 and similarly for $y(0)$. If $u = 1$, the point referred to is (x_2, y_2). As u takes on values between 0 and 1, a curve is traced that goes from the first point to the third of the set. Ordinarily the curve will not pass through the central point of the three. (If they are collinear, the curve is the straight line through them all.) In effect, the points of the second-degree Bezier curve have coordinates that are weighted sums of the coordinates of the three points that are used to define it. From another point of view, one can think of the Bezier equations as weighted sums of three polynomials in u, where the weighting factors are the coordinates of the three points. Applying the general defining equation for $n = 3$, we get the cubic Bezier polynomial that we shall consider in some detail. The properties of other Bezier polynomials are the same as for the cubic. Here is the Bezier cubic:

$$x(u) = (1 - u)^3 x_0 + 3(1 - u)^2 u x_1 + 3(1 - u) u^2 x_2 + u^3 x_3,$$

$$y(u) = (1 - u)^3 y_0 + 3(1 - u)^2 u y_1 + 3(1 - u) u^2 y_2 + u^3 y_3.$$

Observe again that $(x(0), y(0)) = p_0$ and $(x(1), y(1)) = p_3$, and that the curve will not ordinarily go through the intermediate points. As illustrated in the example curves below, changing the intermediate "control" points changes the shape of the curve. The examples are in Figures 3.5(a) through 3.5(e). The first three of these show Bezier curves defined by one group of four points.

Figures 3.5(d) and 3.5(e) demonstrate how cubic Bezier curves can be continued beyond the first set of four points; one just subdivides seven points (p_0 to p_6) into two groups of four, with the central one (p_3) belonging to both sets. Figure 3.5(e) shows that p_2, p_3, and p_4 must be collinear to avoid a discontinuity in the slope at p_3.

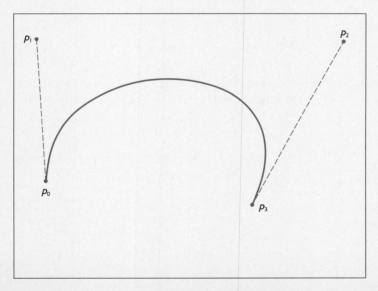

Figure 3.5(a)

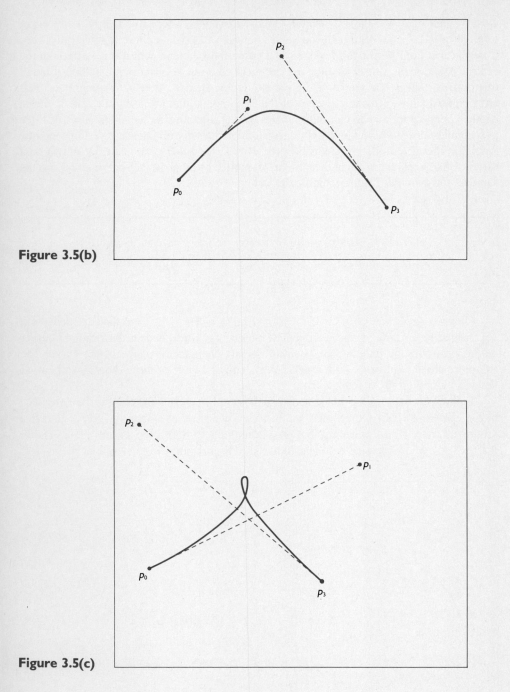

Figure 3.5(b)

Figure 3.5(c)

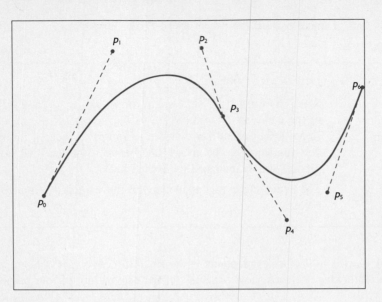

Figure 3.5(d)   Points p_2, p_3, p_4 are not collinear.

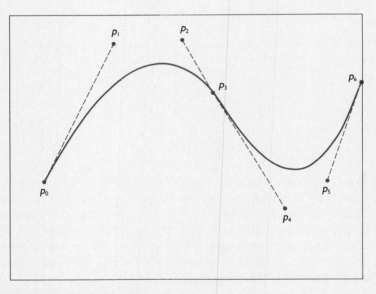

Figure 3.5(e)   Points p_2, p_3, p_4 are collinear.

It is of interest to list the properties of Bezier cubics:

---

1. $P(0) = p_0, \qquad P(1) = p_3.$
2.   Since $dx/du = 3(x_1 - x_0)$ and $dy/du = 3(y_1 - y_0)$ at $u = 0$, the slope of the curve at $u = 0$ is $dy/dx = (y_1 - y_0)/(x_1 - x_0)$, which is the slope of the secant line between p_0 and p_1. Similarly, the slope at $u = 1$ is the same as the secant line between the last two points. In the figures, this is indicated by dashed lines.
3.   The Bezier curve is contained in the convex hull determined by the four points.

The *convex hull* of a set of points is the smallest convex set containing all the points. The following sketches show examples of the convex hull of four points.

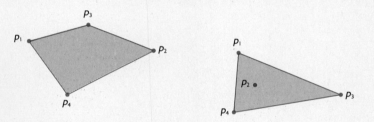

It is often convenient to represent the Bezier curve in matrix form. For Bezier cubics, this is

$$P(u) = [u^3, u^2, u, 1] \begin{bmatrix} -1 & 3 & -3 & 1 \\ 3 & -6 & 3 & 0 \\ -3 & 3 & 0 & 0 \\ 1 & 0 & 0 & 0 \end{bmatrix} \begin{bmatrix} p_0 \\ p_1 \\ p_2 \\ p_3 \end{bmatrix}$$

$$= u^T M_z p.$$

We now discuss B-splines. These curves are like Bezier curves in that they do not ordinarily pass through the given data points. They can be of any degree, but we will concentrate on the cubic form. Cubic B-splines resemble the ordinary cubic splines of the previous section, in that a separate cubic is derived for each pair of points in the set. However, the B-spline need not pass through any of the set that are used in its definition.

We begin the description by stating the formula for a cubic B-spline in terms of parametric equations whose parameter is u:

Given the points $p_i = (x_i, y_i)$, $i = 0, 1, \ldots, n$, the cubic B-spline for the interval (p_i, p_{i+1}), $i = 1, 2, \ldots, n - 1$, is

$$B_i(u) = \sum_{k=-1}^{2} b_k p_{i+k}, \text{ where}$$

$$b_{-1} = (1 - u)^3/6,$$

$$b_0 = u^3/2 - u^2 + 2/3, \tag{3.26}$$
$$b_1 = -u^3/2 + u^2/2 + u/2 + 1/6,$$
$$b_2 = u^3/6, \qquad 0 \le u \le 1.$$

As before, p_i refers to the point (x_i, y_i); it is a two-component vector. The coefficients, the b_k's, serve as a basis and do not change as we move from one pair of points to the next. Observe that they can be considered weighting factors applied to the coordinates of a set of four points. The weighted sum, as u varies from 0 to 1, generates the B-spline curve.

If we write out the equations for x and y from Eq. (3.26), we get

$$x_i(u) = \frac{1}{6}(1-u)^3 x_{i-1} + \frac{1}{6}(3u^3 - 6u^2 + 4)x_i$$
$$+ \frac{1}{6}(-3u^3 + 3u^2 + 3u + 1)x_{i+1} + \frac{1}{6}u^3 x_{i+2};$$

$$y_i(u) = \frac{1}{6}(1-u)^3 y_{i-1} + \frac{1}{6}(3u^3 - 6u^2 + 4)y_i$$
$$+ \frac{1}{6}(-3u^3 + 3u^2 + 3u + 1)y_{i+1} + \frac{1}{6}u^3 y_{i+2}.$$

Note the notation here: $x_i(u)$ is a function (of u) and x_i, y_i are the components of the point p_i.

As we have said, the u-cubics act as weighting factors on the coordinates of the four successive points to generate the curve. For example, at $u = 0$, the weights applied are 1/6, 2/3, 1/6, and 0. At $u = 1$, they are 0, 1/6, 2/3, and 1/6. These values vary throughout the interval from $u = 0$ to $u = 1$. As an exercise, you are asked to graph these factors. This will give you a visual impression of how the weights change with u.

Let us now examine two B-splines determined from a set of exactly four points. Figures 3.6(a) and 3.6(b) show the effect of varying just one of the points. As you would expect, when p_2 is moved upward and to the left, the curve tends to follow; in fact, it is pulled to the opposite side of p_1. You may be surprised to see that the curve is never very close to the two intermediate points, though it begins and ends at positions somewhat adjacent. It will be helpful to think of the curve generated from the defining equation for B_1 as associated with a curve that goes from near p_1 to p_2. It is also helpful to remember that points p_0, p_1, p_2, and p_3 are used to get B_1.

Since a set of four points is required to generate only a portion of the B-spline, that associated with the two inner points, we must consider how to get the B-spline for more than four points as well as how to extend the curve into the region outside of the middle pair. We use a method analogous to the cubic splines of Section 3.9, marching along one point at a time, forming new sets of four. We abandon the first of the old set when we add the new one.

The conditions that we want to impose on the B-spline are exactly the same as for ordinary splines—continuity of the curve and its first and second derivatives. It turns out that the equations for the weighting factors (the u-polynomials, the b_k) are such that these requirements are met. Figure 3.7 shows how three successive parts of a B-spline might look. (We will consider how to fill in the end portions in a moment.)

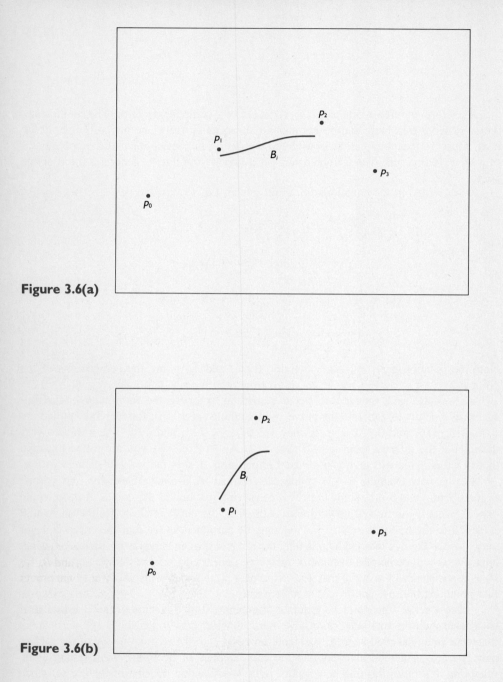

Figure 3.6(a)

Figure 3.6(b)

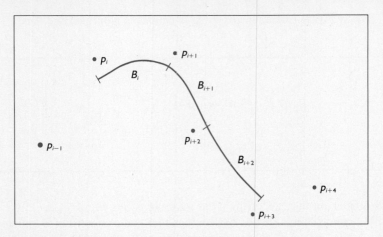

Figure 3.7   Successive B-splines are joined together.

We can summarize the properties of B-splines as

1. Like the cubic splines of Section 3.9, B-splines are pieced together so they agree at their joints in three ways:

a) $B_i(1) = B_{i+1}(0) = (p_i + 4p_{i+1} + p_{i+2})/6.$

b) $B_i'(1) = B_{i+1}'(0) = (-p_i + p_{i+2})/2.$

c) $B_i''(1) = B_{i+1}''(0) = (p_i - 2p_{i+1} + p_{i+2}).$

The subscripts here refer to the portions of the curve and the points in Figure 3.7.

2. The portion of the curve determined by each group of four points is within the convex hull of these points.

Now we consider how to generate the ends of the joined B-spline. If we have points from p_0 to p_n, we already can construct B-splines B_1 through B_{n-2}. We need B_0 and B_{n-1}. Our problem is that, using the procedure already defined, we would need additional points outside the domain of the given points. We probably also want to tie down the curve in some way—having it start and end at the extreme points of the given set seems like a good idea. How can we do this?

First, we can add more points without creating artificiality by making the added points coincide with the given extreme points. If we add not just a single fictitious point at each end of the set, but two at each end, we will find that the new curves not only join properly with the portions already made, but start and end at the extreme points as we wanted. (It looks like we have added two extra portions, but reflection shows these are degenerate, giving only a single point.)

In summary—we add fictitious points p_{-2}, p_{-1}, p_{n+1}, and p_{n+2}, with the first two identical with p_0 and the last two identical with p_n. (There are other methods to handle the starting and ending segments of B-splines that we do not cover.)

The matrix formulation for cubic B-splines is helpful. Here it is:

$$B_i(u) = 1/6[u^3, u^2, u, 1] \begin{bmatrix} -1 & 3 & -3 & 1 \\ 3 & -6 & 3 & 0 \\ -3 & 0 & 3 & 0 \\ 1 & 4 & 1 & 0 \end{bmatrix} \begin{bmatrix} p_{i-1} \\ p_i \\ p_{i+1} \\ p_{i+2} \end{bmatrix} \tag{3.27}$$

$$= u^T M_b p / 6.$$

This applies on the interval [0, 1] and for the points (p_i, p_{i+1}).

We conclude this section by looking at several examples of B-splines. The five parts of Figure 3.8 show B-splines that are defined by the same set of points as the Bezier curves in Figure 3.5. There are significant differences.

B-splines differ from Bezier curves in three ways:

1. For a B-spline, the curve does not begin and end at the extreme points.

2. The slopes of the B-splines do not have any simple relationship to lines drawn between the points.

3. The endpoints of the B-splines are in the vicinity of the two intermediate given points, but neither the x- nor the y-coordinates of these endpoints normally equal the coordinates of the intermediate points.

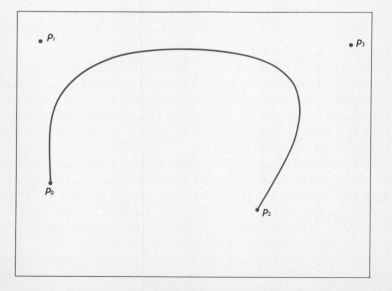

Figure 3.8(a)

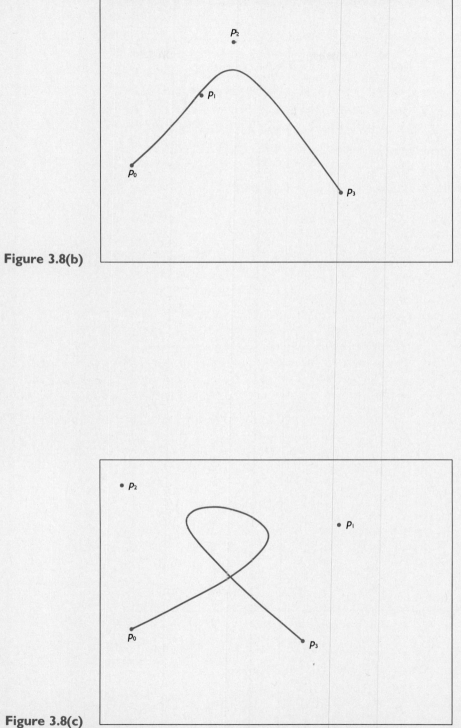

Figure 3.8(b)

Figure 3.8(c)

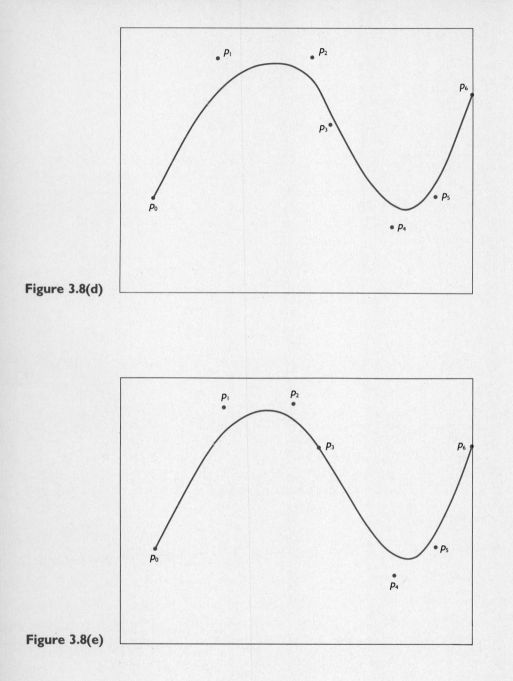

Figure 3.8(d)

Figure 3.8(e)

3.11 POLYNOMIAL APPROXIMATION OF SURFACES

When a function z is a polynomial function* of two variables x and y, say of degree 3 in x and of degree 2 in y, we would have

$$z = f(x, y) = a_0 + a_1 x + a_2 y + a_3 x^2 + a_4 xy + a_5 y^2 + a_6 x^3$$
$$+ a_7 x^2 y + a_8 xy^2 + a_9 x^3 y + a_{10} x^2 y^2 + a_{11} x^3 y^2. \qquad \textbf{(3.28)}$$

Such a function describes a surface; (x, y, z) is a point on it. The functional relation is seen to involve many terms. If we are concerned with four independent variables (three space dimensions plus time, say), even low-degree polynomials would be quite intractable. Except for special purposes, such as when we need an explicit representation, perhaps to permit ready differentiation at an arbitrary point, we can avoid such complications by handling each variable separately. We will treat only this case.

Note the immediate simplification of Eq. (3.28) if we let y take on a constant value, say $y = c$. Combining the y factors with the coefficients, we get

$$z \mid_{y=c} = b_0 + b_1 x + b_2 x^2 + b_3 x^3.$$

This will be our attack in interpolating at the point (a, b) in a table of two variables— hold one variable constant, say $y = y_1$, and the table becomes a single-variable problem. The above methods then apply to give $f(a, y_1)$. If we repeat this at various values of y, $y = y_2, y_3, \ldots, y_n$, we will get a table with x constant at the value $x = a$ and with y varying. We then interpolate at $y = b$.

EXAMPLE   Estimate $f(1.6, 0.33)$ from the values in Table 3.10. Use quadratic interpolation in the x-direction and cubic interpolation for y. We select one of the variables to hold constant, say x. (This choice is arbitrary since we would get the same result (except for differences due to round-off) if we had chosen to hold y constant.) We decide to interpolate for y within the three rows of the table at $x = 1.0$, 1.5, and 2.0 since the desired value at $x = 1.6$ is most nearly centered within this set. We choose y-values of 0.2, 0.3, 0.4, and 0.5 so that $y = 0.33$ is centralized. The shading in Table 3.10 shows the region of fit for our polynomials.

Table 3.10   Tabulation of a function of two variables $z = f(x, y)$

x \ y	0.1	0.2	0.3	0.4	0.5	0.6
0.5	0.165	0.428	0.687	0.942	1.190	1.431
1.0	0.271	0.640	1.003	1.359	1.703	2.035
1.5	0.447	0.990	1.524	2.045	2.549	3.031
2.0	0.738	1.568	2.384	3.177	3.943	4.672
2.5	1.216	2.520	3.800	5.044	6.241	7.379
3.0	2.005	4.090	6.136	8.122	10.030	11.841
3.5	3.306	6.679	9.986	13.196	16.277	19198

*We approximate a nonpolynomial function by a polynomial that agrees with the function, just as we have done with a function of one variable.

We may either use divided differences or derive the interpolated values using difference tables. Let us use the latter method since the data are evenly spaced.

	y	z	Δz	Δ²z	Δ³z
	0.2	0.640			
			0.363		
	0.3	1.003		−0.007	
x = 1.0			0.356		−0.005
	0.4	1.359		−0.012	
			0.344		
	0.5	1.703			
	0.2	0.990			
			0.534		
	0.3	1.524		−0.013	
x = 1.5			0.521		−0.004
	0.4	2.045		−0.017	
			0.504		
	0.5	2.549			
	0.2	1.568			
			0.816		
	0.3	2.384		−0.023	
x = 2.0			0.793		−0.004
	0.4	3.177		−0.027	
			0.766		
	0.5	3.943			

We need the subtables from $y = 0.2$ to $y = 0.5$ since, for a cubic interpolation, four points are required. Using any convenient formula, we arrive at the results:

	x	z	Δz	Δ²z
	1.0	1.1108		
			0.5710	
y = 0.33	1.5	1.6818		0.3717
			0.9427	
	2.0	2.6245		

In the last tabulation we carry one extra decimal to guard against round-off errors. Interpolating again, we get $z = 1.8406$, which we report as $z = 1.841$. ∎

The function tabulated in Table 3.10 is $f(x, y) = e^x \sin y + y - 0.1$, so the true value is $f(1.6, 0.33) = 1.8350$. Our error of -0.006 occurs because quadratic interpolation for x is inadequate in view of the large second difference. In retrospect, it would have been better to use quadratic interpolation for y, since the third differences of the y-subtables are small, and let x take on a third-degree relationship.

It is instructive to observe which of the values in Table 3.10 entered into our computation. The shaded rectangle covers these values. This is the "region of fit" for the interpolating polynomial that we have used. The principle of choosing values so that the point at which the interpolating polynomial is used is centered in the region of fit obviously applies here in exact analogy to the one-way table situation. It also applies to tables of three and four variables in the same way. Of course, the labor of interpolating in such multidimensional cases soon becomes burdensome.

A rectangular region of fit is not the only possibility. We may change the degree of interpolation as we subtabulate the different rows or columns. Intuitively, it would seem best to use higher-degree polynomials for the rows near the interpolating point, decreasing the degree as we get farther away. The coefficient of the error term, when this is done, will be found to be minimized thereby, though for multidimensional interpolating polynomials the error term is quite complex. The region of fit will be diamond-shaped when such tapered degree functions are used.

We may adapt the Lagrangian form of interpolating polynomial to the multidimensional case also. It is perhaps easiest to employ a process similar to the above example. Holding one variable constant, we write a series of Lagrangian polynomials for interpolation at the given value of the other variable, and then combine these values in a final Lagrange form. The net result is a Lagrangian polynomial in which the function factors are replaced by Lagrangian polynomials. The resulting expression for the above example would be

$$
\frac{(y - 0.3)(y - 0.4)(y - 0.5)}{(0.2 - 0.3)(0.2 - 0.4)(0.2 - 0.5)}
$$
$$
\times \left[\frac{(x - 1.5)(x - 2.0)}{(1.0 - 1.5)(1.0 - 2.0)}(0.640) + \frac{(x - 1.0)(x - 2.0)}{(1.5 - 1.0)(1.5 - 2.0)}(0.990) + \frac{(x - 1.0)(x - 1.5)}{(2.0 - 1.0)(2.0 - 1.5)}(1.568) \right]
$$
$$
+ \frac{(y - 0.2)(y - 0.4)(y - 0.5)}{(0.3 - 0.2)(0.3 - 0.4)(0.3 - 0.5)}
$$
$$
\times \left[\frac{(x - 1.5)(x - 2.0)}{(1.0 - 1.5)(1.0 - 2.0)}(1.003) + \frac{(x - 1.0)(x - 2.0)}{(1.5 - 1.0)(1.5 - 2.0)}(1.534) + \frac{(x - 1.0)(x - 1.5)}{(2.0 - 1.0)(2.0 - 1.5)}(2.384) \right]
$$
$$
+ \frac{(y - 0.2)(y - 0.3)(y - 0.5)}{(0.4 - 0.2)(0.4 - 0.3)(0.4 - 0.5)}
$$
$$
\times \left[\frac{(x - 1.5)(x - 2.0)}{(1.0 - 1.5)(1.0 - 2.0)}(1.359) + \frac{(x - 1.0)(x - 2.0)}{(1.5 - 1.0)(1.5 - 2.0)}(2.045) + \frac{(x - 1.0)(x - 1.5)}{(2.0 - 1.0)(2.0 - 1.5)}(3.177) \right]
$$
$$
+ \frac{(y - 0.2)(y - 0.3)(y - 0.4)}{(0.5 - 0.2)(0.5 - 0.3)(0.5 - 0.4)}
$$
$$
\times \left[\frac{(x - 1.5)(x - 2.0)}{(1.0 - 1.5)(1.0 - 2.0)}(1.703) + \frac{(x - 1.0)(x - 2.0)}{(1.5 - 1.0)(1.5 - 2.0)}(2.549) + \frac{(x - 1.0)(x - 1.5)}{(2.0 - 1.0)(2.0 - 1.5)}(3.943) \right].
$$

The equation is easy to write, but its evaluation by hand is laborious. If one is writing a computer program for interpolation in such multivariate situations, the Lagrangian form is recommended. There is a special advantage in that equal spacing in the table is not required. The Lagrangian form is also perhaps the most straightforward way to write out the polynomial as an explicit function.

When the given points are not evenly spaced, the above method using Lagrangian polynomials or the method of divided differences would be used for interpolation. With the latter, exactly the same principle is involved—hold one variable constant while subtables of divided differences are constructed, then combine the interpolated values from these subtables into a new table.

Another alternative is to use cubic splines for interpolation in multivariate cases. There again it is perhaps best to hold one variable constant while constructing one-way splines, then combine the results from these in the second phase. The computational effort would be significant, however.

Interpolating for values of functions of two independent variables can also be thought of as constructing a surface that is defined by the given points. Rather than finding values on a surface that contains the given points, we can construct surfaces that are analogous to Bezier curves and B-spline curves where the surface does not normally contain the given points.

So far we have been able to interpolate simple surfaces where we are given z as a function of x and y. Suppose now we are given a set of points, $p_i = \{(x_i, y_i, z_i), i = 0, \ldots, n\}$, and we wish to fit a surface to those points. This would be the case if we were trying to draw a mountain, an airplane, or a teapot. But first we consider the representation of more general surfaces. Let $p = (x, y, z)$ be any point on the surface. Then the coordinates of each point are represented as the equations

$$x = x(u, v),$$
$$y = y(u, v),$$
$$z = z(u, v),$$

where u, v are the independent variables that range over a given set of values, and x, y, z are the dependent variables. This is a slight change of notation from the first part of this section.

An example of this would be the equations of a sphere of radius r about the origin: $(0, 0, 0)$. Here any point on the surface of the sphere is given by

$$x = r \cos(u) \sin(v),$$
$$y = r \sin(u) \sin(v),$$
$$z = r \cos(v),$$

where u ranges in value from 0 to 2π, and v ranges from 0 to π. Figure 3.9 illustrates this.

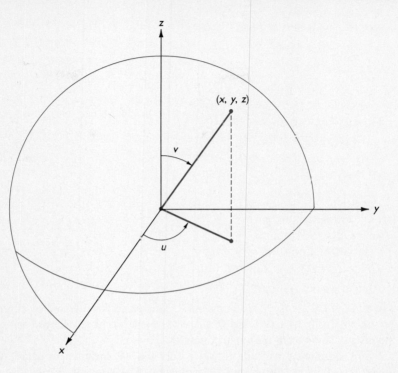

Figure 3.9

We will only describe constructing a B-spline surface. (A most interesting and informative description of Bezier surfaces can be found in Crow (1987).)

From the previous section, we know that a cubic B-spline curve segment starting near the point p_i to near the point p_{i+1} is determined by the four points

where $p_i(u) = (x_i(u), y_i(u))$ in two dimensions, or $p_i(u) = (x_i(u), y_i(u), z_i(u))$ if we had been working in three dimensions. The segment was then extended by introducing p_{i+3}, deleting p_{i-1}, and generating the line for $0 \leq u \leq 1$.

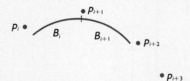

The process is continued until we have B_{n-2}. Finally, the first and last segments are generated by starting with p_0, p_0, p_0, p_1 and ending with p_{n-1}, p_n, p_n, p_n.

In an analogous manner the interpolating B-spline surface patch depends on 16 points, as Figure 3.10 shows.

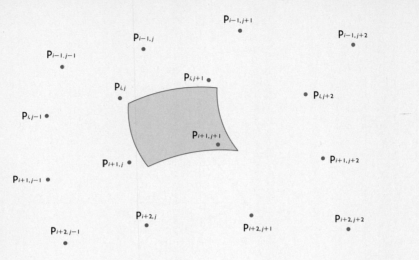

Figure 3.10

Here $p_{i,j} = (x_{i,j}, y_{i,j}, z_{i,j})$, a point in E^3. This patch is generated by computing the points $p_{i,j}(u, v)$, for $0 \le u \le 1$ and $0 \le v \le 1$. Here we have changed the subscripts on the points $p_{i,j}$ so as to fit into matrix notation.

For simplicity, we will consider only the x-coordinate in detail. Comparable formulations hold for the y- and z-coordinates. The simplest formulation for $x_{i,j}(u, v)$ is based on the matrix formulation of Eq. (3.27) and is given by

$$x_{i,j}(u, v) = [u^3, u^2, u, 1]M_b X_{i,j} M_b^T \begin{bmatrix} v^3 \\ v^2 \\ v \\ 1 \end{bmatrix}, \tag{3.29}$$

where $X_{i,j}$ is the 4×4 matrix

$$\begin{bmatrix} x_{i-1,j-1} & x_{i-1,j} & x_{i-1,j+1} & x_{i-1,j+2} \\ x_{i,j-1} & x_{i,j} & x_{i,j+1} & x_{i,j+2} \\ x_{i+1,j-1} & x_{i+1,j} & x_{i+1,j+1} & x_{i+1,j+2} \\ x_{i+2,j-1} & x_{i+2,j} & x_{i+2,j+1} & x_{i+2,j+2} \end{bmatrix},$$

which are just the x-coordinates of the 16 points of Figure 3.10. The matrix M_b is the matrix we saw before in Eq. (3.27).

$$M_b = \begin{bmatrix} -1 & 3 & -3 & 1 \\ 3 & -6 & 3 & 0 \\ -3 & 0 & 3 & 0 \\ 1 & 4 & 1 & 0 \end{bmatrix}$$

The y and z equations are then obtained by merely substituting the corresponding matrices $Y_{i,j}$ and $Z_{i,j}$, which are formed from the y and z components of the 16 points. Since each of these equations is cubic in u and v, they are referred to as *bicubic equations*. The coordinates of the points on a patch are given by

$$x(u, v) = [u^3, u^2, u, 1]M_b X_{i,j} M_b^T [v^3, v^2, v, 1]^T,$$

$$y(u, v) = [u^3, u^2, u, 1]M_b Y_{i,j} M_b^T [v^3, v^2, v, 1]^T,$$

$$z(u, v) = [u^3, u^2, u, 1]M_b Z_{i,j} M_b^T [v^3, v^2, v, 1]^T,$$

as u and v range between 0 and 1. It is easily verified that the weights applied to each of the 16 points are

$$\begin{bmatrix} 1 & 4 & 1 & 0 \\ 4 & 16 & 4 & 0 \\ 1 & 4 & 1 & 0 \\ 0 & 0 & 0 & 0 \end{bmatrix} \text{at } p_{i,j}(u, v) (\text{for } u = 0, v = 0), \text{ and}$$

$$\begin{bmatrix} 0 & 0 & 0 & 0 \\ 0 & 1 & 4 & 1 \\ 0 & 4 & 16 & 4 \\ 0 & 1 & 4 & 1 \end{bmatrix} \text{at } p_{i,j}(u, v) (\text{for } u = 1, v = 1)$$

where each (i, j)th element is the coefficient for the corresponding point in Figure 3.10. In effect, these matrices are templates that overlay the points shown in Figure 3.10.

The surface patch is extended by adding another row or column of points and deleting a corresponding row or column of points. One should verify that the current and previous patches are connected smoothly along the edge where they join. An initial or final patch can be obtained by repeating a corner, as was suggested for the B-spline curve. This will ensure that the patch actually starts or ends at a point. For the surface we would repeat a point nine times instead of three times as was done for the curve.

For a more detailed and informative discussion of interpolating curves and surfaces, the reader should consult Pokorny and Gerald (1989).

3.12 CHAPTER SUMMARY

If you have understood this chapter on interpolation, you are now able to

1. Construct interpolating polynomials for any set of data by (a) use of the Lagrangian formulation or (b) use of divided differences. You can explain why the divided-difference technique is preferred. You can use either to interpolate from a set of given data pairs.

2. Write and use an expression that computes the error of an approximation. You know the difficulty and limitations of such error estimation.

3. Use a table of differences of evenly spaced data to construct a variety of interpolation polynomials (Newton–Gregory forward and backward, Gauss forward and backward, Bessel, and Stirling). You understand that these are merely different ways to create the same polynomial, but you also can select the proper one in situations where one is preferred.

4. Explain why the error expressions for all interpolating polynomials constructed from the same data are the same.

5. Show situations where the error of approximation increases when the degree of the interpolating polynomial is increased and explain why this occurs. You should know techniques to overcome the problem.

6. Replace the error expression with an equivalent $O(h^n)$ and recognize the value of this simplified form for the error.

7. Derive formulas by the symbolic method.

8. Do inverse interpolation from a set of data.

9. Set up the equations for coefficients of cubic splines and get solutions to them that permit interpolation. You should know why splines are an important and often preferred alternative to ordinary interpolating polynomials.

10. Construct Bezier curves and B-splines. By now you know that these have important applications in many fields, especially computer graphics.

11. Do polynomial interpolation in three-dimensional situations.

SELECTED READINGS FOR CHAPTER 3

Bartels, Beatty, and Barsky (1987); de Boor (1978); Newell and Blinn (1987).

3.13 COMPUTER PROGRAMS

The first of the following programs (Program 1, Fig. 3.11) interpolates using the divided-difference method. The program consists of two subprograms: SUBROUTINE DOCOEFF and function EVAL. DOCOEFF returns the coefficients of the interpolating polynomial for the given set of data points. EVAL evaluates the divided-difference polynomial at a given point: $x = u$. Program 1 generates the polynomial for the data given in Table 3.1(b). The coefficients of the polynomial, $p(u)$, as well as the first few points, $(u, p(u))$, are output.

Program 2 (Fig. 3.12) employs the Newton–Gregory interpolating polynomial, of degree up to 9, to interpolate in a function tabulated with uniform x-spacing. The differences are computed as they are needed, and a single array is employed both to compute and to store them, for maximum economy of memory use. The value of x_0 is automatically chosen to put the point of interpolation as near the center of the domain as possible.

Program 3 (Fig. 3.13) uses DOCOEFF and EVAL to find the interpolating polynomial for $f(x) = 1/(1 + 25x^2)$ on the interval $[-1, 1]$ for $n = 11, 13, \ldots, 19$ data points (Section 3.10). On plotting the corresponding polynomials, one will find that for some subintervals the $p_{n-1}(x)$ become more negative. This example was suggested by C. Runge in 1901 and is widely used (see Prenter 1975).

```
      PROGRAM DIVDIF(INPUT,OUTPUT)
C
C     -------------------------------------------------------------
C
C         THIS PROGRAM USES THE DIVIDED DIFFERENCES METHOD TO GET THE
C         INTERPOLATING POLYNCMIAL THAT GOES THROUGH A GIVEN SET OF
C         POINTS.
C
C     -------------------------------------------------------------
C
C          SUBROUTINE DOCOEF: FINDS THE COEFFICIENTS OF THE
C                             INTERPOLATING POLYNOMIAL.
C          FUNCTION EVAL: EVALUATES THE DIVIDED DIFFERENCES POLYNOMIAL
C                         AT A GIVEN VALUE U.
C
C     -------------------------------------------------------------
C
      REAL X(5), Y(5), COEFFS(5), U, EVAL
      INTEGER I, J, N
C
      DATA N/5/
      DATA X/3.2,2.7,1.0,4.8,5.6/
      DATA Y/22.0,17.8,14.2,33.3,51.7/
C
      CALL DOCOEFF(X,Y,COEFFS,N)
C
      PRINT '(///)'
      PRINT *, 'THE COEFFICIENTS FOR THE POLYNOMIAL ARE: '
      PRINT *
      PRINT '(5F9.5)', (COEFFS(I), I = 1,N)
      PRINT '(//)'
      PRINT *, '          U','         P(U)'
      PRINT *
      DO 10 U = 1.0,5.601,0.2
          PRINT '(F11.3,F10.5)', U, EVAL(U, COEFFS,X,N)
10        CONTINUE
C
      PRINT *
      END
C
C     -------------------------------------------------------------
C
      SUBROUTINE DOCOEFF(X,Y,COEFFS,N)
C
C
C     -------------------------------------------------------------
C
C         INPUT: X,Y -  THE GIVEN DATA POINTS
C                N   -  THE NUMBER OF DATA POINTS
C         OUTPUT: COEFFS  -  THE COEFFICIENTS OF P(X)
C
C
C     -------------------------------------------------------------
C
      INTEGER N,  I,J,K
      REAL  X(N),Y(N),COEFFS(N),   TEMP1,TEMP2
C
      DO 10 I = 1,N
          COEFFS(I) = Y(I)
10        CONTINUE
C
```

Figure 3.11 Program 1.

Figure 3.11   (*continued*)

```
        DO 20 J = 2,N
            TEMP1 = COEFFS(J-1)
            DO 30 K = J,N
                TEMP2 = COEFFS(K)
                COEFFS(K) = (COEFFS(K) - TEMP1)/(X(K) - X(K-J+1))
                TEMP1 = TEMP2
30          CONTINUE
20      CONTINUE
C
    RETURN
    END
C
C
C   -------------------------------------------------------------
C
    REAL FUNCTION EVAL(U, COEFFS, X, N)
C
C
C   -------------------------------------------------------------
C
C       INPUT:   U  -  THE X-VALUE
C       OUTPUT:  EVAL  -  THE CORRESPONDING Y-VALUE, I.E. P(U)
C
    REAL U, COEFFS(N), X(N),    SUM
    INTEGER N,  I
C
    SUM = 0.0
C
C   -------------------------------------------------------------
C
C       COMPUTE VALUE BY NESTED MULTIPLICATION FROM HIGHEST TERM
C
C   -------------------------------------------------------------
C
    DO 10 I = N,2,-1
        SUM = (SUM + COEFFS(I)) * (U - X(I-1))
10      CONTINUE
C
    SUM = SUM + COEFFS(1)
    EVAL = SUM
    RETURN
    END



        OUTPUT FOR PROGRAM 1


    THE COEFFICIENTS FOR THE POLYNOMIAL ARE:

    22.00000  8.40000  2.85561  -.52748    .25584



            U       P(U)

        1.000  14.20000
        1.200  12.89775
        1.400  12.25424
        1.600  12.16434
        1.800  12.53272
```

Figure 3.11 (*continued*)

```
2.000   13.27390
2.200   14.31221
2.400   15.58181
2.600   17.02667
2.800   18.60059
3.000   20.26722
3.200   22.00000
3.400   23.78221
3.600   25.60694
3.800   27.47713
4.000   29.40553
4.200   31.41470
4.400   33.53706
4.600   35.81481
4.800   38.30000
5.000   41.05451
5.200   44.15003
5.400   47.66808
5.600   51.70000
```

```
        SUBROUTINE INTERP(X1,XN,H,N,Y,X,M,YOUT)
C
C       -------------------------------------------------------------------
C
C       SUBROUTINE INTERP :
C                          SUBROUTINE TO INTERPOLATE IN A TABLE OF
C       UNIFORMLY SPACED VALUES.
C
C       -------------------------------------------------------------------
C
C       PARAMETERS ARE :
C
C       X1,XN  - BEGINNING AND ENDING X VALUES
C       H      - DELTA X - THE UNIFORM SPACING
C       N      - NUMBER OF ENTRIES IN THE TABLE
C       Y      - ARRAY OF FUNCTION VALUES
C       X      - VALUE AT WHICH Y IS TO BE INTERPOLATED
C       M      - DEGREE OF INTERPOLATING POLYNOMIAL. THE SUBROUTINE WILL
C                HANDLE UP TO 10TH DEGREE, BUT USUALLY THE DEGREE WILL BE
C                LESS THAN 10 TO AVOID ROUND OFF ERRORS.
C       YOUT   - THE INTERPOLATED Y VALUE RETURNED TO THE CALLER
C
C AN ARRAY D IS USED IN THE SUBROUTINE TO HOLD DELTA Y'S.
C
C       -------------------------------------------------------------------
C
        REAL X1,XN,H,Y(N),X,YOUT
        INTEGER N,M,I,J,K
        REAL D(10),FM,FJ,X0,Y0,S,FNUM,DEN,FI
C
C       -------------------------------------------------------------------
C
C FIRST FIND PROPER SUBSCRIPT FOR Y0 SO THAT X IS CENTERED IN THE
C DOMAIN AS WELL AS POSSIBLE. THIS SUBSCRIPT VALUE IS CALLED J.
C
```

Figure 3.12 Program 2.

Figure 3.12 *(continued)*

```
      FM = M + 1
      J = ( X - X1 )/H - FM/2.0 + 2.0
      IF ( X .LE. X1 + FM/2.0*H )  J = 1
      IF ( X .GE. XN - FM/2.0*H )  J = N - M
      FJ = J
      X0 = X1 + ( FJ - 1.0 )*H
      Y0 = Y(J)
C
C     -----------------------------------------------------------------
C
C     COMPUTE THE DIFFERENCES THAT ARE NEEDED
C
      DO 10 I = 1,M
        D(I) = Y(J+1) - Y(J)
        J = J + 1
10    CONTINUE
      IF ( M .GT. 1 ) THEN
        DO 20 J = 2,M
          DO 15 I = J,M
            K = M - I - J
            D(K) = D(K) - D(K-1)
15        CONTINUE
20      CONTINUE
      END IF
C
C     -----------------------------------------------------------------
C
C     COMPUTE S VALUE
C
      S = ( X - X0 ) / H
C
C     COMPUTE INTERPOLATED Y VALUE
C
      YOUT = Y0
      FNUM = S
      DEN = 1.0
      DO 30 I = 1,M
        FI = I
        YOUT = YOUT + FNUM/DEN*D(I)
        FNUM = FNUM * ( S - FI )
        DEN = DEN * ( FI + 1.0 )
30    CONTINUE
      RETURN
      END
```

```
      PROGRAM RUNGE(INPUT,OUTPUT)
C
C     THIS PROGRAM CALLS DOCOEFF AND EVAL AND COMPARES INTERPOLATED
C     VALUES WITH ACTUAL VALUES WITH ACTUAL VALUES  OF F = 1 / (1+25X*X)
C     FOR INCREASING NUMBER OF DATA POINTS.
C
C     -----------------------------------------------------------------
C
```

Figure 3.13 Program 3.

Figure 3.13 (*continued*)

```
          REAL X(20),Y(20),XINT,YOUT,DELX,ERROR,DIFF,A,B,F,H,COEFFS(20)
          INTEGER N,K,J
          DATA A,B / -1.0,1.0 /
C
C       ------------------------------------------------------------
C
C   INITIALIZE VARIABLES AND STORE INFORMATION IN X AND Y ARRAYS
C
          DELX = (B - A) / 64.0
          PRINT 100
          DO 20 K = 11,19,2
            H = (B - A) / (K - 1)
            DO 10 J = 1,K
              X(J) = -1.0 + (J - 1)*H
              Y(J) = F(X(J))
   10     CONTINUE
          ERROR = 0.0
C
C       ------------------------------------------------------------
C
      CALL DOCOEFF (X,Y,COEFFS,K)
C
          DO 15 XINT = -1.0,1.0,DELX
            YOUT = EVAL(XINT,COEFFS,X,K)
            ERROR = MAX( ABS(F(XINT) - YOUT), ERROR )
   15     CONTINUE
          PRINT 200, K-1,ERROR
   20 CONTINUE
      PRINT 300
C
C       ------------------------------------------------------------
C
  100 FORMAT(//T11,'DEGREE',T22,'MAX ERROR FOUND'/)
  200 FORMAT(T12,I3,T24,F11.7)
  300 FORMAT(///)
      STOP
      END
C
C       ------------------------------------------------------------
C
      REAL FUNCTION F(T)
C
      REAL T
      F = 1.0 / (1.0 + 25.0*T*T)
      RETURN
      END



                OUTPUT FOR PROGRAM 3


          DEGREE      MAX ERROR FOUND

            10            1.9132554
            12            3.4710393
            14            7.0075584
            16           14.3699041
            18           29.0384012
```

The fourth program (Fig. 3.14) implements the cubic spline method on the function: $y = x^3 - 8$ (Example 1 in Section 3.9). All four conditions, called IEND, are tested on this function. The program has two subprograms, CUBSPL and SEVAL. SUBROUTINE CUBSPL uses the data points, x,y-pairs to generate the matrix appropriate to the condition, IEND = 1 through 4. The matrix is in tridiagonal form. The subroutine then produces the cubic polynomial coefficients for each subinterval. FUNCTION SEVAL returns the appropriate value for a given $x = u$. SEVAL finds the interval containing u, and uses the polynomial to compute the spline value. For this example we get as output: the matrix in tridiagonal form, the coefficients of the cubic polynomials, the second derivatives, and the spline values compared to the exact function values.

```
      PROGRAM CUBIC(INPUT,OUTPUT)
C
C           THIS PROGRAM FITS DATA WITH A CUBIC INTERPOLATING
C           SPLINE. THERE ARE FOUR KINDS OF SPLINES DEPENDING
C           ON IEND = 1,2,3,OR 4.
C
C           IEND = 1  : LINEAR END CONDITION: S(0) = S(N)
C           IEND = 2  : PARABOLIC END CONDITION: S(0)=S(1),
C                       S(N-1)=S(N)
C           IEND = 3  : CUBIC END CONDITION
C           IEND = 4  : FIRST DERIVATIVE IS KNOWN AT X(0) AND
C                       X(N). THE VALUES ARE STORED IN S(0) AND
C                       S(N) RESPECTIVELY.
C
C
C        THIS EXAMPLE FITS THE APPROPRIATE CUBIC SPLINE
C        FOR THE FOUR POINTS ON THE CURVE DEFINED BY
C        THE FUNCTION:   F(X) = X*X*X - 8
C
      INTEGER N,I,IEND
      REAL X(100),Y(100),A(100),B(100),C(100),S(100)
      DATA N,IEND/5,1/
      DATA X/0,1,2,3,4,95*0.00000000/
      DATA Y/-8,-7,0,19,56,95*0.0/
      DATA S/100*0.0/
       READ *, IEND
      IF (IEND .EQ. 4) THEN
           S(1) = 0.0
           S(N) = 48.000000
           END IF
C
      PRINT '(///)'
      PRINT *, '           OUTPUT FOR IEND = ',IEND
      PRINT '(//)'
C
C
      CALL CUBSPL(X,Y,IEND,N,A,B,C,S)
C
      PRINT 98
      PRINT 99
      PRINT 101, (S(I), I=1,N)
      PRINT 98
      PRINT 200
      DO 31 U = 0.0,4.01,0.25
```

Figure 3.14   Program 4.

Figure 3.14   (*continued*)

```
            TEMP1 = SEVAL(N,U,X,Y,A,B,C)
            TEMP2 = FCN(U)
            PRINT 100, U,TEMP1,TEMP2, TEMP1 - TEMP2
31          CONTINUE
        PRINT 98
98      FORMAT(//T2,60('*')/)
99      FORMAT('  THE SECOND DERIVATIVES S(I) ARE; '/)
100     FORMAT(3X,F4.2,4X,2F10.4,4X,F10.5)
101     FORMAT(10F9.4)
200     FORMAT(T5,' X',T15,'SPLINE(X)',T26,' F(X)',T41,'DIFF'//)
        STOP
        END
C
C
C -----------------------------------------------------------------------
C
C
        SUBROUTINE CUBSPL(X,Y,IEND,N,A,B,C,S)
C
C -------------------------------------------------------------
C
C
C       SUBROUTINE CUBSPL  :
C                           THIS ROUTINE COMPUTES THE MATRIX
C       FOR FINDING THE COEFFICIENTS OF A CUBIC SPLINE THROUGH
C       A SET OF DATA.   THE SYSTEM THEN IS SOLVED TO OBTAIN THE
C       SECOND DERIVATIVE VALUES AND THE COEFFICIENTS OF THE CUBIC
C       SPLINE POLYNOMIALS.
C
C -------------------------------------------------------------
C
C
C       PARAMETERS ARE :
C
C       X,Y : ARRAYS OF X AND Y VALUES TO BE FITTED
C       N   : NUMBER OF POINTS
C       IEND: TYPE OF END CONDITION TO BE USED
C             IEND = 1, LINEAR ENDS, S(1) = S(N) = 0.
C             IEND = 2, PARABOLIC ENDS, S(1) = S(2), S(N) = S(N-1)
C             IEND = 3, CUBIC ENDS, S(1), S(N), ARE EXTRAPOLATED
C             IEND = 4, THE FIRST DERIVATIVES ARE GIVEN AT EITHER END POINT.
C                       FPRIME(X1) IS STORED IN S(1) ON INPUT, AND FPRIME(XN)
C                       IS STORED IN S(N).
C SMATRIX : AUGMENTED MATRIX OF COEFFICIENTS AND R.H.S. FOR
C           FINDING S.
C   A,B,C : ARRAYS OF SPLINE COEFFICIENTS
C
C -------------------------------------------------------------
C
        REAL X(N),Y(N),S(N),A(N),B(N),C(N),SMATRIX(0:20,4)
     +      ,DX1,DY1,DX2,DY2,DXN1,DXN2
        REAL H(20)
        INTEGER N,IEND,NM1,NM2,I,J,FIRST,LAST
C
C -------------------------------------------------------------
C
C       COMPUTE FOR THE N-2 ROWS
C
        NM2 = N - 2
        NM1 = N - 1
        DX1 = X(2) - X(1)
        DY1 = ( Y(2) - Y(1) ) / DX1 * 6.0
        DO 10 I = 1,NM2
           DX2 = X(I + 2) - X(I + 1)
           DY2 = ( Y(I + 2) - Y(I + 1) ) / DX2 * 6.0
```

Figure 3.14 (*continued*)

```
            SMATRIX(I,1) = DX1
            SMATRIX(I,2) = 2.0 * (DX1 + DX2)
            SMATRIX(I,3) = DX2
            SMATRIX(I,4) = DY2 - DY1
            DX1 = DX2
            DY1 = DY2
   10 CONTINUE
      FIRST = 2
      LAST = NM2
C
C------------------------------------------------------------------
C
C ADJUST FIRST AND LAST ROWS APPROPRIATE TO END CONDITION.
C
C
C FOR IEND = 1, NO CHANGE IS NEEDED.
C
      IF ( IEND .EQ.2 ) THEN
C
C FOR IEND = 2, S(1) = S(2), S(N) = S(N-1), PARABOLIC ENDS.
C
   50 SMATRIX(1,2) = SMATRIX(1,2) + X(2) - X(1)
      SMATRIX(NM2,2) = SMATRIX(NM2,2) + X(N) - X(NM1)
C
C         ELSE IF (IEND .EQ. 3 ) THEN
C FOR IEND = 3, CUBIC ENDS, S(1), S(N) ARE EXTRAPOLATED.
C
   80 DX1 = X(2) - X(1)
      DX2 = X(3) - X(2)
      SMATRIX(1,2) = (DX1 + DX2) * (DX1 + 2.0 * DX2) / DX2
      SMATRIX(1,3) = (DX2 * DX2 - DX1 * DX1) / DX2
      DXN2 = X(NM1) - X(NM2)
      DXN1 = X(N) - X(NM1)
      SMATRIX(NM2,1) = (DXN2 * DXN2 - DXN1 * DXN1) / DXN2
      SMATRIX(NM2,2) = (DXN1 + DXN2) *  (DXN1 + 2.0 * DXN2) / DXN2
C
       END IF
      IF (IEND .EQ. 4) THEN
          DX1 = X(2) - X(1)
          DY1 = (Y(2) - Y(1))/DX1
          SMATRIX(0,1) = 1.0
          SMATRIX(0,2) = 2.0*DX1
          SMATRIX(0,3) = DX1
          SMATRIX(0,4) = (DY1 - S(1))*6
          DX1 = X(N) - X(N-1)
          DY1 = (Y(N) - Y(N-1))/DX1
          SMATRIX(NM1,1) = DX1
          SMATRIX(NM1,2) = 2.0*DX1
          SMATRIX(NM1,3) = 0.0
          SMATRIX(NM1,4) = (S(N) - DY1)*6.0
          FIRST = 1
          LAST = N-1
      END IF
C
C         PRINT OUT TRIDIAGONAL MATRIX IN COMPACT FORM
C
C
      DO 11 I = FIRST-1,LAST
   11     PRINT *, (SMATRIX(I,J), J = 1,4)
C
C     ------------------------------------------------------------
C
```

Figure 3.14 (*continued*)

```
C  NOW WE SOLVE THE TRIDIAGONAL SYSTEM.  FIRST REDUCE.
C
      DO 110 I = FIRST,LAST
        SMATRIX(I,1) = SMATRIX(I,1)/SMATRIX(I-1,2)
        SMATRIX(I,2) = SMATRIX(I,2) - SMATRIX(I,1) * SMATRIX(I-1,3)
        SMATRIX(I,4) = SMATRIX(I,4) - SMATRIX(I,1) * SMATRIX(I-1,4)
  110 CONTINUE
C
C  ------------------------------------------------------------
C
C  NOW WE BACK SUBSTITUTE
C
      SMATRIX(LAST,4) = SMATRIX(LAST,4) / SMATRIX(LAST,2)
      DO 120 J = LAST-1,FIRST-1,-1
        SMATRIX(J,4) = ( SMATRIX(J,4) - SMATRIX(J,3) * SMATRIX(J+1,4) )
     +              / SMATRIX(J,2)
  120 CONTINUE
C
C  ------------------------------------------------------------
C
C
C  NOW PUT THE VALUES INTO THE S VECTOR
C
      DO 130 I = FIRST-1,LAST
        S(I+1) = SMATRIX(I,4)
  130 CONTINUE
C
C  ------------------------------------------------------------
C
C  GET S(1) AND S(N) ACCORDING TO END CONDITIONS
C
      IF (IEND .EQ. 1) THEN
C
C  ------------------------------------------------------------
C
C  FOR LINEAR ENDS, S(1) = 0, S(N) = 0.
C
      S(1) = 0.0
      S(N) = 0.0
          ELSE IF (IEND .EQ. 2) THEN
C
C  ------------------------------------------------------------
C
C  FOR PARABOLIC ENDS, S(1) = S(2), S(N) = S(N-1).
C
      S(1) = S(2)
      S(N) = S(N-1)
C
C  ------------------------------------------------------------
C
C  FOR CUBIC ENDS, EXTRAPOLATE TO GET S(1) AND S(N).
C
          ELSE IF (IEND .EQ. 3) THEN
          S(1) = ((DX1 + DX2) * S(2) - DX1 * S(3)) / DX2
          S(N) = ((DXN2 + DXN1) * S(NM1) - DXN1 * S(NM2)) / DXN2
      END IF
C
C            WRITE OUT COEFFICIENTS OF THE POLYNOMIALS
C
      PRINT 99
   99 FORMAT(/T2,60('*')//T10,'THE CUBIC POLYNOMIALS, G(X)'
     + ,' DEFINED ON THE INTERVALS'//)
```

Figure 3.14 *(continued)*

```
          PRINT 101
          DO 200 I = 1,N-1
            H(I) = X(I+1) - X(I)
            A(I) = (S(I+1) - S(I)) / (6 * H(I))
            B(I) = S(I) / 2
            C(I) = ((Y(I+1) - Y(I)) / H(I)) - ((2 * H(I) * S(I) +
     +  H(I)*S(I+1)) / 6)
C
          PRINT 102,I, A(I),B(I),C(I),Y(I)
101       FORMAT(T5,'I',T17,'A',T27,'B',T37,'C',T47,'D'/)
102       FORMAT(T5,I1,T12,F8.4,T22,F8.4,T32,F8.4,T42,F8.4)
    200   CONTINUE
          RETURN
          END
C
C
C
          REAL FUNCTION SEVAL(N,U,X,Y,A,B,C)
          INTEGER N
          REAL U,X(N),Y(N),B(N),C(N),A(N)
C
C     THIS SUBROUTINE EVALUATES THE CUBIC SPLINE FUNCTION
C
C           SEVAL=Y(I)+C(I)*(U-X(I))+B(I)*(U-X(I)**2+A(I)*(U-X(I))**3
C
C           WHERE X(I).LT.U.LT.X(I+1), USING HORNER'S RULE
C
C     IF U.LT.X(1) THEN I=1 IS USED
C                       ELSE I=N IS USED
C
C     INPUT:
C
C     N = THE NUMBER OF DATA POINTS
C     U = THE ABSCISSA AT WHICH THE SPLINE IS TO BE EVALUATED
C     X,Y = THE ARRAYS OF DATA ABSCISSAS AND ORDINATES
C     B,C,A = THE ARRAYS OF SPLINE COEFFICIENTS COMPUTEA BY SPLINE
C
C     IF U IS NOT IN THE SAME INTERVAL AS THE PREVIOUS CALL, THEN A
C     SEARCH IS PERFORMED TO DETERMINE THE PROPER INTERVAL.
C
          INTEGER I,J,K
          REAL DX
          SAVE I
          DATA I/1/
          IF(I.GE.N) THEN
                THEN I=1
          ENDIF
          IF(U.GE.X(N)) THEN
                DX=U-X(N-1)
                SEVAL=Y(N-1)+DX*(C(N-1)+DX*(B(N-1)+DX*A(N-1)))
                RETURN
          END IF
          IF(U.GE.X(I)) THEN
                IF(U.LE.X(I+1)) THEN
                      DX=U-X(I)
                      SEVAL=Y(I)+DX*(C(I)+DX*(B(I)+DX*A(I)))
                      RETURN
                ENDIF
          ENDIF
          I=1
          J=N+1
```

Figure 3.14   (*continued*)

```
10      K=(I+J)/2
        IF(U.LT.X(K)) THEN
             J=K
        ELSE
             I=K
        ENDIF
        IF(J.GT.I+1) THEN
             GOTO 10
        ELSE
C
C       EVALUATE SPLINE
C
             DX=U-X(I)
             SEVAL=Y(I)+DX*(C(I)+DX*(B(I)+DX*A(I)))
             RETURN
        ENDIF
        END
C
C       ----------------------------------------------------------------
C
C                 THE FUNCTION TO BE INTERPOLATED IS DEFINED HERE
C
C       ----------------------------------------------------------------
C
        REAL FUNCTION FCN(X)
        REAL EXP,X
        FCN = X**3 - 8.0
        RETURN
        END

                     OUTPUT FOR PROGRAM 4


        OUTPUT FOR IEND = 1

1. 4. 1. 36.
1. 4. 1. 72.
1. 4. 1. 108.

****************************************************************

        THE CUBIC POLYNCMIALS, G(X) DEFINED ON THE INTERVALS


        I          A          B          C          D

        1       1.0714      .0000    -.0714     -8.0000
        2        .6429     3.2143    3.1429     -7.0000
        3       2.3571     5.1429   11.5000       .0000
        4      -4.0714    12.2143   28.8571     19.0000


****************************************************************

  THE SECOND DERIVATIVES S(I) ARE;

    .0000   6.4286  10.2857  24.4286     .0000


****************************************************************
```

Figure 3.14 (*continued*)

```
     X          SPLINE(X)    F(X)        DIFF

    .00         -8.0000     -8.0000      .00000
    .25         -8.0011     -7.9844     -.01674
    .50         -7.9018     -7.8750     -.02679
    .75         -7.6016     -7.5781     -.02344
   1.00         -7.0000     -7.0000      .00000
   1.25         -6.0033     -6.0469      .04353
   1.50         -4.5446     -4.6250      .08036
   1.75         -2.5636     -2.6406      .07701
   2.00          .0000       .0000       .00000
   2.25         3.2333      3.3906      -.15737
   2.50         7.3304      7.6250      -.29464
   2.75        12.5123     12.7969      -.28460
   3.00        19.0000     19.0000       .00000
   3.25        26.9141     26.3281       .58594
   3.50        35.9732     34.8750      1.09821
   3.75        45.7958     44.7344      1.06138
   4.00        56.0000     56.0000       .00000


*****************************************************************



           OUTPUT FOR IEND = 2



  1.  5.  1.  36.
  1.  4.  1.  72.
  1.  5.  1. 108.

*****************************************************************

       THE CUBIC POLYNOMIALS, G(X) DEFINED ON THE INTERVALS


    I           A           B          C          D

    1          .0000      2.4000     -1.4000     -8.0000
    2         1.2000      2.4000      3.4000     -7.0000
    3         1.2000      6.0000     11.8000      .0000
    4          .0000      9.6000     27.4000     19.0000


*****************************************************************

  THE SECOND DERIVATIVES S(I) ARE;

   4.8000    4.8000   12.0000   19.2000   19.2000


*****************************************************************

     X          SPLINE(X)    F(X)        DIFF


    .00         -8.0000     -8.0000      .00000
    .25         -8.2000     -7.9844     -.21563
```

Figure 3.14 (*continued*)

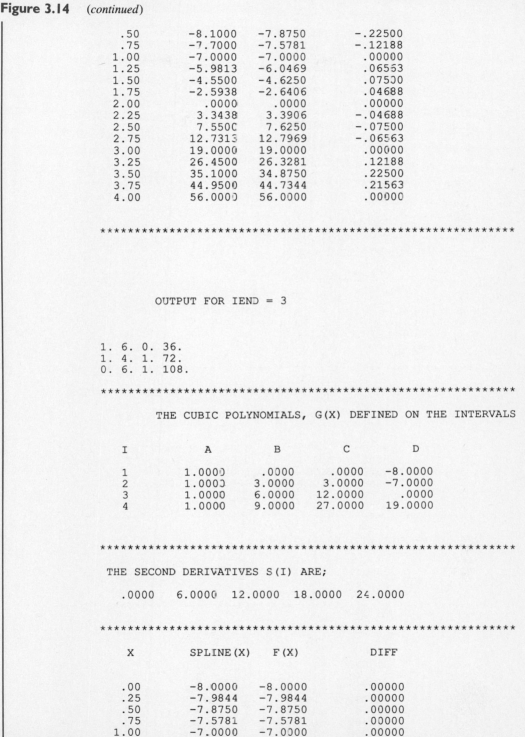

```
           .50        -8.1000        -7.8750         -.22500
           .75        -7.7000        -7.5781         -.12188
          1.00        -7.0000        -7.0000          .00000
          1.25        -5.9813        -6.0469          .06563
          1.50        -4.5500        -4.6250          .07500
          1.75        -2.5938        -2.6406          .04688
          2.00          .0000          .0000          .00000
          2.25         3.3438         3.3906         -.04688
          2.50         7.5500         7.6250         -.07500
          2.75        12.7313        12.7969         -.06563
          3.00        19.0000        19.0000          .00000
          3.25        26.4500        26.3281          .12188
          3.50        35.1000        34.8750          .22500
          3.75        44.9500        44.7344          .21563
          4.00        56.0000        56.0000          .00000


    ************************************************************



                 OUTPUT FOR IEND = 3



    1. 6. 0. 36.
    1. 4. 1. 72.
    0. 6. 1. 108.

    ************************************************************

         THE CUBIC POLYNOMIALS, G(X) DEFINED ON THE INTERVALS


         I          A          B          C          D

         1        1.0000      .0000      .0000    -8.0000
         2        1.0000     3.0000     3.0000    -7.0000
         3        1.0000     6.0000    12.0000      .0000
         4        1.0000     9.0000    27.0000    19.0000



    ************************************************************

      THE SECOND DERIVATIVES S(I) ARE;

        .0000    6.0000   12.0000   18.0000   24.0000


    ************************************************************

         X        SPLINE(X)    F(X)          DIFF


         .00       -8.0000     -8.0000        .00000
         .25       -7.9844     -7.9844        .00000
         .50       -7.8750     -7.8750        .00000
         .75       -7.5781     -7.5781        .00000
        1.00       -7.0000     -7.0000        .00000
        1.25       -6.0469     -6.0469        .00000
```

Figure 3.14 (*continued*)

```
        1.50        -4.6250      -4.6250        .00000
        1.75        -2.6406      -2.6406        .00000
        2.00          .0000        .0000        .00000
        2.25         3.3906       3.3906        .00000
        2.50         7.6250       7.6250        .00000
        2.75        12.7969      12.7969        .00000
        3.00        19.0000      19.0000        .00000
        3.25        26.3281      26.3281        .00000
        3.50        34.8750      34.8750        .00000
        3.75        44.7344      44.7344        .00000
        4.00        56.0000      56.0000        .00000


********************************************************************



            OUTPUT FOR IEND = 4



1.  2.  1.  6.
1.  4.  1.  36.
1.  4.  1.  72.
1.  4.  1.  108.
1.  2.  0.  66.

********************************************************************

        THE CUBIC POLYNOMIALS, G(X) DEFINED ON THE INTERVALS


        I          A          B          C          D

        1        1.0000      .0000       .0000     -8.0000
        2        1.0000     3.0000      3.0000     -7.0000
        3        1.0000     6.0000     12.0000       .0000
        4        1.0000     9.0000     27.0000     19.0000


********************************************************************

    THE SECOND DERIVATIVES S(I) ARE;

    .0000    6.00C0   12.0000   18.0000   24.0000


********************************************************************

        X          SPLINE(X)    F(X)           DIFF


        .00        -8.0000      -8.0000        .00000
        .25        -7.9844      -7.9844        .00000
        .50        -7.8750      -7.8750        .00000
        .75        -7.5781      -7.5781        .00000
        1.00       -7.0000      -7.0000        .00000
        1.25       -6.C469      -6.0469        .00000
        1.50       -4.€250      -4.6250        .00000
        1.75       -2.6406      -2.6406        .00000
```

Figure 3.14   (*continued*)

2.00	.0000	.0000	.00000
2.25	3.3906	3.3906	.00000
2.50	7.6250	7.6250	.00000
2.75	12.7969	12.7969	.00000
3.00	19.000C	19.0000	.00000
3.25	26.3281	26.3281	.00C00
3.50	34.8750	34.8750	.00C00
3.75	44.7344	44.7344	.00C00
4.00	56.0000	56.0000	.00000

```
******************************************************************
```

The fifth and final program (Fig. 3.15) is a Pascal program that can generate a cubic B-spline curve. A graphics package is needed to interface with this program. The curve is actually a set of straight lines. The program allows up to 20 straight lines to make up a segment of the curve determined by any four control points. That information is stored in the variable, lines_per_section. In PROCEDURE set_blending_functions, the values of the four B-spline functions on the interval [0, 1] are evaluated for a fixed number of points. The values are stored in an array called BLEND, which is 4 by N where $N \leq$ 20. After a segment determined by 4 control points (p_i, p_{i+1}, p_{i+2}, p_{i+3}) has been drawn, p_i is dropped (PROCEDURE next_section) and a new point p_{i+4} is added (PROCEDURE put_in_sm). The data points for this program were stored in a file called "bspline2.dat". This Turbo Pascal program uses a graphics package based on the HP AGIOS (Alphanumeric/Graphics Input/Output Subsystem) functions. The output from this program is the B-spline graphs used in this chapter.

```
PROGRAM cubic_B_spline(INPUT,OUTPUT, f);
(*
   This program implements the Algorithms in Chapter 11 of
   S. Harrington but the first and last segments of the curve
   are handled in a simpler manner. Finally, only the cubic
   B-spline functions are generated.

   It is assumed that one has graphics commands such as :

      PROCEDURE drawline(x1,y1,  x2,y2  : INTEGER);
      PROCEDURE plot(x,y  :  INTEGER );
      PROCEDURE graphON;
      PROCEDURE graphOFF;
      PROCEDURE clearGRAPHICS;


   These and others are usually available in a graphics package.
   The one used  in this program as based on the Hewlett-Packard
   AGIOS (ALPHANUMERIC/GRAPHICS INPUT/OUTPUT SUBSYSTEM) functions
   for the HP 150.

*)
```

Figure 3.15   Program 5.

Figure 3.15 (*continued*)

```
CONST
   max_number_of_lines = 20;   (* each curve segment consists
                                     of 20 straight lines       *)
   max_number_of_points = 20;
TYPE
   matrix = ARRAY[1..4,1..max_number_of_lines] OF REAL;
   vector = ARRAY[1..max_number_of_points] OF REAL;
   vector4 = ARRAY[1..4] OF REAL;

VAR
   blend                 : matrix;

   xsm,ysm               (* set of four points to create current
                               segment of B-spline curve         *)
                         : vector4;
   ax,ay                 (* data points as stored in the array  *)
                         : vector;

   x0, y0                : REAL;
   lines_per_section, i,
   number_of_points      : INTEGER;

   f : TEXT;     (* data file that stores the number of
                    points and the data points (x,y) in
                    screen coordinates.                  *)

(*
   INITIALIZE THE BLENDING FUNCTIONS
   -------------------------------------------------------------------------------
   This procedure evaluates the B-spline basis functions at
   the number_of_lines points of u on the interval [0,1]. Since
   these functions are independent of the control points, they
   can be done just once and stored in a 4 x max_number_of_lines
   array.
*)
PROCEDURE set_blending_functions(number_of_lines : INTEGER);
VAR
   i, j : INTEGER;
   u, u_cube, u_square,
   uMINUS1_cube          : REAL;
BEGIN

   FOR i := 1 TO number_of_lines DO                    BEGIN
      u := i/number_of_lines;
      u_square := u*u;  u_cube := u*u_square;
      uMINUS1_cube := (1.0-u)*SQR(1.0-u);

      blend[1,i] := uMINUS1_cube/6.0;
      blend[4,i] := u_cube/6;
      blend[3,i] := (-u_cube/2 + u_square/2 + u/2 + 1.0/6);
      blend[2,i] := (u_cube/2 - u_square + 2/3);


                                                         END;
END;    (* INITIALIZE BLENDING FUNCTIONS  *)
```

Figure 3.15 (*continued*)

```
(*
   PUT_IN_SM
--------------------------------------------------------------------------
   This procedure places a new sample point into the _sm arrays.
*)
PROCEDURE put_in_sm(x,y : REAL);
BEGIN
   xsm[4] := x; ysm[4] := y
END;


(*
   MAKE_CURVE
--------------------------------------------------------------------------
   This procedure fills in a section of the curve.
*)
PROCEDURE make_curve(VAR  b: matrix);
VAR
   i,j      : INTEGER;
   x,y      : REAL;
BEGIN

   FOR j := 1 TO lines_per_section DO                     BEGIN
      x := 0.0; y := 0.0;
      FOR i := 1 TO 4 DO           BEGIN
         x := x + xsm[i]*b[i,j];
         y := y + ysm[i]*b[i,j]     END;
      drawline(ROUND(x0), ROUND(y0), ROUND(x), ROUND(y));
      x0 := x; y0 := y                                    END;

END;     (* make_curve *)


(*
   NEXT_SECTION
--------------------------------------------------------------------------
   After a section of the curve has been drawn, we shift the sample points
   so that our blending functions can be applied to the next section.
*)
PROCEDURE next_section;
VAR
   i : INTEGER;
BEGIN
   FOR i := 1 TO 3 DO           BEGIN
      xsm[i] := xsm[i+1];
      ysm[i] := ysm[i+1]        END
END;


(*
   CURVE_ABS_2
--------------------------------------------------------------------------
   This procedure extends the curve by taking a new sample
   point as its argument and stores it into the _sm arrays.
*)
PROCEDURE curve_abs_2(x,y : REAL);
BEGIN
   put_in_sm(x,y);
   make_curve(blend);
   next_section
END;          (* curve_abs_2  *)
```

Figure 3.15 *(continued)*

```
(*
   START-B-SPLINE
---------------------------------------------------------------------
   We require the first two sample points to start
   the curve.  They are loaded into the _SM arrays.
*)
PROCEDURE start_B_spline(ax,ay : vector
                         (* contains first four points *) );
VAR
   i : INTEGER;

BEGIN
   FOR i := 1 TO 3 DO                 BEGIN
      xsm[i] := ax[1];
      ysm[i] := ay[1];                END;

   xsm[4] := ax[2];  ysm[4] := ay[2];
   make_curve(blend);
   next_section
END;   (* start_B_spline*)


(*
   END_B_SPLINE
---------------------------------------------------------------------
   This procedure terminates the B-spline curve
   by repeating the last point two and three times.
*)
PROCEDURE end_B_spline;
BEGIN
   put_in_sm(ax[number_of_points], ay[number_of_points]);
   make_curve(blend);
   next_section;
   put_in_sm(ax[number_of_points], ay[number_of_points]);
   make_curve(blend)
END;       (*  end_B_spline*)


(*
   INITIALIZE
---------------------------------------------------------------------
   This procedure reads in the control points from a data file.
*)
PROCEDURE initialize;
VAR
   i : INTEGER;
BEGIN

   WRITELN; WRITELN;
   WRITELN(' ':20,' CONTROL POINTS '); WRITELN;
   WRITELN(' X':24,' Y':7); WRITELN;


   READLN(f, number_of_points);
   FOR i := 1 TO number_of_points DO    BEGIN
      READLN(f, ax[i], ay[i]);
      WRITELN(ax[i]:25:0, ay[i]:7:0)     END;
   x0 := ax[1];  y0 := ay[1];

   FOR i := 1 TO number_of_points DO
                                 plotPOINT(TRUNC(ax[i]), TRUNC(ay[i]))
END;
```

Figure 3.15   *(continued)*

```
BEGIN    (* MAIN   *)

    number_of_lines := 10;
    gotoXY(1,1); ClrScr;  graphON; clearGRAPHICS;

    ASSIGN(f,'b:bspline2.dat');
    RESET(f);
    WHILE NOT EOF(f) DO                                    BEGIN

    initialize;

(*
    Create a box about the display
*)
    drawLINE(0,0,0,389);
    drawLINE(0,389, 511,389);
    drawLINE(511,389, 511,0);
    drawLINE(511,0, 0,0);


    set_blending_functions(lines_per_section);
    start_B_spline(ax,ay);

    FOR i := 3 TO number_of_points  DO
        curve_abs_2(ax[i],ay[i]);

    end_B_spline;

    REPEAT UNTIL keyPRESSED;
    clearALPHA;
    clearGRAPHICS                                      END; (* WHILE *)

    graphOFF;
    CLOSE(f)
END.
```

EXERCISES

Section 3.2

1. Write an interpolating polynomial that passes through each point:

x	0.5	1.2	3.1
y	−3.2	1.6	−1.8

    Plot the points and sketch the parabola that passes through them.

▶ 2. Given the four points $(1, 0)$, $(-2, 15)$, $(-1, 0)$, $(2, 9)$, write the Lagrangian form of the cubic that passes through them. Multiply out each term to express in standard form, $ax^3 + bx^2 + cx + d$.

3. Given that ln 2 = 0.69315 and ln 5 = 1.60944, compute the natural logarithms of each integer from 1 to 10. Compare to a five-place table.

▶ 4. If $e^{0.2}$ is estimated by interpolation among the values of $e^0 = 1$, $e^{0.1} = 1.1052$, and $e^{0.3} = 1.3499$, find the maximum and minimum estimates of the error. Compare with the actual error.

5. Repeat Exercise 4, except use extrapolation to get $e^{0.4}$.

Section 3.3

6. Construct the divided-difference table for these points:

x	$f(x)$
−0.2	1.3940
0.5	1.0025
0.1	1.1221
0.7	1.0084
0.0	1.1884

7. Repeat Exercise 2, except use divided differences. Compare the resulting polynomial in standard form with that obtained in Exercise 2.

▶ 8. Use the divided-difference table from Exercise 6 to estimate $f(0.15)$, using:

a) polynomial of degree 2 through the first three points.
b) polynomial of degree 2 through the last three points.
c) polynomial of degree 3 through the first four points.
d) polynomial of degree 3 through the last four points.
e) polynomial of degree 4.

Why are the results different?

9. Repeat Exercise 4, but now use divided differences.

10. Repeat Exercise 5, but now use divided differences.

Section 3.4

11. Complete a difference table for the following data:

x	1.20	1.25	1.30	1.35	1.40	1.45	1.50
$f(x)$	0.1823	0.2231	0.2624	0.3001	0.3365	0.3716	0.4055

▶ 12. In Exercise 11, what degree of polynomial is required to fit exactly to all seven data pairs? What lesser-degree polynomial will nearly fit the data? Justify your answer.

13. Form a difference table for $f(x) = x^3 - 3x^2 + 2x + 1$ for $x = -1(0.2)1$. (Recall that nested multiplication is more efficient, especially on a desk calculator.) Verify from your table the validity of Eq. (3.5).

▶ 14. Use Newton–Gregory forward-interpolating polynomials of degree 3 to estimate $f(0.158)$ and $f(0.636)$, given the following table. In the first polynomial, choose $x_0 = 0.125$. In the second, let $x_0 = 0.375$.

x	$f(x)$	Δf	$\Delta^2 f$	$\Delta^3 f$	$\Delta^4 f$
0.125	0.79168				
		−0.01834			
0.250	0.77334		−0.01129		
		−0.02963		0.00134	
0.375	0.74371		−0.00995		0.00038
		−0.03958		0.00172	
0.500	0.70413		−0.00823		0.00028
		−0.04781		0.00200	
0.625	0.65632		−0.00623		
		−0.05404			
0.750	0.60228				

15. Add one term to the work of Exercise 14 to estimate $f(0.158)$ by a fouth-degree polynomial.

16. Use the data below to find the value of y at $x = 0.58$, using a cubic polynomial that fits the table at x-values of 0.3, 0.5, 0.7, and 0.9.

x	y	Δy	$\Delta^2 y$	$\Delta^3 y$
0.1	0.003			
		0.064		
0.3	0.067		0.017	
		0.081		0.002
0.5	0.148		0.019	
		0.100		0.003
0.7	0.248		0.022	
		0.122		0.004
0.9	0.370		0.026	
		0.148		0.005
1.1	0.518		0.031	
		0.179		
1.3	0.697			

17. What is the minimum-degree polynomial that will exactly fit all seven data pairs of Exercise 16? (Answer is *not* sixth-degree.)

18. Using the x- and y-values given in the table of Exercise 16, construct a divided-difference table up to third differences. How do these values compare to those for Δy, $\Delta^2 y$, $\Delta^3 y$?

Section 3.5

19. Using the data of Exercise 14, write a Newton–Gregory forward polynomial that fits the table at x-values of 0.500, 0.625, and 0.750. Then write the Newton–Gregory backward polynomial that fits the same three points. Demonstrate that these are two different forms of the same polynomial by rewriting both in the form $a_1 x^2 + a_2 x + a_3$

20. Repeat Exercise 14 using Newton–Gregory backward polynomials, but choose $x_0 = 0.500$ in the first and $x_0 = 0.750$ in the second polynomial. Are the same results obtained?

►21.   Using the data of Exercise 14, estimate $f(0.385)$ by the following: Use quadratic polynomials in each case. (a) Newton–Gregory forward, $x_0 = 0.250$; (b) Newton–Gregory backward, $x_0 = 0.500$; (c) Gauss backward, $x_0 = 0.375$; (d) Stirling, $x_0 = 0.375$. Why are all these results identical?

22.   For the data of Exercise 16, write cubic interpolation polynomials that terminate on the third difference whose value is 0.004. Use the formulas to write polynomials of the following forms: (a) Newton–Gregory forward; (b) Newton–Gregory backward; (c) Gauss forward; (d) Gauss backward; (e) Bessel.

►23.   Use each of the polynomials of Exercise 22 to interpolate at $x = 0.92$. For each polynomial, compare in a table the values of y_0, s, $y(0.92)$ that are involved.

Section 3.6

24.   Write the error terms of each polynomial of Exercise 22. Evaluate the coefficients (the s-polynomials).

►25.   What is the error term for linear interpolation? Show that the coefficient is of maximum magnitude at the midpoint between x_0 and x_1.

26.   A table of $\tan x$ with the argument given in radians has $\Delta x = 0.01$. Near $x = \pi/2$, the tangent function increases very rapidly and hence linear interpolation is not very accurate. What degree of interpolating polynomial is needed to interpolate at $x = 1.506$ to three-decimal accuracy?

27.   Estimate the errors for each of the results in Exercise 8. Does this explain the differences in the estimates? The function in Exercise 6 is actually $f(x) = 1/\sin(1 + x)$, and $f(0.15) = 1.0956$. Do the actual errors agree with the estimates of errors?

28.   In Section 3.6, you were asked to explain why the maximum error between $f(x) = 1/(1 + 25x^2)$ and a polynomial that matches $f(x)$ at equispaced points increases when the degree of the polynomial is increased. Why does this occur?

Section 3.7

29.   An operator α is called a *linear* operator if $\alpha(f + g) = \alpha f + \alpha g$ and $\alpha(cf) = c\alpha f$, where $c = $ constant. Show that E, Δ, and ∇ are linear operators based on their definitions by Eq. (3.11).

30.   Two operators, α and β, are said to commute if the result of operating on a function is not changed by changing the order of the operations, that is, $\alpha(\beta f) = \beta(\alpha f)$. Show that E, Δ, and ∇ commute.

31.   Define D as the differentiation operator. Show that D commutes with E, Δ, and ∇.

►32.   Show that

   a) $\Delta[f(x)g(x)] = f(x)\Delta g(x) + g(x + h)\Delta f(x)$.
   b) $\Delta^n x^n = \nabla^n x^n = \nabla^n E^m x^n = n!$ when $h = 1$.

33.   Express $\Delta^3 E^{-2} \nabla^2 y_4$ in terms of y_i entries of the table.

►34.   If $\Delta^n y_s = \nabla^n y_r$, express r in terms of s.

35. $\delta = E^{1/2} - E^{-1/2}$ defines the central-difference operator. Show that

$$\delta^n y_i = \Delta^n E^{-n/2} y_i = \nabla^n E^{n/2} y_i.$$

Section 3.8

36.   In Section 3.8, $y = P_4(x)$ is given in terms of divided differences. Using this, find the x-value

that corresponds to $P_4(x) = 3.0$ by some method from Chapter 1. Then determine the x-value by the method of successive approximations described in Section 3.8.

37. If the x-values are equispaced, inverse interpolation is easier to do using a Gauss forward polynomial rather than with a divided-difference polynomial because the differences remain centered on point of inverse interpolation without having to adjust the range of the polynomial. Use successive approximations based on a Gauss forward polynomial that fits the data of Exercise 11 to find the x-value that corresponds to $f(x) = 0.2852$. (Actually, the data are for $f(x) = \ln x$. Compare your answer to $e^{0.2852}$.)

▶38. For the function $y = x^2$, obviously the points $(1, 1)$, $(2, 4)$, $(3, 9)$ are on its graph. Considering x as a function of y, we have $x = \sqrt{y}$, and for $y = 25$, $x = 5$. But one may also compute x corresponding to $y = 25$ by inverse interpolation from the first three points. Do this by Lagrangian interpolation (extrapolating, of course). What error estimates are available?

39. Repeat Exercise 38, but determine x at $y = 0$ by inverse interpolation. Are you surprised that the interpolated x-value is again 0.6? Sketch the curves for $x = \sqrt{y}$ and for the interpolating parabola.

Section 3.9

40. Consider the function $f(x) = 20/(1 + 5x^2)$ over the interval $[-2, 2]$. Compute and graph polynomials of degree 1, 2, 3, 4, and 5 that agree with $f(x)$ at uniformly spaced points on the interval.

41. Confirm the statement that, for a set of data exactly fitted by a cubic, the values of S at the two ends will be linearly related to the adjacent S-values if the spline curve and the cubic polynomial are the same function. If one changes one end value for S, say S_0, does this change the portions of the spline curve in intervals other than the first?

▶42. Find the coefficient matrix and the right-hand-side vector for fitting a cubic spline to the following data. Use linearity condition on the terminal S-values:

x	y	x	y
0.15	0.3945	1.07	0.2251
0.76	0.2989	1.73	0.0893
0.89	0.2685	2.11	0.0431

43. Solve the set of equations of Exercise 42 (you may wish to utilize a computer program for this), and then determine the constants of the various cubics. The data are the ordinates of the normal probability function. Compare a few interpolated values with tabulated values of the function, say at $x = 0.30, 0.80, 1.50, 2.00$.

44. Fit a natural cubic spline to $f(x) = 20/(1 + 5x^2)$ on the interval $[-2, 2]$. Use five equispaced points on the function $[x = -2(1)2]$. Graph the spline and compare to the graphs of Exercise 40.

45. Repeat Exercise 44, but use end conditions 2 and 3. Compare results to Exercise 44. Repeat again with $f'(x_0) = 0.9$ and $f'(x_n) = -0.9$.

46. A cubic spline fitted to a full period of periodic data will have the first and second derivatives identical at the two endpoints. Develop the equations that give the S-values for this case. Is the matrix tridiagonal?

47. The data for Example 2 in Section 3.9 are actually for a periodic phenomenon, but the

solutions given ignore this fact. Use end conditions appropriate for a periodic function and compare the spline curve so obtained with the results given in Section 3.9.

Section 3.10

48. Show that the matrix forms of the equations for Bezier and B-spline curves are equivalent to the algebraic equations given in Section 3.10.

49. Write the matrix forms of equations for Bezier and B-spline curves of order 4.

50. Prove that the convex hull does enclose all the points for both the Bezier and B-spline curves. *Hint:* use the fact that any point p in the convex hull formed by the points $\{p_0, p_1, \ldots, p_n\}$ can be written as $p = \sum_{i=0}^{n} \alpha_i p_i$ where $\alpha_i \geqq 0$ and $\sum_{i=0}^{n} \alpha_i = 1$.

51. The slope at the ends of the cubic B-spline seems to be the same as that between the adjacent points. Is this true? Does this also hold for B-splines of higher order?

52. A succession of points has been used to construct a set of connected B-spline curves. One of the points then is changed and the curves are recomputed. What part of the combination curve is affected? Is the same kind of effect noticed if the connected curves are Bezier curves? What if the set of points were fitted with a cubic spline? Do the terms *local control* and *global control* apply to the phenomenon you observe?

53. Going to higher degrees of B-splines is a natural extension. What about reducing the degree to give a quadratic B-spline? What assumptions are reasonable in reducing the degree?

54. Compute and then graph the cubic Bezier curves that connect this set of points:

Point	x	y
1	100	100
2	50	150
3	200	150
4	50	50
5	100	200
6	50	100
7	200	100
8	50	50
9	100	200
10	100	100

55. Repeat Exercise 54 for cubic B-splines.

Section 3.11

56. In Section 3.11, the assertion is made that the order of interpolation makes no difference. Demonstrate that this is true by interpolating within the data of Table 3.10 to find the values at $y = 0.33$ (within rows with x constant at 1.0, 1.5, and 2.0), using cubic interpolation formulas that fit the table at $y = 0.2, 0.3, 0.4,$ and 0.5. Then interpolate within these three values to determine $f(1.6, 0.33)$, and compare to the value 1.841 obtained in the text.

▶57. After the example of Section 3.11 was finished, it was observed that a cubic in x and a quadratic in y would have been preferable. Do this to obtain $f(1.6, 0.33)$ and compare to the true value, 1.8350. Use the best "region of fit."

►58. The example of Section 3.11 used a rectangular region of fit when a more nearly circular region would appear to have advantages. Interpolate from the data in Table 3.10 to evaluate $f(1.62, 0.31)$ by a set of polynomials that fits the table at $x = 1.5$ and 2.0 when $y = 0.2$ and when $y = 0.4$, and at $x = 0.5(0.5)2.5$ when $y = 0.3$. Do this by forming a series of difference tables. In this instance it is very awkward to interpolate first with x held constant, but there is no problem if we begin with y constant.

59. Interpolate for $f(3.55, 0.53)$ from the following data, using cubics in both directions.

x \ y	0.1	0.4	0.6	0.9	1.2
1.1	1.100	0.864	0.756	0.637	0.550
3.0	8.182	6.429	5.625	4.737	4.091
3.7	12.445	9.779	8.556	7.205	6.223
5.2	24.582	19.314	16.900	14.232	12.291
6.5	38.409	30.179	26.406	22.237	19.205

60. Find the value at $x = 3.7$, $y = 0.6$ on the B-spline surface patch that is generated from the 16 points in the upper left corner of the data in Exercise 59.

APPLIED PROBLEMS

61. S. H. P. Chen and S. C. Saxena report experimental data for the emittance of tungsten as a function of temperature (*Ind. Eng. Chem. Fund.* **12**, 220 (1973)). Their data are given below. They found that the equation

$$e(T) = 0.02424\left(\frac{T}{303.16}\right)^{1.27591}$$

correlated the data for all temperatures accurately to three digits. What degree of interpolating polynomial is required to match to their correlation at points midway between the tabulated temperatures? Discuss the pros and cons of polynomial interpolation in comparison to using their correlation.

$T,°K$	300	400	500	600	700	800	900	1000	1100
e	0.024	0.035	0.046	0.058	0.067	0.083	0.097	0.111	0.125

$T,°K$	1200	1300	1400	1500	1600	1700	1800	1900	2000
e	0.140	0.155	0.170	0.186	0.202	0.219	0.235	0.252	0.269

62. In studies of radiation-induced polymerization, a source of gamma rays was employed to give measured doses of radiation. However, the dosage varied with position in the apparatus, with these figures being recorded:

Position, in. from base point	0	0.5	1.0	1.5	2.0	3.0	3.5	4.0
Dosage, 10^5 rads/hr	1.90	2.39	2.71	2.98	3.20	3.20	2.98	2.74

For some reason, the reading at 2.5 in. was not reported, but the value of radiation there is needed. Fit interpolating polynomials of various degrees to the data to supply the missing information. What do you think is the best estimate for the dosage level at 2.5 in.?

63. M. S. Selim and R. C. Seagraves studied the kinetics of elution of copper compounds from ion-exchange resins. The normality of the leaching liquid was the most important factor in determining the diffusivity. Their data were obtained at convenient values of normality; we desire a table of D for integer values of normality ($N = 0.0, 1.0, 2.0, 3.0, 4.0, 5.0$). Use their data to construct such a table.

N	$D \times 10^6$, cm²/sec	N	$D \times 10^6$, cm²/sec
0.0521	1.65	0.9863	3.12
0.1028	2.10	1.9739	3.06
0.2036	2.27	2.443	2.92
0.4946	2.76	5.06	2.07

64. When the steady-state heat-flow equation is solved numerically, temperatures $u(x, y)$ are calculated at the nodes of a gridwork constructed in the domain of interest. (This is the content of Chapter 7.) When a certain problem was solved, the values given in the following table were obtained. This procedure does not give the temperatures at points other than the nodes of the grid; if they are desired, one can interpolate to find them. Use the data to estimate the values of the temperature at the points (0.7, 1.2), (1.6, 2.4), and (0.65, 0.82).

x \ y	0.0	0.5	1.0	1.5	2.0	2.5
0.0	0.0	5.00	10.00	15.00	20.00	25.00
0.5	5.00	7.51	10.05	12.70	15.67	20.00
1.0	10.00	10.00	10.00	10.00	10.00	10.00
1.5	15.00	12.51	9.95	7.32	4.33	0.0
2.0	20.00	15.00	10.00	5.00	0.00	−5.00

65. Star S in the Big Dipper (Ursa Major) has a regular variation in its apparent magnitude. Leon Campbell and Laizi Jacchia give data for the mean light curve of this star in their book *The Story of Variable Stars* (Blakeston, 1941). A portion of these data is given here.

Phase	−110	−80	−40	−10	30	80	110
Magnitude	7.98	8.95	10.71	11.70	10.01	8.23	7.86

The data are periodic in that the magnitude for phase = −120 is the same as for phase = +120. The spline functions discussed in Section 3.09 do not allow for periodic behavior. For a periodic function, the slope and second derivatives are the same at the two endpoints. Taking this into account, develop a spline that interpolates the above data.

Other data given by Campbell and Jacchia for the same star are

Phase	−100	−60	−20	20	60	100
Magnitude	8.37	9.40	11.39	10.84	8.53	7.89

How well do interpolants based on your spline function agree with this second set of observations?

66. Develop the matrices to make Eq. (3.29) generate points on a Bezier surface. Show that this passes through the 12 points on the borders of the group of 16 in Figure 3.10. How could a Bezier surface be created that passes through the innermost set of four points?

67. A fictitious chemical experiment produces seven data points:

t	-1	-0.96	-0.86	-0.79	0.22	0.5	0.930
y	-1	-0.151	0.894	0.986	0.895	0.5	-0.306

a) Plot the points and interpolate a smooth curve by intuition.
b) Plot the unique sixth-degree polynomial that interpolates these points.
c) Use a spline program to evaluate enough points to plot this curve.
d) Compare your results with the graph in Fig. 3.16.

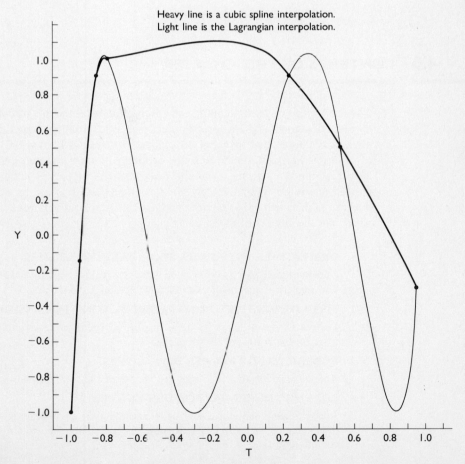

Heavy line is a cubic spline interpolation.
Light line is the Lagrangian interpolation.

Figure 3.16

4

Numerical Differentiation and Numerical Integration

4.0 CONTENTS OF THIS CHAPTER

Chapter 4 tells you how to estimate the derivative or get the integral of a function by numerical methods. Numerical procedures differ from the analytical methods of the calculus in that they can be applied to functions known only as a table of values as well as to functions stated explicitly. Since computers cannot readily be programmed to do the symbolic manipulations of formal integration techniques, nearly all computer programs for integration use these procedures. The material in this chapter has obvious applications in solving differential equations, which are the subject of the next chapter.

4.1 DERIVATIVES, INTEGRALS FROM TABULAR VALUES

Continues the fictional conversation between Rita Laski and Ed Baker to illustrate how numerical integration and differentiation might be required

4.2 FIRST DERIVATIVES FROM INTERPOLATING POLYNOMIALS

Derives several formulas by differentiating an interpolating polynomial; expressions for the errors are also developed

4.3 FORMULAS FOR HIGHER DERIVATIVES

Are developed similarly; in addition, the use of symbolic methods is explained

4.4 LOZENGE DIAGRAMS FOR DERIVATIVES

Are interesting and useful devices for recording the coefficients of formulas for differentiation; they also facilitate the development of alternative forms of formulas for differentiation

4.19 CHAPTER SUMMARY
Lets you evaluate your understanding of the material

4.20 COMPUTER PROGRAMS
Provides several examples of FORTRAN programs for differentiation and integration

4.1 DERIVATIVES, INTEGRALS FROM TABULAR VALUES

Rita and Ed were again at lunch in the cafeteria of Ruscon Engineering.

"How did you make out on the interpolation problem?" Ed asked.

"It went all right," Rita replied. "In fact, we have computer programs now to interpolate either with polynomials or with cubic splines. My boss now wants me to work on some extensions."

"What do you mean, extensions?" Ed asked.

"Well, we need the time derivative of the position. That's the velocity, of course. And some calculations that lead to the fuel consumption require us to integrate a function when we know the values only at discrete times."

"I should think that would follow directly from what you did before," Ed said. "If you have a polynomial that gives you the position as a function of time, can't you just differentiate or integrate that polynomial?"

"Very true," Rita replied. "The catch is that we never really develop the polynomial; we just work with the position values and their differences. My boss warns me also that estimating the errors of derivatives and integrals determined from function values that are known only at discrete points in time is pretty tricky. I suspect that there are ways to tackle these problems that are similar to the methods for interpolation."

Rita is right; there are methods for getting derivatives and integrals from a table of function values, and these resemble methods for interpolation. In this chapter we will explore these methods.

4.2 FIRST DERIVATIVES FROM INTERPOLATING POLYNOMIALS

If a function is reasonably well approximated by an interpolating polynomial, we should expect that the slope of the function should also be approximated by the slope of the polynomial, although we will discover that the error of estimating the slope is greater than the error of estimating the function.

We begin this section with the divided-difference polynomial for approximating a polynomial.

$$
\begin{aligned}
f(x) &= P_n(x) + \text{error} \\
&= f[x_0] + f[x_C, x_1](x - x_0) \\
&\quad + f[x_0, x_1, x_2](x - x_0)(x - x_1) \\
&\quad + \cdots + f[x_0, x_1, \ldots, x_n](x - x_0)(x - x_1) \ldots (x - x_{n-1}) \\
&\quad + \text{error}.
\end{aligned}
\tag{4.1}
$$

As we saw from Section 3.2, the error term of Eq. (4.1) is

$$
\text{Error of } P_n(x) = (x - x_0) \ldots (x - x_n)\frac{f^{(n+1)}(\xi)}{(n+1)!}, \qquad x_0 < \xi < x_n.
\tag{4.2}
$$

Differentiating Eq. (4.1), we get

$$
f'(x) = P_n'(x) = f[x_0, x_1] + f[x_0, x_1, x_2](2x - x_0 - x_1)
$$
$$
+ \cdots + f[x_0, x_1, \ldots, x_n] \sum_{k=0}^{n-1} \frac{(x - x_0) \ldots (x - x_{n-1})}{(x - x_k)}.
\tag{4.3}
$$

Note that

$$
[(x - x_0)(x - x_1) \ldots (x - x_{n-1})]' = \sum_{k=0}^{n-1} \frac{(x - x_0) \ldots (x - x_{n-1})}{(x - x_k)}.
$$

These equations are simplified if we evaluate them at $x = x_0$. Eq. (4.3) becomes

$$
f'(x_0) = f[x_0, x_1] + f[x_0, x_1, x_2](x_0 - x_1)
$$
$$
+ \cdots + f[x_0, x_1, \ldots, x_n](x_0 - x_1)(x_0 - x_2) \ldots (x_0 - x_{n-1}).
\tag{4.4}
$$

Differentiating Eq. (4.2) gives us the error of Eq. (4.4):

$$
\text{Error of } P'(x) = \frac{f^{(n+1)}(\xi)}{(n+1)!} \sum_{k=0}^{n} \frac{(x - x_0) \ldots (x - x_n)}{(x - x_k)}
$$
$$
+ \frac{(x - x_0) \ldots (x - x_n)}{(n+1)!} \frac{d}{dx}[f^{(n+1)}(\xi)]
\tag{4.5}
$$

where ξ depends on x. However, if we set $x = x_0$ in Eq. (4.5), then we get

$$
\text{Error of } P'(x_0) = (x_0 - x_1) \ldots (x_0 - x_n)\frac{f^{(n+1)}(\xi)}{(n+1)!}
$$

Since most of the development of the algorithms presented in Chapters 4 and 5 assume equispaced points, we will make use of the Newton–Gregory forward polynomials and leave the remaining derivations from divided-difference polynomials as exercises.

We begin with a Newton–Gregory forward polynomial:

$$
f(x_s) = P_n(x_s) + \text{error} = f_0 + s\Delta f_0 + \binom{s}{2}\Delta^2 f_0 + \cdots + \binom{s}{n}\Delta^n f_0 + \text{error}.
\tag{4.6}
$$

The error term of Eq. (4.6) is, by our rule for changing the next term,

$$\text{Error of } P_n(x_s) = \binom{s}{n+1} h^{n+1} f^{(n+1)}(\xi), \qquad x_0 < \xi < x_n. \tag{4.7}$$

Differentiating Eq. (4.6), remembering that f_0 and all the Δ-terms are constants (after all, they are just the numbers from the difference table), we have

$$f'(x_s) \doteq P_n'(x_s) = \frac{d}{dx}[P_n(x_s)] = \frac{d}{ds}[P_n(x_s)]\frac{ds}{dx} = \frac{d}{ds}[P_n(x_s)]\frac{1}{h}$$

$$= \frac{1}{h}\left(\Delta f_0 + \frac{1}{2}(s - 1 + s)\Delta^2 f_0 \right.$$

$$\left. + \frac{1}{6}[(s - 1)(s - 2) + s(s - 2) + s(s - 1)]\Delta^3 f_0 + \cdots\right). \tag{4.8}$$

The derivatives of the factorials of s rapidly become algebraically complicated.* We get considerable simplification if we let $s = 0$, giving us the derivative corresponding to x_0 (where $s = 0$), however. If we let $s = 0$, Eq. (4.8) becomes

$$f'(x_0) \doteq \frac{1}{h}\left(\Delta f_0 - \frac{1}{2}\Delta^2 f_0 + \frac{1}{3}\Delta^3 f_0 - \frac{1}{4}\Delta^4 f_0 + \cdots \pm \frac{1}{n}\Delta^n f_0\right). \tag{4.9}$$

In Eq. (4.9) the derivative value, approximating to the derivative of the function, is the derivative of an nth-degree polynomial passing through the point (x_0, f_0) and n additional points to its right, evaluated at $x = x_0$. The error in Eq. (4.9) is obtained by differentiating the error term given in Eq. (4.7):

$$\text{Error of } P_n'(x_s) = h^{n+1}f^{(n+1)}(\xi)\left[\frac{d}{ds}\binom{s}{n+1}\right]\frac{1}{h} + \binom{s}{n+1}h^{n+1}\frac{d}{dx}[f^{(n+1)}(\xi)]. \tag{4.10}$$

The second term here cannot be evaluated, for the way that ξ varies with x is unknown but, when we let $s = 0$, this second term vanishes because

$$\binom{s}{n+1} = \frac{1}{(n+1)!}(s)(s-1)\ldots(s-n) = 0$$

at $s = 0$. Then we need only evaluate the first term of Eq. (4.5). Since

$$\frac{d}{ds}\binom{s}{n+1} = \frac{(s-1)(s-2)\ldots(s-n) + s(s-2)(s-3)\ldots(s-n) + \cdots + s(s-1)\ldots(s-n+1)}{(n+1)!},$$

at $s = 0$, only the first term of the numerator remains, and Eq. (4.10) becomes

---

*In computing the derivatives, we have used the method:

$$\frac{d}{ds}(xyz) = \frac{dx}{ds} \cdot yz + x\frac{dy}{ds}z + xy\frac{dz}{ds}.$$

$$\text{Error of } P'_n(x_0) = h^{n+1} f^{(n+1)}(\xi) \left[(-1)^n \frac{n!}{(n+1)!} \right] \left(\frac{1}{h} \right) \tag{4.11}$$

$$= \frac{(-1)^n}{n+1} h^n f^{(n+1)}(\xi).$$

Note that even though the interpolating polynomial gives the function exactly at $s = 0$, our derivative formula is subject to an error of $O(h^n)$ at that point, unless $f^{(n+1)}(\xi) = 0$. In comparing Eqs. (4.11) and (4.9) we see that the error term of the derivative also can be obtained by changing the $\Delta^n f_0$ in the next term beyond the last one included, into $h^n f^{(n)}(\xi)$ exactly as with our interpolating formulas.

EXAMPLE Use the data in Table 4.1 to estimate the derivative of y at $x = 1.7$. Use $h = 0.2$, and compute using one, two, three, or four terms of the formula.

With one term $\quad y'(1.7) = \dfrac{1}{0.2}(1.212) = 6.060.$

With two terms $\quad y'(1.7) = \dfrac{1}{0.2}\left(1.212 - \dfrac{1}{2}(0.268) \right) = 5.390.$

With three terms $\quad y'(1.7) = \dfrac{1}{0.2}\left(1.212 - \dfrac{1}{2}(0.268) + \dfrac{1}{3}(0.060) \right) = 5.490.$

With four terms $\quad y'(1.7) = \dfrac{1}{0.2}\left(1.212 - \dfrac{1}{2}(0.268) + \dfrac{1}{3}(0.060) - \dfrac{1}{4}(0.012) \right)$

$$= 5.475.$$

Table 4.1

x	y	Δy	$\Delta^2 y$	$\Delta^3 y$	$\Delta^4 y$
1.3	3.669				
		0.813			
1.5	4.482		0.179		
		0.992		0.041	
1.7	5.474		0.220		0.007
		1.212		.0.048	
1.9	6.686		0.268		0.012
		1.480		0.060	
2.1	8.166		0.328		0.012
		1.808		0.072	
2.3	9.974		0.400		
		2.208			
2.5	12.182				

The data tabulated in Table 4.1 are for $y = e^x$, rounded to three decimals.* Since the derivative of e^x is also e^x, we see that the error in the derivative is quite small with four terms, comparing 5.475 to 5.474. We can anticipate this; since the fourth differences in Table 4.1 do not vary greatly, the function is represented reasonably well by a fourth-degree polynomial.

Since we know that $f(x) = e^x$, the estimated errors for the preceding computations can be calculated by Eq. (4.11):

With one term,   error $= \dfrac{(-1)^1}{2}(0.2)f''(\xi)$, $1.7 \le \xi \le 1.9$,

$$= \frac{-0.2}{2}\begin{Bmatrix} e^{1.7} & \text{(min)} \\ e^{1.9} & \text{(max)} \end{Bmatrix} = \begin{Bmatrix} -0.547 & \text{(min)} \\ -0.669 & \text{(max)} \end{Bmatrix}.$$

(Actual error is -0.586.)

With two terms,   error $= \dfrac{(-1)^2}{3}(0.2)^2 f'''(\xi)$, $1.7 \le \xi \le 2.1$,

$$= \frac{0.04}{3}\begin{Bmatrix} e^{1.7} & \text{(min)} \\ e^{2.1} & \text{(max)} \end{Bmatrix} = \begin{Bmatrix} 0.073 & \text{(min)} \\ 0.109 & \text{(max)} \end{Bmatrix}.$$

(Actual error is 0.084.)

With three terms,   error $= \dfrac{(-1)^3}{4}(0.2)^3 f^{iv}(\xi)$, $1.7 \le \xi \le 2.3$,

$$= \frac{-0.008}{4}\begin{Bmatrix} e^{1.7} & \text{(min)} \\ e^{2.3} & \text{(max)} \end{Bmatrix} = \begin{Bmatrix} -0.011 & \text{(min)} \\ -0.020 & \text{(max)} \end{Bmatrix}.$$

(Actual error is -0.016.)

With four terms,   error $= \dfrac{(-1)^4}{5}(0.2)^4 f^{v}(\xi)$, $1.7 \le \xi \le 2.5$,

$$= \frac{0.0016}{5}\begin{Bmatrix} e^{1.7} & \text{(min)} \\ e^{2.5} & \text{(max)} \end{Bmatrix} = \begin{Bmatrix} 0.002 & \text{(min)} \\ 0.004 & \text{(max)} \end{Bmatrix}.$$

(Actual error is -0.001.) ∎

In the case of four terms, the calculated and actual errors do not agree because of round-off errors; Eq.(4.11) accounts only for the truncation error. In the other cases, the round-off is not large enough relative to the truncation error to invalidate the estimates from Eq. (4.11).

The formulas for derivatives given by Eq. (4.9) are called forward-difference approximations because they involve only differences of function values forward from $f(x_0)$. For a forward-difference approximation, the interpolating polynomial that is used does not fit points that are symmetrical about x_0. We have seen in the previous chapter that interpolation is more accurate near the center of the range of fit. A similar effect is observed with derivatives. Consider the first two terms of Eq. (4.8):

$$f'(x) = \frac{1}{h}\left(\Delta f_0 + \frac{1}{2}(s - 1 + s)\Delta^2 f_0\right) + \text{error.} \qquad (4.12)$$

---

*The table incidentally illustrates the effect of rounding errors in the fourth differences. A more accurate table would show fourth differences of 0.0088, 0.0108, 0.0132.

The second-degree interpolating polynomial to which Eq. (4.12) corresponds fits at x_0, x_1, and x_2. Let $s = 1$ in Eq. (4.12) so that we can estimate $f'(x_1)$:

$$f'(x_1) = \frac{1}{h}\left(\Delta f_0 + \frac{1}{2}\Delta^2 f_0\right) + \text{error}.$$

Write out the differences in terms of $f's$:

$$f'(x_1) = \frac{1}{h}\left(f_1 - f_0 + \frac{1}{2}(f_2 - 2f_1 + f_0)\right) = \frac{f_2 - f_0}{2h} + \text{error}. \qquad \textbf{(4.13)}$$

Equation (4.13) can be called a central-difference approximation. The value of x where it is applied is centered in the range of fit of the polynomial—differences of function values on either side of $f(x_1)$ are used.

The error term for Eq. (4.13) can be derived similarly to Eq. (4.11). It is

$$\text{Error in } P_2'(x_1) = -\frac{1}{6}h^2 f'''(\xi), \qquad x_0 \le \xi \le x_2. \qquad \textbf{(4.14)}$$

Note particularly that the power of h in Eq. (4.14) is two, so the error is $O(h^2)$. Contrast this to the first power of h in the error term when only one term of Eq. (4.8) is used, although both computations involve only two functional values. The coefficient in the error term is also significantly smaller. Central-difference formulas are decidedly superior in calculating values for derivatives.

If we apply Eq. (4.13) to the data of Table 4.1 to estimate $f'(1.7)$, we get

$$f'(1.7) \doteq \frac{6.686 - 4.482}{2(0.2)} = 5.510.$$

Estimate of error is $\left\{\begin{array}{ll} -0.030 & \text{(min)} \\ -0.046 & \text{(max)} \end{array}\right\}$.

(Actual error is -0.036.)

Central-difference formulas similar to Eq. (4.13) can be derived using higher-degree polynomials of even order. (The odd-degree polynomials do not have a range of fit that is symmetrical about any of the x-values.) For example, the formula corresponding to a fourth-degree polynomial, expressed in terms of function values rather than differences, and related to the point x_0, is

$$f'(x_0) = \frac{1}{h}\frac{f_{-2} - 8f_{-1} + 8f_1 - f_2}{12}, \qquad \text{Error} = \frac{1}{30}h^4 f^v(\xi), \qquad x_{-2} \le \xi \le x_2.$$

In many applications it is preferable to use the simpler formula of Eq. (4.13) and control the error by making h small. Observe, however, that the error of the derivative from an nth-degree polynomial is $O(h^n)$, while the error of interpolation is $O(h^{n+1})$.

4.3 FORMULAS FOR HIGHER DERIVATIVES

We can get formulas for higher derivatives by further differentiating Eq. (4.8), and differentiating Eq. (4.10) will give the error terms. For example,

$$f''(x_s) \doteq P_n''(x_s) = \frac{1}{h^2} \frac{d^2}{ds^2} P_n(x_s)$$

$$\doteq \frac{1}{h^2} \Big(\frac{1}{2}(2)\Delta^2 f_0 + \frac{1}{6}[(s-2) + (s-1) + (s-2) + s \tag{4.15}$$

$$+ (s-1) + s]\Delta^3 f_0 + \cdots \Big).$$

At $s = 0$,

$$f''(x_0) \doteq \frac{1}{h^2}(\Delta^2 f_0 - \Delta^3 f_0 + \cdots).$$

There is no simple pattern for the coefficients, and the error terms have complicated coefficients as well. The situation gets more complex with still higher derivatives, so we would like a simpler method. We have seen before that symbolic methods have given interpolation formulas with relative ease. It is true for derivative formulas also:

$$E = 1 + \Delta,$$

$$y_s = E^s y_0,$$

$$y_s' = \frac{d}{dx}(E^s y_0) = \frac{1}{h} \frac{d}{ds}(E^s y_0) = \frac{1}{h}(\ln E)E^s y_0. \tag{4.16}$$

At $s = 0$,

$$y_0' = \frac{1}{h}(\ln E)y_0 = \frac{1}{h}\ln(1 + \Delta)y_0$$

$$= \frac{1}{h}\Big(\Delta y_0 - \frac{1}{2}\Delta^2 y_0 + \frac{1}{3}\Delta^3 y_0 - \frac{1}{4}\Delta^4 y_0 + \cdots\Big). \tag{4.17}$$

In writing Eq. (4.17), we have used the Maclaurin expansion for $\ln(1 + \Delta)$. Note that Eq. (4.17) is identical to Eq. (4.9), which we derived before by nonsymbolic means.

If we use the symbol D for the derivative operator,

$$Dy_0 = \frac{1}{h}\ln(1 + \Delta)y_0. \tag{4.18}$$

We abstract the equivalence between operators from Eq. (4.18):

$$D = \frac{1}{h}\ln(1 + \Delta).$$

$$(4.19)$$

The value of the symbolic method is that algebraic operations on operator relations give valid results. Let us raise each side of Eq. (4.19) to powers:

$$D^2 = \frac{1}{h^2}\ln^2(1 + \Delta),$$

$$D^3 = \frac{1}{h^3}\ln^3(1 + \Delta),$$

$$D^n = \frac{1}{h^n}\ln^n(1 + \Delta).$$

$$(4.20)$$

Equations (4.20) show that we can get formulas for higher derivatives by multiplying the series in Eq. (4.17) by itself. The second derivative is given by:

$$D^2 y_0 = \frac{1}{h^2}\left(\Delta - \frac{1}{2}\Delta^2 + \frac{1}{3}\Delta^3 - \frac{1}{4}\Delta^4 + \cdots\right)^2 y_0$$

$$= \frac{1}{h^2}\left(\Delta^2 - \Delta^3 + \frac{11}{12}\Delta^4 - \frac{5}{6}\Delta^5 + \cdots\right)y_0;$$

$$y'' = \frac{1}{h^2}\left(\Delta^2 y_0 - \Delta^3 y_0 + \frac{11}{12}\Delta^4 y_0 - \frac{5}{6}\Delta^5 y_0 + \cdots\right).$$

$$(4.21)$$

Formulas for derivatives of higher order can be similarly obtained by multiplication of the series in Eq. (4.17).

Equations (4.20) lead to an important generalization for the estimation of derivatives from forward-difference formulas. The first term of all the formulas in terms of differences has only Δ^n. Hence, as a first approximation,*

$$D^n y = \frac{1}{h^n}\Delta^n y + \text{error } O(h).$$

$$(4.22)$$

E X A M P L E Use formula (4.21) to estimate $y''(1.7)$ from Table 4.1 using terms through Δ^3. Also estimate the error.

$$y''(1.7) = \frac{1}{(0.2)^2}(0.268 - 0.060)$$

$$= 5.200 \qquad \text{(compare to 5.474, exact answer)}.$$

$$\text{Error} = \frac{1}{h^2}\left(\frac{11}{12}h^4 y^{iv}(\xi)\right), \qquad 1.7 \le \xi \le 2.1,$$

*This was used in the previous chapter in developing the "next-term rule" for the error of an interpolation formula.

by converting the next term in the formula. Since $y = e^x$, $y^{iv} = e^x$.

$$\text{Max value of error} = \frac{11}{12}(0.2)^2(e^{2.1}) = 0.298,$$

$$\text{Min value of error} = \frac{11}{12}(0.2)^2(e^{1.7}) = 0.201.$$

Compare to actual error of 0.274. ∎

In the absence of knowledge of the function, we could have used the fourth differences to estimate $h^4 f^{iv}$, as shown by Eq. (4.22). If the differences vary greatly, this is hazardous, however. In this instance we would estimate the error as about 0.23 if we used 0.010 (the average value of $\Delta^4 y$) as an estimate of $h^4 f^{iv}(\xi)$.

Central differences for higher derivatives will have more accuracy than the forward-difference formula represented by Eq. (4.21). We prefer to develop these through the lozenge diagrams of the next section.

4.4 LOZENGE DIAGRAMS FOR DERIVATIVES

Since there are a variety of equivalent forms of the interpolating polynomial, we may obtain a variety of forms of derivative formulas by differentiating the various polynomials. They are interrelated, of course, as we have seen for interpolation formulas. It is convenient to exhibit this interrelation by a so-called lozenge diagram, as in Fig. 4.1.

Examination of the lozenge diagram in Fig. 4.1 shows it to be a difference table in symbolic form (similar to Table 3.3) with numerical values interspersed.

These values are used as coefficients of the differences of f in forming the derivative formula and are selected by tracing a pattern through the difference table.

Certain rules must be followed:

1. One begins at the column of f's at f_0. The point of beginning dictates the subscripting, and hence the value of x_0 must be consistent with the subscripts of the f's.

2. One proceeds from left to right, either diagonally upward or downward to the next difference column, or alternatively horizontally. A term is added for every column crossed.

3. The term that we add is the entry in the lozenge that is traversed, multiplied by the entry above if the last step was diagonally downward, by the entry below if the last step was diagonally upward, or by the average of entries above and below if it was horizontal. The coefficient of the f-term is always zero, as the interspersed figures indicate.

There is a special property of lozenge diagrams that makes them easy to construct. The diagram can be considered to be of two parts, the differences of the function and the interspersed coefficients, one superimposed on the other. Consider the array of coefficients of Fig. 4.1, shown here as Fig. 4.2.

Compare the columns in pairs, beginning with columns A and B. Observe that the values in A are the differences of the values in B. Now compare columns B and C; column

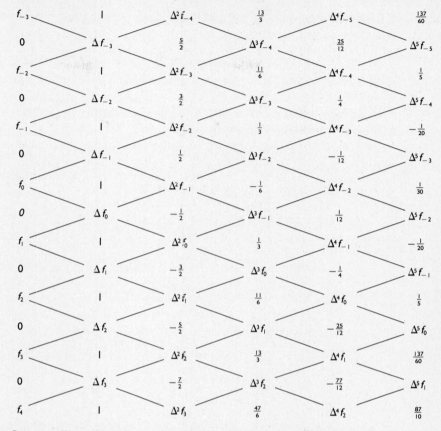

Figure 4.1   Lozenge diagram for $f'(x)$. (*Note*: All formulas must be multiplied by $1/h$.)

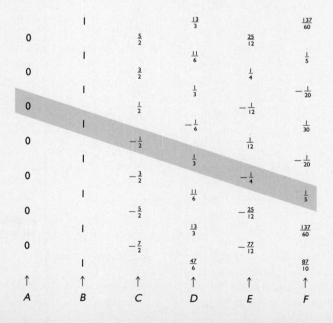

Figure 4.2

B is just the differences of the values in column C, with each value subtracted from the one above it. The same thing is true for columns C and D, columns D and E, and so on. In other words, the array of coefficients is a difference table, but written from right to left and upside down. This means that one line of coefficients, which we get from Eq. (4.9), suffices to create the entire array of coefficients! In Fig. (4.2), this set of coefficients is shaded.

Equation (4.21) serves to start a lozenge diagram for the second derivative at $x = x_0$ in a similar fashion. The result is Fig. 4.3.

We use the lozenge diagrams to write formulas for differentiation of a tabulated function by following a path through the difference table. Any path through the table may be chosen; the coefficeints are then given by the rules discussed above. Suppose we use a horizontal path beginning at f_0:

$$\left.\frac{df}{dx}\right|_{x=x_0} = \frac{1}{h}\left[\frac{\Delta f_0 + \Delta f_{-1}}{2} + (0)\Delta^2 f_{-1} + \left(-\frac{1}{6}\right)\frac{\Delta^3 f_{-1} + \Delta^3 f_{-2}}{2} + \cdots\right]$$

$$= \frac{1}{h}\left[\frac{(f_1 - f_0) + (f_0 - f_{-1})}{2} + O(h^3)\right] = \frac{f_1 - f_{-1}}{2h} + O(h^2). \qquad (4.23)$$

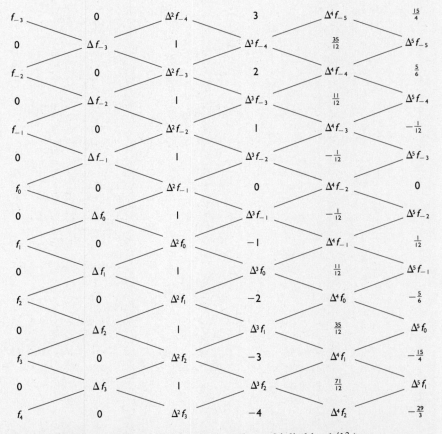

Figure 4.3 Lozenge diagram for $f''(x)$. (*Note*: All formulas must be multiplied by $1/h^2$.)

Using Fig. 4.3 to find the second derivative, again following a horizontal path, we obtain

$$\frac{d^2f}{dx^2}\Bigg|_{x=x_0} = \frac{1}{h^2}\left[\Delta^2 f_{-1} + 0 + \left(-\frac{1}{12}\right)\Delta^4 f_{-2} + \cdots\right]$$

$$= \frac{1}{h^2}[(f_1 - 2f_0 + f_{-1}) + O(h^4)]$$

$$= \frac{f_1 - 2f_0 + f_{-1}}{h^2} + O(h^2). \tag{4.24}$$

Observe that following a horizontal path leads to a central-difference approximation. Following a diagonally downward path gives a formula based on a Newton-Gregory forward polynomial.

Equations (4.23) and (4.24) are formulas of special importance, Eq. (4.23) being the same as Eq. (4.13). Note that even though they are only one-term formulas, they have errors of $O(h^2)$, in contrast to the $O(h)$ error for Eqs. (4.8) and (4.21), when only one term is used. As previously stated, these $O(h^2)$ formulas are known as *central-difference approximations*. Their good error terms occur because the point at $x = x_0$ is at the center of the range of x-values at which the interpolating polynomial matches the table. Central-difference formulas of higher order can be developed by continuing further on our horizontal path.

The efficiency of the central-difference approximations for second derivatives, as compared to the forward-difference approximation, is illustrated by the following example.

Using the data of Table 4.1, we compute $f''(1.7)$.

Using forward differences, Eq. (4.21),

$$f''(1.7) = \frac{0.268}{(0.2)^2} = 6.700,$$

$$\text{Error} = -1.226.$$

Using central differences, Eq. (4.24),

$$f''(1.7) = \frac{0.220}{(0.2)^2} = 5.500,$$

$$\text{Error} = -0.026.$$

The central-difference expression is much more accurate. The estimated errors for the above computations are -1.095 (min) to -1.633 (max) for forward differences, and -0.015 (min) to -0.022 (max) for central differences. (Note that, in the second case, round-off causes the actual error to fall outside these limits.)

Figure 4.3 gives a formula of $O(h^4)$ error for the second derivative if we extend the horizontal path until we reach the column of fourth differences:

$$\frac{d^2f}{dx^2}\Bigg|_{x=x_0} = \frac{1}{h^2}\left[\Delta^2 f_{-1} - \frac{1}{12}\Delta^4 f_{-2}\right] + O(h^4).$$

Rewriting this in terms of the f-values, we get an equivalent expression:

$$\left.\frac{d^2f}{dx^2}\right|_{x=x_0} = \frac{-f_{-2} + 16f_{-1} - 30f_0 + 16f_1 - f_2}{12h^2} + O(h^4).$$

Applying these formulas to the data of Table 4.1, we compute $f''(1.7)$ as

$$f''(1.7) = \frac{1}{(0.2)^2}\left[0.220 - \frac{1}{12}(0.007)\right] = 5.485$$

$$= \frac{-8.166 + 16(6.686) - 30(5.474) + 16(4.482) - 3.669}{12(0.2)^2} = 5.485.$$

(Compare the result to the exact value: 5.474. The error, -0.011, is much larger than the truncation error (-0.0049 to -0.00108) due to round-off.)

The central-difference approximations extend directly to higher derivatives as well. Lozenge diagrams can be constructed for these, but if we need only the first nonzero term, a simple formula will suffice. For the even-order derivatives it is

$$f^{(n)}(x)|_{x=x_0} = \frac{1}{h^n}\Delta^n f_{-n/2} + O(h^2), \quad n \text{ even.}$$

The difference required is always found in the table on the horizontal line through x_0 and f_0. If the difference table has not been constructed, we would express it in terms of the f-values.

For odd-order derivatives, the expression involves the average of differences found just above and just below the horizontal line through x_0:

$$f^{(n)}(x)|_{x=x_0} = \frac{\Delta^n f_{(-n+1)/2} + \Delta^n f_{(-n-1)/2}}{2h^n} + O(h^2), \quad n \text{ odd.} \tag{4.25}$$

4.5 EXTRAPOLATION TECHNIQUES

There is an alternative way of successively improving the accuracy of our estimates of the derivative that is equivalent to adding terms as given by the lozenge diagram. This is the technique of *successive extrapolations*. It has special utility in computer programs for differentiation of arbitrary functions.

We illustrate the method first for a function known only from a table, such as Table 4.2. We desire the derivative $f'(x)$ at $x = 2.5$. The central-difference formula, Eq. (4.13) or Eq. (4.23), gives

$$f'(x)|_{x=2.5} = \frac{f_1 - f_{-1}}{2h} = \frac{0.25337 - 0.31729}{2(0.1)} = -0.3196.$$

This estimate has an error $O(h^2)$.

Table 4.2

x	f(x)	x	f(x)
2.0	0.42298	2.6	0.25337
2.1	0.40051	2.7	0.22008
2.2	0.37507	2.8	0.18649
2.3	0.34718	2.9	0.15290
2.4	0.31729	3.0	0.11963
2.5	0.28587		

We can also compute the derivative by using values at $x = 2.7$ and $x = 2.3$, for which the spacing is 0.2, and get

$$f'(x)|_{x=2.5} \doteq \frac{0.22008 - 0.34718}{2(0.2)} = -0.3178.$$

The error here is also $O(h^2)$, but now h is twice as large as in the previous calculation.

In order to combine these two estimates and extrapolate to a more accurate estimate, we examine the nature of the error term of Eq. (4.23) in greater detail. Using the lozenge diagram, Fig. 4.1, we have

$$f_0' = \frac{1}{h}\left(\frac{\Delta f_{-1} + \Delta f_0}{2} - \frac{1}{6}\frac{\Delta^3 f_{-2} + \Delta^3 f_{-1}}{2} + \frac{1}{30}\frac{\Delta^5 f_{-3} + \Delta^5 f_{-2}}{2} - \cdots\right). \quad \textbf{(4.26)}$$

If we truncate after the first term, we get an error of $O(h^2)$, but Eq. (4.25) shows that

$$f_0''' = \frac{\Delta^3 f_{-2} + \Delta^3 f_{-1}}{2h^3} + O(h^2).$$

Equivalently, we may write

$$-\frac{1}{6}\frac{\Delta^3 f_{-2} + \Delta^3 f_{-1}}{2} = -\frac{1}{6}h^3 f_0''' + O(h^5) = Ch^3 + O(h^5).$$

In this expression C is a constant; we merely recognize that the third derivative of f evaluated at x_0 has a fixed value. Equation (4.26) can hence be written as

$$f_0' = \frac{1}{h}\frac{\Delta f_{-1} + \Delta f_0}{2} + Ch^2 + O(h^4) + \frac{1}{30}h^4 f^v(\xi)$$

$$= \frac{1}{h}\frac{\Delta f_{-1} + \Delta f_0}{2} + Ch^2 + O(h^4).$$

We see that we make only an order h^4 error in assuming that the error of Eq. (4.23) is proportional to h^2. We then combine the two estimates of the derivative, -0.3196 with $h = 0.1$ and -0.3178 with $h = 0.2$, as follows:

$$f_0'(\text{exact}) = -0.3178 + Ch^2 + O(h^4), \quad \text{based on less accurate value,}$$

$$\doteq -0.3178 + C(0.2)^2, \quad \text{if we neglect the } O(h^4) \text{ term.}$$

$$f_0'(\text{exact}) = -0.3196 + Ch^2 + O(h^4), \quad \text{based on more accurate value,}$$

$$\doteq -0.3196 + C(0.1)^2, \quad \text{if we neglect the } O(h^4) \text{ term.}$$

We solve for the unknown value of f_0' (exact) by eliminating the terms in C between the two expressions,

$$f_0'(\text{exact}) \doteq -0.3196 + \frac{1}{3}[-0.3196 - (-0.3178)]$$

$$\doteq -0.3203.$$

The method can be applied to any two estimates, each with $O(h^n)$ errors and with values of h varying by 2 to 1. The rule is

$$\text{Exact value} \doteq \text{more accurate} \tag{4.27}$$

$$+ \left(\frac{1}{2^n - 1}\right)(\text{more accurate} - \text{less accurate}).$$

We show that the relationship is not perfect by using the $\doteq$ sign; this lack of exactness is the result of our neglect of the $O(h^{n+2})$ terms in the two expressions.

This extrapolation method can be extended since the first-order extrapolations with error $O(h^4)$ can be shown to have errors of the form $Ch^4 + O(h^6)$. (Observe that this follows from the alternately zero coefficients on the horizontal path through Fig. 4.1, so this principle extends to all orders of extrapolation. Errors due to rounding of the original data eventually dominate, however, so the extrapolation technique must be abandoned at some point.)

For the data in Table 4.2 we can extend to one more order of extrapolation. We arrange the data in tabular form.

h	Initial estimate	First-order extrapolation	Second-order extrapolation
0.1	-0.31960		
		-0.32022	
0.2	-0.31775		-0.32020
		-0.32050	
0.4	-0.30951		

The last entry was calculated as

$$-0.32022 + \frac{1}{2^4 - 1}(-0.32022 + 0.32050) = -0.32020.$$

An extra guard figure was carried to minimize round-off in the computations.

Since the lozenge for second derivatives, Fig. 4.3, also shows alternate coefficients to be zero, approximations for the second derivative can be extrapolated in exactly the same way, the order of the error increasing to h^4, h^6, h^8, and so on, at each step.

It is left as an exercise for the student to show that this extrapolation method is the equivalent to passing a higher-degree interpolating polynomial through the data (with the point $x = x_0$ at the center of the region of fit for each of these), and then finding the derivative by central-difference formulas.

The application of this extrapolation method in a computer program is important because successive estimates, with h decreasing, have smaller and smaller truncation errors, and hence provide a built-in measure of accuracy. The method is generally employed to differentiate a function of known form, rather than a tabulated function. The basic scheme is to compute the derivative at two values of h, one being one-half the value of the other, and to extrapolate as in the above example. If a significant change is made by the extrapolation, compared to the more accurate of the original estimates, h is halved again and a second-order extrapolation is made. Again if a significant improvement results, another computation is made with h halved once more and a third-order extrapolation is made. The process is continued until the improvement is less than some criterion.

Since round-off affects calculation by computers just as it does hand computations, eventually these errors will control the accuracy of successive values, so the student is cautioned against indiscriminate use of extrapolation. It is generally good practice to put a limit to the number of repetitions of the process. This is discussed in the next section.

4.6   ROUND-OFF AND ACCURACY OF DERIVATIVES

It is important to realize the effect of round-off error on the accuracy of derivatives computed by finite-difference formulas. As we have seen, decreasing the value of h or increasing the degree of the interpolating polynomial increases the accuracy of our derivative formulas. (These procedures reduce the truncation error.)

As h is reduced, however, we are required to subtract function values that are more nearly the same value, and this, we have seen, incurs a large error due to round-off. The increase in round-off error as h gets smaller causes the best accuracy to occur at some intermediate point. The data in Fig. 4.4(a) and (b) illustrate this phenomenon. The test was made with both single-precision arithmetic (giving 6–7 significant figure accuracy) and double-precision (14–15 figures). The function was $f(x) = e^x$, at $x = 0$. The true value of $f'(0)$ is 1.0, of course. Central-difference approximations were employed.

In the single-precision test, the optimum value of h was only 0.01 for the first derivative, and the best accuracy of the second derivative was at $h = 0.1$, a surprisingly large value. When the round-off errors were reduced by using double-precision, essentially exact results were obtained for the first derivative throughout the range of h values. For the second derivative, $h = 0.1 \times 10^{-2}$ or $h = 0.1 \times 10^{-3}$ gives the best accuracy. Note how large the error grows; it is especially marked for second derivatives when h becomes

H	DX		DDX	
0.1E 00	0.1001663E	01	0.1000761E	01
0.1E−01	0.1000001E	01	0.9959934E	00
0.1E−02	0.9999569E	00	0.8940693E	00
0.1E−03	0.9959930E	00	−0.8344640E	02
0.1E−04	0.9775158E	00	−0.4768363E	04
0.1E−05	0.9834763E	00	−0.5960462E	05
0.1E−06	0.5960463E	00	−0.1192093E	08
0.1E−07	0.2980230E	01	−0.5960458E	09
0.1E−08	0.2980229E	02	−0.5960459E	11
0.1E−09	0.2980229E	03	−0.5960456E	13

a) Results with single precision

H	DX		DDX	
0.1E 00	0.1001667E	01	0.1000834E	01
0.1E−01	0.1000016E	01	0.1000008E	01
0.1E−02	0.1000000E	01	0.1000000E	01
0.1E−03	0.1000000E	01	0.9999999E	00
0.1E−04	0.1000000E	01	0.9999989E	00
0.1E−05	0.1000000E	01	0.1000031E	01
0.1E−06	0.9999999E	01	0.9894834E	00
0.1E−07	0.9999999E	00	−0.4163322E	00
0.1E−08	0.9999999E	00	−0.4163319E	02
0.1E−09	0.1000000E	01	0.2775545E	04

b) Results with double precision

Figure 4.4   Derivatives computed by central-difference formulas $f(x) = e^x$, at $x = 0$.

much smaller than the optimum. Higher derivatives than the second will be expected to show an even more exaggerated influence of round-off.

Different models of computers vary widely in the number of digits of accuracy they provide, due to the different number of bits provided for the storage of floating-point values and also, to a lesser degree, due to the format of their floating-point representations. For example, a CDC Cyber provides 60 bits for floating-point values, and gives the equivalent of about 14 decimal digits of accuracy. Another factor that affects the accuracy of functional values is the algorithm used in their evaluation. Chapter 10 gives some insight into this. Different software packages that are used with computer operating systems employ different algorithms, which are of varying accuracy. Further, as Chapter 10 will show, the error of approximating the function differs with the value of the argument.

No analytical way of predicting the effects of round-off error seems to exist. The best we can do is to warn budding numerical analysts so they will be properly wary when computing derivatives numerically. In the words of many authorities, *the process of differentiating is basically an unstable process*—meaning that small errors made during the process cause greatly magnified errors in the final result.

Differentiation of "noisy" data encounters a similar problem. If the data being differentiated are from experimental tests, or are observations subject to errors of measurement, the errors so influence the derivative values calculated by numerical procedures that they may be meaningless. The usual recommendation is to smooth the data first, using methods that are discussed in Chapters 3 and 10. Passing a cubic spline through the points and then getting the derivative of this approximation to the data has become quite popular. A least-squares curve may also be used. The strategy involved is straight-forward—we don't try to represent the function by one that fits exactly to the data points, since this fits to the errors as well as to the trend of the information. Rather, we approximate with a smoother curve that we hope is closer to the truth than the data themselves. The problem, of course, is how much smoothing should be done. One can go too far and "smooth" beyond the point where only errors are eliminated.

A final situation should be mentioned. Some functions, or data from a series of tests, are inherently "rough." By this we mean that the function values change rapidly; a graph would show sharp local variations. When the derivative values of the function incur rapid changes, a sampling of the information may not reflect them. In this instance, the data indicate a smoother function than actually exists. Unless enough data are at hand to show the local variations, valid values of the derivatives just cannot be obtained. The only solution is more data, especially near the "rough" spots. And then we are beset by problems of accuracy of the data!

For convenience, we collect formulas for computing derivatives here.

Formulas for Computing Derivatives

Formulas for the first derivative:

$$f'(x_0) = \frac{f_1 - f_0}{h} + O(h).$$

$$f'(x_0) = \frac{f_1 - f_{-1}}{2h} + O(h^2). \qquad \text{(Central difference)}$$

$$f'(x_0) = \frac{-f_2 + 4f_1 - 3f_0}{2h} + O(h^2).$$

$$f'(x_0) = \frac{-f_2 + 8f_1 - 8f_{-1} + f_{-2}}{12h} + O(h^4). \qquad \text{(Central difference)}$$

Formulas for the second derivative:

$$f''(x_0) = \frac{f_2 - 2f_1 + f_0}{h^2} + O(h).$$

$$f''(x_0) = \frac{f_1 - 2f_0 + f_{-1}}{h^2} + O(h^2). \qquad \text{(Central difference)}$$

$$f''(x_0) = \frac{-f_3 + 4f_2 - 5f_1 + 2f_0}{h^2} + O(h^2).$$

$$f''(x_0) = \frac{-f_2 + 16f_1 - 30f_0 + 16f_{-1} - f_{-2}}{12h^2} + O(h^4). \qquad \text{(Central difference)}$$

Formulas for the third derivative:

$$f'''(x_0) = \frac{f_3 - 3f_2 + 3f_1 - f_0}{h^3} + O(h).$$

$$f'''(x_0) = \frac{f_2 - 2f_1 + 2f_{-1} - f_{-2}}{2h^3} + O(h^2). \qquad \text{(Averaged differences)}$$

Formulas for the fourth derivative:

$$f^{iv}(x_0) = \frac{f_4 - 4f_3 + 6f_2 - 4f_1 + f_0}{h^4} + O(h).$$

$$f^{iv}(x_0) = \frac{f_2 - 4f_1 + 6f_0 - 4f_{-1} + f_{-2}}{h^4} + O(h^2). \qquad \text{(Central difference)}$$

Perhaps an easier and more intuitive approach to deriving the above formulas is through a Taylor series expansion about the point x_0. Suppose we have the points $(x_{-2}, f_{-2}), \ldots, (x_2, f_2)$, where $f_i = f(x_i)$ and the points are evenly spaced, that is, $x_i = x_0 + i * h$, $i = -2, \ldots, 2$. If we consider the series expansion about x_0 for h, we get

$$f(x_1) = f_1 = f_0 + hf_0' + \frac{h^2}{2}f_0'' + \frac{h^3}{6}f_0''' + \frac{h^4}{24}f_0^{iv} + \cdots; \qquad (4.28)$$

$$f(x_{-1}) = f_{-1} = f_0 - hf_0' + \frac{h^2}{2}f_0'' - \frac{h^3}{6}f_0''' + \frac{h^4}{24}f_0^{iv} - \cdots. \qquad (4.29)$$

Subtracting these we find that

$$f_0' = \frac{f_1 - f_{-1}}{2h} + O(h^2);$$

adding the same we get

$$f_0'' = \frac{f_1 - 2f_0 + f_{-1}}{h^2} + O(h^2).$$

If we take a Taylor series expansion for $x = \pm 2h$, we obtain

$$f_2 = f_0 + 2hf_0' + 2h^2f_0'' + \frac{4}{3}h^3f_0''' + \frac{2}{3}h^4f_0^{iv} + \cdots; \qquad (4.30)$$

$$f_{-2} = f_0 - 2hf_0' + 2h^2f_0'' - \frac{4}{3}h^3f_0''' + \frac{2}{3}h^4f_0^{iv} - \cdots. \qquad (4.31)$$

If we combine these new ones with the previous, it is not hard to see that

$$f_0' = \frac{-f_2 + 8f_1 - 8f_{-1} + f_{-2}}{12h} + O(h^4),$$

$$f_0'' = \frac{-f_2 + 16f_1 - 30f_0 + 16f_{-1} - f_{-2}}{12h^2} + O(h^4).$$

Using this same approach, we can generate our own formulas for special needs. For instance, suppose our data points are unevenly spaced as in the following table:

x	$f(x)$
0.90	2.4596
1.00	2.7183
1.11	3.0344

The expansion about x_0 would be

$$f(x_0 + t) = f_0 + tf_0' + \frac{t^2}{2}f_0'' + \frac{t^3}{6}f_0''' + \cdots ; \tag{4.32}$$

$$f(x_0 - s) = f_0 - sf_0' + \frac{s^2}{2}f_0'' - \frac{s^3}{6}f_0''' + \cdots . \tag{4.33}$$

From these we can derive the equation

$$f_0' = \frac{\left(\frac{1}{t^2}f(x_0 + t) - \frac{1}{s^2}f(x_0 - s)\right)}{\left(\frac{1}{t} + \frac{1}{s}\right)} - \left(\frac{1}{t} - \frac{1}{s}\right)f_0. \tag{4.34}$$

Using the values in the table with $t = 0.11$ and $s = 0.10$, we get from Eq. (4.34)

$$f_0' = \frac{\left(\frac{1}{0.0121}(3.0344) - \frac{1}{0.01}(2.4596)\right)}{\left(\frac{1}{0.11} + \frac{1}{0.10}\right)} - \left(\frac{1}{0.11} - \frac{1}{0.10}\right)(2.7183)$$

$$= 2.7235.$$

Since these values are based on $f(x) = e^x$, the exact answer is 2.7183 at $x = 1.0$.

4.7 NEWTON–COTES INTEGRATION FORMULAS

The usual strategy in developing formulas for numerical integration is similar to that for numerical differentiation. We pass a polynomial through points of the function, and then integrate this polynomial approximation to the function. This permits us to integrate a function known only as a table of values. When the values are equispaced, our familiar Newton–Gregory forward polynomial is a convenient starting point, so

$$\int_a^b f(x)\, dx \doteq \int_a^b P_n(x_s)\, dx. \tag{4.35}$$

The formula we get from Eq. (4.35) will not be exact because the polynomial is not identical with $f(x)$. We get an expression for the error by integrating the error term of $P_n(x_s)$:

$$\text{Error} = \int_a^b \binom{s}{n+1} h^{n+1} f^{(n+1)}(\xi)\, dx.$$

There are various ways that we can employ Eq. (4.35). The interval of integration (a, b) can match the range of fit of the polynomial, (x_0, x_n). In this case, we get the Newton–Cotes formulas; these are a set of integration rules corresponding to the varying degrees of the interpolating polynomial. The first three, with the degree of the polynomial 1, 2, or 3, are particularly important, and we discuss them at length in the following sections.

If the degree of the polynomial is too high, errors due to round-off and local irregularities can cause a problem. This explains why it is only the lower-degree Newton–Cotes formulas that are often used.

The range of the polynomial and the interval of integration do not have to be the same. If the interval of integration extends outside the range of fit, however, we are extrapolating, and this incurs larger errors. If we only desire to get the integral of a function whose values are known, we will normally avoid extrapolation. As we will see in the next chapter, however, integrating the polynomial outside its range of fit leads to some important methods for solving differential equations. Using an interval of integration that is a subset of the points at which the polynomial agrees with the function also has special application in solving differential equations numerically.

The utility of numerical integration extends beyond the need to integrate a function known only as a table of values. Most computer programs for integration of functions whose form is known use these numerical techniques rather than the analytical methods of the calculus. While it is possible to write a program to do symbol manipulation and hence to perform analytical integration for simple forms of elementary functions, these routines lack generality. Furthermore, in a large number of cases no closed form for the integral exists. Numerical integration applies regardless of the complexity of the integrand or the existence of a closed form for the integral.

Let us now develop our three important Newton–Cotes formulas. During the integration, we will need to change the variable of integration from x to s since our polynomials are expressed in terms of s. Observe that $dx = h\, ds$.

For $n = 1$,

$$\int_{x_0}^{x_1} f(x)\, dx \doteq \int_{x_0}^{x_1} (f_0 + s\Delta f_0)\, dx = h \int_{s=0}^{s=1} (f_0 + s\Delta f_0)\, ds$$

$$= hf_0 s \Big]_0^1 + h\Delta f_0 \frac{s^2}{2} \Big]_0^1 = h\left(f_0 + \frac{1}{2}\Delta f_0\right)$$

$$= \frac{h}{2}[(2f_0 + (f_1 - f_0)] = \frac{h}{2}(f_0 + f_1). \tag{4.36}$$

$$\text{Error} = \int_{x_0}^{x_1} \frac{s(s-1)}{2} h^2 f''(\xi) \, dx = h^3 f''(\xi_1) \int_0^1 \frac{s^2 - s}{2} \, ds$$

$$= h^3 f''(\xi_1) \left(\frac{s^3}{6} - \frac{s^2}{4} \right) \Big]_0^1 = -\frac{1}{12} h^3 f''(\xi_1), \qquad x_0 \le \xi_1 \le x_1. \qquad (4.37)$$

In getting the error term of Eq. (4.37), we have used the mean-value theorem for integrals, which is that

$$\int_a^b f(x) g(x) \, dx = f(\xi) \int_a^b g(x) \, dx, \qquad a \le \xi \le b,$$

provided that $g(x)$ does not change sign in the interval (a, b). This applies to Eq. (4.37) because $s(s-1)$ is always less than or equal to zero on $(0, 1)$. The value of ξ_1 is not necessarily the same as ξ, though both fall in the interval (x_0, x_1).

For $n = 2$,

$$\int_{x_0}^{x_2} f(x) \, dx \doteq \int_{x_0}^{x_2} \left(f_0 + s \Delta f_0 + \frac{s(s-1)}{2} \Delta^2 f_0 \right) dx$$

$$= h \int_0^2 \left(f_0 + s \Delta f_0 + \frac{s(s-1)}{2} \Delta^2 f_0 \right) ds$$

$$= h f_0 s \Big]_0^2 + h \Delta f_0 \frac{s^2}{2} \Big]_0^2 + h \Delta^2 f_0 \left(\frac{s^3}{6} - \frac{s^2}{4} \right) \Big]_0^2$$

$$= h \left(2 f_0 + 2 \Delta f_0 + \frac{1}{3} \Delta^2 f_0 \right) = \frac{h}{3} (f_0 + 4 f_1 + f_2). \qquad (4.38)$$

When we integrate the next term to get the error, we find

$$\int_{x_0}^{x_2} \frac{s(s-1)(s-2)}{6} \Delta^3 f_0 \, dx = 0,$$

so, to get the error, we must integrate the succeeding term:

$$\text{Error} = \int_{x_0}^{x_2} \frac{s(s-1)(s-2)(s-3)}{24} h^4 f^{iv}(\xi) \, dx$$

$$= -\frac{1}{90} h^5 f^{iv}(\xi_1), \qquad x_0 \le \xi_1 \le x_2. \qquad (4.39)$$

We omit the details of the integration in Eq. (4.39).

Similarly, for $n = 3$, we find

$$\int_{x_0}^{x_3} f(x) \, dx = \int_{x_0}^{x_3} P_3(x_s) \, dx = \frac{3h}{8} (f_0 + 3 f_1 + 3 f_2 + f_3). \qquad (4.40)$$

$$\text{Error} = -\frac{3}{80} h^5 f^{iv}(\xi_1), \qquad x_0 \le \xi_1 \le x_3. \qquad (4.41)$$

In summary, the basic Newton–Cotes formulas are

$$\int_{x_0}^{x_1} f(x)\, dx = \frac{h}{2}(f_0 + f_1) - \frac{1}{12}h^3 f''(\xi)$$

$$\int_{x_0}^{x_2} f(x)\, dx = \frac{h}{3}(f_0 + 4f_1 + f_2) - \frac{1}{90}h^5 f^{\text{iv}}(\xi)$$

$$\int_{x_0}^{x_3} f(x)\, dx = \frac{3h}{8}(f_0 + 3f_1 + 3f_2 + f_3) - \frac{3}{80}h^5 f^{\text{iv}}(\xi)$$

An important item to observe is that the error terms for both $n = 2$ and $n = 3$ are $O(h^5)$. This means that the error of integration using a quadratic is similar to the integral using a cubic; it is a consequence of finding the integral of the next term equal to zero in deriving the error term of Eq. (4.39). Note also that the coefficient in Eq. (4.39)— $\left(-\frac{1}{90}\right)$—is smaller than that in Eq. (4.41)—$\left(-\frac{3}{80}\right)$. The formula based on a quadratic is unexpectedly accurate.

This phenomenon is true of all the even-order Newton–Cotes formulas; each has an order of h in its error term the same as for the formula of next higher order. This suggests that the even-order rules are especially useful.

4.8   THE TRAPEZOIDAL RULE

The first of the Newton–Cotes formulas, based on approximating $f(x)$ on (x_0, x_1) by a straight line, is also called the *trapezoidal rule*. We have derived it by integrating $P_1(x_s)$, but the familiar and simple trapezoidal rule can also be considered to be an adaptation of the definition of the definite integral as a sum. To evaluate $\int_a^b f(x)\, dx$, we subdivide the interval from a to b into n subintervals, as in Fig. 4.5. The area under the curve in each subinterval is approximated by the trapezoid formed by replacing the curve by its secant line drawn between the endpoints of the curve. The integral is then approximated by the sum of all the trapezoidal areas. There is no necessity to make the subintervals equal in width, but our formula is simpler if this is done. Let h be the constant Δx. Since the area of a trapezoid is its average height times the base, for each subinterval,

$$\int_{x_i}^{x_{i+1}} f(x)\, dx \doteq \frac{f(x_i) + f(x_{i+1})}{2}(\Delta x) = \frac{h}{2}(f_i + f_{i+1}), \qquad (4.42)$$

and for $[a, b]$ subdivided into subintervals of size h,

$$\int_a^b f(x)\, dx \doteq \sum_{i=1}^{n} \frac{h}{2}(f_i + f_{i+1}) = \frac{h}{2}(f_1 + f_2 + f_2 + f_3 + \cdots + f_n + f_{n+1});$$

$$\int_a^b f(x)\, dx = \frac{h}{2}(f_1 + 2f_2 + 2f_3 + \cdots + 2f_n + f_{n+1}). \qquad (4.43)$$

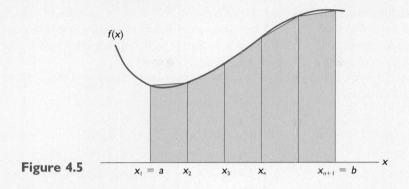

Figure 4.5

Equation (4.42) is identical to Eq. (4.36). Equation (4.43) is called the *extended trapezoidal rule*: it lets us apply the formula over an extended region where $f(x)$ is far from linear by applying the procedure to subintervals in which it can be approximated by linear segments. The formula is beautifully simple, and its applicability to unequally spaced values is useful in finding the integral of an experimentally determined function. It is obvious from Fig. 4.5 that the method is subject to large errors unless the subintervals are small, for replacing a curve by a straight line is hardly accurate.

EXAMPLE Suppose we wished to integrate the function tabulated in Table 4.3 over the interval from $x = 1.8$ to $x = 3.4$. The extended trapezoidal rule gives

$$\int_{1.8}^{3.4} f(x)\,dx \doteq \frac{0.2}{6}[6.050 + 2(7.389) + 2(9.025) + 2(11.023) + 2(13.464)$$

$$+ 2(16.445) + 2(20.086) + 2(24.533) + 29.964] = 23.9944.$$

The data in Table 4.3 are for $f(x) = e^x$, so the true value of the integral is $e^{3.4} - e^{1.8} = 23.9144$. We are off in the second decimal.

We have previously derived the error of the trapezoidal rule as Eq. (4.37). We repeat it here:

$$\begin{array}{c}\text{Local error of}\\ \text{trapezoidal rule}\end{array} = -\frac{1}{12}h^3 f''(\xi_1), \qquad x_0 \leq \xi_1 \leq x_1.$$

Table 4.3

x	$f(x)$	x	$f(x)$
1.6	4.953	2.8	16.445
1.8	6.050	3.0	20.086
2.0	7.389	3.2	24.533
2.2	9.025	3.4	29.964
2.4	11.023	3.6	36.598
2.6	13.464	3.8	44.701

This error, it should be emphasized, is the error of only a single step, and is hence called the *local error*. We normally apply the trapezoidal formula to a series of subintervals to get the integral over a large interval from $x = a$ to $x = b$. We are interested in the total error, which is called the *global error*.

To develop the formula for global error of the trapezoidal rule, we note that it is the sum of the local errors:

$$\text{Global error} = -\frac{1}{12}h^3[f''(\xi_1) + f''(\xi_2) + \cdots + f''(\xi_n)]. \tag{4.44}$$

In Eq. (4.44) each of the values of ξ_i is found in the n successive subintervals. If we assume that $f''(x)$ is continuous on (a, b), there is some value of x in (a, b), say $x = \xi$, at which the value of the sum in Eq. (4.44) is equal to $n \cdot f''(\xi)$. Since $nh = b - a$, the global error becomes

$$\begin{array}{l}\text{Global error of} \\ \text{trapezoidal rule}\end{array} = -\frac{1}{12}h^3 n f''(\xi) = \frac{-(b - a)}{12}h^2 f''(\xi) = O(h^2). \tag{4.45}$$

The fact that the global error is $O(h^2)$ while the local error is $O(h^3)$ is reasonable since, for example, if h is halved the number of subintervals is doubled, so we add together twice as many errors.

When the function $f(x)$ is known, Eq. (4.45) permits us to estimate the error of numerical integration by the trapezoidal rule. In applying this equation we bracket the error by calculating with the maximum and the minimum values of $f''(x)$ on the interval $[a, b]$.

For the example above, our error expression gives these estimates:

$$\text{Error} = -\frac{1}{12}h^3 n f''(\xi), \qquad 1.8 \le \xi \le 3.4,$$

$$= -\frac{1}{12}(0.2)^3(8)\begin{Bmatrix} e^{1.8} & (\text{min}) \\ e^{3.4} & (\text{max}) \end{Bmatrix} = \begin{Bmatrix} -0.0323 & (\text{min}) \\ -0.1598 & (\text{max}) \end{Bmatrix}.$$

Alternatively,

$$\text{Error} = -\frac{1}{12}(0.2)^2(3.4 - 1.8)\begin{Bmatrix} e^{1.8} & (\text{min}) \\ e^{3.4} & (\text{max}) \end{Bmatrix} = \begin{Bmatrix} -0.0323 \\ -0.1598 \end{Bmatrix}.$$

The actual error was -0.080. ∎

If we did not know the form of the function for which we have tabulated values, we would estimate $h^2 f''(\xi)$ from the second differences.

4.9 ROMBERG INTEGRATION

There is a way to improve on the accuracy of the simple trapezoidal rule when the function is known at equispaced intervals, or when it is explicitly known so that it can be computed as desired. This method is known as *Romberg integration*, or as *extrapolation to the limit*. It is based on the same principle as that used in Section 4.5 for the extrapolation of derivative values. We illustrate the method by a reconsideration of the data in Table 4.3. We saw that the trapezoidal rule gave a value of 23.9944 for $\int_{1.8}^{3.4} f(x) \, dx$, which differs from the true value of 23.9144. Purely for illustrative purposes, we do not desire to take advantage of our knowledge of the function, so we do not make h smaller to improve our estimate of the integral.

Even if we should not make the interval smaller, we can certainly make it larger, by using only every other value, taking $h = 0.4$. If we do this, we get:

$$I(h = 0.4) \doteq \frac{0.4}{2}[6.050 + 2(9.025) + 2(13.464) + 2(20.086) + 29.964]$$

$$= 24.2328.$$

This value is even more in error than before, which, of course, we expect with h larger; but we can take advantage of this to nearly eliminate the error of our former computation if we assume that errors of $O(h^2)$ can be interpreted as proportional to h^2, say error $= Ch^2$. Taking C to be a constant applies strictly only in the limit, of course, and at $h = 0.2$ and $h = 0.4$ we are not very near zero. It can be shown,* however, that we make an error of only $O(h^4)$ by assuming that the errors are proportional to h^2. We then have, for the two computations above,

$$\text{True value} = \text{computed value} + \text{error},$$

$$\text{True value} = 23.9944 + C(0.2)^2,$$

$$\text{True value} = 24.2328 + C(0.4)^2. \qquad \textbf{(4.46)}$$

In Eq. (4.46) there are two unknowns, the "true value" and the proportionality constant C, but we can certainly solve for these with two equations. We get, subtracting the second equation from four times the first,

$$\text{True value} \doteq 23.9944 + \frac{1}{3}(23.9944 - 24.2328)$$

$$\doteq 23.9149 \quad \text{(versus 23.9144, exact value)}.$$

We have extrapolated from two inexact values to an improved one. (What we have called "true value" is not exact because it used the erroneous assumption that $O(h^2)$ was the same as Ch^2.)

---

*See, for example, Davis and Rabinowitz (1967), who show that the error can be expressed as $C_1h^2 + C_2h^4 + C_3h^6 + \cdots$.

We can make a further improvement by extrapolation again, if we have two such estimates of $O(h^4)$ accuracy, by assuming that their errors can be stated as Ch^4. Such a second estimate can be obtained from the data in Table 4.3 by backing off once more, and computing the area with $h = 0.8$. If we do so we get $I(h = 0.8) = 25.1768$. We use this with $I(h = 0.4) = 24.2328$, as shown below.

h	Trapezoidal rule estimate $O(h^2)$ errors	Extrapolated value $O(h^4)$ errors
0.2	23.9944	
		23.9149
0.4	24.2328	
		23.9181
0.8	25.1768	

As before, we set up relationships:

$$\text{True value} = 23.9149 + C(0.2)^4,$$
$$\text{True value} = 23.9181 + C(0.4)^4.$$

Solving, we get

$$\text{True value} = 23.9149 + \frac{1}{15}(23.9149 - 23.9181)$$
$$= 23.9147.$$

This is now as close as can be determined to the exact value, 23.9144. One further order of extrapolation could be made, for we could start with $h = 1.6$ and, by combining with $h = 0.8$ and $h = 0.4$, get another second-order extrapolation. These second-order extrapolations can be shown to be of $O(h^6)$ accuracy and can be combined to an estimate of $O(h^8)$. No further improvement results from this, however, due to the influence of round-off errors in the original data. With only three-decimal accuracy there, it is not surprising to find errors in the fourth decimal of the integrals.

The number of successive improvements that can be made, and hence the order of the error that can be attained, obviously depends on the number of functional values that are available, or that are computed when $f(x)$ is a known function. The number of successive improvements that *should* be made may be less than the number that can, in principle, be made, because of the effects of round-off error.

We summarize our extrapolation rule, to be used for improving two values for which h varies $2:1$ (it is identical to Eq. (4.27), which was derived for derivatives):

$$\text{Improved value} = \text{more accurate} + \left(\frac{1}{2^n - 1}\right)(\text{more accurate} - \text{less accurate}). \quad \textbf{(4.47)}$$

The exponent n in the coefficient is the exponent on h in the error term $O(h^n)$ that applies to the two values used for extrapolation. One always knows that the more accurate value is the one computed with smaller h.

Actually, one hardly ever improves the estimate of an integral by comparing results from a small h to those with a larger h; instead the reverse is done. The previous discussion is primarily motivational and to give a basis for the following.

When this technique of extrapolation is applied to integrals of known functions so that values can be computed for smaller values of h, it is known as Romberg integration. It is quite popular in computer programs,* often being available as a stock subroutine in the computer library. The technique is to compute the integral for an arbitrarily chosen h, and then to compute it again with h halved. If the two values for the integral differ by more than some tolerance value, the value is improved by extrapolation, and a second integration is computed with h halved again. This is combined with the previous value of the integral to give a second extrapolated value that is compared to the first. If necessary, these two are combined to give a higher-order extrapolation. The method is continued until a pair of extrapolated values of the integral agree satisfactorily.

The Romberg method is applicable to a wide class of functions—to any Reimann-integrable function, in fact. Smoothness and continuity are not required.

4.10   SIMPSON'S RULES

The Newton–Cotes formulas based on quadratic and cubic polynomials are also widely used. In their composite forms they are known as *Simpson's rules*. There are two of these.

The second-degree Newton–Cotes formula integrates a quadratic over $2\Delta x$ intervals that are of uniform width. (Such intervals are often called *panels*.) We repeat the following equation from Section 4.7:

$$\int_{x_0}^{x_2} f(x)\, dx = \frac{h}{3}(f_0 + 4f_1 + f_2) - \frac{h^5}{90} f^{iv}(\xi), \qquad x_0 \le \xi \le x_2. \tag{4.48}$$

This is the very popular Simpson's $\frac{1}{3}$ rule, which has a local error of $O(h^5)$. If we apply this to a succession of pairs of panels to evaluate $\int_a^b f(x)\, dx$, we get

$$\int_a^b f(x)\, dx = \frac{h}{3}(f_1 + 4f_2 + 2f_3 + 4f_4 + 2f_5 + \cdots + 2f_{n-1} + 4f_n + f_{n+1})$$
$$- \frac{(b - a)}{180} h^4 f^{iv}(\xi), \qquad x_1 \le \xi \le x_{n+1}. \tag{4.49}$$

---

*A subroutine that performs Romberg integration is given at the end of this chapter.

In Eq. (4.49) we have subscripted the x-values so that $x_1 = a$ and $x_{n+1} = b$. In getting the global error we have used the fact that $\Sigma f^{iv}(\xi_i) = nf^{iv}(\xi)$, which imposes a continuity condition on the fourth derivative. Here n, the number of times that the local rule was used, is only one-half the number of Δx steps, because we integrate over two panels each time; this causes the denominator to change to 180 from 90. Note that Simpson's $\frac{1}{3}$ rule has a global error of $O(h^4)$. Its use requires that we subdivide into an even number of panels.

EXAMPLE   Apply Simpson's $\frac{1}{3}$ rule to the data of Table 4.3 to evaluate $\int_{1.8}^{3.4} f(x)\ dx$.
Using $h = 0.2$, we get

$$\int_{1.8}^{3.4} f(x)\ dx \doteq \frac{0.2}{3}[6.050 + 4(7.389) + 2(9.025) + 4(11.023)$$
$$+ 2(13.464) + 4(16.445) + 2(20.086)$$
$$+ 4(24.533) + 29.964]$$
$$= 23.9149.$$

It is no coincidence that this value is the same as the extrapolated value from the trapezoidal-rule estimates at $h = 0.2$ and $h = 0.4$. The student should demonstrate that these two procedures always give identical results.

$$\text{Error} = \frac{-(3.4 - 1.8)}{180}(0.2)^4 \begin{Bmatrix} e^{1.8} & (\min) \\ e^{3.4} & (\max) \end{Bmatrix} = \begin{Bmatrix} -8.6 \times 10^{-5} \\ -4.3 \times 10^{-4} \end{Bmatrix}.$$

The actual error is -5×10^{-4}; round-off has magnified it.   ∎

An extrapolation can also be made for this example. Using $h = 0.4$, we get

$$\int_{1.8}^{3.4} f(x)\ dx \doteq \frac{0.2}{3}[6.050 + 4(9.025) + 2(13.464) + 4(20.086) + 29.964]$$
$$= 23.9181.$$

Similarly, with $h = 0.8$, we get 23.9653. Our results can be summarized in a table as we did in the previous section:

h	Simpson's rule estimate $O(h^4)$ errors	Extrapolated value $O(h^6)$ errors
0.2	23.9149	
		23.9147
0.4	23.9181	
		23.9166
0.8	23.9653	

The values in the last column are obtained by the extrapolations:

$$\text{True value} \doteq 23.9149 + \frac{1}{15}(23.9149 - 23.9181) = 23.9147,$$

$$\text{True value} \doteq 23.9181 + \frac{1}{15}(23.9181 - 23.9653) = 23.9166.$$

A further extrapolation of these last two values gives

$$\text{True value} \doteq 23.9147 + \frac{1}{63}(23.9147 - 23.9166) = 23.9147.$$

No improvement is obtained due to round-off errors and the short precision used in Table 4.3. Had we used 10 digits of precision throughout, this last result would have been

$$\text{True value} \doteq 23.914456 + \frac{1}{63}(23.914456 - 23.914644)$$

$$= 23.914453.$$

$$\text{Exact value} = e^{3.4} - e^{1.8} \doteq 23.914453.$$

Since the error term for Simpson's rule can be written as $Ch^4 + Ch^6 + Ch^8 + \cdots$, we can do a Romberg-like set of successive extrapolations, as we have shown in this example.

The third Newton–Cotes formula that finds frequent use is obtained by integrating a cubic interpolating polynomial over its range of fit. We have already derived it as Eq. (4.40) and the error term as Eq. (4.41). The local rule is

$$\int_{x_0}^{x_3} f(x)\,dx = \frac{3h}{8}(f_0 + 3f_1 + 3f_2 + f_3) - \frac{3}{80}h^5 f^{\text{iv}}(\xi), \qquad x_0 \le \xi \le x_3.$$

The value of the coefficient $3h/8$ gives it its common name, *Simpson's $\frac{3}{8}$ rule*. When applied over the interval from $x_1 = a$ to $x_{n+1} = b$ (the interval must be divided into a multiple of three panels), we get

$$\int_a^b f(x)\,dx = \frac{3h}{8}(f_1 - 3f_2 + 3f_3 + 2f_4 + 3f_5 + 3f_6 + \cdots + 2f_{n-2} + 3f_{n-1}$$

$$+ 3f_n + f_{n+1}) - \frac{(b-a)}{80}h^4 f^{\text{iv}}(\xi), \qquad x_1 \le \xi \le x_{n+1}. \tag{4.50}$$

The global rule has a coefficient in the error term of $\frac{1}{80}$, reduced from $\frac{3}{80}$ in the local error, because we apply the local rule $n/3$ times when there are n panels.

As we have previously seen, the order of the global error, $O(h^4)$, is no higher than for Simpson's $\frac{1}{3}$ rule and the coefficient is larger. One might wonder whether the $\frac{3}{8}$ rule should ever be used.

As we have seen, Simpson's $\frac{1}{3}$ rule requires that the interval be subdivided into an even number of panels. When applied to tabulated data, this is often impossible. In such cases, one can combine the two rules, applying the $\frac{1}{3}$ rule as far as possible and completing the integration by picking up three panels with the $\frac{3}{8}$ rule. An alternative method for an odd number of panels is to integrate over one panel by the trapezoidal rule. This involves a penalty of an $O(h^2)$ error over that subinterval, however, and is somewhat less accurate.

EXAMPLE Use the data of Table 4.3 to find the integral from $x = 1.6$ to $x = 3.4$. There are nine panels, so we cannot apply the $\frac{1}{3}$ rule directly. We have several choices to handle this problem: apply the trapezoidal rule; apply Simpson's $\frac{3}{8}$ rule; or combine Simpson's $\frac{1}{3}$ rule with either one of the others. We investigate each choice.

With the trapeziodal rule:

$$\int_{1.6}^{3.4} f(x) \, dx = 25.0947, \qquad \text{error} = -0.0836.$$

With Simpson's $\frac{3}{8}$ rule:

$$\int_{1.6}^{3.4} f(x) \, dx = 25.0119, \qquad \text{error} = -0.0008.$$

With Simpson's $\frac{1}{3}$ rule from 1.8 to 3.4, and the trapezoidal rule from 1.6 to 1.8:

$$\int_{1.6}^{3.4} f(x) \, dx = 25.0152, \qquad \text{error} = -0.0041.$$

With Simpson's $\frac{1}{3}$ rule from 2.2 to 3.4, and the $\frac{3}{8}$ rule from 1.6 to 2.2:

$$\int_{1.6}^{3.4} f(x) \, dx = 25.0115, \qquad \text{error} = -0.0004. \quad \blacksquare$$

The results are as we would anticipate: The best accuracy is obtained when the $\frac{1}{3}$ and $\frac{3}{8}$ rules are combined. We still have a choice to make when we combine the rules: at which three panels should we use the $\frac{3}{8}$ rule? A consideration of the error term of the $\frac{3}{8}$ rule gives the answer: apply it where its error is least. This means that we should choose the three panels where the fourth derivative of the function is smallest. For the data of this example, we know the function $[f(x) = e^x]$ so the choice we made is best. The choice will not always be obvious when we know $f(x)$ only as the table of values. The differences will be minor, however, and round-off errors distort the picture, so this point is of more theoretical than practical importance.

For convenience, we collect here the Newton–Cotes integration formulas. In a later section we will consider adaptive integration methods using these formulas. In addition, we will consider ways to find the step size, h, so that accuracy is obtained without unduly increasing the number of function evaluations.

Formulas for integration   (Uniform spacing, $\Delta x = h$)

Trapezoidal Rule:

$$\int_a^b f(x)\,dx = \frac{h}{2}(f_1 + 2f_2 + 2f_3 + \cdots + 2f_n + f_{n+1})$$

$$-\frac{(b-a)}{12}h^2 f''(\xi), \qquad a \le \xi \le b.$$

Simpson's $\frac{1}{3}$ Rule:

$$\int_a^b f(x)\,dx = \frac{h}{3}(f_1 + 4f_2 + 2f_3 + 4f_4 + 2f_5 + \cdots + 4f_n + f_{n+1})$$

$$-\frac{(b-a)}{180}h^4 f^{iv}(\xi), \qquad a \le \xi \le b.$$

(Requires an even number of panels.)

Simpson's $\frac{3}{8}$ Rule:

$$\int_a^b f(x)\,dx = \frac{3h}{8}(f_1 + 3f_2 + 3f_3 + 2f_4 + 3f_5 + 3f_6 + \cdots + 3f_n + f_{n+1})$$

$$-\frac{(b-a)}{80}h^4 f^{iv}(\xi), \qquad a \le \xi \le b.$$

(Requires a number of panels divisible by 3.)

4.11   OTHER WAYS TO DERIVE INTEGRATION FORMULAS

It is interesting to examine other ways of deriving integration formulas than by integrating the interpolating polynomial. One way is to use symbolic methods, similar to those of Section 4.3 and Section 3.7.

In terms of the stepping operator E,

$$f(x_s) = f_s = E^s f_0.$$

Multiplying by $dx = h\,ds$ and integrating from x_0 to x_1 ($s = 0$ to $s = 1$), we obtain

$$\int_{x_2}^{x_1} f(x)\,dx = h \int_0^1 E^s f_0\,ds = \left[\frac{hE^s}{\ln E}f_0\right]_0^1 = \frac{h(E-1)}{\ln E}f_0.$$

Let $E = 1 + \Delta$ and expand $\ln(1 + \Delta)$ as a power series:

$$\ln(1 + \Delta) = \Delta - \frac{1}{2}\Delta^2 + \frac{1}{3}\Delta^3 - \frac{1}{4}\Delta^4 + \cdots.$$

On dividing Δ by this series, we get

$$\int_{x_0}^{x_1} f(x) \, dx = \frac{h\Delta}{\Delta - \frac{1}{2}\Delta^2 + \frac{1}{3}\Delta^3 - \frac{1}{4}\Delta^4 + \cdots} f_0$$

$$= h(f_0 + \frac{1}{2}\Delta f_0 - \frac{1}{12}\Delta^2 f_0 + \frac{1}{24}\Delta^3 f_0 - \cdots). \qquad (4.51)$$

The coefficients are considerably easier to get by this technique than by the term-by-term integration of Section 4.7.

Equation (4.51) is not a Newton–Cotes formula unless we use only the first two terms, making the interval of integration and the range of fit of the polynomial agree. When n terms are used, it represents a polynomial of degree n, fitting from x_0 to x_n, but integrated only from x_0 to x_1. This is especially useful in connection with differential-equation methods.

We can develop the formula for an nth-degree interpolating polynomial integrated over two panels, from x_0 to x_2, in a similar fashion:

$$\int_{x_0}^{x_2} f(x) \, dx = \left[\frac{hE^s}{\ln E} f_0\right]_0^2 = \frac{h(E^2 - 1)}{\ln E} f_0 = \frac{h(E + 1)(E - 1)}{\ln E} f_0.$$

Again letting $E = 1 + \Delta$ so $E - 1 = \Delta$ and $E + 1 = 2 + \Delta$, and dividing Δ by the series for $\ln (1 + \Delta)$, we get

$$\int_{x_0}^{x_2} f(x) \, dx = h(2 + \Delta)(f_0 + \frac{1}{2}\Delta f_0 - \frac{1}{12}\Delta^2 f_0 + \frac{1}{24}\Delta^3 f_0 - \cdots)$$

$$= h(2f_0 + \Delta f_0 - \frac{1}{6}\Delta^2 f_0 + \frac{1}{12}\Delta^3 f_0 - \cdots$$

$$+ \Delta f_0 + \frac{1}{2}\Delta^2 f_0 - \frac{1}{12}\Delta^3 f_0 + \cdots)$$

$$= h(2f_0 + 2\Delta f_0 + \frac{1}{3}\Delta^2 f_0 + 0 - \cdots). \qquad (4.52)$$

The method can obviously be extended. One reason why we might desire formulas such as Eqs. (4.51) and (4.52) is to use them to construct lozenge diagrams similar to Figs. 4.1 and 4.3. These would permit integration over m panels based on polynomials of degree n. The diagrams are easy to construct if we remember that the coefficients themselves form a difference table that is interlaced with the function differences. Since they are of rather special application, we do not exhibit them here.

We now present still another interesting method of deriving formulas that can be applied to a variety of situations including the development of integration formulas. It may be called the method of *undetermined coefficients*.* We express the formula as a sum of $n + 1$ terms with unknown coefficients, and then evaluate the coefficients by requiring that the formula be exact for all polynomials of degree n or less. We illustrate

---

*This method is described in more detail in the Appendix, and several examples are presented.

it here by finding Simpson's $\frac{1}{3}$ rule by this other technique. Express the integral as a weighted sum of three equispaced function values:

$$\int_{-1}^{1} f(x)\, dx = af(-1) + bf(0) + cf(+1). \tag{4.53}$$

The symmetrical interval of integration simplifies the arithmetic. We stipulate that the function is to be evaluated at three equally spaced intervals, the two end values and the midpoint. Since the formula contains three terms, we can require it to be correct for all polynomials of degree 2 or less. If that is true, it certainly must be true for the three special cases of $f(x) = x^2$, $f(x) = x$, and $f(x) = 1$. We rewrite Eq. (4.53) three times, applying each definition of $f(x)$ in turn:

$$f(x) = 1: \qquad \int_{-1}^{1} dx = 2 = a(1) + b(1) + c(1) = a + b + c;$$

$$f(x) = x: \qquad \int_{-1}^{1} x\, dx = 0 = a(-1) + b(0) + c(1) = -a + c;$$

$$f(x) = x^2: \qquad \int_{-1}^{1} x^2\, dx = \frac{2}{3} = a(1) + b(0) + c(1) = a + c.$$

Solving the three equations simultaneously gives $a = \frac{1}{3}$, $b = \frac{4}{3}$, $c = \frac{1}{3}$. Here the spacing between points was unity; obviously the integral is proportional to $\Delta x = h$. We then get Simpson's $\frac{1}{3}$ rule:

$$\int_{-h}^{h} f(x)\, dx = h\left[\frac{1}{3}f(-h) + \frac{4}{3}f(0) + \frac{1}{3}f(h)\right].$$

4.12 GAUSSIAN QUADRATURE

Our previous formulas for numerical integration were all predicated on evenly spaced x-values; this means the x-values were predetermined. With a formula of three terms, then, there were three parameters, the coefficients (weighting factors) applied to each of the functional values. A formula with three parameters corresponds to a polynomial of the second degree, one less than the number of parameters. Gauss observed that if we remove the requirement that the function be evaluated at predetermined x-values, a three-term formula will contain six parameters (the three x-values are now unknowns, plus the three weights) and should correspond to an interpolating polynomial of degree 5. Formulas based on this principle are called *Gaussian quadrature formulas*. They can be applied only when $f(x)$ is explicitly known, so that it can be evaluated at any desired value of x.

We will determine the parameters in the simple case of a two-term formula containing four unknown parameters:

$$\int_{-1}^{1} f(t) \doteq af(t_1) + bf(t_2).$$

The method is the same as that illustrated in the previous section, by determining unknown parameters. We use a symmetrical interval of integration to simplify the arithmetic, and call our variable t. (This agrees with the notation of most authors. Since the variable of integration is only a dummy variable, its name is unimportant.) Our formula is to be valid for any polynomial of degree 3; hence it will hold if $f(t) = t^3$, $f(t) = t^2$, $f(t) = t$, and $f(t) = 1$:

$$f(t) = t^3: \qquad \int_{-1}^{1} t^3 \, dt = 0 = at_1^3 + bt_2^3;$$

$$f(t) = t^2: \qquad \int_{-1}^{1} t^2 \, dt = \frac{2}{3} = at_1^2 + bt_2^2;$$

$$f(t) = t: \qquad \int_{-1}^{1} t \, dt = 0 = at_1 + bt_2;$$

$$f(t) = 1: \qquad \int_{-1}^{1} dt = 2 = a + b. \qquad (4.54)$$

Multiplying the third equation by t_1^2, and subtracting from the first, we have

$$0 = 0 + b\left[t_2^3 - t_2 t_1^2\right] = b(t_2)(t_2 - t_1)(t_2 + t_1). \qquad (4.55)$$

We can satisfy Eq. (4.55) by either $b = 0$, $t_2 = 0$, $t_1 = t_2$ or $t_2 = -t_1$. Only the last of these possibilities is satisfactory, the others being invalid, or else reduce our formula to only a single term, so we choose $t_1 = -t_2$. We then find that

$$a = b = 1,$$

$$t_2 = -t_1 = \sqrt{\frac{1}{3}} = 0.5773,$$

$$\int_{-1}^{1} f(t) \, dt \doteq f(-0.5773) + f(0.5773).$$

It is remarkable that adding these two values of the function gives the exact value for the integral of any cubic polynomial over the interval from -1 to 1.

Suppose our limits of integration are from a to b, and not -1 to 1 for which we derived this formula. To use the tabulated Gaussian quadrature parameters, we must change the interval of integration to $(-1, 1)$ by a change of variable. We replace the given variable by another to which it is linearly related according to the following scheme: If we let

$$x = \frac{(b - a)t + b + a}{2} \qquad \text{so that } dx = \left(\frac{b - a}{2}\right) dt,$$

then

$$\int_{a}^{b} f(x) \, dx = \frac{b - a}{2} \int_{-1}^{1} f\left(\frac{(b - a)t + b + a}{2}\right) dt.$$

E X A M P L E   Evaluate $I = \int_{0}^{\pi^2} \sin x \, dx$. (Obviously, $I = 1.0$, so we can readily see the error of our estimate.)

To use the two-term Gaussian formula, we must change the variable of integration to make the limits of integration from -1 to 1.

Let

$$x = \frac{(\pi/2)t + \pi/2}{2}, \quad \text{so } dx = (\pi/4)\, dt.$$

Observe that when $t = -1$, $x = 0$; when $t = 1$, $x = \pi/2$.) Then

$$I = \frac{\pi}{4} \int_{-1}^{1} \sin\left(\frac{\pi t + \pi}{4}\right) dt.$$

The Gaussian formula calculates the value of the new integral as a weighted sum of two values of the integrand, at $t = -0.5773$ and at $t = 0.5773$. Hence,

$$I = \frac{\pi}{4}[(1.0)(\sin(0.10566\pi) + (1.0)(\sin(0.39434\pi)]$$

$$= 0.99847.$$

The error is 1.53×10^{-3}. ∎

The power of the Gaussian method derives from the fact that we need only two functional evaluations. If we had used the trapezoidal rule, which also requires only two evaluations, our estimate would have been $(\pi/4)(0.0 + 1.0) = 0.7854$, an answer quite far from the mark. Simpson's $\frac{1}{3}$ rule requires three functional evaluations and gives $I = 1.0023$, with an error of -2.3×10^{-3}, somewhat greater than for Gaussian quadrature.

Gaussian quadrature can be extended beyond two terms. The formula is then given by

$$\int_{-1}^{1} f(t)\, dt \doteq \sum_{i=1}^{n} w_i f(t_i), \quad \text{for } n \text{ points.} \tag{4.56}$$

This formula is *exact* for functions $f(t)$ that are polynomials of degree $2n - 1$ or less! Moreover, by extending the method we used previously for the 2-point formula, for each n we obtain a system of $2n$ equations:

$$w_1 t_1^k + \cdots + w_n t_n^k = \begin{cases} 0, & \text{for } k = 1, 3, 5, \ldots, 2n - 1; \\ \dfrac{2}{k + 1}, & \text{for } k = 0, 2, 4, \ldots, 2n - 2. \end{cases}$$

This approach is obvious. However, this set of equations we get by writing $f(t)$ as a succession of polynomials is not easily solved. We wish to indicate an approach that is easier than the methods for a nonlinear system that we used in Chapter. 2.

It turns out that the t_i's for a given n are the roots of the nth-degree Legendre polynomial. The Legendre polynomials are defined by recursion:

$$(n + 1)L_{n+1}(x) - (2n + 1)xL_n(x) + nL_{n-1}(x) = 0,$$

$$\text{with } L_0(x) = 1, L_1(x) = x.$$

Then $L_2(x)$ is

$$L_2(x) = \frac{3xL_1(x) - (1)L_0(x)}{2} = \frac{3}{2}x^2 - \frac{1}{2};$$

here roots are $\pm\sqrt{\frac{1}{3}} = \pm 0.5773$, precisely the t-values for the two-term formula. By using the recursion relation, we find

$$L_3(x) = \frac{5x^3 - 3x}{2},$$

$$L_4(x) = \frac{35x^4 - 30x^2 + 3}{8}, \quad \text{and so on.}$$

The methods of Chapter 1 allow us to find the roots of these polynomials. After these have been determined, the set of equations analogous to Eqs. (4. 54) can easily be solved for the weighting factors because the equations are linear with respect to these unknowns.

Table 4.4 lists the zeros of Legendre polynomials up to degree 5, giving values that we need for Gaussian quadrature where the basic polynomial is up to degree 9. For example, $L_3(x)$ has zeros at $x = 0, +0.77459667$, and -0.77459667.

Before continuing with an example of the use of Gaussian quadrature, it is of interest to summarize the properties of Legendre polynomials.

1. The Legendre polynomials are *orthogonal* over the interval $[-1, 1]$. By this we mean that

$$\int_{-1}^{1} L_n(x) \, L_m(x) \, dx \begin{cases} = 0 \text{ if } n \neq m; \\ > 0 \text{ if } n = m. \end{cases}$$

This is a property of several other important functions, such as $\{\cos(nx), n = 0, 1, \ldots\}$. Here we have

$$\int_0^{2\pi} \cos(mx) \cos(nx) \, dx \begin{cases} = 0 \text{ if } n \neq m; \\ > 0 \text{ if } n = m. \end{cases}$$

In this case we say that these functions are orthogonal over the interval $[0, 2\pi]$.

2. Any polynomial of degree n can be written as a sum of the Legendre polynomials:

$$P_n(x) = \sum_{i=0}^{n} c_i L_i(x).$$

3. The n roots of $L_n(x) = 0$ lie in the interval $[-1, 1]$.

Using these properties, we are able to show that Eq. (4.56) is exact for polynomials of degree $2n - 1$ or less.

The weighting factors and t-values for Gaussian quadrature have been tabulated. (Love (1966) gives values for up to 200-term formulas.) We are content to give a few of the values in Table 4.4.

We illustrate the three-term formula with an example.

EXAMPLE Evaluate $I = \int_{0.2}^{1.5} e^{-x^2}\, dx$ using the three-term Gaussian formula.

$$x = \frac{(1.5 - 0.2)t + 1.5 + 0.2}{2} = 0.65t + 0.85.$$

Then

$$I = \frac{1.5 - 0.2}{2} \int_{-1}^{1} e^{-(0.65t+0.85)^2}\, dt$$

$$= 0.65[0.555 \ldots e^{-[0.65(-0.774\ldots)+0.85]^2} + 0.888 \ldots e^{-[0.65(0.0)+0.85]^2}$$

$$+ 0.555 \ldots e^{-[0.65(0.774\ldots)+0.85]^2}]$$

$$= 0.6586 \quad \text{(compare to correct value 0.65882).} \quad \blacksquare$$

Table 4.4 Values for Gaussian quadrature

Number of terms	Values of t	Weighting factor	Valid up to degree
2	−0.57735027	1.0	3
	0.57735027	1.0	
3	−0.77459667	0.55555555	5
	0.0	0.88888889	
	0.77459667	0.55555555	
4	−0.86113631	0.34785485	7
	−0.33998104	0.65214515	
	0.33998104	0.65214515	
	0.86113631	0.34785485	
5	−0.90617975	0.23692689	9
	−0.53846931	0.47862867	
	0.0	0.56888889	
	0.53846931	0.47862867	
	0.90617975	0.23692689	

4.13   IMPROPER INTEGRALS AND INDEFINITE INTEGRALS

Our previous examples have been for integrals with finite upper and lower limits and with bounded integrands. Sometimes we want to evaluate integrals such as

$$I_1 = \int_0^\infty x e^{-x} \, dx; \tag{4.57}$$

$$I_2 = \int_0^2 \frac{dx}{\sqrt{x}}; \tag{4.58}$$

$$I_3 = \int_0^t x\sqrt{x^2 + 1} \, dx. \tag{4.59}$$

Equation (4.57) has an infinite upper limit; Eq. (4.58) has an infinite value for the integrand at the lower limit, $x = 0$; Eq. (4.59) is an indefinite integral whose value is a function of t.

Sometimes we can transform such integrals into the simpler forms that we have discussed above, by a change of variable. For example, we can break I_1 into the sum of two integrals:

$$I_1 = \int_0^1 x e^{-x} \, dx + \int_1^\infty x e^{-x} \, dx,$$

and then change the variable in the second term by substituting $y = 1/x$, so that

$$\int_1^\infty x e^{-x} \, dx = \int_1^0 \frac{1}{y} e^{-1/y} \left(-\frac{dy}{y^2} \right) = \int_0^1 \frac{e^{-1/y}}{y^3} \, dy.$$

In the transformed integral, the value of the integrand at $y = 0$ is indeterminate $(0/0)$, but this causes us no trouble because the limit of the integrand is zero as $y \to 0$.

Where the integral exists, as in Eq. (4.57), we can apply our numerical methods directly, however. We just evaluate

$$\int_0^A x e^{-x} \, dx$$

for increasing values of A, and use the limiting value of these results as the value of the integral with infinite upper limit. Observe that this is just an adaptation of the usual calculus definition of the value of an integral with an infinite upper limit. One must be on guard in doing this when the convergence of the integral is unknown, for round-off error may give an upper limit to the sequence of results even though the integral actually diverges.

EXAMPLE   Table 4.5 gives results for the integral in Eq. (4.57). They show how the value of the integral reaches a limiting value as the value of A increases. (A computer program that implements Romberg integration was used.)

The value of the integral in Eq. (4.58) can be found similarly by finding the value of I_2 with a variable lower limit that approaches closer and closer to zero, the point where the integrand is undefined. Equation (4.59) is also amenable to this same approach; we let t take on a series of values and produce a table that relates the value of I_3 to values of t.

Table 4.5   Values for $I = \int_0^A xe^{-x}\,dx$, $A \to \infty$

A	I	
1	0.26424	
10	0.99950	
100	1.00001	
1000	1.00001	
10000	1.00001	
(∞)	(1.00000)	(analytical result)

Equation (4.59) is a simple special case of a situation we will treat at length in the next chapter, which deals with differential equations.

4.14 ADAPTIVE INTEGRATION

In Sections 4.8 and 4.10, we presented the trapezoidal rule and Simpson's rules for finding the integral of $f(x)$ over a fixed interval $[a, b]$ using a uniform value for Δx. When $f(x)$ is a known function, one can choose the value for $\Delta x = h$ arbitrarily. The problem is that we do not know what value to choose for h to attain a desired accuracy.

One way to select the value for h is that of Section 4.9. We can start with two panels, $h = h_1 = (b - a)/2$, and apply one of the formulas. Then we let $h_2 = h_1/2$ and apply the formula again, now with four panels, and compare the two results. If the new value is sufficiently close, we terminate and apply a Romberg extrapolation to further reduce the error. If the second result is not close enough to the first, we again halve h and repeat the procedure. We continue in the same way until the last result is close enough to its predecessor.

We illustrate this obvious procedure with an example:

EXAMPLE   Integrate $f(x) = 1/x^2$ over the interval $[0.2, 1]$. Use a tolerance value of 0.02 to terminate the halving of $h = \Delta x$. From calculus, we know that the exact answer is 4.0.

In order to compare this to an alternative approach, we introduce a notation that will be helpful throughout this section.

Define: $I(f) = \displaystyle\int_a^b f(x)\,dx$, the exact value.

$S_n(f) =$ the computed value using Simpson's $\frac{1}{3}$ rule with $\Delta x = h_n$.

If we use this notation, the extended Simpson's rule becomes

$$I(f) = S_n(f) - \frac{(b - a)}{180}h_n^4 f^{\mathrm{iv}}(\xi), \qquad a < \xi < b.$$

Using this with $h_1 = (1.0 - 0.2)/2 = 0.4$, we compute $S_1(f)$. We continue halving h, $h_{n+1} = h_n/2$, computing its corresponding $S_{n+1}(f)$ until $|S_{n+1}(f) - S_n(f)| < 0.02$, the tolerance value. The following table shows the results.

n	h_n	S_n	$\|S_{n+1}(f) - S_n(f)\|$
1	0.4	4.948148	
			0.761111
2	0.2	4.187037	
			0.162819
3	0.1	4.024281	
			0.022054
4	0.05	4.002164	
			0.002010
5	0.025	4.000154	

From the table we see that, at $n = 5$, we have met the tolerance criterion since $|S(f) - S(f)| < 0.02$. A Romberg extrapolation gives

$$RS(f) = S_5(f) + \frac{S_5(f) - S_4(f)}{15} = 4.00002.$$

(We use the notation $RS(f)$ to represent the Romberg extrapolation from Simpson's rule.)

∎

The disadvantage of this technique is that the value of h is the same over the entire interval of integration, while the behavior of $f(x)$ may not require this. Consider Fig. 4.6. It is obvious that, in the subinterval $[c, b]$, h can be much larger than in subinterval $[a, c]$. One could subdivide $[a, b]$ by personal intervention after examining the curve of $f(x)$. We prefer to avoid such intervention, of course.

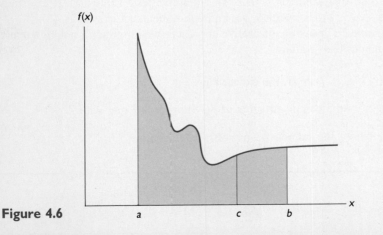

Figure 4.6

Adaptive integration automatically allows for different h's on different subintervals of $[a, b]$, choosing values adequate for a specified accuracy. In adaptive integration, we do not specify how the total interval $[a, b]$ is to be broken up; the point c could occur anywhere within it. Since the place to subdivide $[a, b]$ is not known a priori, we use something like a binary search to find it. Actually, the total interval $[a, b]$ may be broken into many subintervals. This depends on the tolerance value and the behavior of $f(x)$.

The strategy is to first divide $[a, b]$ in half, compare the two results, and if the tolerance criterion is exceeded, focus our attention only on the right half of $[a, b]$, setting aside the results for the left half for later use. We then divide the right half of $[a, b]$ in half and again compare results (now using a criterion half as great to compensate for the reduced interval). We repeat until the terminating condition is met. Then we focus attention on the portions of the interval of integration that were set aside, working from right to left.

We repeat the above example to illustrate the procedure, again using a tolerance value of 0.02.

Because we are working with different subintervals of $[a, b]$ we modify the above notation slightly:

$S_n(a_i, b_i)$ is the computed value from Simpson's rule for $h_n = (b - a)/2^n$, where $a \leq a_i < b_i \leq b$.

The example begins just as before. Take $h_1 = (1.0 - 0.2)/2 = 0.4$. Then

$$S_1(0.2, 1.0) = 4.94814815,$$
$$S_2(0.2, 1.0) = S_2(0.2, 0.6) + S_2(0.6, 1.0)$$
$$= 3.51851852 + 0.66851852 = 4.18703704,$$
$$h_2 = 0.2.$$

Since $|S_2(0.2, 1.0) - S_1(0.2, 1.0)| = 0.7611111 > 0.02$, we must continue. We set aside all the information we have obtained in evaluating $S_2(0.2, 0.6)$ for the moment. Focusing on $[0.6, 1.0]$, we compute

$$S_3(0.6, 1.0) = S_3(0.6, 0.8) + S_3(0.8, 1.0)$$
$$= 0.41678477 + 0.25002572 = 0.66681049,$$
$$h_3 = 0.1.$$

Since $|S_3(0.6, 1.0) - S_2(0.6, 1.0)| = 0.00171803 < 0.02/2 = 0.01$, we meet the criterion. We use a Romberg extrapolation to get

$$RS(0.6, 1.0) = 0.66681049 + \frac{0.66681049 - 0.66851852}{15}$$
$$= 0.66669662.$$

We are finished with the subinterval [0.6, 1.0] and turn our attention to the interval [0.2, 0.6].

$$S_3(0.2, 0.6) = S_3(0.2, 0.4) + S_3(0.4, 0.6)$$
$$= 2.52314815 + 0.83425926 = 3.35740741,$$
$$h_3 = 0.1.$$

Now $|S_3(0.2, 0.6) - S_2(0.2, 0.6)| = 0.16111 > 0.02/2$. We must continue with $h_4 = 0.05$ and restrict ourselves to [0.4, 0.6]. We find that

$$S_4(0.4, 0.6) = S_4(0.4, 0.5) + S_4(0.5, 0.6)$$
$$= 0.50005144 + 0.33334864 = 0.83340008,$$
$$h_4 = 0.05.$$

Since $|S_4(0.4, 0.6) - S_3(0.4, 0.6)| = 0.00085918 < 0.02/4$, we use a Romberg extrapolation to get

$$RS(0.4, 0.6) = 0.83334280.$$

We move on to the last set-aside interval, [0.2, 0.4]. We just summarize the results:

$$RS(0.3, 0.4) = 0.83334924 + \frac{0.83334924 - 0.83356954}{15}$$
$$= 0.83333455, \text{ and}$$
$$RS(0.2, 0.3) = 1.66680016 + \frac{1.66680016 - 1.66851852}{15}$$
$$= 1.66668560.$$

For the last two operations, we met a local tolerance of 0.02/8. Summing all the RS values, we get

$$RS(0.2, 1.0) = RS(0.2, 0.3) + RS(0.3, 0.4) + RS(0.4, 0.6) + RS(0.6, 1.0)$$
$$= 4.00005957.$$

The result should be compared to $I(0.2, 1.0) = 4.0$, the exact value. Using adaptive integration instead of the first simplistic procedure, we reduced the number of function evaluations from 33 to 17.

This diagram shows how the total interval was subdivided, with varying step sizes.

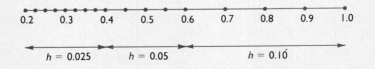

Each point on the line represents a function evaluation using the adaptive Simpson integration. For more complicated functions, the adaptive method can be extremely efficient in reducing the total number of function evaluations. However, one must store the set-aside information for later use. Before we explain how this can be done in a computer, let us review the information we had to set aside in this example. There were three times when we had to set aside several values—$S_2(0.2, 0.6)$, $S_3(0.2, 0.4)$, and $S_4(0.2, 0.3)$. We surely don't want to recompute function values, so there are seven quantities to remember for each; we use a seven-column array for these:

Left point	$f(x)$ at left	$f(x)$ at middle	$f(x)$ at right	Step size	Local tolerance	$S(a, c)$
0.2	25.0	6.25	2.7778	0.2	0.01	3.51851852
0.2	25.0	11.1111	6.25	0.1	0.005	2.52314815
0.2	25.0	16.0	11.1111	0.05	0.0025	1.66851852

The first row was stored from $S_2(0.2, 0.6)$; after computing $RS(0.6, 1.0)$ it was retrieved. The second row is from $S_3(0.2, 0.4)$; it was retrieved after computing $RS(0.4, 0.6)$. The third row is from $S_4(0.2, 0.3)$ and was retrieved after computing $RS(0.3, 0.4)$.

For a more complicated function we would have more rows but still seven columns. Observe that we always retrieve the last row stored and store the next row at the bottom of the array.

The following algorithm implements the adaptive integration presented.

Algorithm for Computing $I(f) = \int_a^b f(x)\, dx$

    Set: value = 0.0,

    Evaluate: $h_1 = (b - a)/2$, $c = a + h_1$, $Fa = f(a)$,
              $Fc = f(c)$, $Fb = f(b)$, $Sab = S_1(a, b)$.

    STORE(a, Fa, Fc, Fb, h_1, tol, Sab).

REPEAT

    RETRIEVE(a, Fa, Fc, Fb, h_1, tol, Sab)

    Evaluate: $h_2 = h_1/2$, $d = a + h_2$, $e = a + 3h_2$, $Fd = f(d)$,
        $Fe = f(e)$,
        $Sac = S_2(a, c)$, $Scb = S_2(c, b)$, $S_2(a, b) = Sac + Scb$.

    IF $|S_2(a, b) - S_1(a, b)| <$ tol THEN

        Compute $RS(a, b)$

        value = value + $RS(a, b)$

    ELSE

        $h_1 = h_2$, tol = tol$/2$
        STORE(a, Fa, Fd, Fc, h_1, tol, Sac)
        STORE(c, Fc, Fe, Fb, h_1, tol, Scb)

UNTIL TABLE is EMPTY.

$I(f) =$ value.

(Note that in the FORTRAN implementation of this algorithm we will use the well-known structure from computer science known as a *stack*. Our table of seven columns is just an $n \times 7$ two-dimensional stack. To store a set of values, we do an operation called PUSH; to retrieve a set, we do a POP.)

4.15   APPLICATIONS OF CUBIC SPLINE FUNCTIONS

In addition to their obvious use for interpolation, splines (Chapter 3) can be used for finding derivatives and integrals of functions, even when the function is known only as a table of values. The smoothness of splines, because of the requirement that each portion have the same first and second derivatives as its neighbor, can give improved accuracy in some cases.

For the cubic spline that approximates $f(x)$, we can write, for the interval $x_i \leq x \leq x_{i+1}$,

$$f(x) = a_i(x - x_i)^3 + b_i(x - x_i)^2 + c_i(x - x_i) + d_i,$$

where the coefficients are determined as in Section 3.9. The method outlined in that section computes S_i and S_{i+1}, the values of the second derivative at each end of the subinterval. From these S-values and the values of $f(x)$, we compute the coefficients of the cubic:

$$a_i = \frac{S_{i+1} - S_i}{6(x_{i+1} - x_i)},$$

$$b_i = \frac{S_i}{2},$$

$$c_i = \frac{f(x_{i+1}) - f(x_i)}{x_{i+1} - x_i} - \frac{2(x_{i+1} - x_i)S_i + (x_{i+1} - x_i)S_{i+1}}{6},$$

$$d_i = f(x_i).$$

Approximating the first and second derivatives is straightforward; we estimate these as the values of the derivatives of the cubic:

$$f'(x) \doteq 3a_i(x - x_i)^2 + 2b_i(x - x_i) + c_i,$$
$$f''(x) \doteq 6a_i(x - x_i) + 2b_i.$$

At the $n + 1$ points x_i where the function is known and the spline matches $f(x)$, these formulas are particularly simple:

$$f'(x_i) \doteq c_i,$$
$$f''(x_i) \doteq 2b_i.$$

(We note that a cubic spline is not useful for approximating derivatives of order higher than the second. A higher degree of spline function would be required for these.)

Approximating the integral of $f(x)$ over the n intervals where $f(x)$ is approximated by the spline is similarly straightforward:

$$\int_{x_1}^{x_{n+1}} f(x)\, dx = \int_{x_1}^{x_{n+1}} P_3(x)\, dx$$

$$= \sum_{i=1}^{n} \left[\frac{a_i}{4}(x - x_i)^4 + \frac{b_i}{3}(x - x_i)^3 + \frac{c_i}{2}(x - x_i)^2 + d_i(x - x_i) \right]_{x_i}^{x_{i+1}}$$

$$= \sum_{i=1}^{n} \left[\frac{a_i}{4}(x_{i+1} - x_i)^4 + \frac{b_i}{3}(x_{i+1} - x_i)^3 + \frac{c_i}{2}(x_{i+1} - x_i)^2 + d_i(x_{i+1} - x_i) \right].$$

If the intervals are all the same size, ($h = x_{i+1} - x_i$), this equation becomes

$$\int_{x_1}^{x_{n+1}} f(x)\, dx = \frac{h^4}{4} \sum_{i=1}^{n} a_i + \frac{h^3}{3} \sum_{i=1}^{n} b_i + \frac{h^2}{2} \sum_{i=1}^{n} c_i + h \sum_{i=1}^{n} d_i.$$

We illustrate the use of splines to compute derivatives and integrals by a simple example.

EXAMPLE   Compute the integral and derivatives of $f(x) = \sin \pi x$ over the interval $0 \le x \le 1$ from the spline that fits at $x = 0$, 0.25, 0.5, 0.75, and 1.0. (See Table 4.6.) We use end condition 1: $S_1 = 0$, $S_5 = 0$. Solving for the coefficients of the cubic spline, we get the results shown in Table 4.7.

The derivatives can now be computed:

At $x = 0.25$: $f'(x) = 2.2164$ $f''(x) = -7.344$

(exact = 2.2214);   (exact = −6.979);

At $x = 0.50$: $f'(x) = 0$ $f''(x) = -10.387$

(exact = 0);   (exact = −9.870);

At $x = 0.75$: $f'(x) = -2.2164$ $f''(x) = -7.344$

(exact = −2.2214);   (exact = −6.979).

The errors of the first derivatives computed from the cubic spline are smaller than those from a fourth-degree interpolating polynomial; the errors of the second derivatives from the spline are somewhat larger than from the polynomial.

Table 4.6

i, point number	x	$f(x)$
1	0	0
2	0.25	0.7071
3	0.50	1.0000
4	0.75	0.7071
5	1.0	0

Table 4.7

i	x	S_i	a_i	b_i	c_i	d_i
1	0	0	−4.8960	0	3.1340	0
2	0.25	−7.3440	−2.0288	−3.6720	2.2164	0.7071
3	0.50	−10.3872	2.0288	−5.1936	0	1.0
4	0.75	−7.3440	4.8960	−3.6720	−2.2164	0.7071

We readily compute the integral:

$$\int_0^1 f(x)\,dx = \frac{(0.25)^4}{4}(0) + \frac{(0.25)^3}{3}(-12.5376) + \frac{(0.25)^2}{2}(3.1340)$$

$$+ \; 0.25(2.4142)$$

$$= 0.6362 \qquad (\text{exact} = 0.6366;\ \text{error} = +0.0004).$$

For this example, the accuracy is significantly better than that obtained with Simpson's $\frac{1}{3}$ rule, which gives 0.6381, error = −0.0015. ∎

4.16 MULTIPLE INTEGRALS

We consider first the case when the limits of integration are constants. In the calculus we learned that a double integral may be evaluated as an iterated integral; in other words, we may write

$$\int\!\!\int_A f(x, y)\,dA = \int_a^b \left(\int_c^d f(x, y)\,dy \right) dx = \int_c^d \left(\int_a^b f(x, y)\,dx \right) dy. \qquad (4.60)$$

In Eq. (4.60) the rectangular region A is bounded by the lines

$$x = a, \qquad x = b, \qquad y = c, \qquad y = d.$$

In computing the iterated integrals, we hold x constant while integrating with respect to y (vice versa in the second case).

Adapting the numerical integration formulas developed earlier in this chapter for integration with respect to just one independent variable is quite straightforward. Recall that any of the integration formulas are just a linear combination of values of the function, evaluated at varying values of the independent variable. In other words, a quadrature formula is just a weighted sum of certain functional values. The inner integral is written then as a weighted sum of function values with one variable held constant. We then add together a weighted sum of these sums. If the function is known only at the nodes of a rectangular grid through the region, we are constrained to use these values. The Newton–Cotes formulas are a convenient set to employ. There is no reason why the same formula must be used in each direction, although it is often particularly convenient to do so.

EXAMPLE   We illustrate this technique by evaluating the integral of the function of Table 4.8 over the rectangular region bounded by

$$x = 1.5, \quad x = 3.0, \quad y = 0.2, \quad y = 0.6.$$

Let us use the trapezoidal rule in the x-direction and Simpson's $\frac{1}{3}$ rule in the y-direction. (Since the number of panels in the x-direction is not even, Simpson's $\frac{1}{3}$ rule does not apply readily.) It is immaterial which integral we evaluate first. Suppose we start with y constant:

$$y = 0.2: \quad \int_{1.5}^{3.0} f(x, y)\, dx = \int_{1.5}^{3.0} f(x, 0.2)\, dx = \frac{h}{2}(f_1 + 2f_2 + 2f_3 + f_4)$$

$$= \frac{0.5}{2}[0.990 + 2(1.568) + 2(2.520) + 4.090]$$

$$= 3.3140;$$

$$y = 0.3: \quad \int_{1.5}^{3.0} f(x, 0.3)\, dx = \frac{0.5}{2}[1.524 + 2(2.384) + 2(3.800) + 6.136]$$

$$= 5.0070.$$

Similarly, at

$$y = 0.4, \quad I = 6.6522;$$
$$y = 0.5, \quad I = 8.2368;$$
$$y = 0.6, \quad I = 9.7435.$$

We now sum these in the y-direction according to Simpson's rule:

$$f(x, y)\, dx = \frac{0.1}{3}[3.3140 + 4(5.0070) + 2(6.6522) + 4(8.2368) + 9.7435]$$

$$= 2.6446. \quad \blacksquare$$

Table 4.8   Tabulation of a function of two variables, $u = f(x, y)$

x \ y	0.1	0.2	0.3	0.4	0.5	0.6
0.5	0.165	0.428	0.687	0.942	1.190	1.431
1.0	0.271	0.640	1.003	1.359	1.703	2.035
1.5	0.447	0.990	1.524	2.045	2.549	3.031
2.0	0.738	1.568	2.384	3.177	3.943	4.672
2.5	1.216	2.520	3.800	5.044	6.241	7.379
3.0	2.005	4.090	6.136	8.122	10.030	11.841
3.5	3.306	6.679	9.986	13.196	16.277	19.198

(In this example our answer does not check well with the analytical value of 2.5944 because the x-intervals are large. We could improve our estimate somewhat by fitting a higher-degree polynomial than the first to provide the integration formula. We can even use values outside the range of integration for this, using the techniques of Section 4.11 to derive the formulas.)

The previous example shows that double integration by numerical means reduces to a double summation of weighted function values. The calculations we have just made could be written in the form

$$\int f(x, y)\, dx\, dy = \sum_{j=1}^{m} v_j \sum_{i=1}^{n} w_i f_{ij}$$

$$= \frac{\Delta y}{3} \frac{\Delta x}{2} [(f_{1,1} + 2f_{2,1} + 2f_{3,1} + f_{4,1})$$

$$+ 4(f_{1,2} + 2f_{2,2} + 2f_{3,2} + f_{4,2})$$

$$+ \cdots + (f_{1,5} + 2f_{2,5} + 2f_{3,5} + f_{4,5})].$$

It is convenient to write this in a pictorial operator form, in which the weighting factors are displayed in an array that is a map to the location of the functional values to which they are applied.

$$\int f(x, y)\, dx\, dy = \frac{\Delta y}{3} \frac{\Delta x}{2} \begin{pmatrix} 1 & 4 & 2 & 4 & 1 \\ 2 & 8 & 4 & 8 & 2 \\ 2 & 8 & 4 & 8 & 2 \\ 1 & 4 & 2 & 4 & 1 \end{pmatrix} f_{i,j}. \tag{4.61}$$

We interpret the numbers in the array of Eq. (4.61) in this manner: We use the values 1, 4, 2, 4, and 1 as weighting factors for functional values in the top row of the portion of Table 4.8 that we integrate over (values where $x = 1.5$ and y varies from 0.2 to 0.6.) Similarly, the second column of the array in Eq. (4.61) represents weighting factors that are applied to a column of function values where $y = 0.4$ and x varies from 1.5 to 3.0. Observe that the values in the pictorial operator of Eq. (4.61) follow immediately from the Newton–Cotes coefficients for single-variable integration.

Other combinations of Newton–Cotes formulas give similar results. It is probably easiest for hand calculation to use these pictorial integration operators. Pictorial integration is readily adapted to any desired combination of integration formulas. Except for the difficulty of representation beyond two dimensions, this operator technique also applies to triple and quadruple integrals.

There is an alternative representation to such pictorial operators that is easier to translate into a computer program. We also derive it somewhat differently. Consider the numerical integration formula for one variable

$$\int_{-1}^{1} f(x)\, dx \doteq \sum_{i=1}^{n} a_i f(x_i). \tag{4.62}$$

We have seen in Section 4.12 that such formulas can be made exact if $f(x)$ is any polynomial of a certain degree. Assume that Eq. (4.62) holds for polynomials up to degree s.*

---

*For Newton–Cotes formulas, $s = n - 1$ for n even and $s = n$ for n odd. For Gaussian quadrature formulas, $s = 2n - 1$, and the x_i will be unevenly spaced.

We now consider the multiple integral formula

$$\int_{-1}^{1} \int_{-1}^{1} \int_{-1}^{1} f(x, y, z) \, dx \, dy \, dz \stackrel{?}{=} \sum_{i=1}^{n} \sum_{j=1}^{n} \sum_{k=1}^{n} a_i a_j a_k f(x_i, y_j, z_k). \qquad (4.63)$$

We wish to show that Eq. (4.63) is exact for all polynomials in x, y, and z up to degree s. Such a polynomial is a linear combination of terms of the form $x^{\alpha} y^{\beta} z^{\gamma}$, where α, β, and γ are nonnegative integers whose sum is equal to s or less. If we can prove that Eq. (4.63) holds for the general term of this form, it will then hold for the polynomial.

To do this we assume that

$$f(x, y, z) = x^{\alpha} y^{\beta} z^{\gamma}.$$

Then, since the limits are constants and the integrand is factorable,

$$I = \int_{-1}^{1} \int_{-1}^{1} \int_{-1}^{1} x^{\alpha} y^{\beta} z^{\gamma} \, dx \, dy \, dz$$

$$= \left(\int_{-1}^{1} x^{\alpha} \, dx \right) \left(\int_{-1}^{1} y^{\beta} \, dy \right) \left(\int_{-1}^{1} z^{\gamma} \, dz \right).$$

Replacing each term according to Eq. (4.62), we get,

$$I = \left(\sum_{i=1}^{n} a_i x_i^{\alpha} \right) \left(\sum_{j=1}^{n} a_j y_j^{\beta} \right) \left(\sum_{k=1}^{n} a_k z_k^{\gamma} \right) = \sum_{i=1}^{n} a_i x_i^{\alpha} \sum_{j=1}^{n} a_j y_j^{\beta} \sum_{k=1}^{n} a_k z_k^{\gamma}. \qquad (4.64)$$

We need now an elementary rule about the product of summations. We illustrate it for a simple case. We assert that

$$\left(\sum_{i=1}^{3} u_i \right) \left(\sum_{j=1}^{2} v_j \right) = \sum_{i=1}^{3} \left(\sum_{j=1}^{2} u_i v_j \right)$$

$$= \sum_{i=1}^{3} \sum_{j=1}^{2} u_i v_j.$$

The last equality is purely notational. We prove the first by expanding both sides:

$$\left(\sum_{i=1}^{3} u_i \right) \left(\sum_{j=1}^{2} v_j \right) = \sum_{i=1}^{3} u_i \sum_{j=1}^{2} v_j$$

$$= (u_1 + u_2 + u_3)(v_1 + v_2)$$

$$= u_1 v_1 + u_1 v_2 + u_2 v_1 + u_2 v_2 + u_3 v_1 + u_3 v_2;$$

$$\sum_{i=1}^{3} \sum_{j=1}^{2} u_i v_j = (u_1 v_1 + u_1 v_2) + (u_2 v_1 + u_2 v_2) + (u_3 v_1 + u_3 v_2).$$

On removing parentheses we see the two sides are the same. Using this principle, we can write Eq. (4.64) in the form

$$I = \sum_{i=1}^{n} \sum_{j=1}^{n} \sum_{k=1}^{n} a_i a_j a_k x_i^{\alpha} y_j^{\beta} z_k^{\gamma}, \qquad (4.65)$$

which shows that the questioned equality of Eq. (4.63) is valid, and we can write a program for a triple integral by three nested DO loops. The coefficients a_i are chosen from any numerical integration formula. If the three one-variable formulas corresponding to Eq. (4.63) are not identical, an obvious modification of Eq. (4.65) applies. In some cases a change of variable is needed to correspond to Eq. (4.62).

If we are evaluating a multiple integral numerically where the integrand is a known function, our choice of the form of Eq. (4.62) is wider. Of higher efficiency than the Newton–Cotes formulas is Gaussian quadrature. Since it also fits the pattern of Eq. (4.62) the formula of Eq. (4.65) applies. We illustrate this with a simple example.

EXAMPLE   Evaluate

$$I = \int_0^1 \int_{-1}^0 \int_{-1}^1 yze^x \, dx \, dy \, dz$$

by Gaussian quadrature using a three-term formula for x and two-term formulas for y and z. We first make the changes of variables to adjust the limits for y and z to $(-1, 1)$:

$$y = \frac{1}{2}(u + 1), \qquad dy = \frac{1}{2} \, du;$$

$$z = \frac{1}{2}(v - 1), \qquad dz = \frac{1}{2} \, dv.$$

Our integral becomes

$$I = \frac{1}{16} \int_{-1}^1 \int_{-1}^1 \int_{-1}^1 (u + 1)(v - 1)e^x \, dx \, du \, dv.$$

The two- and three-point Gaussian formulas are, from Section 4.12:

$$\int_{-1}^1 f(x) \, dx = (1)f(-0.5774) + (1)f(0.5774),$$

$$\int_{-1}^1 f(x) \, dx = \left(\frac{5}{9}\right)f(-0.7746) + \left(\frac{8}{9}\right)f(0) + \left(\frac{5}{9}\right)f(0.7746).$$

The integral is then

$$I = \frac{1}{16} \sum_{i=1}^2 \sum_{j=1}^2 \sum_{k=1}^3 a_i a_j b_k (u_i + 1)(v_j - 1)e^{x_k},$$

$$a_1 = 1, \qquad a_2 = 1,$$

$$b_1 = \frac{5}{9}, \qquad b_2 = \frac{8}{9}, \qquad b_3 = \frac{5}{9},$$

and values of u, v, and x as above.

A few representative terms of the sum are

$$I = \frac{1}{16}\left[(1)(1)\left(\frac{5}{9}\right)(-0.5774 + 1)(-0.5774 - 1)e^{-0.7446} \right.$$

$$+ (1)(1)\left(\frac{8}{9}\right)(-0.5774 + 1)(-0.5774 - 1)e^0$$

$$+ (1)(1)\left(\frac{5}{9}\right)(-0.5774 + 1)(-0.5774 - 1)e^{0.7746}$$

$$+ (1)(1)\left(\frac{5}{9}\right)(0.5774 + 1)(-0.5774 - 1)e^{-0.7746}$$

$$+ \cdots \bigg].$$

On evaluating, we get $I = -0.58758$. The analytical value is

$$-\frac{1}{4}(e - e^{-1}) = -0.58760. \quad \blacksquare$$

4.17 ERRORS IN MULTIPLE INTEGRATION AND EXTRAPOLATIONS

The error term of a one-variable quadrature formula is an additive one just like the other terms in the linear combination (although of special form). It would seem reasonable that it would go through the multiple summations in a similar fashion, so we should expect error terms for multiple integration that are analogous to the one-dimensional case. We illustrate that this is true for double integration using the trapezoidal rule in both directions, with uniform spacings, choosing n intervals in the x-direction and m in the y-direction.

From Section 4.8 we have

$$\text{Error of } \int_a^b f(x) \, dx = -\frac{b - a}{12}h^2 f''(\xi) = O(h^2),$$

$$h = \Delta x = \frac{b - a}{n}.$$

In developing Romberg integration, we found that the error term could be written as

$$\text{Error} = O(h^2) = Ah^2 + O(h^4) \doteq Ah^2 + Bh^4,$$

where A is a constant and the value of B depends on a fourth derivative of the function. Appending this error term to the trapezoidal rule, we get (equivalent to Eq. (4.62))

$$\int_a^b f(x) \, dx \bigg|_{y=y_j} = \frac{h}{2}(f_{0,j} + 2f_{1,j} + 2f_{2,j} + \cdots + f_{n,j}] + A_j h^2 + B_j h^4.$$

Summing these in the y-direction and retaining only the error terms, we have

$$\int_c^d \int_a^b f(x, y) \, dx \, dy = \frac{k}{2}\frac{h}{2}\sum_{i=0}^{n} \sum_{j=0}^{m} a_i a_j f_{i,j} + \frac{k}{2}(A_0 + 2A_1 + 2A_2 + \cdots + A_m)h^2$$

$$+ \frac{k}{2}(B_0 + 2B_1 + 2B_2 + \cdots + B_m)h^4 + \bar{A}k^2 + \bar{B}k^4, \qquad (4.66)$$

$$k = \Delta y = \frac{d - c}{m}.$$

In Eq. (4.66), $\bar{A}$ and $\bar{B}$ are the coefficients of the error term for y. The coefficients A and B for the error terms in the x-direction may be different for each of the $(m + 1)$ y-values.

but each of the sums in parentheses in Eq. (4.66) is $2n$ times some average value of A or B, so the error terms become

$$\text{Error} = \frac{k}{2}(nA_{av})h^2 + \frac{k}{2}(nB_{av})h^4 + \bar{A}k^2 + \bar{B}k^4. \tag{4.67}$$

Since both Δx and Δy are constant, we may take $\Delta y = k = \alpha \Delta x = \alpha h$, where $\alpha = \Delta y/\Delta x$, and Eq. (4.67) can be written, with $nh = (b - a)$,

$$\text{Error} = \left(\frac{b-a}{2}A_{av}\alpha\right)h^2 + \left(\frac{b-a}{2}B_{av}\alpha\right)h^4 + A\alpha^2h^2 + B\alpha^4h^4$$

$$= K_1h^2 + K_2h^4.$$

Here K_2 will depend on fourth-order partial derivatives. This confirms our expectation that the error term of double integration by numerical means is of the same form as for single integration.

Since this is true, a Romberg integration may be applied to multiple integration, whereby we extrapolate to an $O(h^4)$ estimate from two trapezoidal computations at a 2-to-1 interval ratio. From two such $O(h^4)$ computations we may extrapolate to one of $O(h^6)$ error.

4.18 MULTIPLE INTEGRATION WITH VARIABLE LIMITS

If the limits of integration are not constant, so that the region in the x,y-plane upon which the integrand $f(x, y)$ is to be summed is not rectangular, we must modify the above procedure. Let us consider a simple example to illustrate the method. Evaluate

$$\int \int f(x, y) \, dy \, dx$$

over the region bounded by the lines $x = 0$, $x = 1$, $y = 0$, and the curve $y = x^2 + 1$. The region is sketched in Fig. 4.7. If we draw vertical lines spaced at $\Delta x = 0.2$ apart, shown as dashed lines in Fig. 4.7, it is obvious that we can approximate the inner integral at constant x-values along any one of the vertical lines (including $x = 0$ and $x = 1$). If we use the trapezoidal rule with five panels for each of these, we get the series of sums

$$S_1 = \frac{h_1}{2}(f_a + 2f_b + 2f_c + 2f_d + 2f_e + f_f),$$

$$S_2 = \frac{h_2}{2}(f_g + 2f_h + 2f_i + 2f_j + 2f_k + f_l),$$

$$S_3 = \frac{h_3}{2}(f_m + 2f_n + \cdots),$$

$$\vdots$$

$$S_6 = \frac{h_6}{2}(f_u + 2f_v + 2f_w + 2f_x + 2f_y + f_z).$$

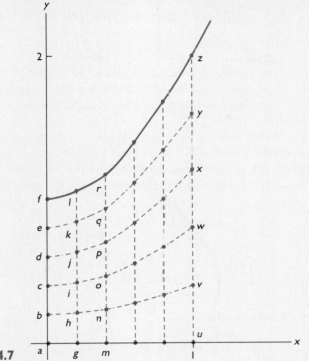

Figure 4.7

The subscripts here indicate the values of the function at the points so labeled in Fig. 4.7. The values of the h_i are not the same in the above equations, but in each they are the vertical distances divided by 5. The combination of these sums to give an estimate of the double integral will then be

$$\text{Integral} = \frac{0.2}{2}(S_1 + 2S_2 + 2S_3 + 2S_4 + 2S_5 + S_6).$$

To be even more specific, suppose $f(x, y) = xy$. Then

$$S_1 = \frac{1.0/5}{2}(0 + 0 + 0 + 0 + 0 + 0) = 0,$$

$$S_2 = \frac{1.04/5}{2}(0 + 0.0832 + 0.1664 + 0.2496 + 0.3328 + 0.208) = 0.1082,$$

$$S_3 = \frac{1.16/5}{2}(0 + 0.1856 + 0.3712 + 0.5568 + 0.7428 + 0.464) = 0.2692,$$

$$S_4 = \frac{1.36/5}{2}(0 + 0.3264 + 0.6528 + 0.9792 + 1.3056 + 0.816) = 0.5549,$$

$$S_5 = \frac{1.64/5}{2}(0 + 0.5248 + 1.0496 + 1.5744 + 2.0992 + 1.312) = 1.0758,$$

$$S_6 = \frac{2.0/5}{2}(0 + 0.8 + 1.6 + 2.4 + 3.2 + 2.0) = 2.0;$$

$$\text{Integral} = \frac{0.2}{2}(0 + 0.2164 + 0.5384 + 1.1098 + 2.1516 + 2.0)$$

$$= 0.6016 \quad \text{versus analytical value of } 0.583333.$$

The extension of this to more complicated regions and the adaptation to the use of Simpson's rule should be obvious. If the functions that define the region are not single-valued, one must divide the region into subregions to avoid the problem, but this must also be done to integrate analytically.

The previous calculations were not very accurate because the trapezoidal rule has relatively large errors. Gaussian quadrature should be an improvement, even using fewer points within the region. Let us use 3-point quadrature in the x-direction and 4-point quadrature in the y-direction. As in Section 4.12, we must change the limits of integration:

$$\int_0^1 \int_0^{x^2+1} x\, y\, dy\, dx$$

to

$$\frac{1}{4} \int_{-1}^1 \int_{-1}^1 \frac{s+1}{2} \left[\frac{(x^2(s) + 1)^2 t + (x^2(s) + 1)^2}{2} \right] dt\, ds$$

in which we made the following substitutions:

$$x = \frac{s+1}{2} \qquad y = \frac{(x^2(s) + 1)^2 t + (x^2(s) + 1)^2}{2}.$$

The integral is approximated by the sum

$$\sum_{i=1}^3 \sum_{j=1}^4 w_i W_j f(s_i, t_j),$$

where the w_i's, W_j's, s_i's, and t_j's are the values taken from Table 4.4. Using that table, we set $w_1 = 0.55555555$, $w_3 = w_1$, and $w_2 = 0.88888889$; we set $s_1 = -0.77459667$, $s_3 = -s_1$, and $s_2 = 0.0$. The values for the W_j's and t_j's are obtained in the same way. For each fixed i, $i = 1, 2, 3$, let S_i be the corresponding value obtained using Gaussian quadrature for a fixed s_i, where $S_i = \sum_{j=1}^4 W_j f(s_i, t_j)$.

The following intermediate values are easily verified:

$$S_1 = (0.00279158 + 0.02487506 + 0.05050174 + 0.03741447) = 0.11558285,$$

$$S_2 = (0.01886891 + 0.16813600 + 0.34135240 + 0.25289269) = 0.78125000,$$

$$S_3 = (0.06845742 + 0.61000649 + 1.23844492 + 0.91750833) = 2.83441716.$$

We sum these as follows:

$$\frac{w_1 S_1 + w_2 S_2 + w_3 S_3}{4} = 0.58333334$$

which agrees with the exact answer to seven places. In this case we used only 12 evaluations of the function (exceptionally simple to do here, but usually more costly) compared to the 36 used with the trapezoidal rule.

To keep track of the intermediate computations, it is convenient to use a template such as

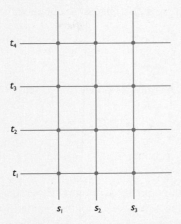

and the S_i's are computed along the verticals.

4.19 CHAPTER SUMMARY

If you are not able to do all of the following, we advise you to restudy portions of this chapter. You should now be able to:

1. Develop formulas for differentiation and integration from an interpolating polynomial.

2. Derive an expression for the error terms for the formulas developed in the preceding question.

3. Use a lozenge diagram to write several formulas for differentiation and distinguish whether they are based on a forward, backward, or other type of interpolating polynomial.

4. Derive formulas by the symbolic method and from Taylor series.

5. Apply formulas to obtain derivatives and integrals and estimate the error in the result.

6. Use extrapolation techniques to get improved estimates for the values of the derivative and integral. You should be able to explain why these methods work.

7. Explain how round-off errors influence the accuracy of the results and why differentiation is more susceptible to such errors than integration.

8. Evaluate an integral using Gaussian quadrature. You should be able to develop a Legendre polynomial of degree n from the recursion formula and to tell what is meant by *orthogonal function*.

9. Apply the various integration procedures to improper and indefinite integrals and to multiple integrals.

10. Understand the principle of adaptive integration well enough to apply it to a simple example.

11. Use a cubic spline fitted to a set of data to obtain the derivative and integral. You should be able to explain why a spline may give improved accuracy.

12. Utilize the FORTRAN programs given in this chapter to obtain values for derivatives and integrals.

SELECTED READINGS FOR CHAPTER 4

Atkinson (1978); Conte and de Boor (1980); Forsythe, Malcolm, and Moler (1977); Shampine and Allen (1973).

4.20 COMPUTER PROGRAMS

We conclude this chapter with a series of computer programs. Programs 1 (Fig. 4.9) and 2 (Fig. 4.10) perform differentiation. The first is a subroutine named DERIV that finds the first derivative of tabular data with equispaced x-values. Parabolas are passed through groups of three points and the derivative is computed as the slope of the parabola at the point in question. For each interior point, this amounts to using a central-difference formula for the derivative of $f(x)$. The endpoints use three-point formulas for the derivative, but the point at which the derivative is estimated is not a central point. The lozenge diagram of Fig. 4.1 is the source of these formulas; they have been rewritten in terms of function values in the routine. The subroutine is passed the values of the function in an array, the number of points, and the uniform x-interval. An array holding the derivative values is returned.

When the tabular data given below were passed to the subroutine, the results in Fig. 4.8 were obtained. The data are recordings of the speed of a car at 10-second intervals:

Time, sec	0	10	20	30	40	50	60	70	80	90	100
Speed, mph	48	38	31	33	36	41	51	43	35	29	28

Program 1 Output
Function Values and Derivative Values

Point #	F	DF
1	0.480000E 02	−0.115000E 01
2	0.380000E 02	−0.850000E 00
3	0.310000E 02	−0.250000E 00
4	0.330000E 02	0.250000E 00
5	0.360000E 02	0.400000E 00
6	0.410000E 02	0.750000E 00
7	0.510000E 02	0.100000E 00
8	0.430000E 02	−0.800000E 00
9	0.350000E 02	−0.700000E 00
10	0.290000E 02	−0.350000E 00
11	0.280000E 02	0.150000E 00

Figure 4.8

```
      SUBROUTINE DERIV(F,DF,N,H)
C
C   ------------------------------------------------------------------
C
C     SUBROUTINE DERIV :
C
C                      THIS SUBROUTINE FINDS THE FIRST DERIVATIVE OF
C     EQUISPACED DATA. CENTRAL DIFFERENCE FORMULAS ARE USED FOR ALL POINTS
C     EXCEPT THE FIRST AND THE LAST. FOR THESE, A QUADRATIC IS PASSED
C     THROUGH THREE SUCCESSIVE POINTS TO OBTAIN THE DERIVATIVE.
C
C   ------------------------------------------------------------------
C
C     PARAMETERS ARE :
C
C     F    - THE ARRAY OF FUNCTION VALUES
C     DF   - AN ARRAY THAT HOLDS THE DERIVATIVE VALUES, RETURNED TO CALLER
C     N    - THE NUMBER OF POINTS
C     H    - THE UNIFORM SPACING BETWEEN X VALUES
C
C   ------------------------------------------------------------------
C
      REAL F(N),DF(N),H
      INTEGER N,I,NM1
C
C   ------------------------------------------------------------------
C
C COMPUTE THE DERIVATIVES AT X(2) THROUGH X(N-1)
C
      NM1 = N - 1
      DO 10 I = 2,NM1
        DF(I) = ( F(I+1) - F(I-1) ) / 2.0 / H
   10 CONTINUE
C
C   ------------------------------------------------------------------
C
C NOW COMPUTE THE DERIVATIVE AT X(1)
C
      DF(1) = ( 2.0*F(2) - 1.5*F(1) - 0.5*F(3) ) / H
C
C AND GET IT AT X(N)
C
      DF(N) = ( 1.5*F(N) - 2.0*F(N-1) + 0.5*F(N-2) ) / H
      RETURN
      END
```

Figure 4.9 Program 1.

In view of the discussion of "'noisy data" in Section 4.6, what can you say about the accuracy of these results? How could the accuracy of the acceleration determination be improved?

The second program (DER2), in Fig. 4.10, is a subroutine that finds the first and second derivatives of a function whose form is known. The name of the function is passed as a parameter; it must therefore be declared EXTERNAL in the calling program. Central-difference formulas are used, values of $f(x)$ being used at the distance H on either side of the point where the derivatives are desired. This routine and a modified version that employed double-precision arithmetic were used to create the tables in Section 4.6 that

```
          SUBROUTINE DER2(FCT,X,H,DX,DDX)
C
C      --------------------------------------------------------------------
C
C      SUBROUTINE DER2 :
C                        THIS SUBROUTINE COMPUTES THE FIRST AND SECOND
C      DERIVATIVES OF THE FUNCTION FCT AT THE POINT X. CENTRAL DIFFERENCE
C      FORMULAS ARE USED, EMPLOYING A STEP SIZE EQUAL TO H.
C
C      --------------------------------------------------------------------
C
C      PARAMETERS ARE :
C
C      FCT      - FUNCTIONS WHOSE DERIVATIVES ARE DESIRED. MUST BE DECLARED
C                 EXTERNAL IN CALLING PROGRAM.
C      X        - POINT AT WHICH DERIVATIVES ARE TO BE CALCULATED.
C      H        - STEP SIZE TO BE USED.
C      DX       - RETURNS THE FIRST DERIVATIVE TO THE CALLER.
C      DDX      - RETURNS THE SECOND DERIVATIVE TO THE CALLER.
C
C      --------------------------------------------------------------------
C
          REAL FCT,X,H,DX,DDX,F0,FR,FL
C
          F0 = FCT(X)
          FR = FCT(X+H)
          FL = FCT(X-H)
          DX = ( FR - FL ) / 2.0 / H
          DDX = ( FR - 2.0*F0 + FL ) / H / H
          RETURN
          END
```

Figure 4.10   Program 2.

illustrate how the best accuracy in computing derivatives occurs at a step size where the sum of round-off and truncation errors is minimized.

Four integration programs are presented. Program 3 (Fig. 4.11) is a subroutine named SIMPS, designed to integrate uniformly spaced data that are known from a table. A combination of Simpson's $\frac{1}{3}$ and $\frac{3}{8}$ rules are used when there is an odd number of intervals. If there is an even number, only the $\frac{1}{3}$ rule is used, which normally gives better accuracy. An array containing the function values is passed, together with $H = \Delta x$ and the number of points. The program first checks to see whether the number of panels is odd or even. If it is odd, one application of the $\frac{3}{8}$ rule is performed to integrate over the first three panels, and the $\frac{1}{3}$ rule is applied over all the others. The parameter RESULT returns the value of the integral. When this subroutine is applied to the velocity data tabulated with Program 1, the result obtained is 3726.67 mile-sec/hr. If we unscramble the mixed time units by dividing by 3600, we find that the distance traveled, as indicated by the speedometer, is 1.035 mi.

Program 4 (Fig. 4.12) is a subroutine that applies the Romberg method to integrate a function. The name of the function is passed as a parameter (thus it must be declared EXTERNAL), along with the limits of integration and a tolerance value to control the error. RESULT returns the final value to the caller, but intermediate results are printed out by the subroutine. The value of H is halved at each stage of the integration until TOL

```
C
C
      SUBROUTINE SIMPS(F,N,H,RESULT)
C
C     ----------------------------------------------------------------------
C
C     SUBROUTINE SIMPS :
C                        THIS ROUTINE PERFORMS SIMPSON'S RULE INTEGRATION
C     OF A FUNCTION DEFINED BY A TABLE OF EQUISPACED VALUES.
C
C     ----------------------------------------------------------------------
C
C     PARAMETERS ARE :
C
C     F      - ARRAY OF VALUES OF THE FUNCTION
C     N      - NUMBER OF POINTS
C     H      - THE UNIFORM SPACING BETWEEN X VALUES
C     RESULT - ESTIMATE OF THE INTEGRAL THAT IS RETURNED TO THE CALLER
C
C     ----------------------------------------------------------------------
C
      REAL F(1024),H,RESULT
      INTEGER N,NPANEL,NHALF,NBEGIN,NEND
C
C     ----------------------------------------------------------------------
C
C CHECK TO SEE IF NUMBER OF PANELS IS EVEN. NUMBER OF PANELS IS N - 1.
C
      NPANEL = N - 1
      NHALF = NPANEL / 2
      NBEGIN = 1
      RESULT = 0.0
      IF ( (NPANEL - 2*NHALF) .NE. 0 ) THEN
C
C NUMBER OF PANELS IS ODD. USE 3/8 RULE ON FIRST THREE, 1/3 RULE ON REST.
C
         RESULT = 3.0*H/8.0*( F(1) + 3.0*F(2) + 3.0*F(3) + F(4) )
         NBEGIN = 4
         IF ( N .EQ. 4 ) RETURN
      END IF
C
C     ----------------------------------------------------------------------
C
C APPLY 1/3 RULE - ADD IN FIRST, SECOND AND LAST VALUES.
C
      RESULT = RESULT + H/3.0*( F(NBEGIN) + 4.0*F(NBEGIN+1) + F(N) )
      NBEGIN = NBEGIN + 2
      IF ( NBEGIN .EQ. N ) RETURN
C
C     ----------------------------------------------------------------------
C
C THE PATTERN AFTER NBEGIN+2  IS REPETITIVE. GET NEND, THE PLACE TO STOP.
C
      NEND = N - 2
      DO 10 I = NBEGIN,NEND,2
   10 RESULT = RESULT + H/3.0*( 2.0*F(I) + 4.0*F(I+1) )
      RETURN
      END
```

Figure 4.11   Program 3.

```
      PROGRAM PROMB(INPUT,OUTPUT)
C
C     ------------------------------------------------------------------------
C
      REAL FCT,TOL,TOTAL,A,B,RESULT
      INTEGER I
      EXTERNAL FCT
      DATA TOL,TOTAL,A,B / 0.0001,0.0,0.0,1.0 /
C
C     ------------------------------------------------------------------------
C
      DO 10 I = 1,5
        CALL ROMB(FCT,A,B,TOL,RESULT)
        TOTAL = TOTAL + RESULT
        PRINT 200, B,RESULT,TOTAL
        IF ( ABS(RESULT) .LT. 0.001 ) STOP
        A = B
        B = B*10
   10 CONTINUE
C
  200 FORMAT(/' B = ',F7.0,' INCREMENT IS ',E15.7,' INTEGRAL IS ',
     +        E15.7//)
      STOP
      END
C
C     ------------------------------------------------------------------------
C
      REAL FUNCTION FCT(X)
C
      REAL X
      FCT = X * EXP(-X)
      RETURN
      END
C
C     ------------------------------------------------------------------------
C
      SUBROUTINE ROMB(FCN,A,B,TOL,RESULT)
C
C     ------------------------------------------------------------------------
C
C
C     SUBROUTINE ROMB :
C                         SUBROUTINE FOR ROMBERG INTEGRATION. PROGRAM BEGINS
C     WITH TRAPEZOIDAL INTEGRATION WITH 10 SUBINTERVALS. INTERVALS ARE THEN
C     HALVED AND RESULTS ARE EXTRAPOLATED UP TO EIGHTH ORDER. MAXIMUM
C     NUMBER OF SUBINTERVALS USED IN PROGRAM IS 2560.
C
C     ------------------------------------------------------------------------
C
C
C     PARAMETERS ARE :
C
C     FCN    - FUNCTION THAT COMPUTES F(X), DECLARED EXTERNAL IN MAIN
C     A,B    - INITIAL AND FINAL X VALUES
C     TOL    - TOLERANCE VALUE USED TO TERMINATE ITERATIONS
C     RESULT - RETURNS VALUE OF INTEGRAL TO CALLER
C     TRAP   - DOUBLY SUBSCRIPTED ARRAY THAT HOLDS INTERMEDIATE VALUES
C                FOR COMPARISONS AND EXTRAPOLATION
C     KFLAG  - FLAG USED INTERNALLY TO SIGNAL NON-CONVERGENCE. WHEN
C                KFLAG=0, MEANS NON-CONVERGENT, =1 MEANS ALL OK.
C
C     ------------------------------------------------------------------------
C
```

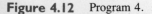

Figure 4.12   Program 4.

Figure 4.12   (*continued*)

```
        REAL FCN,A,B,TOL,H,SUM,X.TRAP(9,9)
        INTEGER I,L,K,KFLAG
C
C       ----------------------------------------------------------------
C
C  SET FLAG AT 1 INITIALLY
C
        KFLAG = 1
C
C  COMPUTE FIRST INTEGRAL WITH 10 SUBINTERVALS AND USING TRAP RULE
C
        H = (B - A) / 10.0
        SUM = FCN(A) + FCN(B)
        X = A
        DO 10 I = 2,10
           X = X + H
           SUM = SUM + FCN(X)*2.0
     10 CONTINUE
        TRAP(1,1) = H / 2.0 * SUM
C
C       ----------------------------------------------------------------
C
C  RECOMPUTE INTEGRAL WITH H HALVED, EXTRAPOLATE AND TEST. REPEAT
C  UP TO EIGHT TIMES.
C
        DO 20 I = 1,8
           H = H / 2.0
           X = A + H
           K = 10*2**I
           DO 30 J=2,K,2
              SUM = SUM + FCN(X)*2.0
              X = X + H + H
     30    CONTINUE
           TRAP(1,I+1) = H / 2.0 * SUM
           DO 40 L = 1,I
              TRAP(L+1,I+1) = TRAP(L,I+1) + 1.0/(4.0**L - 1.0) *
     +                      (TRAP(L,I+1) - TRAP(L,I))
     40    CONTINUE
           IF ( ABS(TRAP(I+1,I+1) - TRAP(I,I+1)) - TOL ) 50,50,20
     20 CONTINUE
C
C       ----------------------------------------------------------------
C
C  IF TOLERANCE NOT MET AFTER 8 EXTRAPOLATIONS, PRINT NOTE & SET KFLAG = 0.
C
        KFLAG = 0
        PRINT 200
C
C       ----------------------------------------------------------------
C
C  PRINT INTERMEDIATE RESULTS
C
     50 I = I + 1
        DO 70 L = 1,I
           PRINT 203, (TRAP(J,L),J=1,L)
     70 CONTINUE
        IF ( KFLAG .EQ. 0 ) STOP
        RESULT = TRAP(I,I)
    200 FORMAT(/' TOLERANCE NOT MET. CALCULATED VALUES WERE ')
    203 FORMAT(1X,8F12.6)
        RETURN
        END
```

Figure 4.12   (*continued*)

```
                    OUTPUT FOR PROGRAM 4

     .263408
     .264033     .264241
     .264189     .264241      .264241

 B =      1. INCREMENT IS      .2642411E+00 INTEGRAL IS      .2642411E+00


     .735878
     .735294     .735100
     .735260     .735249      .735259

 B =     10. INCREMENT IS      .7352591E+00 INTEGRAL IS      .9995002E+00


     .002044
     .001055     .000725
     .000661     .000530      .000517

 B =    100. INCREMENT IS      .5171925E-03 INTEGRAL IS      .1000017E+01
```

is met on that stage. The program avoids recomputing function values that have been used in prior stages.

A main program and a function are also shown with Program 4, illustrating how the subroutine can be used to find an integral with infinite upper limit. The integral being evaluated is

$$\int_0^\infty xe^{-x}\, dx,$$

whose analytical result is unity. The routine, with TOL = 0.0001, finds a result of almost perfect accuracy. In the program, the upper limit is increased progressively until a negligibly small increment is computed.

The fifth program (see Fig. 4.13) is based on the adaptive Simpson method of Section 4.14. The example executed by this program is the very same one in that section. The integrand is evaluated in FUNCTION F(X). SUBROUTINE ADAPT follows the algorithm presented in Section 4.14 very closely. A Romberg refinement is made once the integral value estimates, S1AB and S2AB, are sufficiently close. Moreover, as the limits of integration are reduced, we also reduce the stopping criterion, TOL, by the same factor. The PUSH and POP subroutines make the implementation of the algorithm straightforward.

The sixth program (DBLINT) (see Fig. 4.14) integrates a function of two independent variables over a rectangular region. The Romberg method again is used in the program, and the name of the function is passed as a parameter to provide greater flexibility. This program is similar to Program 4, except that a series of integrals in the *x*-direction is combined to integrate in the *y*-direction; a function subprogram is called to do this. Program 6 is less efficient than Program 4 in that some function values are recomputed as the step size is reduced.

```
      PROGRAM PADAPT(INPUT,OUTPUT)
C
C     ------------------------------------------------------------------
C
C          THIS PROGRAM IMPLEMENTS THE ADAPTIVE SIMPSON INTEGRATION
C          PRESENTED IN SECTION 8
C
C     ------------------------------------------------------------------
C
      REAL F, A,B,TOL, ANSWER
      INTEGER  MAXCNT, FCNCNT
      EXTERNAL F
      COMMON MAXCNT, FCNCNT
C
      DATA A,B,TOL/0.2, 1.0, 0.050/
C
C     ------------------------------------------------------------------
C
C          MAXCNT: MAXIMUN NUMBER OF FUNCTION CALLS
C          FCNCNT: ACTUAL NUMBER OF FUNCTION CALLS
C
C     ------------------------------------------------------------------
C
      MAXCNT = 500
      FCNCNT = 0
C
C
      CALL ADAPT(F, A,B,TOL,  ANSWER)
C
C         PRINT ANSWER = INTEGRAL(F) FROM A TO B
C
      PRINT 100, ANSWER
       PRINT 200, FCNCNT
100   FORMAT(///,10X,'THE INTEGRAL VALUE IS ', F10.6/)
200   FORMAT(10X,'THE NUMBER OF FUNCTION CALLS = ', I4///)
      STOP
      END
C
C          SUBROUTINE ADAPT
C
      SUBROUTINE ADAPT(F,  A,B,TOL,  ANSWER)
C
C     ------------------------------------------------------------------
C
C     THIS SUBROUTINE USES THE ADAPTIVE SIMPSON METHOD AS PRESENTED
C     IN SECTION 8. THE VARIABLES USED HERE ARE DEFINED THE SAME AS
C     GIVEN IN THE ALGORITHM.
C
C     ------------------------------------------------------------------
C
      REAL A,B,ANSWER,TOL,              .
     +     H1,H2,C,D,E,FA,FB,FC,FD,FE,
     +     S1AB,S2AB,S2AC,S2CB, STORE
      REAL STACK(7,100)
      INTEGER COUNT, TOP , MAXCNT,FCNCNT
C
      COMMON MAXCNT, FCNCNT
      COMMON /SUB/STACK, TOP
C
      STORE = 0.0
```

Figure 4.13   Program 5.

Figure 4.13 (*continued*)

```
            COUNT = 0
            H1 = (B-A)/2.0
            C = A + H1
            FA = F(A)
            FB = F(B)
            FC = F(C)
            S1AB = H1*(FA + 4.0*FC + FB)/3.0
            TOP = 0
            CALL PUSH(A,FA,FC,FB,H1,TOL,S1AB)
C
10          CONTINUE
                COUNT = COUNT + 1
                CALL POP(A,FA,FC,FB,H1,TOL,S1AB)
                H2 = H1/2.0
                D = A + H2
                E = A + 3.0*H2
                C = A + H1
                B = A + 2.0*H1
                FD = F(D)
                FE = F(E)
                S2AC = H2*(FA + 4.0*FD + FC)/3.0
                S2CB = H2*(FC + 4.0*FE + FB)/3.0
C
                S2AB = S2AC + S2CB
C
                IF (ABS(S2AB - S1AB) .LT. TOL )  THEN
                    STORE = STORE + S2AB + (S2AB - S1AB)/15.0
                ELSE
                    H1 = H2
                    TOL = TOL/2
                    CALL PUSH(A,FA,FD,FC,H1,TOL,S2AC)
                    CALL PUSH(C,FC,FE,FB,H1,TOL,S2CB)
                END IF
C
                IF (TOP .GT. 0  .AND. COUNT .LT. MAXCNT) GO TO 10
C
C               RETURN ANSWER
C
            ANSWER = STORE
            RETURN
            END
C
C   ----------------------------------------------------------------
C
C           SUBROUTINE PUSH WHICH STORES THE VALUES IN A STACK
C
C   ----------------------------------------------------------------
C
        SUBROUTINE PUSH(LPT,LVAL,MIDVAL,RTVAL,STPSZE, TOL,INTGRL)
C
        REAL LPT,LVAL,MIDVAL,RTVAL,STPSZE,TOL,INTGRL
        REAL STACK(7,100)
        INTEGER TOP
        COMMON /SUB/ STACK,TOP
C
            TOP = TOP + 1
            STACK(1,TOP) = LPT
            STACK(2,TOP) = LVAL
            STACK(3,TOP) = MIDVAL
            STACK(4,TOP) = RTVAL
            STACK(5,TOP) = STPSZE
            STACK(6,TOP) = TOL
            STACK(7,TOP) = INTGRL
```

Figure 4.13   (*continued*)

```
      RETURN
      END
C
C     ----------------------------------------------------------------
C
C        SUBROUTINE PUSH RETURNS THE STORED VALUES FROM THE STACK
C
C     ----------------------------------------------------------------
C

      SUBROUTINE POP(LPT,LVAL,MIDVAL,RTVAL,STPSZE, TOL,INTGRL)
C
      REAL LPT,LVAL,MIDVAL,RTVAL,STPSZE,TOL,INTGRL
      REAL STACK(7,100)
      INTEGER TOP
      COMMON /SUB/ STACK,TOP
C
         LPT = STACK(1,TOP)
         LVAL = STACK(2,TOP)
         MIDVAL = STACK(3,TOP)
         RTVAL = STACK(4,TOP)
         STPSZE = STACK(5,TOP)
         TOL = STACK(6,TOP)
         INTGRL = STACK(7,TOP)
         TOP = TOP - 1
      RETURN
      END
C
C     ----------------------------------------------------------------
C
C        THE INTEGRAND IS DEFINED
C
C     ----------------------------------------------------------------
C
      REAL FUNCTION F(X)
      REAL X
      INTEGER MAXCNT, FCNCNT
      COMMON MAXCNT, FCNCNT
         FCNCNT = FCNCNT + 1
         F = 1.0/(X*X)
      RETURN
      END

            OUTPUT FOR PROGRAM 5

      THE INTEGRAL VALUE IS    4.000060

      THE NUMBER OF FUNCTION CALLS =    17
```

The seventh and final program of this section (see Fig. 4.15) is one that implements the Gaussian quadrature method. For simplicity, the subprogram GAUSSQ can solve only for 2, 3, or 4 points. The choice is determined by the value given to the variable— NTERMS—which can only handle values for 2, 3, or 4. It would be easy to add additional T and W values in the IF (NTERMS.EQ.) . . . statement for additional points. Although we use only 4 points in this example program, the solution compares very well with the exact integral, which is 0.6588234. It would be quite easy to extend this program to two dimensions. The integrand is defined in FUNCTION F(X), and only the limits of integration, A, B, and the number of terms, NTERMS, need to be input.

```
      SUBROUTINE DBLINT(FCT,XA,XB,YA,YB,TOL,RESULT)
C
C ----------------------------------------------------------------------
C
C     SUBROUTINE DBLINT :
C                         THIS ROUTINE COMPUTES THE INTEGRAL OF A
C     FUNCTION OF TWO VARIABLES. THE ROMBERG METHOD IS USED. INITIALLY,
C     FOUR SUBDIVISIONS ARE USED. THESE ARE HALVED UNTIL THE TOLERANCE
C     IS MET, WITH A MAXIMUM NUMBER OF SUBDIVIDINGS OF FIVE.
C
C ----------------------------------------------------------------------
C
C     PARAMETERS ARE :
C
C     FCT    - FUNCTION SUBPROGRAM TO COMPUTE F(X,Y). DECLARED EXTERNAL
C                IN CALLING PROGRAM.
C     XA,XB  - LOWER AND UPPER LIMITS FOR X.
C     YA,YB  - LOWER AND UPPER LIMITS FOR Y.
C     TOL    - TOLERANCE TO TERMINATE INTEGRATION. WHEN NOT MET, A
C                MESSAGE IS PRINTED AND LAST VALUE RETURNED.
C     RESULT - RETURNS VALUE OF INTEGRAL TO CALLER.
C     ARRAY  - DOUBLY SUBSCRIPTED ARRAY TO HOLD INTERMEDIATE VALUES
C                FOR COMPARISON AND EXTRAPOLATION.
C
C A FUNCTION SUBPROGRAM NAMED SUMROW, IS CALLED TO COMPUTE SUMS
C ACROSS ONE ROW OF THE REGION.
C
C ----------------------------------------------------------------------
C
      REAL FCT,XA,XB,YA,YB,TOL,RESULT
      INTEGER I,N,J,K
      REAL ARRAY(6,6),DELX,DELY,SUMROW,Y
      EXTERNAL FCT
C
C ----------------------------------------------------------------------
C
C INITIALIZE DEL VALUES AND SUM TOP AND BOTTOM ROWS
C
      DELX = (XB - XA) / 4.0
      DELY = (YB - YA) / 4.0
      N = 4
      ARRAY(1,1) = SUMROW(FCT,XA,XB,YA,DELX,N) +
     +             SUMROW(FCT,XA,XB,YB,DELX,N)
C
C ----------------------------------------------------------------------
C
C GET THE SUMS FOR INTERMEDIATE ROWS
C
      Y = YA
      DO 10 I = 2,N
         Y = Y + DELY
         ARRAY(1,1) = ARRAY(1,1) + 2.0*SUMROW(FCT,XA,XB,Y,DELX,N)
   10 CONTINUE
      ARRAY(1,1) = ARRAY(1,1) * DELX * DELY / 4.0
C
C ----------------------------------------------------------------------
C
C NOW HALVE THE VALUES OF DELX AND DELY, RECOMPUTE THE INTEGRAL, AND
C EXTRAPOLATE, THEN TEST TO SEE IF TOLERANCE IS MET. REPEAT UP TO
C FIVE TIMES.
C
```

Figure 4.14 Program 6.

Figure 4.14   (*continued*)

```
      DO 40 J = 1,5
        DELX = DELX / 2.0
        DELY = DELY / 2.0
        N = 2*N
C
C     ----------------------------------------------------------------
C
C     DO TOP AND BOTTOM ROWS FIRST
C
        ARRAY(J+1,1) = SUMROW(FCT,XA,XB,YA,DELX,N) +
     +                 SUMROW(FCT,XA,XB,YB,DELX,N)
C
C     THEN THE INTERMEDIATE ROWS
C
        Y = YA
        DO 20 I = 2,N
          Y = Y + DELY
          ARRAY(J+1,1) = ARRAY(J+1,1) + 2.0*SUMROW(FCT,XA,XB,Y,DELX,N)
  20    CONTINUE
        ARRAY(J+1,1) = ARRAY(J+1,1) * DELX * DELY / 4.0
C
C     ----------------------------------------------------------------
C
C     NOW WE EXTRAPOLATE
C
        DO 30 K = 1,J
          ARRAY(J+1,K+1) = ARRAY(J+1,K) + 1.0 / (4.0**K - 1.0) *
     +                   ( ARRAY(J+1,K) - ARRAY(J,K) )
  30    CONTINUE
        IF ( ABS(ARRAY(J+1,J+1) - ARRAY(J+1,J)) - TOL ) 50,50,40
  40  CONTINUE
C
C     ----------------------------------------------------------------
C
C     WE HAVE A NORMAL TERMINATION OF LOOP 40 ONLY WHEN THE TOLERANCE IS
C     NOT MET, SO PRINT MESSAGE AND RETURN.
C
      PRINT 201, TOL
      RESULT = ARRAY(6,6)
      RETURN
  50  RESULT = ARRAY(J+1,J+1)
      RETURN
C
 201  FORMAT(/' TOLERANCE OF ',E14.7,' NOT MET AFTER FIVE ',
     +        'EXTRAPOLATIONS.'/)
      END
C
C     ----------------------------------------------------------------
C
      REAL FUNCTION SUMROW(FCT,XA,XB,Y,DELX,N)
C
C     ----------------------------------------------------------------
C
C     FUNCTION SUMROW :
C                      THIS FUNCTION COMPUTES THE WEIGHTED SUM FOR
C     TRAPEZOIDAL RULE INTEGRATION ACROSS ONE ROW OF A REGION, FROM
C     XA TO XB WITH INTERVALS OF DELX, WHERE THE VALUE OF Y IS Y.
C
C     ----------------------------------------------------------------
C
C     PARAMETERS ARE :
C
```

Figure 4.14 (*continued*)

```
C     FCT    - EXTERNAL FUNCTION THAT COMPUTES F(X,Y)
C     XA,XB  - LIMITS FOR X VALUES
C     Y      - VALUE OF Y
C     DELX   - STEP SIZE FOR X
C     N      - NUMBER OF INTERVALS
C
C     ----------------------------------------------------------------
C
C GET FIRST AND LAST VALUES TO START.
C
      SUMROW = FCT(XA,Y) + FCT(XB,Y)
C
C NOW ADD IN THE INTERMEDIATE VALUES.
C
      X = XA
      DO 10 I = 2,N
        X = X + DELX
        SUMROW = SUMROW + 2.0 * FCT(X,Y)
   10 CONTINUE
      RETURN
      END
```

```
      PROGRAM GAUSQDT(INPUT,OUTPUT)
C
C     ----------------------------------------------------------------
C
C     THIS PROGRAM IMPLEMENTS THE GAUSSIAN QUADRATURE METHOD AS PRESENTED
C     IN SECTION 4.12. IT CAN HANDLE ONLY 2,3,4 TERMS DEPENDING ON THE
C     PARAMETER, NTERMS.
C
C     THIS PROGRAM SOLVES THE EXAMPLE IN SECTION 4.12.
C
C     ----------------------------------------------------------------
C
      REAL A,B, GAUSSQ
      EXTERNAL F
C
      DATA A,B,NTERMS/0.2,1.5,4/
C
      PRINT '(//)'
      PRINT 100 , GAUSSQ(F,A,B,NTERMS)
  100 FORMAT('THE VALUE OF THE INTEGRAL IS: ', F10.7///)
C
      STOP
      END
C
C              FUNCTION GAUSSQ(A,B,NTERMS)
C
      REAL FUNCTION GAUSSQ(F,A,B,NTERMS)
C
C     ----------------------------------------------------------------
C
C     INPUT:  A,B  - INTERVAL [A,B] OVER WHICH FUNCTION IS TO BE
C                    INTEGRATED.
C             NTERMS - NUMBER OF TERMS TO BE USED
C             F  -  THE INTEGRAND FUNCTION
```

Figure 4.15 Program 7.

Figure 4.15   (*continued*)

```
C
C        OUTPUT: GAUSSQ  -  THE VALUE OF THE INTEGRAL
C
C    ----------------------------------------------------------------
C
C        LOCAL VARIABLES:
C
C            BPLUSA:   (B + A)/2
C            BLESSA:   (B - A)/2
C            T:        THE VECTOR OF T-VALUES IN [-1,1]
C            W:        THE VECTOR OF WEIGHTS
C
C    ----------------------------------------------------------------
C
        REAL A,B, T(4), W(4), BPLUSA, BLESSA, SUM
        INTEGER NTERMS,  I,J
C
        IF (NTERMS .EQ. 2) THEN
              T(1) = -0.5773502691
              T(2) = -    T(1)
              W(1) = 1.0
              W(2) = 1.0
           ELSE IF (NTERMS .EQ. 3) THEN
              T(1) = -0.7745966691
              T(2) = 0.0
              T(3) = -T(1)
              W(1) = 0.5555555556
              W(2) = 0.8888888839
              W(3) = W(1)
           ELSE IF (NTERMS .EQ. 4) THEN
              T(1) = -0.8611363115
              T(2) = -0.3399810435
              T(3) = -T(2)
              T(4) = -T(1)
              W(1) = 0.3478548451
              W(2) = 0.6521451549
              W(3) = W(2)
              W(4) = W(1)
           END IF
C
C    ----------------------------------------------------------------
C
C        COMPUTE INTEGRAL FROM ABOVE VALUES
C
C    ----------------------------------------------------------------
C
        BLESSA = (B - A)/2.0
        BPLUSA = (B + A)/2.0
        SUM = 0.0
        DO 10 I = 1, NTERMS
            SUM = SUM + W(I)*F(BLESSA*T(I) + BPLUSA)
10      CONTINUE
C
        GAUSSQ = BLESSA * SUM
        RETURN
        END
C
C    ----------------------------------------------------------------
C
C        DEFINE FUNCTION TO BE INTEGRATED
C
```

Figure 4.15   (*continued*)

```
C   ------------------------------------------------------------------
C
      REAL FUNCTION F(X)
C
      REAL X
      F = 1.0/EXP(X*X)
      RETURN
      END


             OUTPUT FOR PROGRAM 7

      THE VALUE OF THE INTEGRAL IS:      .6588291
```

EXERCISES

Section 4.2

▶ 1. The following table is for $(1 + \log x)$. Determine estimates of $d(1 + \log x)/dx$ at $x = 0.15$, 0.19, and 0.23, using (a) one term, (b) two terms, and (c) three terms of Eq. (4.9). By comparing to the analytical values, determine the errors of each estimate.

x	$1 + \log x$	Δ	Δ^2	Δ^3
0.15	0.1761			
		0.0543		
0.17	0.2304		−0.0059	
		0.0484		0.0009
0.19	0.2788		−0.0050	
		0.0434		0.0011
0.21	0.3222		−0.0039	
		0.0395		0.0006
0.23	0.3617		−0.0033	
		0.0362		0.0005
0.25	0.3979		−0.0027	
		0.0335		0.0002
0.27	0.4314		−0.0025	
		0.0310		0.0005
0.29	0.4624		−0.0020	
		0.0290		
0.31	0.4914			

▶ 2. Write expressions for the errors in each computation of Exercise 1, by properly interpreting Eq. (4.11). From these expressions find upper and lower bounds for each of the computations.

3. If one wished the derivative of $(1 + \log x)$ at $x = 0.31$, the table in Exercise 1 would be inadequate for this, using Eq. (4.9), because the necessary forward differences cannot be

computed. A formula in terms of backward differences can be derived, however, by differentiating the Newton–Gregory backward-interpolation polynomial. Do this, and use three terms to obtain $d(1 + \log x)/dx$ at $x = 0.31$ from the data given in the table of Exercise 1.

4. Derive the error term for the formula of Exercise 3, and show that the "next-term rule" applies.

5. A central-difference approximation for derivatives is more accurate than a forward- or backward-difference approximation, as discussed in Section 4.2. Repeat Exercise 1(b), but use Eq. (4.13) and compare the errors.

6. Find bounds on the errors of Exercise 5 and compare the actual errors to the bounds.

7. Exercises 1 through 6 used values from a table with equispaced x-values. The data below are not equispaced, so Eq. (4.4) is needed to compute the derivative and errors are given by Eq. (4.5). Use these to compute

   a) $f'(1.4)$ with one, two, three terms of Eq. (4.4).
   b) $f'(2.1)$ with one, two, three terms of Eq. (4.4).
   c) Estimate the errors in parts (a) and (b).

x	1.4	1.8	1.9	2.1
$f(x)$	2.8867	3.4017	3.5355	3.7947

Section 4.3

8. The Newton–Gregory backward-interpolating formula can be developed by expanding the symbolic relation

$$f_s = E^s f_0 = (1 - \nabla)^{-s} f_0.$$

Using this representation of the interpolation formula, derive the derivative formula required in Exercise 3, by the symbolic technique.

▶ 9. Suppose the nature of the function tabulated in Exercise 1 were unknown. How could estimates of the errors of each computation in that problem be determined? Compare with the error bounds determined in Exercise 2.

Section 4.4

10. If one begins with Stirling's interpolation formula (Section 3.5), differentiates, and then evaluates at $x = x_0$, one obtains a derivative formula in terms of central differences. Show that, when this is done, the coefficients are the same as those given by a horizontal path through the lozenge diagram, Fig. 4.1, starting at f_0.

11. If one differentiates Stirling's formula twice and then sets $s = 0$, a central-difference formula for the second derivative results. Show that the coefficients so obtained match those on a horizontal path from f_0 in Fig. 4.3.

12. Make a lozenge diagram, similar to Figs. 4.1 and 4.3, for the third derivative.

13. If the difference table of Exercise 1 is extended to the right, differences up to the eighth order can be calculated. The lozenge diagram, Fig. 4.1, can also be extended to a similar order of differences. Discuss whether the accuracy of estimates of the derivative of the logarithm function will be improved by such extensions.

14. An alternative way of deriving error terms for derivative formulas is through Taylor series expansions. For example, the error term of Eq. (4.23) can be obtained by expanding $f_1 = f(x_0 + h)$ and $f_{-1} = f(x_0 - h)$ each about the point $x = x_0$, and then combining the two series. Show that the error term of Eq. (4.23) is $-\frac{1}{6}h^2 f'''(\xi)$.

15. For unevenly spaced data, a derivative expression analogous to Eq. (4.23) is

$$f'(x)\Big|_{x=x_0} \doteq \frac{f(x_1) - f(x_{-1})}{x_1 - x_{-1}}.$$

In this case $x_1 - x_0 \neq x_0 - x_{-1}$. Take $x_1 - x_0 = h$ and $x_0 - x_{-1} = \alpha h$, $\alpha \neq 1$. Show by the Taylor series method that the error is now $O(h)$ and not $O(h^2)$ unless $\alpha = 1$.

16. Show that the results of Exercise 15 are exactly what is given by divided-difference formulas.

Section 4.5

▶17. Use the extrapolation method to determine $f'(0.23)$ for the data of Exercise 1, to an $O(h^6)$ extrapolation.

18. Show that the first-order extrapolation for $f'(x_0)$ with Δx-values differing by $2:1$ is the same as the formula

$$f_0' = \frac{1}{h}\left(\frac{\Delta f_{-1} + \Delta f_0}{2} - \frac{1}{6}\frac{\Delta^3 f_{-2} + \Delta^3 f_{-1}}{2}\right),$$

where h is the smaller of the two Δx-values. Note that this comes directly from the lozenge diagram.

19. Show that a second-order extrapolation for $f'(x_0)$ is the equivalent of using central differences through $\Delta^5 f$.

20. Consider whether extrapolation similar to that of Section 4.5 can be performed when derivatives for unevenly spaced data are estimated through divided differences. You may want to use Taylor series expansions as in Exercise 14. If you succeed in getting a formula, apply it to get $f'(1.8)$ from the data in Exercise 7.

Section 4.6

21. The data below are for $\tan \theta$ in the neighborhood of $\theta = 1$ radian.

▶ a) Truncating the data to four significant digits, compute the derivative at $\theta = 1.0$, using a central-difference formula (Eq. (4.8)) but with varying step size. At what point is the greatest accuracy obtained?

▶ b) Repeat part (a) but with the data rounded to four digits.

c) Repeat with the data as given below, which are rounded to six digits.

θ	$\tan \theta$	θ	$\tan \theta$
0.9000	1.26016	1.0001	1.55775
0.9900	1.52368	1.0010	1.56084
0.9990	1.55399	1.0100	1.59221
0.9999	1.55706	1.1000	1.96476
1.0000	1.55741		

22. Repeat Exercise 21, but use forward- and backward-difference formulas.

23. Repeat Exercise 21, but use a central-difference formula for second derivatives.

24. Use Program 2 (Fig. 4.10) to determine the optimum step size for most accurate computation of first and second derivatives for a number of functions, including $\sin x$, $\cos x$, $\sinh x$, $\cosh x$, $\ln x$. If your computer has double precision in FORTRAN (also if extended precision is available), modify the program to investigate for these cases.

Section 4.7

25. Equations (4.36) through (4.41) are Newton–Cotes formulas, and their error terms are derived from forward-difference interpolating polynomials. Rederive these, beginning with backward-difference formulas.

▶26. Prove the assertion made at the end of Section 4.7 that all Newton–Cotes integration formulas of even order have an order of h in the error term one greater than would normally be expected.

27. Derive the Newton–Cotes formulas of orders 4 and 5.

28. Beginning with the divided-difference form of the interpolating polynomial, rederive the Newton–Cotes formulas of orders 1, 2, and 3. (Assume, of course, that $(x_{i+1} - x_i) = h$ for all i.)

29. Continue Exercise 28 to derive the error terms.

Section 4.8

▶30. The following values of a function are given.

x	$f(x)$	x	$f(x)$
1.0	1.543	1.5	2.352
1.1	1.668	1.6	2.577
1.2	1.811	1.7	2.828
1.3	1.971	1.8	3.107
1.4	2.151		

Find $\int_{1.0}^{1.8} f(x)\, dx$, using the trapezoidal rule with

a) $h = 0.1$ b) $h = 0.2$ c) $h = 0.4$

31. The function tabulated in Exercise 30 is $\cosh x$. What are the errors of the computations in parts (a), (b), (c)? How closely are the errors proportional to h^2? What other errors besides that of Eq. (4.45) are present?

32. In Section 4.8, the value of $\int_{1.8}^{3.4} e^x\, dx$ is computed as 23.9944 using the trapezoidal rule with $h = 0.2$, and is in error by -0.08. Equation (4.45) shows this to fall within the predicted bounds for the error. Suppose we did not know what $f(x)$ was; estimate the error using second-order differences of the data in Table 4.3.

▶33. If one wished to compute $\int_{1.8}^{3.4} e^x\, dx$ using the trapezoidal rule, and wished to be certain of five-decimal accuracy (error $\leqslant 0.000005$), how small must h be?

34. Find $\int_0^2 f(x)\, dx$:

x	$f(x)$	x	$f(x)$
0	1.0000	1.08	0.3396
0.12	0.8869	1.43	0.2393
0.53	0.5886	2.00	0.1353
0.87	0.4190		

Section 4.9

35. Extrapolate the individual answers of Exercise 30 to get estimates of improved accuracy. What are the orders of the errors of these extrapolations?

▶36. Use extrapolation to the limit to evaluate $\int_0^1 f(x)\, dx$, getting a result with error $O(h^6)$:

x	$f(x)$	x	$f(x)$
0	0.3989	0.50	0.3521
0.25	0.3867	0.75	0.3011
		1.00	0.2420

37. Use Romberg integration (successive extrapolations with h halved each time) to evaluate $\int_1^2 dx/x$. Carry six decimals and continue until no change in the fifth place occurs. Compare to the analytical value $\ln 2 = 0.69315$.

Section 4.10

38. Repeat Exercise 30 using Simpson's $\frac{1}{3}$ rule with (a) $h = 0.1$, (b) $h = 0.2$, (c) $h = 0.4$.

39. Use the error expression of Eq. (4.49) to estimate the maximum and minimum errors to be expected in Exercise 38. ($f(x)$ in Exercise 30 is $\cosh x$.)

▶40. Evaluate $\int_0^{1.0} e^x\, dx$ by Simpson's $\frac{1}{3}$ rule, choosing h small enough to guarantee five-decimal accuracy. How large can h be?

41. In the text it is stated that the results of using Simpson's $\frac{1}{3}$ rule and the extrapolated result from two applications of the trapezoidal rule (with h changed by 2:1) are identical. Show that this will always be true.

42. Compute

$$\int_0^1 \frac{\sin x}{x}\, dx$$

using Simpson's $\frac{1}{3}$ rule with $h = 0.5$ and with $h = 0.25$. Extrapolate the results. What is the order of the error for the extrapolated result?

43. Repeat Exercise 40, but use Simpson's $\frac{3}{8}$ rule.

▶44. Apply a combination of the two Simpson rules to evaluate the integral from $x = 3$ to $x = 6.5$. Note that this is a technique to handle an odd number of panels:

x	$f(x)$	x	$f(x)$
3.0	0.33906	5.0	−0.32758
3.5	0.13738	5.5	−0.34144
4.0	−0.06604	6.0	−0.27668
4.5	−0.23106	6.5	−0.15384

45. The fact that Simpson's $\frac{1}{3}$ rule has an error term dependent on the third derivative and not the second (as is also true for the $\frac{3}{8}$ rule) means that these integrations are exact for any cubic polynomial. However, the $\frac{1}{3}$ rule is based on fitting a quadratic through three equispaced points. The implication of this is that the area under any cubic, from $x = a$ to $x = b$, is

exactly the same as the area under the parabola that intersects the cubic at the two endpoints, and also at $x = \frac{1}{2}(a + b)$. Prove this.

Section 4.11

▶46. Use the symbolic method to determine a formula for three-panel integration, using a polynomial of degree m.

47. Integration from x_0 to x_1, as in Eq. (4.51), is an arbitrary choice taken for convenience only. If the limits were taken from x_{-1} to x_0, a similar formula for one-panel integration would result, but different coefficients would be obtained. Carry out this computation.

48. Perform a computation similar to Exercise 47, but get a two-panel integration formula from x_{-1} to x_1.

49. Use the method of undetermined coefficients to derive (a) the trapezoidal rule, (b) Simpson's $\frac{3}{8}$ rule.

50. Use the method of undetermined coefficients to derive the central-difference formulas, Eqs. (4.23) and (4.24), for $f'(x_0)$ and $f''(x_0)$.

Section 4.12

▶51. Evaluate

$$\int_0^1 \frac{\sin x}{x}\, dx$$

by a three-term Gaussian quadrature formula.

52. By computing with Gaussian quadrature formulas of increasing complexity, determine how many terms are needed to evaluate $\int_{1.8}^{3.4} e^x\, dx$ to five-decimal accuracy. (The exact value of the integral is 23.9144526.)

53. An n-term Gaussian quadrature formula assumes that a polynomial of degree $2n - 1$ is used to fit the function between $x = a$ and $x = b$. Does this mean that the error term is the same as for the Newton–Cotes integration formulas based on polynomials of $(2n - 1)$ degree?

Section 4.13

54. As discussed in Section 4.13, an integral with infinite limits can be approximated numerically if it is convergent. For example, the exponential integral $Ei(x)$ can be evaluated by taking the upper limit U sufficiently large in

$$Ei(x) = \int_x^\infty \frac{e^{-v}}{v}\, dv \doteq \int_x^U \frac{e^{-v}}{v}\, dv.$$

We know that U is "sufficiently large" when the additional contributions of making U larger are negligible. Estimate $Ei(0.5)$ using the trapezoidal rule. Note that one can use larger subintervals as v increases. Compare to the tabular value of 0.5598.

55. Apply the method of Section 4.13 to show that

$$\int_{-\infty}^\infty \frac{1}{\sqrt{2\pi}} e^{-x^2/2}\, dx = 1.$$

(The values of the integrand are available as ordinates of the standard normal curve.)

▶56. Evaluate these integrals

a) $\displaystyle\int_0^2 \frac{dx}{\sqrt{x}}$ b) $\displaystyle\int_0^1 \frac{dx}{\sqrt{1 - x}}$ c) $\displaystyle\int_0^\infty \frac{x\, dx}{e^x + 1}$ (Analytical value is $\pi^2/12$)

57. In integrating $\int_a^b f(x)\,dx$, Gaussian quadrature does not require the value of $f(x)$ at the endpoints. Therefore that technique has special utility when there are singularities at the endpoints of the interval of integration. Apply this to Exercise 56 (a) and (b).

▶58. It often helps to break up an integral into parts. Consider

$$\int_0^\infty \frac{dx}{e^x + e^{-x}}.$$

At $x = 10$, e^{-x} is only 4×10^{-5}, while e^x is 22,026.5; for $x \geq 10$, the integral can be considered to be just $\int dx/e^x$. Evaluate the integral numerically as

$$\int_0^\infty \frac{dx}{e^x + e^{-x}} = \int_0^{10} \frac{dx}{e^x + e^{-x}} + \int_{10}^\infty \frac{dx}{e^x}.$$

59. One must be cautious in integrating on a computer. $\int_0^1 dx/x$ surely diverges, but you will probably not find this to be true when calculating numerically because the small values of delta x near $x = 0$ may cancel the contribution to the total. It will depend on how one defines the step size as x gets small. See what you get as a result of numerically evaluating this integral.

Section 4.14

60. a) Use a calculator to verify the results in the example of Section 4.14.
    b) Redo the example of Section 4.14 using an adaptive trapezoidal method. How many function evaluations would be required? Compare this number with the number required if you did not use an adaptive method.

61. Most programs will compute the appropriate step size h in the manner described in the discussion of adaptive integration. However, this could lead to significant errors. For instance,

$$\int_0^{\pi/2} \sin^2(16x)\,dx = \frac{\pi}{4},$$

but it is easy to see that $S_1(0, \pi/2) = S_2(0, \pi/2) = 0$, where $h_1 = (\pi/4)$ and $h_2 = (\pi/8)$. How should one solve this problem correctly using the adaptive method described in this section? (It is interesting to see that the integration function on the HP-15C avoids this error.)

Section 4.15

▶62. Consider this table of data (which are obviously for $f(x) = 1/x$).

x	1.0	1.5	2.0	2.5	3.0
$f(x)$	1.000	0.667	0.500	0.400	0.333

Find values for $f'(x)$ and $f''(x)$ at $x = 1.5$, 2.0, and 2.5 from the cubic spline functions that approximate $f(x)$. Compare to analytical values for $f'(x)$ and $f''(x)$ to determine their errors. Also compare to derivative values computed from central-difference formulas.

63. The comparison in Exercise 62 between the spline and central-difference values may favor the spline method because it is based on approximation by a cubic while the central-difference formulas approximate $f(x)$ with a quadratic. Suppose we fit the function by polynomials of degree 3 and 4. How do estimates of $f'(x)$ ad $f''(x)$ from these approximations compare to those from the cubic splines?

64. The natural spline in Exercise 62 is handicapped by setting the second derivatives to zero at the endpoints, while the correct values are 2 and 2/27. Repeat Exercise 63 for a spline that has the correct values for its second derivative at $x = 1$ and $x = 3$.

▶65. Find $\int_0^2 \operatorname{sech} x \, dx$ by integrating the natural cubic spline that fits at five equispaced points on [0, 2]. Compare to the analytical value; also compare to the Simpson's $\frac{1}{3}$ rule value.

66. In Exercise 65, the natural spline assumes that the second derivatives at the extremes of the interval are zero. How does the value of the integral change when conditions 2, 3, and 4 are employed?

67. The best values to use in a cubic spline for the values of its second derivatives at the endpoints should be the correct values of $f''(x)$. Repeat Exercise 65, using the correct values for $f''(x)$ where $f(x) = \operatorname{sech} x$.

Section 4.16

68. In connection with the first example of Section 4.16, it was stated that it is immaterial which integral of a double integral (with constant limits) we integrate first. Confirm this by evaluating $\int_{1.5}^{3.0}\int_{0.2}^{0.6} f(x, y) \, dy \, dx$, with $f(x, y)$ given by Table 4.8, performing the integration first with respect to y.

▶69. Write the pictorial operators similar to that portrayed in Eq. (4.61) that result by using:

a) Simpson's $\frac{1}{3}$ rule in the y-direction and the trapezoidal rule in the x-direction;
b) Simpson's $\frac{1}{3}$ rule in both directions;
c) Simpson's $\frac{3}{8}$ rule in both directions.
d) What conditions are placed on the number of panels in each direction by the methods employed in parts (a), (b), and (c)?

70. Since Simpson's $\frac{1}{3}$ rule is accurate for a cubic, evaluation of the triple integral below using this rule should be exact. Confirm this by evaluating both numerically and analytically:

$$\int_0^1 \int_0^1 \int_0^1 x^3 y z^2 \, dx \, dy \, dz.$$

Use Eq. (4.65) adapted to this case.

71. Draw a pictorial operator (three dimensions) to represent the formula used in Exercise 70. It is perhaps easiest to do this with three widely separated planes on which one indicates the coefficients (see illustration).

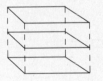

▶72. Evaluate

$$\int_{0.1}^{0.7} \int_{-.2}^{0.6} e^x \sin y \, dy \, dx$$

a) Using the trapezoidal rule in both directions, $\Delta x = \Delta y = 0.1$.
b) Using Simpson's $\frac{1}{3}$ rule in both directions, $\Delta x = \Delta y = 0.1$.
c) Using Gaussian quadrature, three-term formulas in both directions.

Section 4.18

73. Solve Exercise 72 by extrapolating from the trapezoidal-rule evaluations using $\Delta x = \Delta y = 0.2$, and $\Delta x = \Delta y = 0.1$. This should give the same result as part (b) of Exercise 72. Does it?

74. Integrate $\int_0^1\int_0^1 (x^2 + y^2)\, dx\, dy$ using the trapezoidal rule with varying Δx and Δy. Show that the errors decrease proportionately to h^2.

Section 4.19

▶75. Integrate $\int\int \sin x \sin y\, dx\, dy$ over the region defined by that portion of the unit circle that lies in the first quadrant. Integrate first with respect to x holding y constant with $\Delta x = 0.25$. Subdivide the vertical lines into four panels.

   a) Use the trapezoidal rule.   b) Use Simpson's $\frac{1}{3}$ rule.

76. The order of integration in multiple integration may usually be changed. Evaluate the integral of Exercise 75 by integrating first with respect to y, holding x constant. The integral then becomes

$$\int_0^1 \int_0^{\sqrt{1-x^2}} \sin x \sin y\, dy\, dx.$$

▶77. Integrate the function $e^{-x^2 y^2}$ over the region bounded by the two parabolas $y = x^2$ and $y = 2x^2 - 1$. Note that the integrand is an even function and that the region is symmetrical about the y-axis, so that the integral over half the region may be evaluated and then doubled. Choose reasonable values for Δx and Δy.

78. a) Use Gaussian quadrature to solve Exercise 75. Use three-point formulas.
   b) Solve Exercise 77 using four-point Gaussian quadrature formulas.

APPLIED PROBLEMS AND PROJECTS

79. Differential thermal analysis is a specialized technique that can be used to determine transition temperatures and the thermodynamics of chemical reactions. It has special application in the study of minerals and clays. Vold (*Anal. Chem.* **21**, 683 (1949)) describes the technique. In this method, the temperature of a sample of the material being studied is compared to the temperature of an inert reference material when both are heated simultaneously under identical conditions. The furnace housing the two materials is normally heated so that its temperature T_f increases (approximately) linearly with time (t), and the difference in temperatures (ΔT) between the sample and the reference is recorded. Some typical data are

t, min	0	1	2	3	4	5	6	7
ΔT, °F	0.00	0.34	1.86	4.32	8.07	13.12	16.80	18.95
T_f, °F	86.2	87.8	89.4	91.0	92.7	94.3	95.9	97.5

t	8	9	10	11	12	13	14	15	16
ΔT	18.07	16.69	15.26	13.86	12.58	11.40	10.33	8.95	6.46
T_f	99.2	100.8	102.3	103.9	105.5	107.1	108.6	110.2	111.8

t	17	18	19	20	21	22	23	24	25
ΔT	4.65	3.37	2.40	1.76	1.26	0.88	0.63	0.42	0.30
T_f	113.5	115.1	116.8	118.4	120.0	121.6	123.2	124.9	126.5

The ΔT values increase to a maximum, then decrease. This is due to the heat evolved in an exothermic reaction. One item of interest is the time (and furnace temperature) when the reaction is complete. Vold shows that the logarithm of ΔT should decrease linearly after the reaction is over; while the chemical reaction is occurring, the data depart from this linear relation. She used a graphical method to find this point. Perform numerical computations to find, from the above data, the time and the furnace temperature when the reaction terminates. Compare the merits of doing it graphically or numerically.

80. The temperature difference data in Problem 79 can be used to compute the heat of reaction. To do this, the integral of the values of ΔT are required, from the point where the reaction begins (which is at the point where ΔT becomes nonzero) to the time when the reaction ceases, as found in Exercise 79. Determine the value of the required integral. Which of the methods of this chapter should give the best value for the integral?

81. *Fugacity* is a term used by engineers to describe the available work from an isothermal process. For an ideal gas, the fugacity f is equal to its pressure P, but for real gases,

$$\ln \frac{f}{P} = \int_0^P \frac{C - 1}{P} \, dp,$$

where C is the experimentally determined *compressibility factor*. For methane, values of C are

P (atm)	C	P (atm)	C
1	0.9940	80	0.3429
10	0.9370	120	0.4259
20	0.8683	160	0.5252
40	0.7043	250	0.7468
60	0.4515	400	1.0980

Write a program that reads in the P and C values and uses them to compute and print f corresponding to each pressure given in the table. Assume that the value of C varies linearly between the tabulated values (a more precise assumption would fit a polynomial to the tabulated C values). The value of C approaches 1.0 as P approaches 0.

82. The stress developed in a rectangular bar when it is twisted can be computed if one knows the values of a torsion function U that satisfies a certain partial-differential equation. Chapter 7 describes a numerical method that can determine values of U. To compute the stress, it is necessary to integrate $\iint U \, dx \, dy$ over the rectangular region for which the data given below apply. (You may be able to simplify the integration because of the symmetry in the data.)

x \ y	0.0	0.2	0.4	0.6	0.8	1.0	1.2
0.0	0	0	0	0	0	0	0
0.2	0	2.043	3.048	3.354	3.048	2.043	0
0.4	0	3.123	4.794	5.319	4.794	3.123	0
0.6	0	3.657	5.686	6.335	5.686	3.657	0
0.8	0	3.818	5.960	6.647	5.960	3.818	0
1.0	0	3.657	5.686	6.336	5.686	3.657	0
1.2	0	3.123	4.794	5.319	4.794	3.123	0
1.4	0	2.043	3.048	3.354	3.048	2.043	0
1.6	0	0	0	0	0	0	0

83. Make a critical comparison of the accuracy of Newton–Cotes integration formulas compared to Gaussian quadrature. Test the formulas for a variety of functions for which you can calculate the integrals analytically. Select some functions that are smooth, some that have sharp changes in value, and some with periodic behavior.

84. Write a general-purpose subroutine that performs Gaussian quadrature. You should have the subroutine change the limits of the integration appropriately and call a function subprogram to compute function values, with the name of the function subprogram passed as an argument. It should also receive as an argument the degree of the formula to be employed.

85. The data in Exercise 62 exhibit round-off in the second and in the last entries. By recalculating the spline function, with more and less accurate values of $f(1.5)$ and $f(3.0)$, determine how the values of S are affected by the precision of the data. Also calculate how the precision affects the estimates of $f'(x)$ and $f''(x)$. Compare the effects of the precision of the original data, using a cubic spline, with the effects of precision on the values of $f'(x)$ and $f''(x)$ when central-difference formulas are used.

5

Numerical Solution of Ordinary Differential Equations

5.0 CONTENTS OF THIS CHAPTER

Ordinary differential equations describe many physical situations: spring–mass systems, resistor–capacitor–inductance circuits, bending of beams, chemical reactions, pendulums, and so on. Their prominence in applied mathematics is due to the fact that most scientific laws are more readily expressed in terms of rates of change. For example,

$$\frac{du}{dt} = -0.27(u - 60)^{5/4}$$

is an equation describing (approximately) the rate of change of temperature u of a body losing heat by natural convection with constant-temperature surroundings. This is termed a first-order differential equation because the highest-order derivative is the first.

If the equation contains derivatives of nth order, it is said to be an nth-order differential equation. For example, a second-order equation describing the oscillation of a weight acted upon by a spring, with resistance to motion proportional to the square of the velocity, might be

$$\frac{d^2x}{dt^2} + 4\left(\frac{dx}{dt}\right)^2 + 0.6x = 0,$$

where x is the displacement and t is time.

The solution to a differential equation is the function that satisfies the differential equation and that also satisfies certain initial conditions on the function. In solving a differential equation analytically, one usually finds a general solution containing arbitrary constants and then evaluates the arbitrary constants so that the expression agrees with the initial conditions. For an nth-order equation, n independent initial conditions must usually be known. The analytical methods are limited to certain special forms of equations; elementary courses normally treat only linear equations with constant coefficients when the degree of the equation is higher than the first. Neither of the above examples is linear.

Numerical methods have no such limitations to only standard forms. We obtain the solution as a tabulation of the values of the function at various values of the independent variable, however, and not as a functional relationship. We must also pay a price for our ability to solve practically any equation in that we must recompute the entire table if the initial conditions are changed.

Our procedure will be to explore several methods of solving first-order equations, and then to show how these same methods can be applied to systems of simultaneous first-order equations and to higher-order differential equations. We will use for our typical first-order equation the form

$$\frac{dy}{dx} = f(x, y),$$

$$y(x_0) = y_0.$$

5.1 POPULATION CHARACTERISTICS OF FIELD MICE

Outlines a typical problem from the field of biology/ecology that involves a differential equation.

5.2 TAYLOR-SERIES METHOD

Is a straightforward adaptation of classical calculus to develop the solution as an infinite series. The catch is that a computer usually cannot be programmed to construct the terms and one doesn't know how many terms should be used.

5.3 EULER AND MODIFIED EULER METHODS

Are simple to use but subject to error unless the step size Δx is made very small.

5.4 RUNGE–KUTTA METHODS

Are very popular because of their good efficiency; these are used in most computer programs for differential equations. These are single-step methods, as are the Euler methods. In this section we compare the methods presented so far.

5.5 MULTISTEP METHODS

Are even more efficient than the previous methods but cannot be used at the beginning of the interval of integration; they can be employed after several steps with a single-step method. These methods have a built-in ability to monitor the error in the solution.

5.6 MILNE'S METHOD

Is a multistep method that appears very attractive until one finds that it may be unstable.

5.7 ADAMS–MOULTON METHOD

Is a multistep method that does not suffer from the fault of instability. This section compares the methods of Adams–Moulton and Milne for accuracy and stability through an example.

5.8 CONVERGENCE CRITERIA

Impose additional limitations on the step size beyond the requirement for accuracy; here you are exposed to some theory.

5.9 ERRORS AND ERROR PROPAGATION

Examines the various sources of error in the context of solving differential equations; the important concept of propagated error is examined in a simple case, again giving you a taste of theoretical numerical analysis.

5.10 SYSTEMS OF EQUATIONS AND HIGHER-ORDER EQUATIONS

Are the usual situations in applied problems; fortunately our methods are readily extended to cover them. The various methods are again applied to the same example to show how they compare.

5.11 COMPARISON OF METHODS

Summarizes the various methods and makes a critical comparison of their strengths and weaknesses. The somewhat esoteric notion of "stiff" equations is examined.

5.12 CHAPTER SUMMARY

Is our usual review section.

5.13 COMPUTER PROGRAMS

Gives programs for modified Euler and fourth-order Runge–Kutta procedures that can be applied to single equations, systems, or higher-order equations.

5.1 POPULATION CHARACTERISTICS OF FIELD MICE

An ecologist has been studying the effects of the environment on the population of field mice. Her research shows that the number of mice born each month is proportional to the number of females in the group, and that the fraction of females is normally constant in any group. This implies that the number of births per month is proportional to the total population.

She has located a test plot for further research, which is a restricted area of semiarid land. She has constructed barriers around the plot so mice cannot enter or leave. Under the conditions of the experiment, the food supply is limited, and it is found that the death rate is affected as a result, with mice dying of starvation at a rate proportional to some power of the population. (She also hypothesizes that when the mother is undernourished, the babies have less chance for survival, and that starving males tend to attack each other, but these factors are only speculation.)

The net result of this scientific analysis is the following equation, with N being the number of mice at time t (with t expressed in months.) She has come to you for help in solving the equation; her calculus doesn't seem to apply.

$$\frac{dN}{dt} = aN - BN^{1.7}, \quad \text{with } B \text{ given by Table 5.1.}$$

As the season progresses, the amount of vegetation varies. She accounts for this change in the food supply by using a "constant" B that varies with the season.

If 100 mice were initially released into the test plot and if $a = 0.9$, estimate the number of mice as a function of t, for $t = 0$ to $t = 8$.

In this chapter we will find how such a problem can be solved by numerical methods.

Table 5.1

t	B	t	B
0	0.0070	5	0.0013
1	0.0036	6	0.0028
2	0.0011	7	0.0043
3	0.0001	8	0.0056
4	0.0004		

5.2 TAYLOR-SERIES METHOD

The first method we discuss is not strictly a numerical method, but it is sometimes used in conjunction with the numerical schemes, is of general applicability, and serves as an introduction to the other techniques we will study. Consider the example problem

$$\frac{dy}{dx} = x + y, \quad y(0) = 1. \tag{5.1}$$

(This particularly simple example is chosen to illustrate the method so that you can readily check the computational work. The analytical solution, $y = 2e^x - x - 1$, is obtained immediately by application of the standard methods, and will be compared with our numerical results to show exactly the error at any step.)

We develop the relation between y and x by finding the coefficients of the Taylor series in which we expand y about the point $x = x_0$:

$$y(x) = y(x_0) + y'(x_0)(x - x_0) + \frac{y''(x_0)}{2!}(x - x_0)^2 + \frac{y'''(x_0)}{3!}(x - x_0)^3 + \cdots$$

If we let $x - x_0 = h$, we can write the series as

$$y(x) = y(x_0) + y'(x_0)h + \frac{y''(x_0)}{2}h^2 + \frac{y'''(x_0)}{6}h^3 + \cdots$$

Since $y(x_0)$ is our initial condition, the first term is known from the initial condition $y(0) = 1$. (Since the expansion is about the point $x = 0$, our Taylor series is actually a Maclaurin series in this example.)

We get the coefficient of the second term by substituting $x = 0$, $y = 1$ into the equation for the first derivative, Eq. (5.1):

$$y'(x_0) = y'(0) = 0 + 1 = 1.$$

We get equations for the second- and higher-order derivatives by successively differentiating the equation for the first derivative. Each of these is evaluated corresponding to $x = 0$ to get the various coefficents:

$$y''(x) = 1 + y', \qquad y''(0) = 1 + 1 = 2,$$
$$y'''(x) = y'', \qquad y'''(0) = 2,$$
$$y^{iv}(x) = y''', \qquad y^{iv}(0) = 2,$$
$$\text{and so on.} \qquad y^{(n)}(0) = 2.$$

(Getting the derivatives is deceptively easy in this example. You should compare this to the function $f(x, y) = x/y$, with $y(1) = 1$.)

We then write our series solution for y, letting $x = h$ be the value at which we wish to determine y:

$$y(h) = 1 + h + h^2 + \frac{1}{3}h^3 + \frac{1}{12}h^4 + \text{error}.$$

The solution to our differential equation, $dy/dx = x + y$, $y(0) = 1$, is then given by Table 5.2.

As we computed the last two entries, there was some doubt in our minds as to their accuracy without using more terms of the Taylor series, because the successive terms were decreasing less and less rapidly. We need more terms than we have calculated to get four-decimal-place accuracy.

The error when a convergent Taylor series is truncated is simple to express. We merely take the next term and evaluate the derivative at the point $x = \xi$, $0 < \xi < h$,

Table 5.2

x	y	y, analytical
0	1.000	1.0000
0.1	1.1103	1.1103
0.2	1.2428	1.2428
0.3	1.3997	1.3997
0.4	1.5835 (?)	1.5836
0.5	1.7969 (?)	1.7974

instead of at the point $x = x_0$. This is exactly what we did to write error terms in the previous chapters. The error term of the Taylor series after the h^4 term is

$$\text{Error} = \frac{y^{(v)}(\xi)}{5!}h^5, \qquad 0 < \xi < h.$$

However, this cannot be computed because evaluating the derivative at $x = \xi$ is impossible with ξ unknown, and even bounding it in the interval $[0, h]$ is impossible because the derivatives are known only at $x = 0$ and not at $x = h$.

Numerical analysis is sometimes termed an art instead of a science because, in situations like this, the number of Taylor-series terms to be included is a matter of judgment and experience. We normally truncate the Taylor series when the contribution of the last term is negligible to the number of decimal places to which we are working. However, this is correct only when the succeeding terms become small rapidly enough—in some cases the sum of the many neglected small terms is significant.

The Taylor series is easily applied to a higher-order equation. For example, if we are given

$$y'' = 3 + x - y^2, \qquad y(0) = 1, \qquad y'(0) = -2,$$

we can find the derivative terms in the Taylor series as follows:

$y(0)$, $y'(0)$ are given by the initial conditions.

$y''(0)$,   results by substitution into the differential equation from $y(0)$ and $y'(0)$.

$y'''(0)$, $y^{iv}(0)$, . . . are found by differentiating the equation for the previous order of derivative and substituting previously computed values.

5.3   EULER AND MODIFIED EULER METHODS

As we have seen, the Taylor-series method is awkward to apply if the various derivatives are complicated, and the error is difficult to determine. An even more significant criticism in this computer age is that taking derivatives of arbitrary functions cannot easily be written into a computer program. We look for an approach that is not subject to these disadvantages.

One thing we do know about the Taylor series. The error will be small if the step size h (the interval beyond x_0 where we evaluate the series) is small. In fact, if it is small enough, only a few terms are needed for good accuracy. The Euler method may be thought of as following this idea to the extreme for first-order differential equations. Suppose we choose h small enough that we may truncate after the first-derivative term. Then

$$y(x_0 + h) = y(x_0) + hy'(x_0) + \frac{y''(\xi)}{2}h^2, \qquad x_0 < \xi < x_0 + h.$$

We have written the usual form of the error term for the truncated Taylor series.

In using this equation, the value of $y(x_0)$ is given by the initial condition and $y'(x_0)$ is evaluated from $f(x_0, y_0)$, given by the differential equation, $dy/dx = f(x, y)$. It will of course be necessary to use this method iteratively, advancing the solution to $x = x_0 + 2h$ after $y(x_0 + h)$ has been found, then to $x = x_0 + 3h$, and so on. Adopting a subscript notation for the successive y-values and representing the error by the order relation, we may write the algorithm for the Euler method as

$$y_{n+1} = y_n + hy'_n + O(h^2)\text{error.*} \qquad\qquad \textbf{(5.2)}$$

As an example, we apply this to the simple equation

$$\frac{dy}{dx} = x + y, \qquad y(0) = 1,$$

where the computations can be done mentally. It is convenient to arrange the work as in Table 5.3. Take $h = 0.02$.

Table 5.3

x_n	y_n	y'_n	hy'_n
0	1.0000	1.0000	0.0200
0.02	1.0200	1.0400	0.0208
0.04	1.0408	1.0808	0.0216
0.06	1.0624	1.1224	0.0224
0.08	1.0848	1.1648	0.0233
0.10	1.1081		

(1.1103 analytical, error is 0.0022)

*This is the order of error for one step only, the "local error." As detailed in a later section, over many steps the error becomes $O(h)$. Such accumulated error is termed the *global error*.

Each of the y_n values is computed using Eq. (5.2), adding hy'_n and y_n of the previous line. Comparing the last result to the analytical answer $y(0.10) = 1.1103$, we see that there is only two-decimal-place accuracy, even though we have advanced the solution only five steps. To gain four-decimal accuracy, we must reduce the error at least 22-fold. Since the global error is about proportional to h, we will need to reduce the step about 22-fold to < 0.004.

The trouble with this most simple method is its lack of accuracy, requiring an extremely small step size. Figure 5.1 suggests how we might improve this method with little additional effort.

In the simple Euler method, we use the slope at the beginning of the interval, y'_n, to determine the increment to the function, but this is always wrong. After all, if the slope of the function were constant, the solution is an obvious linear relation. We need to use an average slope over the interval if we hope to estimate the change in y with precision.

Suppose we do this, using the arithmetic average of the slopes at the beginning and end of the interval:

$$y_{n+1} = y_n + h\frac{y'_n + y'_{n+1}}{2}. \tag{5.3}$$

This will surely give an improved estimate for y at x_{n+1}. However, we are unable to employ Eq. (5.3) immediately, because, since the derivative is a function of both x and y, we cannot evaluate y'_{n+1} with y_{n+1} unknown. The modified Euler method surmounts the difficulty by estimating, or "predicting," a value of y_{n+1} by the simple Euler relation, Eq. (5.2), and uses this value to compute y'_{n+1}, giving an improved estimate ("corrected" value) for y_{n+1}. Since the value of y'_{n+1} was computed using the predicted value, of less than perfect accuracy, one is tempted to recorrect the y_{n+1} value as many times as will make a significant difference. (If more than two or three recorrections are required, it is more efficient to reduce the step size.)

We will illustrate the modified Euler method, which we also call the *Euler predictor–corrector method,* on the same problem previously treated. Table 5.4 is convenient.

$$\frac{dy}{dx} = x + y, \qquad y(0) = 1, \qquad h = 0.02.$$

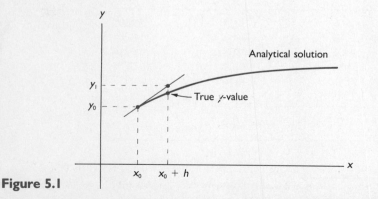

Figure 5.1

Table 5.4

x_n	y_n	y_n'	hy_n'	y_{n+1}	y_{n+1}'	y_{av}'	hy_{av}'
0	1.0000	1.0000	0.0200	1.0200	1.0400	1.0200	0.0204
				1.0204*	1.0404	1.0202	0.0204
0.02	1.0204	1.0404	0.0208	1.0412	1.0812	1.0608	0.0212
				1.0416	1.0816	1.0610	0.0212
0.04	1.0416	1.0816	0.0216	1.0632	1.1232	1.1024	0.0220
				1.0636	1.1236	1.1026	0.0221
				1.0637	1.1237	1.1027	0.0221
0.06	1.0637	1.1237	0.0225	1.0862	1.1662	1.1449	0.0229
				1.0866	1.1666	1.1451	0.0229
0.08	1.0866	1.1666	0.0233	1.1099	1.2099	1.1883	0.0238
				1.1104	1.2104	1.1885	0.0238
0.10	1.1104						

(1.1103 analytical value)

*It is convenient to use this column for both the predicted and corrected values of y_{n+1}. The first entry in any x-row is the predicted value. Corrected and recorrected values follow. In this example, it was necessary to recorrect only where $x = 0.04$. All the other calculations indicate no need to recorrect the values of y_{n+1}. We will discuss in a later section whether one should recorrect the first corrected value of y_{n+1} in this type of method. In general, it is better to make h small and not recorrect.

In this table, we tabulate the corrected values of y_{n+1} in the same column as the predicted ones. y_{av}' is the mean of y_n' and the last value of y_{n+1}'. The answer agrees within 1 in the fourth decimal place. We have done more work than in the simple Euler method, but certainly not the 22 times more that would have been needed with that method to attain four-decimal-place accuracy. Hence the modified method is more efficient.

We can find the error of the modified Euler method by comparing with the Taylor series.

$$y_{n+1} = y_n + y_n'h + \frac{1}{2}y_n''h^2 + \frac{y'''(\xi)}{6}h^3, \qquad x_n < \xi < x_n + h.$$

Replace the second derivative by the forward-difference approximation for y'', $(y_{n+1}' - y_n')/h$, which has error of $O(h)$, and write the error term as $O(h^3)$:

$$y_{n+1} = y_n + h\left(y_n' + \frac{1}{2}\left[\frac{y_{n+1}' - y_n'}{h} + O(h)\right]h\right) + O(h^3),$$

$$y_{n+1} = y_n + h\left(y_n' + \frac{1}{2}y_{n+1}' - \frac{1}{2}y_n'\right) + O(h^3),$$

$$y_{n+1} = y_n + h\left(\frac{y_n' + y_{n+1}'}{2}\right) + O(h^3).$$

This shows that the error of one step of the modified Euler method is $O(h^3)$. This is the "local error." There is an accumulation of errors from step to step, so that the error over the whole range of application, the so-called global error, is $O(h^2)$. This seems intuitively reasonable since the number of steps into which the interval is subdivided is proportional to $1/h$; hence the order of error is reduced to $O(h^2)$ on the continuing application. We treat the accumulation of errors more fully in a later section.

5.4 RUNGE–KUTTA METHODS

A further advance in efficiency (that is, obtaining the most accuracy per unit of computational effort) can be secured with a group of methods due to the German mathematicians Runge and Kutta. The fourth-order Runge–Kutta methods are widely used in computer solutions to differential equations. The development of this technique is algebraically complicated.

To convey some idea of how the Runge–Kutta methods are developed, we show the derivation of a second-order method. We write the increment to y as a weighted average of two estimates of Δy, k_1 and k_2. For the equation $dy/dx = f(x, y)$,

$$
\begin{aligned}
y_{n+1} &= y_n + ak_1 + bk_2, \\
k_1 &= hf(x_n, y_n), \\
k_2 &= hf(x_n + \alpha h, y_n + \beta k_1).
\end{aligned}
\tag{5.4}
$$

We can think of the values k_1 and k_2 as estimates of the change in y when x advances by h because they are the product of the change in x and a value for the slope of the curve, dy/dx. The Runge–Kutta methods always use as the first estimate of Δy the simple Euler estimate; the other estimate is taken with x and y stepped up by the fractions α and β of h and of the earlier estimate of Δy, k_1. Our problem is to devise a scheme of choosing the four parameters, a, b, α, β. We do this by making Eq. (5.4) agree as well as possible with the Taylor-series expansion, in which the y-derivatives are written in terms of f, from $dy/dx = f(x, y)$,

$$
y_{n+1} = y_n + hf(x_n, y_n) + (h^2/2)f'(x_n, y_n) + \cdots\cdots
$$

An equivalent form, since $df/dx = f_x + f_y\, dy/dx = f_x + f_y f$, is

$$
y_{n+1} = y_n + hf_n + h^2\!\left(\frac{1}{2}f_x + \frac{1}{2}f_y f\right)_n.
\tag{5.5}
$$

(All the derivatives in Eq. (5.5) are evaluated at the point (x_n, y_n).) We now rewrite Eq. (5.4) by substituting the definitions of k_1 and k_2:

$$
y_{n+1} = y_n + ahf(x_n, y_n) + bhf[x_n + \alpha h, y_n + \beta hf(x_n, y_n)].
\tag{5.6}
$$

To make the last term of Eq. (5.6) comparable to Eq. (5.5) we expand $f(x, y)$ in a Taylor series in terms of x_n, y_n, remembering that f is a function of two variables,* retaining only first derivative terms.

$$f[x_n + \alpha h, y_n + \beta h f(x_n, y_n)] \doteq (f + f_x \alpha h + f_y \beta h f)_n. \tag{5.7}$$

On the right side of both Eqs. (5.5) and (5.7), f and its partial derivatives are all to be evaluated at (x_n, y_n).

Substituting from Eq. (5.7) into Eq. (5.6), we have

$$y_{n+1} = y_n + ahf_n + bh(f + f_x \alpha h + f_y \beta h f)_n,$$

or, rearranging,

$$y_{n+1} = y_n + (a + b)hf_n + h^2(\alpha b f_x + \beta b f_y f)_n. \tag{5.8}$$

Equation (5.8) will be identical to Eq. (5.5) if

$$a + b = 1, \qquad \alpha b = \frac{1}{2}, \qquad \beta b = \frac{1}{2}.$$

Note that there are only three equations that need to be satisfied by the four unknowns. We can choose one value arbitrarily (with minor restrictions); hence we have a set of second-order methods. For example, taking $a = \frac{2}{3}$, we have $b = \frac{1}{3}$, $\alpha = \frac{3}{2}$, $\beta = \frac{3}{2}$. Other choices give other sets of parameters that agree with the Taylor-series expansion. If one takes $a = \frac{1}{2}$, the other variables are $b = \frac{1}{2}$, $\alpha = 1$, $\beta = 1$. This last set of parameters gives the modified Euler algorithm that we have previously discussed; the modified Euler method is a special case of a second-order Runge–Kutta method.

Fourth-order Runge–Kutta methods are most widely used and are derived in similar fashion. Greater complexity results from having to compare terms through h^4, and gives a set of 11 equations in 13 unknowns. The set of 11 equations can be solved with 2 unknowns being chosen arbitrarily. The most commonly used set of values leads to the algorithm

$$y_{n+1} = y_n + \frac{1}{6}(k_1 + 2k_2 + 2k_3 + k_4),$$

$$k_1 = hf(x_n, y_n),$$

$$k_2 = hf(x_n + \frac{1}{2}h, y_n + \frac{1}{2}k_1),$$

$$k_3 = hf(x_n + \frac{1}{2}h, y_n + \frac{1}{2}k_2),$$

$$k_4 = hf(x_n + h, y_n + k_3).$$

---

*Appendix A will remind readers of this expansion.

As an example, we again solve $dy/dx = x + y$, $y(0) = 1$, this time taking $h = 0.1$:

$$k_1 = 0.1(0 + 1) = 0.10000,$$

$$k_2 = 0.1(0.05 + 1.05) = 0.11000,$$

$$k_3 = 0.1(0.05 + 1.055) = 0.11050,$$

$$k_4 = 0.1(0.10 + 1.1105) = 0.12105,$$

$$y(0.1) = 1.0000 + \frac{1}{6}(0.10000 + 0.2200 + 0.22100 + 0.12105)$$

$$= 1.11034.$$

This agrees to five decimals with the analytical result, and illustrates a further gain in accuracy with less effort than required by our example of Section 5.2.

The local error term for the fourth-order Runge–Kutta is $O(h^5)$; the global error would be $O(h^4)$. It is computationally more efficient than the modified Euler method because, while four evaluations of the function are required per step rather than two, the steps can be manyfold larger for the same accuracy.

It is easy to see why the Runge–Kutta technique is so popular. Since going from second to fourth order was so beneficial, we may wonder if we should use a still higher order of formula. Higher-order (fifth, sixth, and so on) Runge–Kutta formulas have been developed and can be used to advantage in determining a suitable size, h, as we will see.

A standard way to determine whether the Runge–Kutta values are sufficiently accurate is to recompute the value at the end of each interval with the step size cut in half. If this makes a change of negligible magnitude, the results are accepted; if not, the step is halved again until the results are satisfactory. This is very expensive, however, because seven additional sets of function evaluations are made just to determine the accuracy. (We may not have to make a new computation after the last one to ascertain its accuracy, of course. The knowledge that the method is of $O(h^5)$ accuracy permits us to anticipate the change that an additional computation would make if we have repeated once already.) There have been several schemes proposed to minimize the effort to determine the error in a Runge–Kutta computation. They demand some additional effort, but fewer than seven additional function evaluations, which, plus the original four, total eleven.

The Runge–Kutta–Fehlberg method is one of the most popular of these methods at the present time. Six functional evaluations are required, but we then have an estimate of the error as well:

$$k_1 = h \cdot f(x_n, y_n),$$

$$k_2 = h \cdot f(x_n + h/4, y_n + k_1/4),$$

$$k_3 = h \cdot f(x_n + 3h/8, y_n + 3k_1/32 + 9k_2/32),$$

$$k_4 = h \cdot f(x_n + 12h/13, y_n + 1932k_1/2197$$
$$- 7200k_2/2197 + 7296k_3/2197),$$

$$k_5 = h \cdot f(x_n + h, y_n + 439k_1/216 - 8k_2$$
$$+ 3680k_3/513 - 845k_4/4104),$$
$$k_6 = h \cdot f(x_n + h/2, y_n - 8k_1/27 + 2k_2$$
$$- 3544k_3/2565 + 1859k_4/4104 - 11k_5/40);$$
$$\hat{y}_{n+1} = y_n + (25k_1/216 + 1408k_3/2565$$
$$+ 2197k_4/4104 - k_5/5), \text{ with global error } O(h^4),$$
$$y_{n+1} = y_n + (16k_1/135 + 6656k_3/12{,}825 + 28{,}561k_4/56{,}430$$
$$- 9k_5/50 + 2k_6/55) \text{ with global error } O(h^5);$$
$$E = k_1/360 - 128k_3/4275 - 2197k_4/75{,}240$$
$$+ k_5/50 + 2k_6/55.$$

The basis for the Runge–Kutta–Fehlberg scheme is to compute two Runge–Kutta estimates for the new value of $\hat{y}_{n+1}$ but of different orders of errors. Thus, instead of comparing estimates of y_{n+1} for h and $h/2$, we compare the estimates $\hat{y}_{n+1}$ and y_{n+1} using fourth- and fifth-order (global) Runge–Kutta formulas. Moreover, both equations make use of the same k's, so only six function evaluations are needed versus the previous eleven. In addition, one can increase or decrease h depending on the value of E at each step. As our estimate for the new y_{n+1} we use the fifth-order (global) estimate.

As an example, we once more solve $dy/dx = x + y$, $y(0) = 1$, again taking $h = 0.1$:

$$k_1 = 0.1(0 + 1) = 0.1,$$
$$k_2 = 0.1(0.025 + 1.025) = 0.105,$$
$$k_3 = 0.1(0.0375 + 1.0389) = 0.10764,$$
$$k_4 = 0.1(0.09231 + 1.101295) = 0.119360,$$
$$k_5 = 0.1(0.1 + 1.110824) = 0.121082,$$
$$k_6 = 0.1(0.05 + 1.052415) = 0.110242;$$
$$\hat{y}_{n+1} = 1 + (0.01157 + 0.05909 + 0.06390 - 0.02422)$$
$$= 1.11034197,$$
$$y_{n+1} = 1 + (0.1185 + 0.05586 + 0.06041 - 0.02180 + 0.00040)$$
$$= 1.110341834.$$
$$E = 0.00000014 \text{ and the exact value is: } 1.110341836!$$

This agrees to eight decimal places (the actual computation was done in ten-digit arithmetic) with the analytical result, with only two more function evaluations. Moreover, we have the value E to adjust our step size for the next iteration.

Let us summarize and compare (see Table 5.5) the four numerical methods we have studied for solving $y' = f(x, y)$.

Table 5.5

Method	Estimate of slope over x-interval	Global error	Local error	Evaluations of $f(x, y)$ per step
Euler	Initial value	$O(h)$	$O(h^2)$	1
Modified Euler	Arithmetic average of initial and final predicted slope	$O(h^2)$	$O(h^3)$	2
Runge–Kutta (fourth-order)	Weighted average of four values	$O(h^4)$	$O(h^5)$	4
Runge–Kutta–Fehlberg	Weighted average of six values	$O(h^5)$	$O(h^6)$	6

Table 5.6

Method	Step size	Result	Error	Number of function evaluations
Euler	0.02	1.1081	0.0022	5
Modified Euler	0.02	1.1104	0.0001	12
Runge–Kutta (fourth-order)	0.1	1.11034	0.000001	4
Runge–Kutta–Fehlberg	0.1	1.11034183	0.000000002	6

In the example problem, $y' = x + y$, $y(0) = 1$, we obtained the results shown in Table 5.6 for the y-value at $x = 0.1$.

For completeness we mention the Runge–Kutta–Merson method. This method computes the Δy in the next step; an estimate of the local error is then available from a weighted sum of the individual estimates:

$$k_1 = h \cdot f(x_n, y_n),$$

$$k_2 = h \cdot f(x_n + h/3, y_n + k_1/3),$$

$$k_3 = h \cdot f(x_n + h/3, y_n + k_1/6 + k_2/6),$$

$$k_4 = h \cdot f(x_n + h/2, y_n + k_1/8 + 3k_3/8),$$

$$k_5 = h \cdot f(x_n + h, y_n + k_1/2 - 3k_3/2 + 2k_4);$$

$$y_{n+1} = y_n + (k_1 + 4k_4 + k_5)/6 + O(h^5);$$

$$E \doteq \frac{1}{30}(2k_1 - 9k_3 + 8k_4 - k_5).$$

Finally, there is an IMSL subroutine DVERK that uses Runge–Kutta formulas of orders 5 and 6 that were developed by J. H. Verner. This method uses eight function evaluations to find the two estimates of y_{n+1}.

5.5   MULTISTEP METHODS

Runge–Kutta-type methods (which include Euler and modified Euler as special cases) are called single-step methods because they use only the information from the last step computed. In this they have the ability to perform the next step with a different step size and are ideal for beginning the solution where only the initial conditions are available. After the solution has begun, however, there is additional information available about the function (and its derivative) if we are wise enough to retain it in the memory of the computer. A multistep method is one that takes advantage of this fact.

The principle behind a multistep method is to utilize the past values of y and/or y' to construct a polynomial that approximates the derivative function, and extrapolate this into the next interval. Most methods use equispaced past values to make the construction of the polynomial easy. The Adams method is typical. The number of past points that are used sets the degree of the polynomial and is therefore responsible for the truncation error. The order of the method is equal to the power of h in the global error term of the formula, which is also equal to one more than the degree of the polynomial.

To derive the relations for the Adams method, we write the differential equation $dy/dx = f(x, y)$ in the form

$$dy = f(x, y)\, dx,$$

and we integrate between x_n and x_{n+1}:

$$\int_{x_n}^{x_{n+1}} dy = y_{n+1} - y_n = \int_{x_n}^{x_{n+1}} f(x, y)\, dx.$$

In order to integrate the term on the right, we approximate $f(x, y)$ as a polynomial in x, deriving this by making it fit at several past points. If we use three past points, the approximating polynomial will be a quadratic. If we use four points, it will be a cubic. The more points we use, the better the accuracy (until round-off interferes, of course).

Suppose we fit a second-degree polynomial through three past points, writing this as a Newton–Gregory backward polynomial:

$$\int_{x_n}^{x_{n+1}} dy = y_{n+1} - y_n = \int_{x_n}^{x_{n+1}} \left(f_n + s\Delta f_{n-1} + \frac{(s + 1)s}{2} \Delta^2 f_{n-2} + \text{error} \right) dx$$

$$= \int_{s=0}^{s=1} \left(f_n + s\Delta f_{n-1} + \frac{(s + 1)s}{2} \Delta^2 f_{n-2} \right) h\, ds$$

$$+ \int_{s=0}^{s=1} \frac{s(s + 1)(s + 2)}{6} h^3 f'''(\xi) h\, ds.$$

In the preceding, we have changed the variable to s and identified x_n as x_0. The interval of integration becomes $s = 0$ to $s = 1$. Performing the integration, we get

$$y_{n+1} - y_n = h\left(f_n + \frac{1}{2}\Delta f_{n-1} + \frac{5}{12}\Delta^2 f_{n-2} \right) + O(h^4). \tag{5.9}$$

While it is perfectly possible to use Eq. (5.9) as it stands, constructing a difference table is not necessary. If we expand the differences of f in terms of the values of f_n, f_{n-1}, and f_{n-2}, we get a more useful formula:

$$y_{n+1} = y_n + h\left[f_n + \frac{f_n - f_{n-1}}{2} + \frac{5(f_n - 2f_{n-1} + f_{n-2})}{12}\right]$$

$$= y_n + \frac{h}{12}[23f_n - 16f_{n-1} + 5f_{n-2}] + O(h^4). \qquad (5.10)$$

Observe that Eq. (5.10) resembles the single-step formulas of the previous sections in that the increment to y is a weighted sum of the derivatives times the step size, but differs in that past values are used rather than estimates in the forward direction.

E X A M P L E   We illustrate the use of Eq. (5.10) to calculate $y(0.6)$ for $dy/dx = x + y$, $y(0) = 1$. We compute good values for $y(0.2)$ and $y(0.4)$ using a single-step method. In this case we obtain these values using the Runge–Kutta–Fehlberg method with $h = 0.2$. These values are given in Table 5.7.

Table 5.7

x	y	$f(x, y)$
0.0	1.00000	1.00000
0.2	1.24281	1.44281
0.4	1.58365	1.98365

Then from Eq. (5.10):

$$y(0.6) = 1.58365 + \frac{0.2}{12}[23(1.98365) - 16(1.44281) + 5(1.0)] = 2.04260.$$

Comparing our result with the exact solution (2.04424), we find that the computed value has an error of 0.00164. We can reduce the size of the error by doing the calculations with a smaller step size of 0.1. We use the Runge–Kutta–Fehlberg method once again to obtain the values in Table 5.8. Using the values at $x = 0.3$, $x = 0.4$, $x = 0.5$ from Table 5.8, we repeat the computations:

$$y(0.6) = 1.79744 + \frac{0.1}{12}[23(2.29744) - 16(1.98364) + 5(1.69972)] = 2.04412,$$

which has an error of 0.00012 compared with the exact answer of 2.04424.

Table 5.8

x	y	$f(x, y)$
0.0	1.00000	1.00000
0.1	1.11034	1.21034
0.2	1.24281	1.44281
0.3	1.39972	1.69972
0.4	1.58364	1.98364
0.5	1.79744	2.29744

In Section 5.7 we will see that we can reduce the error by using more past points for fitting the polynomial. In fact, when the derivation is done for four points using the Newton–Gregory backward polynomial, the following results are obtained:

$$y_{n+1} = y_n + h\left(f_n + \frac{1}{2}\Delta f_{n-1} + \frac{5}{12}\Delta^2 f_{n-2} + \frac{3}{8}\Delta^3 f_{n-3}\right) + O(h^5),$$

or

$$y_{n+1} = y_n + \frac{h}{24}[55f_n - 59f_{n-1} + 37f_{n-2} - 9f_{n-3}] + O(h^5).$$

We repeat the above example with $h = 0.1$ to compute $y(0.6)$, now using the values in Table 5.8 for $x = 0.2$, $x = 0.3$, $x = 0.4$, $x = 0.5$, and computing

$$y(0.6) = 1.79744 + \frac{0.1}{24}[55(2.29744) - 59(1.98364) + 37(1.69972) - 9(1.44281)]$$

$$= 2.04423.$$

The error of this computation has been reduced to 0.00001. We summarize the results of these two formulas in Table 5.9.

Table 5.9

Number of points used	Estimate of $y(0.6)$	Error $(h = 0.1)$
3	2.04412	0.00012
4	2.04423	0.00001

5.6   MILNE'S METHOD

The method of Milne is a multistep method that first predicts a value for y_{n+1} by extrapolating values for the derivative. It differs from the Adams method in that it integrates over more than one interval. The past values that we require may have been computed by the Range–Kutta method, or possibly by the Taylor-series method. In the Milne method, we suppose that four equispaced starting values of y are known, at the points x_n, x_{n-1}, x_{n-2}, and x_{n-3}. We can employ quadrature formulas to integrate as follows:

$$\frac{dy}{dx} = f(x, y),$$

$$\int_{x_{n-3}}^{x_{n+1}} \left(\frac{dy}{dx}\right) dx = \int_{x_{n-3}}^{x_{n+1}} f(x, y) \, dx = \int_{x_{n-3}}^{x_{n+1}} P_2(x) \, dx; \tag{5.11}$$

so

$$y_{n+1} - y_{n-3} = \frac{4h}{3}(2f_n - f_{n-1} + 2f_{n-2}) + \frac{28}{90}h^5 y^v(\xi_1), \qquad x_{n-3} < \xi_1 < x_{n+1}.$$

We integrate the function $f(x, y)$ by replacing it with a quadratic interpolating polynomial that fits at the three points, where $x = x_n, x_{n-1}$, and x_{n-2}, and integrating according to the methods of Chapter 4.* Note that we extrapolate in the integration by one panel both to the left and to the right of the region of fit. Hence the error is larger, because of the extrapolation, than it would be with only interpolation.

With the value of y_{n+1} one can calculate f_{n+1} reasonably accurately. In Milne's method, we use Eq. (5.11) as a predictor formula and then correct with

$$\int_{x_{n-1}}^{x_{n+1}} \left(\frac{dy}{dx}\right) dx = \int_{x_{n-1}}^{x_{n+1}} f(x, y) \, dx = \int_{x_{n-1}}^{x_{n+1}} P_2(x) \, dx, \tag{5.12}$$

$$y_{n+1,c} - y_{n-1} = \frac{h}{3}(f_{n+1} + 4f_n + f_{n-1}) - \frac{h^5}{90}y^v(\xi_2), \qquad x_{n-1} < \xi_2 < x_{n+1}.$$

In Eq. (5.12), the polynomial P_2 is not identical to that in Eq. (5.11) because they do not fit the function at the same points. In Eq. (5.12) the polynomial fits at x_{n+1}, x_n, and x_{n-1}. Note the changed range of integration and the smaller coefficient in the error term of Eq. (5.12) because the polynomial is not extrapolated; f_{n+1} is calculated using y_{n+1} from the predictor formula. The integration formula is the familiar Simpson's

---

*The formula can also be derived by the method of undetermined coefficients, as shown in the Appendix.

Table 5.10

	x	y	$f(x, y) = x + y$	
From Taylor series	0	1.00000	1.00000	
	0.1	1.11034	1.21034	
	0.2	1.24280	1.44280	
	0.3	1.39971	1.69971	
	0.4	(1.58364)	(1.98364)	Predictor value
		1.58364	1.98364	Corrector value
	0.5	(1.79742)	(2.29742)	Predictor value
		1.79743	2.29743	Corrector value

Analytical value at $x = 0.5$ is 1.79744

$\frac{1}{3}$ rule, since we integrate a quadratic over two panels within the region of fit. In this method we do not try to recorrect y_{n+1}.

We illustrate with our familiar simple problem, $dy/dx = x + y$, $y(0) = 1$. From Section 5.1, we take the four values calculated by Taylor series and carry five decimals (see Table 5.10).

For this example, the first predictor and corrector values agree, and the next set differ by only one in the fifth decimal place. The corrected value at $x = 0.5$ is in error by only one in the fifth decimal place. This discrepancy is mostly due to round-off error in the computed values. In this case, the value of h could have been chosen larger. With the set of values available, h cannot be increased without additional computations, but if we had seven equally spaced values, we could double h by taking only every other one and still have four values to move ahead from.

Normally, the values of y_{n+1} from the predictor and the corrector do not agree. Consideration of the error terms of Eqs. (5.11) and (5.12) suggests that the true value should usually lie between the two values and closer to the corrector value. While ξ_1 and ξ_2 are not necessarily the same value, they lie in similar intervals. If one assumes that the values of $y^{v}(\xi_1)$ and $y^{v}(\xi_2)$ are equal, the error in the corrector formula is $\frac{1}{28}$ times the error in the predictor formula. Hence the difference between the predictor and corrector formula is about 29 times the error in the corrected value. This is frequently used as a criterion of accuracy for Milne's method.* The ease with which we can monitor the error is a particular advantage of predictor–corrector methods. We are able to know immediately whether the step size is too big to give the desired degree of accuracy. This is in strong contrast to the Runge–Kutta method of the previous section (but not of Runge–Kutta–Fehlberg).

Milne's method is simple and has a good error term, $O(h^5)$ for local error. It is subject to an instability problem in certain cases, however, in which the errors do not tend to zero as h is made smaller. This unexpected phenomenon is discussed below. Because of possible instability, another method, a modification of that of Adams, is more widely used than Milne's method.

---

*A further criterion as to whether or not the single correction that is normally used in Milne's method is adequate is discussed in Section 5.8.

It is sufficient to show the instability of Milne's method for one simple case. We will be able to draw the necessary conclusions from this. Consider the differential equation $dy/dx = Ay$, where A is a constant. The general solution is $y = ce^{Ax}$. If $y(x_0) = y_0$ is the initial condition that the solution must satisfy, $c = y_0 e^{-Ax_0}$. Hence, letting y_n be the value of the function when $x = x_n$, the analytical solution is

$$y_n = y_0 e^{A(x_n - x_0)}. \tag{5.13}$$

If we solve the differential equation by the method of Milne, we have, from the corrector formula,

$$y_{n+1} = y_{n-1} + \frac{h}{3}(y'_{n+1} + 4y'_n + y'_{n-1}).$$

Letting $y'_n = Ay_n$, from the original differential equation, and rearranging, we get

$$y_{n+1} = y_{n-1} + \frac{h}{3}(Ay_{n+1} + 4Ay_n + Ay_{n-1}),$$

$$\left(1 - \frac{hA}{3}\right)y_{n+1} - \frac{4hA}{3}y_n - \left(1 + \frac{hA}{3}\right)y_{n-1} = 0. \tag{5.14}$$

We would like to solve this equation for y_n in terms of y_0 to compare to Eq. (5.13). Equation (5.14) is a second-order linear difference equation that can be solved in a manner analogous to that for differential equations. The solution is

$$y_n = C_1 Z_1^n + C_2 Z_2^n, \tag{5.15}$$

where Z_1, Z_2 are the roots of the quadratic

$$\left(1 - \frac{hA}{3}\right)Z^2 - \frac{4hA}{3}Z - \left(1 + \frac{hA}{3}\right) = 0. \tag{5.16}$$

(The reader should check that Eq. (5.15) is a solution of Eq. (5.14) by direct substitution.) For simplification, let $hA/3 = r$; the roots of Eq. (5.16) are then

$$Z_1 = \frac{2r + \sqrt{3r^2 + 1}}{1 - r},$$

$$Z_2 = \frac{2r - \sqrt{3r^2 + 1}}{1 - r}. \tag{5.17}$$

We are interested in comparing the behavior of Eqs. (5.15) and (5.13) as the step size h becomes small. As $h \to 0$, $r \to 0$, and $r^2 \to 0$ even faster. Neglecting the $3r^2$ terms in comparison to the constant 1 under the radical in Eq. (5.17) gives

$$Z_1 \doteq \frac{2r + 1}{1 - r} = 1 + 3r + O(r^2) = 1 + Ah + O(h^2),$$

$$Z_2 \doteq \frac{2r - 1}{1 - r} = -1 + r + O(r^2) = -\left(1 - \frac{Ah}{3}\right) + O(h^2). \tag{5.18}$$

The last results are obtained by dividing the fractions. We now compare Eq. (5.18) with the Maclaurin series of the exponential function,

$$e^{hA} = 1 + hA + O(h^2),$$

$$e^{-hA/3} = 1 - \frac{hA}{3} + O(h^2).$$

We see that, for $h \to 0$,

$$Z_1 \doteq e^{-hA}, \qquad Z_2 \doteq e^{-hA/3}.$$

Hence the Milne solution is represented by

$$y_n = C_1(e^{hA})^n + C_2(e^{-hA/3})^n = C_1 e^{A(x_n - x_0)} + C_2 e^{-A(x_n - x_0)/3}. \qquad \textbf{(5.19)}$$

In Eq. (5.19), we have used $x_n - x_0 = nh$. The solution consists of two parts. The first term obviously matches with the analytical solution, Eq. (5.13). The second term, called a *parasitic term*, will die out as x_n increases if A is a positive constant, but if A is negative, it will grow exponentially with x_n. Note that we get this peculiar behavior independent of h; smaller step size is of no benefit in eliminating the error.

A numerical example that demonstrates the instability of Milne's method is given in the next section. Such a demonstration by a numerical example is less conclusive than the analytical approach just given, but is much easier to grasp.

5.7   ADAMS–MOULTON METHOD

A method that does not have the same instability problem as the Milne method, but is about as efficient, is the Adams–Moulton method. It also assumes a set of starting values already calculated by some other technique. Here we take a cubic through four points, from x_{n-3} to x_n, and integrate over one step, from x_n to x_{n+1}. This is the same as the method described in Section 5.5. For $dy/dx = f(x, y)$,

$$\int_{x_n}^{x_{n+1}} \left(\frac{dy}{dx}\right) dx = \int_{x_n}^{x_{n+1}} f(x, y)\, dx = \int_{xn}^{x_{n+1}} P_3(x)\, dx,$$

$$y_{n+1} - y_n = h\left(f_n + \frac{1}{2}\Delta f_{n-1} + \frac{5}{12}\Delta^2 f_{n-2} + \frac{3}{8}\Delta^3 f_{n-3}\right) + \frac{251}{720}h^5 y^v(\xi),$$

$$x_{n-3} < \xi < x_{n+1}. \qquad \textbf{(5.20)}$$

The integration formula can be derived by the methods of Chapter 4 or by the method of undetermined coefficients. Alternatively, it can be developed by writing $P_3(x)$ as a Newton–Gregory backward interpolating polynomial (fitting at x_n, x_{n-1}, x_{n-2}, and x_{n-3}) and integrating. In the Adams–Moulton method we continue by correcting y_{n+1} before calculating the next step. Using Eq. (5.20) as a predictor formula, we can compute a nearly correct value of f_{n+1}. If we now approximate $f(x, y)$ as a cubic that fits over the range from x_{n-2} to x_{n+1}, and integrate from x_n to x_{n+1}, we will not be extrapolating the polynomial, and will have a more favorable error term. The result is

$$
y_{n+1} - y_n = h\left(f_{n+1} - \frac{1}{2}\Delta f_n - \frac{1}{12}\Delta^2 f_{n-1} - \frac{1}{24}\Delta^3 f_{n-2}\right) - \frac{19}{720}h^5 y^v(\xi_1),
$$
$$
x_{n-2} < \xi_1 < x_{n+1}. \tag{5.21}
$$

The Adams–Moulton method consists of using Eq. (5.20) as a predictor and Eq. (5.21) as a corrector. We illustrate it with the same example as before, $dy/dx = x + y$, $y(0) = 1$. A difference table (Table 5.11) is computed to assist us (though in a computer program we would use a different scheme: see Eqs. (5.22)).

By the predictor formula,

$$
y(0.4) = 1.39971 + 0.1\left[1.69971 + \frac{1}{2}(0.25691) + \frac{5}{12}(0.02445) + \frac{3}{8}(0.00233)\right]
$$
$$
= 1.58363.
$$

Table 5.11

	x	y	f	Δf	$\Delta^2 f$	$\Delta^3 f$
	0	1.00000	1.00000			
				0.21034		
	0.1	1.11034	1.21034		0.02212	
Starting				0.23246		0.00233
values	0.2	1.24280	1.44280		0.02445	
				0.25691		(0.00256)
	0.3	1.39971	1.69971		(0.02701)	0.00257
				(0.28392)	0.02702	(0.00283)
Predicted 0.4		(1.58363)	(1.98363)	0.28393	(0.02985)	
Corrected		1.58364	1.98364	(0.31378)		
Predicted 0.5		(1.79742)	(2.29742)			
Corrected		1.79743				

Analytical value at $x = 0.5$ is 1.79744.

Then f at $x = 0.4$ is computed and the difference table is calculated. The corrector formula then gives

$$y(0.4) = 1.39971 + 0.1\left[1.98363 - \frac{1}{2}(0.28392) - \frac{1}{12}(0.02701) - \frac{1}{24}(0.00256)\right]$$

$$= 1.58364.$$

The computations are continued in the same manner to get $y(0.5)$. The corrected value almost agrees to five decimals with the predicted value. Comparing error terms of Eqs. (5.20) and (5.21), and assuming that the two fifth-derivative values are equal, we see that the true value should lie between the predicted and corrected values, with the error in the corrected value being about

$$\frac{19}{251 + 19} \quad \text{or} \quad \frac{1}{14.2}$$

times the difference between the predicted and corrected values. A frequently used criterion for accuracy of the Adams–Moulton method with four starting values is that the corrected value is not in error by more than 1 in the last place if the difference between predicted and corrected values is less than 14 in the last decimal place.* If this degree of accuracy is inadequate, we know that h is too large.

Equations (5.20) and (5.21) are not well suited to computer utilization because calculating and storing the difference tables is wasteful of both time and memory space. Each of these differences is a linear function of the various f-values, however. Expressing the differences in terms of the f's gives an alternative form for the Adams–Moulton method that is usually employed in a computer program.

Predictor: $\displaystyle y_{n+1} = y_n + \frac{h}{24}(55f_n - 59f_{n-1} + 37f_{n-2} - 9f_{n-3}).$

Corrector: $\displaystyle y_{n+1} = y_n + \frac{h}{24}(9f_{n+1} + 19f_n - 5f_{n-1} + f_{n-2}).$ **(5.22)**

These same formulas are derived directly by the method of undetermined coefficients as Eqs. (B.28) and (B.29) of Appendix B.

These equations are better suited to machine calculation and to digital computer programming. Without such calculation aids, the large coefficients and alternating signs lead to large round-off errors unless extra guard figures are carried.

Adams–Moulton formulas employing more or less than four starting values can be derived in analogous fashion. The fourth-order formulas we have given are widely used, especially in combination with Runge–Kutta, because both kinds of methods then have local errors of $O(h^5)$.

---

*The convergence criterion of Section 5.8 should also be met.

When the predicted and corrected values agree to as many decimals as the desired accuracy, we can save computational effort by increasing the step size. As mentioned in connection with Milne's method, we can conveniently double the step size, after we have seven equispaced values, by omitting every second one. When the difference between predicted and corrected values reaches or exceeds the accuracy criterion, we should decrease step size. If we interpolate two additional y-values with a fourth-degree polynomial, where the error will be $O(h^5)$, consistent with the rest of our work, we can readily halve the step size.* Convenient formulas for this are

$$y_{n-1/2} = \frac{1}{128}[35y_n + 140y_{n-1} - 70y_{n-2} + 28y_{n-3} - 5y_{n-4}],$$

$$y_{n-3/2} = \frac{1}{64}[-y_n + 24y_{n-1} + 54y_{n-2} - 16y_{n-3} + 3y_{n-4}]. \tag{5.23}$$

Use of these values with y_n, y_{n-1} gives four values of the function at intervals of $\Delta x = h/2$.

The efficiency of both the Adams–Moulton and Milne methods is about twice that of the Runge–Kutta–Fehlberg and Runge–Kutta methods. Only two function evaluations are needed per step for the former methods, while six or four are required with the single-step alternatives. All have similar error terms. Change of step size with the multistep methods is considerably more awkward, however.

We have stressed that the advantage of the Adams–Moulton method over that of Milne is that it is a stable method rather than an unstable one. An analytical proof of the stability of Adams–Moulton for the equation $y' = Ay$ is similar to the analysis of Section 5.6—an equation results with parasitic terms that do die out as h gets small regardless of the sign of A. Such a treatment is not entirely satisfying because we can't prove stability by examining only one case (proving that an assertion is *not* true is always easier because we need to find only one counterexample).

It is much clearer to compare the stability of Adams–Moulton with Milne through a numerical example. The table in Fig. 5.2 presents results from a computer program that solved $y' = -10y$, $y(0) = 1$ (for which the analytical solution is $y = e^{-10x}$) over the interval $x = 0$ to $x = 2$. The first part of the table is computed with $h = 0.04$. In the second part, for which only partial output is given, $h = 0.004$.

Observe that the results with the Milne method grow to have very large relative errors. In fact, near $x = 2.0$, the relative error is practically 100%. The oscillatory behavior of the solution is also characteristic of instability. Even with the smaller h, the Milne method blows up. The errors are even greater near $x = 2.0$ in spite of the smaller step size. (The Milne solution when x is small is considerably more accurate when $h = 0.004$. Results are not shown for this.)

The results by the Adams–Moulton method do not show this anomalous behavior. The relative error, while growing to some degree, still stays manageable (8% at $x = 2.0$ with $h = 0.04$, $< 0.1\%$ at $x = 2.0$ with $h = 0.004$). The expected decrease of error

---

*An alternative but computationally more expensive way to get intermediate points would be to compute them with Runge–Kutta formulas.

	Solution Using Milne Method			Solution Using Adams–Moulton Method	
Values with $h = 0.04$					
x	y	Error		y	Error
0.000000E 00	0.100000E 01	0.000000E 00		0.100000E 01	0.000000E 00
0.400000E−01	0.670320E 00	0.000000E 00		0.670320E 00	0.000000E 00
0.800000E−01	0.449329E 00	0.000000E 00		0.449329E 00	0.000000E 00
0.120000E 00	0.301195E 00	0.000000E 00		0.301195E 00	0.000000E 00
0.160000E 00	0.201667E 00	0.229836E−03		0.201552E 00	0.344634E−03
0.200000E 00	0.135271E 00	0.640750E−04		0.134873E 00	0.462234E−03
0.240000E 00	0.904572E−01	0.260949E−03		0.902753E−01	0.442863E−03
0.280000E 00	0.607594E−01	0.507981E−04		0.604129E−01	0.397373E−03
0.320000E 00	0.405497E−01	0.212613E−03		0.404273E−01	0.335068E−03
0.360000E 00	0.273069E−01	0.169165E−04		0.270547E−01	0.269119E−03
0.400000E 00	0.181599E−01	0.155795E−03		0.181052E−01	0.210475E−03
0.440000E 00	0.122874E−01	−0.100434E−04		0.121160E−01	0.161357E−03
0.480000E 00	0.811849E−02	0.111297E−03		0.810813E−02	0.121657E−03
0.520000E 00	0.554193E−02	−0.253431E−04		0.542601E−02	0.905804E−04
0.560000E 00	0.361742E−02	0.804588E−04		0.363111E−02	0.667726E−04
0.599999E 00	0.251041E−02	−0.316396E−04		0.242996E−02	0.488090E−04
0.639999E 00	0.160174E−02	0.598307E−04		0.162614E−02	0.354277E−04
0.679999E 00	0.114626E−02	−0.324817E−04		0.108822E−02	0.255615E−04
0.719999E 00	0.700667E−03	0.459237E−04		0.728243E−03	0.183482E−04
0.759999E 00	0.530960E−03	−0.305050E−04		0.487344E−03	0.131114E−04
0.799999E 00	0.299215E−03	0.362506E−04		0.326133E−03	0.933232E−05
0.839999E 00	0.252197E−03	−0.273281E−04		0.218250E−03	0.661935E−05
0.879999E 00	0.121497E−03	0.292376E−04		0.146054E−03	0.468052E−05
0.919999E 00	0.124879E−03	−0.238383E−04		0.977399E−04	0.330036E−05
0.959999E 00	0.437888E−04	0.239405E−04		0.654081E−04	0.232129E−05
0.999999E 00	0.658748E−04	−0.204745E−04		0.437714E−04	0.162897E−05
0.104000E 01	0.106330E−04	0.197998E−04		0.292920E−04	0.114074E−05
0.108000E 01	0.378257E−04	−0.174260E−04		0.196024E−04	0.797314E−06
0.112000E 01	−0.280455E−05	0.164789E−04		0.131180E−04	0.556327E−06
0.116000E 01	0.239183E−04	−0.147521E−04		0.877864E−05	0.387549E−06
0.120000E 01	−0.762395E−05	0.137682E−04		0.587471E−05	0.269573E−06
0.124000E 01	0.165678E−04	−0.124492E−04		0.393138E−05	0.187252E−06
0.128000E 01	−0.876950E−05	0.115303E−04		0.263090E−05	0.129902E−06
0.132000E 01	0.123371E−04	−0.104865E−04		0.176061E−05	0.900100E−07
0.136000E 01	−0.842895E−05	0.966946E−05		0.117821E−05	0.622986E−07
0.140000E 01	0.965523E−05	−0.882369E−05		0.788466E−06	0.430742E−07
0.144000E 01	−0.755810E−05	0.811550E−05		0.527645E−06	0.297529E−07
0.148000E 01	0.779361E−05	−0.741997E−05		0.353103E−06	0.205325E−07
0.152000E 01	−0.656401E−05	0.681446E−05		0.236298E−06	0.141571E−07

Figure 5.2 Comparison of relative error in results obtained by the Milne and Adams–Moulton methods.

Figure 5.2   (*continued*)

		Solution Using Milne Method	Solution Using Adams–Moulton Method	

Values with $h = 0.04$ continued

x		y	Error	y	Error
0.156000E	01	0.640522E−05	−0.623733E−05	0.158132E−06	0.975331E−08
0.160000E	01	−0.561102E−05	0.572355E−05	0.105823E−06	0.671417E−08
0.164000E	01	0.531755E−05	−0.524211E−05	0.708171E−07	0.461910E−08
0.168000E	01	−0.475746E−05	0.480803E−05	0.473911E−07	0.317556E−08
0.172000E	01	0.443906E−05	−0.440517E−05	0.317144E−07	0.218119E−08
0.176000E	01	−0.401659E−05	0.403931E−05	0.212234E−07	0.149759E−08
0.180000E	01	0.371683E−05	−0.370160E−05	0.142028E−07	0.102763E−08
0.184000E	01	−0.338345E−05	0.339366E−05	0.950460E−08	0.704578E−09
0.188000E	01	0.311713E−05	−0.311028E−05	0.636053E−08	0.482931E−09
0.192000E	01	−0.284671E−05	0.285129E−05	0.425650E−08	0.330768E−09
0.196000E	01	0.261645E−05	−0.261337E−05	0.284847E−08	0.226487E−09
0.200000E	01	−0.239358E−05	0.239564E−05	0.190621E−08	0.155008E−09

Values Near $x = 0.2$ with $h = 0.004$

0.191192E	01	−0.734619E−05	0.735117E−05	0.496895E−08	0.425615E−11
0.191592E	01	0.744385E−05	−0.743908E−05	0.477411E−08	0.405009E−11
0.191992E	01	−0.752346E−05	0.752805E−05	0.458691E−08	0.392220E−11
0.192392E	01	0.762249E−05	−0.761808E−05	0.440706E−08	0.379785E−11
0.192792E	01	−0.770495E−05	0.770919E−05	0.423425E−08	0.368061E−11
0.193192E	01	0.780545E−05	−0.780138E−05	0.406822E−08	0.350298E−11
0.193592E	01	−0.789078E−05	0.789469E−05	0.390870E−08	0.339284E−11
0.193992E	01	0.799286E−05	−0.798910E−05	0.375544E−08	0.328626E−11
0.194392E	01	−0.808104E−05	0.808465E−05	0.360819E−08	0.318456E−11
0.194792E	01	0.818480E−05	−0.818133E−05	0.346671E−08	0.303002E−11
0.195192E	01	−0.827585E−05	0.827918E−05	0.333078E−08	0.293365E−11
0.195592E	01	0.838139E−05	−0.837819E−05	0.320017E−08	0.284017E−11
0.195992E	01	−0.847532E−05	0.847839E−05	0.307469E−08	0.274958E−11
0.196392E	01	0.858274E−05	−0.857979E−05	0.295413E−08	0.261657E−11
0.196792E	01	−0.867956E−05	0.868240E−05	0.283830E−08	0.253331E−11
0.197192E	01	0.878896E−05	−0.878623E−05	0.272701E−08	0.245226E−11
0.197592E	01	−0.888869E−05	0.889132E−05	0.262008E−08	0.237388E−11
0.197992E	01	0.900017E−05	−0.899765E−05	0.251735E−08	0.225930E−11
0.198392E	01	−0.910284E−05	0.910526E−05	0.241864E−08	0.218714E−11
0.198792E	01	0.921648E−05	−0.921415E−05	0.232380E−08	0.211720E−11
0.199192E	01	−0.932212E−05	0.932435E−05	0.223268E−08	0.204925E−11
0.199592E	01	0.943801E−05	−0.943586E−05	0.214514E−08	0.195066E−11
0.199992E	01	−0.954665E−05	0.954871E−05	0.206103E−08	0.188805E−11

with decrease in step size is realized. The oscillation of values we notice in the results by Milne is absent. In sum, we conclude that the Adams–Moulton method gives good results, particularly at the smaller value of h, while the results of the Milne method are hopeless.

It is worth remarking that we usually don't have analytical answers to compare with our numerical results. Observing oscillatory behavior in itself does not mean instability, because the correct solution may *be* oscillatory. For this reason, the usual practice in numerical analysis is to entirely avoid methods that are sometimes unstable, even though they might be more accurate in some instances.

5.8 CONVERGENCE CRITERIA

In Section 5.3, we recorrected in the modified Euler method until no further change in y_{n+1} resulted. Usually this requires one more calculation than would otherwise be needed if we could predict whether the recorrection would make a significant change. In the methods of Milne and Adams–Moulton, we usually do not recorrect, but use a value of h small enough that this is unnecessary. We now look for a criterion to show how small h should be in the Adams–Moulton method, for $dy/dx = f(x, y)$, so that recorrections are not necessary. Let

$$
y_p = \text{value of } y_{n+1} \text{ from predictor formula,}
$$

$$
y_c = \text{value of } y_{n+1} \text{ from corrector formula,}
$$

$$
y_{cc}, y_{ccc}, \text{ etc.} = \text{values of } y_{n+1} \text{ if successive recorrections are made,}
$$

$$
y_\infty = \text{value to which successive recorrections converge,}
$$

$$
D = y_c - y_p.
$$

The change of y_c by recorrecting would be

$$
\begin{aligned}
y_{cc} - y_c &= \left(y_n + \frac{h}{24}(9y_c' + 19y_n' - 5y_{n-1}' + y_{n-2}') \right) \\
&\quad - \left(y_n + \frac{h}{24}(9y_p' + 19y_n' - 5y_{n-1}' + y_{n-2}') \right) \\
&= \frac{9h}{24}(y_c' - y_p').
\end{aligned} \tag{5.24}
$$

In Eq. (5.24) we have used the subscript p or c to denote which y-value is used in evaluating the derivative at $x = x_{n+1}$. We now manipulate the difference $(y_c' - y_p')$:

$$
y_c' - y_p' = f(x_{n+1}, y_c) - f(x_{n+1}, y_p) = \frac{f(x_{n+1}, y_c) - f(x_{n+1}, y_p)}{(y_c - y_p)}(y_c - y_p)
$$

$$
= f_y(\xi_1)D, \qquad \xi_1 \text{ between } y_c \text{ and } y_p, \text{ with } D = y_c - y_p.
$$

Hence,

$$y_{cc} - y_c = \frac{9hD}{24} f_y(\xi_1)$$

is the difference on recorrecting. If recorrected again, the result is

$$
\begin{aligned}
y_{ccc} - y_{cc} &= \frac{9h}{24}(y'_{cc} - y'_c) \\
&= \frac{9h}{24} f_y(\xi_2) \cdot (y_{cc} - y_c) \\
&= \frac{9h}{24} f_y(\xi_2) \left[\frac{9hD}{24} f_y(\xi_1) \right] \\
&= \left(\frac{9h}{24} \right)^2 [f_y(\xi)]^2 D, \qquad \xi_2 \text{ between } y_c \text{ and } y_{cc}.
\end{aligned}
\tag{5.25}
$$

In Eq. (5.25), we need to impose the restrictions that f be continuous and $f_y(\xi_2)$ have the same sign; ξ lies between the extremes of y_p, y_c, and y_{cc}.

On further recorrections we will have a similar relation. We get y_∞ by adding all the corrections of y_p together:

$$
\begin{aligned}
y_\infty &= y_p + (y_c - y_p) + (y_{cc} - y_c) + (y_{ccc} - y_{cc}) + \cdots \\
&= y_p + D + \frac{9hf_y(\xi)}{24}D + \left(\frac{9hf_y(\xi)}{24} \right)^2 D + \left(\frac{9hf_y(\xi)}{24} \right)^3 D + \cdots.
\end{aligned}
$$

The increment to y_p is a geometric series; so, if the ratio is less than unity, which is necessary if a geometric series is to have a sum,

$$y_\infty = y_p + \frac{D}{1 - r}, \qquad r = \frac{9hf_y(\xi)}{24}, \qquad \xi \text{ between } y_p \text{ and } y_\infty.$$

Hence, unless

$$|r| = \frac{h|f_y(\xi)|}{24/9} \doteq \frac{h|f_y(x_n, y_n)|}{24/9} < 1,$$

the successive recorrections diverge. Our first convergence criterion is

$$h < \frac{24/9}{|f_y(x_n, y_n)|}. \tag{5.26}$$

If we wish to have y_c and y_∞ the same to within one in the Nth decimal place, then

$$y_\infty - y_c = \left(y_p + \frac{D}{1 - r} \right) - (y_p + D) = \frac{rD}{1 - r} < 10^{-N}.$$

If $r \ll 1$, the fraction

$$\frac{r}{1 - r} \doteq r;$$

and a second convergence criterion, which ensures that the first corrected value is adequate (that is, it will not be changed in the Nth decimal place by further corrections) is

$$D \cdot 10^N < \left| \frac{1}{r} \right| \doteq \frac{24/9}{h|f_y(x_n, y_n)|}. \tag{5.27}$$

For the Adams–Moulton method we have the three criteria given below. If all are met, the corrected value should be good to N decimals:

$$
\text{Convergence criteria:} \begin{cases} h < \dfrac{24/9}{|f_y|}, \\[2ex] D \cdot 10^N < \dfrac{24/9}{h|f_n|}; \end{cases}
$$
$$\text{Accuracy criterion: } D \cdot 10^N < 14.2.$$

Similar criteria for the Milne method are derived in the same way. They are

$$
\text{Convergence criteria:} \begin{cases} h < \dfrac{3}{|f_y|}, \\[2ex] D \cdot 10^N < \dfrac{3}{h|f_y|}; \end{cases}
$$
$$\text{Accuracy criterion: } D \cdot 10^N < 29. \tag{5.28}$$

These criteria are for a single first-order equation only. A similar analysis for a system is much more complicated.

We illustrate the use of these criteria with an example. Given the equation $dy/dx = \sin x - 3y$, $y(0) = 1$. In the neighborhood of the point $(1, 0.3)$, what maximum value of h is permitted if we wish to compute by (1) Milne's method, (2) the Adams–Moulton method and get accuracy to five decimals? How close must the predictor and corrector values be?

1. $f_y = -3$, so $h < \dfrac{3}{|-3|} = 1$

to ensure convergence if we were to recorrect y_c in the Milne method. This requirement is not severe—we certainly would choose Δx smaller than this to give information throughout the range of x values from 0 to 1. Suppose h were taken as 0.2. Then

$$D < \frac{3}{h|f_y| \, 10^N} = \frac{3}{0.2\,|-3|\,10^5} = 5.0 \times 10^{-5}.$$

The difference between y_p and y_c cannot exceed this value; if it does, recorrections will be needed. This is more severe than $D < 29 \times 10^{-5}$ required for accuracy to one in the fifth decimal for y_c. We should monitor a Milne program in order to be sure that the difference between y_p and y_c does not exceed 5×10^{-5}. If it should exceed this value, we would need to reduce h so that this criterion could be met.

2. For Adams–Moulton, we calculate

$$h < \frac{24/9}{|-3|} = 0.89.$$

If $h = 0.2$,

$$D < \frac{24/9}{0.2 \, |-3| \, 10^5} = 4.4 \times 10^{-5}.$$

Again, this difference in y_p and y_c is more severe than $D < 14.2 \times 10^{-5}$ and should be used to control the value of h.

5.9 ERRORS AND ERROR PROPAGATION

Our previous error analyses have examined the error of a single step only, the so-called *local truncation error* of the methods. Since all practical applications of numerical methods to differential equations involve many steps, the accumulation of these errors, termed the *global truncation error*, is important. We remember that there are several sources of error in a numerical calculation in addition to the truncation error.

ORIGINAL DATA ERRORS

If the initial conditions are not known exactly (or must be expressed inexactly as a terminated decimal number), the solution will be affected to a greater or lesser degree depending on the sensitivity of the equation. Highly sensitive equations are said to be subject to *inherent instability*.

ROUND-OFF ERRORS

Since we can carry only a finite number of decimal places, our computations are subject to inaccuracy from this source, no matter whether we round or whether we chop off. Carrying more decimal places in the intermediate calculations than we require in the final answer is the normal practice to minimize this, but in lengthy calculations this is a source of error that is extremely difficult to analyze and control. Furthermore, in a computer program, if we use double precision, we require a longer execution time and also more storage to hold the more precise values. If these values are for a large array, the memory space needed may exceed that available to the program. This type of error is especially acute when two nearly equal quantities are subtracted. Both floating-point and fixed-point calculations in computers are subject to round-off errors.

TRUNCATION ERRORS OF THE METHOD

These are the types of error we have been discussing, because we use truncated series for approximation in our work, when an infinite series is needed for exactness. The choice of method is our best control here, with suitable selection of h.

In addition to these three types of error, when we solve differential equations numerically we must worry about the propagation of previous errors through the subsequent steps. Since we use the end values at each step as the starting values for the next one, it is as if incorrect original data were distorting the later values. (Round-off would almost always produce error even if our method were exact.) This effect we now examine, but only for the very simple case of the Euler method. This will show how such error studies are made, as well as suggesting how difficult the analysis of more practical methods is.

We consider the first-order equation $dy/dx = f(x, y)$, $y(x_0) = y_0$. Let

$$Y_n = \text{calculated value at } x_n,$$

$$y_n = \text{true value at } x_n,$$

$$e_n = y_n - Y_n = \text{error in } Y_n; \ y_n = Y_n + e_n.$$

By the Euler algorithm,

$$Y_{n+1} = Y_n + hf(x_n, Y_n).$$

By Taylor series,

$$y_{n+1} = y_n + hf(x_r, y_n) + \frac{h^2}{2}y''(\xi_n), \qquad x_n < \xi_n < x_n + h,$$

$$e_{n+1} = y_{n+1} - Y_{n+1} = y_n - Y_n + h[f(x_n, y_n) - f(x_n, Y_n)] + \frac{h^2}{2}y''(\xi_n)$$

$$= e_n + h\frac{f(x_n, y_n) - f(x_n, Y_n)}{y_n - Y_n}(y_n - Y_n) + \frac{h^2}{2}y''(\xi_n)$$

$$= e_n + hf_y(x_n, \eta_n)e_n + \frac{h^2}{2}y''(\xi_n), \qquad \eta_n \text{ between } y_n, Y_n. \tag{5.29}$$

In Eq. (5.29) we have used the mean-value theorem, imposing continuity and existence conditions on $f(x, y)$ and f_y. We suppose, in addition, that the magnitude of f_y is bounded by the positive constant K in the region of x, y-space in which we are interested.* Hence,

$$e_{n+1} \leq (1 + hK)e_n + \frac{1}{2}h^2y''(\xi_n). \tag{5.30}$$

Here $y(x_0) = y_0$ is our initial condition, which we assume free of error. Since $Y_0 = y_0$, $e_0 = 0$:

$$e_1 \leq (1 + hK)e_0 + \frac{1}{2}h^2y''(\xi_0) = \frac{1}{2}h^2y''(\xi_0),$$

$$e_2 \leq (1 + hK)\left[\frac{1}{2}h^2y''(\xi_0)\right] + \frac{1}{2}h^2y''(\xi_1) = \frac{1}{2}h^2[(1 + hK)y''(\xi_0) + y''(\xi_1)].$$

---

*This is essentially the same as the Lipschitz condition, which will guarantee existence and uniqueness of a solution.

Similarly,

$$e_3 \leq \frac{1}{2}h^2[(1 + hK)^2 y''(\xi_0) + (1 + hK)y''(\xi_1) + y''(\xi_2)],$$

$$e_n \leq \frac{1}{2}h^2[(1 + hK)^{n-1}y''(\xi_0) + (1 + hK)^{n-2}y''(\xi_1) + \cdots + y''(\xi_{n-1})].$$

If $f_y \leq K$ is positive, the truncation error at every step is propagated to every later step after being amplified by the factor $(1 + hf_y)$ each time. Note that as $h \to 0$, the error at any point is just the sum of all the previous errors. If the f_y are negative and of magnitude such that $|hf_y| < 2$, the errors are propagated with diminishing effect.

We now show that the accumulated error after n steps is $O(h)$; that is, the global error of the simple Euler method is $O(h)$. We assume, in addition to the above, that y'' is bounded, $|y''(x)| < M, M > 0$. Equation (5.30) becomes, after taking absolute values,

$$|e_{n+1}| \leq (1 + hK)|e_n| + \frac{1}{2}h^2 M.$$

Compare to the second-order difference equation

$$Z_{n+1} = (1 + hK)Z_n + \frac{1}{2}h^2 M, \tag{5.31a}$$

$$Z_0 = 0.$$

Obviously the values of Z_n are at least equal to the magnitudes of $|e_n|$. The solution to Eq. (5.31a) is (check by direct substitution)

$$Z_n = \frac{hM}{2K}(1 + hK)^n - \frac{hM}{2K}.$$

The Maclaurin expansion of e^{hK} is

$$e^{hK} = 1 + hK + \frac{(hK)^2}{2} + \frac{(hK)^3}{6} + \cdots,$$

so that

$$1 + hK < e^{hK} \quad (K > 0),$$

$$Z_n < \frac{hM}{2k}(e^{hK})^n - \frac{hM}{2K} = \frac{hM}{2K}(e^{nhK} - 1)$$

$$= \frac{hM}{2K}(e^{(x_n - x_0)K} - 1) = O(h). \tag{5.31b}$$

It follows that the global error e_n is $O(h)$. (This result can be derived without difference equations. See Exercise 42.)

5.10 SYSTEMS OF EQUATIONS AND HIGHER-ORDER EQUATIONS

We have so far treated only the case of a first-order differential equation. Most differential equations that are the mathematical model for a physical problem are of higher order, or even a *set* of simultaneous higher-order differential equations. For example,

$$\frac{w}{g} \frac{d^2x}{dt^2} + b\frac{dx}{dt} + kx = f(x, t)$$

represents a vibrating system in which a linear spring with spring constant k restores a displaced mass of weight w against a resisting force whose resistance is b times the velocity. The function $f(x, t)$ is an external forcing function acting on the mass.

An analogous second-order equation describes the flow of electricity in a circuit containing inductance, capacitance, and resistance. The external forcing function in this case represents the applied electromotive force. Compound spring–mass systems and electrical networks can be simulated by a system of such second-order equations.

We first show how a higher-order differential equation can be reduced to a system of simultaneous first-order equations. We then show that these can be solved by an application of the methods previously studied. We treat here initial-value problems only, for which n values of the functions or derivatives (with n equal to the order of the system) are all specified at the same (initial) value of the independent variable. When some of the conditions are specified at one value of the independent variable and others at a second value, we call it a *boundary-value problem*. Methods of solving these are discussed in the next chapter.

By solving for the second derivative, we can normally express a second-order equation as

$$\frac{d^2x}{dt^2} = f\left(t, x, \frac{dx}{dt}\right), \qquad x(t_0) = x_0, \qquad x'(t_0) = x'_0. \tag{5.32}$$

The initial value of the function x and its derivative are generally specified. We convert this to a pair of first-order equations by the simple expedient of defining the derivative as a second function. Then, since $d^2x/dt^2 = (d/dt)(dx/dt)$,

$$\frac{dx}{dt} = y, \qquad x(t_0) = x_0,$$

$$\frac{dy}{dt} = f(t, x, y), \qquad y(t_0) = x'_0.$$

This pair of first-order equations is equivalent to the original Eq. (5.32). For even higher orders, each of the lower derivatives is defined as a new function, giving a set of n first-order equations that correspond to an nth-order differential equation. For a system of higher-order equations, each is similarly converted, so that a larger set of first-order equations results. Thus the nth-order differential equation

$$y^{(n)} = f(x, y, y', \dots, y^{(n-1)}),$$

$$y(x_0) = A_1, \qquad y'(x_0) = A_2, \qquad \dots, \qquad y^{(n-1)}(x_0) = A_n$$

is converted into a system of n first-order differential equations by letting $y_1 = y$ and

$$y_1' = y_2$$
$$y_2' = y_3$$
$$\vdots$$
$$y_{n-1}' = y_n$$
$$y_n' = f(x, y_1, y_2, \ldots, y_n) \quad \text{with initial conditions}$$
$$y_1(x_0) = A_1, \quad y_2(x_0) = A_2, \quad \ldots, \quad y_n(x_0) = A_n.$$

We now illustrate the application of the various methods to the pair of first-order equations

$$\frac{dx}{dt} = xy + t, \quad x(0) = 1,$$

$$\frac{dy}{dt} = ty + x, \quad y(0) = -1. \tag{5.33}$$

TAYLOR-SERIES METHOD

We need the various derivatives x', x'', x''', $\ldots$, y', y'', y''', $\ldots$, all evaluated at $t = 0$:

$$x' = xy + t, \qquad\qquad x'(0) = (1)(-1) + 0 = -1,$$
$$y' = ty + x, \qquad\qquad y'(0) = (0)(-1) + 1 = 1,$$
$$x'' = xy' + x'y + 1, \qquad x''(0) = (1)(1) + (-1)(-1) + 1 = 3,$$
$$y'' = y + ty' + x', \qquad y''(0) = -1 + (0)(1) - 1 = -2,$$
$$x''' = x'y' + xy'' + x''y + x'y', \qquad x'''(0) = -7,$$
$$y''' = y' + y' + ty'' + x'', \qquad y'''(0) = 5,$$

and so on; and so on;

$$x(t) = 1 - t + \frac{3}{2}t^2 - \frac{7}{6}t^3 + \frac{27}{24}t^4 - \frac{124}{120}t^5 + \cdots,$$

$$y(t) = -1 + t - t^2 + \frac{5}{6}t^3 - \frac{13}{24}t^4 + \frac{47}{120}t^5 + \cdots. \tag{5.34}$$

At $t = 0.1$, $x = 0.9139$ and $y = -0.9092$.

Equations (5.34) are the solution to the set (5.33). Note that we need to alternate between the functions in getting the derivatives; for example, we cannot get $x''(0)$ until $y'(0)$ is known; we cannot get $y'''(0)$ until $x''(0)$ is known. After we have obtained the coefficients of the Taylor-series expansions in Eq. (5.34), we can evaluate x and y at any value of t, but the error will depend on how many terms we employ.

EULER PREDICTOR–CORRECTOR

We apply the predictor to each equation; then the corrector can be used. Again note that we work alternately with the two functions.

Take $h = 0.1$. Let p and c subscripts indicate predicted and corrected values, respectively:

$$x_p(0.1) = 1 + 0.1[(1)(-1) + 0] = 0.9,$$

$$y_p(0.1) = -1 + 0.1[(0)(-1) + 1] = -0.9,$$

$$x_c(0.1) = 1 + 0.1\left(\frac{-1 + [(0.9)(-0.9) + 0.1]}{2}\right) = 0.9145,$$

$$y_c(0.1) = -1 - 0.1\left(\frac{1 + [(0.1)(-0.9) + 0.9145]}{2}\right) = -0.9088.$$

In computing $x_c(0.1)$, we used the x_p and y_p. In computing $y_c(0.1)$ after $x_c(0.1)$ is known, we have a choice between x_p and x_c. There is an intuitive feel that one should use x_c, with the idea that one should always use the best available values. This does not always expedite convergence, probably due to compensating errors. Here we have used the best values to date. Recorrecting in the obvious manner gives

$$x(0.1) = 0.9135,$$

$$y(0.1) = -0.9089.$$

We can now advance the solution another step if desired, by using the computed values at $t = 0.1$ as the starting values. From this point we can advance one more step, and so on for any value of t. The errors will be the combination of local truncation error at each step plus the propagated error resulting from the use of inexact starting values.

RUNGE–KUTTA–FEHLBERG

Again there is an alternation between the x and y calculations. In applying this method, one always uses the previous k-value in incrementing the function values and the value of h to increment the independent variable. As in the previous calculations, we oscillate between computations for x and for y; for example, we do $k_{1,x}$, then $k_{1,y}$, before doing $k_{2,x}$, and so on.

Keeping in mind that the equations are

$$\frac{dx}{dt} = f(t, x, y) = xy + t, \qquad x(0) = 1,$$

$$\frac{dy}{dt} = g(t, x, y) = ty + x, \qquad y(0) = -1,$$

the k-values for x and y are

for x:

$$k_{1,x} = hf(0, 1, -1)$$
$$= 0.1[(1)(-1) + 0]$$
$$= -0.1;$$

$$k_{2,x} = hf(0.025, 0.975, -0.975)$$
$$= 0.1[(0.975)(-0.975) + 0.025]$$
$$= -0.092562;$$

$$k_{3,x} = hf(0.038, 0.965, -0.964)$$
$$= 0.1[(0.965)(-0.964) + 0.038]$$
$$= -0.089226;$$

$$k_{4,x} = hf(0.092, 0.919, -0.915)$$
$$= 0.1[(0.919)(-0.915) + 0.092]$$
$$= -0.074892;$$

$$k_{5,x} = hf(0.1, 0.913, -0.908)$$
$$= 0.1[(0.913)(-0.908) + 0.1]$$
$$= -0.072904;$$

$$k_{6,x} = hf(0.05, 0.954, -0.953)$$
$$= 0.1[(0.954)(-0.953) + 0.05]$$
$$= -0.085868.$$

for y:

$$k_{1,y} = hg(0, 1, -1)$$
$$= 0.1[(0)(-1) + 1]$$
$$= 0.1;$$

$$k_{2,y} = hg(0.025, 0.975, -0.975)$$
$$= 0.1[(0.025)(-0.975) + 0.975]$$
$$= 0.095062;$$

$$k_{3,y} = hg(0.038, 0.965, -0.964)$$
$$= 0.1[(0.038)(-0.964) + 0.095]$$
$$= 0.092845;$$

$$k_{4,y} = hg(0.092, 0.919, -0.915)$$
$$= 0.1[(0.092)(-0.915) + 0.919]$$
$$= 0.083461;$$

$$k_{5,y} = hg(0.1, 0.913, -0.908)$$
$$= 0.1[(0.1)(-0.908) + 0.913]$$
$$= 0.082178;$$

$$k_{6,y} = hg(0.05, 0.954, -0.953)$$
$$= 0.1[(0.05)(-0.953) + 0.954]$$
$$= 0.090628.$$

Then using the fifth-order formula, we get

$$x(0.1) = 1 + (-0.01185 - 0.046307 - 0.037905 + 0.013123 - 0.003122)$$
$$= 0.913936;$$

$$y(0.1) = -1 + (0.01185 + 0.048185 + 0.042242 - 0.014792 + 0.003296)$$
$$= -0.909217.$$

Extending the Taylor-series solution even further shows that the Runge–Kutta–Fehlberg values are correct to more than five decimals, while the modified Euler values are correct to only three, so $h = 0.1$ may be too large for that method.

Advancing the solution by the Runge–Kutta–Fehlberg method will again involve using the computed values of x and y as the initial values for another step. The errors here will be much less than those for the Euler predictor–corrector method.

ADAMS–MOULTON

After getting four starting values, we proceed with the algorithm of Eq. (5.22), again alternately computing x and then y (see Table 5.12).

In the computations we first get predicted values of x and y:

$$x(0.1) = 0.9330 + \frac{0.025}{24}[55(-0.7929) - 59(-0.8582) + 37(-0.9271) - 9(-1.0)]$$

$$= 0.913937;$$

$$y(0.1) = -0.9303 + \frac{0.025}{24}[55(0.8632) - 59(0.9060) + 37(0.9515) - 9(1.0)]$$

$$= -0.909217.$$

After getting x' and y' at $t = 0.1$, using $x(0.1)$ and $y(0.1)$, we then correct:

$$x(0.1) = 0.9330 + \frac{0.025}{24}[9(-0.7310) + 19(-0.7929) - 5(-0.8582) + (-0.9271)]$$

$$= 0.913936;$$

$$y(0.1) = -0.9303 + \frac{0.025}{24}[9(0.8230) + 19(0.8632) - 5(0.9060) + (0.9515)]$$

$$= -0.909217.$$

The close agreement of predicted and corrected values indicates six-decimal-place accuracy.

In this method, as we advance the solution to larger values of t, the comparison between predictor and corrector values tells us whether the step size needs to be changed.

MILNE

This method can be applied to a system of first-order equations exactly analogously to the application of Adams–Moulton. We do not illustrate it because it is subject to instability.

Table 5.12

	t	x	x'	t	y	y'
Starting values	0.000	1.0	−1.0	0.00	−1.0	1.0
	0.025	0.9759	−0.9271	0.025	−0.9756	0.9515
	0.050	0.9536	−0.8582	0.050	−0.9524	0.9060
	0.075	0.9330	−0.7929	0.075	−0.9303	0.8632
Predicted	0.10	(0.9139)	(−0.7310)	0.10	(−0.9092)	(0.8230)
Corrected		0.9139			−0.9092	

5.11 COMPARISON OF METHODS

It is appropriate that we summarize the various methods that have been discussed in this chapter and compare them. Table 5.13 compares the accuracy, effort required, stability, and other features of the methods.

The data in Table 5.13 lead us to draw the usual conclusion about the best scheme for solving a differential equation of higher order, or a system of N first-order equations: We begin with a fourth-order Runge–Kutta to get a total of four values for each of the functions (this also allows us to compute four values for each of the derivatives), and then advance the solution with Adams–Moulton. At each step after employing Adams–Moulton, we check* the accuracy and adjust the step size when appropriate.

There is still a problem during the starting phase when Runge–Kutta is being used. For instance, when a Runge–Kutta–Fehlberg method (Section 5.4) is used to start the solution for a multistep method that needs equispaced function values, an additional restriction is imposed; the step size must be uniform, which may mean that closer spaced values may need to be computed than are required to meet the accuracy criterion alone.

A method due to Hamming has been widely accepted. It is available through subroutines in some FORTRAN libraries. It begins the solution with a fourth-order Runge–Kutta, and then continues with a predictor–corrector. The equations employed are

$$y_{i+1,p} = y_{i-3} + \frac{4h}{3}(2f_i - f_{i-1} + 2f_{i-2}),$$

$$y_{i+1,m} = y_{i+1,p} - \frac{112}{121}(y_{i,p} - y_{i,c}),$$

$$y_{i+1,c} = \frac{1}{8}[9y_i - y_{i-2} + 3h(f_{i+1,m} + 2f_i - f_{i-1})],$$

$$y_{i+1} = y_{i+1,c} + \frac{9}{121}(y_{i+1,p} - y_{i+1,c}). \qquad \textbf{(5.35)}$$

The predictor equation is the Milne predictor. Before correcting, the estimate of y_{i+1} is modified using the difference between the predicted and corrected values in the *previous* interval (this is omitted in the first interval, since these are not available). A corrector formula is used that depends on two previous y-values, though heavily weighted in favor of the last one. Finally, an adjustment is made based on the error estimate computed from the difference between predicted and corrected values. Note that, while two additional equations are employed in each step compared to the predictor–corrector methods previously described, only two evaluations of the derivative function are needed, the same as before. For many applications, it is the evaluation of the derivative function that is costly in computer time—the two extra algebraic steps don't count for much.

---

*Ordinarily a weighted average of the N errors is monitored. Alternatively, the maximum error is controlled.

Table 5.13   Comparison of methods for differential equations

Method	Type	Local error	Global error	Function evaluation/step	Stability	Ease of changing step size	Recommended?
Modified Euler	Single-step	$O(h^3)$	$O(h^2)$	2	Good	Good	No
Fourth-order Runge–Kutta	Single-step	$O(h^5)$	$O(h^4)$	4	Good	Good	Yes
Runge–Kutta–Fehlberg	Single-step	$O(h^6)$	$O(h^5)$	6	Good	Good	Yes
Milne	Multistep	$O(h^5)$	$O(h^4)$	2	Poor	Poor	No
Adams–Moulton	Multistep	$O(h^5)$	$O(h^4)$	2	Good	Poor	Yes

The special merits of Hamming's method are stability combined with good accuracy. Like Milne's and Adams', the method has a local error of $O(h^5)$ and a global error of $O(h^4)$.

Gear (1967) has proposed a predictor–corrector method that has an $O(h^6)$ local error but uses only three previous steps rather than the four previous steps employed by Adams–Moulton and Milne. It obtains its high order of error by using recorrected values of the function and derivative values. The formulas are

$$y_{n+1,p} = -18y_n + 9y_{n-1,c_1} + 10y_{n-2,c_2} + 9hy_n' + 18hy_{n-1}' + 3hy_{n-2}',$$

$$hy_{n+1,p}' = -57y_n + 24y_{n-1,c_1} + 33y_{n-2,c_2} + 24hy_n' + 57hy_{n-1}' + 10hy_{n-2}',$$

$$F = hy_{n+1,p}' - hf(x_{n+1}, y_{n+1,p}),$$

$$y_{n+1} = y_{n+1,p} - \frac{95}{288}F,$$

$$y_{n,c_1} = y_n + \frac{3}{160}F,$$

$$y_{n-1,c_2} = y_{n-1,c_1} - \frac{11}{1440}F,$$

$$hy_{n+1}' = hy_{n+1,p}' - F.$$

This method is stable and is applicable to systems of first-order differential equations. In addition, Gear (1971) has a listing and a complete description of a subroutine called DIFSUB, which includes both the Adams predictor–corrector method and Gear's stiff methods. This subroutine is also the basis of an IMSL subroutine DGEAR, which contains Adams methods up to order 12 and methods for stiff differential equations.

A stiff equation results from phenomena with widely differing time scales. For example, the general solution of a differential equation may involve sums or differences of terms of the form ae^{ct}, be^{dt}, where both c and d are negative but c is much smaller than d. In such cases, using a small value for the step size can introduce enough round-off errors to cause instability.

An example is the following:

$$x' = 1195x - 1995y, \quad x(0) = 2,$$
$$y' = 1197x - 1997y, \quad y(0) = -2. \tag{5.36}$$

The analytical solution of Eq. (5.36) is

$$x(t) = 10e^{-2t} - 8e^{-800t}, \quad y(t) = 6e^{-2t} - 8e^{-800t}.$$

Observe that the exponents are all negative and of very different magnitude, qualifying this as a stiff equation. Suppose we solve Eq. (5.36) by the simple Euler method with $h = 0.1$, applying just one step. The iterations are

$$x_{i+1} = x_i + hf(x_i, y_i) = x_i + 0.1(1195x_i - 1995y_i),$$
$$y_{i+1} = y_i + hg(x_i, y_i) = y_i + 0.1(1197x_i - 1997y_i).$$

This gives $x(0.1) = 640$, $y(0.1) = 636$, while the exact values are $x(0.1) = 8.187$ and $y(0.1) = 4.912$. Such a result is typical (though here exaggerated) for stiff equations.

One solution to this problem is to use an implicit method rather than an explicit one. All the methods so far discussed have been explicit, meaning that the new values, x_{i+1} and y_{i+1}, are computed in terms of the previous ones, x_i and y_i. The implicit form of Euler's method is

$$x_{i+1} = x_i + hf(x_{i+1}, y_{i+1}),$$
$$y_{i+1} = y_i + h(x_{i+1}, y_{i+1}). \tag{5.37}$$

If the derivative functions $f(x, y)$ and $g(x, y)$ are nonlinear, this is difficult to solve. However, in Eq. (5.36), they are linear. Solving Eq. (5.36) by use of Eq. (5.37), we have

$$x_{i+1} = x_i + 0.1(1195x_{i+1} - 1995y_{i+1}),$$
$$y_{i+1} = y_i + 0.1(1197x_{i+1} - 1997y_{i+1}).$$

Since the system is linear, we can write

$$\begin{bmatrix} x_{i+1} \\ y_{i+1} \end{bmatrix} = \begin{bmatrix} (1 - 1195(0.1)) & 1995(0.1) \\ -1197 & (1 + 1997(0.1)) \end{bmatrix}^{-1} \begin{bmatrix} x_i \\ y_i \end{bmatrix}$$

which has the solution $x(0.1) = 8.23$, $y(0.1) = 4.90$, reasonably close to the analytical values.

In summary, our results for the solution of Eq. (5.36) are

	$x(0.1)$	$y(0.1)$
Exact	8.19	4.91
Euler		
Explicit	640	636
Implicit	8.23	4.90

If the step size is very small, we can get good results from the simpler Euler after the first step. With $h = 0.0001$, the table of results becomes

	$x(0.0001)$	$y(0.0001)$
Exact	2.61	-1.39
Euler		
Explicit	2.64	-1.36
Implicit	2.60	-1.41

but this would require 1000 steps to reach $t = 0.1$, and round-off errors would be large.

If we anticipate some material from the next chapter, we can give a better description of stiffness as well as indicate the derivation of the general solution to Eq. (5.36). We rewrite Eq. (5.36) in matrix form:

$$\begin{bmatrix} x \\ y \end{bmatrix}' = A \begin{bmatrix} x \\ y \end{bmatrix}, \text{ where } A = \begin{bmatrix} 1195 & -1995 \\ 1197 & -1997 \end{bmatrix}.$$

The general solution, in matrix form, is

$$\begin{bmatrix} x \\ y \end{bmatrix} = ae^{-2t}v_1 + ce^{-800t}v_2,$$

where

$$v_1 = \begin{bmatrix} 5 \\ 3 \end{bmatrix} \text{ and } v_2 = \begin{bmatrix} 1 \\ 1 \end{bmatrix}.$$

You can easily verify that $Av_1 = -2v_1$ and $Av_2 = -800v_2$. In Chapter 6 we will see that this means that v_1 is an eigenvector of A and that -2 is the corresponding eigenvalue. Similarly, v_2 is an eigenvector of A with the corresponding eigenvalue of -800. (In Chapter 6 you will learn how to find the eigenvectors and eigenvalues of a matrix.)

A stiff equation can be defined in terms of the eigenvalues of the matrix A that represents the right-hand sides of the system of differential equations. When the eigenvalues of A have real parts that are negative and differ widely in magnitude as in this example, the system is stiff. In the case of a nonlinear system

$$\begin{bmatrix} x_1 \\ x_2 \\ \vdots \\ x_n \end{bmatrix}' = \begin{bmatrix} f_1(x_1, x_2, \ldots, x_n) \\ f_2(x_1, x_2, \ldots, x_n) \\ \vdots \\ f_n(x_1, x_2, \ldots, x_n) \end{bmatrix},$$

one must consider the Jacobian matrix whose terms are $\partial f_i / \partial x_j$. See Gear (1971) for more information.

5.12 CHAPTER SUMMARY

These are things you should be able to do if you understand the material in Chapter 5:

1. Use the Taylor-series procedure to solve a first- or second-order differential equation. You should be able to explain why this method is not generally used.

2. Solve first-order equations and systems of first-order equations with all of these methods:

   a. Simple Euler
   b. Modified Euler
   c. Fourth-order Runge–Kutta
   d. Runge–Kutta–Fehlberg
   e. Adams–Moulton (applied after starting the solution with another method)

3. Rewrite a higher-order differential equation or system of higher-order equations as an equivalent system of first-order equations.

4. Explain these terms: *efficiency*, *accuracy*, *stability*, and *convergence* as applied to the procedures of this chapter. In particular, you should be able to explain why Milne's method is rejected.

5. Compare all of these methods for accuracy, efficiency, stability, ability to monitor errors, and ease of changing step size.

6. Outline the arguments used to show whether errors grow as they propagate and whether a method is stable.

7. Explain what is meant by the term *stiff equation* and give examples of explicit and implicit methods.

8. Utilize computer programs to obtain the solution to a given differential equation or system of equations.

SELECTED READINGS FOR CHAPTER 5

Gear (1971); Press, Flannery, Teukolsky, and Vetterling (1986); Stoer and Bulirsch (1980).

5.13 COMPUTER PROGRAMS

Program 1 (Fig. 5.3) solves a second-order differential equation by the modified Euler method, after reducing it to a pair of first-order equations. The computer output is the solution to the equation that represents a damped spring system,

$$\frac{W}{g}\frac{d^2x}{dt^2} + b\frac{dx}{dt} + kx = a \sin \omega t.$$

In the program, each of the constants in the differential equation has the same name except that k is called XK and ω is called C. In the program, X0 and T0 are the initial values of displacement and time. The variable y is used for dx/dt, and Y0 is its initial

```
      PROGRAM PEULER(INPUT,OUTPUT)
C
C     PROGRAM FOR EULER PREDICTOR-CORRECTOR SOLUTION TO A SECOND
C ORDER DIFFERENTIAL EQUATION OR TWO SIMULTANEOUS FIRST ORDER
C EQUATIONS.
C
C ----------------------------------------------------------------
C
      REAL X,Y,T,A,B,C,XK,W,X0.Y0,T0,TEND,DT,TOL,YP,DX0,DY0
     +     ,Z,XC,YC,TOUT
      INTEGER I
C
C ----------------------------------------------------------------
C
C DEFINE THE DERIVATIVE FUNCTIONS. FOR A SECOND ORDER EQUATION,
C ALWAYS USE FCN1 = VARIABLE THAT REPLACES THE FIRST DERIVATIVE.
C
      FCN1(X,Y,T) = Y
      FCN2(X,Y,T) = ( A*SIN(C*T) - B*Y - XK*X ) * 32.1725 / W
C
C ----------------------------------------------------------------
C
C READ THE CONSTANTS OF THE PROBLEM
C
C      READ*, W,B,XK,A,C
C      READ*, X0,Y0,T0,TEND,DT,TOL
      DATA W,B,XK,A,C/10,5,6,20,2/
      DATA X0,Y0,T0,TEND,DT,TCL/1,-2,0,2,0.05,0.01/
C
      PRINT 200
      PRINT 201, T0,X0,Y0
C
C ----------------------------------------------------------------
C
C COMPUTE BY PREDICTOR FORMULAS
C
    5 IF ( T0 .LT. TEND ) THEN
         DX0 = FCN1(X0,Y0,T0)
         XP = X0 + DT*DX0
         DY0 = FCN2(X0,Y0,T0)
         YP = Y0 + DT*DY0
C
C ----------------------------------------------------------------
C
C NOW CORRECT. REPEAT CORRECTIONS UP TO FIVE TIMES.
C
         DO 10 I = 1,5
            XC = X0 + DT*( DX0 + FCN1(XP,YP,T0+DT) ) / 2.0
            YC = Y0 + DT*( DY0 + FCN2(XC,YP,T0+DT) ) / 2.0
            Z = ( ABS(XC - XP) + ABS(YC - YP) ) / 2.0
            IF ( Z .LE. TOL ) GO TO 20
            XP = XC
            YP = YC
   10    CONTINUE
         TOUT = T0 + DT
         PRINT 202, TOUT
C
C ----------------------------------------------------------------
C
C RESET VARIABLES. PRINT A NEW LINE AND GO ON TO NEXT STEP.
C
```

Figure 5.3 Program 1.

Figure 5.3   (*continued*)

```
  20   T0 = T0 + DT
       PRINT 201, T0,XC,YC
         X0 = XC
         Y0 = YC
         GO TO 5
       END IF
C
 200 FORMAT(///2X,'TIME',7X,'DISTANCE',15X,'VELOCITY'/)
 201 FORMAT(1X,F5.2,6X,F10.7,12X,F13.7)
 202 FORMAT(1X,'AT T = ',F5.2,' TOLERANCE NOT MET WITH FIVE',
     +         ' RECORRECTIONS ')
       STOP
       END


                OUTPUT FOR PROGRAM 1


    TIME         DISTANCE                 VELOCITY

    .00         1.0000000               -2.0000000
    .05          .9152416               -1.4002323
    .10          .8583095                -.8601237
    .15          .8270768                -.3760873
    .20          .8189385                 .0622915
    .25          .8316736                 .4585103
    .30          .8631817                 .8132604
    .35          .9113776                1.1261918
    .40          .9741545                1.3966751
    .45         1.0493767                1.6241366
    .50         1.1348858                1.8082081
    .55         1.2285121                1.9487964
    .60         1.3280888                2.0461160
    .65         1.4314684                2.1007043
    .70         1.5365380                2.1134275
    .75         1.6412367                2.0854794
    .80         1.7435708                2.0183761
    .85         1.8416304                1.9139465
    .90         1.9336045                1.7743193
    .95         2.0177954                1.6019060
   1.00         2.0926322                1.3993814
   1.05         2.1566839                1.1696611
   1.10         2.2086704                 .9158759
   1.15         2.2477825                 .6352084
   1.20         2.2725406                 .3426194
   1.25         2.2822756                 .0372065
   1.30         2.2764537                -.2771621
   1.35         2.2547013                -.5971912
   1.40         2.2168205                -.9193613
   1.45         2.1627919               -1.2400556
   1.50         2.0927802               -1.5556447
   1.55         2.0071316               -1.8625485
   1.60         1.9063695               -2.1572853
   1.65         1.7915245               -2.4420739
   1.70         1.6629711               -2.7060522
   1.75         1.5217995               -2.9474093
   1.80         1.3691976               -3.1639028
   1.85         1.2064641               -3.3532707
   1.90         1.0350061               -3.5134185
   1.95          .8563286               -3.6425097
   2.00          .6720204               -3.7390133
```

value. The solution is computed from T0 to TEND, using a step size equal to DT. TOL is a value used to determine whether recorrections should be continued or not; they terminate when the change on recorrection is less than TOL. The parameters and initial conditions are read in. The value of the function is recorrected until the difference in successive values is less than a certain tolerance. If the tolerance is not met in five recorrections, the value is printed together with a message to indicate that the tolerance was not met.

Program 2 (Fig. 5.4) uses a subroutine, RK4TH, to solve a single first-order differential equation of the form $dx/dt = f(x, t)$, using the Runge–Kutta method. Parameters are passed to specify the initial value of the function, X0, and the initial value and the increment for the independent variable, T0 and H. The subroutine returns X, the value of the function at the end of the interval. An external function calculates values of the derivative function.

The ecologist's problem concerning the growth in population of field mice, described at the beginning of this chapter, is solved in Program 2 using $\Delta t = 0.2$. In this problem one of the constants in the differential equation, B, is known as a function of time only through a table of values. To find the proper values of this constant, a subroutine that does interpolation is called. (This subroutine, INTERP, is the same one presented earlier, in Chapter 3.) The values of the two constants needed by the function are passed through COMMON. (They cannot be passed as parameters because the subroutine RK4TH was written with a call to the function that did not include them as parameters.)

The output from Program 2 shows how the mouse population is expected to vary. Of course, only whole numbers of mice are possible; our mathematical model considers population as a *continuous* variable. We are not surprised to see that the population reaches a maximum as the food supply diminishes with time.

The third program is PRKSYST (Fig. 5.5), which solves a system of N simultaneous first-order equations by the Runge–Kutta–Fehlberg method. Parameters passed to the subroutine include the name of a function that computes each of the N derivative values, the initial value and the step size for the independent variable, an array of function values at the beginning of the interval, and the number of equations. Values of the functions at the end of the interval are returned in an array. A doubly dimensioned array (declared of size $6 \times N$ in the main program) is employed to hold intermediate values.

This routine can be used to solve an nth-order differential equation, or a system of higher-order equations, by reducing each of them to a set of simultaneous first-order equations.

```
      PROGRAM PRK4TH(INPUT,OUTPUT)
C
C     MAIN PROGRAM TO SOLVE THE FIELD MICE PROBLEM. IT EMPLOYS TWO
C  SUBROUTINES. ROUTINE RK4TH SOLVES THE DIFFERENTIAL EQUATION AND
C  ROUTINE INTERP INTERPOLATES IN A TABLE OF VALUES FOR ONE OF THE
C  PARAMETERS. THE PARAMETERS OF THE DIFFERENTIAL EQUATION ARE
C  PASSED TO A FUNCTION SUBPROGRAM THROUGH COMMON. THIS FUNCTION
C  IS AN EXTERNAL SUBPROGRAM BECAUSE IT IS PASSED AS A PARAMETER.
C
C  ------------------------------------------------------------------
C
```

Figure 5.4   Program 2.

Figure 5.4 (*continued*)

```
          REAL A,BOUT,B(9),D(10),FCT,T0,X0,H,TF,X
          INTEGER I
          EXTERNAL FCT
          COMMON /PARAM/A,BOUT
          A = 0.9
C
C    ------------------------------------------------------------------
C
C    READ IN THE VALUES OF THE CONSTANT B, WHICH ARE DEFINED ONLY AS
C    A TABLE OF VALUES FOR VARIOUS T VALUES.
C
          READ*, ( B(I), I = 1,9 )
          READ*, T0,X0,H,TF
C
C    ------------------------------------------------------------------
C
C    WRITE HEADING AND INITIAL VALUES
C
          PRINT 200
          PRINT 201, T0,X0
C
C    ------------------------------------------------------------------
C
C    CALL THE SUBROUTINE RK4TH FOR THE NEXT VALUE AND PRINT IT, BUT
C    MUST GET PROPERLY INTERPOLATED B VALUE FROM INTERP.
C
        5 CALL INTERP(0.C,8.0,1.0,9,B,T0,3,BOUT)
          CALL RK4TH(FCT,T0,H,X0,X)
C
C    ADVANCE VARIABLES
C
          T0 = T0 + H
          X0 = X
          PRINT 201, T0,X
C
C    ------------------------------------------------------------------
C
C    ARE WE DONE ? TEST X VALUE TO SEE.
C
          IF ( T0 .LT. TF ) GO TO 5
C
      200 FORMAT(///'SOLUTION TO A DIFFERENTIAL EQUATION',/
         +         '          BY RK4TH METHOD',
         +        //5X,'T',13X,'N',//)
      201 FORMAT(1X,F6.1,1X,F14.1)
          STOP
          END
C
C    ------------------------------------------------------------------
C
          REAL FUNCTION FCT(XN,T)
C
C    THIS FUNCTION COMPUTES VALUES FOR THE DIFFERENTIAL EQUATION.
C
C    ------------------------------------------------------------------
C
          REAL A,BOUT,XN,T
          COMMON /PARAM/A,BOUT
          FCT = A*XN - BOUT*XN**1.7
          RETURN
          END
```

Figure 5.4 (*continued*)

```
C
C     ------------------------------------------------------------------
C
      SUBROUTINE RK4TH(FCN,T0,H,X0,X)
C
C     ------------------------------------------------------------------
C
C     SUBROUTINE RK4TH :
C                         THIS SUBROUTINE ADVANCES THE SOLUTION OF A
C     FIRST ORDER DIFFERENTIAL EQUATION OF THE FORM DX/DT = F(X,T),
C     USING THE RUNGE-KUTTA FOURTH ORDER METHOD.
C
C     ------------------------------------------------------------------
C
C     PARAMETERS ARE :
C
C     FCN     - FUNCTION SUBPROGRAM TO COMPUTE DX/DT = F(X,T). THIS
C                 MUST BE DECLARED EXTERNAL IN THE CALLING PROGRAM.
C     X0,T0   - X AND T VALUES AT THE BEGINNING OF THE INTERVAL.
C     H       - STEP SIZE, THE VALUE OF DELTA T.
C     X       - X VALUE AT THE END OF THE INTERVAL AS RETURNED.
C
C     ------------------------------------------------------------------
C
      REAL FCN,T0,H,X0,X,XK1,XK2,XK3,XK4
C
      XK1 = H * FCN(X0,T0)
      XK2 = H * FCN(X0+XK1/2.0,T0+H/2.0)
      XK3 = H * FCN(X0+XK2/2.0,T0+H/2.0)
      XK4 = H * FCN(X0+XK3,T0+H)
      X = X0 + (XK1 + 2.0*XK2 + 2.0*XK3 + XK4) / 6.0
      RETURN
      END
C
C     ------------------------------------------------------------------
C
      SUBROUTINE INTERP(X1,XN,H,N,Y,X,M,YOUT)
C
C     ------------------------------------------------------------------
C
C     SUBROUTINE INTERP :
C                         SUBROUTINE TO INTERPOLATE IN A TABLE OF
C     UNIFORMLY SPACED VALUES.
C
C     ------------------------------------------------------------------
C
C     PARAMETERS ARE :
C
C     X1,XN   - BEGINNING AND ENDING X VALUES
C     H       - DELTA X - THE UNIFORM SPACING
C     N       - NUMBER OF ENTRIES IN THE TABLE
C     Y       - ARRAY OF FUNCTION VALUES
C     X       - VALUE AT WHICH Y IS TO BE INTERPOLATED
C     M       - DEGREE OF INTERPOLATING POLYNOMIAL. THE SUBROUTINE WILL
C                 HANDLE UP TO 10TH DEGREE, BUT USUALLY THE DEGREE WILL BE
C                 LESS THAN 10 TO AVOID ROUND OFF ERRORS.
C     YOUT    - THE INTERPOLATED Y VALUE RETURNED TO THE CALLER
C
C     AN ARRAY D IS USED IN THE SUBROUTINE TO HOLD DELTA Y'S.
C
C     ------------------------------------------------------------------
```

Figure 5.4 (*continued*)

```
      C
            REAL X1,XN,H,Y(N),X,YOUT
            INTEGER N,M,I,J,K
            REAL D(10),FM,FJ,X0,Y0,S,FNUM,DEN,FI
      C
      C     ----------------------------------------------------------------
      C
      C     FIRST FIND PROPER SUBSCRIPT FOR Y0 SO THAT X IS CENTERED IN THE
      C     DOMAIN AS WELL AS POSSIBLE. THIS SUBSCRIPT VALUE IS CALLED J.
      C
            FM = M + 1
            J = ( X - X1 )/H - FM/2.0 + 2.0
            IF ( X .LE. X1 + FM/2.0*H )  J = 1
            IF ( X .GE. XN - FM/2.0*H )  J = N - M
            FJ = J
            X0 = X1 + ( FJ - 1.0 )*H
            Y0 = Y(J)
      C
      C     ----------------------------------------------------------------
      C
      C     COMPUTE THE DIFFERENCES THAT ARE NEEDED
      C
            DO 10 I = 1,M
               D(I) = Y(J+1) - Y(J)
               J = J + 1
         10 CONTINUE
            IF ( M .GT. 1 ) THEN
               DO 20 J = 2,M
                  DO 15 I = J,M
                     K = M - I + J
                     D(K) = D(K) - D(K-1)
         15       CONTINUE
         20    CONTINUE
            END IF
      C
      C     ----------------------------------------------------------------
      C
      C     COMPUTE S VALUE
      C
            S = ( X - X0 ) / H
      C
      C     COMPUTE INTERPOLATED Y VALUE
      C
            YOUT = Y0
            FNUM = S
            DEN = 1.0
            DO 30 I = 1,M
               FI = I
               YOUT = YOUT + FNUM/DEN*D(I)
               FNUM = FNUM * ( S - FI )
               DEN = DEN * ( FI + 1.0 )
         30 CONTINUE
            RETURN
            END


         OUTPUT FOR PROGRAM 2


   SOLUTION TO A DIFFERENTIAL EQUATION
         BY RK4TH METHOD
```

Figure 5.4   (*continued*)

```
        T                N


      0.0              100.0
       .2              115.4
       .4              133.2
       .6              153.8
       .8              177.8
      1.0              205.8
      1.2              238.8
      1.4              277.6
      1.6              323.6
      1.8              378.4
      2.0              444.0
      2.2              522.8
      2.4              617.6
      2.6              731.9
      2.8              870.0
      3.0             1036.8
      3.2             1237.8
      3.4             1478.3
      3.6             1763.5
      3.8             2097.1
      4.0             2480.5
      4.2             2910.5
      4.4             3382.0
      4.6             3879.9
      4.8             4379.2
      5.0             4846.2
      5.2             5241.5
      5.4             5518.6
      5.6             5651.1
      5.8             5633.4
      6.0             5480.9
      6.2             5224.1
      6.4             4903.9
      6.6             4551.8
      6.8             4192.6
      7.0             3843.7
```

```
C         PROGRAM PRKSYST(INPUT,OUTPUT)
C
C             THIS PROGRAM USES SUBROUTINE RKFSYS TO SOLVE THE SYSTEM
C             OF EQUATIONS:
C                 X' = F(T,X,Y) = Y
C                 Y' = G(T,X,Y) = X + T
C             WHERE X,Y ARE THE DEPENDENT VARIABLES AND T IS THE
C             INDEPENDENT VARIABLE.
C
C         WE COMPARE THE COMPUTED SOLUTIONS WITH THE EXACT ONES:
C             X(T) = COSH(T) - T
C             Y(T) = SINH(T) - 1
C
```

Figure 5.5   Program 3.

Figure 5.5 (*continued*)

```
         REAL T0,H,XEND(10),F(10),X0(10),
              TFINAL,TPRINT, TOL
         INTEGER N,I,J
C
C        ----------------------------------------------------------------
C
         COMMON TOL, TFINAL
         EXTERNAL DERIVS
         DATA N,X0/2,1,-1,8*0.0/
         DATA XEND/1,-1,8*0.0/
         DATA T0/0.0/
         DATA H/0.1/
C
C
C        ----------------------------------------------------------------
C
         PRINT 200
         TOL = 0.0001
         TFINAL = 1.5
         T0 = 0.0
10       IF (T0 .LT. TFINAL+0.001) THEN
             PRINT 100, T0,XEND(1),XEND(2),COSH(T0)-T0,SINH(T0)-1.0
             CALL RKFSYS(DERIVS,T0,H,X0,XEND,F,N)
             X0(1) = XEND(1)
             X0(2) = XEND(2)
             GO TO 10
         ENDIF
C
100      FORMAT(5X,F5.3,2X,5F15.7)
200      FORMAT(///T5,'TIME',T20,'X(TIME)',T35,'Y(TIME)'
       +          ,T50,'X-EXACT',T65,'Y-EXACT'/)
         STOP
         END
C
C        ----------------------------------------------------------------
C
         SUBROUTINE RKFSYS(DERIVS,T0,H,X0,XEND,F,N)
C
C        ------------------------------------------------------------------
C
C
C        SUBROUTINE RKFSYS :
C                         THIS SUBROUTINE SOLVES A SYSTEM OF N FIRST
C        ORDER DIFFERENTIAL EQUATIONS BY THE RUNGE-KUTTA-FEHLBERG
C        METHOD. THE EQUATIONS ARE OF THE FORM:
C
C                         DX1/DT = F1(X,T) , DX2/DT = F2(X,T),ETC.
C                         DX2/DT = F2(X,T),ETC.
C
C        ------------------------------------------------------------------
C
C        PARAMETERS ARE :
C
C        DERIVS - A SUBROUTINE THAT COMPUTES VALUES OF THE N DERIVATIVES.
C                 IT MUST BE DECLARED EXTERNAL BY THE CALLER. IT IS
C                 INVOKED BY THE STATEMENT :
C                                     CALL DERIVS(X,T,F,N)
C        T0     - THE INITIAL VALUE OF INDEPENDENT VARIABLE
C        H      - THE INCREMENT TO T, THE STEP SIZE
C        X0     - THE ARRAY THAT HOLDS THE INITIAL VALUES OF THE FUNCTIONS
C        XEND   - AN ARRAY THAT RETURNS THE FINAL VALUES OF THE FUNCTIONS
C        XWRK   - AN ARRAY USED TO HOLD INTERMEDIATE VALUES DURING THE
C                 COMPUTATION. IT MUST BE DIMENSIONED OF SIZE 4 X N IN
```

Figure 5.5   (*continued*)

```
C                    THE MAIN PROGRAM.
C       N       - THE NUMBER OF EQUATIONS IN THE SYSTEM BEING SOLVED
C       F       - AN ARRAY THAT HOLDS VALUES OF THE DERIVATIVES
C
C       -----------------------------------------------------------------
C
        REAL X0(N),XEND(N),XWRK(6,10),F(N),H,T0 , TOL
        INTEGER I,N
C
C           LOCAL VARIABLES
C
        REAL ERROR, SUM, ABS, MAX, STOREH, TEND
C
        COMMON TOL, TFINAL
C
C
C           INITIALIZE FOR INTERVAL [T0, T0+H]
C
        TEND = T0 + H
        STOREH = H
C
C           CHECK TO SEE IF WE ARE FINISHED
C
1       IF (T0 .GE. TEND) THEN
            H = STOREH
            RETURN
C
C
                                            ELSE
C
C
C       -----------------------------------------------------------------
C
C   GET FIRST ESTIMATE OF THE DELTA X'S
C
        CALL DERIVS(X0,T0,F,N)
        DO 10 I = 1,N
           XWRK(1,I) = H * F(I)
           XEND(I) = X0(I) + XWRK(1,I)/4.0
    10 CONTINUE
C
C       -----------------------------------------------------------------
C
C   GET THE SECOND ESTIMATE. THE XEND VECTOR HOLDS THE X VALUES.
C
        CALL DERIVS(XEND,T0+H/4.0,F,N)
        DO 20 I = 1,N
           XWRK(2,I) = H * F(I)
           XEND(I) = X0(I) + (XWRK(1,I)*3.0 + XWRK(2,I)*9.0)/32.0
    20 CONTINUE
C
C       -----------------------------------------------------------------
C
C   REPEAT FOR THIRD ESTIMATE
C
        CALL DERIVS(XEND,T0+3.0*H/8.0,F,N)
        DO 30 I = 1,N
           XWRK(3,I) = H * F(I)
           XEND(I) = X0(I) + (XWRK(1,I)*1932.0 - XWRK(2,I)*7200.0
                   + XWRK(3,I)*7296.0)/2197.0
    30 CONTINUE
C
```

Figure 5.5 (*continued*)

```
C     --------------------------------------------------------------------
C
C     NOW GET FOURTH ESTIMATE
C
      CALL DERIVS(XEND,T0+12.0*H/13.0,F,N)
      DO 40 I = 1,N
         XWRK(4,I) = H * F(I)
         XEND(I) = X0(I) + (439.0*XWRK(1,I)/216.0 - 8.0*XWRK(2,I)
     .              + 3680.0*XWRK(3,I)/513.0
     .              - 845.0*XWRK(4,I)/4104.0 )
   40 CONTINUE
C
C     --------------------------------------------------------------------
C
C     NOW GET FIFTH ESTIMATE
C
      CALL DERIVS(XEND,T0+H,F,N)
      DO 50 I = 1,N
         XWRK(5,I) = H*F(I)
         XEND(I) = X0(I) - 8.0*XWRK(1,I)/27.0 + 2.0*XWRK(2,I)
     .              - 3544.0*XWRK(3,I)/2565.0 + 1859.0*XWRK(4,I)/4104.0
     .              - 11.0*XWRK(5,I)/40.0
   50 CONTINUE
C
C     --------------------------------------------------------------------
C
C     NOW GET SIXTH ESTIMATE
C
      CALL DERIVS(XEND,T0+H/2.0,F,N)
      DO 60 I = 1,N
         XWRK(6,I) = H*F(I)
   60 CONTINUE
C
C     --------------------------------------------------------------------
C
C          WE ESTIMATE THE ERROR BY COMPUTING THE DIFFERENCE BETWEEN
C          THE FOURTH AND FIFTH ORDER EQUATIONS.
C
      SUM = 0.0
      ERROR = 0.0
      DO 70 I = 1,N
         SUM = ABS(XWRK(1,I)/360.0 - 128.0*XWRK(3,I)/4275.0
     .              - 2197.0*XWRK(4,I)/75240.0
     .              + XWRK(5,I)/50.0 + 2.0*XWRK(6,I)/55.0)
         ERROR = AMAX1(ERROR,SUM)
   70 CONTINUE
C
C
      IF (ERROR .LT. TOL) THEN
C
C     --------------------------------------------------------------------
C
C     WE COMPUTE THE X AT THE END OF THE INTERVAL FROM A WEIGHTED AVERAGE
C     OF THE SIX ESTIMATES, THEN RETURN.
C
         DO 80 I = 1,N
            XEND(I) = X0(I) + 16.0*XWRK(1,I)/135.0
     .                         + 6656.0*XWRK(3,I)/12825.0
     .                         + 28561.0*XWRK(4,I)/56430.0
     .                         - 9.0*XWRK(5,I)/50.0
     .                         + 2.0*XWRK(6,I)/55.0
            X0(I) = XEND(I)
```

Figure 5.5   (*continued*)

```
80              CONTINUE
        T0 = T0 + H
C
        END IF
C
        IF (ERROR .GT. TOL) THEN
            H = H/2.0
        ENDIF
C
        IF (ERROR .LT. H*TOL/10.0) THEN
            H = H*2.0
        ENDIF
C
        IF (T0 + H .GT. TEND) THEN
            H = TEND - T0
        ENDIF
C
C                                               ENDIF
C
C
        GO TO 1
        END
C
C       -----------------------------------------------------------------
C
C           DEFINE THE FUNCTIONS OF THE PROBLEM IN TERMS OF THE
C           F'S AND THE X'S.
C
C       -----------------------------------------------------------------
C
        SUBROUTINE DERIVS(X,T,F,N)
        REAL X(N),T,F(N)
        INTEGER N
C
C           X' = Y        F(1) = X(2)
C           Y' = X + T    F(2) = X(1) + T
C
        F(1) = X(2)
        F(2) = X(1) + T
        RETURN
        END
```

                    OUTPUT FOR PROGRAM 3

TIME	X(TIME)	Y(TIME)	X-EXACT	Y-EXACT
.000	1.0000000	-1.0000000	1.0000000	-1.0000000
.100	.9050042	-.8998333	.9050042	-.8998332
.200	.8200668	-.7986640	.8200668	-.7986640
.300	.7453385	-.6954797	.7453385	-.6954797
.400	.6810724	-.5892477	.6810724	-.5892477
.500	.6276260	-.4789047	.6276260	-.4789047
.600	.5854652	-.3633464	.5854652	-.3633464
.700	.5551690	-.2414163	.5551690	-.2414163
.800	.5374349	-.1118940	.5374349	-.1118940
.900	.5330864	.0265167	.5330864	.0265167
1.000	.5430806	.1752012	.5430806	.1752012
1.100	.5685185	.3356475	.5685186	.3356475
1.200	.6106555	.5094613	.6106556	.5094614
1.300	.6709142	.6983824	.6709142	.6983824
1.400	.7508984	.9043015	.7508985	.9043015
1.500	.8524096	1.1292794	.8524096	1.1292795

EXERCISES

Section 5.2

▶ 1. a) Solve the differential equation

$$\frac{dy}{dx} = x + y + xy, \qquad y(0) = 1$$

by Taylor-series expansion to get the value of y at $x = 0.1$ and at $x = 0.5$. Use terms through x^5.

b) Do the same for

$$\frac{dy}{dx} = -y - 2x, \qquad y(0) = -1.$$

(Analytical solution is $y(x) = -3e^{-x} - 2x + 2$.)

c) Do the same for

$$\frac{dy}{dx} = \frac{5x}{y} - xy, \qquad y(0) = 2.$$

(Analytical solution is $y(x) = \sqrt{5 - e^{-x^2}}$.)

2. The general solution to a differential equation normally defines a family of curves. For the differential equation

$$\frac{dy}{dx} = x^2 y^2,$$

find, using the Taylor-series method, the particular curve that passes through $(1, 0)$. Also find the curve through $(0, 1)$. Compare to the analytical solutions.

▶ 3. Use the Taylor-series method to get y at $x = 0.2(0.2)0.6$,* given that

$$y'' = xy, \qquad y(0) = 1, \qquad y'(0) = 1.$$

4. A spring system has resistance to motion proportional to the square of the velocity, and its motion is described by

$$\frac{d^2x}{dt^2} + 0.1\left(\frac{dx}{dt}\right)^2 + 0.6x = 0.$$

If the spring is released from a point that is a unit distance above its equilibrium point, $x(0) = 1$, $x'(0) = 0$, use the Taylor-series method to write a series expression for the displacement as a function of time, including terms up to t^6.

Section 5.3

▶ 5. Use the simple Euler method to solve for $y(0.1)$ from

$$\frac{dy}{dx} = x + y + xy, \qquad y(0) = 1,$$

with $h = 0.01$. Comparing your result to the value determined by Taylor series in Exercise 1, estimate how small h would need to be to obtain four-decimal accuracy.

---

*This notation means for $x = 0.2$ through $x = 0.6$ with increments of 0.2.

6. Solve the differential equation

$$\frac{dy}{dx} = \frac{x}{y}, \qquad y(0) = 1,$$

by the simple Euler method with $h = 0.1$, to get $y(1)$. Then repeat with $h = 0.2$ to get another estimate of $y(1)$. Extrapolate these results, assuming errors are proportional to step size, and then compare them to the analytical result. (Analytical result is $y^2 = 1 + x^2$.)

▶ 7. Repeat Exercise 5, but with the modified Euler method with $h = 0.025$, so that the solution is obtained after four steps. Comparing with the result of Exercise 5, about how much less effort is it to solve this problem to four decimals with the modified Euler method in comparison to the simple Euler method?

8. Find the solution to

$$\frac{dy}{dt} = y^2 + t^2, \qquad y(1) = 0, \qquad \text{at } t = 2,$$

by the modified Euler method, using $h = 0.1$. Repeat with $h = 0.05$. From the two results, estimate the accuracy of the second computation.

▶ 9. Solve $y' = \sin x + y$, $y(0) = 2$, by the modified Euler method to get y at $x = 0.1(0.1)0.5$.

10. A sky diver jumps from a plane, and during the time before the parachute opens, the air resistance is proportional to the $\frac{3}{2}$ power of the diver's velocity. If it is known that the maximum rate of fall under these conditions is 80 mph, determine the diver's velocity during the first 2 sec of fall using the modified Euler method with $\Delta t = 0.2$. Neglect horizontal drift and assume an initial velocity of zero.

Section 5.4

11. Solve Exercise 5 by the Runge–Kutta method but with $h = 0.1$ so that the solution is obtained in only one step. Carry five decimals, and compare the accuracy and amount of work required with this method against the simple and modified Euler techniques in Exercises 5 and 7.

12. Solve Exercise 8 by the Runge–Kutta method, using $h = 0.2$, 0.1, and 0.05.

▶ 13. Determine y at $x = 0.2(0.2)0.6$ by the Runge–Kutta technique, given that

$$\frac{dy}{dx} = \frac{1}{x + y}, \qquad y(0) = 2.$$

14. Using the conditions of Exercise 10, determine how long it takes for the jumper to reach 90% of his or her maximum velocity, by integrating the equation using the Runge–Kutta technique with $\Delta t = 0.5$ until the velocity exceeds this value, and then interpolating. Then use numerical integration on the velocity values to determine the distance the diver falls in attaining $0.9v_{max}$.

15. It is not easy to know the accuracy with which the function has been determined by either the Euler methods or the Runge–Kutta method. A possible way to measure accuracy is to repeat the problem with a smaller step size, and compare results. If the two computations agree to n decimal places, one then assumes the values are correct to that many places. Repeat Exercise 14 with $\Delta t = 0.3$, which should give a global error about one-eighth as large, and by comparing results, determine the accuracy in Exercise 14. (Why do we expect to reduce the error eightfold by this change in Δt?)

16. Write a FORTRAN program to implement the Runge–Kutta–Fehlberg algorithm for solving an initial-value differential equation of the form $y' = f(x, y)$, $y(x_0) = A$.

17. Solve Exercises 1, 6, and 9 using the Runge–Kutta–Fehlberg algorithm.

▶18. Solve $y' = 2x - y$, $y(0) = -1$, by the Runge–Kutta–Fehlberg algorithm for $x = 0.2(0.2)1.0$. (The exact answer is $y(x) = e^{-x} + 2x - 2$.)

19. a) Exercise 15 suggests a way to estimate the accuracy of a classical Runge–Kutta computation. Runge–Kutta–Fehlberg has an advantage over the classical fourth-order procedure because it gives an error estimate without additional computations. Use Runge–Kutta–Fehlberg in Exercise 14 with a step size of $t = 0.5$. Compare the estimate of accuracy with what you obtained in Exercise 15.

    b) Repeat part (a) but now use Runge–Kutta–Merson.

Section 5.5

▶20. The equation

$$\frac{dy}{dx} = f(x, y) = 2x(y - 1), \qquad y(0) = 0,$$

has initial values as follows:

x	y	f
0	0	0
0.1	−0.01005	−0.20201
0.2	−0.04081	−0.41632
0.3	−0.09417	−0.65650
0.4	−0.17351	−0.93881

a) Using the Adams procedure described in Section 5.5, compute $y(0.5)$ by fitting a quadratic through the last three values of $f(x, y)$. Compare to the exact answer: -0.28403.

b) Repeat, using the last four points to fit a cubic.

c) Repeat again, but use five points to fit a quartic.

21. For the differential equation

$$\frac{dy}{dt} = y - t^2, \qquad y(0) = 1,$$

starting values are known:

$$y(0.2) = 1.2186, \qquad y(0.4) = 1.4682, \qquad y(0.6) = 1.7379.$$

Use the Adams method, fitting cubics with the last four (y, t) values and advance the solution to $t = 1.2$. Compare to the analytical solution.

22. For the equation

$$\frac{dy}{dt} = t^2 - t, \qquad y(1) = 0,$$

the analytical solution is easy to find:

$$y = \frac{t^3}{3} - \frac{t^2}{2} + \frac{1}{6}.$$

If we use three points in the Adams method, what error would we expect in the numerical solution? Confirm your expectation by performing the computations.

23. Continue the results of Exercise 13 using Adams' method, employing cubics to extrapolate the y'-values.

Section 5.6

24. For the differential equation,

$$\frac{dy}{dx} = y - x^2, \qquad y(0) = 1,$$

starting values are known:

$$y(0.2) = 1.2186, \qquad y(0.4) = 1.4682, \qquad y(0.6) = 1.7379.$$

Use the Milne method to advance the solution to $x = 1.2$. Carry four decimals and compare to the analytical solution.

▶25. For the differential equation $dy/dx = x/y$, the following values are given:

x	y
0	$\sqrt{1}$
1	$\sqrt{2}$
2	$\sqrt{5}$
3	$\sqrt{10}$

To how many decimal places will Milne's method give the value at $x = 4$? How many decimal places must be carried in the starting values of y to ensure this accuracy?

26. For the equation $y' = y \sin \pi x$, $y(0) = 1$, get starting values by the Runge–Kutta–Fehlberg method for $x = 0.2(0.2)0.6$, and advance the solution to $x = 1.0$ by Milne's method.

▶27. Continue the results of Exercise 13 to $x = 2.0$ by the method of Milne. If you find that the corrector formula reproduces the predictor values, double the value of h after sufficient values are available.

28. Check that y_n as defined by Eq. (5.15) is a solution of the difference equation, Eq. (5.14).

29. Perform the long division to show that

$$Z_1 = 1 + Ah + O(h^2), \qquad Z_2 = -\left(1 - \frac{Ah}{3}\right) + O(h^2),$$

as given in Eq. (5.18).

Section 5.7

30. Express the differences in Eqs. (5.20) and (5.21) in terms of functional values to show that Eqs. (5.22) are equivalent.

31. Solve Exercise 21 using the Adams–Moulton method.

32. Repeat Exercise 26 using the Adams–Moulton method.

▶33. For the equation

$$\frac{dy}{dx} = x^3 + y^2, \qquad y(0) = 0,$$

using $h = 0.2$, compute three new values by the Runge–Kutta method (four decimals). Then advance to $x = 1.4$ using the Adams–Moulton method. If you find the accuracy criterion is not met, use Eqs. (5.23) to interpolate additional values so that four-place accuracy is maintained.

34. Derive the interpolation formulas given in Eq. (5.23).

Section 5.8

▶35. Given the linear differential equation $dy/dx = y \sin x$.
  a) What is the maximum value of h that ensures convergence of the Adams–Moulton method when continuing applications of the corrector formula are made?
  b) If an h one-tenth of this maximum value is used, how close must the predictor and corrector values be so that recorrections are not required?
  c) In terms of the maximum h in part (a), what size of h is implied in the accuracy criterion, $D \cdot 10^n < 14.2$?

36. Repeat Exercise 35 for the differential equation

$$\frac{dy}{dx} = x^3 + y^2, \qquad y(0) = 0,$$

in the neighborhood of the point $(1.0, 0.15)$.

37. Repeat Exercise 35, using the Milne method. For part (c), the accuracy criterion is $D \cdot 10^n < 29$.

38. Derive Eq. (5.28).

39. Derive convergence criteria similar to Eqs. (5.26) and (5.27) for the Euler predictor–corrector method. Why can one not derive an accuracy criterion similar to those for the methods of Milne and Adams–Moulton?

Section 5.9

40. Estimate the propagated error at each step when the equation

$$\frac{dy}{dx} = x + y, \qquad y(0) = 1,$$

is solved by the simple Euler method with $h = 0.02$, for $x = 0(0.02)0.1$. (The equation is solved by this method in Section 5.3.) Compare to the actual errors.

▶41. Follow the propagated error between $x = 1$ and $x = 1.6$ when the simple Euler method is used to solve

$$\frac{dy}{dx} = xy^2, \qquad y(1) = 1.$$

Take $h = 0.1$. Compare to the actual errors at each step. The analytical solution is $y = 2/(3 - x^2)$.

42. We can derive the global error (Eq. 5.31b) for Euler's method without making use of the second-order difference equation. With the same assumptions about M and K and using the fact that the series

$$1 + s + s^2 + \cdots + s^n = \frac{s^{n+1} - 1}{s - 1},$$

show that

$$e_n \leq \frac{hM}{2K}(e^{(x_n-x_0)K} - 1).$$

(*Hint:* Let $s = 1 + hK$.)

Section 5.10

43. The mathematical model of an electrical circuit is given by the equation

$$0.5\frac{d^2Q}{dt^2} + 6\frac{dQ}{dt} + 50Q = 24 \sin 10t,$$

with $Q = 0$ and $I = dQ/dt = 0$ at $t = 0$. Express as a pair of first-order equations.

▶44. In the theory of beams it is shown that the radius of curvature at any point is proportional to the bending moment:

$$EI\frac{y''}{[1 + (y')^2]^{3/2}} = M(x),$$

where y is the deflection of the neutral axis. In the usual approach, $(y')^2$ is neglected in comparison to unity, but if the beam has appreciable curvature, this is invalid. For the cantilever beam for which $y(0) = y'(0) = 0$, express the equation as a pair of simultaneous first-order equations.

45. The motion of the compound spring system as sketched in Fig. 5.6 is given by the solution of the pair of simultaneous equations.

$$m_1\frac{d^2y_1}{dt^2} = -k_1y_1 - k_2(y_1 - y_2), \qquad m_2\frac{d^2y_2}{dt^2} = k_2(y_1 - y_2),$$

where y_1 and y_2 are the displacements of the two masses from their equilibrium positions. The initial conditions are

$$y_1(0) = A, \qquad y_1'(0) = B, \qquad y_2(0) = C, \qquad y_2'(0) = D.$$

Express as a set of first-order equations.

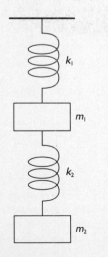

Figure 5.6

▶46. Solve the pair of simultaneous equations

$$dx/dt = xy + t, \qquad x(0) = 0, \qquad dx/dt = x - t, \qquad y(0) = 1,$$

by the modified Euler method for $t = 0.2(0.2)0.6$. (Carry three decimals rounded.) Recorrect until reproduced to three decimals.

47. Advance the solution of Exercise 46 to $x = 1.0$ $(h = 0.2)$ by the Adams–Moulton method.

48. Repeat Exercise 47 but use Milne's method.

▶49. Find y at $x = 0.6$, given that

$$y'' = yy', \qquad y(0) = 1, \qquad y'(0) = -1.$$

Begin the solution by the Taylor-series method, getting

$$y(0.1), \qquad y(0.2), \qquad y(0.3).$$

Then advance to $x = 0.6$ employing the Adams–Moulton technique with $h = 0.1$ on the equivalent set of first-order equations.

50. Express the third-order equation

$$y''' + ty'' - ty' - 2y = t, \qquad y(0) = y''(0) = 0, \qquad y'(0) = 1,$$

as a set of first-order equations and solve at $t = 0.2, 0.4, 0.6$ by the Runge–Kutta method $(h = 0.2)$.

51. Using the Adams–Moulton method with $h = 0.5$, advance the solution of Exercise 50 to $t = 1.0$. Estimate the accuracy of the value of y at $t = 1.0$.

Section 5.11

▶52. For a resonant spring system with a periodic forcing function, the differential equation is

$$\frac{d^2x}{dt^2} + 64x = 16 \cos 8t, \qquad x(0) = x'(0) = 0.$$

Determine the displacement at $t = 0.1(0.1)0.8$ by the method of Eq. (5.35), getting the starting values by any other method of this chapter. Compare to the analytical solution $t \sin 8t$.

53. Apply Gear's formula for solving a differential equation, as given in Section 5.11, to the problem of Exercise 52.

54. For the first-order equation

$$\frac{dy}{dt} + 12y = \frac{3}{2}t^2 - 12t + 6t^3 - 1, \qquad y(0) = a,$$

it is not difficult to verify that the solution is

$$y = ae^{-12t} + \frac{t^3}{2} - t.$$

If $a = 0$ (so $y(0) = 0$), the solution reduces to a simple cubic in t. Still, accumulated round-off errors act as if the initial value of y is not exactly zero, causing the exponential term to appear when it should not. This makes the problem a stiff equation. Assume that $y(0) = 0$ and demonstrate that you do not get the analytical answer when you integrate from $t = 0$ to $t = 2$ using the Runge–Kutta fourth-order method with a step size of 0.005.

APPLIED PROBLEMS AND PROJECTS

55. In finding the deflection of a beam, the y' term is often neglected (see Exercise 44). (The analytical method is easy if y' is neglected in the equation.) What is the difference in the calculated values of the maximum deflection when the nonlinear relationship (Exercise 44) is used in comparison to the simpler linear equation? For light loads, the error is negligible, of course; only for heavier ones is there a need to use the nonlinear equation. At what value of the load is the error equal to 1% of the true value?

56. Write a program that solves a first-order initial-value problem by the Adams–Moulton method, calling the RK4TH subroutine of the text to obtain starting values. As each new point is computed, monitor the accuracy by comparing the error estimate to a tolerance parameter, printing out a message if the tolerance is not met, but continuing the solution anyway. This might best be done by incorporating the Adams–Moulton method in a subroutine that advances the solution one step.

57. The equation $y' = 1 + y^2$, $y(0)$, has the analytical solution $y = \tan x$. The tangent function is infinite at $x = \pi/2$. Use the program you wrote in Problem 56 to solve this equation between $x = 0$ and $x = 1.6$ with a step size of 0.1. Compare the results of the program with the analytical solution.

58. What is the behavior of the Runge–Kutta fourth-order method when used over $x = 0$ to $x = 1.6$ on the equation of Problem 57? How does it compare with the multistep methods? How does its behavior change with step size?

59. Enlarge on Problem 56 by modifying your program so that the step size is halved if the error estimate exceeds the tolerance value. Use Eq. (5.23) to interpolate for the new values needed when this is done. If the error estimate is very small, say $\frac{1}{20}$ times the tolerance, your program should double the step size. (Some programs that provide for doubling the step size keep track of how many uniformly spaced values are available and defer the doubling until there are seven. You may wish to do this, but it is a tricky bit of programming. Alternatively, one could always compute two additional values after discovering that the step size can be increased; this guarantees that there are at least seven equispaced values. Another method would be to extrapolate to obtain a value at t_{n-5} when we find, at t_{n+1}, that the error is $\frac{1}{20}$ of the tolerance.)

60. In an electrical circuit (Fig. 5.7) containing resistance, inductance, and capacitance (and every circuit does), the voltage drop across the resistance is iR (i is current in amperes, R is resistance in ohms), across the inductance it is $L(di/dt)$ (L is inductance in henries), and across the capacitance it is q/C (q is charge in the capacitor in coulombs, C is capacitance in farads). We then can write, for the voltage difference between points A and B,

$$V_{AB} = L\frac{di}{dt} + Ri + \frac{q}{C}.$$

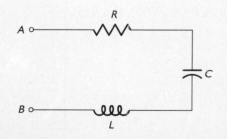

Figure 5.7

Differentiating with respect to t and remembering that $dq/dt = i$, we have a second-order differential equation:

$$L\frac{d^2i}{dt^2} + R\frac{di}{dt} + \frac{1}{C}i = \frac{dV}{dt}.$$

If the voltage V_{AB} (which has previously been 0 V) is suddenly brought to 15 V (let us say, by connecting a battery across the terminals) and maintained steadily at 15 V (so $dV/dt = 0$), current will flow through the circuit. Use an appropriate numerical method to determine how the current varies with time between 0 and 0.1 sec if $C = 1000\ \mu f$, $L = 50$ mH, and $R = 4.7$ ohms; use Δt of 0.002 sec. Also determine how the voltage builds up across the capacitor during this time. You may want to compare the computations with the analytical solution.

61. Repeat Problem 60 but let the voltage source be a 60-Hz sinusoidal input:

$$V_{AB} = 15\sin(120\pi t).$$

How closely does the voltage across the capacitor resemble a sine wave during the last full cycle of voltage variation?

62. After the voltages have stabilized in Problem 60 (15 V across the capacitor), the battery is shorted so that the capacitor discharges through the resistance and inductor. Follow the current and the capacitor voltages for 0.1 sec, again with $\Delta t = 0.002$ sec. The oscillations of decreasing amplitude are called *damped oscillations*. If the calculations are repeated but with the resistance value increased, the oscillations will be damped out more quickly; at $R = 14.14$ ohms the oscillations should disappear; this is called *critical damping*. Perform numerical computations with values of R increasing from 4.7 to 22 ohms to confirm that critical damping occurs at 14.14 ohms.

63. In Chapter 2, electrical networks were discussed, but these were simple ones with resistance only. A more realistic situation is to have RLC circuits combined into a network. Using either the current or voltage laws of Kirchhoff, we can set up a set of simultaneous equations, but now these are simultaneous differential equations. For example, for the circuit in Fig. 5.8, which has two voltage sources, we have

$$(L_1 + L_3)\frac{d^2i_1}{dt^2} + (R_1 + R_3)\frac{di_1}{dt} + (C_1 + C_3)i_1 - L_3\frac{d^2i_2}{dt^2} - R_3\frac{di_2}{dt} - C_3i_2 = e_1(t),$$

$$(L_2 + L_3)\frac{d^2i_2}{dt^2} + (R_2 + R_3)\frac{di_2}{dt} + (C_2 + C_3)i_2 - L_3\frac{d^2i_1}{dt^2} - R_3\frac{di_1}{dt} - C_3i_1 = e_2(t).$$

In the equations, i_1 and i_2 represent the currents in each of the loops. Solve the equations for i_1 and i_2 between $t = 0$ and $t = 0.2$ sec, if

$$e_1(t) = 100\sin(120\pi t), \qquad e_2(t) = 0,$$

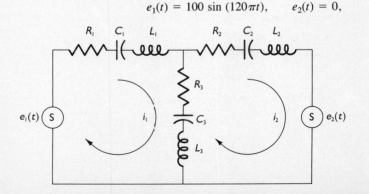

Figure 5.8

given that

$$R_1 = 22 \text{ ohms}, \qquad C_1 = C_2 = C_3 = 10 \ \mu\text{f},$$
$$R_2 = 4.7 \text{ ohms}, \qquad L_1 = 2.5 \text{ mH},$$
$$R_3 = 47 \text{ ohms}, \qquad L_2 = L_3 = 0.5 \text{ mH}.$$

64. A Foucault pendulum is one free to swing in both the x- and y-directions. It is frequently displayed in science museums to exhibit the rotation of the earth, which causes the pendulum to swing in directions that continuously vary. The equations of motion are

$$\ddot{x} - 2\omega \sin \psi \dot{y} + k^2 x = 0,$$
$$\ddot{y} + 2\omega \sin \psi \dot{x} + k^2 y = 0,$$

when damping is absent (or compensated for). In these equations the dots over the variable represent differentiation with respect to time. Here ω is the angular velocity of the earth's rotation ($7.29 \times 10^{-5} \text{ sec}^{-1}$), ψ is the latitude, $k^2 = g/\ell$ where ℓ is the length of the pendulum. How long will it take a 10-m-long pendulum to rotate its plane of swing by $45°$ at the latitude where you live? How long if located in Quebec, Canada?

65. Condon and Odishaw (1967) discuss Duffing's equation for the flux ϕ in a transformer. This nonlinear differential equation is

$$\ddot{\phi} + \omega_0^2 \phi + b\phi^3 = \frac{\omega}{N} E \cos \omega t.$$

In this equation, $E \sin \omega t$ is the sinusoidal source voltage and N is the number of turns in the primary winding, while ω_0 and b are parameters of the transformer design. Make a plot of ϕ versus t (and compare to the source voltage) if $E = 165$, $\omega = 120\pi$, $N = 600$, $\omega_0^2 = 83$, and $b = 0.14$. For approximate calculations, the nonlinear term $b\phi^3$ is sometimes neglected. Evaluate your results to determine whether this makes a significant error in the results.

66. Ethylene oxide is an important raw material for the manufacture of organic chemicals. It is produced by reacting ethylene and oxygen together over a silver catalyst. Laboratory studies have been reported by Wan (1952).

It is planned to use this process commercially by passing the gaseous mixture through tubes filled with catalyst. Wan's studies show that the reaction rate varies with pressure, temperature, and concentrations of ethylene and oxygen, according to this equation:

$$r = 1.7 \times 10^6 e^{-9716/T} \left(\frac{P}{14.7} \right) C_E^{0.328} C_O^{0.672},$$

where

r = reaction rate (units of ethylene oxide formed per lb of catalyst per hr),

T = temperature, °K (°C + 273),

P = absolute pressure (lb/in²),

C_E = concentration of ethylene,

C_O = concentration of oxygen.

Under the planned conditions, the reaction will occur, as the gas flows through the tube, according to the equation

$$\frac{dx}{dL} = 6.42r,$$

where

$$x = \text{fraction of ethylene converted to ethylene oxide,}$$

$$L = \text{length of reactor tube (ft).}$$

The reaction is strongly exothermic, so that it is necessary to cool the tubular reactor to prevent overheating. (Excessively high temperatures produce undesirable side reactions.) The reactor will be cooled by surrounding the catalyst tubes with boiling coolant under pressure so that the tube walls are kept at 225°C. This will remove heat proportional to the temperature difference between the gas and the boiling water. Of course, heat is generated by the reaction. The net effect can be expressed by this equation for the temperature change per foot of tube, where B is a design parameter:

$$\frac{dT}{dL} = 24{,}320r - B(T - 225).$$

For preliminary computations, it has been agreed that we can neglect the change in pressure as the gases flow through the tubes; we will use the average pressure of $P = 22$ lb/in^2 absolute. We will also neglect the difference between the catalyst temperature (which should be used to find the reaction rate) and the gas temperature. You are to compute the length of tubes required for 65% conversion of ethylene if the inlet temperature is 250°C. Oxygen is consumed in proportion to the ethylene converted; material balances show that the concentrations of ethylene and oxygen vary with x, the fraction of ethylene converted, as follows:

$$C_E = \frac{1 - x}{4 - 0.375x},$$

$$C_O = \frac{1 - 1.125x}{4 - 0.375x},$$

The design parameter B will be determined by the diameter of tubes that contain the catalyst. (The number of tubes in parallel will be chosen to accommodate the quantities of materials flowing through the reactor.) The tube size will be chosen to control the maximum temperature of the reaction, as set by the minimum allowable value of B. If the tubes are too large in diameter (for which the value of B is small), the temperatures will run wild. If the tubes are too *small* (giving a large value to B), so much heat is lost that the reaction tends to be quenched. In your studies, vary B to find the least value that will keep the maximum temperature below 300°C. Permissible values for the parameter B are from 1.0 to 10.0.

In addition to finding how long the tubes must be, we need to know how the temperature varies with x and with the distance along the tubes. In order to have some indication of the controllability of the process, you are also asked to determine how much the outlet temperature will change for a 1°C change in the inlet temperature, using the value of B as determined above.

6

Boundary-Value Problems and Characteristic-Value Problems

6.0 CONTENTS OF THIS CHAPTER

In all differential equations of order greater than 1, two or more values must be known to evaluate the constants in the particular function that satisfies the differential equation. In the problems discussed in Chapter 5, these several values were all specified at the same value of the independent variable, generally at the start or *initial value*. For that reason, such problems are termed *initial-value problems*.

For an important class of problems, the several values of the function or its derivatives are not all known at the same point, but rather at two different values of the independent variable. Because these values of the independent variable are usually the endpoints (or *boundaries*) of some domain of interest, problems with this type of conditions are classed as *boundary-value problems*. Determining the deflection of a simply supported beam is a typical example, where the conditions specified are the deflections and second derivatives of the elastic curve at the supports. Heat-flow problems fall in this class when the temperatures or temperature gradients are given at two points. A special case of the boundary-value problem occurs in vibration problems.

We will study three different methods for solving boundary-value problems. In addition, material on a special class of boundary-value problems, characteristic-value problems, is presented.

6.1   THE "SHOOTING" METHOD

Applies the techniques of Chapter 5 to boundary-value problems by assuming values needed to make it into an initial-value problem. This involves a degree of trial and error, but, for a linear problem, no more than two trials are required.

6.2   SOLUTION THROUGH A SET OF EQUATIONS

Uses finite-difference approximations for the derivatives, allowing one to write a set of equations whose unknowns are values for the independent variable at several points within the domain. The solution to this system gives a solution to the original boundary-value problem.

6.3   DERIVATIVE BOUNDARY CONDITIONS

Require a modification of the method of Section 6.2; this is straightforward, but it does require an artificial extension of the domain.

6.4   RAYLEIGH–RITZ METHODS

Use techniques based on the calculus of variations. An approximating function is found by optimizing a functional, an entity related to the original differential equation.

6.5   CHARACTERISTIC-VALUE PROBLEMS

Are an important special class of boundary-value problems that have a solution only for certain characteristic values of a parameter, also called *eigenvalues*. In this section, you learn about both eigenvalues and eigenvectors, essential matrix-related quantities that have many important applications.

6.6   EIGENVALUES OF A MATRIX BY ITERATION

Describes a method for getting eigenvalues and the associated eigenvectors that is well adapted to computers.

6.7   EIGENVALUES/EIGENVECTORS IN ITERATIVE METHODS

Illustrates how the use of eigenvalues can provide a theoretical basis for explaining when an iterative method will converge and how rapidly.

6.8   CHAPTER SUMMARY

Is provided to allow you to test your knowledge of the essential points of the chapter.

6.9   COMPUTER PROGRAMS

Are given that implement the shooting method for boundary-value problems and the power method for eigenvalues/eigenvectors.

6.1   THE "SHOOTING METHOD"

Suppose we wish to solve the second-order boundary-value problem

$$\frac{d^2x}{dt^2} - \left(1 - \frac{t}{5}\right)x = t, \qquad x(1) = 2, \qquad x(3) = -1. \qquad (6.1)$$

Note that if $x'(1)$ were given in addition to $x(1)$, this would be an initial-value problem. There is a way to adapt our previous methods to this problem, as illustrated in Fig. 6.1. We know the value of x at $t = 1$, and at $t = 3$, as given by the dots. The curve

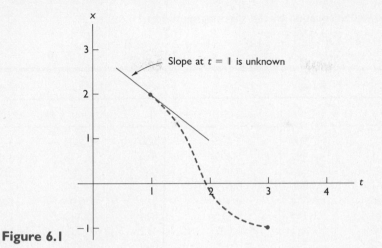

Figure 6.1

that represents x between these two points is desired. We anticipate that some such curve, such as the dotted line, exists;* its slope and curvature are interrelated to x and t by the differential Eq. (6.1).

If we assume the slope of the curve at $t = 1$, say $x'(1) = -1.5$, we could solve the equation as an initial-value problem using this assumed value. The test of our assumption is whether we calculate x at $t = 3$ to match the known value, $x(3) = -1$. In Table 6.1 we show the results of this computation employing the computer program of Section 5.13 (Fig. 5.3), modified for this equation, using the modified Euler method with $\Delta t = 0.2$. Since the result of this gives $x(3) = 4.811$, and not the desired $x(3) = -1$, we assume another value for $x'(1)$ and repeat. Since the calculated value of $x(3)$ is too high, we assume a smaller value for the slope, say $x'(1) = -3.0$. The second computation gives $x(3) = 0.453$, which is better but still too high. After these two trials, we linearly interpolate (here extrapolate) for a third trial. At $x'(1) = -3.500$, we get the correct value of $x(3)$. In some problems more attempts are needed to get the correct solution.

The method we have illustrated is called the *shooting method* because it resembles an artillery problem. One sets the elevation of the gun and fires a preliminary round at the target. After successive shots have straddled the target, one zeroes in on it by using intermediate values of the gun's elevation. This corresponds to our using assumed values of the initial slope and interpolating based on how close we come to $x(3)$.

You may wonder whether it was only a lucky accident that gave correct results on our third trial, using the initial slope $x'(1) = -3.500$, extrapolated from our earlier guesses. This desirable result will always be true when the boundary-value problem is *linear*, as in this case. (A differential equation is linear when the coefficients of each derivative term and the function are not functions of x. They may be functions of t, as in this example, however.) The reason for this is that, if there are two different functions of x that satisfy the differential equation (but agreeing with different boundary conditions, of course), then a linear combination of them is also a solution. This is quite easy to show.

---

*The existence of a solution to a boundary-value problem cannot be taken for granted, however. For example, the problem $y'' + y = 0$, $y(0) = 1$, $y(\pi) = 0$ has no solution.

Table 6.1 Solving a second-order equation by the shooting method:

$$x'' = t + \left(1 - \frac{t}{5}\right)x, \qquad x(1) = 2, \qquad x(3) = -1$$

Time	Distance	Velocity	Distance	Velocity	Distance	Velocity
$\Delta t = 0.2$: Assume $x'(1) = -1.5$			Assume $x'(1) = -3.0$		Assume $x'(1) = -3.500$	
1.00	2.000	−1.500	2.000	−3.000	2.000	−3.500
1.20	1.751	−0.987	1.499	−2.510	1.348	−3.018
1.40	1.605	−0.478	0.991	−2.068	0.787	−2.599
1.60	1.561	0.043	0.619	−1.655	0.305	−2.221
1.80	1.625	0.594	0.328	−1.252	−0.104	−1.867
2.00	1.803	1.186	0.118	−0.844	−0.443	−1.521
2.20	2.105	1.832	−0.007	−0.417	−0.712	−1.167
2.40	2.542	2.542	−0.045	0.040	−0.908	−0.794
2.60	3.128	3.324	0.013	0.539	−1.026	−0.391
2.80	3.880	4.185	0.175	1.087	−1.060	0.054
3.00	4.811	5.128	0.453	1.693	−1.000	0.547

Suppose that $x_1(t)$ satisfies $x'' + Fx' + Gx = H$, where F, G, and H are functions of t only. Suppose also that $x_2(t)$ is a solution. Then

$$y = \frac{c_1 x_1 + c_2 x_2}{c_1 + c_2}$$

will be a solution. We show that this will always be a solution in the following manner:
Since x_1 and x_2 are solutions,

$$x_1'' + Fx_1' + Gx_1 = H,$$

and

$$x_2'' + Fx_2' + Gx_2 = H.$$

Substituting y into the differential equation, with

$$y' = \frac{c_1 x_1' + c_2 x_2'}{c_1 + c_2}, \qquad y'' = \frac{c_1 x_1'' + c_2 x_2''}{c_1 + c_2},$$

we get

$$\frac{c_1 x_1'' + c_2 x_2''}{c_1 + c_2} + F\frac{c_1 x_1' + c_2 x_2'}{c_1 + c_2} + G\frac{c_1 x_1 + c_2 x_2}{c_1 + c_2}$$

$$= \frac{c_1 x_1'' + c_1 Fx_1' + c_1 Gx_1}{c_1 + c_2} + \frac{c_2 x_2'' + c_2 Fx_2' + c_2 Gx_2}{c_1 + c_2}$$

$$= \frac{c_1 H}{c_1 + c_2} + \frac{c_2 H}{c_1 + c_2} = H.$$

With two functions x_1 and x_2 both agreeing with $x(1) = 2.0$ at the left boundary and differing at $t = 3.0$, we can take a linear combination, so the correct value of $x(3.0)$ results.

Let G1 = first guess at initial slope;

Let G2 = second guess at initial slope;

Let R1 = first result at endpoint (using G1);

Let R2 = second result at endpoint (using G2).

With D = the desired value at the endpoint, we can write

$$\text{Extrapolated estimate for initial slope} = G1 + \frac{G2 - G1}{R2 - R1}(D - R1).$$

In the above example, we get

$$-1.5 + \frac{(-3.0) - (-1.5)}{0.453 - 4.811}(-1.0 - 4.811) = -3.500.$$

The consequence of having $(c_1x_1 + c_2x_2)/(c_1 + c_2)$ as a solution of the differential equation goes further than just allowing us to interpolate from the two initial guesses of $x'(1)$ to predict exactly the correct value required for $x'(1)$ that is needed in order to give agreement with the right-hand boundary condition. This relationship is true *throughout* the interval as well. This means that each of the values of x could be calculated by taking the proper combination of the values obtained from the earlier calculations, with this combination exactly the same one as that which gives the correct value of x at $t = 3$. (The true curve for x versus t will be intermediate between two curves that bracket the final condition.)

In our example,

$$c_1x_1 + c_2x_2 = \text{true values}.$$

At left end,

$$t = 1: \quad c_1(2.0) + c_2(2.0) = 2.0.$$

At right end,

$$t = 3: \quad c_1(4.811) + c_2(0.453) = -1.0.$$

From the first equation, $c_2 = 1 - c_1$, and from the second, $c_1 = -0.3334$, giving $c_2 = 1.3334$. We can calculate the correct values of x using these values. For example,

$$\text{at } t = 2.0: \quad -0.3334(1.803) + 1.3334(0.118) = -0.443.$$

We therefore conclude that a second-order linear boundary-value problem can always be solved by doing only two computations using two different guesses for the initial slope.

With higher-order linear boundary-value problems, the situation is somewhat more difficult, although with a third-order equation, we can choose the starting point at that boundary where two of the three conditions are specified. (If this means working backward from the right-hand end, Δt will be negative.) For a fourth-order problem, with two conditions specified at each boundary, we seem to have a double trial-and-error situation. However, a combination of solutions also is a solution in this case, permitting us to get the correct solution by combining four trial computations. In these higher-order problems, some of the conditions we must match are values of the derivatives, requiring us to estimate the derivatives of our computed functions using approximation methods of Chapter 4, and these are more subject to errors.

When the differential equation is nonlinear (the coefficients F, G, and/or H in $x'' + Fx' + Gx = H$ are functions of x or x'), the interpolation between two assumed values for the initial slope does not usually give the correct value. Consider

$$x'' - \left(1 - \frac{t}{5}\right)(x')(x) = t, \quad x(1) = 2, \quad x(3) = -1.$$

The presence of the $(x')(x)$ term makes it nonlinear. In this case, the values in Table 6.2 result, using the modified Euler method with $\Delta t = 0.2$.

Table 6.2 Values for a nonlinear boundary-value problem by the shooting method:

$$x'' = t + \left(1 - \frac{t}{5}\right)xx', \qquad x(1) = 2, \qquad x(3) = -1$$

Assumed value for $x'(1)$	Calculated value for $x(3)$
-1.5	-0.016
-3.0	-2.085
-2.213*	-1.271
-2.0	-0.972
-1.8	-0.642
-2.017*	-0.998
-2.01	-0.987
-2.02	-1.002
-2.018*	-1.000

*Interpolated from two previous values.

Only when the assumed values are very near to the correct initial slope does one get the correct value by interpolating linearly. (If the computer program that solves the differential equation is available through a time-sharing system, the above trial-and-error search can be conducted quickly and conveniently. This was the case with the data in Table 6.2.)

Note that we are faced with a root-finding problem, in that the value of $x(3)$ is a function of $x'(1)$. The methods of Chapter 1 then apply, although Newton's method requires the *slope* of this function, which is not readily available. Probably Muller's method, applied after three estimates have been made, would be particularly beneficial.

After we know the correct value for $x'(1)$, we solve the problem as an initial-value problem, getting the results shown in Table 6.3.

A Summary of the Shooting Method

To solve a boundary-value problem, create an initial-value problem by assuming a sufficient number of initial values. (Choose the right-hand end of the interval as the "initial point," and use a negative value of h if this requires fewer assumed conditions.) Solve this initial-value problem and compare the computed values with the given conditions at the other boundary.

Repeat the solution with varying values of the assumed conditions until agreement is attained at the other boundary. Methods for solving nonlinear systems can be employed to estimate the new assumed values for the successive iterations in an efficient manner.

The shooting method is often quite laborious. Especially with problems of fourth and higher order, the necessity to assume two or more conditions at the starting point (and match with the same number of conditions at the end) is slow and tedious. Although it tends to use considerable computer time, the NLSYST program of Chapter 2 is a convenient tool for such higher-order problems.

Table 6.3

t	x	x'
1.000	2.000	−2.018
1.200	1.557	−2.407
1.400	1.053	−2.632
1.600	0.527	−2.625
1.800	0.026	−2.383
2.000	−0.408	−1.959
2.200	−0.747	−1.431
2.400	−0.977	−0.867
2.600	−1.094	−0.307
2.800	−1.101	0.237
3.000	−1.000	0.775

6.2 SOLUTION THROUGH A SET OF EQUATIONS

We have seen in Chapter 4 how the derivatives of a function can be approximated by finite-difference quotients. If we replace the derivatives in a differential equation by such expressions, we convert it to a difference equation whose solution is an approximation to the solution of the differential equation. This method is sometimes preferred over the shooting method discussed earlier. Consider the same linear example as before:

$$\frac{d^2x}{dt^2} - \left(1 - \frac{t}{5}\right)x = t, \qquad x(1) = 2, \qquad x(3) = -1. \tag{6.2}$$

Central-difference approximations to derivatives are more accurate than forward or backward approximations ($O(h^2)$ versus $O(h)$), so we replace the derivatives with

$$\left.\frac{dx}{dt}\right|_{t=t_i} = \frac{x_{i+1} - x_{i-1}}{2h} + O(h^2),$$

$$\left.\frac{d^2x}{dt^2}\right|_{t=t_i} = \frac{x_{i+1} - 2x_i + x_{i-1}}{h^2} + O(h^2).$$

The quantity h is the constant difference in t-values. Substituting these equivalences into Eq. (6.2), and rearranging, we get

$$\frac{x_{i+1} - 2x_i + x_{i-1}}{h^2} - \left(1 - \frac{t_i}{5}\right)x_i = t_i,$$

$$x_{i-1} - \left[2 + h^2\left(1 - \frac{t_i}{5}\right)\right]x_i + x_{i+1} = h^2 t_i. \tag{6.3}$$

In Eqs. (6.3), we have replaced x with x_i and t with t_i, since these values correspond to the point at which the difference quotients represent the derivatives.* Our problem now reduces to solving Eqs. (6.3) at points in the interval from $t = 1$ to $t = 3$. Let us subdivide the interval into a number of equal subintervals. For example, if $h = \Delta t = \frac{1}{2}$, the points

$$t_1 = 1, \qquad t_2 = 1.5, \qquad t_3 = 2, \qquad t_4 = 2.5, \qquad \text{and} \qquad t_5 = 3$$

subdivide the interval into four subintervals.

---

*Our procedure is to subdivide the interval from the initial to the final value of t into equal subintervals, replacing the differential equation with a difference equation at each of the discrete points where the function is unknown.

We write the difference equation in (6.3) for each of these values of t at which x is unknown, giving

$$\text{at } t = t_2 = 1.5: \quad x_1 - \left[2 + \left(\frac{1}{2}\right)\left(\frac{1}{2}\right)\left(1 - \frac{1.5}{5}\right)\right]x_2 + x_3 = \left(\frac{1}{2}\right)\left(\frac{1}{2}\right)(1.5);$$

$$\text{at } t = t_3 = 2.0: \quad x_2 - \left[2 + \left(\frac{1}{2}\right)\left(\frac{1}{2}\right)\left(1 - \frac{2.0}{5}\right)\right]x_3 + x_4 = \left(\frac{1}{2}\right)\left(\frac{1}{2}\right)(2.0);$$

$$\text{at } t = t_4 = 2.5: \quad x_3 - \left[2 + \left(\frac{1}{2}\right)\left(\frac{1}{2}\right)\left(1 - \frac{2.5}{5}\right)\right]x_4 + x_5 = \left(\frac{1}{2}\right)\left(\frac{1}{2}\right)(2.5).$$

We know $x_1 = 2$ and $x_5 = -1$; hence the difference equation is not written corresponding to $t = 1$ or $t = 3$. Substituting the values $x_1 = 2$ and $x_5 = -1$ and simplifying, we have, in matrix form,

$$\begin{bmatrix} -2.175 & 1 & 0 \\ 1 & -2.150 & 1 \\ 0 & 1 & -2.125 \end{bmatrix}\begin{bmatrix} x_2 \\ x_3 \\ x_4 \end{bmatrix} = \begin{bmatrix} -1.625 \\ 0.5 \\ 1.625 \end{bmatrix}.$$

Solving, we get

$$x_2 = 0.552, \quad x_3 = -0.424, \quad \text{and} \quad x_4 = -0.964.$$

These are reasonably close to the values obtained by the shooting method. We would expect some significant error because our step size h is large and the finite differences will be poor approximations to the derivatives. If we take $h = 0.2$ (10 subdivisions) and write the approximating equations, we get

$$\begin{bmatrix} -2.0304 & 1 & 0 & 0 & 0 & 0 & 0 & 0 & 0 \\ 1 & -2.0288 & 1 & 0 & 0 & 0 & 0 & 0 & 0 \\ 0 & 1 & -2.0272 & 1 & 0 & 0 & 0 & 0 & 0 \\ 0 & 0 & 1 & -2.0256 & 1 & 0 & 0 & 0 & 0 \\ 0 & 0 & 0 & 1 & -2.0240 & 1 & 0 & 0 & 0 \\ 0 & 0 & 0 & 0 & 1 & -2.0224 & 1 & 0 & 0 \\ 0 & 0 & 0 & 0 & 0 & 1 & -2.0208 & 1 & 0 \\ 0 & 0 & 0 & 0 & 0 & 0 & 1 & -2.0192 & 1 \\ 0 & 0 & 0 & 0 & 0 & 0 & 0 & 1 & -2.0176 \end{bmatrix} x = \begin{bmatrix} -1.952 \\ 0.056 \\ 0.064 \\ 0.072 \\ 0.080 \\ 0.088 \\ 0.096 \\ 0.104 \\ 1.112 \end{bmatrix}. \quad (6.4)$$

We observe that the system is tridiagonal and therefore speedy to solve and also economical of memory space to store the coefficients. This will be true even if the value of N is very large, because we only use x_{i-1}, x_i, x_{i+1} in any equation to replace x' or x''. This is one reason why the finite-difference method is widely used to solve second-order linear boundary-value problems.

Table 6.4   Results for linear boundary-value problem:

$$x'' = t + \left(1 - \frac{t}{5}\right)x, \quad x(1) = 2, \quad x(3) = -1$$

t	Values from finite-difference method	Values from shooting method
1.0	2.000	2.000
1.2	1.351	1.348
1.4	0.792	0.787
1.6	0.311	0.305
1.8	−0.097	−0.104
2.0	−0.436	−0.443
2.2	−0.705	−0.712
2.4	−0.903	−0.908
2.6	−1.022	−1.026
2.8	−1.058	−1.060
3.0	−1.000	−1.000

If we use the subroutine TRIDG from Chapter 2 to compute the solution to Eqs. (6.4), we get the values shown in Table 6.4, where we also display the results by the shooting method from Section 6.1 for comparison.

The values obtained with $\Delta t = 0.2$ are more accurate. We might anticipate that the error would decrease proportionately to h^2 since the approximations to the derivatives are of $O(h^2)$ error. If we use the values from the shooting method as the exact values for comparison, the errors are reduced only about threefold (comparing the values at $t = 2$) instead of sixfold. However, the shooting-method calculations are themselves imperfect, and this adds an element of doubt to our observation. To observe how the errors decrease when Δt decreases, we look at another example, one where the exact solution is known.

E X A M P L E   Solve

$$\frac{d^2y}{dx^2} = y, \quad y(1) = 1.1752, \quad y(3) = 10.0179.$$

(The analytical solution is $y = \sinh x$, to which we can compare our estimates of the function.)

While it is not strictly necessary, it is common to normalize the function to the interval $(0,1)$.* This we can do by the change of variable

$$x = (b - a)t + a,$$

---

*Perhaps the most important reason is to make a computer program more general. Normalizing is really a scaling of the interval so we have an immediate understanding of whether a given value of h is "small" or "large." Also, when we compute the LU equivalent of the coefficient matrix to solve the system, it is then of more general applicability.

where (a, b) is the original interval to be normalized. Letting $x = 2t + 1$, we write

$$\frac{dy}{dx} = \frac{dy}{dt}\frac{dt}{dx} = \frac{1}{2}\frac{dy}{dt},$$

$$\frac{d^2y}{dt^2} = \frac{d}{dt}\left(\frac{1}{2}\frac{dy}{dt}\right)\frac{dt}{dx} = \frac{1}{2^2}\frac{d^2y}{dt^2},$$

and the problem becomes

$$\frac{d^2y}{dt^2} = 4y, \qquad y(0) = 1.1752, \qquad y(1) = 10.0179.$$

Replacing the second derivative by the central-difference approximation, and thus converting the differential equation to a difference equation, we have

$$\frac{y_{i+1} - 2y_i + y_{i-1}}{h^2} = 4y_i, \qquad i = 1, 2, 3, \ldots, n.$$

Subdividing the interval $(0, 1)$ into four parts, $h = 0.25$, and writing the difference equation at the three internal points where y is unknown, we have the set of equations

$$\begin{bmatrix} -2.25 & 1 & 0 \\ 1 & -2.25 & 1 \\ 0 & 1 & -2.25 \end{bmatrix}\begin{bmatrix} y_2 \\ y_3 \\ y_4 \end{bmatrix} = \begin{bmatrix} -1.1752 \\ 0 \\ -10.0179 \end{bmatrix}. \tag{6.5}$$

In Eqs. (6.5) we have used $y_1 = 1.1752$, $y_5 = 10.0179$. Again, a tridiagonal matrix of coefficients occurs. The solution to this set of equations is given in Table 6.5.

Table 6.5

t	x	Calculated y	Exact value, sinh x	Error
0	1.0	(1.1752)	1.1752	—
0.25	1.5	2.1467	2.1293	−0.0174
0.50	2.0	3.6549	3.6269	−0.0280
0.75	2.5	6.0768	6.0502	−0.0266
1.0	3.0	(10.0179)	10.0179	—

If we recalculate this problem using varying values of h, we get the results shown in Table 6.6 for y at $x = 2.0$ (equivalent to $t = 0.5$).

We observe an approximate fourfold decrease in the error when the step size is halved, meaning an $O(h^2)$ error does exist. When we know how errors vary with step size, we can perform an extrapolation similar to that for Romberg integration in Chapter 3, using results for which the step size is halved:

$$\begin{array}{c} \text{Improved} \\ \text{value} \end{array} = \begin{array}{c} \text{More} \\ \text{accurate} \end{array} + \frac{1}{3}\left[\left(\begin{array}{c} \text{More} \\ \text{accurate} \end{array}\right) - \left(\begin{array}{c} \text{Less} \\ \text{accurate} \end{array}\right)\right].$$

Table 6.6

Number of equations	$h = \Delta t$ for normalized interval	Result	Error
1	0.50	3.7310	−0.1041
3	0.25	3.6549	−0.0280
7	0.125	3.6340	−0.0071
∞	0	3.6269	(exact)

In this instance we get $3.6340 + \frac{1}{3}(3.6340 - 3.6549) = 3.6270$, when results with $\Delta t = \frac{1}{4}$ and $\Delta t = \frac{1}{8}$ are combined. ∎

When the differential equation underlying a boundary-value problem is nonlinear, this method of finite differences runs into a problem in that the resulting system of equations is nonlinear. Consider the nonlinear problem solved by the shooting method in Section 6.1:

$$\frac{d^2x}{dt^2} - \left(1 - \frac{t}{5}\right)\left(\frac{dx}{dt}\right)(x) = t, \qquad x(1) = 2, \qquad x(3) = -1.$$

When finite differences are substituted for the derivatives, we obtain an equivalent set of difference equations as follows:

$$x_{i-1} - \left[2 + \frac{h}{2}\left(1 - \frac{t_i}{5}\right)(x_{i+1} - x_{i-1})\right]x_i + x_{i+1} = h^2 t_i, \qquad i = 1, 2, 3, \ldots, n, \quad \textbf{(6.6)}$$

which is to be written at each point t_i at which x is unknown. In the middle term, products of the x's occur, exhibiting nonlinearity in the difference equations as a consequence of the nonlinearity in the differential equation. The standard elimination method fails for these equations.

In such cases, iteration techniques can be employed. Suppose we obtain some estimates of the x-vector, the solution to the equations. We could use these values as a means of approximating the coefficient of x_i in Eq. (6.6). (Since the values are multiplied by $h/2$, we might hope that errors of the estimate would be diluted in their effect, especially if h is fairly small, in comparison to the number 2, which should dominate in the coefficient of x_i.)

When this was done, using $h = 0.2$ and the initial x_i values estimated from $x = 3.5 - 1.5t$ (a linear relation between $x(1) = 2$ and $x(3) = -1$), the values in Table 6.7 resulted.

Further iterations after the eighth did not change the values. In Table 6.7, we have listed the results by the shooting method for comparison. In both techniques, an iterative process was used; in the finite-difference method, we iterate on the approximations to x, while in the shooting method, we use successive estimates to the initial slope. For this

Table 6.7   Successive approximations to finite-difference equations for

$$x'' - \left(1 - \frac{t}{5}\right)(x')(x) = t, \qquad x(1) = 2, \qquad x(3) = -1$$

		Approximations to x-vector				
t	Initial estimate	After 1 iteration	After 2 iterations	After 4 iterations	After 8 iterations	Results from shooting method
1.0	2.000	(2.000)	(2.000)	(2.000)	(2.000)	(2.000)
1.2	1.700	1.431	1.601	1.558	1.552	1.557
1.4	1.400	0.845	1.110	1.048	1.039	1.053
1.6	1.100	0.279	0.583	0.513	0.503	0.527
1.8	0.800	−0.235	0.077	0.006	−0.005	0.026
2.0	0.500	−0.668	−0.362	−0.430	−0.440	−0.408
2.2	0.200	−0.997	−0.704	−0.766	−0.775	−0.747
2.4	−0.100	−0.204	−0.937	−0.990	−0.998	−0.977
2.6	−0.400	−1.278	−1.061	−1.101	−1.107	−1.094
2.8	−0.700	−1.211	−0.080	−1.104	−1.107	−1.101
3.0	−1.000	(−1.000)	(−1.000)	(−1.000)	(−1.000)	(−1.000)

example, about the same number of iterations were required, and roughly the same accuracy was achieved. (The correct solution is intermediate between the results for the two methods shown in Table 6.7.)

The choice between the finite-difference method and the shooting method for nonlinear boundary-value problems is not clearcut. The choice depends on whether a reasonably good initial estimate for the x-vector is available and how strongly nonlinear the equation is. When the nonlinearity in the set of equations is not diluted in its effect and when no good initial estimate is available, iterative methods applied to the algebraic finite-difference equations may not even converge. Although the shooting method often involves a somewhat greater computational effort, it is usually more certain of giving a solution.

6.3   DERIVATIVE BOUNDARY CONDITIONS

The conditions that the solution of a differential equation must satisfy need not necessarily be just the value of the function. In many applied problems, some derivative of the function may be known at the boundaries of an interval. In the more general case a linear combination of the function and its derivatives is specified. The finite-difference procedure needs modification for this type of boundary conditions.* We illustrate by an example:

---

*With the shooting method, derivative boundary conditions do not require any change in procedure, but we will need to use finite-difference approximations to know when the results match with specified derivatives at the far end.

$$\frac{d^2y}{dx^2} = y,$$

$$y'(1) = 1.1752,$$

$$y'(3) = 10.0179.$$

(This problem has the same differential equation as the example of Section 6.2 but, with the values specified for the derivative at $x = 1$ and $x = 3$, it now has the analytical solution $y = \cosh x$.)

We begin just as before. We change variables to make the interval $(0, 1)$ by letting $x = 2t + 1$:

$$\frac{d^2y}{dt^2} = 4y.$$

We now replace the derivative by a central-difference approximation and write the difference equation at each point where y is unknown. With $h = 0.25$,

$$
\begin{aligned}
t = 0: \qquad & y_\ell - 2.25y_1 + y_2 = 0, \\
t = 0.25: \qquad & y_1 - 2.25y_2 + y_3 = 0, \\
t = 0.50: \qquad & y_2 - 2.25y_3 + y_4 = 0, \\
t = 0.75: \qquad & y_3 - 2.25y_4 + y_5 = 0, \\
t = 1.00: \qquad & y_4 - 2.25y_5 + y_r = 0.
\end{aligned}
\tag{6.7}
$$

Two more equations are required than in Section 6.2 because y is unknown at $t = 0$ and $t = 1$ as well as at the interior points. These two equations involve the values y_ℓ, y_r, points one space to the left and to the right of the interval $(0, 1)$. We assume that the domain of the differential equation can be so extended.

Our problem is now that Eqs. (6.7) contain seven unknowns, and we have only five equations. The boundary conditions, however, have not yet been involved. Let us also express these as difference quotients, preferring central-difference approximations of $O(h^2)$ error as used in replacing derivatives in the original equation:

$$\left.\frac{dy}{dx}\right|_{x=1} = \left.\frac{dy}{dt}\frac{dt}{dx}\right|_{t=0} \doteq \left(\frac{1}{2}\right)\frac{y_2 - y_\ell}{2h} = 1.1752, \qquad y_\ell = y_2 - (4)(0.25)(1.1752),$$

$$\left.\frac{dy}{dx}\right|_{x=3} = \left.\frac{dy}{dt}\frac{dt}{dx}\right|_{t=1} \doteq \left(\frac{1}{2}\right)\frac{y_r - y_4}{2h} = 10.0179, \qquad y_r = y_4 + (4)(0.25)(10.0179). \tag{6.8}$$

The relations of (6.8), when substituted into (6.7), reduce the number of unknowns to five, and we solve the equations in the usual way. The solution is given in Table 6.8.

Table 6.8

t	x	y	cosh x	Error
0	1.0	1.5522	1.5431	−0.0091
0.25	1.5	2.3338	2.3524	0.0186
0.50	2.0	3.6989	3.7622	0.0633
0.75	2.5	5.9887	6.1323	0.1436
1.00	3.0	9.7757	10.0677	0.2920

We observe here that the errors are much greater than for the previous example, being very large at $x = 3.0$. The explanation is that our approximation for the derivative is poor at large x-values. Although the central-difference approximations are $O(h^2)$, the third derivative appearing in the error term is large in magnitude. In the previous example, knowing the function at the endpoints eliminated the errors there.

Repeating the calculations with the step size cut in half reduces the errors about fourfold. For example, at $t = 0.5$, we get 3.7459 (error 0.0163 versus 0.0633); and at $t = 1.0$, we get 9.9921 (error 0.0756 versus 0.2920). Halving h again gives another fourfold reduction in errors ($y = 10.0486$ at $t = 1.0$, error of 0.0191). An extrapolation based on errors of $O(h^2)$ will give nearly perfect results.

It is appropriate to summarize the finite-difference method for solving a boundary-value problem.

The Finite-Difference Method for Boundary-Value Problems

To solve a boundary-value problem, replace the differential equation with a finite-difference equation by replacing the derivatives with central-difference quotients. Subdivide the interval into a suitable number of equal subintervals, and write the difference equation at each point where the value of the function is unknown. When the boundary values involve derivatives, this will require that the domain be extended beyond the interval. Utilize the derivative boundary conditions to write difference quotients that permit the elimination of the fictitious points outside the interval.

Solve the system of equations so created to obtain approximate values for the solution of the differential equation at discrete points on the interval. If the original differential equation is nonlinear, the system of equations will also be nonlinear. In such situations, the shooting method will normally be preferred.

We could, of course, use more accurate finite-difference approximations to the derivatives, not only for the boundary values, but for the equation itself. The disadvantage of doing this is that the system of equations is then not tridiagonal, and the solution is more difficult to arrive at.

If the more general form of boundary condition applies, $ay + by' = c$, equations similar to (6.8) result, and after the boundary conditions have been approximated and the exterior y-values eliminated from the set of equations, the solution of the problem proceeds as before.

Equations of order higher than the second will involve approximations for the third or higher derivatives. Central-difference formulas will involve points more than h away from the point where the derivative is being approximated, and may be unsymmetrical in the case of the odd derivatives. Probably the method of undetermined coefficients is the easiest way to derive the necessary formulas (see Appendix B). Again, nontridiagonal systems result. Fortunately, most of the important physical problems are simulated by equations of order 2.

Besides the two methods presented in this chapter there are others that are commonly referred to as Rayleigh–Ritz, collocation, and finite-element methods. These approximate the solution to the boundary-value problem, $y(x)$, by writing it as a linear combination $\Sigma c_i \phi_i(x)$, $i = 1, 2, \ldots, n$, where the ϕ_i's are specially chosen functions. These methods have been found to be very effective for certain kinds of problems. We will introduce you to one of these methods in the next section.

6.4   RAYLEIGH–RITZ METHODS

In addition to the previous two methods for boundary-value problems, you should know something about the Rayleigh–Ritz method. It is based on an elegant branch of mathematics, the calculus of variations. In this method we solve a boundary-value problem by approximating the solution with a finite linear combination of simple basis functions that are chosen to fulfill certain criteria, including meeting the boundary conditions.

The calculus of variations seeks to optimize (often minimize) a special class of functions called *functionals*. The usual form for the functional (in problems of one independent variable) is

$$I[y] = \int_a^b F\left(x, y, \frac{dy}{dx}\right) dx. \tag{6.9}$$

Observe that $I[y]$ is not a function of x because x disappears when the definite integral is evaluated. The argument y of $I[y]$ is not a simple variable but a function, $y = y(x)$. The square brackets in $I[y]$ emphasize this fact. A functional can be thought of as a "function of functions." The value of the right-hand side of Eq. (6.9) will change as the function $y(x)$ is varied, but when $y(x)$ is fixed, it evaluates to a scalar quantity (a constant).

Let us illustrate this concept by a very simple example where the solution is obvious in advance—find the function $y(x)$ that minimizes the distance between two points. While we know what $y(x)$ must be, let's pretend we don't. The figure suggests that we are to chose from among the set of curves $y_i(x)$ of which $y_1(x)$, $y_2(x)$, and $y_3(x)$ are representative.

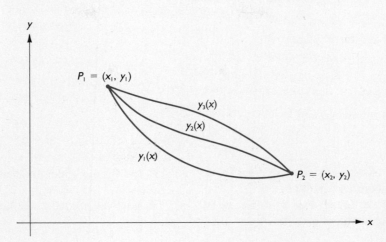

In this simple case, the functional is the integral of the distance along any of these curves:

$$I[y] = \int_{x_1}^{x_2} \sqrt{(dx)^2 + (dy)^2} = \int_{x_1}^{x_2} \sqrt{1 + \left(\frac{dy}{dx}\right)^2}\, dx.$$

To minimize $I[y]$, just as in calculus, we set its derivative to zero. There are certain restrictions on all the curves $y_i(x)$. Obviously each must pass through the points (x_1, y_1) and (x_2, y_2). In addition, for the optimal trajectory, the Euler–Lagrange equation must be satisfied:

$$\frac{d}{dx}\left[\frac{\partial}{\partial y'} F(x, y, y')\right] = \frac{\partial}{\partial y} F(x, y, y'). \qquad (6.10)$$

Applying this to the functional for shortest distance, we have

$$F(x, y, y') = (1 + (y')^2)^{1/2},$$

$$\frac{\partial F}{\partial y} = 0,$$

$$\frac{\partial F}{\partial y'} = \frac{1}{2}(1 + (y')^2)^{-1/2}\,(2y'),$$

$$\frac{d}{dx}\frac{\partial F}{\partial y'} = \frac{d}{dx}\left(\frac{y'}{\sqrt{1 + (y')^2}}\right) = 0.$$

From this, it follows that

$$\frac{y'}{\sqrt{1 + (y')^2}} = c.$$

Solving for y' gives

$$y' = \sqrt{\frac{c^2}{1 - c^2}} = \text{a constant} = b,$$

and, on integrating,

$$y = bx + a.$$

As stated above, $y(x)$ must pass through P_1 and P_2; this condition is used to evaluate the constants a and b.

Let us advance to a less trivial case. Consider the second-order linear boundary-value problem over $[a, b]$,

$$y'' + Q(x)y = F(x), \qquad y(a) = 0, \qquad y(b) = 0. \tag{6.11}$$

The equation is subject to homogeneous Dirichlet conditions. It turns out that the functional that corresponds to Eq. (6.11) is

$$I[u] = \int_a^b \left[\left(\frac{du}{dx}\right)^2 - Qu^2 + 2Fu \right] dx. \tag{6.12}$$

We can transform Eq. (6.12) into Eq. (6.11) through the Euler–Lagrange conditions, so optimizing Eq. (6.12) gives the solution to Eq. (6.11). Observe carefully what benefits result from operating with the functional rather than the original equation—we have now only first-order derivatives instead of second-order ones. This not only simplifies the mathematics but permits finding solutions even when there are discontinuities that cause y to not have sufficiently high derivatives. (There is another approach that shows the equivalence—multiply Eq. (6.11) by the function u and then use integration by parts on the first term.) If the differential equation has boundary conditions that involve the derivative of y, the functional in Eq. (6.12) must be modified.

If we know the solution to our differential equation, substituting it for u in Eq. (6.12) will make $I[u]$ a minimum. When the answer isn't known, perhaps we can approximate it by some (almost) arbitrary function and see if we can minimize the functional by a suitable choice of the parameters in the approximation. The Rayleigh–Ritz method is based on this idea. We assume that u can be approximated by a linear sum of functions:

$$u = c_1 v_1 + c_2 v_2 + \cdots + c_n v_n = \sum_{i=1}^{n} c_i v_i.$$

We substitute u and du/dx from this into Eq. (6.12) to get

$$I(c_1, c_2, \ldots, c_n) = \int_a^b \left[\left(\frac{d}{dx} \sum c_i v_i\right)^2 - Q(\sum c_i v_i)^2 + 2F \sum c_i v_i \right] dx.$$

At this point, observe that I is an ordinary function of the unknown c_i's; our notation reflects this. To get the optimum, we differentiate partially with respect to each c_i in turn, setting each to zero. This gives, when the integrals are evaluated, a linear system in the c_i's that we solve by the usual methods.

E X A M P L E   Solve $y'' = 6x - 6,$ subject to $y(0) = 0$, $y(1) = 0$.

The functional is

$$I[u] = \int_0^1 \left[\left(\frac{du}{dx} \right)^2 + 2(6x - 6)u \right] dx.$$

Assume $u = a + bx + cx^2$ (here $v_1 = 1$, $v_2 = x$, $v_3 = x^2$). To satisfy the boundary conditions, we take $a = 0$, $b = -c$, so

$$u = c(x)\,(x - 1) = c(x^2 - x),$$

$$\frac{du}{dx} = c(2x - 1),$$

$$\left(\frac{du}{dx} \right)^2 = c^2(2x - 1)^2.$$

The second term of the integrand is

$$2(6x - 6)(u) = 2(6x - 6)(c)(x^2 - x)$$
$$= 12c(x^3 - 2x^2 + x).$$

Now we have

$$I(c) = \int_0^1 c^2(2x - 1)^2\, dx + \int_0^1 12c(x^3 - 2x^2 + x)\, dx.$$

Differentiate with respect to c and set to zero:

$$\frac{\partial I}{\partial c} = 0 = c \int_0^1 (2x - 1)^2(2)\, dx + 12 \int_0^1 (x^3 - 2x^2 + x)\, dx.$$

On evaluating the integrals we find that

$$\frac{2}{3}c = -1, \qquad c = -\frac{3}{2},$$

$$u = -\frac{3}{2}(x)(x - 1).$$

Our example is very easy because there is only one unknown coefficient. If there were N of these, a set of N linear equations in the N coefficients would result from setting the several partial derivatives of I to zero.

Table 6.9 compares this result to the analytical answer, $y = 2x - 3x^2 + x^3$.

Table 6.9

x	Exact	$u(x) = (3/2)(x)(x - 1)$
0	0	0
0.2	0.288	0.240
0.4	0.384	0.360
0.6	0.336	0.360
0.8	0.192	0.240
1.0	0	0

We leave it as an exercise for you to show that using a cubic polynomial for u gives a perfect match to the analytical answer. ∎

The difficulty with this classical form of the Rayleigh–Ritz method is that the proper form of approximating function is often not easy to find; polynomials are commonly used but may be unsuitable. The trouble lies in trying to fit a complex function over the entire interval $[a, b]$. But we have previously seen how we can overcome this problem—even low-degree polynomials can approximate any function if applied over a small enough interval. We now examine the application of Rayleigh–Ritz to subintervals of $[a, b]$.

What we do here is equivalent to interpolating between values of the solution to the differential equation at distinct points within $[a, b]$. If we knew values for y at these points, we could interpolate to get y at points between; if we assume a linear relation, we can get y in the subinterval $[x_i, x_{i+1}]$ from y_i and y_{i+1} by defining

$$y(x) = N_i y_i + N_{i+1} y_{i+1}, \qquad \text{with}$$

$$N_i(x) = \begin{cases} (x - x_{i-1})/h_{i-1} & \text{on } [x_{i-1}, x_i] \\ (x_{i+1} - x)/h_i & \text{on } [x_i, x_{i+1}] \\ 0 & \text{elsewhere,} \end{cases}$$

where $h_i = |x_{i+1} - x_i|$. In effect, y within the interval is a weighted average of the values at the ends. We will take h as a constant in the following.

Here is a summary of some of the properties of these weighting functions:

1. $N_i(x_j) = \begin{cases} 1 \text{ if } i = j \\ 0 \text{ otherwise (if } i \neq j), \end{cases}$

2. $N_i(x) * N_j(x) = 0$ for $|i - j| > 1$ (intervals other than those that have x_i as an endpoint),

3. $N_i'(x) * N_j'(x) = 0$ for $|i - j| > 1$,

4. $N_i(x) * N_{i+1}(x) = \begin{cases} (x_{i+1} - x)(x - x_i)/h^2 & \text{on } [x_i, x_{i+1}] \\ 0 & \text{elsewhere,} \end{cases}$

5. $N_i'(x) * N_{i+1}'(x) = \begin{cases} -1/h^2 & \text{on } [x_i, x_{i+1}] \\ 0 & \text{elsewhere.} \end{cases}$

This sketch shows the appearance of the $N_i(x)$:

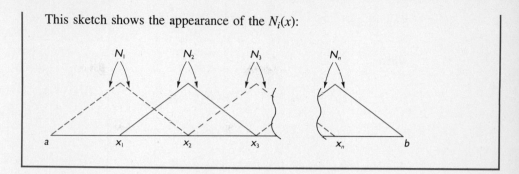

We will approximate the solution to the differential equation by substituting into the functional for $u(x)$:

$$u(x) = \sum_{i=1}^{n} c_i N_i(x).$$

When we do so and minimize by setting derivatives with respect to c_i equal to zero, the values of the c_i's will be approximations to y at the nodes within $[a, b]$.

EXAMPLE Solve $y'' - (1 - x/5)y = x$, $y(1) = 2$ and $y(3) = -1$. (This is the same problem as in Sections 6.1 and 6.2, but the variables have been renamed.)

Subdivide the interval $[1, 3]$ into four equal subintervals, making $h = 0.5$. The $N_i(x)$ functions look like this:

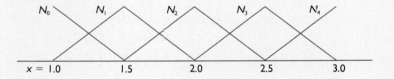

(Observe that N_0 and N_4 are different from the others. This is required to handle the nonhomogeneous end conditions.) We approximate the function with

$$y(x) = c_0 N_0(x) + c_1 N_1(x) + c_2 N_2(x) + c_3 N_3(x) + c_4 N_4(x)$$

(c_0 will be 2 and c_4 will be -1 to satisfy the end conditions, but we postpone this substitution.) On substituting into the functional, we have

$$I(c_0, \ldots, c_4) = \int_1^3 [(c_0 N_0' + c_1 N_1' + c_2 N_2' + c_3 N_3' + c_4 N_4')^2$$
$$+ \left(1 - \frac{x}{5}\right)(c_0 N_0 + c_1 N_1 + c_2 N_2 + c_3 N_3 + c_4 N_4)^2$$
$$+ 2x(c_0 N_0 + c_1 N_1 + c_2 N_2 + c_3 N_3 + c_4 N_4)] \; dx$$

We will set the derivatives of I with respect to c_i, $i = 1, 2, 3$ to zero. We do not do this for c_0 and c_4 because they are fixed. We also substitute the known values for c_0 and c_4.

This gives

$$Ac = b,$$

where

$$A_{ki} = \int_1^3 \left[N_k'(x)N_i'(x) + \left(1 - \frac{x}{5}\right)N_k(x)N_i(x) \right] dx,$$

$$b_k = -\int_1^3 xN_k(x)\, dx.$$

Some example terms that you should verify are

$$A(1, 1) = 4 + 4\int_{1.0}^{1.5} \left(1 - \frac{x}{5}\right)(x - 1)^2\, dx + 4\int_{1.5}^{2.0} \left(1 - \frac{x}{5}\right)(2 - x)^2\, dx,$$

$$A(1, 2) = -2 + 4\int_{1.5}^{2.0} \left(1 - \frac{x}{5}\right)(2 - x)(x - 1.5)\, dx,$$

$$b(1) = -2\int_{1.0}^{1.5} x(x - 1)\, dx - 2\int_{1.5}^{2.0} x(2 - x)\, dx.$$

We used Simpson's method to evaluate the integrals, getting

$$A = \begin{bmatrix} 4.2333 & -1.9458 & 0.0 \\ -1.9458 & 4.1999 & -1.9542 \\ 0.0 & -1.9541 & 4.1667 \end{bmatrix}, \quad b = \begin{bmatrix} 3.125 \\ -1.0 \\ -3.2125 \end{bmatrix},$$

which has the solution

$$c = \begin{bmatrix} -0.5323 \\ -0.4480 \\ -0.9811 \end{bmatrix}.$$

This follows from the fact that we set $c_0 = 2$ and $c_4 = -1$ in the system

$$\begin{bmatrix} -1.9375 & 4.2333 & -1.9458 & 0.0 & 0.0 \\ 0.0 & -1.9458 & 4.1999 & -1.9542 & 0.0 \\ 0.0 & 0.0 & -1.9541 & 4.1667 & -1.9625 \end{bmatrix} \begin{bmatrix} 2.0 \\ c_1 \\ c_2 \\ c_3 \\ -1.0 \end{bmatrix} = \begin{bmatrix} -0.75 \\ -1.0 \\ -1.25 \end{bmatrix}.$$

Now we have an approximation to the solution of our differential equation:

$$y(x) = 2N_0(x) - 0.5323N_1(x) - 0.4480N_2(x) - 0.9811N_3(x) - N_4(x).$$

The nodal values are quite obvious:

$$y(1.5) = 0.5323,$$
$$y(2.0) = -0.4480,$$
$$y(2.5) = -0.9811,$$

but we can also use the equation for y to compute for any point in $[1, 3]$. ∎

Table 6.10

| x | Rayleigh–Ritz method | | Runge–Kutta–Fehlberg |
| | y | y | y |
(h =	0.50	0.25	0.25)
1.25	(1.2262)*	1.2033	1.2011
1.50	0.5323	0.5381	0.5400
1.75	(0.0422)*	−0.0064	−0.0041
2.00	−0.4480	−0.4408	−0.4385
2.25	(−0.7146)*	−0.7659	−0.7637
2.50	−0.9811	−0.9758	−0.9741
2.75	(−0.9906)*	−1.0593	−1.0583

*Interpolated.

We repeated this same application of Rayleigh–Ritz to subintervals of [1, 3] but with $h = 0.25$. This produced nine weighting functions instead of five and allowed the computation of more accurate values for the solution. Table 6.10 summarizes the results and compares them to those from Runge–Kutta–Fehlberg.

What we have done above is really an application of the finite-element method. Basically, this breaks up the region of a boundary-value problem into subintervals and applies a variational method within the subintervals. There are alternative ways to derive the equations that produce the linear system of equations whose solutions are the approximations to the solution of the boundary-value problem at the nodes. One of these is the Galerkin method, which, for problems of the same type as our example, results in exactly the same set of equations.

6.5 CHARACTERISTIC-VALUE PROBLEMS

Problems in the fields of elasticity and vibration (including applications of the wave equations of modern physics) fall into a special class of boundary-value problems known as *characteristic-value problems*. (Certain problems in statistics also reduce to such problems.) We discuss only the most elementary forms of characteristic-value problems here.

Consider the homogeneous* second-order equation with homogeneous boundary conditions:

$$\frac{d^2y}{dx^2} + k^2y = 0, \quad y(0) = 0, \quad y(1) = 0, \tag{6.13}$$

*Homogeneous here means that all the terms are alike in being functions of y or its derivatives.

where k^2 is a parameter. We first solve this equation nonnumerically to show that there is a solution for only certain particular, or "characteristic," values of the parameter. The general solution is

$$y = a \sin kx + b \cos kx,$$

which can easily be verified by substituting into the differential equation; the solution contains the two arbitrary constants a and b, because the differential equation is second-order. The constants a and b are to be determined to make the general solution agree with the boundary conditions.

At $x = 0$, $y = 0 = a \sin(0) + b \cos(0) = b$. Then b must be zero. At $x = 1$, $y = 0 = a \sin(k)$; we may have either $a = 0$ or $\sin(k) = 0$ to satisfy this condition. The former leads to $y(x) = 0$, which we call the *trivial solution*. This function, y everywhere zero, will always be a solution to any homogeneous differential equation with homogeneous boundary conditions. (The trivial solution is usually of no interest.) To get a nontrivial solution, we must choose the other alternative, $\sin(k) = 0$, which is satisfied only for certain values of k, the characteristic values of the system. The solution to Eq. (6.9) must then be

$$k = \pm n\pi, \qquad n = 1, 2, \ldots, \qquad y = a \sin n\pi x. \tag{6.14}$$

Note that the arbitrary constant a can have any value and still permit the function y to meet the boundary conditions, so that the solution is determined only to within a multiplicative constant. In Fig. 6.2 we sketch several of the solutions as given by Eq. (6.14).

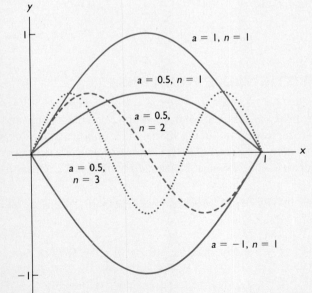

$a = 1, n = 1$

$a = 0.5, n = 1$

$a = 0.5,$
$n = 2$

$a = 0.5,$
$n = 3$

$a = -1, n = 1$

Figure 6.2

The information of interest in characteristic-value problems are these characteristic values, or *eigenvalues*, for the system. If we are dealing with a vibration problem, these give the natural frequencies of the system, which are especially important because, with external loads applied at or near these frequencies, resonance will cause an amplification of motion so that failure is likely. In the field of elasticity, there is an eigenfunction corresponding to each eigenvalue, and these determine the possible shapes of the elastic curve when the system is in equilibrium. Often the smallest nonzero value of the parameter is especially important; this gives the fundamental frequency of the system.

To illustrate the use of numerical methods, we will solve Eq. (6.13) again. Our attack is the previous one of replacing the differential equation by a difference equation, and writing this at all points into which the x-interval has been subdivided and where y is unknown.* Replacing the derivative in (6.13) by a central-difference approximation, we have

$$\frac{y_{i+1} - 2y_i + y_{i-1}}{h^2} + k^2 y_i = 0. \tag{6.15}$$

Letting $h = 0.2$, and writing out Eq. (6.15) at each of the four interior points, we get

$$y_1 - (2 - 0.04k^2)y_2 + y_3 = 0,$$
$$y_2 - (2 - 0.04k^2)y_3 + y_4 = 0,$$
$$y_3 - (2 - 0.04k^2)y_4 + y_5 = 0,$$
$$y_4 - (2 - 0.04k^2)y_5 + y_6 = 0. \tag{6.16}$$

The boundary conditions give $y_1 = y_6 = 0$. Making this substitution and writing in matrix form, after multiplying all equations by -1, we get

$$\begin{bmatrix} 2 - 0.04k^2 & -1 & 0 & 0 \\ -1 & 2 - 0.04k^2 & -1 & 0 \\ 0 & -1 & 2 - 0.04k^2 & -1 \\ 0 & 0 & -1 & 2 - 0.04k^2 \end{bmatrix} \begin{bmatrix} y_2 \\ y_3 \\ y_4 \\ y_5 \end{bmatrix} = 0. \tag{6.17}$$

Note that we can write this as the matrix equation $(A - \lambda I)y = 0$, where

$$A = \begin{bmatrix} 2 & -1 & 0 & 0 \\ -1 & 2 & -1 & 0 \\ 0 & -1 & 2 & -1 \\ 0 & 0 & -1 & 2 \end{bmatrix}, \qquad \lambda = 0.04k^2,$$

and I is the identity matrix of order 4. We will consider the problem in this form in the next section.

---

*There is an alternative trial-and-error approach, analogous to the shooting method, but it generally involves more work.

The solution to our characteristic-value problem, Eq. (6.13), reduces to solving the set of equations in (6.16) or (6.17).

Such a set of homogeneous linear equations has a nontrivial solution only if the determinant of the coefficient matrix is equal to zero. Hence

$$
\text{Det}
\begin{bmatrix}
2 - 0.04k^2 & -1 & 0 & 0 \\
-1 & 2 - 0.04k^2 & -1 & 0 \\
0 & -1 & 2 - 0.04k^2 & -1 \\
0 & 0 & -1 & 2 - 0.04k^2
\end{bmatrix} = 0.
$$

Expanding this determinant will give an eighth-degree polynomial in k (this is called the characteristic polynomial of matrix A); the roots of this polynomial will be approximations to the characteristic values of the system. Letting $2 - 0.04k^2 = Z$ makes the expansion simpler, and we get

$$Z^4 - 3Z^2 + 1 = 0.$$

The roots of this biquadratic are $Z^2 = (3 \pm \sqrt{5})/2$, or

$$Z = 1.618, \quad -1.618, \quad 0.618, \quad -0.618.$$

We will get an estimate of the principal eigenvalues from $2 - 0.04k^2 = 1.618$, giving $k = \pm 3.09$ (compare to $\pm \pi$). The next eigenvalues are obtained from

$$2 - 0.04k^2 = 0.618, \qquad \text{giving } k = \pm 5.88 \text{ (compare to } \pm 2\pi\text{);}$$

$$2 - 0.04k^2 = -0.618, \qquad \text{giving } k = \pm 8.09 \text{ (compare to } \pm 3\pi\text{);}$$

$$2 - 0.04k^2 = -1.618, \qquad \text{giving } k = \pm 9.51 \text{ (compare to } \pm 4\pi\text{).}$$

Our estimates of the characteristic values get progressively worse. Fortunately the smallest values of k are of principal interest in many applied problems. To improve accuracy, we will need to write our difference equation using smaller values of h. The work soon becomes inordinately long, and we look for a simpler method to find the eigenvalues of a matrix. This is the subject of the next section.

Another reason, beyond the amount of computational effort, that argues against expanding the determinant and then finding the roots of the characteristic equation is accuracy. Finding the zeros of a polynomial of high degree is a process that is often subject to large errors due to round-off. The process is very ill-conditioned.

6.66 EIGENVALUES OF A MATRIX BY ITERATION

We have seen that particular values of the parameter occurring in a characteristic-value problem are of special interest, and the numerical solution involves finding the eigenvalues of the coefficient matrix of the set of difference equations. To do this, we need to find the values of λ that satisfy the matrix equation

$$Ax = \lambda x$$

where A is a square matrix. In effect, we look for a certain vector x that gives a scalar multiple of itself when multiplied by the matrix A. For most vectors chosen at random, this will not be true. If A is $n \times n$ and nonsingular, there are usually exactly n different vectors x which, when multiplied by the matrix A, give a scalar multiple of x. These vectors are called *eigenvectors* or *characteristic vectors*. The multipliers λ, which are n in number, are termed the eigenvalues or characteristic values of the matrix A. As we have seen in the previous section, one method is to solve the equivalent matrix equation

$$(A - \lambda I)x = 0$$

by determining the values of λ that make the determinant of $(A - \lambda I)$ equal to zero. This involves expanding the determinant to give a polynomial in λ, the *characteristic equation*, whose roots must be evaluated. This tells us, of course, that there will be n eigenvalues for the system. but they will not necessarily all be different in value. In fact, it is possible that some eigenvalues will be complex-valued. Solving for the roots of the characteristic equation is laborious.* This section develops an alternative iterative procedure that involves less effort.

Consider a simple 3×3 system:

$$\begin{bmatrix} 10 & 0 & 0 \\ 1 & -3 & -7 \\ 0 & 2 & 6 \end{bmatrix} \begin{bmatrix} x_1 \\ x_2 \\ x_3 \end{bmatrix} = \lambda \begin{bmatrix} x_1 \\ x_2 \\ x_3 \end{bmatrix},$$

$$\begin{bmatrix} 10 - \lambda & 0 & 0 \\ 1 & -3 - \lambda & -7 \\ 0 & 2 & 6 - \lambda \end{bmatrix} \begin{bmatrix} x_1 \\ x_2 \\ x_3 \end{bmatrix} = \begin{bmatrix} 0 \\ 0 \\ 0 \end{bmatrix}. \tag{6.18}$$

We wish to find the values of λ and vectors x that satisfy Eq. (6.18). Let us first see what they are.

The characteristic equation is

$$(10 - \lambda)(-3 - \lambda)(6 - \lambda) + 14(10 - \lambda) = 0, \quad (10 - \lambda)(\lambda - 4)(\lambda + 1) = 0.$$

The roots are

$$\lambda_1 = 10, \qquad \lambda_2 = 4, \qquad \text{and} \qquad \lambda_3 = -1.$$

We can find the eigenvectors corresponding to these from the three sets of equations:

$$\begin{bmatrix} 10 & 0 & 0 \\ 1 & -3 & -7 \\ 0 & 2 & 6 \end{bmatrix} \begin{bmatrix} x_1 \\ x_2 \\ x_3 \end{bmatrix} = 10 \begin{bmatrix} x_1 \\ x_2 \\ x_3 \end{bmatrix}, \quad \begin{cases} 0x_1 & = 0 \\ x_1 - 13x_2 - 7x_3 = 0, \\ 2x_2 - 4x_3 = 0 \end{cases} \quad \begin{cases} x_1 = 1 \\ x_2 = \frac{2}{33}. \\ x_3 = \frac{1}{33} \end{cases}$$

$$\begin{bmatrix} 10 & 0 & 0 \\ 1 & -3 & -7 \\ 0 & 2 & 6 \end{bmatrix} \begin{bmatrix} x_1 \\ x_2 \\ x_3 \end{bmatrix} = 4 \begin{bmatrix} x_1 \\ x_2 \\ x_3 \end{bmatrix}, \quad \begin{cases} 6x_1 & = 0 \\ x_1 - 7x_2 - 7x_3 = 0, \\ 2x_2 + 2x_3 = 0 \end{cases} \quad \begin{cases} x_1 = 0 \\ x_2 = -1. \\ x_3 = 1 \end{cases}$$

---

*The labor involved is not only in finding the roots of the characteristic polynomial but in evaluating the coefficients in the first place.

$$\begin{bmatrix} 10 & 0 & 0 \\ 1 & -3 & -7 \\ 0 & 2 & 6 \end{bmatrix} \begin{bmatrix} x_1 \\ x_2 \\ x_3 \end{bmatrix} = -1 \begin{bmatrix} x_1 \\ x_2 \\ x_3 \end{bmatrix}, \quad \begin{cases} 11x_1 & = 0 \\ x_1 - 2x_2 - 7x_3 = 0, \\ 2x_2 + 7x_3 = 0 \end{cases} \quad \begin{cases} x_1 = 0 \\ x_2 = 1 \\ x_3 = -\frac{2}{7} \end{cases}.$$

We see that any nonzero value can be taken for x_1 in the first vector, for x_2 in the second, and for x_3 in the third. Our eigenvectors are then any multiple of

$$\begin{bmatrix} 1 \\ \frac{2}{33} \\ \frac{1}{33} \end{bmatrix}, \quad \begin{bmatrix} 0 \\ -1 \\ 1 \end{bmatrix}, \quad \text{and} \quad \begin{bmatrix} 0 \\ 1 \\ -\frac{2}{7} \end{bmatrix}.$$

In the matrix of Eq. (6.18) the sum of the eigenvalues 10, 4, -1 is equal to the sum of the diagonal elements of the given matrix. This is no coincidence; in fact, for any matrix A, the sum of its eigenvalues is equal to the trace of A, tr(A), that was defined in Section 2.2.

In Chapter 5 we considered initial-value differential equations. The eigenvalues and eigenvectors of a matrix are important in the solution of a system of differential equations with constant coefficients. Consider the system

$$x' = 10x,$$
$$y' = x - 3y - 7z,$$
$$z' = 2y + 6z,$$

where $x(0) = 1$, $y(0) = -1$, $z(0) = 2$. We may rewrite these equations as

$$X'(t) = \begin{bmatrix} 10 & 0 & 0 \\ 1 & -3 & -7 \\ 0 & 2 & 6 \end{bmatrix} X(t), \quad X(0) = \begin{bmatrix} 1 \\ -1 \\ 2 \end{bmatrix}.$$

The general solution to this problem is

$$X(t) = Ae^{10t} \begin{bmatrix} 1 \\ \frac{2}{33} \\ \frac{1}{33} \end{bmatrix} + Be^{4t} \begin{bmatrix} 0 \\ -1 \\ 1 \end{bmatrix} + Ce^{-t} \begin{bmatrix} 0 \\ 1 \\ -\frac{2}{7} \end{bmatrix};$$

that is,

$$x(t) = Ae^{10t}$$

$$y(t) = \frac{2}{33}Ae^{10t} - Be^{4t} + Ce^{-t}$$

$$z(t) = \frac{1}{33}Ae^{10t} + Be^{4t} - \frac{2}{7}Ce^{-t}$$

where

$$1 = A$$

$$-1 = \frac{2}{33}A - B + C$$

$$2 = \frac{1}{33}A + B - \frac{2}{7}C \quad \text{at } t = 0.$$

Unfortunately, we cannot find the eigenvalues of a general matrix by simply reducing it to triangular form by Gaussian elimination, as one at first might hope. The reason is that row reduction changes the eigenvalues. For example, consider the matrices A and B:

$$A = \begin{bmatrix} 2 & 1 \\ 1 & 2 \end{bmatrix}, \quad B = \begin{bmatrix} 2 & 1 \\ 0 & \frac{3}{2} \end{bmatrix}.$$

Matrix B is derived from A by row reduction: it has eigenvalues 2 and $\frac{3}{2}$, while the eigenvalues of A are 3 and 1.

We now illustrate an iterative procedure for determining the eigenvalue of largest magnitude of Eq. (6.18). This is called the *power method*. We start with an arbitrary vector and multiply it by the matrix repeatedly. Using

$$\begin{bmatrix} 1 \\ 0 \\ 0 \end{bmatrix},$$

we have the results below. At each step, we normalize the vector by making its largest component equal to unity.

$$\begin{bmatrix} 10 & 0 & 0 \\ 1 & -3 & -7 \\ 0 & 2 & 6 \end{bmatrix}\begin{bmatrix} 1 \\ 0 \\ 0 \end{bmatrix} = \begin{bmatrix} 10 \\ 1 \\ 0 \end{bmatrix} = 10\begin{bmatrix} 1 \\ 0.1 \\ 0 \end{bmatrix},$$

$$\begin{bmatrix} 10 & 0 & 0 \\ 1 & -3 & -7 \\ 0 & 2 & 6 \end{bmatrix}\begin{bmatrix} 1 \\ 0.1 \\ 0 \end{bmatrix} = \begin{bmatrix} 10 \\ 0.7 \\ 0.2 \end{bmatrix} = 10\begin{bmatrix} 1 \\ 0.07 \\ 0.02 \end{bmatrix},$$

$$\begin{bmatrix} 10 & 0 & 0 \\ 1 & -3 & -7 \\ 0 & 2 & 6 \end{bmatrix}\begin{bmatrix} 1 \\ 0.07 \\ 0.02 \end{bmatrix} = \begin{bmatrix} 10 \\ 0.65 \\ 0.26 \end{bmatrix} = 10\begin{bmatrix} 1 \\ 0.065 \\ 0.026 \end{bmatrix}.$$

$$\begin{bmatrix} 10 & 0 & 0 \\ 1 & -3 & -7 \\ 0 & 2 & 6 \end{bmatrix}\begin{bmatrix} 1 \\ 0.065 \\ 0.026 \end{bmatrix} = \begin{bmatrix} 10 \\ 0.623 \\ 0.286 \end{bmatrix} = 10\begin{bmatrix} 1 \\ 0.0623 \\ 0.0286 \end{bmatrix},$$

$$\begin{bmatrix} 10 & 0 & 0 \\ 1 & -3 & -7 \\ 0 & 2 & 6 \end{bmatrix}\begin{bmatrix} 1 \\ 0.0623 \\ 0.0286 \end{bmatrix} = \begin{bmatrix} 10 \\ 0.6129 \\ 0.2962 \end{bmatrix} = 10\begin{bmatrix} 1 \\ 0.06129 \\ 0.02962 \end{bmatrix},$$

We see that the successive values of the product vector approach closer and closer to multiples of the eigenvector

$$\begin{bmatrix} 1 \\ \frac{2}{33} \\ \frac{1}{33} \end{bmatrix}.$$

The normalization factors approach the value of the largest eigenvalue.

Before leaving this example, we again observe that the

$$\text{tr}\begin{pmatrix} 10 & 0 & 0 \\ 1 & -3 & -7 \\ 0 & 2 & 6 \end{pmatrix} = 13 = 10 + 4 - 1.$$

In programming, one should use this fact to keep a check on the accuracy of the eigenvalues found.

The method is slow to converge if the magnitudes of the largest eigenvalues are nearly the same.

With more realistically sized matrices, the iteration takes many steps and requires a computer for its practical execution. The method will converge only when the eigenvalue largest in modulus is uniquely determined.

The power method works because the eigenvectors are a set of *basis vectors*. By this we mean that they are a set of linearly independent vectors that *span the space*; that is, any n-component vector can be written as a unique linear combination of them. Let $v^{(0)}$ by any vector and $x_1, x_2, \ldots, x_n$ be eigenvectors. Then

$$v^{(0)} = c_1 x_1 + c_2 x_2 + \cdots + c_n x_n.$$

If we multiply $v^{(0)}$ by the matrix A, we have (since the x_i are eigenvectors with corresponding eigenvalues λ_i),

$$v^{(1)} = Av^{(0)} = c_1 A x_1 + c_2 A x_2 + \cdots + c_n A x_n$$
$$= c_1 \lambda_1 x_1 + c_2 \lambda_2 x_2 + \cdots + c_n \lambda_n x_n. \tag{6.19}$$

Upon repeated multiplication by A we get, after m times,

$$v^{(m)} = A^m v^{(0)} = c_1 \lambda_1^m x_1 + c_2 \lambda_2^m x_2 + \cdots + c_n \lambda_n^m x_n. \tag{6.20}$$

If one eigenvalue, say λ_1, is larger in magnitude than all the rest, the values of λ_i^m, $i \neq 1$, will be negligibly small in comparison to λ_1^m when m is large and

$$A^m v \rightarrow c_1 \lambda^m x_1, \quad \text{or} \quad A^m v \rightarrow \text{Multiple of eigenvector } x_1,$$

with the normalization factor of λ_1, provided that $c_1 \neq 0$. This is the principle behind the power method.

In applying Eq. (6.20) to find the dominant eigenvalue (the eigenvalue whose magnitude exceeds all others), we normalize at each step; this in effect is just scaling the c's after each multiplication, and will not change the relation. If some other eigenvalue than λ_1 were the dominant one, we would merely renumber the eigenvalues. It is easy to see why the rate of convergence depends strongly on the ratio of the two eigenvalues of largest magnitude.

Obviously, if we are able to choose the starting vector $v^{(0)}$ close to the eigenvector x_1, we will converge more rapidly; the coefficients c_i, $i \neq 1$, in Eq. (6.19) will be small in this case. We usually have no knowledge of this eigenvector, so we start with an arbitrary vector with all components equal to unity. This choice could be a bad one. If the value of c_1 in Eq. (6.19) is zero for this arbitrary choice, the process may never converge. Actually, the process often does converge in spite of this; round-off errors incurred as we repeatedly multiply by A produce components in the direction of x_1 so that we do converge, although slowly.

A special advantage of the power method is that the eigenvector that corresponds to the dominant eigenvalue is generated at the same time. For most methods of determining eigenvalues, a separate computation is needed to obtain the eigenvector. A disadvantage, of course, is that it gives only one eigenvalue. Sometimes the largest eigenvalue is the most important, but if it is not, we must modify the method.

In some problems, the most important eigenvalue and eigenvector is that where the eigenvalue is of least magnitude. We can find this one if we apply the power method to the inverse of A. This is because the inverse matrix has a set of eigenvalues that are the reciprocals of the eigenvalues of A. This is readily shown:

Given that $Ax = \lambda x$.

Multiply by A^{-1}:

$$A^{-1}Ax = A^{-1}\lambda x = \lambda A^{-1}x.$$

From this we see that

$$x = \lambda A^{-1}x \quad \text{or} \quad A^{-1}x = \frac{1}{\lambda}x.$$

(It is inefficient to actually invert A before applying the power method to get the eigenvalue. We use the LU equivalent of A, obtainable at much less effort, and solve for $v^{(n+1)}$ from

$$LUv^{(n+1)} = Av^{(n+1)} = v^{(n)}, \quad \text{equivalent to} \quad A^{-1}v^{(n)} = v^{(n+1)}. \quad \textbf{(6.21)}$$

The solution of Eq. (6.21) when the LU is known requires the same amount of effort as a matrix multiplication.)

The power method is apt to be slowly convergent. We can accelerate it if we know an approximate value for the eigenvalue. This is based on the result of shifting the eigenvalues. Observe that subtracting a constant from the diagonal elements of A gives a system whose eigenvalues are those of A with the same constant subtracted:

Given $Ax = \lambda x$.

Subtract $sIx = sx$ from both sides.

$$Ax - sIx = \lambda x - sx,$$

$$(A - sI)x = (\lambda - s)x.$$

The above relationship can be applied in two ways. Suppose we wish to determine the value of an eigenvalue near to some number s. We shift the eigenvalues by subtracting s from the diagonal elements; there is then an eigenvalue very near to zero in the shifted matrix. We use the power method on the inverse of the shifted matrix. This is often rapidly convergent because the reciprocal of the very small value is very large, and is usually much larger than the next largest one (for the shifted inverse system). After we obtain it, we reverse the transformations to obtain the desired value for the original matrix. This process is called the *inverse power method*. It is illustrated in the example below.

Another application of shifting is to determine the eigenvalue at the other extreme after the dominant one has been computed. Suppose a matrix has eigenvalues whose values are 8, 4, 1, and -5. After applying the power method to get the one at 8, we shift by subtracting 8. The eigenvalues of the shifted system are then 0, -4, -7, and -13. The power method will get the dominant one, at -13. When we add 8 back to reverse the effect of the shifting, we obtain -5, the eigenvalue of the original matrix at the opposite extreme from 8.

We illustrate the use of the power method and its several variations with the 3×3 matrix:

$$A = \begin{bmatrix} 4 & -1 & 1 \\ 1 & 1 & 1 \\ -2 & 0 & -6 \end{bmatrix}.$$

If we begin with the initial vector $(1, 1, 1)^T$, the regular power method gives the results shown in Table 6.11. The iterations were terminated when the normalization factor changed by less than 0.0001 from the previous value.

After the dominant eigenvalue at -5.76849 was determined, the matrix was shifted by subtracting this value from the elements on the diagonal. For this shifted matrix, the power method gave, after 32 iterations, a normalization factor of 9.23759 with a vector of $(1, 0.31963, -0.21121)^T$. We then know that the original matrix had an eigenvalue at 3.4691 (from $9.23759 + (-5.76849)$). (The corresponding eigenvector is the same as that determined by the power method for the shifted matrix.)

Table 6.11

Iteration number	Normalization factor	Resultant vector
1	-8	$(-0.5, -0.375, 1)^T$
2	-5	$(0.125, -0.025, 1)^T$
3	-6.25	$(-0.244, -0.176, 1)^T$
.	.	.
.	.	.
.	.	.
22	-5.76849	$(-0.1157, -0.1306, 1)^T$

Applying the power method to A^{-1} (through application of the LU equivalent, of course) gave, after nine iterations, the value of 0.76989 with a vector of $(0.4121, 1, -0.1129)^T$. Now we know that the original matrix A had its smallest eigenvalue at $1/0.76989 = 1.29923$. For this 3×3 system, we have computed all its eigenvalues.

The dominant eigenvalue came with some slowness; 22 iterations were needed. If we shift by -5.77 and apply the method to the inverse, only four iterations are required to obtain a value of 672.438 and a vector of $(-0.1157, -0.1306, 1)^T$. This is the same vector as that obtained in the first application of the method. We get the dominant eigenvalue for A by adding -5.77 to the reciprocal of 672.438:

$$\frac{1}{672.438} + (-5.77) = -5.76851.$$

Many iterations have been saved.

The key to the inverse power method is finding an approximate value for the desired eigenvalue. Sometimes the physical problem itself can suggest a value. Applying the standard power method for only a few iterations frequently gives an approximate value (the many successive iterations are needed to refine the value). Gerschgorin's theorems can give good estimates if there is strong diagonal dominance. Let D_i be a circle whose center is a_{ii} and whose radius is

$$\sum |a_{ij}|, \quad j = 1, 2, \ldots, n \quad \text{and} \quad j \neq i.$$

Then Gerschgorin I says that every eigenvalue of A must lie in the union of those circles. Gerschgorin II says that, if k of these circles do not touch the other $n - k$ circles, then exactly k eigenvalues (counting multiplicities) lie in the union of those k circles. For the previous example we have

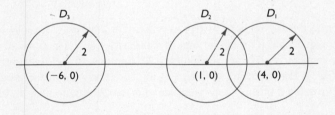

The dominant eigenvalue, -5.76851, was found to lie in D_3, and the other two are in $D_1 \cup D_2$.

When a matrix has two eigenvalues of equal largest magnitude,* the power method as described will fail. In the case of two largest eigenvalues of opposite sign, the normalization factor does not converge but oscillates. For example, when the power method is applied to

$$A = \begin{bmatrix} -10 & -7 & 7 \\ 5 & 5 & -4 \\ -7 & -5 & 6 \end{bmatrix},$$

---

*This may occur if the eigenvalues are complex; for a matrix with real coefficients, complex eigenvalues will come in complex conjugate pairs.

which has eigenvalues of 4, -4, and 1, the power method gives the results shown in Table 6.12 and the normalizations continue indefinitely to be -10 and -1.6. In such a case, the magnitude of the largest eigenvalue(s) is given by taking the square root of the product of the alternating normalization factors; in this example, -10 and -1.6 have a product, 16, whose square roots are 4 and -4.

In case all the eigenvalues of a large system are required, the QR algorithm is the usual method. This method is based on several properties of eigenvalues that we already know. For instance, let $Ax = \lambda x$ where λ is an eigenvalue of A and x an eigenvector corresponding to λ. From this it is easy to show that if A is nonsingular, then λ^{-1} is an eigenvalue of A^{-1}. Similarly, we have seen that for this same matrix A we have $(\lambda - \alpha)$ as an eigenvalue of $(A - \alpha I)$ and $(\lambda - \alpha)^{-1}$ as an eigenvalue of $(A - \alpha I)^{-1}$. In these last two matrices we just subtract α from the diagonal elements of A. There is only one new property to add to this list—namely, that for any nonsingular matrix B:

$$BAB^{-1}x = \lambda x.$$

This follows from the property of determinants that says that $\det(A - \lambda I) = \det(BAB^{-1} - \lambda I)$. From this we can show that all the eigenvalues of A and BAB^{-1} are the same. This property is used again and again in the QR method. In fact, two matrices A and C are said to be *similar* if we have $BAB^{-1} = C$.

There is an important class of matrices called the *Hessenberg matrices*. They are of the form $H(i, j) = 0$ if $i > (j + 1)$. A few examples will clarify this definition.

$$H_1 = \begin{bmatrix} 4 & 3 & 1 \\ 5 & 3 & 4 \\ 0 & 3 & -9 \end{bmatrix}, \quad \text{and} \quad H_2 = \begin{bmatrix} 7 & 8 & -8 & 3 & -3 \\ 6 & 9 & -2 & -1 & 29 \\ 0 & 9 & 5 & 9 & 1 \\ 0 & 0 & 2 & 12 & 9 \\ 0 & 0 & 0 & -9 & 21 \end{bmatrix}.$$

We can describe the Hessenberg matrix as "almost triangular." The QR method takes a matrix and makes it into a triangular one by similarity transformations so that the eigenvalues of the original matrix are maintained. The details of the algorithm can be found in Stewart (1973). However, we will give a simple description of the method along with an example. Program 3 of this chapter implements the QR method for finding eigenvalues of a matrix.

Suppose we start with a matrix A, which we denote as A_0.

Table 6.12

Iteration number	Normalization factor	Resultant vector
1	-10	$(1, -0.6, 0.6)^T$
2	-1.6	$(1, 0.25, 0.25)^T$
3	-10	$(1, -0.525, 0.675)^T$
4	-1.6	$(1, 0.2031, 0.2031)^T$

Step 1. Let $A_1 = PA_0P^{-1}$ where A_1 is a Hessenberg matrix. The matrix P is just the product of elementary matrices, called *Householder reflectors*. These reflectors have a very useful property in that, if B is such a reflector, then $B = B^T = B^{-1}$. These matrices can be generated easily. Take any nonzero vector x; then $B = I - \alpha xx^T$ is a reflector, provided $\alpha = 2/(x^Tx)$. (*Note:* xx^T is an "outer product" since x is $n \times 1$, and xx^T is $n \times n$.)

For instance, let A_0 be a 4×4 matrix,

$$A_0 = \begin{bmatrix} * & * & * & * \\ * & * & * & * \\ * & * & * & * \\ * & * & * & * \end{bmatrix} ; \text{ then reflectors } B_1B_2 \text{ can be}$$

found so that

$$B_1A_0B_1^T = \begin{bmatrix} * & * & * & * \\ * & * & * & * \\ 0 & * & * & * \\ 0 & * & * & * \end{bmatrix} \quad \text{and}$$

$$B_2B_1A_0B_1^TB_2^T = \begin{bmatrix} * & * & * & * \\ * & * & * & * \\ 0 & * & * & * \\ 0 & 0 & * & * \end{bmatrix}. \text{ Then let } P = B_2B_1.$$

This first step produces a matrix that is almost upper-triangular.

Step 2. Now we perform an iterative procedure on the matrix from step 1 as it is transformed into one with more zeros below the main diagonal until it becomes upper-triangular. Here we have for $i = 1, 2, 3, \ldots$.

$A_i = Q_iR_i$, where $Q_i^TQ_i = I$ and R_i is upper-triangular. Then set

$A_{i+1} = R_iQ_i$. It is easy to verify that A_i, A_{i+1}, are similar, since

$A_{i+1} = Q_i^TA_iQ_i$. Once all the elements below the main diagonal are zero (less than TOL in absolute value) we stop. Actually the process is speeded up by shifting the matrix as we go along. By modifying the preceding formulation, we have

$$A_i - \alpha_iI = Q_iR_i,$$

$$A_{i+1} = R_iQ_i + \alpha_iI.$$

In this case A_i, A_{i-1}, are still similar and so have the same eigenvalues.

We illustrate these steps by seeing the output from Program 3. Let A_0 be the matrix

$$A_0 = \begin{bmatrix} 7 & 8 & 6 & 6 \\ 1 & 6 & -1 & -2 \\ 1 & -2 & 5 & -2 \\ 3 & 4 & 3 & 4 \end{bmatrix},$$

whose eigenvalues are $\{10, 1, 4, 7\}$. After step 1, we have

$$A_1 = \begin{bmatrix} 7.00 & -9.65 & -6.15 & -2.24 \\ -3.32 & 4.82 & 4.44 & 1.23 \\ 0.00 & -3.24 & 3.17 & -0.81 \\ 0.00 & 0.00 & 0.40 & 7.01 \end{bmatrix},$$

a Hessenberg matrix. In step 2 we find that after eight iterations we obtain

$$A_3 = \begin{bmatrix} 10.00 & 9.83 & 4.9 & -3.27 \\ 0.00 & 1.00 & 1.83 & 2.72 \\ 0.00 & 0.00 & 4.00 & -1.70 \\ 0.00 & 0.00 & 0.00 & 7.00 \end{bmatrix}.$$

The intermediate matrices are given after Program 3 in Figure 6.5. It should be noted that, although the matrices are different, trace $(A_i) = 22$, $i = 0, 1, \ldots, 9$, since we have mentioned previously that the trace of a matrix is equal to the sum of its eigenvalues. This property turns out to be a very useful check on the implementation of the method on the computer.

The EISPACK programs and IMSL contain subroutines for finding eigenvalues and eigenvectors that are based on the QR method. In particular, these will handle the complex case. In addition, the LINPACK library uses QR in the singular-value-decomposition of a (possibly rectangular) matrix. This latter is applied to least-squares problems (Chapter 10). There are other variations on this theme. For symmetric matrices, the Jacobi method has been widely used.

6.7 EIGENVALUES/EIGENVECTORS IN ITERATIVE METHODS

As an illustration of the importance of eigenvalues in numerical analysis, we conclude this chapter by discussing their application to the convergence of iterative methods for solving a set of linear equations.

In Section 2.10, we considered an example system of equations, $Ax = b$, where

$$A = \begin{bmatrix} 8 & 1 & -1 \\ 1 & -7 & 2 \\ 2 & 1 & 9 \end{bmatrix}, \qquad b = \begin{bmatrix} 8 \\ -4 \\ 12 \end{bmatrix}.$$

In Section 2.10, we showed that both the Jacobi and Gauss–Seidel methods can be written in the form

$$x^{(n+1)} = Gx^{(n)} = b' - Bx^{(n)}. \tag{6.22}$$

For the Jacobi method applied to the example,

$$B = \begin{bmatrix} 0.0 & 0.125 & -0.125 \\ -0.143 & 0.0 & -0.286 \\ 0.222 & 0.111 & 0.0 \end{bmatrix}, \tag{6.23}$$

and for Gauss–Seidel,

$$B = \begin{bmatrix} 0.0 & 0.125 & -0.125 \\ 0.0 & 0.018 & -0.304 \\ 0.0 & -0.030 & 0.062 \end{bmatrix}. \tag{6.24}$$

We now will show that whether an iterative method converges is determined by the eigenvalues of its B matrix, and, further, that the rate of convergence is also related to their magnitude.

The general formulation for iteration is $x^{(n+1)} = b' - Bx^{(n)}$. If an iterative method converges, then $x^{(n)} \to \bar{x}$, where the symbol $\bar{x}$ represents the solution. Hence $A\bar{x} = b$. Eq. (6.22) becomes, for $x = \bar{x}$,

$$\bar{x} = b' - B\bar{x}.$$

Let $e^{(n)}$ be the error of the nth iteration:

$$e^{(n)} = x^{(n)} - \bar{x}.$$

When the method converges (and it certainly does if the matrix A is diagonally dominant), $e^{(n)} \to 0$, the zero vector, as n increases. Using Eg. (6.22), we have

$$e^{(n+1)} = -Be^{(n)} = B^2 e^{(n-1)} = \cdots = (-B)^{n+1} e^{(0)}.$$

If it should be that $B^n \to 0$, the zero matrix, obviously $e^{(n)} \to 0$. In linear algebra it is shown that any square matrix B can be written as $B = UDU^{-1}$. If the eigenvalues of B are distinct, then D is a diagonal matrix with the eigenvalues of B on its diagonal. (If some of the eigenvalues of B are repeated, then D may be triangular rather than diagonal. Our argument holds in either case.)

We have

$$B = UDU^{-1}, \qquad B^2 = UD^2U^{-1}, \qquad B^3 = UD^3U^{-1}, \qquad \ldots, \qquad B^n = UD^nU^{-1}.$$

If all the eigenvalues of B (which are the elements on the diagonal of D) have magnitudes less than 1, it is clear that D^n will approach the zero matrix. This implies that $B^n \to 0$-matrix, $e^{(n)} \to 0$-vector, and $x^{(n)} \to \bar{x}$. In other words, if all the eigenvalues of B, the iteration matrix, are less than 1 in magnitude, the iteration will always converge (even if the system is not diagonally dominant). It is also easy to see that the rate of convergence is determined by the largest eigenvalue in magnitude of B, since this controls how rapidly B^n approaches the zero matrix.

We now look at the eigenvalues of the B matrices in Eqs. (6.23) and (6.24). For the Jacobi matrix (6.23), we find that two are complex and one is real:

$$\lambda_1 = 0.036 + 0.285i, \qquad \lambda_2 = 0.036 - 0.285i, \qquad \lambda_3 = -0.072.$$

The largest magnitude is $(0.036^2 + 0.285^2)^{1/2} = 0.287$. Since all of its eigenvalues are less than 1 in magnitude, Jacobi converges for the example problem. (This confirms the demonstration in Section 2.10.)

For the Gauss–Seidel B matrix (Eq. (6.24)), all the eigenvalues are real:

$$\lambda_1 = 0, \qquad \lambda_2 = -0.058, \qquad \lambda_3 = 0.137,$$

of which the largest magnitude obviously is 0.137. Therefore the Gauss–Seidel also converges (again confirming the demonstration). It converges faster than Jacobi because its eigenvalue of largest magnitude is smaller. The more superficial argument about the relative rates of convergence given in Section 2.10 is now put on a sound basis.

We used the IMSL subroutine EIGRF to obtain these eigenvalues. EIGRF is useful to find the eigenvalues and eigenvectors of a matrix whose components are all real numbers.

6.8 CHAPTER SUMMARY

Can you do all of the following? You can if you really understand Chapter 6.

1. Solve boundary-value problems using either the shooting method or through a set of equations.

2. Explain when no more than two trials are needed when using the shooting method. You can outline an efficient procedure to use when more than two trials are required.

3. Apply the Rayleigh–Ritz method to a second-order differential equation with homogeneous boundary conditions.

4. Explain what is meant by the term *characteristic-value problem* and solve simple examples of them.

5. Find the largest and smallest eigenvalues of a matrix and the associated eigenvectors.

6. Outline the arguments that relate convergence of an iterative method to the eigenvalues of a particular matrix.

7. Use computer programs that employ the shooting method for a boundary-value problem and the power method for eigenvalues. You should be able to adapt the former to handle derivative end conditions.

SELECTED READINGS FOR CHAPTER 6

Burnett (1987); Stewart (1973); Vichnevetsky (1981); Wilkinson (1965).

6.9 COMPUTER PROGRAMS

Program 1 (Fig. 6.3) is an example of how a boundary-value problem can be solved by a computer. In it, a second-order boundary-value problem is solved, using SUBROUTINE RKSYST to integrate the differential equation. Two preliminary solutions are obtained, employing two guesses for the initial slope of the function. From these, an improved estimate of the initial slope is computed by interpolation; the process of interpolation based on the last two values is iterated until the calculated value for the function at the end of the interval closely approximates the desired value. The program is a straightforward implementation of the technique of Section 6.1.

If one wishes to program the difference-equation method, the differential equation would be approximated by a system of finite-difference equations and the solution to this system would be obtained by the subroutines of Chapter 2. In case the boundary-value problem is nonlinear, some iterative method normally would be required.

A subroutine POWER, which finds the largest eigenvalue of an $N \times N$ matrix (together with its corresponding eigenvector), is given in Program 2 (Fig. 6.4). To use it, the matrix and an initial estimate of the eigenvector are passed as parameters. When no estimate of the eigenvector is known, one usually sends a vector that has each element equal to unity, although there are some matrices for which this will not produce convergence. In such cases, one should call the subroutine with a different starting vector, although it must be remembered that the power method fails for some matrices.

In subroutine POWER, the starting vector is multiplied successively by the matrix in each iteration, with the product of each multiplication being normalized. When the method converges, the normalized product vectors approach the eigenvector and the normalization factors approach the eigenvalue. The subroutine terminates iterations when successive values of the normalization factor differ by less than a stated tolerance value. Normalization is done by dividing each element of the vector by the largest element, so the normalized vectors always have unity as their largest element.

Program 3 (Fig. 6.5) implements the QR algorithm given in Stewart (1973), namely, Algorithms 1.1 and 3.5. This program can find the real eigenvalues of a matrix. The input values are A, the real matrix; N, the size of the matrix; TOL, the value for determining when a subdiagonal element is close enough to 0; and MAXIT, the maximum number of iterations. SUBROUTINE HESSN returns the matrix that is similar to the original matrix and that is in Hessenberg form. SUBROUTINE QR is the main one that generates a sequence of matrices that converge to a triangular one or reaches MAXIT. For an $N \times N$ matrix, an additional two rows are used to store values. With some modifications one can generate the eigenvectors as well or make use of SUBROUTINE POWER and shifting. For output we have (1) the original matrix, (2) the Hessenberg matrix similar to the original matrix, (3) the iterates on the Hessenberg matrix as it converges through similarity matrices to an upper-triangular matrix whose diagonal elements are the eigenvalues of the original matrix, and (4) the eigenvalues listed. One can verify that the trace for each matrix equals 22 since similar matrices have the same eigenvalues.

```
      PROGRAM BNDARY(INPUT,OUTPUT)
C
C     THIS PROGRAM SOLVES A NON-LINEAR SECOND ORDER BOUNDARY VALUE
C PROBLEM BY THE SHOOTING METHOD. IT EMPLOYS RKSYST TO SOLVE THE
C EQUATIONS WITH TWO ASSUMED VALUES OF THE INITIAL SLOPE.
C
C SUBROUTINE RKSYST IS BASED ON THE SIMPLE FOURTH ORDER RUNGE-
C KUTTA METHOD. IT WOULD BE EASY TO REPLACE IT WITH SUBROUTINE
C RKFSYS IN CHAPTER 5.
C
C ------------------------------------------------------------------------
C
C     THE EQUATION IS : D2X/DT2 + (3X)(DX/DT) + X = T   X(1)=2,X(3)=7
C
C     IT IS REWRITTEN AS TWO FIRST ORDER EQUATIONS :
C
C       DX1/DT = X2             X1(1) = 2
C       DX2/DT = T - X1 - 3(X1)(X2)        X2(1) = ?
C
C     WHERE X1 = X AND X2 = DX/DT
C
C ------------------------------------------------------------------------
C
C     THE PROGRAM INTERPOLATES FROM THE INITIAL RESULTS TO FIND A
C BETTER ESTIMATE OF THE UNKNOWN INITIAL VALUE OF DX/DT = X2(1).
C ITERATIONS ARE CONTINUED UNTIL SUCCESSIVE VALUES OF X(3) EQUAL 7
C WITHIN A SPECIFIED TOLERANCE. IN THE PROGRAM, VARIABLES X1 & X2 ARE
C THE FIRST AND SECOND COMPONENTS OF THE VECTOR X0.
C
C ------------------------------------------------------------------------
C
      REAL X0(2),XEND(2),XWRK(4,2),F(2),T0,TSTART,H,G1,G2
     +    ,TOL,XSTART,D,R1,R2
      INTEGER I,ITER
      EXTERNAL DERIVS
C
C ------------------------------------------------------------------------
C
C INITIALIZE THE VARIABLES WITH A DATA STATEMENT
C
      DATA H,N,G1,G2,TOL/0.1,2,-1.0,1.0,.001/
      DATA TSTART,XSTART,D/1.0,2.0,7.0/
C
C ------------------------------------------------------------------------
C
C DO THE INTEGRATION ONCE.
C
      X0(1) = XSTART
      T0 = TSTART
      X0(2) = G1
      DO 10 I = 1,20
        CALL RKSYST(DERIVS,T0,H,X0,XEND,XWRK,F,N)
C
C STEP UP THE VARIABLES FOR THE NEXT INTERVAL.
C
        X0(1) = XEND(1)
        X0(2) = XEND(2)
        T0 = T0 + H
   10 CONTINUE
C
C ------------------------------------------------------------------------
```

Figure 6.3 Program 1.

Figure 6.3   (*continued*)

```
C
C  SAVE THE FIRST RESULT FOR EXTRAPOLATING.
C
       R1 = X0(1)
C
C  DO THE INTEGRATION AGAIN WITH THE SECOND ESTIMATE OF DX/DT.
C
       T0 = TSTART
       X0(1) = XSTART
       X0(2) = G2
       DO 20 I = 1,20
          CALL RKSYST(DERIVS,T0,H,X0,XEND,XWRK,F,N)
          X0(1) = XEND(1)
          X0(2) = XEND(2)
          T0 = T0 + H
   20 CONTINUE
C
C  -----------------------------------------------------------------------
C
C  SAVE THE SECOND CALCULATED VALUE.
C
       R2 = X0(1)
C
C  -----------------------------------------------------------------------
C
C  NOW WE EXTRAPOLATE. THEN REPEAT THE ABOVE CALCULATIONS A MAXIMUM
C  OF 20 TIMES, OR UNTIL WE MATCH DESIRED VALUE OF X1(3).
C
       DO 40 ITER = 1,20
          PRINT 205, G1,G2,  R1,R2
          IF ( ABS(R2-D) .LE. TOL ) GO TO 99
          T0 = TSTART
          X0(1) = XSTART
          X0(2) = G1 + (G2 - G1)/(R2 - R1)*(D - R1)
          G1 = G2
          G2 = X0(2)
          R1 = R2
          DO 30 I = 1,20
             CALL RKSYST(DERIVS,T0,H,X0,XEND,XWRK,F,N)
             X0(1) = XEND(1)
             X0(2) = XEND(2)
             T0 = T0 + H
   30     CONTINUE
          R2 = X0(1)
   40 CONTINUE
C
C  -----------------------------------------------------------------------
C
C  WHEN WE HAVE A NORMAL TERMINATION OF THE LOOP, WE DIDN'T CONVERGE.
C  PRINT A MESSAGE, THEN THE FINAL VALUE.
C
       PRINT 200
       GO TO 100
C
C  WE DID CONVERGE IF WE COME HERE. PRINT MESSAGE AND VALUE.
C
   99 PRINT 201, ITER
  100 PRINT 202, R2,G2
C
C  -----------------------------------------------------------------------
C
C  NOW RECOMPUTE AND WRITE OUT THE LAST SET OF COMPUTATIONS.
```

Figure 6.3 *(continued)*

```
C
         T0 = TSTART
         X0(1) = XSTART
         X0(2) = G2
         PRINT 203, T0,X0
         DO 50 I = 1,20
           CALL RKSYST(DERIVS,T0,H,X0,XEND,XWRK,F,N)
           X0(1) = XEND(1)
           X0(2) = XEND(2)
           T0 = T0 + H
           PRINT 204, T0,X0
    50 CONTINUE
C
C  ----------------------------------------------------------------------
C
  200 FORMAT(/' WE DID NOT MEET TOLERANCE CRITERION IN 20 ',
      +        'ITERATIONS. FINAL VALUE AT END OF INTERVAL WAS ')
  201 FORMAT(/' TOLERANCE CRITERION MET IN ',I3,' ITERATIONS.',
      +        ' FINAL VALUE AT END OF INTERVAL WAS ')
  202 FORMAT(/20X,F7.4,' USING INITIAL SLOPE VALUE OF ',F7.4)
  203 FORMAT(/' LAST COMPUTATIONS WERE '//6X,'TIME',5X,
      +        'X1 VALUE' , 5X,
      +        'X2 VALUE'//7X,F3.1,6X,F7.4,6X,F7.4)
  204 FORMAT(7X,F3.1,6X,F7.4,6X,F7.4)
  205 FORMAT(/' GUESSES ARE: ',2F8.4,';   FINAL VALUES ARE: ',2F8.4)
C
         STOP
         END
C
C  ----------------------------------------------------------------------
C
         SUBROUTINE DERIVS(X,T,F,N)
C
C  ----------------------------------------------------------------------
C
C
C     SUBROUTINE DERIVS :
C                          THIS SUBROUTINE COMPUTES DERIVATIVE VALUES
C     FOR THE R-K ROUTINE.
C
C  ----------------------------------------------------------------------
C
         REAL X(N),F(N),T
         INTEGER N
C
C  ----------------------------------------------------------------------
C
         F(1) = X(2)
         F(2) = T - X(1) - 3.0*X(1)*X(2)
         RETURN
         END
         SUBROUTINE RKSYST(DERIVS,T0,H,X0,XEND,XWRK,F,N)
C
C  ----------------------------------------------------------------------
C
C
C     SUBROUTINE RKSYST :
C                          THIS SUBROUTINE SOLVES A SYSTEM OF N FIRST
C     ORDER DIFFERENTIAL EQUATIONS BY THE RUNGE-KUTTA METHOD. THE
C     EQUATIONS ARE OF THE FORM  DX1/DT = F1(X,T) , DX2/DT = F2(X,T),ETC.
C
C  ----------------------------------------------------------------------
C
C     PARAMETERS ARE :
```

Figure 6.3 (*continued*)

```
C
C          DERIVS - A SUBROUTINE THAT COMPUTES VALUES OF THE N DERIVATIVES.
C                   IT MUST BE DECLARED EXTERNAL BY THE CALLER. IT IS
C                   INVOKED BY THE STATEMENT :
C                                          CALL DERIVS(X,T,F,N)
C          T0     - THE INITIAL VALUE OF INDEPENDENT VARIABLE
C          H      - THE INCREMENT TO T, THE STEP SIZE
C          X0     - THE ARRAY THAT HOLDS THE INITIAL VALUES OF THE FUNCTIONS
C          XEND   - AN ARRAY THAT RETURNS THE FINAL VALUES OF THE FUNCTIONS
C          XWRK   - AN ARRAY USED TO HOLD INTERMEDIATE VALUES DURING THE
C                   COMPUTATION. IT MUST BE DIMENSIONED OF SIZE 4 X N IN
C                   THE MAIN PROGRAM.
C          N      - THE NUMBER OF EQUATIONS IN THE SYSTEM BEING SOLVED
C          F      - AN ARRAY THAT HOLDS VALUES OF THE DERIVATIVES
C
C     -------------------------------------------------------------------
C
      REAL X0(N),XEND(N),XWRK(4,N),F(N),H,T0
      INTEGER I,N
C
C     -------------------------------------------------------------------
C
C GET FIRST ESTIMATE OF THE DELTA X'S
C
      CALL DERIVS(X0,T0,F,N)
      DO 10 I = 1,N
         XWRK(1,I) = H * F(I)
         XEND(I) = X0(I) + XWRK(1,I)/2.0
   10 CONTINUE
C
C     -------------------------------------------------------------------
C
C GET THE SECOND ESTIMATE. THE XEND VECTOR HOLDS THE X VALUES.
C
      CALL DERIVS(XEND,T0+H/2.0,F,N)
      DO 20 I = 1,N
         XWRK(2,I) = H * F(I)
         XEND(I) = X0(I) + XWRK(2,I)/2.0
   20 CONTINUE
C
C     -------------------------------------------------------------------
C
C REPEAT FOR THIRD ESTIMATE
C
      CALL DERIVS(XEND,T0+H/2.0,F,N)
      DO 30 I = 1,N
         XWRK(3,I) = H * F(I)
         XEND(I) = X0(I) + XWRK(3,I)
   30 CONTINUE
C
C     -------------------------------------------------------------------
C
C NOW GET LAST ESTIMATE
C
      CALL DERIVS(XEND,T0+H,F,N)
      DO 40 I = 1,N
         XWRK(4,I) = H * F(I)
   40 CONTINUE
C
C     -------------------------------------------------------------------
C
C WE COMPUTE THE X AT THE END OF THE INTERVAL FROM A WEIGHTED AVERAGE
C OF THE FOUR ESTIMATES, THEN RETURN.
```

Figure 6.3 (*continued*)

```
C
      DO 50 I = 1,N
         XEND(I) = X0(I) + ( XWRK(1,I) + 2.0*XWRK(2,I) +
     +            2.0*XWRK(3,I) + XWRK(4,I) ) / 6.0
   50 CONTINUE
      RETURN
      END
C
C         DEFINE THE FUNCTIONS OF THE PROBLEM IN TERMS OF THE
C         F'S AND THE X'S.
C
```

```
           OUTPUT FOR PROGRAM 1


GUESSES ARE:  -1.0000  1.0000;   FINAL VALUES ARE:   1.8623  2.1153

GUESSES ARE:   1.0000 39.6079;   FINAL VALUES ARE:   2.1153  5.1279

GUESSES ARE:  39.6079 63.5995;   FINAL VALUES ARE:   5.1279  6.3780

GUESSES ARE:  63.5995 75.5366;   FINAL VALUES ARE:   6.3780  6.9346

GUESSES ARE:  75.5366 76.9387;   FINAL VALUES ARE:   6.9346  6.9986

GUESSES ARE:  76.9387 76.9694;   FINAL VALUES ARE:   6.9986  7.0000

TOLERANCE CRITERION MET IN   6 ITERATIONS. FINAL VALUE AT END OF INTERVAL WAS
           7.0000 USING INITIAL SLOPE VALUE OF 76.9694

LAST COMPUTATIONS WERE

      TIME    X1 VALUE    X2 VALUE

      1.0      2.0000      76.9694
      1.1      6.6453      14.9031
      1.2      7.1277       5.8486
      1.3      7.2761       2.3699
      1.4      7.3204        .8684
      1.5      7.3237        .2186
      1.6      7.3101      -.0579
      1.7      7.2895      -.1726
      1.8      7.2663      -.2178
      1.9      7.2422      -.2338
      2.0      7.2180      -.2374
      2.1      7.1941      -.2360
      2.2      7.1706      -.2324
      2.3      7.1475      -.2280
      2.4      7.1249      -.2232
      2.5      7.1028      -.2183
```

Figure 6.3   (*continued*)

```
        2.6        7.0813        -.2133
        2.7        7.0602        -.2082
        2.8        7.0396        -.2032
        2.9        7.0195        -.1980
        3.0        7.0000        -.1929
```

```
      SUBROUTINE POWER(A,X,C,TOL,NLIMIT,N,NDIM,XWRK)
C
C    ----------------------------------------------------------------------
C
C    SUBROUTINE POWER :
C                       THIS SUBROUTINE COMPUTES THE LARGEST EIGEN-
C    VALUE AND ITS CORRESPONDING EIGENVECTOR BY THE POWER METHOD.
C
C    ----------------------------------------------------------------------
C
C    PARAMETERS ARE :
C
C    A      - AN N X N MATRIX WHOSE EIGENS ARE BEING DETERMINED
C    X      - ESTIMATE FOR THE EIGENVECTOR USED TO BEGIN ITERATIONS.
C             IF NO APPROXIMATION TO THIS IS KNOWN, THE USUAL CHOICE
C             IS A VECTOR WITH ALL COMPONENTS EQUAL TO UNITY. X ALSO
C             RETURNS THE FINAL EIGENVECTOR TO THE CALLER.
C    C      - RETURNS THE VALUE OF THE EIGENVALUE
C    TOL    - TOLERANCE VALUE USED TO DETERMINE CONVERGENCE. ITERATIONS
C             CONTINUE UNTIL SUCCESSIVE ESTIMATES OF THE EIGENVALUE
C             ARE THE SAME WITHIN TOL IN VALUE.
C    NLIMIT - LIMIT TO THE NUMBER OF ITERATIONS IF NOT CONVERGENT
C    N      - SIZE OF THE MATRIX AND THE VECTOR X
C    NDIM   - FIRST DIMENSION OF MATRIX A IN THE CALLING PROGRAM
C    XWRK   - VECTOR USED TO STORE INTERMEDIATE VALUES. MUST BE
C             DIMENSIONED TO HOLD AT LEAST N ELEMENTS IN THE
C             MAIN PROGRAM.
C
C    ----------------------------------------------------------------------
C
      REAL A(NDIM,N),X(N),XWRK(N),C,TOL,SAVE
      INTEGER N,NDIM,NLIMIT,IROW,ITER,JCOL
C
C    ----------------------------------------------------------------------
C
C BEGIN THE ITERATIONS. GET THE PRODUCT OF A AND X, THEN NORMALIZE.
C THE NORMALIZATION FACTORS SHOULD CONVERGE TO THE EIGENVALUE, C, ON
C REPEATED MULTIPLICATIONS. WE STORE THE CURRENT VALUE OF C FOR
C COMPARISON WITH THE NEXT VALUE TO TEST CONVERGENCE. TO BEGIN, WE
C MAKE SAVE = 0.
C
      SAVE = 0.0
      DO 50 ITER = 1,NLIMIT
        DO 20 IROW = 1,N
          XWRK(IROW) = 0.0
          DO 10 JCOL = 1,N
            XWRK(IROW) = XWRK(IROW) + A(IROW,JCOL)*X(JCOL)
   10     CONTINUE
   20   CONTINUE
```

Figure 6.4   Program 2.

Figure 6.4 (*continued*)

```
C
C    -----------------------------------------------------------------------
C
C    FIND THE LARGEST ELEMENT OF THE PRODUCT VECTOR FOR NORMALIZING.
C
        C = 0.0
        DO 30 IROW = 1,N
          IF ( ABS(C) .LT. ABS(XWRK(IROW)) )  C = XWRK(IROW)
30      CONTINUE
C
C    -----------------------------------------------------------------------
C
C    NOW NORMALIZE THE PRODUCT VECTOR AND PUT INTO X FOR NEXT ITERATION.
C
        DO 40 IROW = 1,N
          X(IROW) = XWRK(IROW) / C
40      CONTINUE
C
C    -----------------------------------------------------------------------
C
C    SEE IF TOLERANCE IS MET. IF SO, WE ARE DONE. IF NOT, WE CONTINUE
C    THE ITERATIONS.
C
        IF ( ABS(C - SAVE) .LE. TOL ) RETURN
        SAVE = C
50   CONTINUE
C
C    -----------------------------------------------------------------------
C
C    IF WE DO NOT MEET THE TOLERANCE FOR CONVERGENCE, PRINT A MESSAGE
C    AND RETURN LAST VALUES CALCULATED.
C
      PRINT 200, NLIMIT
200   FORMAT(/' CONVERGENCE NOT REACHED IN ',I5,' ITERATIONS')
      RETURN
      END
```

```
      PROGRAM PQR(INPUT,OUTPUT)
C    ---------------------------------------------------------------
C
C        THIS PROGRAM IMPLEMENTS THE QR ALGORITHM.
C          INPUT: A,N,TOL
C            A - THE MATRIX WHOSE EIGENVALUES WE WANT.
C            N - THE SIZE OF THE MATRIX, A
C            TOL - THE VALUE ON THE SUBDIAGONAL ELEMENT WE SHALL TAKE
C                  AS 0
C
C        TWO SUBROUTINES ARE CALLED:
C          SUBROUTINE HESSN - PRODUCES THE HESSENBERG MATRIX
C          SUBROUTINE QR    - THE ACTUAL QR METHOD IS APPLIED TO THE
C                             HESSENBERG MATRIX
C
C        NOTE: THIS PROGRAM AS GIVEN HERE JUST PRODUCES THE REAL
C              EIGENVALUES OF THE INPUT MATRIX: A.
```

Figure 6.5 Program 3.

Figure 6.5   (*continued*)

```
C
C            ALGORITHMS 1.1 AND 3.5 OF STEWART ARE IMPLEMENTED IN THE
C            SUBROUTINES.
C
C     ----------------------------------------------------------------
C
      REAL A(12,10), TOL
      INTEGER N,I,J
C
      DATA N,NDIM,TOL/4,12,1.0E-6/
      DATA MAXIT/50/
      DATA A/7,1,1,3,8*0,8,6,-2,4,8*0, 6,-1,5,3,8*0,
     +       6,-2,-2,4,80*0/
C
C
C            ECHO ORIGINAL MATRIX
C
      PRINT 100
      CALL PRINTA(A,NDIM,N)
      PRINT 200
      CALL HESSN(A,NDIM,N)
      PRINT 300
      CALL PRINTA(A,NDIM,N)
      PRINT 200
      CALL QR(A,NDIM,N,MAXIT,TOL)
      PRINT 400
      CALL PRINTA(A,NDIM,N)
C
C            PRINT OUT THE EIGENVALUES
C
      PRINT 500
      PRINT '(8F10.4)', (A(I,I), I = 1,N)
C
100      FORMAT(///T10,'THE ORIGINAL MATRIX IS: '//)
200      FORMAT(///)
300      FORMAT(///T10,'THE HESSENBERG MATRIX IS: '//)
400      FORMAT(///T10,'THE FINAL MATRIX IS: '//)
500      FORMAT(////T10,'THE EIGENVALUES ARE: '/)
      STOP
      END
C
C     -----------------------------------------------------------------
C
C     PRINT OUT THE N X N MATRIX
C
C     -----------------------------------------------------------------
C
      SUBROUTINE PRINTA(A,NDIM,N)
      REAL A(NDIM,N)
      INTEGER NDIM,N,    I,J
C
      DO 10 I = 1,N
          PRINT 100, (A(I,J), J = 1,N)
10        CONTINUE
100      FORMAT(8F10.4)
      RETURN
      END
C
      SUBROUTINE HESSN(A,NDIM,N)
      REAL A(NDIM,N), V(10), SIGMA, ETA, SUM, PI
      INTEGER N,NLESS1, NLESS2, K, KPLUS1, I, J
C
C     -----------------------------------------------------------------
```

Figure 6.5   *(continued)*

```
C
C          SUBROUTINE HESSN
C             INPUT : A THE ORIGINAL MATRIX A
C             OUTPUT : THE MATRIX SIMILAR TO A BUT IN HESSENBERG FORM.
C
C          SIMPLE REFLECTORS ARE GENERATED, V(1), V(2), ... , V(N-2) SO
C          THAT A IS TRANSFORMED TO:
C
C             V(N-2)*...*V(2)*V(1)*A*V(1)*V(2)*...*V(N-2)
C
C    ------------------------------------------------------------------
C
       NLESS1 = N-1
       NPLUS1 = N+1
       NPLUS2 = N+2
       NLESS2 = N-2
       DO 10 K = 1,NLESS2
           KPLUS1 = K+1
           ETA = ABS(A(KPLUS1,K))
C
C
           DO 20 I = KPLUS1,N
           IF (ETA .LT. ABS(A(I,K))) ETA = ABS(A(I,K))
20         CONTINUE
           IF (ETA .NE. 0.0) THEN
           SUM = 0.0
           DO 30 I = KPLUS1,N
               V(I) = A(I,K)/ETA
               A(I,K) = V(I)
               SUM = SUM + V(I)**2
30             CONTINUE
C
           IF (V(KPLUS1) .NE. 0.0) THEN
               SIGMA= SQRT(SUM)*V(KPLUS1)/ABS(V(KPLUS1))
               ENDIF
           V(KPLUS1) = V(KPLUS1) + SIGMA
           PI = SIGMA*V(KPLUS1)
           A(NPLUS1,K) = PI
C
C    ------------------------------------------------------------------
C
C       PREMULTIPLY BY THE APPROPRIATE V
C
C    ------------------------------------------------------------------
C
           DO 40 J = K+1,N
               SUM = 0.0
               DO 50 I = KPLUS1,N
                   SUM = SUM + V(I)*A(I,J)
50                 CONTINUE
               RHO = SUM/PI
             DO 60 I = KPLUS1,N
                 A(I,J) = A(I,J) - RHO*V(I)
60               CONTINUE
40         CONTINUE
C
C    ------------------------------------------------------------------
C
C          POSTMULTIPLY BY THE SAME V REFLECTOR
C
C    ------------------------------------------------------------------
C
```

Figure 6.5   (*continued*)

```
            DO 80 I = 1,N
               SUM = 0.0
               DO 90 J = K+1,N
                  SUM = SUM + A(I,J)*V(J)
90             CONTINUE
               RHO = SUM/PI
               DO 100 J = KPLUS1,N
                  A(I,J) = A(I,J) - RHO*V(J)
100            CONTINUE
80          CONTINUE
C
C
C

            A(NPLUS2,K) = V(KPLUS1)
            A(KPLUS1,K) = -ETA*SIGMA
         ELSE
            A(NPLUS1,K) = 0.0
         ENDIF
10       CONTINUE
C
C -------------------------------------------------------------
C
C          ZERO OUT ENTRIES BELOW SUB-DIAGONAL
C
C -------------------------------------------------------------
C
         DO 110 J = 1,NLESS2
            DO 120 I = J+2,N
               A(I,J) = 0.0
120         CONTINUE
110      CONTINUE
         RETURN
         END


         SUBROUTINE QR(A,NDIM,N,MAXIT,TOL)
         REAL A(NDIM,N),GAMMA(10), ETA, ALFA,BETA,DELTA,
     +      KAPPA, SIGMA(10)
C
         INTEGER NDIM,N,NSTORE,  MAXIT,KPLUS1,KLESS1,I,J,K
C
         NSTORE = N
         DO 10 I = 1,MAXIT
C
C -------------------------------------------------------------
C
C          WE CHOOSE OUR CURRENT SHIFT VALUE SO THAT WE
C          ARE WORKING ON: A - K*I
C
C -------------------------------------------------------------
C
            KAPPA = A(N,N)
            A(1,1) = A(1,1) - KAPPA
C
C
            DO 20 K = 1,N
               KPLUS1 = K+1
               IF (K.NE.N) THEN
C
C -------------------------------------------------------------
C
C          DETERMINE ROTATION PARAMETERS: GAMMA, SIGMA, XNU
```

Figure 6.5 (*continued*)

```
C
C      ------------------------------------------------------------------
C
               ETA = AMAX1( ABS(A(K,K)),ABS(A(KPLUS1,K)) )
               ALFA = A(K,K)/ETA
               BETA = A(KPLUS1,K)/ETA
               DELTA = SQRT(ALFA*ALFA + BETA*BETA)
               GAMMA(K) = ALFA/DELTA
               SIGMA(K) = BETA/DELTA
               XNU = ETA*DELTA
C
C

               A(K,K) = XNU
               A(KPLUS1,K) = 0.0
               A(KPLUS1,KPLUS1) = A(KPLUS1,KPLUS1) - KAPPA
C
C      ------------------------------------------------------------------
C
C
C         PREMULTIPLY ON RIGHT
C
C      ------------------------------------------------------------------
C
            DO 30 J = KPLUS1,N
            AKJ = A(K,J)
            A(K,J) = GAMMA(K)*A(K,J) + SIGMA(K)*A(KPLUS1,J)
            A(KPLUS1,J) = GAMMA(K)*A(KPLUS1,J) - SIGMA(K)*AKJ
 30         CONTINUE
          ENDIF
C
C      ------------------------------------------------------------------
C
C
C         POSTMULTIPLY
C
C      ------------------------------------------------------------------
C
          IF (K .NE. 1) THEN
            DO 40 J = 1,K
               KLESS1 = K-1
               AIKL1 = A(J,KLESS1)
               A(J,KLESS1) = GAMMA(KLESS1)*A(J,KLESS1)
     +                        + SIGMA(KLESS1)*A(J,K)
               A(J,K) = GAMMA(KLESS1)*A(J,K) - SIGMA(KLESS1)*AIKL1
 40         CONTINUE
          A(KLESS1,KLESS1) = A(KLESS1,KLESS1) + KAPPA
          ENDIF
C
 20      CONTINUE
         A(N,N) = A(N,N) + KAPPA
C
C      ------------------------------------------------------------------
C
C
C         THE MATRIX SIZE IS REDUCED BY 1 AFTER EACH ITERATION IN
C         ORDER TO BE MORE EFFICIENT.
C
C         THE EIGENVALUES OF THE MATRIX APPEAR ON THE DIAGONAL ENTRIES OF
C         A STARTING AT N.
C
C      ------------------------------------------------------------------
C
         IF (ABS(A(N,N-1)) .LT.TOL) N = N-1
         IF (N .EQ. 1) THEN
          N = NSTORE
          RETURN
```

Figure 6.5   (*continued*)

```
              END IF
C
C   ----------------------------------------------------------------
C
C             PRINT OUT INTERMEDIATE MATRIX
C
C   ----------------------------------------------------------------
C
              PRINT 100, I
              CALL PRINTA(A,NDIM,NSTORE)
100           FORMAT(///T10,'AT ITERATION ',I2,4X,'THE MATRIX IS: '//)
10            CONTINUE
         RETURN
         END

              OUTPUT FOR PROGRAM 3


         THE ORIGINAL MATRIX IS:


     7.0000      8.0000      6.0000      6.0000
     1.0000      6.0000     -1.0000     -2.0000
     1.0000     -2.0000      5.0000     -2.0000
     3.0000      4.0000      3.0000      4.0000



         THE HESSENBERG MATRIX IS:


     7.0000     -9.6484     -6.1545     -2.2431
    -3.3166      4.8182      4.4403      1.2343
      .0000     -3.2423      3.1692      -.8097
      .0000       .0000       .3963      7.0126



         AT ITERATION  1    THE MATRIX IS:


     4.7690     -1.7275      4.8182     -2.3993
   -10.1705      4.7939     -5.9875      3.8708
      .0000      -.5489      5.4359       .4239
      .0000       .0000      -.0027      7.0012



         AT ITERATION  2    THE MATRIX IS:


     2.3015    -10.7577      -.4193     -3.2668
    -1.3015      9.2441      4.2663      2.7199
      .0000     -1.6038      3.4545     -1.6958
      .0000       .0000       .0000      7.0000



         AT ITERATION  3    THE MATRIX IS:


      .2056       .9400      2.8150     -3.2668
    -8.9815     10.9236     -4.1503      2.7199
      .0000       .0561      3.8708     -1.6958
      .0000       .0000       .0000      7.0000
```

Figure 6.5   (*continued*)

```
            AT ITERATION  4    THE MATRIX IS:


    6.5804   -11.4904    -3.1399    -3.2668
   -1.6623     4.4183     4.1916     2.7199
     .0000     -.0041     4.0013    -1.6958
     .0000      .0000      .0000     7.0000



            AT ITERATION  5    THE MATRIX IS:


   11.9350     6.6492     4.9054    -3.2668
   -3.1823     -.9350     1.8256     2.7199
     .0000      .0000     4.0000    -1.6958
     .0000      .0000      .0000     7.0000



            AT ITERATION  6    THE MATRIX IS:


   10.3857     9.4484     4.9054    -3.2668
    -.3831      .6143     1.8256     2.7199
     .0000      .0000     4.0000    -1.6958
     .0000      .0000      .0000     7.0000



            AT ITERATION  7    THE MATRIX IS:


   10.0158     9.8170     4.9054    -3.2668
    -.0145      .9842     1.8256     2.7199
     .0000      .0000     4.0000    -1.6958
     .0000      .0000      .0000     7.0000



            AT ITERATION  8    THE MATRIX IS:


   10.0000     9.8315     4.9054    -3.2668
     .0000     1.0000     1.8256     2.7199
     .0000      .0000     4.0000    -1.6958
     .0000      .0000      .0000     7.0000



         THE FINAL MATRIX IS:


   10.0000     9.8315     4.9054    -3.2668
     .0000     1.0000     1.8256     2.7199
     .0000      .0000     4.0000    -1.6958
     .0000      .0000      .0000     7.0000



         THE EIGENVALUES ARE:

   10.0000     1.0000     4.0000     7.0000
```

EXERCISES

Section 6.1

▶ 1. Solve the boundary-value problem

$$y'' + xy' - 3y = 4.2x,$$
$$y(0) = 0, \, y(1) = 1.9,$$

by the shooting method, assuming two values of the initial slope, which is near unity. Use $h = 0.25$ and use the modified Euler method, carrying three decimals. Compare to the analytical solution $y = x^3 + 0.9x$.

2. Show that the solution to Exercise 1 found by integrating with an initial slope interpolated from the results of the first two trials can be obtained by interpolating from the trial solutions themselves.

3. In Exercise 1, the exact (analytical) value of the initial slope is 0.9. Using $h = 0.25$, you probably find that $y(1) = 1.9$ is achieved with a different value of $y'(0)$. Explain this discrepancy.

4. The shooting method can work backward through the interval as well as forward. Solve Exercise 1 by moving backward from $x = 1$ (Δx is then negative), and use an assumed value for the slope at $x = 1$.

▶ 5. Use the shooting method to solve

$$y'' - yy' = e^x,$$
$$y(0) = 1, \, y(1) = -1.$$

Use the computer programs of Chapter 5, or another one that you have written, to solve the problem with assumed values of the initial slope. Use $h = 0.2$.

6. Exercise 5 solves a nonlinear boundary-value problem. The solution will not be exact because the solution of a differential equation by a numerical method has errors that depend on the step size. If the step size is reduced, the errors will decrease. Vary the step size in Exercise 5 until you believe that the value of y at $x = 0.5$ is accurate to within 5×10^{-5}. Justify your belief that you have attained this accuracy.

7. Use the shooting method to solve

$$\frac{d^3y}{dt^3} + 10\frac{dy}{dt} - 5y^3 = t^2 - ty,$$

$$y(0) = 0, \quad y(2) = -1, \quad y''(2) = 0.$$

Section 6.2

8. Given the boundary problem

$$\frac{d^2y}{d\theta^2} + y = 0, \quad y(0) = 0, \quad y\left(\frac{\pi}{2}\right) = 1.$$

a) Normalize to the interval [0, 1] by an appropriate change of variable.
b) Solve the normalized problem by the finite-difference method, replacing the derivative by a difference quotient of error $O(h^2)$, and then solving the set of equations. Use $h = 0.25$.
▶ c) Solve the original equation (without normalizing) by finite difference. Take $\Delta\theta = \frac{\pi}{8}$.
d) Compare your solutions to the analytical solution $y = \sin\theta$.
e) Repeat (b) with $h = 0.5$, and extrapolate the value at the midpoint.

9. Solve the boundary-value problem

$$y'' + xy' - xy = 2x, \qquad y(0) = 1, \qquad y(1) = 0.$$

Take $h = 0.2$.

10. Solve Exercise 1 by replacing the derivatives with finite-difference quotients.

11. Solve Exercise 5 by rewriting the differential equation as a set of difference equations.

12. Solve Exercise 7 through a set of difference equations.

Section 6.3

▶13. Repeat Exercise 8, except for the boundary-value problem with derivative conditions:

$$\frac{d^2y}{d\theta^2} + y = 0, \qquad y'(0) = 0, \qquad y'\left(\frac{\pi}{2}\right) = 1.$$

In part (d), compare to $y = -\cos\theta$.

14. Solve

$$\frac{d^2y}{d\theta^2} + y = 0, \qquad y'(0) + y(0) = 1, \qquad y'\left(\frac{\pi}{2}\right) + y\left(\frac{\pi}{2}\right) = 0.$$

Subdivide the interval into four equal parts.

15. The most general form of conditions that can be specified for a second-order boundary-value problem is a linear combination of the function and its derivative at both ends of the interval. Set up the equations to solve

$$y'' - xy' + x^2y = x^3, \qquad \begin{cases} y(0) + y'(0) - y(1) - y'(1) = 1, \\ y(0) + y'(0) + y(1) + y'(1) = 2. \end{cases}$$

Use $h = \frac{1}{3}$.

16. Solve the third-order boundary-value problem

$$y''' - y' = e^x, \qquad y(0) = 0, \qquad y(1) = 1, \qquad y'(1) = 0.$$

Use $h = 0.2$. Approximate the third derivative in terms of the average third central difference

$$y_0''' = \frac{y_2 - 2y_1 + 2y_{-1} - y_{-2}}{2h^3} + O(h^2)$$

at $x = 0.4, 0.6, 0.8$. Using the derivative condition will eliminate the assumed function value at $x = 1.2$, but we are short one equation. Obtain this by writing the equation at $x = 0.2$ using the unsymmetrical approximation for y''':

$$y_0''' = \frac{-y_3 + 6y_2 - 12y_1 + 10y_0 - 3y_{-1}}{2h^3} + O(h^2).$$

▶17. Solve the fourth-order problem

$$y^{iv} - y''' + y = x^2, \qquad y(0) = 0, \qquad y'(0) = 0, \qquad y(1) = 2, \qquad y'(1) = 0.$$

Use symmetrical expressions for the derivatives.

18. Solve the differential equation in Exercise 17 subject to the conditions

$$y(0) = y'(0) = y''(0) = 0, \qquad y(1) = 2.$$

Unsymmetrical expressions may not be required!

19. Solve the nonlinear problem

$$y'' = 2 - \frac{4y^2}{\sin^2 x}, \qquad y'(1) = 0.9093, \qquad y(2) = 0.8268,$$

by a set of difference equations. Compare to the analytical solution, $y = \sin^2 x$.

Section 6.4

20. Show that the integrand of Eq. (6.12) becomes Eq. (6.11) by use of the Euler–Lagrange condition. This means that Eq. (6.12) is the functional for any second-order boundary value problem of the form

$$y'' + Q(x)y = F(x),$$

when the boundary conditions are homogeneous: $y(a) = 0$, $y(b) = 0$.

▶21. Use the Rayleigh–Ritz method to approximate the solution of

$$y'' = 2x, \qquad y(0) = 0, \qquad y(1) = 0,$$

using a quadratic in x as the approximating function. Compare to the analytical solution by graphing both the approximate and analytical solutions.

22. a) Repeat Exercise 21 except use, for the approximating function,

$$ax(x - 1) + bx^2(x - 1).$$

Show that this reproduces the analytical solutions.

▶ b) Suppose the boundary conditions for part (a) are $y(0) = 1$, $y(1) = 3$. How can you modify part (a) to give the solution?

c) Another form of a cubic approximation that meets the boundary conditions is

$$ax(x - 1) + bx(x - 1)^2.$$

Is the analytical solution reproduced when this is used?

23. Show that the use of a cubic approximating function in the first example of Section 6.4 reproduces the analytical solution.

24. Solve Exercise 21 by the piecewise application of Rayleigh–Ritz using first two, then four intervals. Compare these solutions to the analytical answer by graphing them.

25. Solve by the piecewise Rayleigh–Ritz method, using five intervals:

$$y'' + \frac{y}{2 - x^2} = 1 + x^2, \qquad y(1) = 2, \qquad y(4) = -2.$$

Compare to solutions by the shooting method (use Runge–Kutta–Fehlberg) and by finite-difference approximations of the derivative.

Section 6.5

26. Consider the characteristic-value problem with k restricted to real values:

$$\frac{d^2y}{dx^2} - k^2y = 0, \qquad y(0) = 0, \qquad y(1) = 0.$$

a) Show analytically that there is no solution except the trivial solution $y = 0$.

b) Show, by setting up the set of difference equations corresponding to the differential equation with $h = 0.2$, that there are no real values of k for which a solution to the difference equations exists.

▶27. Consider the characteristic-value problem

$$y'' + 2y' + k^2 y = 0, \qquad y(0) = y(1) = 0.$$

Find the principal eigenvalue and compare to $k = \pm\sqrt{1 + \pi^2} = \pm3.297$.

a) Use $h = \frac{1}{2}$.

b) Use $h = \frac{1}{3}$.

c) Use $h = \frac{1}{4}$.

d) Assuming errors are proportional to h^2, extrapolate from parts (a) and (c), then from parts (b) and (c), to get improved values for the principal eigenvalue.

28. Find the principal eigenvalue of

$$y'' + k^2 x^2 y = 0, \qquad y(0) = y(1) = 0.$$

29. Using the principal value, $k = 3.297$, in Exercise 27, find y as a function of x over the interval $[0, 1]$. This function is the corresponding eigenfunction.

▶30. The second eigenvalue in Exercise 27 is $k = \sqrt{1 + 4\pi^2} = 6.3623$. Find the corresponding eigenfunction.

Section 6.6

31. Find the dominant eigenvalue and the corresponding eigenvector by the power method:

a) $\begin{bmatrix} 2 & 0 \\ 1 & 11 \end{bmatrix}$ ▶b) $\begin{bmatrix} 1 & 2 \\ 5 & 4 \end{bmatrix}$ c) $\begin{bmatrix} 1 & 1 \\ 1 & -1 \end{bmatrix}$

d) $\begin{bmatrix} 4 & 1 & 0 \\ 0 & 2 & 1 \\ 0 & 0 & -1 \end{bmatrix}$ ▶e) $\begin{bmatrix} 1 & 1 & 2 \\ 0 & 1 & 3 \\ 1 & 1 & 1 \end{bmatrix}$

(In part (c) the two eigenvalues are equal but of opposite sign.)

32. Let

$$A = \begin{bmatrix} -5 & 2 & 0 \\ 2 & -9 & -1 \\ 3 & -1 & 7 \end{bmatrix}, \qquad B = \begin{bmatrix} -6 + 2i & -1 & -2i \\ -3 & 7 + i & 0 \\ 1 & -2 & 4 - i \end{bmatrix}.$$

a) Locate the eigenvalues of these matrices using Gerschgorin's theorems.

b) From part (a) determine whether either of the matrices is nonsingular.

33. Use iteration to find all the eigenvalues of the matrices for Exercise 27(a), (b), (c).

34. Invert the matrices in Exercise 31, then use the power method to get the smallest eigenvalues of the original matrices. Repeat, except avoid the inversion by using the LU decomposition of the matrix. Compare the effort in the two cases.

▶35. Find all the eigenvalues of matrix A in Exercise 32. Then invert the matrix and again find the eigenvalues. Show that the two sets of eigenvalues are reciprocals of each other.

36. After finding the dominant eigenvalue in Exercise 35, subtract that value from each of the diagonal elements and use the power method. Compare the value obtained with the second largest eigenvalue as determined in Exercise 35.

37. Using your results from Exercises 31 and 34, find the general solutions to the following systems of differential equations.

a) $\dot{x} = 2x,$ b) $\dot{x} = x + 2y,$
   $\dot{y} = x + 11y.$ $\dot{y} = 5x + 4y.$

Section 6.7

38. Let

$$T = \begin{bmatrix} \lambda & 0 \\ x & \lambda \end{bmatrix}.$$

Show that

$$T^n = \begin{bmatrix} \lambda^n & 0 \\ n\lambda^{n-1}x & \lambda^n \end{bmatrix}.$$

From this infer that $T^n \to$ 0-matrix if $|\lambda| < 1$.

39. Let

$$T = \begin{bmatrix} \lambda_1 & 0 & 0 \\ a & \lambda_2 & 0 \\ b & c & \lambda_3 \end{bmatrix} \quad \text{and}$$

$$\bar{T} = \begin{bmatrix} \lambda & 0 & 0 \\ a & \lambda & 0 \\ b & c & \lambda \end{bmatrix} \quad \text{where}$$

$$\lambda = \text{Max} \{ |\lambda_1|, |\lambda_2|, |\lambda_3| \}.$$

Write $\bar{T} = L + D$, where

$$L = \begin{bmatrix} 0 & 0 & 0 \\ a & 0 & 0 \\ b & c & 0 \end{bmatrix}, \quad D = \begin{bmatrix} \lambda & 0 & 0 \\ 0 & \lambda & 0 \\ 0 & 0 & \lambda \end{bmatrix}.$$

a) Show that $L^3 =$ 0-matrix.
b) Since $LD = DL$, show that

$$\bar{T}^n = (L + D)^n$$
$$= \binom{n}{2}L^2 D^{n-2} + \binom{n}{1}LD^{n-1} + D^n,$$

using the binomial formula. From this we see that $\bar{T}^n \to$ 0-matrix and also T^n.
c) Prove this for any general triangular matrix T, where $|T_{i,i}| < 1$ for $i = 1, 2, \ldots, n$.

40. Suppose, for $Ax = b$, that

$$A = \begin{bmatrix} 4 & 3 & 2 \\ 2 & 3 & 4 \\ 2 & 4 & a \end{bmatrix}.$$

What is the smallest value of a for which iteration will converge

a) using Jacobi?
b) using Gauss–Seidel?

APPLIED PROBLEMS AND PROJECTS

41. If a cantilever beam of length L, which bends due to a uniform load of w lb/ft, is also subject to an axial force P at its free end (see Fig. 6.6), the equation of its elastic curve is

$$EI\frac{dy^2}{dx^2} = Py - \frac{wx^2}{2}.$$

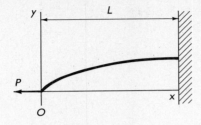

Figure 6.6

For this equation, the origin O has been taken at the free end. At the point $x = L$, $dy/dx = 0$; at $(0, 0)$, $y = 0$. Solve this boundary-value problem by the shooting method for a $2'' \times 4'' \times 10$-ft wooden beam for which $E = 12 \times 10^5$ lb/in^2. Find y versus x when the beam has the $4''$ dimension vertical with $w = 25$ lb/ft and a tension force of $P = 500$ lb. Also solve for the deflections if the beam is turned so that the $4''$ dimension is horizontal.

42. Solve Problem 41 by replacing the differential equation by difference equations. Compare the solutions by each method.

43. In Problem 41, y'' is used although the radius of curvature should be employed in the equation:

$$EI\frac{y''}{[1 + (y')^2]^{3/2}} = Py - \frac{wx^2}{2}.$$

If the deflection of the beam is small, the difference is negligible, but in the second part of Problem 41 at least, this is not true. Furthermore, if there is considerable bending of the beam, the horizontal distance from the origin to the wall is less than L, the original length of the beam. Solve Problem 41 taking these factors into account, and determine by how much the deflections differ from those previously calculated.

44. A cylindrical pipe has a hot fluid flowing through it. Since the pressure is very high, the walls of the pipe are thick. For such a situation, the differential equation that relates temperatures in the metal wall to radial distance is

$$r\frac{d^2u}{dr^2} + \frac{du}{dr} = 0,$$

where

$r =$ radial distance from the centerline,
$u =$ temperature.

Solve for the temperatures within a pipe whose inner radius is 1 cm and whose outer radius is 2 cm if the fluid is at 540°C and the temperature of the outer circumference is 20°C.

45. The pipe in Problem 44 is insulated to reduce the heat loss. The insulation used has properties such that the gradient du/dr at the outer circumference is proportional to the difference in temperatures from the outer wall to the surroundings:

$$\left.\frac{du}{dr}\right|_{r=2} = 0.083[u(2) - 20].$$

Solve Problem 44 with this boundary condition.

46. When air flows at a velocity of u_∞ past a flat plate placed horizontally to the flow, experiments show that the air velocity at the surface of the plate is zero and that a boundary layer within which the velocities are less than u_∞ builds up as flow progresses along the plate (Fig. 6.7).

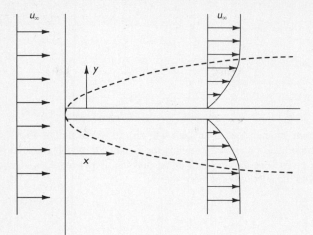

Figure 6.7

This is a typical problem of interest to aeronautical engineers. To solve for the velocities within the boundary layer, a stream function f is defined for which

$$\frac{d^3f}{dz^3} + \frac{f}{2}\frac{d^2f}{dz^2} = 0,$$

where z is a dimensionless distance involving x, y, u_∞, and the viscosity of air. This is a boundary-value problem because $f = 0$ and $f' = 0$ at $z = 0$, while $f' = 1$ at $z = \infty$. In solving this problem, one does not use $z = \infty$ as a boundary; rather, the integration is taken to a value of z large enough so that there is an insignificant change when it is increased. Solve this problem to find f and df/dz as functions of z.

47. A simple spring–mass system obeys the equation

$$\frac{d^2y}{dt^2} + \alpha^2 y = 0,$$

where the positive constant α^2 equals k/m, the ratio of the spring constant to the mass. If it is known that $y = 0$ at $t = 0$ and again at $t = 1.26$ sec (so that its period is 2.52 sec), this is a characteristic-value problem that has a solution for only certain values of α. These characteristic values are discussed in Section 6.5 for this equation. Suppose, however, that the spring is not an ideal one with constant k, but that the spring force and elongation vary with y according to the equation $k = b(1 - y^{0.1})$. If it is still true that $y = 0$ at $t = 0$ and also at $t = 1.26$, is this a characteristic-value problem requiring that the ratio k/m be at certain fixed values in order to have a nontrivial solution? If it is, find these values.

48. A shaft rotates in fixed bearings (Fig. 6.8) so that at these points $y = 0$ and $dy/dx = 0$. The governing equation is

$$EI\frac{d^4y}{dx^4} = \frac{w}{g}\omega^2 y,$$

where E is Young's modulus, I is the moment of inertia of the cross-sectional area of the shaft, y is the displacement from the horizontal, x is the distance measured from the point midway between the bearings, w is the weight of the shaft per unit length, g is the acceleration of gravity, $2L$ is the length of the shaft between the bearings, and ω is the rotational speed.

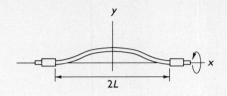

At certain critical speeds, the shaft will distort from a straight line. These critical speeds are very important to know in designing the machine of which the shaft is a part. For parameter values given here in consistent units, find the first two values of critical rotational speeds. Also find the curve $y(x)$ at these speeds.

$$E = 2 \times 10^6 \text{ lb/in}^2 \qquad g = 386.4 \text{ in/sec}^2$$
$$I = 0.46 \text{ in}^4 \qquad L = 16 \text{ in}$$
$$w = 0.025 \text{ lb/in}$$

7

Numerical Solution of Partial-Differential Equations

7.0 CONTENTS OF THIS CHAPTER

Chapter 7 introduces you to the important topic of partial-differential equations in which there are two or more independent variables. The class of equations we study here finds application in steady-state heat flow, fluid flow, electrical potential distributions, and so on. We apply two numerical methods to these problems: finite differences and finite elements.

7.1 EQUILIBRIUM TEMPERATURES IN A HEATED SLAB

Presents a typical two-dimensional problem that illustrates one practical application of an elliptic partial-differential equation.

7.2 EQUATION FOR STEADY-STATE HEAT FLOW

Derives the equation that governs this situation, introduces you to some terminology, and shows how partial-differential equations are classified.

7.3 REPRESENTATION AS A DIFFERENCE EQUATION

Utilizes finite-difference approximations (Chapter 4) to change from a differential equation to a difference equation.

7.4 LAPLACE'S EQUATION ON A RECTANGULAR REGION

Illustrates how the finite-difference method leads to a system of algebraic equations that can be solved to give approximate values for the potential (temperature, concentration, voltage, or other potential quantity) at points within a rectangular region. The question of accuracy is explored.

7.5 ITERATIVE METHODS FOR LAPLACE'S EQUATION

Applies iterative techniques (Chapter 3) to solve the system of equations that is often too large for elimination methods. You are introduced to the S.O.R. method, a means to accelerate the convergence of the iterations.

7.6 THE POISSON EQUATION

Is another form of elliptic equation that applies to problems such as torsion, membrane displacement, and so on. These can also be handled with the finite-difference technique.

7.7 EIGENVALUES AND S.O.R.

Applies matrices and their properties to the analysis of acceleration via S.O.R.

7.8 DERIVATIVE BOUNDARY CONDITIONS

Extends the finite-difference method to handle a larger class of boundary conditions.

7.9 IRREGULAR REGIONS AND NONRECTANGULAR GRIDS

Tackles real-world situations that do not fit neatly into the simpler approach taken in previous sections of the chapter.

7.10 THE LAPLACIAN OPERATOR IN THREE DIMENSIONS

Shows that, in principle, the finite-difference method extends readily to three-dimensional problems but points out that the size of the system to be solved may easily exceed the available computing power.

7.11 MATRIX PATTERNS, SPARSENESS, AND THE A.D.I. METHOD

Examines the structure of the matrices that result from finite-difference approximations and introduces you to a newer method that offers significant advantages over the previous technique.

7.12 FINITE-ELEMENT METHOD

Takes an entirely new approach to solving elliptic partial-differential equations. It offers an alternative to finite differences that is often preferred, especially when the region is irregular.

7.13 CHAPTER SUMMARY

Lets you systematically review the topics of Chapter 7.

7.14 COMPUTER PROGRAMS

Implements the finite-difference method using both S.O.R. and A.D.I. techniques.

7.1 EQUILIBRIUM TEMPERATURES IN A HEATED SLAB

Many important scientific and engineering problems fall into the field of partial-differential equations. Here is a situation that is typical, in which the temperatures are a function of the coordinates of position of the point in question.

A piece of metal is 12 in. × 3 in. × 6 ft. Three feet of the slab is kept inside a furnace but half of the slab protrudes (see Fig. 7.1). In order to decrease heat losses to the air, the protruding half is covered with a 1-in. thickness of insulation. If the furnace is maintained at 950°F, will all points of the metal reach a temperature of 800°F or higher,

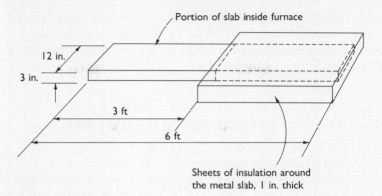

Portion of slab inside furnace

12 in.

3 in.

3 ft

6 ft

Sheets of insulation around
the metal slab, 1 in. thick

Figure 7.1

in spite of heat loss through the insulation? Such a question might arise in heat-treating the slab when the only furnace available to heat the metal is too small to contain the whole slab. The sketch in Fig. 7.1 will explain the problem.

You will recognize that this is a boundary-value problem: The temperature is known for the part of the slab inside the furnace, and something is known about the temperature gradients for the insulated part. The problem differs from those discussed in the previous chapter in that more than one independent variable is involved in specifying the coordinates of points within the slab. (Specifically, these variables are the x-, y-, and z-coordinates.)

This chapter presents methods to solve this problem. These methods are generally referred to as finite-difference methods. However, since the 1950s an equally important method has been developed for solving elliptic partial-differential equations, namely, the finite-element method. This method is very suitable for computer coding and for handling very complex figures and irregularities. The finite-element method breaks the complex region into simple subregions such as triangles and pyramids and applies solution techniques to these simpler parts (see Vichnevetsky, 1981).

There are well-known codes that use either or both of these methods. Among them are ELLPACK, which uses a variety of methods including those presented in this chapter, and TWODEPEP (IMSL), which uses the finite-element method for two-dimensional problems. However, in this chapter we consider mainly finite-difference methods; we do show how finite elements can be used.

Many physical phenomena are a function of more than one independent variable and must be represented by a partial-differential equation, by which term we mean an equation involving partial derivatives. Most scientific problems have mathematical models that are second-order equations, with the highest order of derivative being the second. If u is a function of the two independent variables x and y, there are three second-order partial derivatives:

$$\frac{\partial^2 u}{\partial x^2}, \qquad \frac{\partial^2 u}{\partial x\, \partial y}, \qquad \frac{\partial^2 u}{\partial y^2}.$$

(We will treat only functions for which the order of differentiation is unimportant, so $\partial^2 u/\partial y\, \partial x = \partial^2 u/\partial x\, \partial y$.) Depending on the values of the coefficients of the second-derivative terms, partial-differential equations are classified as elliptic, parabolic, or hyperbolic. The most important distinctions to us are the kinds of problems and the nature of boundary conditions that lead to one type of equation or another. Steady-state potential distribution problems fall in the class of elliptic equations, which we discuss first. The equilibrium temperatures attained by the metal piece protruding from the furnace is such a potential distribution problem.

7.2 EQUATION FOR STEADY-STATE HEAT FLOW

We derive the relationship for temperature u as a function of the two space variables x and y for the equilibrium temperature distribution on a flat plate (Fig. 7.2). Consider the element of the plate whose surface area is $dx\,dy$. We assume that heat flows only in the x- and y-directions and not in the perpendicular direction. If the plate is very thin, or if the upper and lower surfaces are both well insulated, the physical situation will agree with our assumption. Let t be the thickness of the plate.

Heat flows at a rate proportional to the cross-sectional area, to the temperature gradient ($\partial u/\partial x$ or $\partial u/\partial y$), and to the thermal conductivity k, which we will assume constant at all points. The flow of heat is from high to low temperature, of course, meaning opposite to the direction of increasing gradient. We use a minus sign in the equation to account for this:

Rate of heat flow into element at $x = x_0$, in the x-direction:

$$-k(t\,dy)\frac{\partial u}{\partial x}.$$

The gradient at $x_0 + dx$ is the gradient at x_0 plus the increment in gradient over the distance dx:

Gradient at $x_0 + dx$:

$$\frac{\partial u}{\partial x} + \frac{\partial}{\partial x}\left(\frac{\partial u}{\partial x}\right)\,dx.$$

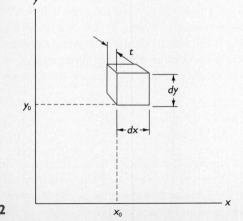

Figure 7.2

Rate of heat flow out of element at $x = x_0 + dx$:

$$-k(t\,dy)\left[\frac{\partial u}{\partial x} + \frac{\partial}{\partial x}\left(\frac{\partial u}{\partial x}\right)dx\right].$$

Net rate of heat flow into element in x-direction:

$$-k(t\,dy)\left[\frac{\partial u}{\partial x} - \left(\frac{\partial u}{\partial x} + \frac{\partial}{\partial x}\left(\frac{\partial u}{\partial x}\right)dx\right)\right] = kt\,dx\,dy\,\frac{\partial^2 u}{\partial x^2}.$$

Similarly, in the y-direction we have the following.

Net rate of heat flow into element in y-direction:

$$-k(t\,dx)\left[\frac{\partial u}{\partial y} - \left(\frac{\partial u}{\partial y} + \frac{\partial}{\partial y}\left(\frac{\partial u}{\partial y}\right)dy\right)\right] = kt\,dx\,dy\,\frac{\partial^2 u}{\partial y^2}.$$

The total heat flowing into the elemental volume by conduction is the sum of these net flows in the x- and y-directions. If there were heat generated within the element, this would be added to the heat entering by conduction. The sum must be equal to the rate of heat lost by other mechanisms, or else there will be a buildup of heat within the element, causing its temperature to increase with time. In this chapter we will consider only the case where the temperatures do not change with time, the *steady-state* case.

If there is equilibrium as to temperature distribution (that is, steady state), the total rate of heat flow into the element plus heat generated must be zero. Hence,

$$kt(dx)(dy)\left(\frac{\partial^2 u}{\partial x^2} + \frac{\partial^2 u}{\partial y^2}\right) + Qt(dx)(dy) = 0,$$

where Q is the rate of heat generation per unit volume. (Obviously, if there is heat generation, there must be heat flow *from* the element by conduction that just balances it.) Q will often be a function of x and y.

If $Q = 0$, we have

$$kt\,dx\,dy\left(\frac{\partial^2 u}{\partial x^2} + \frac{\partial^2 u}{\partial y^2}\right) = 0,$$

$$\frac{\partial^2 u}{\partial x^2} + \frac{\partial^2 u}{\partial y^2} = \nabla^2 u = 0. \qquad (7.1)$$

The operator

$$\nabla^2 = \left(\frac{\partial^2}{\partial x^2} + \frac{\partial^2}{\partial y^2}\right)$$

is called the *Laplacian*, and Eq. (7.1) is called *Laplace's equation*. For three-dimensional heat-flow problems, we would have, analogously,

$$\nabla^2 u = \left(\frac{\partial^2}{\partial x^2} + \frac{\partial^2}{\partial y^2} + \frac{\partial^2}{\partial z^2}\right)u = 0. \qquad (7.2)$$

Equation (7.2), which has been derived with reference to heat flow, applies as well to steady-state diffusion problems (where u is now concentration of material) and to electrical-potential distribution (where u is electromotive force); in fact, Laplace's equation holds for the steady-state distribution of the potential of any quantity where the rate of flow is proportional to the gradient, and where the proportionality constant does not vary with position or the value of u. In our examples, we will generally use terminology corresponding to heat flow as being more closely related to the average student's everyday experience.

With Laplace's equation, we assume that no heat is being generated at points in the plate, as by electric heaters embedded in it, nor does any removal take place, as by cooling coils or other means. In the presence of such "sources" or "sinks," we would not equate the net flow into the element to zero, as in Eq. (7.1), but the net flow into the element of volume would equal the net rate of heat removal from the element, if steady state applies. Assuming this removal rate to be a function of the location of the element in the xy-plane, $f(x, y)$, we would have, with Q equal to the rate of heat generation per unit area,

$$\nabla^2 u = -\frac{1}{k}Q(x, y) = f(x, y). \qquad (7.3)$$

This equation is called *Poisson's equation*. Our numerical methods of solving elliptic differential equations apply equally well to both Laplace's and Poisson's equations. Analytical methods, however, find the solution of Poisson's equations or even Laplace's equations with complicated boundary conditions considerably more difficult. These two equations include most of the physical applications of elliptic partial-differential equations.

The distinction as to elliptic, parabolic, or hyperbolic for second-order partial-differential equations depends on the coefficients of the second-derivative terms. We can write any such second-order equation (in two independent variables) as

$$A\frac{\partial^2 u}{\partial x^2} + B\frac{\partial^2 u}{\partial x \, \partial y} + C\frac{\partial^2 u}{\partial y^2} + D\left(x, y, u, \frac{\partial u}{\partial x}, \frac{\partial u}{\partial y}\right) = 0.$$

Depending on the value of $B^2 - 4AC$, we classify the equation as

$$\text{Elliptic, if } B^2 - 4AC < 0;$$

$$\text{Parabolic, if } B^2 - 4AC = 0;$$

$$\text{Hyperbolic, if } B^2 - 4AC > 0.$$

If the coefficients A, B, and C are functions of x, y, and/or u, the equation may change from one classification to another at various points in the domain.

For Laplace's and Poisson's equations, $B = 0$, $A = C = 1$, so these are always elliptic. We study the other types in later chapters.

Equation (7.2) or Eq. (7.3) describes how u varies within the interior of a closed region. The function u is determined on the boundaries of the region by boundary conditions. The boundary conditions may specify the value of u at all points on the boundary, or some combination of the potential and the normal derivative.

7.3 REPRESENTATION AS A DIFFERENCE EQUATION

One scheme for solving all kinds of partial-differential equations is to replace the derivatives by difference quotients, converting the equation to a difference equation.* We then write the difference equation corresponding to each point at the intersections (nodes) of a gridwork that subdivides the region of interest at which the function values are unknown. Solving these equations simultaneously gives values for the function at each node that approximate the true values. We begin with the two-dimensional case.

Approximating derivatives by difference quotients was the subject matter of Chapter 4. They may also be derived by the method of undetermined coefficients (Appendix B). We rederive the few relations we need independently.

Let $h = \Delta x =$ equal spacing of gridwork in the x-direction (see Fig. 7.3). We assume that the function $f(x)$ has a continuous fourth derivative. By Taylor series,

$$f(x_n + h) = f(x_n) + f'(x_n)h + \frac{f''(x_n)}{2}h^2 + \frac{f'''(x_n)}{6}h^3 + \frac{f^{iv}(\xi_1)}{24}h^4,$$

$$x_n < \xi_1 < x_n + h,$$

$$f(x_n - h) = f(x_n) - f'(x_n)h + \frac{f''(x_n)}{2}h^2 - \frac{f'''(x_n)}{6}h^3 + \frac{f^{iv}(\xi_2)}{24}h^4,$$

$$x_n - h < \xi_2 < x_n.$$

It follows that

$$\frac{f(x_n + h) - 2f(x_n) + f(x_n - h)}{h^2} = f''(x_n) + \frac{f^{iv}(\xi)}{12}h^2, \qquad x_n - h < \xi < x_n + h.$$

---

*This is exactly the same as the method of solving a boundary-value problem through replacing it by difference equations, as discussed in Chapter 6.

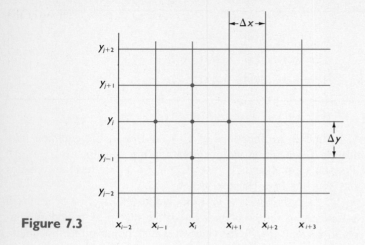

Figure 7.3

A subscript notation is convenient:

$$\frac{f_{n+1} - 2f_n + f_{n-1}}{h^2} = f''_n + O(h^2).$$

(7.4)

In Eq. (7 4) the subscripts on f indicate the x-values at which it is evaluated. The order relation $O(h^2)$ signifies that the error approaches proportionality to h^2 as $h \to 0$. Similarly, the first derivative is approximated,

$$\frac{f(x_n + h) - f(x_n - h)}{2h} = f'(x_n) + \frac{f'''(\xi)}{6}h^2, \qquad x_n - h < \xi < x_n + h,$$

$$\frac{f_{n+1} - f_{n-1}}{2h} = f'_n + O(h^2).$$

(7.5)

(The first derivative could also be approximated by the forward or backward difference, but this would have an error of $O(h)$. We prefer the more accurate central-difference approximation.)

When f is a function of both x and y, we get the second partial derivative with respect to x, $\partial^2 u/\partial x^2$, by holding y constant and evaluating the function at three points where x equals x_n, $x_n + h$, and $x_n - h$. The partial derivative $\partial^2 u/\partial y^2$ is similarly computed, holding x constant. We require that fourth derivatives with respect to both variables exist.

To solve the Laplace equation on a region in the xy-plane, we subdivide the region with equispaced lines parallel to the x- and y-axes. Consider a portion of the region near (x_i, y_j). (See Fig. 7.3.) We wish to approximate

$$\nabla^2 u = \frac{\partial^2 u}{\partial x^2} + \frac{\partial^2 u}{\partial y^2} = 0.$$

Replacing the derivatives by difference quotients that approximate the derivatives at the point (x_i, y_j), we get

$$\nabla^2 u(x_i, y_j) = \frac{u(x_{i+1}, y_j) - 2u(x_i, y_j) + u(x_{i-1}, y_j)}{(\Delta x)^2}$$
$$+ \frac{u(x_i, y_{j+1}) - 2u(x_i, y_j) + u(x_i, y_{j-1})}{(\Delta y)^2} = 0.$$

It is convenient to let double subscripts on u indicate the x- and y-values:

$$\nabla^2 u_{i,j} = \frac{u_{i+1,j} - 2u_{i,j} + u_{i-1,j}}{(\Delta x)^2} + \frac{u_{i,j+1} - 2u_{i,j} + u_{i,j-1}}{(\Delta y)^2} = 0.$$

It is common to take $\Delta x = \Delta y = h$, resulting in considerable simplification, so that

$$\nabla^2 u_{i,j} = \frac{1}{h^2}[u_{i+1,j} + u_{i-1,j} + u_{i,j+1} + u_{i,j-1} - 4u_{i,j}] = 0. \quad \textbf{(7.6a)}$$

Note that five points are involved in the relationship of Eq. (7.6a), points to the right, left, above, and below the central point (x_i, y_j). It is convenient to represent the relationship pictorially, where the linear combination of u's is represented symbolically. Equation (7.6a) becomes

$$\nabla^2 u_{i,j} = \frac{1}{h^2}\left\{ \begin{matrix} & 1 & \\ 1 & -4 & 1 \\ & 1 & \end{matrix} \right\} u_{i,j} = 0. \quad \textbf{(7.6b)}$$

The representation of the Laplacian operator by the pictorial operator that represents the linear combination of u-values is fundamental to the finite-difference method for elliptic partial-differential equations. The approximation has $O(h^2)$ error, provided that u is sufficiently smooth. This formula is referred to as the five-point or five-point star formula.

In addition to the five-point formula just given, one can derive the nine-point formula for Laplace's equation by similar methods to get

$$\nabla^2 u_{i,j} = \frac{1}{6h^2}\left[\begin{matrix} 1 & 4 & 1 \\ 4 & -20 & 4 \\ 1 & 4 & 1 \end{matrix} \right] u_{i,j} = 0. \quad \textbf{(7.6c)}$$

In this case the approximation has $O(h^6)$ error, provided that u is sufficiently smooth.

7.4 LAPLACE'S EQUATION ON A RECTANGULAR REGION

A typical steady-state heat-flow problem is the following: A thin steel plate is a 10×20 cm rectangle. If one of the 10-cm edges is held at 100°C and the other three edges are held at 0°C, what are the steady-state temperatures at interior points? For steel, $k = 0.16$ cal/sec $\cdot$ cm^2 $\cdot$ °C/cm. We can state the problem mathematically in this way:

Find $u(x, y)$ such that

$$\frac{\partial^2 u}{\partial x^2} + \frac{\partial^2 u}{\partial y^2} = 0, \tag{7.7}$$

with

$$u(x, 0) = 0,$$

$$u(x, 10) = 0,$$

$$u(0, y) = 0,$$

$$u(20, y) = 100.$$

In this statement of the problem, we imagine one corner of the plate at the origin, with boundary conditions as sketched in Fig. 7.4.

A problem with temperatures known on each boundary is said to have *Dirichlet boundary conditions*. This problem can be solved analytically, giving $u(x, y)$ as an infinite Fourier series, but we will use it to illustrate the numerical technique. We replace the differential equation by a difference equation, as described in the previous section:

$$\frac{1}{h^2}[u_{i+1,j} + u_{i-1,j} + u_{i,j+1} + u_{i,j-1} - 4u_{ij}] = 0,$$

which relates the temperature at the point (x_i, y_j) to the temperatures at four neighboring points, each the distance h away from (x_i, y_j). An approximation to the solution of Eq. (7.7) results when we select a set of such points and find the solution to the set of difference equations that result. Suppose we choose $h = 5$ cm. (In normal practice we would choose a smaller mesh size.) Figure 7.5 shows the notation we use.

The system of equations is

$$\frac{1}{5^2}(0 + 0 + u_2 + 0 - 4u_1) = 0,$$

$$\frac{1}{5^2}(u_1 + 0 + u_3 + 0 - 4u_2) = 0,$$

$$\frac{1}{5^2}(u_2 + 0 + 100 + 0 - 4u_3) = 0. \tag{7.8}$$

The pictorial operator of Section 7.3,

$$\nabla^2 u_{ij} = \frac{1}{h^2}\left\{ \begin{matrix} & 1 & \\ 1 & -4 & 1 \\ & 1 & \end{matrix} \right\} u_{ij} = 0,$$

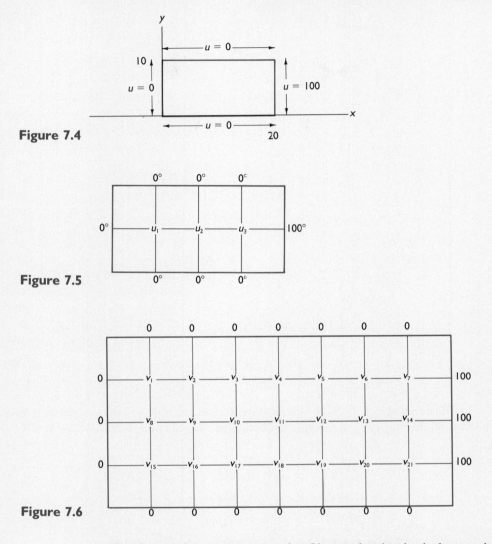

Figure 7.4

Figure 7.5

Figure 7.6

makes the equations very easy to write. Observe that the physical properties of the material do not enter the relation,* and that the constant $1/h^2 = 1/5^2$ can be divided out of Eqs. (7.8).

The solution to the set of equations is easy when there are only three of them:

$$u_1 = 1.786, \quad u_2 = 7.143, \quad u_3 = 26.786.$$

Of course, with such a coarse grid as that shown in Fig. 7.5, we don't expect high accuracy. Redoing the problem with a smaller value for h should improve the precision. Figure 7.6 and Eqs. (7.9a) result when we make $h = 2.5$ cm. (The equations are expressed in matrix form. We use v instead of u for the temperature variable in order to distinguish it from the previous solution.)

---

*If the conductivity varied at different regions of the plate, this would not be true.

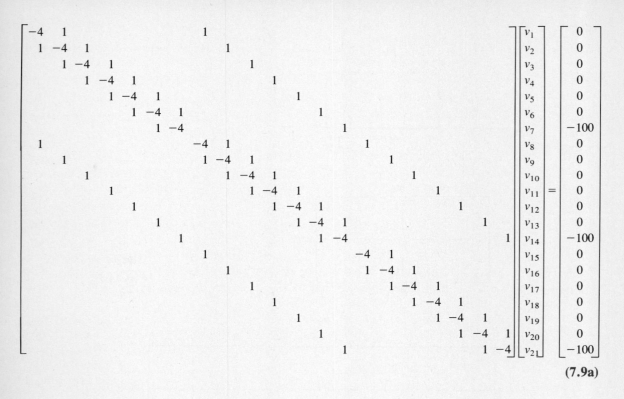

$$(7.9a)$$

We can solve Eqs. (7.9a) by elimination. It turns out that the largest element always occurs on the diagonal, so pivoting isn't required. The solution vector is

$$v_1 = 0.3530, \qquad v_8 = 0.4988, \qquad v_{15} = 0.3530,$$

$$v_2 = 0.9132, \qquad v_9 = 1.2894, \qquad v_{16} = 0.9132,$$

$$v_3 = 2.0103, \qquad v_{10} = 2.8323, \qquad v_{17} = 2.0103,$$

$$v_4 = 4.2957, \qquad v_{11} = 6.0193, \qquad v_{18} = 4.2957,$$

$$v_5 = 9.1531, \qquad v_{12} = 12.6537, \qquad v_{19} = 9.1531,$$

$$v_6 = 19.6631, \qquad v_{13} = 26.2893, \qquad v_{20} = 19.6631,$$

$$v_7 = 43.2101, \qquad v_{14} = 53.1774, \qquad v_{21} = 43.2101.$$

Symmetry in the system is apparent by inspection, and we could have reduced the number of equations by taking this into account. The regular pattern of coefficients in the matrix of Eqs. (7.9a) would have been lost, however.

We now are able to examine the effect of grid size h. Three points are identical in the two sets of equations. We show these results side by side in the first two columns of Table 7.1.

Table 7.1

Point (Fig. 7.5)	From Eq. (7.8), with $h = 5$ cm		From Eq. (7.9a), with $h = 2.5$ cm		From Eq. (7.9b), with $h = 2.5$ cm		Analytical value
	Value	Error	Value	Error	Value	Error	
u_1	1.786	-0.692	1.289	-0.195	1.0938	$+0.0005$	1.0943
u_2	7.143	-1.655	6.019	-0.531	5.4853	$+0.0032$	5.4885
u_3	26.786	-0.692	26.289	-0.195	26.0938	$+0.0006$	26.0944

The answers with $h = 2.5$ cm show significant improvement, as expected. In fact, since the error in approximating the derivatives by central-difference quotients is of $O(h^2)$, we might anticipate that the errors in the solution through sets of difference equations would also vary as h^2; this is roughly true. When we have such knowledge of the effect of h, an extrapolation can be made. When this is done, temperatures within 3% of the analytical values are obtained.

Another way to reduce the error is to use the nine-point formula for the Laplacian, Eq. (7.6c). Doing so gives the matrix equation of Eq. (7.9b); here we assumed values for the corner points equal to the average of adjacent boundary points. The solution to this gives much better accuracy. Table 7.1 shows three of these values.

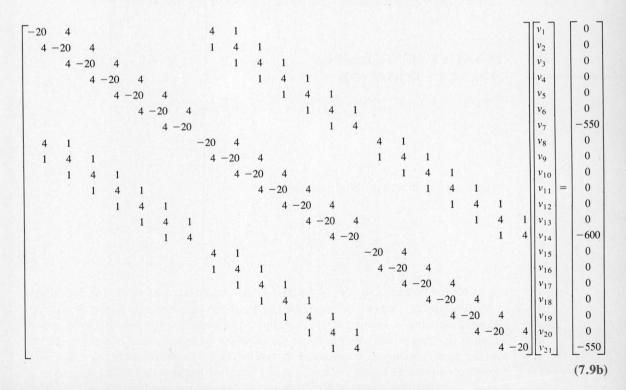

$$(7.9b)$$

In principle, then, we can use the numerical method to solve the steady-state heat-flow equation to any desired accuracy; all we need to do is to continue to make h smaller. In practice, however, this procedure runs into severe difficulties. It is very apparent that the number of equations increases inordinately fast. With $h = 1.25$, we would have 105 discrete interior points; with $h = 0.625$, we have 465, and so on. Storing a matrix with 105 rows and 105 columns would require 105^2 or about 11,000 words (44,000 bytes) of computer memory; with 465 equations we would need 216,000 words (865,000 bytes). Few computer systems allow us such a generous partition, and overlaying memory space from disk storage would be extremely time-consuming. There must be a better way, you are thinking, and there is. Along with memory requirements, we worry about execution times.

The remedy to the problem of overlarge matrices is to take advantage of the sparseness exhibited by Eqs. (7.9a). That system of 21 equations has nonzero terms in only 85 out of 441 elements, or 19%. The larger matrices that result when h is smaller have an even smaller proportion of nonzero terms (4% and 1% with 105 and 465 equations, respectively). Unfortunately, the systems are not tridiagonal. If they were, we could work with only the nonzero terms, as we found in Chapter 2. With banded matrices such as we have here, none of the zero elements outside the bands changes, but the elements between the side bands do fill up with nonzeros. The preferred method for large numbers of points in our mesh is *iteration*, which we discuss next.

7.5 ITERATIVE METHODS FOR LAPLACE'S EQUATION

When the steady-state heat-flow equation in two dimensions,

$$\nabla^2 u = \frac{\partial^2 u}{\partial x^2} + \frac{\partial^2 u}{\partial y^2} = 0,$$

is approximated by the pictorial relation

$$\nabla^2 u = \frac{1}{h^2} \left\{ \begin{matrix} & 1 & \\ 1 & -4 & 1 \\ & 1 & \end{matrix} \right\} u_{ij} = 0,$$

at a set of grid points (x_i, y_i), a set of simultaneous linear equations results, as Eqs. (7.8) and (7.9) illustrate. Proper ordering of the equations gives a *diagonally dominant* system. (Diagonal dominance means that the magnitude of the coefficient on the diagonal is not less than the sum of the magnitudes of the other coefficients in the same row.) As we saw in Chapter 2, such systems can be solved by Gauss–Seidel iteration.

When applied to Laplace's equation, the iterative technique is called *Liebmann's method*. We illustrate this procedure with the same problem as in the last section. The difference equations, when $n = 5$ cm, are

$$-4u_1 + u_2 \qquad = 0,$$
$$u_1 - 4u_2 + u_2 = 0,$$
$$u_2 - 4u_3 = -100. \qquad (7.10)$$

The first step in applying Liebmann's method is to rearrange these equations into a new form by solving each equation for the variable on the diagonal:

$$u_1 = \frac{u_2}{4},$$

$$u_2 = \frac{u_1 + u_3}{4},$$

$$u_3 = \frac{u_2 + 100}{4}. \qquad (7.11)$$

Observe that the pictorial operator can give this rearrangement directly:

$$\left\{ \begin{array}{ccc} & 1 & \\ 1 & -4 & 1 \\ & 1 & \end{array} \right\} u_{ij} = 0 \rightarrow u_{ij} = \frac{u_{i+1,j} + u_{i-1,j} + u_{i,j+1} + u_{i,j-1}}{4}. \qquad (7.12)$$

We solve Eqs. (7.10) by beginning with some initial approximation $\bar{u}$ to u, the solution vector, substituting values from this approximation into the right-hand sides of Eqs. (7.11), and generating new components of $\bar{u}$ that converge to the solution. (Convergence is usually faster if each new component of $\bar{u}$ is used as soon as it is available. The former procedure is called the Jacobi method and the latter is called Gauss–Seidel.)

The better the beginning approximation, the more rapid the convergence, but, because the system is diagonally dominant, even a poor initial vector will eventually converge. Inspection of the problem underlying the equations lets us guess at reasonably good starting values: we know that u_1 will be small, that u_2 will be about four times as large as u_1, and that u_3 will be a little more than one fourth of 100. Suppose we begin with $u_3 = 30$, $u_2 = 7.5$, $u_1 = 2$. Performing the iterations gives the values shown in Table 7.2.

The iterations converge as anticipated and to the same values as in the previous section when we solved the equations by elimination. We terminate the iterations when all components of $\bar{u}$ converge to constant values. For only three equations, iteration is surely more computational effort than elimination, but for large systems there are important advantages.

The most obvious advantage is a greatly reduced demand for computer memory space. In our example problem, if h is 1.25 cm, there are 105 interior points at grid intersections. To store the full 105×105 matrix, we need 44,000 bytes, not even taking into account the right-hand sides and the solution vector. With Liebmann's method, we need to store only the 105 values of u_{ij} plus 44 boundary values; this requires about 600 bytes. The difference in storage requirements is even more striking when $h = 0.625$ cm. The 865,000 bytes for the full matrix is reduced to a very moderate 2300 bytes.

Table 7.2

u_1	u_2	u_3
	Initial values	
2	7.5	30
1.875	7.969	26.992
1.992	7.246	26.812
1.812	7.156	26.789
1.789	7.144	26.786
1.786	7.143	26.786
1.786	7.143	26.786

With large systems there is also a very significant reduction in execution time as well, although the magnitude of this depends on how good a starting vector we can supply, and very strongly on the precision we require. It is customary to terminate the iterations when the largest change in any component of $\bar{u}$ meets a supplied tolerance value. This tolerance should be chosen with consideration for the importance of precision in the final answers (many engineering applications do not need extreme accuracy) and only after determining whether the parameters of the problem (boundary conditions, dimensions, material properties, and so on) are themselves known with great accuracy.

Since the speed of convergence depends on the initial value of $\bar{u}$, it is frequently worthwhile to spend some effort in obtaining a good initial estimate. Some guidelines can be helpful. It is obvious, since the temperature at each point is the *average* of temperatures at surrounding points, that no interior point can have a temperature *greater* than the hottest boundary point, or a temperature *lower* than the coldest. Sometimes the simple expedient of setting all interior points equal to the mean of all the boundary temperatures is used to start iterations. Interior points near cold edges have low temperatures and those near hot edges have higher temperatures. This may let us make good guesses from which to begin iterations. A more refined technique solves the problem with a coarse mesh, getting a quick solution; these values then are used as a basis to fill in initial values at intermediate points.

When our sample problem was solved with $h = 2.5$ (21 interior points), employing the average of boundary values as the starting temperatures, these results were obtained after 21 iterations. (The iterations were stopped when the largest change in any $\bar{u}$ component was less than 0.001.)

$$u_1 = 0.354, \qquad u_8 = 0.500, \qquad u_{15} = 0.354,$$
$$u_2 = 0.914, \qquad u_9 = 1.291, \qquad u_{16} = 0.914,$$
$$u_3 = 2.012, \qquad u_{10} = 2.834, \qquad u_{17} = 2.011,$$
$$u_4 = 4.297, \qquad u_{11} = 6.021, \qquad u_{18} = 4.296,$$
$$u_5 = 9.154, \qquad u_{12} = 12.655, \qquad u_{19} = 9.154,$$
$$u_6 = 19.664, \qquad u_{13} = 26.290, \qquad u_{20} = 19.663,$$
$$u_7 = 43.210, \qquad u_{14} = 53.178, \qquad u_{21} = 43.210.$$

Comparison with the values in Section 7.4 shows that the same results are obtained as when the equations are solved by elimination.

Liebmann's method essentially takes care of the problem of memory requirement for the steady-state heat-flow problem (though three-dimensional problems can still be demanding). The chief drawback in Liebmann's method is slow convergence, which is especially acute when there are a larger number of points, because then each iteration is lengthy and, in addition, more iterations are required to reach a given tolerance.

The relaxation method of Southwell, as discussed in Section 2.11, would be a way of attaining faster convergence in the iterative method. (In fact, Southwell developed his method to solve potential problems.) As pointed out in the discussion of that section, relaxation is not adapted to computer solution of sets of equations. Based on Southwell's technique, the use of an overrelaxation factor can give significantly faster convergence, however. Since we handle each equation in a standard and repetitive order, this method is called *successive overrelaxation*, frequently abbreviated as S.O.R.

To show how successive overrelaxation can be applied to Laplace's equation, we begin with Eq. (7.12), adding superscripts to show that a new value is computed from previous iterates,

$$u_{ij}^{(k+1)} = \frac{u_{i+1,j}^{(k)} + u_{i-1,j}^{(k+1)} + u_{i,j+1}^{(k)} + u_{i,j-1}^{(k+1)}}{4}.$$

We now both add and subtract $u_{ij}^{(k)}$ on the right-hand side, getting

$$u_{ij}^{(k+1)} = u_{ij}^{(k)} + \left[\frac{u_{i+1,j}^{(k)} + u_{i-1,j}^{(k+1)} + u_{i,j+1}^{(k)} + u_{i,j-1}^{(k+1)} - 4u_{ij}^{(k)}}{4} \right]. \qquad (7.13)$$

(The numerator term will be zero when final values, after convergence, are used. The term in brackets is what Southwell called the "residual," which he "relaxed" to zero by changes in the temperature at (x_i, y_j).)

We can consider the bracketed term in Eq. (7.13) to be an adjustment to the old value $u_{ij}^{(k)}$, to give the new and improved value $u_{ij}^{(k+1)}$. If, instead of adding just the bracketed term, we add a larger value (thus "overrelaxing"), we get faster convergence. We modify Eq. (7.13) by including an overrelaxation factor ω to get the new iterating relation

$$u_{ij}^{(k+1)} = u_{ij}^{(k)} + \omega \left[\frac{u_{i+1,j}^{(k)} + u_{i-1,j}^{(k+1)} + u_{i,j+1}^{(k)} + u_{i,j-1}^{(k+1)} - 4u_{ij}^{(k)}}{4} \right]. \qquad (7.14)$$

Maximum acceleration is obtained for some optimum value of ω. This optimum value will always lie between 1.0 and 2.0.

Table 7.3 shows some results that demonstrate how S.O.R. can speed up the convergence in our example problem for $h = 2.5$ (21 interior points) and for $h = 1.25$ (105 interior points.)

Table 7.3 Effect* of overrelaxation factor on speed of convergence

h = 2.5 cm (21 interior points)		h = 1.25 cm (105 interior points)	
ω	Number of iterations	ω	Number of iterations
1.00	20	1.00	70
1.10	15	1.10	58
1.20	13	1.20	46
1.30	12	1.30	35
1.40	15	1.40	29
1.50	18	1.50	26
1.60	23	1.60	28
		1.70	36
ω_{opt} at about 1.3		ω_{opt} at about 1.5	

*The example problem using S.O.R. with iterations was continued until maximum change in any component of $\bar{u}$ was less than 0.001.

The optimum value for ω is not always predictable in advance. There are methods of using the results of the first few iterations to find a value of ω that is near optimum.* We do not discuss these here. For a rectangular region having constant boundary conditions (Dirichlet conditions), a reasonable estimate of the optimum ω can be determined as the smaller root of the quadratic equation

$$\left(\cos \frac{\pi}{p} + \cos \frac{\pi}{q}\right)^2 \omega^2 - 16\omega + 16 = 0, \qquad (7.15)$$

where p and q are the number of mesh divisions on each side of the rectangular region. Solving the above equation for ω gives

$$\omega_{opt} = \frac{4}{2 + \sqrt{4 - c^2}},$$

with

$$c = \left(\cos \frac{\pi}{p} + \cos \frac{\pi}{q}\right).$$

For our example problem, Eq. (7.15) predicts $\omega_{opt} = 1.267$ for $h = 2.5$ and $\omega_{opt} = 1.532$ for $h = 1.25$. We see from Table 7.3 that there is good agreement with actual results.

To show the range of values for ω_{opt} that are customary in steady-state problems, we show in Table 7.4 solutions to Eq. (7.15) for several values of p and q where $p = q$ (a square region).

---

*Hageman and Young (1981) present details. We also consider S.O.R. and the optimum ω as an eigenvalue problem in Section 7.7.

Table 7.4

Value of $p = q$	ω_{opt}
2	1.000
3	1.072
5	1.260
10	1.528
20	1.729
100	1.939
∞	2.000

This section has used a somewhat intuitive approach to S.O.R. and values for ω, the overrelaxation factor. Actually this is an eigenvalue problem, as you will see in Section 7.7.

7.6 THE POISSON EQUATION

The methods of the previous section are readily applied to Poisson's equation. We illustrate with an analysis of torsion in a rectangular bar subject to twisting. The torsion function ϕ satisfies the Poisson equation:

$$\nabla^2\phi + 2 = 0, \qquad \phi = 0 \text{ on boundary.}^* \qquad (7.16)$$

The tangential stresses are proportional to the partial derivatives of ϕ for a twisted prismatic bar of constant cross section. Let us find ϕ over the cross section of a rectangular bar 6×8 in in size.

Subdivide the cross section into 2-in squares, so that there are six interior points, as in Fig. 7.7. In terms of difference quotients, Eq. (7.16) becomes

$$\frac{1}{h^2}\left\{\begin{matrix} & 1 & \\ 1 & -4 & 1 \\ & 1 & \end{matrix}\right\}\phi_{i,j} + 2 = 0,$$

or

$$\frac{1}{4}\left\{\begin{matrix} & 1 & \\ 1 & -4 & 1 \\ & 1 & \end{matrix}\right\}\phi_{i,j} + 2 = 0. \qquad (7.17)$$

(The function ϕ is dimensional; with our choice of h, ϕ will have square inches as units.)

---

*The second term in Eq. (7.16)—here equal to 2—actually should be multiplied by the angular twist per unit length and by the modulus of rigidity for the material.

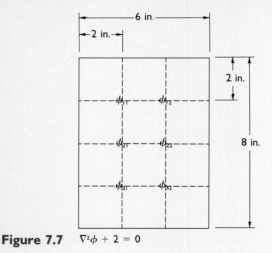

Figure 7.7 $\nabla^2\phi + 2 = 0$

The set of equations, when (7.17) is applied at each interior point, is

$$0 \quad + 0 \quad + \phi_{12} + \phi_{21} - 4\phi_{11} + 8 = 0,$$
$$\phi_{11} + 0 \quad + 0 \quad + \phi_{22} - 4\phi_{12} + 8 = 0,$$
$$0 \quad + \phi_{11} + \phi_{22} + \phi_{31} - 4\phi_{21} + 8 = 0,$$
$$\phi_{21} + \phi_{12} + 0 \quad + \phi_{32} - 4\phi_{22} + 8 = 0,$$
$$0 \quad + \phi_{21} + \phi_{32} + 0 \quad - 4\phi_{31} + 8 = 0,$$
$$\phi_{31} + \phi_{22} + 0 \quad + 0 \quad - 4\phi_{32} + 8 = 0. \tag{7.18}$$

Symmetry considerations show that $\phi_{11} = \phi_{12} = \phi_{31} = \phi_{32}$ and $\phi_{21} = \phi_{22}$, so only two unknowns are left in (7.18) after substitutions:

$$\phi_{21} - 3\phi_{11} + 8 = 0,$$
$$2\phi_{11} - 3\phi_{21} + 8 = 0.$$

Obviously these would be solved by elimination; $\phi_{11} = 4.56$, $\phi_{21} = 5.72$.

Now let us solve the problem with a 1-in-square mesh using iteration. To use Liebmann's method, we need initial estimates of ϕ. As shown in Fig. 7.8(a), we estimate these, using our previous results as a guide. We converge in 25 iterations (tolerance = 0.001) to the values shown in Fig. 7.8(b), employing Eq. (7.19):*

$$\phi_{ij} = \frac{1}{4}(\phi_{i+1,j} + \phi_{i-1,j} + \phi_{i,j+1} + \phi_{i,j-1} + 2). \tag{7.19}$$

---

*Equation (7.19) differs from Eq. (7.12), the Liebmann method for Laplace's equation, only in that the $f(x, y)$ term of Poisson's equation must be included.

1.4	2.6	3.0	2.6	1.4	2.042	3.047	3.353	3.047	2.043
2.5	**4.6**	5.0	**4.6**	2.5	3.123	4.794	5.319	4.794	3.123
3.0	5.2	5.5	5.2	3.0	3.657	5.686	6.335	5.686	3.657
3.5	**5.7**	6.0	**5.7**	3.5	3.181	5.960	6.647	5.960	3.818
3.0	5.2	5.5	5.2	3.0	3.657	5.686	6.335	5.686	3.657
2.5	**4.6**	5.0	**4.6**	2.5	3.124	4.794	5.319	4.794	3.124
1.4	2.6	3.0	2.6	1.4	2.043	3.048	3.354	3.048	2.043

(a) Initial values of ϕ for iteration   (b) Final value for ϕ by iterative methods

Figure 7.8   Torsion in a rectangular bar; mesh size of 1 in.

As we have seen before, applying an overrelaxation factor should speed up the convergence. Accordingly, we rewrite Eq. (7.19) with an overrelaxation factor:

$$\phi_{ij}^{(k+1)} = \phi_{ij}^{(k)} + \frac{\omega}{4}\left(\phi_{i+1,j}^{(k)} + \phi_{i-1,j}^{(k+1)} + \phi_{i,j+1}^{(k)} + \phi_{i,j-1}^{(k+1)} - 4\phi_{ij}^{(k)} + 2\right). \quad (7.20)$$

The optimum overrelaxation factor calculated from Eq. (7.15) is 1.383. Using this in Eq. (7.20), we converge in 13 iterations to the same set of values as before and as tabulated in Fig. 7.8(b). Using S.O.R., rather than the standard Liebmann's method, cuts down the number of iterations by nearly 50%.

The nine-point formula for the Poisson equation,

$$\nabla^2\phi + f = 0,$$

becomes

$$\frac{1}{5h^2}\begin{Bmatrix} 1 & 4 & 1 \\ 4 & -20 & 4 \\ 1 & 4 & 1 \end{Bmatrix}\phi_{i,j} + f = 0,$$

provided that f is a function of x and y only. The truncation error in this case is still $O(h^6)$. Equations (7.17) and (7.18) become

$$\frac{1}{24}\begin{Bmatrix} 1 & 4 & 1 \\ 4 & -20 & 4 \\ 1 & 4 & 1 \end{Bmatrix}\phi_{i,j} + 2 = 0,$$

and the set of equations becomes

$$
\begin{aligned}
-20\phi_{11} + 4\phi_{12} + 4\phi_{21} + \phi_{22} && + 48 &= 0, \\
4\phi_{11} - 20\phi_{12} + \phi_{21} + 4\phi_{22} && + 48 &= 0, \\
4\phi_{11} + \phi_{12} - 20\phi_{21} + 4\phi_{22} + 4\phi_{31} + \phi_{32} && &= 0, \\
\phi_{11} + 4\phi_{12} + 4\phi_{21} - 20\phi_{22} + \phi_{31} + 4\phi_{32} + 48 &= 0, \\
4\phi_{21} + \phi_{22} - 20\phi_{31} + 4\phi_{32} + 48 &= 0, \\
\phi_{21} + 4\phi_{22} + 4\phi_{31} - 20\phi_{32} + 48 &= 0.
\end{aligned}
$$

7.7   EIGENVALUES AND S.O.R.

The S.O.R. method used in Sections 7.5 and 7.6 is actually a special case of the relaxation method of Section 2.11. In fact, Eqs. (7.14) and (7.20) are just special cases of Eq. (2.28), which we repeat here:

$$
x_i^{(k+1)} = x_i^{(k)} + \frac{\omega}{a_{ii}}\left(b_i - \sum_{j=1}^{i-1} a_{ij}x_j^{(k+1)} - \sum_{j=i}^{n} a_{ij}x_j^{(k)}\right),
$$
$$
i = 1, 2, \ldots, n. \tag{7.21}
$$

Since this equation reduces to Gauss–Seidel for $\omega = 1$, we will see in this section that the purpose of introducing ω is to reduce the eigenvalues of the iteration matrix in order to accelerate convergence (as explained in Section 6.7).

Let us review the argument of Section 6.7 about speed of convergence. We prefer to work with matrices rather than the individual equations. We start with $Ax = b$, where A is $n \times n$. If we rewrite A as $L + D + U$, then

$$
Ax = (L + D + U)x = b, \text{ or}
$$
$$
(L + D)x = b - Ux.
$$

We can put this into an iterative form:

$$
(L + D)x^{(k+1)} = b - Ux^{(k)},
$$
$$
Dx^{(k+1)} = b - Lx^{(k+1)} - Ux^{(k)}, \tag{7.22}
$$

$$
x^{(k+1)} = D^{-1}b - D^{-1}Lx^{(k+1)} - D^{-1}Ux^{(k)},
$$

giving the Gauss–Seidel equation, Eq. (2.23). In using this, we begin with an initial approximation $x^{(0)}$. We see that Eq. (7.22) can be written as

$$x^{(k+1)} = b' - Bx^{(k)}, \qquad b' = (L + D)^{-1}b, \qquad B = (L + D)^{-1}U.$$

Suppose that $\bar{x}$ is the solution to $Ax = b$. Then $\bar{x}$ certainly satisfies

$$\bar{x} = b' - B\bar{x}.$$

Subtracting this from Eq. (7.23), we get

$$x^{(k+1)} - \bar{x} = e^{(k+1)} = -B(x^{(k)} - \bar{x}) = -Be^{(k)},$$

where $e^{(i)} = x^{(i)} - \bar{x}$. This means that

$$e^{(k+1)} = -Be^{(k)} = B^2 e^{(k-1)} = \cdots = (-B)^{k+1} e^{(0)}$$

with $e^{(0)} = x^{(0)} - \bar{x}$, the error in the initial approximation to the solution. Now, assuming that we can write $e^{(0)}$ in terms of the eigenvectors of B, where $Bv_i = \lambda_i v_i$,

$$e^{(0)} = c_1 v_1 + c_2 v_2 + \cdots + c_n v_n.$$

Now

$$B^k e^{(0)} = c_1 B^k v_1 + c_2 B^k v_2 + \cdots + c_n B^k v_n$$
$$= c_1 \lambda_1^k v_1 + c_2 \lambda_2^k v_2 + \cdots + c_n \lambda_n^k v_n,$$

and we see that, if all the eigenvalues of B are less than 1 in magnitude, the iteration will converge and that the rate of convergence is faster when the eigenvalue of greatest magnitude is small. We can speed up the convergence if we can find a way to reduce the eigenvalue of maximum magnitude of B.

Consider now a simple example that shows how the relaxation factor can reduce the eigenvalues of the iteration matrix and hence accelerate convergence. Let $Ax = b$ be

$$\begin{bmatrix} 2 & 1 \\ 1 & 3 \end{bmatrix} x = \begin{bmatrix} 6 \\ -2 \end{bmatrix} \qquad \text{(solution is } \{4 \quad -2\}).^*$$

First, let us relate Eq. (7.21) (single-equation form for overrelaxation) to Eq. (7.22) (the matrix form for iteration). Let $A = L + D + U$ and write

$$\omega(b - Ax) = 0 = \omega(b - (L + D + U)x). \tag{7.23}$$

(Actually, $b - (L + D + U)x$ is the residual in Southwell's relaxation method of Section 2.11.) Obviously, we can write

$$Dx = Dx.$$

Add (7.23) to both sides:

$$Dx + 0 = Dx - \omega Lx - \omega Dx - \omega Ux + \omega b.$$

---

*In this section, braces indicate a column vector.

Rearrange to give the iterative form:

$$(D + \omega L)x^{(k+1)} = [(1 - \omega)D - \omega U]x^{(k)} + \omega b, \text{ or}$$

$$x^{(k+1)} = (D - \omega L)^{-1}[(1 - \omega)D - \omega U]x^{(k)} + \omega(D + \omega L)^{-1}b. \quad (7.24)$$

This is the matrix formulation of Eq. (7.21) for overrelaxation.

Let us apply this to the simple problem, $Ax = b$, where

$$A = \begin{bmatrix} 2 & 1 \\ 1 & 3 \end{bmatrix}, b = \begin{bmatrix} 6 \\ -2 \end{bmatrix}, D = \begin{bmatrix} 2 & 0 \\ 0 & 3 \end{bmatrix}, L = \begin{bmatrix} 0 & 0 \\ 1 & 0 \end{bmatrix}, U = \begin{bmatrix} 0 & 1 \\ 0 & 0 \end{bmatrix}.$$

In the form of Eq. (7.24), this is

$$\underbrace{\begin{bmatrix} 2 & 0 \\ \omega & 3 \end{bmatrix}}_{(D + \omega L)} x^{(k+1)} = \underbrace{\begin{bmatrix} 2(1 - \omega) & -\omega \\ 0 & 3(1 - \omega) \end{bmatrix}}_{(D - \omega D - \omega U)} x^{(k)} + \omega \underbrace{\begin{bmatrix} 6 \\ -2 \end{bmatrix}}_{\omega b}.$$

The inverse of $(D + \omega L)$ is not difficult to compute. Doing so and multiplying the matrices gives

$$x^{(k+1)} = \begin{bmatrix} 1 - \omega & -\omega/2 \\ \omega(\omega - 1)/3 & (\omega^2/6 - \omega + 1) \end{bmatrix} x^{(k)} + \begin{bmatrix} 3\omega \\ -\omega^2 - 2\omega/3 \end{bmatrix}. \quad (7.25)$$

(You should verify that, for $\omega = 1$, this reduces to the Gauss–Seidel matrix.) In fact, this is

$$\begin{bmatrix} 0 & -\frac{1}{2} \\ 0 & \frac{1}{6} \end{bmatrix} \quad \text{whose eigenvalues are } \{0 \quad \tfrac{1}{6}\}.$$

We wish to determine the optimum value ω_{opt} of the overrelaxation factor in the matrix of Eq. (7.25) that reduces the largest magnitude eigenvalue of the matrix as much as possible. To do this, recall four properties of matrices:

1. $\det(AB) = \det(A)\det(B)$.
2. $\det(A) = $ product of its eigenvalues.
3. $\text{trace}(A) = $ sum of diagonal elements $=$ sum of its eigenvalues.
4. The eigenvectors of a matrix A form a basis for R^n if the matrix can be diagonalized. We used this earlier.

Applying these to the matrix of Eq. (7.25), we have

$$\lambda_1 \lambda_2 = (\omega - 1)^2,$$

where $0 < \omega < 2$. This follows because we want all the eigenvalues to be less than 1 in absolute value. Moreover, since we want the largest eigenvalue as small as possible, we take

$$\lambda_1 = \lambda_2 = \omega - 1.$$

By the trace property, it follows that

$$\lambda_1 + \lambda_2 = \frac{\omega^2}{6} - 2\omega + 2 = 2(\omega - 1).$$

This quadratic equation has one solution $\omega_{opt} = 1.0455488$, which falls in the range from 0 to 2.

To summarize what we have done: We wanted to solve $Ax = b$ by iteration. If we had used Gauss–Seidel, our iteration matrix would be

$$\begin{bmatrix} 0 & -\frac{1}{2} \\ 0 & \frac{1}{6} \end{bmatrix} \quad \text{with eigenvalues 0 and } \tfrac{1}{6}.$$

Starting with $x^{(0)} = \{0 \quad 0\}$, succeeding iterates are $\{3 \quad -1.6675\}$, $\{3.8333 \quad -1.9444\}$, $\{3.9722 \quad -1.9907\}$, $\rightarrow \{4 \quad -2\}$.

However, by using Eq. (7.25) with $\omega = 1.0455488$, the iteration matrix becomes

$$\begin{bmatrix} -0.0455 & -0.5228 \\ 0.0159 & 0.1366 \end{bmatrix} \quad \text{with eigenvalues 0.0455 and 0.0455.}$$

Starting again with $\{0 \quad 0\}$, we get for $x^{(1)}$, $\{3.1366 \quad -1.7902\}$, for $x^{(2)}$, $\{3.9296 \quad -1.9850\}$, then $\{3.9953 \quad -1.9991\}$. Reducing the largest-magnitude eigenvalue from $\frac{1}{6}$ to 0.0455 gives faster convergence. If the reduction in magnitude had been greater, a more significant speeding up would have been observed.

You will do another example as an exercise. For larger matrices, a different approach to finding the optimal ω_{opt} would be necessary. Our purpose here is only to interrelate eigenvalues of the relaxation matrices to the convergence that is obtained. We discussed one of these other methods for eigenvalues in Chapter 6, the power method. We also described the QR method in Chapter 6; this is based on similarity transformations. The selected references listed at the end of this chapter will give you further insight.

7.8   DERIVATIVE BOUNDARY CONDITIONS

In the previous examples, we solved for the steady-state temperatures at interior points of a rectangular plate, with the boundary temperatures being specified. In many problems, instead of knowing the boundary temperatures, we know the temperature gradient in the direction normal to the boundary, as for example when heat is being lost from the surface by radiation and conduction.

EXAMPLE   Consider a rectangular plate within which heat is also being uniformly generated at each point at a rate of Q cal/cm^3 · sec. The plate is steel and is 4 × 8 cm, 1 cm thick. k = 0.16 cal/sec · cm^2 · °C/cm. For this situation, Poisson's equation holds in the form

$$\nabla^2 u = -\frac{Q}{k} = \frac{1}{h^2}\begin{Bmatrix} & 1 & \\ 1 & -4 & 1 \\ & 1 & \end{Bmatrix} u_{i,j}. \qquad (7.26)$$

Suppose that $Q = 5$ for our example. The top and bottom faces are perfectly insulated so that no heat is lost, while the upper and lower edges lose heat so that

$$\frac{\partial u}{\partial y} = -15°C/cm$$

in the outward direction, and the right and left edges are held at a constant temperature $u = 20°C$. We will find the steady-state temperatures at points in the plate.

In Fig. 7.9 we sketch the plate, with a gridwork to give 21 interior points with h = 1 cm. In this problem, the upper and lower edge temperatures are also unknown, increasing the total number of points at which u is to be determined to 35. Because of symmetry, the equations can be written in terms of 12 quantities; these are indicated on the diagram.

We now write Eq. (7.26) at each unknown point, which is our general procedure in elliptic partial-differential equations, except that at the upper and lower edges we cannot form the five-point combination—there are not enough points to form the star. We get around this by the device of extending our network to a row of exterior points. We utilize

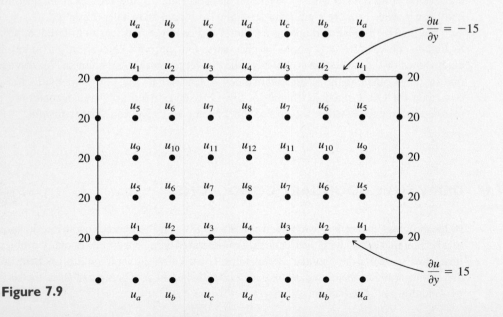

Figure 7.9

these fictitious exterior points to include in the set of equations those whose central point is on the upper or lower edge:

$$(20 + u_a + u_2 + u_5 - 4u_1) = -5/0.16,$$
$$(u_1 + u_b + u_3 + u_6 - 4u_2) = -5/0.16,$$
$$(u_2 + u_c + u_4 + u_7 - 4u_3) = -5/0.16,$$
$$(u_3 + u_d + u_3 + u_8 - 4u_4) = -5/0.16,$$
$$(20 + u_1 + u_6 + u_9 - 4u_5) = -5/0.16,$$
$$(u_5 + u_2 + u_7 + u_{10} - 4u_6) = -5/0.16,$$
$$(u_6 + u_3 + u_8 + u_{11} - 4u_7) = -5/0.16,$$
$$(u_7 + u_4 + u_7 + u_{12} - 4u_8) = -5/0.16,$$
$$(20 + u_5 + u_{10} + u_5 - 4u_9) = -5/0.16,$$
$$(u_9 + u_6 + u_{11} + u_6 - 4u_{10}) = -5/0.16,$$
$$(u_{10} + u_7 + u_{12} + u_7 - 4u_{11}) = -5/0.16,$$
$$(u_{11} + u_8 + u_{11} + u_8 - 4u_{12}) = -5/0.16. \tag{7.27}$$

It would appear that we have not helped ourselves by the fictitious points outside the plate. We have made it possible to write twelve equations, but four new unknowns have been introduced. However, we have not yet utilized the gradient conditions. Write the derivative conditions as central-difference quotients, as discussed in Chapter 4:

$$-\left(\frac{\partial u}{\partial y}\right)_1 = \frac{u_5 - u_a}{2(1)} = 15, \qquad u_a = u_5 - 30,$$

$$-\left(\frac{\partial u}{\partial y}\right)_2 = \frac{u_6 - u_b}{2(1)} = 15, \qquad u_b = u_6 - 30,$$

$$-\left(\frac{\partial u}{\partial y}\right)_3 = \frac{u_7 - u_c}{2(1)} = 15, \qquad u_c = u_7 - 30,$$

$$-\left(\frac{\partial u}{\partial y}\right)_4 = \frac{u_8 - u_d}{2(1)} = 15, \qquad u_d = u_8 - 30. \tag{7.28}$$

We write these approximations for the gradient condition, choosing the proper order of points so that the outward normal is negative, since heat is flowing outwardly. Using a central-difference approximation of error $O(h^2)$ makes these compatible with our other difference-quotient approximations. Each of the difference quotients allows us to write the fictitious temperature values in terms of points within the rectangle.

We solve the set of equations in (7.27) after eliminating the fictitious points by using Eq. (7.28). Elimination, iteration (Liebmann's method), or relaxation may be used. The solution of the equations is left as an exercise. ∎

We can apply the same method to the more general boundary condition,

$$au - b\frac{\partial u}{\partial x} = c,$$

where a, b, and c are constants, in an obvious application of the relationships of Eq. (7.28).

7.9 IRREGULAR REGIONS AND NONRECTANGULAR GRIDS

When the boundary of the region is not such that the network can be drawn to have the boundary coincide with the nodes of the mesh, we must proceed differently at points near the boundary. Consider the general case of a group of five points whose spacing is nonuniform, arranged in an unequal-armed star. We represent each distance by $\theta_i h$, where θ_i is the fraction of the standard spacing h that the particular distance represents (Fig. 7.10). Along the line from u_1 to u_0 to u_3, we may approximate the first derivatives:

$$\left(\frac{\partial u}{\partial x}\right)_{1-0} \doteq \frac{u_0 - u_1}{\theta_1 h}; \qquad \left(\frac{\partial u}{\partial x}\right)_{0-3} \doteq \frac{u_3 - u_0}{\theta_3 h}.$$

Since

$$\frac{\partial^2 u}{\partial x^2} = \frac{\partial}{\partial x}\left(\frac{\partial u}{\partial x}\right),$$

we have

$$\frac{\partial^2 u}{\partial x^2} \doteq \frac{(u_3 - u_0)/\theta_3 h - (u_0 - u_1)/\theta_1 h}{\frac{1}{2}(\theta_1 + \theta_3)h} = \frac{2}{h^2}\left[\frac{u_1 - u_0}{\theta_1(\theta_1 + \theta_3)} + \frac{u_3 - u_0}{\theta_3(\theta_1 + \theta_3)}\right]. \qquad (7.29)$$

Similarly,

$$\frac{\partial^2 u}{\partial y^2} \doteq \frac{2}{h^2}\left[\frac{u_2 - u_0}{\theta_2(\theta_2 + \theta_4)} + \frac{u_4 - u_0}{\theta_4(\theta_2 + \theta_4)}\right]. \qquad (7.30)$$

The expressions in Eqs. (7.29) and (7.30) have errors $O(h)$, which introduce larger errors in the computations than for points that are arranged in an equal-armed star.

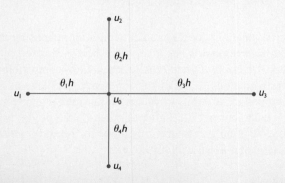

Figure 7.10

Combining, we get

$$\nabla^2 u = \frac{\partial^2 u}{\partial x^2} + \frac{\partial^2 u}{\partial y^2}$$

$$= \frac{2}{h^2} \left[\frac{u_1}{\theta_1(\theta_1 + \theta_3)} + \frac{u_2}{\theta_2(\theta_2 + \theta_4)} + \frac{u_3}{\theta_3(\theta_1 + \theta_3)} + \frac{u_4}{\theta_4(\theta_2 + \theta_4)} \right.$$

$$\left. - \left(\frac{1}{\theta_1 \theta_3} + \frac{1}{\theta_2 \theta_4} \right) u_0 \right]. \tag{7.31}$$

We use the operator of Eq. (7.31) for points adjacent to boundary points when the boundary points do not coincide with the mesh, instead of our standard operator. If the boundary conditions involve normal derivatives, great complications arise, especially for curved boundaries. The finite-element method of Section 7.12 offers less difficulty.

Let us illustrate this procedure with an example. A semicircular plate of radius a has the temperature at the base (the straight side) held at $0°$ while the circumference is held at $c°$. We desire the steady-state temperatures. The analytical solution to this problem is given by the infinite series

$$u(r, \theta) = \frac{4c}{\pi} \sum_{n=1}^{\infty} \frac{1}{2n - 1} \left(\frac{r}{a} \right)^{2n-1} \sin (2n - 1)\theta, \tag{7.32}$$

where (r, θ) are the polar coordinates of a point on the plate. (For points near the circumference, several hundred terms are needed in order to compute temperatures with any degree of accuracy.) There is right-to-left symmetry in the temperatures.

The finite-difference method superimposes a gridwork on the plate. Suppose we take these values for the parameters of the problem: $a = 1$, $c = 100$. With $h = 0.2$, the diagram of Fig. 7.11 results; there are 17 unknowns after we utilize the left-to-right symmetry. For points u_2, u_3, u_{12}, and u_{17}, the grid does not coincide with the boundary. It is easy to find by analytic geometry that the short arms at u_2 and u_{17} have a length of $0.8990h$ and those at u_3 and u_{12} have a length of $0.5826h$. Applying Eq. (7.31) we get

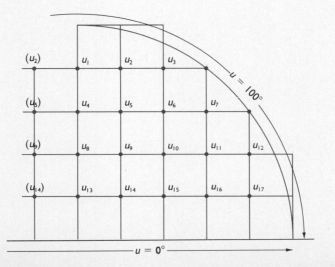

Figure 7.11 $u = 0°$

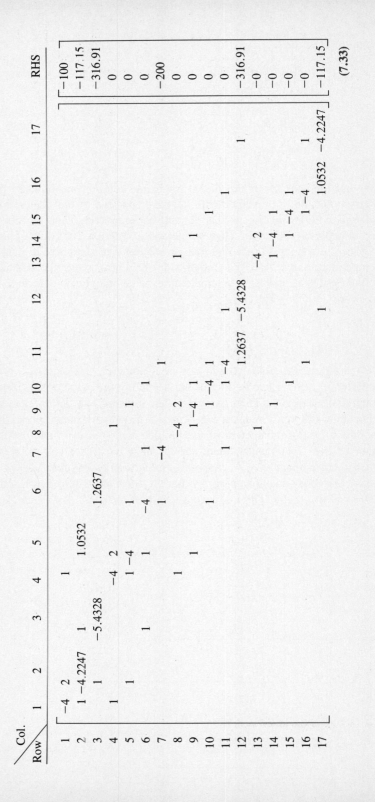

(7.33)

these pictorial operators:

At point 2:
$$\frac{1}{(0.2)^2}\left\{1 \quad \begin{matrix} 1.1715 \\ -4.2247 \\ 1.0532 \end{matrix} \quad 1\right\}u_2 = 0;$$

At point 3:
$$\frac{1}{(0.2)^2}\left\{1 \quad \begin{matrix} 2.1691 \\ -5.4328 \\ 1.2637 \end{matrix} \quad 1\right\}u_3 = 0;$$

At point 12:
$$\frac{1}{(0.2)^2}\left\{1.2637 \quad \begin{matrix} 1 \\ -5.4328 \\ 1 \end{matrix} \quad 2.1691\right\}u_{12} = 0;$$

At point 17:
$$\frac{1}{(0.2)^2}\left\{1.0532 \quad \begin{matrix} 1 \\ -4.2247 \\ 1 \end{matrix} \quad 1.1715\right\}u_{17} = 0.$$

(When computing these operators, it is useful to observe that the central number is the negative of the sum of the other four values. This helps check for accuracy.)

At all the other points, the standard operator applies. The matrix for the set of equations is given in (7.33).

Observe how sparse matrix (7.33) is. For a larger number of equations, the density of nonzero values will be even less, since no row ever has more than five nonzero values.

When the equations of (7.33) are solved, the values in Table 7.5 result. The analytical results given by Eq. (7.32) are included for comparison, using values of r and θ to coincide with the grid points. In general, the agreement is good. One might expect greater

Table 7.5 Comparison of results for example by finite differences, and analytical solution

x	y	Point number	u, finite differences	u, series solution, Eq. (7.27)
0	0.8	1	86.053	85.906
0.2	0.8	2	87.548	87.417
0.4	0.8	3	92.124	92.094
0	0.6	4	69.116	68.807
0.2	0.6	5	70.773	70.482
0.4	0.6	6	75.994	75.772
0.6	0.6	7	85.471	85.405
0	0.4	8	48.864	48.448
0.2	0.4	9	50.436	50.000
0.4	0.4	10	55.606	55.151
0.6	0.4	11	65.891	65.593
0.8	0.4	12	84.189	84.195
0	0.2	13	25.466	25.133
0.2	0.2	14	26.501	26.109
0.4	0.2	15	30.102	29.527
0.6	0.2	16	38.300	37.436
0.8	0.2	17	57.206	57.006

errors at points where the less accurate unequal-arm operators were used. This is not the case for this example, however.

If our mesh of points is chosen very fine (as we would do to get high accuracy), there is an even simpler way to handle irregular boundaries. One uses the closest mesh point as the boundary point, thus in effect perturbing the actual region to one that coincides with the network, and then the standard operator of Eq. (7.26) applies everywhere. This introduces some error, of course, but often its effect is no worse than the $O(h)$ operator of (7.31). It is also usually easier to program a computer solution using this perturbed-region technique.

Sometimes the region, while not fit by a rectangular mesh, can be fit with nodes in a different arrangement. It is occasionally useful to have a finite-difference approximation to the Laplacian for an equispaced triangular network of points. To derive this, we need the formulas for a rotation of axes from analytical geometry. The point (x, y), written in terms of its coordinates with respect to the pair of x',y'-axes that are rotated an angle θ from the x,y-system, is (see Fig. 7.12):

$$x = x' \cos \theta - y' \sin \theta,$$
$$y = x' \sin \theta + y' \cos \theta. \tag{7.34}$$

We first compute $\partial u/\partial x'$:

$$\frac{\partial u}{\partial x'} = \frac{\partial u}{\partial x}\frac{\partial x}{\partial x'} + \frac{\partial u}{\partial y}\frac{\partial y}{\partial x'} = \frac{\partial u}{\partial x}\cos \theta + \frac{\partial u}{\partial y}\sin \theta = u_x\cos \theta + u_y\sin \theta.$$

Then we get

$$\frac{\partial^2 u}{\partial x'^2} = \cos \theta(u_{xx}\cos \theta + u_{xy}\sin \theta) + \sin \theta(u_{yx}\cos \theta + u_{yy}\sin \theta)$$

$$= u_{xx}\cos^2\theta + 2u_{xy}\sin \theta \cos \theta + u_{yy}\sin^2\theta.$$

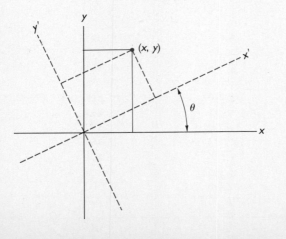

Figure 7.12

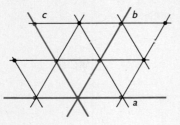

Figure 7.13

For an equispaced triangular network, the connections between nodes make angles of 0°, 60°, and 120° with the horizontal (Fig. 7.13). Call these the a-, b-, and c-axes.

For $\theta = 0°$:
$$\frac{\partial^2 u}{\partial a^2} = u_{xx};$$

For $\theta = 60°$:
$$\frac{\partial^2 u}{\partial b^2} = \frac{1}{4}u_{xx} + \frac{\sqrt{3}}{2}u_{xy} + \frac{3}{4}u_{yy};$$

For $\theta = 120°$:
$$\frac{\partial^2 u}{\partial c^2} = \frac{1}{4}u_{xx} - \frac{\sqrt{3}}{2}u_{xy} + \frac{3}{4}u_{yy}.$$

Adding, we obtain

$$\frac{\partial^2 u}{\partial a^2} + \frac{\partial^2 u}{\partial b^2} + \frac{\partial^2 u}{\partial c^2} = \frac{3}{2}(u_{xx} + 0 + u_{yy}) = \frac{3}{2}\nabla^2 u.$$

Laplace's equation, $\nabla^2 u = 0$, can be represented by

$$\frac{2}{3}\left(\frac{\partial^2 u}{\partial a^2} + \frac{\partial^2 u}{\partial b^2} + \frac{\partial^2 u}{\partial c^2}\right) = 0.$$

Using finite-difference quotients to approximate the partial derivatives gives a pictorial operator for a triangular network:

$$\nabla^2 u = \frac{2}{3h^2}\left\{\begin{matrix} & 1 & & 1 & \\ 1 & & -6 & & 1 \\ & 1 & & 1 & \end{matrix}\right\}u.$$

Note that, for Laplace's equation, the potential at every point is the arithmetic average of the potentials at its six equidistant neighbors. We observe this rule of averages even in the three-dimensional situation below.

For circular regions, one may derive a finite-difference approximation to the Laplacian in polar coordinates. Consider the group of points that are the nodes of a polar-coordinate network (Fig. 7.14):

$$\nabla^2 u = \frac{\partial^2 u}{\partial r^2} + \frac{1}{r}\frac{\partial u}{\partial r} + \frac{1}{r^2}\frac{\partial^2 u}{\partial \theta^2}$$

$$= \frac{u_3 - 2u_0 + u_1}{(\Delta r)^2} + \frac{1}{r_0}\frac{u_3 - u_1}{2\Delta r} + \frac{1}{r_0^2}\frac{u_2 - 2u_0 + u_4}{(\Delta \theta)^2}.$$

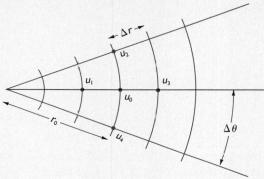

Figure 7.14

No "standard" operator can be written for this, so finding the coefficients and solving the set of equations by iteration when the problem is in polar coordinates is awkward, but it helps to group the various coefficients of each term.

Calling u_0 the central point, u_1 the point nearer the center, u_3 the point farther from the center, and u_2 and u_4 the points on either side at the same radial distance from the center, as shown in Fig. 7.14, we get, for Laplace's equation

$$\nabla^2 u = \frac{1}{(\Delta r)^2}\left[\left(1 - \frac{\Delta r}{2r_0}\right)u_1 + \left(1 + \frac{\Delta r}{2r_0}\right)u_3 + \left(\frac{\Delta r}{r_0\Delta\theta}\right)^2 u_2 \right.$$

$$\left. + \left(\frac{\Delta r}{r_0\Delta\theta}\right)^2 u_4 - \left(2 + 2\left(\frac{\Delta r}{r_0\Delta\theta}\right)^2\right)u_0\right] = 0. \qquad (7.35)$$

Let us illustrate the Laplacian in polar coordinates by again solving the example presented at the beginning of this section. In this we computed the steady-state temperatures on a semicircular plate of radius equal to one, with its base held at $0°$ and its circumference at $100°$. We superimpose a polar-coordinate system; with $\Delta r = 0.2$ and $\Delta\theta = \pi/8$, Fig. 7.15 results. There are 16 u-values to be determined, after we allow for left-to-right symmetry.

When we compute the coefficients for Eq. (7.35) applied to each of the 16 points, the matrix representation of matrix (7.36) results. Table 7.6 compares the results by finite differences with those computed from the infinite-series solution, Eq. (7.32).

The agreement in Table 7.6 is about as good as that observed in Table 7.5 with a rectangular grid. However, one point, no. 4, stands out in Table 7.6 as having an unusually large error. Large errors often occur in finite-difference solutions near a discontinuity in the boundary conditions, and this could possibly be the explanation. On the other hand, this phenomenon is absent in Table 7.5, possibly due to a cancellation of errors.

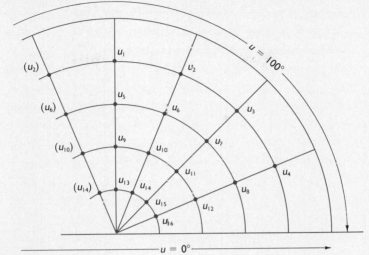

Figure 7.15

Table 7.6   Results for example by finite differences in polar coordinates, and analytical solution

r	θ	Point number	u, finite differences	u, series solution Eq. (7.32)
0.8	$\pi/2$	1	85.661	85.906
0.8	$3\pi/8$	2	84.425	84.792
0.8	$\pi/4$	3	79.459	80.394
0.8	$\pi/8$	4	63.679	66.186
0.6	$\pi/2$	5	68.371	68.808
0.6	$3\pi/8$	6	66.126	66.673
0.6	$\pi/4$	7	58.058	58.862
0.6	$\pi/8$	8	39.165	39.624
0.4	$\pi/2$	9	48.048	48.448
0.4	$3\pi/8$	10	45.543	45.938
0.4	$\pi/4$	11	37.457	37.730
0.4	$\pi/8$	12	22.374	22.252
0.2	$\pi/2$	13	25.003	25.133
0.2	$3\pi/8$	14	23.301	23.393
0.2	$\pi/4$	15	18.245	18.241
0.2	$\pi/8$	16	10.147	10.066

We observe another sparse matrix in (7.36), this time with a band structure. The structure will depend on the ordering of the points, however.

Col. Row	1	2	3	4	5	6	7	8	9	10	11	12	13	14	15	16	RHS
1	-2.8106	0.8106			0.875												-112.5
2	0.4053	-2.8106	0.4053			0.875											-112.5
3		0.4053	-2.8106	0.4053			0.875										-112.5
4			0.4053	-2.8106				0.875									-112.5
5	1.1667				-3.4410	1.4410			0.8333								0
6		1.1667			0.7205	-3.4410	0.7205			0.8333							0
7			1.1667			0.7205	-3.4410	0.7205			0.8333						0
8				1.667			0.7205	-3.4410				0.8333					0
9					1.25				-5.2423	3.2423			0.75				0
10						1.25			1.6211	-5.2423	1.6211			0.75			0
11							1.25			1.6211	-5.2423	1.6211			0.75		0
12								1.25			1.6211	-5.2423				0.75	0
13									1.5				-14.9691	12.9691			0
14										1.5			6.4855	-14.9691	6.4855		0
15											1.5			6.4855	-14.9691	6.4855	0
16												1.5			6.4855	-14.9691	0

$$(7.36)$$

7.10 THE LAPLACIAN OPERATOR IN THREE DIMENSIONS

Writing a difference equation to approximate the three-dimensional Laplacian is straightforward. We use triple subscripts to indicate the spatial position of points, and take the mesh distance the same in each direction:

$$\nabla^2 u = \frac{\partial^2 u}{\partial x^2} + \frac{\partial^2 u}{\partial y^2} + \frac{\partial^2 u}{\partial z^2}$$

$$= \frac{u_{i+1,j,k} - 2u_{i,j,k} + u_{i-1,j,k}}{h^2}$$

$$+ \frac{u_{i,j+1,k} - 2u_{i,j,k} + u_{i,j-1,k}}{h^2} + \frac{u_{i,j,k+1} - 2u_{i,j,k} + u_{i,j,k-1}}{h^2}$$

$$\nabla_u^2 = \frac{1}{h^2}\left\{\begin{array}{c} 1 \\ | \quad 1 \\ 1 - -6 - 1 \\ 1 \quad | \\ 1 \end{array}\right\} u_{i,j,k}.$$

We see again, for Laplace's equation $\nabla^2 u = 0$, that the potential u is the arithmetic average of its six nearest neighboring values. The set of equations for this case is more extensive and more tedious to solve, but in principle the methods are unchanged. In hand calculations using iterative methods, keeping track of the successive values of u is awkward. An isometric projection of the points is generally recommended over the use of superimposed sheets of paper. In a computer, triple subscripting solves the problem, but large storage requirements are imposed by problems encountered in practice.

In three-dimensional problems, it is easy to exceed the memory space of even large computer systems. For example, if the volume under consideration has 100 points on a side, a total of $100^3 = 1,000,000$ points are involved. If we were to represent the coefficients of the one million equations in a square matrix, no computer system could hold all the values in real memory. Virtual memory systems could perhaps accommodate that many values, but the speed of computation would be excruciatingly slow.

Of course, there are at most seven nonzero coefficients in any of these one million equations, but even 8,000,000 numerical values (the right-hand sides are needed too) will exceed the memory capacity. Obviously, the problem must be cut down in size. However, even only 30 points on each side of a cubical volume results in 27,000 equations, with about 216,000 nonzero numerical values. While such a number of coefficients can be stored in large real memories or in virtual memory systems, lots of computer time is involved in iterating through the large set of equations. Unfortunately, many more iterations are required for convergence where the number of equations is great. It is little wonder that three-dimensional problems, while perfectly tractable in principle, are terribly expensive to solve.

7.11 MATRIX PATTERNS, SPARSENESS, AND THE A.D.I. METHOD

All of the examples that have been presented in this chapter illustrate the fact that the coefficient matrix is sparse when an elliptic partial-differential equation is solved by the finite-difference method. Especially in a three-dimensional case, the number of nonzero coefficients is a small fraction of the total.

The relative sparseness increases as the number of equations increases. In the two-dimensional example of Section 7.4, the 21×21 coefficient matrix of Eq. (7.9) has 81% of its positions filled with zeros. If symmetry were taken into account, there would have been a 14×14 coefficient matrix with 71% zeros. Decreasing the value of h so as to have 105 points within the region gives rise to a coefficient matrix with 96% zeros. For a $30 \times 30 \times 30$ three-dimensional case, there would be 90,176 nonzero coefficients, but this is only 0.012% of the 729×10^6 values in the coefficient matrix!

In Chapter 2 it was shown that iterative methods are usually preferred for sparse matrices unless they have a tridiagonal structure. The reason for this is that elimination does not preserve the sparseness (unless the matrix is tridiagonal), so that one cannot work only with nonzero terms. This is illustrated by Fig. 7.16. The original matrix in (a) is shown in partially triangularized form in (b). The shaded portion of the matrix, which original had 50% zeros, has lost all but one of these.

Frequently the coefficient matrix has a band structure, as illustrated by Eqs. (7.33) and (7.36). Equation (7.36) illustrates a special regularity for the nonzero elements; the ordering of the points must be carefully done to attain this. A band structure is worth working for, however, since elimination does not introduce nonzero terms outside of the limits defined by the original bands. Zeros in the gaps between the parallel lines are not preserved, though, so the "tightest" possible bandedness is preferred. Sometimes it is possible to order the points so that a pentadiagonal matrix results.

$$
\begin{bmatrix}
3 & 1 & 0 & 0 & 1 & 0 & 0 & 2 \\
1 & 2 & 1 & 0 & 0 & 1 & 0 & 0 \\
0 & 2 & 3 & 1 & 0 & 0 & 1 & 0 \\
1 & 0 & 0 & 2 & 0 & 0 & 1 & 1 \\
0 & 1 & 1 & 3 & 0 & 0 & 1 & 0 \\
0 & 2 & 0 & 1 & 0 & 2 & 0 & 3 \\
2 & 0 & 0 & 3 & 0 & 1 & 1 & 0 \\
0 & 1 & 0 & 1 & 0 & 0 & 2 & 3
\end{bmatrix}
\qquad
\begin{bmatrix}
3 & 1 & 0 & 0 & 1 & 0 & 0 & 2 \\
0 & 1.677 & 1 & 0 & -0.333 & 1 & 0 & -0.666 \\
0 & 0 & 1.8 & 1 & 0.4 & -1.2 & 1 & 0.8 \\
0 & 0 & 0.6 & 2 & -0.4 & 0.2 & 1 & 0.2 \\
0 & 0 & 0.4 & 2.4 & 0.2 & -0.6 & 1 & 0.4 \\
0 & 0 & -1.2 & 1 & 0.4 & 0.8 & 0 & 3.8 \\
0 & 0 & 0.4 & 3 & -0.8 & 1.4 & 1 & -1.6 \\
0 & 0 & -0.6 & 1 & 0.2 & -0.6 & 2 & 3.5
\end{bmatrix}
$$

Figure 7.16   (a)   (b)

The best of the band structures is tridiagonal, with corresponding economy of storage and speed of solution, as discussed in Chapter 2. A method for the steady-state heat equation called the *alternating-direction-implicit* (A.D.I.) method results in tridiagonal matrices and is of growing popularity. It was initially developed for unsteady-state problems, which are the subject of Chapter 8.

In the A.D.I. method, we write the finite-difference approximation to Laplace's equation in this fashion:

$$\nabla^2 u = \frac{u_{i,j-1}^{(n)} - 2u_{i,j}^{(n)} + u_{i,j+1}^{(n)}}{(\Delta x)^2} + \frac{u_{i-1,j}^{(n+1)} - 2u_{i,j}^{(n+1)} + u_{i+1,j}^{(n+1)}}{(\Delta y)^2} = 0. \quad (7.37)$$

The superscripts indicate the iteration number. Note that when we are calculating the $(n + 1)$st iterate, we use for the approximation $\dfrac{\partial^2 u}{\partial x^2}$, a set of values in a horizontal row that are already known, the results of the nth iteration. By writing the equations in the order of the points in each column, we obtain a tridiagonal matrix when the known values of $u_{i,j-1}^{(n)}$, $u_{i,j}^{(n)}$, and $u_{i,j+1}^{(n)}$ are carried over to the right-hand sides.

The solution for steady-state values of $u_{i,j}$ is iterative, with the $(k + 1)$st and $(k + 2)$nd values computed from*

$$u_{i,j}^{(k+1)} = u_{i,j}^{(k)} + \rho(u_{i-1,j}^{(k+1)} - 2u_{i,j}^{(k+1)} + u_{i+1,j}^{(k+1)}) + \rho(u_{i,j-1}^{(k)} - 2u_{i,j}^{(k)} + u_{i,j+1}^{(k)}), \quad (7.38)$$

followed by

$$\begin{aligned} u_{i,j}^{(k+2)} = u_{i,j}^{(k+1)} &+ \rho(u_{i-1,j}^{(k+1)} - 2u_{i,j}^{(k+1)} + u_{i+1,j}^{(k+1)}) \\ &+ \rho(u_{i,j-1}^{(k+2)} - 2u_{i,j}^{(k+2)} + u_{i,j+1}^{(k+2)}). \end{aligned} \quad (7.39)$$

The relations are implicit but, by properly ordering the equations, we obtain tri-diagonal coefficient matrices. This requires that Eq. (7.38) proceed down each column of points in turn, while Eq. (7.39) processes them row-wise. (This is the reason the method is called the *alternating direction implicit*). The proper choice of the parameter ρ can accelerate the convergence; for fastest convergence of large systems, a sequence of values of ρ is employed.[†]

The iterations begin with an initial estimate of the $u^{(0)}$ vector. The iterations are terminated when successive calculations agree within a given tolerance. Since there is a bias in the calculated results in the first, third, fifth, . . . iterates, one usually ignores these and records the results of only the even-numbered iterations.

Except for the additional work of recalculating right-hand sides, this method has a minimum of computational effort, equivalent to S.O.R., but converging faster. The coefficient matrices are always the same, so the reduction step need be performed only once, if one uses the LU equivalent or some similar procedure.

EXAMPLE   A square plate conducts only in the x- and y-directions. Two adjoining edges are kept at $0°$ and the other edges are kept at $100°$. A gridwork is drawn with four interior nodes (Fig. 7.17). Use the A.D.I. method to estimate the temperatures at these four nodes.

---

*The basis for these iteration relations will become clear in the next chapter when we discuss the A.D.I. method for parabolic partial-differential equations. The method applied to elliptic equations is really a special case of that technique.

[†]Birkhoff, Varga, and Young (1962, pp. 189–273) give an extensive survey of A.D.I. methods for elliptic equations and present many examples.

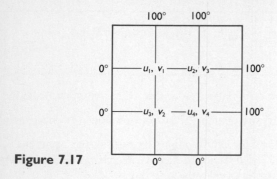

Figure 7.17

It will be convenient to use v for the temperatures when ordering the points columnwise and u for the temperatures when ordering the points row-wise. Equations (7.38) and (7.39) become, in matrix form,

$$\begin{bmatrix} (1/\rho) + 2 & -1 & & \\ -1 & (1/\rho) + 2 & 0 & \\ & 0 & (1/\rho) + 2 & -1 \\ & & -1 & (1/\rho) + 2 \end{bmatrix}\begin{bmatrix} v_1 \\ v_2 \\ v_3 \\ v_4 \end{bmatrix} = \begin{bmatrix} 100 + 0 & + [(1/\rho) - 2]u_1 + u_2 \\ 0 + 0 & + [(1/\rho) - 2]u_3 + u_4 \\ 100 + u_1 & + [(1/\rho) - 2]u_2 + 100 \\ 0 + u_3 & + [(1/\rho) - 2]u_4 + 100 \end{bmatrix},$$

$$\begin{bmatrix} (1/\rho) + 2 & -1 & & \\ -1 & (1/\rho) + 2 & 0 & \\ & 0 & (1/\rho) + 2 & -1 \\ & & -1 & (1/\rho) + 2 \end{bmatrix}\begin{bmatrix} u_1 \\ u_2 \\ u_3 \\ u_4 \end{bmatrix} = \begin{bmatrix} 0 + 100 + [(1/\rho) - 2]v_1 + v_2 \\ 100 + 100 + [(1/\rho) - 2]v_3 + v_4 \\ 0 + v_1 + [(1/\rho) - 2]v_2 + 0 \\ 100 + v_3 + [(1/\rho) - 2]v_4 + 0 \end{bmatrix}.$$

The results of successive iterations, beginning with $u^{(0)} = (0, 0, 0, 0)$ and $\rho = 0.9$, are shown in Table 7.7.

Table 7.7

Iteration number	u_1	u_2	u_3	u_4	v_1	v_2	v_3	v_4
1	0.00	0.00	0.00	0.00	35.85	11.52	83.21	58.89
2	49.86	75.47	24.26	49.86				
3					50.27	25.26	74.73	49.71
4	50.00	74.98	25.01	50.00				
5					49.99	24.99	75.01	50.01
6	50.00	75.00	25.00	50.00				
7					50.00	25.00	75.00	50.00

For this example, the optimum value of ρ is about 1.00. Figure 7.18 shows how the errors after six iterations vary with ρ.

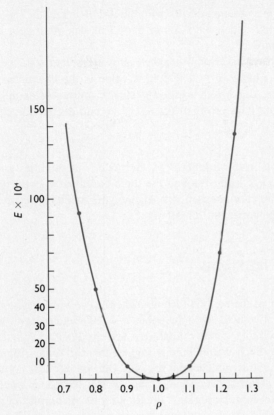

Figure 7.18

With the alternating-direction-implicit method, the errors show a most interesting pattern as the iterations proceed. Table 7.8 gives the average error at each iteration. Note that the error after an odd number of iterations is only slightly reduced over that of the previous iteration. This is true generally for A.D.I. methods. It is as if there is a bias in the values from the odd iterations that is compensated for in the succeeding even iterate.

Table 7.8 Reduction in errors with iteration number—comparison of methods in example problem

Iteration number	Average error		
	A.D.I.	Liebmann's	S.O.R.
1	11.185	17.970	15.055
2	0.3725	5.272	2.930
3	0.2725	1.318	0.2675
4	0.0075	0.330	0.0275
5	0.0065	0.082	0.0232
6	0.0002	0.020	0.0022

A very important phenomenon occurs with the A.D.I. method. The equations break up into independent subsets, here into sets of two. This can often help in the computer solution of very large problems.

Table 7.8 also compares the errors when the example problem is solved by Liebmann's method and by S.O.R., using $\omega = 1.072$ (this value is the calculated ω_{opt} from Eq. (7.15)). In this example, the A.D.I. method is clearly preferable as judged by the rate of convergence, particularly as errors of the A.D.I. would decrease even faster if ρ had been taken at 1.0.   ■

For elliptic problems in three dimensions, the A.D.I. method has great advantages. The directions alternate cyclically through the three coordinates, with new values of u being used in only one direction for each step, keeping the coefficient matrices tridiagonal. The references should be consulted for details.

7.12   FINITE-ELEMENT METHOD

In Section 7.9, we applied the method of finite differences, with some agony, to regions whose boundary does not coincide with the points of a rectangular mesh. We also showed how one can derive operators to solve Laplace's and the Poisson equations for meshes that are not rectangular. Such manipulations suggest that the finite-difference method is ill-suited for other than regular rectangular regions. But the world is not always regular!

The finite-element method, which we discuss in this section in an elementary way, is much better adapted to irregular regions. It uses a series of complex computations, but computers take care of that. There are many commercial packages, such as NASTRAN and CADAM, that handle a variety of elliptical differential equations.

You have already had an introduction to the finite-element method. In Chapter 6, we solved ordinary boundary-value problems by minimizing a quadratic functional. In that procedure, we did not use the differential equation itself to get the solution; the function that minimized the functional was the solution to the differential equation. We also saw that these so-called variational methods can be applied piecewise within the domain of the problem. This is called a finite-element method because the domain is broken into a number of finite elements. Such methods are of great and growing popularity in solving partial-differential equations. Finite elements are especially useful when the domain of the boundary-value problem is two- or three-dimensional, particularly for irregular regions. Using finite elements also facilitates local mesh refinement in those parts of the region where the variables of interest vary rapidly or where discontinuities occur.

In breaking a two-dimensional area into subregions, there is a wide choice of elements that span the region. Rectangles can be used but don't fit as well to irregular boundaries as triangles do. Other polygonal shapes are possible and even elements with curved boundaries, but these are considerably more complex to use. Triangles are very popular and we will discuss only these.* A mixture of triangles and quadrilaterals is most commonly used.

---

*In three dimensions, the corresponding choice for the elements is a tetrahedron, for the same reasons.

In close analogy to the piecewise application of Rayleigh–Ritz in Chapter 6, we do the following:

1. Find the quadratic functional that corresponds to the differential equation. This is well known for a large class of problems.

2. Subdivide the region into subregions (we use triangles) that span the region of the problem.

3. Write relations that interpolate values of the function (the solution) at the nodes (vertices) to give values of the function at points within the element. The interpolation relations are chosen to be zero outside the element so there is a purely local effect. We use a sum of these relations, weighted by the nodal values, as an approximation to the solution to the problem.

4. Substitute this weighted sum into the quadratic functional and minimize with respect to each unknown weighting factor by setting derivatives to zero. The quadratic functional breaks into a sum of integrals over each element. This leads to a set of linear equations that we can solve to give the solution to our original partial-differential equation. There is an alternative approach, the Galerkin method, that arrives at the same end result.

We will examine each of these four steps in turn. While we will work only in two dimensions, everything extends readily to three.

1. Find Functional   The second-order linear boundary-value problem that will be our model is

$$\partial^2 u / \partial x^2 + \partial^2 u / \partial y^2 + Qu = F \text{ on region } R, \tag{7.40}$$

with boundary conditions

$$u = u_0 \text{ on } L_1, \qquad \partial u / \partial n + \alpha u = \beta \text{ on } L_2.$$

In the above, Q, F, u, α, β are all functions of x and y. The boundary of R is divided into portions (L_1) where u is known (Dirichlet condition) and portions (L_2) where a mixed boundary condition holds. $\partial u / \partial n$ is the derivative of u normal to L_2.

Using methods similar to those of Section 6.4, the quadratic functional corresponding to Eq. (7.40) can be developed. It is

$$\int \int_R \left[\left(\frac{\partial u}{\partial x} \right)^2 + \left(\frac{\partial u}{\partial y} \right)^2 - Qu^2 + 2Fu \right] dx \, dy + \int_{L_2} [\alpha u^2 + 2\beta u] \, dL. \tag{7.41}$$

2. Divide into Elements   As we have said, we elect to use triangular elements. The choice of where to locate the vertices of the triangles is, in part, an art. In general, one puts many vertices in areas where the function is expected to change rapidly (corresponding to making a smaller mesh size with finite differences). It is a good idea to make the sides run in the direction of largest gradient. Along a curved or irregular boundary, the sides of the elements should closely approximate the boundary. Every vertex must be a vertex of all triangles that touch at that point. As you probably expect, the more elements, the smaller the error in the solution.

The chore of defining the coordinates of the nodes (element vertices) is facilitated by computer programs, especially when one has access to a graphics terminal. At that device, the numerical analyst can point to locations where he or she wants a node and its coordinates are automatically recorded. There are routines that can divide any given region into triangles, but these usually don't have the expertise of an experienced engineer.

3. Interpolating Relations   More sophisticated interpolating relations are often used, but we will assume that, within any single triangular element, u is a linear function of the u-values at the vertices. Note that assuming a linear relation within an element does not imply that u is linear over the whole region R. Figure 7.19(a) illustrates the region R subdivided into elements. Each node within R is shared by two or more triangular elements. One of the elements is shown; its upper facet represents u within the element.

Figure 7.20(a) shows a typical element in plan view with nodes numbered 1, 2, 3 in a local numbering system. We select the nodes (vertices) in counterclockwise order. (It is important to be consistent in the direction of ordering.) The coordinates of the nodes are (x_i, y_i), $i = 1, 2, 3$.

The linear relation within the element can be written

$$u(x, y) = N_1 u_1 + N_2 u_2 + N_3 u_3 = (N_1 \quad N_2 \quad N_3) \begin{bmatrix} u_1 \\ u_2 \\ u_3 \end{bmatrix} = N\{u\}. \qquad (7.42)$$

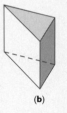

Figure 7.19   (a)   (b)

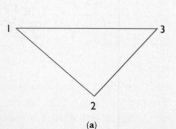

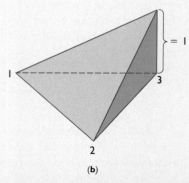

Figure 7.20   (a)   (b)

The N_i are called *shape functions* or *pyramid functions*. (The shape function for node 3 looks like an unsymmetrical pyramid of height = 1 with its apex exactly above node 3. Its base is the element. It is sketched in Figure 7.20(b). The shape functions for nodes 1 and 2 are similar.) It will be convenient to work with vectors and matrices as shown in the last part of Eq. (7.42). Obviously each N_i depends on the location (x, y) within the element where the value of u is to be computed, as well as on the location of the nodes (x_i, y_i), $i = 1, 2, 3$. We now develop the expression for the N_i.

If $u(x, y)$ varies linearly with position within the element, an alternative way to write the linear relation is

$$u(x, y) = c_1 + c_2x + c_3y,$$

which must agree with the nodal values when $(x, y) = (x_i, y_i)$. Hence

$$u_1 = c_1 + c_2x_1 + c_3y_1,$$
$$u_2 = c_1 + c_2x_2 + c_3y_2,$$
$$u_3 = c_1 + c_2x_3 + c_3y_3.$$

This is a system of equations

$$M\{c\} = \{u\} \qquad \text{(The curly brackets indicate a column vector.)}$$

where

$$M = \begin{bmatrix} 1 & x_1 & y_1 \\ 1 & x_2 & y_2 \\ 1 & x_3 & y_3 \end{bmatrix}, \; c = (c_1 \quad c_2 \quad c_3), \; u = (u_1 \quad u_2 \quad u_3). \tag{7.43}$$

Solving for c:

$$\{c\} = M^{-1}\{u\}.$$

The inverse of M is not difficult to find:

$$M^{-1} = \frac{1}{2A}\begin{bmatrix} (x_2y_3 - x_3y_2) & (x_3y_1 - x_1y_3) & (x_1y_2 - x_2y_1) \\ (y_2 - y_3) & (y_3 - y_1) & (y_1 - y_2) \\ (x_3 - x_2) & (x_1 - x_3) & (x_2 - x_1) \end{bmatrix}, \tag{7.44}$$

with $2A = \det(M)$, which turns out to be the sum of the elements in row 1 of Eq. (7.44) within the brackets. A is the area of the triangular element.* You should verify that $M^{-1}M = I$ to ensure that Eq. (7.44) truly gives the inverse matrix.

To apply the interpolating function to the minimizing of the quadratic functional (Eq. (7.41)), we prefer to write u in terms of the shape functions (Eq. (7.42)). This is easy to do. We have

$$u(x, y) = c_1 + c_2x + c_3y = (1 \quad x \quad y)\{c\} = (1 \quad x \quad y)M^{-1}\{u\}.$$

---

*That $A = \frac{1}{2}\det(M)$ is shown in most books on vectors where the cross product is explained.

But, in terms of N (from Eq. (7.42)),

$$u(x, y) = N\{u\}.$$

Comparing the two expression gives

$$N = (1 \quad x \quad y)M^{-1} \tag{7.45}$$

where M^{-1} is given by Eq. (7.44). This concludes step 3. Before we go on to step 4, we show an example to clarify step 3.

E X A M P L E   Consider the triangular element with nodes 1, 2, 3 in *ccw* order (see Figure 7.21):

Node	x	y	u
1	0	0	100
2	2	0	200
3	0	1	300

Find c, N, and $u(0.8, 0.4)$.

Before we use the above formulations, we can solve by inspection. Point (a) is at (0, 0.4), so u there is 180 by linear interpolation between nodes 1 and 3. Similarly, u at (b) is 240. The point (0.8, 0.4) is $\frac{2}{3}$ of the distance between (a) and (b), so $u(0.8, 0.4) = 180 + \frac{2}{3} (240 - 180) = 220$. We get the same result by interpolating between points (c) and (d), and between node 1 and (e).

To get c we first compute

$$M = \begin{bmatrix} 1 & 0 & 0 \\ 1 & 2 & 0 \\ 1 & 0 & 1 \end{bmatrix}, \qquad M^{-1} = \begin{bmatrix} 1 & 0 & 0 \\ -0.5 & 0.5 & 0 \\ -1 & 0 & 1 \end{bmatrix},$$

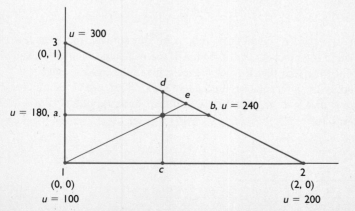

Figure 7.21

by substituting x_i, y_i, $i = 1, 2, 3$ in Eqs. (7.43) and (7.44). So

$$\{c\} = M^{-1}\{u\} = \begin{bmatrix} 1 & 0 & 0 \\ -0.5 & 0.5 & 0 \\ -1 & 0 & 1 \end{bmatrix} \begin{bmatrix} 100 \\ 200 \\ 300 \end{bmatrix} = \begin{bmatrix} 100 \\ 50 \\ 200 \end{bmatrix},$$

giving $u(x, y) = 100 + 50x + 200y$. (You should confirm that this gives the correct values for each u_i.)

From Eq. (7.45):

$$N = (1 \quad x \quad y)M^{-1} = (1 - 0.5x - y \quad 0.5x \quad y).$$

In nonmatrix form:

$$u(x, y) = N\{u\} = (1 - 0.5x - y)u_1 + (0.5x)u_2 + (y)u_3.$$

Observe that the coefficients of each N_i can be read directly from the columns of M^{-1}.

At this point we know how to write $u(x, y)$ within the single triangular element (e) as $u(x, y) = N^{(e)}\{u^{(e)}\}$. (The superscript (e) tells which element is being considered when this is necessary.) We now stipulate that every $N^{(e)} \equiv 0$ everywhere outside of element (e). Therefore we can write, for any point (x, y) within region R,

$$u(x, y) = \sum_{\text{all elements}} N^{(e)} \{u^{(e)}\}.$$

This implies that $u(x, y)$ is a surface composed of joined planar triangular facets. ∎

4. Substitute into Functional and Minimize The development of step 4 is rather involved, but it flows logically. We are to substitute, for u in Eq. (7.41),

$$u(x, y) = \sum_{\text{all elements}} N^{(e)} \{u^{(e)}\}, \tag{7.46}$$

where the elements of each $N^{(e)}$ are obtained in step 3. The $u^{(e)}$ are not yet known except where specified by Dirichlet boundary conditions. Our objective in step 4 is to obtain a system of linear equations of the form

$$K\{u\} = \{b\}$$

that we will solve for the unknown u. We develop values in the rows of K and b by setting $\partial I / \partial u_i = 0$ for each unknown u_i, where I is the functional (Eq. (7.41)).

When the relation in (7.46) is substituted into the functional, we get

$$I[u] = \int \int_R \left[\left(\frac{\partial}{\partial x} \sum \bar{N}\{\bar{u}\} \right)^2 + \left(\frac{\partial}{\partial y} \sum \bar{N}\{\bar{u}\} \right)^2 - Q \left(\sum \bar{N}\{\bar{u}\} \right)^2 + 2F \sum \bar{N}\{\bar{u}\} \right] dx \, dy$$

$$+ \int_{L_2} \left[\alpha \left(\sum \bar{N}\{\bar{u}\} \right)^2 + 2\beta \sum \bar{N}\{\bar{u}\} \right] dL.$$

Because all $N^{(e)} \equiv 0$ outside of element (e), this breaks into a sum of integrals over each individual element:

$$I[u] = \sum \left[\int\!\!\int \int^{(e)} \left(\frac{\partial}{\partial x} N^{(e)}\{u^{(e)}\} \right)^2 dx\, dy + \int\!\!\int \int^{(e)} \left(\frac{\partial}{\partial y} N^{(e)}\{u^{(e)}\} \right)^2 dx\, dy \right.$$

$$\boxed{1} \qquad\qquad\qquad\qquad\qquad \boxed{2}$$

$$+ \int\!\!\int \int^{(e)} -Q(N^{(e)}\{u^{(e)}\})^2 \, dx\, dy + \int\!\!\int \int^{(e)} 2FN^{(e)}\{u^{(e)}\} \, dx\, dy$$

$$\boxed{3} \qquad\qquad\qquad\qquad\qquad \boxed{4}$$

$$\left. + \int_{L_2} \alpha\,(N^{(e)}\{u^{(e)}\})^2 \, dL + \int_{L_2} 2\beta N^{(e)}\{u^{(e)}\} \, dL \right]. \qquad (7.47)$$

$$\boxed{5} \qquad\qquad\qquad\qquad \boxed{6}$$

The separate integrals are interconnected through the nodes that are shared by adjacent elements. What this means is that several members of the summation may contribute to the same elements of K and b. We will treat each part, 1 through 6, of Eq. (7.47) in turn.

The development of the partial derivatives of Eq. (7.47) will be clearer if we follow a specific example. To keep things simpler, the example has only two elements and four nodes, and u is unknown at only two of these nodes.

EXAMPLE Solve

$$\frac{\partial^2 u}{\partial x^2} + \frac{\partial^2 u}{\partial y^2} - \frac{u}{2} = -\frac{3}{2}$$

over the region and with boundary conditions as shown in Figure 7.22. In comparing to

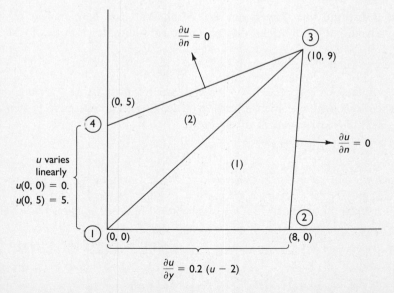

Figure 7.22

Eq. (7.40), we have $Q = -\frac{1}{2}$, $F = -\frac{3}{2}$, $\alpha = -0.2$, $\beta = -0.4$. Using the procedure of step 3 to get $N^{(1)}$ and $N^{(2)}$:

$$M^{(1)} = \begin{array}{c} ① \\ ② \\ ③ \end{array}\begin{bmatrix} 1 & 0 & 0 \\ 1 & 8 & 0 \\ 1 & 10 & 9 \end{bmatrix}, \quad M^{-1} = \begin{bmatrix} 1 & 0 & 0 \\ -0.125 & 0.125 & 0 \\ 0.028 & -0.139 & 0.111 \end{bmatrix} \begin{array}{l} \\ \leftarrow \partial u/\partial x \\ \leftarrow \partial u/\partial y, \end{array}$$
$$\qquad\qquad\qquad\qquad\qquad\qquad ① \qquad ② \qquad ③$$

$$M^{(2)} = \begin{array}{c} ① \\ ③ \\ ④ \end{array}\begin{bmatrix} 1 & 0 & 0 \\ 1 & 10 & 9 \\ 1 & 0 & 5 \end{bmatrix}, \quad M^{-1} = \begin{bmatrix} 1 & 0 & 0 \\ 0.080 & 0.100 & -0.180 \\ -0.200 & 0 & 0.200 \end{bmatrix} \begin{array}{l} \\ \leftarrow \partial u/\partial x \\ \leftarrow \partial u/\partial y, \end{array}$$
$$\qquad\qquad\qquad\qquad\qquad\qquad ① \qquad ③ \qquad ④$$

$$A^{(1)} = 36, \qquad A^{(2)} = 25.$$

The circled numbers indicate the nodes that are related to the rows or columns. The row indicators in the inverses will be explained later.

We then have

$$u(x, y) = \begin{cases} (1 - 0.125x + 0.028y)u_1 + (0.125x - 0.139y)u_2 + (0.111y)u_3 \text{ in } (1), \\ (1 - 0.080x - 0.200y)u_1 + (0.100x)u_3 + (0.180x + 0.200y)u_4 \text{ in } (2). \end{cases}$$

$$(7.48)$$

The sum breaks into disjoint functions because $N^{(e)} \equiv 0$ outside of (e). We could simplify the preceding by substituting the known values for u_1 and u_4 (known from the boundary conditions), but we defer.

Now consider each part of Eq. (7.47) in turn.

For part $\boxed{1}$ of Eq. (7.47):

Using Eq. (7.48),

$$\frac{\partial u^{(1)}}{\partial x} = -0.125u_1 + 0.125u_2 + (0)u_3,$$

$$\frac{\partial u^{(2)}}{\partial x} = 0.080u_1 + 0.100u_3 - 0.180u_4.$$

Observe that these come directly from the second row of the inverse matrices as indicated by the arrows.

Squaring these and taking partial derivatives $\partial/\partial u_2$ and $\partial/\partial u_3$ (we don't take derivatives with respect to u_1 and u_4 because these are fixed by the boundary conditions; they don't vary),

$$\frac{\partial}{\partial u_2}: \qquad 2(-0.125u_1 + 0.125u_2)(0.125) \int\int^{(1)} dA$$

$$= 2(0.125)(-0.125u_1 + 0.125u_2)A^{(1)};$$

$$\frac{\partial}{\partial u_3}: \qquad 2(-0.125u_1 + 0.125u_2)(0) \int\int^{(1)} dA$$

$$+ \ 2(0.080u_1 + 0.100u_3 - 0.180u_4)(0.100) \int\int^{(2)} dA$$

$$= \ 0 + 2(0.100)(0.080u_1 + 0.100u_3 - 0.180u_4)A^{(2)}.$$

Again, observe that these could be written directly from the inverse matrices. Note that there is no term from (2) in $\partial/\partial u_2$ because node 2 is not in element (2). The zero occurs in $\partial/\partial u_3$ because the coefficient of u_3 is zero in $N^{(1)}$.

In the preceding, $u_1 = 0$, $u_4 = 5$, $A^{(1)} = 36$, $A^{(2)} = 25$, so we have these contributions to rows 1 and 2 of K and b:

To row 1:	$1.125u_2,$
To row 2:	$0.500u_3 - 4.5.$

For part $\boxed{2}$ of Eq. (7.47):

This is very similar to part $\boxed{1}$, and we know now how to write the result directly from the inverse matrices:

$$\frac{\partial}{\partial u_2}: \qquad 2(0.028u_1 - 0.139u_2 - 0.111u_3)(-0.139)A^{(1)};$$

$$\frac{\partial}{\partial u_3}: \qquad 2(0.028u_1 - 0.139u_2 - 0.111u_3)(0.111)A^{(1)} + 0.$$

Again, terms vanish for the same reasons as in $\boxed{1}$. Substituting $u_1 = 0$, $A^{(1)} = 36$, we get these contributions to elements of K:

To row 1:	$1.391u_2 - 1.111u_3,$
To row 2:	$-1.111u_2 + \ \ 0.887u_3.$

For part $\boxed{3}$ of Eq. (7.47):

Consider a general term for element (e) whose nodes are r, s, t:

$$\int\int^{(e)} -Qu^2 \, dx \, dy = \int\int^{(e)} -Q(N_r u_r + N_s u_s + N_t u_t)^2 \, dx \, dy$$

$$= \int\int^{(e)} -Q(N_r^2 u_r^2 + N_s^2 u_s^2 + N_t^2 u_t^2$$

$$+ \ 2N_r N_s u_r u_s + 2N_r N_t u_r u_t + 2N_s N_t u_s u_t) \, dx \, dy.$$

If we take the partial derivative $\partial/\partial u_s$:

$$\frac{\partial}{\partial u_s}(\quad) = \int\int^{(e)} -Q(2N_s^2 u_s + 2N_s N_r u_r + 2N_s N_t u_t) \, dx \, dy,$$

we see that we must evaluate integrals of the form

$$2u_j \int\int^{(e)} -QN_s N_j \, dA, \qquad j = r, s, t.$$

If Q is constant over the element (e), this is not hard to evaluate, because

$$-Q \int\int^{(e)} N_k N_j \, dx \, dy = \begin{cases} -QA^{(e)}/6 & \text{if } j = k, \\ -QA^{(e)}/12 & \text{if } j \neq k.^* \end{cases}$$

Usually, for a Q varying with x and y, one takes an average value for $Q(x, y)$ throughout the element, often the average of the values of Q at the vertices.

Returning now to our specific example where $Q = -\frac{1}{2}$, we have

$$\frac{\partial}{\partial u_2}: \qquad \frac{1}{2} \int\int^{(1)} 2(N_2 N_1 u_1 + N_2^2 u_2 + N_2 N_3 u_3) \, dx \, dy = u_1 \frac{A^{(1)}}{12} + u_2 \frac{A^{(1)}}{6} + u_3 \frac{A^{(1)}}{12};$$

$$\frac{\partial}{\partial u_3}: \qquad \frac{1}{12} \int\int^{(1)} 2(N_3 N_1 u_1 + N_3 N_2 u_2 + N_3^2 u_3) \, dx \, dy$$

$$+ \frac{1}{12} \int\int^{(2)} 2(N_3 N_1 u_1 + N_3^2 u_3 + N_3 N_4 u_4) \, dx \, dy$$

$$= u_1 \frac{A^{(1)}}{12} + u_2 \frac{A^{(1)}}{12} + u_3 \frac{A^{(1)}}{6} + u_1 \frac{A^{(2)}}{12} + u_3 \frac{A^{(2)}}{6} + u_4 \frac{A^{(2)}}{12}.$$

Substituting the known values of $u_1 = 0$, $u_4 = 5$, $A^{(1)} = 36$, $A^{(2)} = 25$, we get these contributions to K and b:

To row 1:	$6u_2 + 3u_3$,
To row 2:	$3u_2 + 10.166u_3 + 10.417$.

For part $\boxed{4}$ of Eq. (7.47):

Consider again the general case of element (e) with nodes r, s, t:

$$\int\int^{(e)} 2Fu \, dx \, dy = \int\int^{(e)} 2F(N_r u_r + N_s u_s + N_t u_t) \, dx \, dy.$$

---

*This comes from the integral of products of powers of shape functions over a triangular element:

$$\int\int^{(e)} N_r^a N_s^b N_t^c \, dA = \frac{2a! \, b! \, c!}{(a + b + c + 2)!} A^{(e)}, \qquad A^{(e)} = \text{area of element}.$$

Deriving this is not easy!

If we differentiate with respect to u_s,

$$\frac{\partial}{\partial u_s}(\quad) = \int\int^{(e)} 2FN_s \, dx \, dy.$$

This is simple if F is a constant. When it is not, one can take F equal to an average value over element (e), and

$$\frac{\partial}{\partial u_s}(\quad) = 2F\int\int^{(e)} N_s \, dx \, dy = 2F\frac{A^{(e)}}{3}.$$

This comes from the fact that the volume of a pyramid is equal to one-third the height times the area of the base. (We get the same result from the general formula given earlier.)

For our specific example, $F = -\frac{3}{2}$, so we get

$$\frac{\partial}{\partial u_2}: \quad 2\left(-\frac{3}{2}\right)\int\int^{(1)} N_2 \, dx \, dy = -3\frac{A^{(1)}}{3};$$

$$\frac{\partial}{\partial u_3}: \quad 2\left(-\frac{3}{2}\right)\int\int^{(1)} N_3 \, dx \, dy + 2\left(-\frac{3}{2}\right)\int\int^{(2)} N_3 \, dx \, dy = -3\frac{A^{(1)}}{3} - 3\frac{A^{(2)}}{3}.$$

Substituting $A^{(1)} = 36$ and $A^{(2)} = 25$, we get contributions

To row 1:	-36,
To row 2:	$-36 - 25 = -61$.

Observe that term $\boxed{4}$ never contributes to K, only to b.

For part $\boxed{5}$ of Eq. (7.47):

This integral (of $\alpha u^2 \, dL$ over L_2) applies only to those elements that have a side on L_2. An element may have two sides on L_2. If so, this is repeated for each side.

We attack this the same way as in part $\boxed{3}$, which this parallels. Consider the general element (e) with nodes r, s, t. Suppose that side s-t is on L_2. Again, if α is a constant, or if an average value is used,

$$\int_{L_2} \alpha \, u^2 \, dL = \alpha \int_{\text{node } s}^{\text{node } t} (N_r u_r + N_s u_s + N_t u_t)^2 \, dL.$$

As in part $\boxed{3}$, we get, on differentiating

$$\frac{\partial}{\partial u_s}(\quad) = \alpha \int_s^t (2N_s^2 u_s + 2N_s N_r u_r + 2N_s N_t u_t) \, dL.$$

We need to evaluate integrals of the form

$$2u_j \alpha \int N_s N_j \, dL, \quad j = r, s, t.$$

It is not hard to show that

$$\alpha \int N_j N_k \, dL = \begin{cases} \alpha L/3 & \text{if } j = k, \\ \alpha L/6 & \text{if } j \neq k, \end{cases}$$

where L = the length of the side from node s to node t. (Actually, on this side $N_r = 0$, N_s varies 1 to 0, and N_t varies 0 to 1.)

For our specific example, only element (1) has a side (from node 1 to node 2) on L_2 and $\alpha = -0.2$. Hence these contributions to K result:

$$\frac{\partial}{\partial u_2}: \qquad -0.2 \int_{x_1}^{x_2} 2(N_2 N_1 u_1 + N_2^2 u_2 + N_2 N_3 u_3) \, dx$$

$$= u_1 \frac{-0.4(x_2 - x_1)}{6} + u_2 \frac{-0.4(x_2 - x_1)}{3} + u_3(0);$$

$$\frac{\partial}{\partial u_3}: \qquad -0.2 \int_{x_1}^{x_2} 2(N_3 N_1 u_1 + N_3 N_2 u_2 + N_3^2 u_3) \, dx$$

$$= 0 \quad (\text{because } N_3 = 0 \text{ on } L_2).$$

Substituting $u_1 = 0$, $x_2 - x_1 = 8$, we get the contributions

To row 1:	$-1.066 u_2$,
To row 2:	none.

For part $\boxed{6}$ of Eq. (7.47):

The integral (of $\beta u \, dL$ over L_2) also applies only to elements that have a side on L_2. This part parallels part $\boxed{4}$: We attack it in the same way. Consider the general element (e) with nodes r, s, t. Suppose that side s-t is on L_2. We will again take $\beta = $ constant.

As in part $\boxed{4}$,

$$\frac{\partial}{\partial u_s}(\quad) = 2\beta \int_{\text{node } s}^{\text{node } t} N_s \, dL = 2\beta \frac{L}{2},$$

where L is the length of the side s-t, because N_s is a triangle with side s-t as its base and a height of 1.

In our specific example, $\beta = -0.4$, $L = 8$, so we get contributions to b:

$\dfrac{\partial}{\partial u_2}:$	$2(-0.4)\dfrac{8}{2} = -3.2$, to row 1.
$\dfrac{\partial}{\partial u_3}:$	none, because $N_3 = 0$ on the side 1–2.

Table 7.9 assembles the separate contributions.

Table 7.9

Source →	1	2	3	4	5	6	Total
To k_{11}:	1.125	1.391	6		−1.066		7.45
k_{12}:		−1.111	3				1.889
k_{21}:		−1.111	3				1.889
k_{22}:	0.500	0.887	10.166				11.553
b_1:				−36		−3.2	−39.2
b_2:	−4.5		10.417	−61			−55.083

Now we have the system from which we solve for u:

$$\begin{bmatrix} 7.45 & 1.889 \\ 1.889 & 11.553 \end{bmatrix} \begin{bmatrix} u_2 \\ u_3 \end{bmatrix} = \begin{bmatrix} 39.2 \\ 46.083 \end{bmatrix},$$

which gives $u_2 = 4.230$ and $u_3 = 4.075$. (Note carefully that the signs for b change from those given in the assembly display. In that display they were in terms of $Ku + b = 0$, so we had to change signs to put them on the right-hand side.)

All this seems very tedious when done by hand, but everything is quite straightforward and not too difficult to put into a computer program. The most difficult part is to guide the user in inputting the coordinates of the nodes, the boundary conditions, and the coefficients of the equations. Care is needed in the design of the data structure so that elements, nodes, and sides can be interrelated. ∎

7.13 CHAPTER SUMMARY

If you fully understand Chapter 7, you should be able to do the following.

1. Explain how to analyze heat flow to derive the partial-differential equation that governs this and similar potential flow situations.

2. Tell in which class a given differential equation falls.

3. Set up the system of equations that results from replacing derivatives with finite-difference approximations and solve by either elimination or iteration. You should be able to handle a variety of boundary conditions.

4. List the advantages of iteration over elimination for solving large systems and explain why overrelaxation speeds up convergence.

5. Show how the finite-difference method can be adapted to nonrectangular regions.

6. Discuss the problems encountered in three-dimensional situations.

7. Use the A.D.I. procedure to solve problems and tell why it is advantageous.

8. Solve a small problem by the finite-element method and compare it to the finite-difference method.

9. Utilize the computer programs of this chapter to get the numerical solution to typical examples.

SELECTED READINGS FOR CHAPTER 7

Allaire (1985); Burnett (1987); Davis (1980); Smith (1978); Vichnevetsky (1981).

7.14   COMPUTER PROGRAMS

Since Laplace's equation is a special case of Poisson's equation, we present two computer programs for the latter, in two dimensions. They can be applied to Laplace's equation by defining $f(x, y) = 0$.

The first (Fig. 7.23) is a program that generates the coefficient matrix for the Laplacian on a rectangular region. The parameters passed to the subroutine LAPMTX include the matrix A into which the coefficients are placed, NDIM, the first dimension of the matrix A in the calling program, and N and M, the number of intervals in the x- and y-directions. Only nonzero values are inserted in the subroutine, so that the main program must set all the values of A to zero. Finally the parameter, METHOD, determines whether the five-point (Eq. 7.6b) or the nine-point (Eq. 7.6c) is used. This is determined by setting METHOD equal to either 5 or 9 in the calling arguments. This particular program implements the example in Section 7.4 and produces (Eqs. 7.9a) and (Eqs. 7.9b) for METHOD = 5 and 9 respectively.

SUBROUTINE LAPMTX is used in connection with a linear-equation solver such as subroutines LUD and SOLVE of Chapter 2; the main program must establish the right-hand-side values based on the boundary conditions and the $f(x, y)$ values, combining them with the matrix produced by LAPMTX and sending them to the linear-equation solver to obtain the solution vector, which will then be the calculated steady-state potential values at each node of the grid. The difficulty with this technique is the large memory requirements that may be needed to store the values of matrix A.

Program 2 (Fig. 7.24) illustrates how successive overrelaxation can be used on problems involving Poisson's equation. The program as shown is applied to Laplace's equation by defining the $f(x, y)$ term in Poisson's equation as zero.

The constants of the problem are set by DATA statements and DO loops; these would be modified for a different problem. As shown, they apply to the example problem of Section 7.5 in which the steady-state temperatures within a rectangular region are determined with 105 interior points. The value of the overrelaxation factor was varied to generate the data of Table 7.3. The program can be readily adapted to other problems with a rectangular region and Dirichlet conditions.

Program 2 begins the iterations by setting all interior points at the average potential of the boundary points. If better starting values are known, they might be read in rather than computing an average potential. If $f(x, y)$ is variable over the region, appropriate

```
      PROGRAM LAPLACE(INPUT,OUTPUT,TAPE6=OUTPUT)
C
C       THIS PROGRAM GENERATES THE COEFFICIENT MATRIX FOR THE LAPLACIAN ON
C       RECTANGULAR REGION THAT HAS N POINTS IN THE X-DIRECTION AND M POINTS
C       IN THE Y-DIRECTION.
C
C       THIS PROGRAM CAN GENERATE A MATRIX BASED ON EITHER THE 5-POINT OR THE
C       9-POINT METHOD PRESENTED IN SECTION 7.4
C       THIS PROGRAM SOLVES THE EXAMPLE PROBLEM IN SECTION 7.4 USING BOTH
C       THE 5-POINT AND 9-POINT METHODS OF FINITE DIFFERENCES WITH H TAKEN
C       AS H = 0.125
C
C
C       ---------------------------------------------------------------------
C
      REAL A(40,40), RHS(40), H
      INTEGER N,M, MAXSIZ,MSIZE,METHOD,   I,J
C
C
      DATA A,RHS/1640*0.0/
      DATA MAXSIZ/40/
      DATA N,M/7,3/
      DATA NDIM/40/
      PRINT *, ' ENTER A ''5'' FOR THE 5-POINT OR '
      PRINT *, '        A ''9'' FOR THE 9-POINT. '
C
1     READ *, METHOD
      IF ( (METHOD .NE. 5) .AND. (METHOD .NE. 9) ) THEN
          PRINT *, '    ENTER ONLY A ''5'' OR A ''9'' '
          GO TO 1
          END IF
C
      CALL LAPMTX(A,NDIM,N,M,METHOD)
C
C       ---------------------------------------------------------------------
C
C
C       PRINT OUT MATRIX
C
C
C       ---------------------------------------------------------------------
C
      MSIZE = M * N
      DO 10 I = 1, MSIZE
          WRITE(6,'(40I5)') (INT(A(I,J)), J = 1,MSIZE)
10        CONTINUE
C
      STOP
      END
C
C       ---------------------------------------------------------------------
C
C
C                       SUBROUTINE LAPMTX
C
C       ---------------------------------------------------------------------
C
      SUBROUTINE LAPMTX(A,NDIM,N,M, METHOD)
C       SUBROUTINE LAPMTX :
C                       THIS SUBROUTINE FORMS THE COEFFICIENT MATRIX
C       FOR THE LAPLACIAN ON A RECTANGULAR REGION THAT HAS N POINTS IN THE
C       X DIRECTION AND M POINTS IN THE Y DIRECTION. IT ASSUMES THAT THE
C       MATRIX ALREADY HAS ZEROS EVERYWHERE.
C
```

Figure 7.23   Program 1.

Figure 7.23   (*continued*)

```
C     ------------------------------------------------------------------
C
C     PARAMETERS ARE :
C
C     A      - MATRIX THAT RETURNS COEFFICIENTS, MUST CONTAIN ZEROS TO START.
C     NDIM   - FIRST DIMENSION OF A IN THE MAIN PROGRAM
C     N      - NUMBER OF GRID POINTS IN THE X DIRECTION
C     M      - NUMBER OF GRID POINTS IN THE Y DIRECTION
C     METHOD -   1 IMPLIES FIVE POINT FORMULA
C                2 IMPLIES NINE POINT FORMULA
C
C     ------------------------------------------------------------------
C
      REAL A(NDIM,NDIM)
      INTEGER NDIM,N,M,NL1,NP1,MSIZE,MSLN, METHOD, I
C
C     ------------------------------------------------------------------
C
C     FIRST WE COMPUTE SOME CONSTANTS
C
      NL1 = N - 1
      NP1 = N + 1
      MSIZE = N * M
      MSLN = MSIZE - N
C
C     ------------------------------------------------------------------
C
C          FIVE POINT METHOD IS DONE HERE
C
C     ------------------------------------------------------------------
C
      IF (METHOD .EQ. 5 ) THEN
C
C     ------------------------------------------------------------------
C
C     PUT -4 ON THE DIAGONAL
C
      DO 10 I = 1,MSIZE
         A(I,I) = -4.0
   10 CONTINUE
C
C     ------------------------------------------------------------------
C
C     NOW PUT 1 ABOVE AND BELOW THE DIAGONAL
C
      DO 15 I = 2,MSIZE
         A(I-1,I) = 1.0
         A(I,I-1) = 1.0
   15 CONTINUE
C
C     ------------------------------------------------------------------
C
C     NOW REPLACE SOME ONES WITH ZEROS
C
      DO 20 I = N,MSLN,N
         A(I,I+1) = 0.0
         A(I+1,I) = 0.0
   20 CONTINUE
C
C     ------------------------------------------------------------------
C
C     PUT IN THE SIDE BANDS AND GO BACK
C
```

Figure 7.23   (*continued*)

```
      DO 30 I = NP1,MSIZE
        A(I,I-N) = 1.0
        A(I-N,I) = 1.0
   30 CONTINUE
C
      ELSE
C
C ------------------------------------------------------------------------
C
C     PUT -20 ON THE DIAGONAL
C
C ------------------------------------------------------------------------
C
        DO 40 I = 1,MSIZE
          A(I,I) = -20.0
   40     CONTINUE
C
C ------------------------------------------------------------------------
C
C     PUT 4 ABOVE AND BELOW THE MAIN DIAGONAL
C
C ------------------------------------------------------------------------
C
        DO 50 I = 2, MSIZE
          A(I-1,I) = 4.0
          A(I,I-1) = 4.0
   50     CONTINUE
C
C ------------------------------------------------------------------------
C
C     REPLACE SOME 4'S WITH 0'S
C
C ------------------------------------------------------------------------
C
      DO 60 I = N,MSLN,N
          A(I,I+1) = 0.0
          A(I+1,I) = 0.0
   60     CONTINUE
C
C ------------------------------------------------------------------------
C
C     PUT IN THE SIDE BANDS
C
C ------------------------------------------------------------------------
C
      DO 70 I = NP1,MSIZE
          A(I,I-N) = 4.0
          A(I-N,I) = 4.0
          A(I,I-N+1) = 1.0
          A(I-N+1,I) = 1.0
          A(I+1,I-N) = 1.0
          A(I-N,I+1) = 1.0
   70     CONTINUE
C
        DO 80 I = NP1,MSIZE,N
          A(I-1,I+N) = 0.0
          A(I,I+N-1) = 0.0
          A(I+N-1,I) = 0.0
          A(I+N,I-1) = 0.0
   80     CONTINUE
C
      ENDIF
```

Figure 7.23   (*continued*)

```
C
      RETURN
      END
                    OUTPUT FOR PROGRAM 1
                    FIVE POINT METHOD
-4    1    0    0    0    0    0    1    0    0    0    0
 0    0    0    0    0
 1   -4    1    0    0    0    0    0    1    0    0    0
 0    0    0    0    0
 0    1   -4    1    0    0    0    0    0    1    0    0
 0    0    0    0    0
 0    0    1   -4    1    0    0    0    0    0    1    0
 0    0    0    0    0
 0    0    0    1   -4    1    0    0    0    0    0    1
 0    0    0    0    0
 0    0    0    0    1   -4    1    0    0    0    0    0
 0    0    0    0    0
 0    0    0    0    0    1   -4    0    0    0    0    0
 0    0    0    0    0
 1    0    0    0    0    0    0   -4    1    0    0    0
 0    0    0    0    0
 0    1    0    0    0    0    0    1   -4    1    0    0
 0    0    0    0    0
 0    0    1    0    0    0    0    0    1   -4    1    0
 1    0    0    0    0
 0    0    0    1    0    0    0    0    0    1   -4    1
 0    1    0    0    0
 0    0    0    0    1    0    0    0    0    0    1   -4
 0    0    1    0    0
 0    0    0    0    0    1    0    0    0    0    0    1
 0    0    0    1    0
 0    0    0    0    0    0    1    0    0    0    0    0
 0    0    0    0    1
 0    0    0    0    0    0    0    1    0    0    0    0
 0    0    0    0    0
 0    0    0    0    0    0    0    0    1    0    0    0
 1    0    0    0    0
 0    0    0    0    0    0    0    0    0    1    0    0
-4    1    0    0    0
 0    0    0    0    0    0    0    0    0    0    1    0
 1   -4    1    0    0
 0    0    0    0    0    0    0    0    0    0    0    1
 0    1   -4    1    0
 0    0    0    0    0    0    0    0    0    0    0    0
 0    0    1   -4    1
 0    0    0    0    0    0    0    0    0    0    0    0
 0    0    0    1   -4

                    NINE POINT METHOD
-20    4    0    0    0    0    C    4    1    0    0    0    0    0
  0    0    0    0    0
  4  -20    4    0    0    0    C    1    4    1    0    0    0    0
  0    0    0    0    0
  0    4  -20    4    0    0    C    0    1    4    1    0    0    0
  0    0    0    0    0
  0    0    4  -20    4    0    C    0    0    1    4    1    0    0
  0    0    0    0    0
  0    0    0    4  -20    4    0    0    0    0    1    4    1    0
  0    0    0    0    0
```

Figure 7.23 (*continued*)

```
   0    0    0    0    4   -20    4    0    0    0    0    1    4    1
   0    0    0    0    0
   0    0    0    0    0    4  -20    0    0    0    0    0    1    4
   0    0    0    0    0
   4    1    0    0    0    0    0  -20    4    0    0    0    0    0
   0    0    0    0    0
   1    4    1    0    0    0    0    4  -20    4    0    0    0    0
   1    0    0    0    0
   0    1    4    1    0    0    0    0    4  -20    4    0    0    0
   4    1    0    0    0
   0    0    1    4    1    0    0    0    0    4  -20    4    0    0
   1    4    1    0    0
   0    0    0    1    4    1    0    0    0    0    4  -20    4    0
   0    1    4    1    0
   0    0    0    0    1    4    1    0    0    0    0    4  -20    4
   0    0    1    4    1
   0    0    0    0    0    1    4    0    0    0    0    0    4  -20
   0    0    0    1    4
   0    0    0    0    0    0    0    4    1    0    0    0    0    0
   0    0    0    0    0
   0    0    0    0    0    0    0    1    4    1    0    0    0    0
   4    0    0    0    0
   0    0    0    0    0    0    0    0    1    4    1    0    0    0
 -20    4    0    0    0
   0    0    0    0    0    0    0    0    0    1    4    1    0    0
   4  -20    4    0    0
   0    0   -0    0    0    0    0    0    0    0    1    4    1    0
   0    4  -20    4    0
   0    0    0    0    0    0    0    0    0    0    0    1    4    1
   0    0    4  -20    4
   0    0    0    0   -0    0    0    0    0    0    0    0    1    4
   0    0    0    4  -20
```

Figure 7.24 Program 2.

```
        PROGRAM POISSON(INPUT,OUTPUT)
C
C     A PROGRAM TO SOLVE POISSON'S EQUATION ON A RECTANGULAR REGION.
C THE ACCELERATED LIEBMANN'S ( OVER/RELAXATION ) METHOD IS USED.
C
C     -----------------------------------------------------------------
C
C     PARAMETERS ARE :
C
C     NWIDE    - NUMBER OF NODES IN THE X DIRECTION
C     NHIGH    - NUMBER OF NODES IN THE Y DIRECTION
C     F(X,Y)   - R.H.S. FUNCTION FOR POISSON EQUATION
C     TOL      - TOLERANCE TO STOP ITERATIONS
C     W        - OVER-RELAXATION FACTOR
C     H        - MESH SIZE
C
C     -----------------------------------------------------------------
C
        REAL U(100,100),SUM,UAVG,RESID,CHGMAX,TOL,W,H
        INTEGER NWIDE,NHIGH,NHP1,NWP1,I,J
C
C     -----------------------------------------------------------------
C
```

Figure 7.24   (*continued*)

```
C   DEFINE THE FUNCTION WITH AN ARITHMETIC STATEMENT FUNCTION AND
C   INITIALIZE DATA AND VARIABLES.
C
      F(X,Y) = 0.0
      DATA NWIDE,NHIGH,TOL,W,H/16,8,.001,1.4,1.25/
      NHP1 = NHIGH + 1
      NWP1 = NWIDE + 1
C
C   -----------------------------------------------------------------------
C
C   USE DO LOOPS TO ESTABLISH BOUNDARY VALUES
C
      DO 1 I = 1,NHP1
         U(I,1) = 0.0
         U(I,NWP1) = 100.0
    1 CONTINUE
      DO 2 I = 2,NWIDE
         U(1,I) = 0.0
         U(NHP1,I) = 0.0
    2 CONTINUE
C
C   -----------------------------------------------------------------------
C
C   COMPUTE MEAN VALUE OF THE BOUNDARY VALUES TO USE FOR INITIAL VALUE
C   OF INTERIOR POINTS
C
    5 SUM = 0.0
      DO 10 I = 1,NHP1
         SUM = SUM + U(I,1) + U(I,NWP1)
   10 CONTINUE
      DO 20 I = 2,NWIDE
         SUM = SUM + U(1,I) + U(NHP1,I)
   20 CONTINUE
      UAVG = SUM / FLOAT(2*NWP1 + 2*(NHIGH - 1))
      X = 0.0
      Y = 0.0
      DO 30 I = 2,NHIGH
         DO 30 J = 2,NWIDE
            U(I,J) = UAVG + H*H*F(X,Y)
   30 CONTINUE
C
C   -----------------------------------------------------------------------
C
C   PRINT A HEADING AND BEGIN THE ITERATIONS. LIMIT TO 100 ITERATIONS OR
C   UNTIL THE MAX CHANGE IN U VALUES IS .LE. TOL.
C
      PRINT 199, W
      DO 50 KNT = 1,100
         CHGMAX = 0.0
         DO 40 I = 2,NHIGH
            Y = (I-1) * H
            DO 35 J = 2,NWIDE
            X = (J-1) * H
            RESID = W/4.0*( U(I+1,J) + U(I-1,J) + U(I,J+1) + U(I,J-1) -
     +              4.0*U(I,J) + H*H*F(X,Y) )
            IF ( CHGMAX .LT. ABS(RESID) ) CHGMAX = ABS(RESID)
            U(I,J) = U(I,J) + RESID
   35    CONTINUE
   40    CONTINUE
         IF ( CHGMAX .LT. TOL ) GO TO 55
   50 CONTINUE
   55 PRINT 200, KNT,CHGMAX
      DO 45 I = 1,NHP1
```

Figure 7.24 (*continued*)

```
        PRINT 201, ( U(I,J), J = 1,NWP1 )
     45 CONTINUE
        W = W + 0.1
        IF ( W .LT. 1.8 ) GO TO 5
C
C   --------------------------------------------------------------------------
C
    199 FORMAT(///' ITERATIONS WITH OVER-RELAXATION FACTOR OF ',F5.2)
    200 FORMAT(/' AFTER ITERATION NO. ',I3,' MAX CHANGE IN U = ',
       +        F8.4,' U MATRIX IS '/)
    201 FORMAT(1X,9F8.2)
        STOP
        END


              OUTPUT FOR PROGRAM 2

   ITERATIONS WITH OVER-RELAXATION FACTOR OF  1.40

   AFTER ITERATION NO.  29 MAX CHANGE IN U =     .0008 U MATRIX IS

      .00     .00     .00     .00     .00     .00     .00     .00     .00
      .00     .00     .00     .00     .00     .00     .00  100.00
      .00     .08     .17     .28     .44     .66     .99    1.46    2.16
     3.20    4.75    7.10   10.73   16.64   27.10   48.34  100.00
      .00     .14     .31     .52     .81    1.22    1.82    2.70    3.99
     5.90    8.72   12.91   19.19   28.72   43.41   66.25  100.00
      .00     .19     .40     .68    1.05    1.59    2.38    3.52    5.20
     7.68   11.31   16.63   24.40   35.64   51.58   73.24  100.00
      .00     .20     .43     .73    1.14    1.73    2.57    3.81    5.63
     8.30   12.20   17.90   26.14   37.87   54.04   75.13  100.00
      .00     .19     .40     .68    1.06    1.60    2.38    3.52    5.21
     7.68   11.31   16.63   24.40   35.64   51.58   73.24  100.00
      .00     .14     .31     .52     .81    1.22    1.82    2.70    3.99
     5.90    8.72   12.91   19.19   28.72   43.41   66.25  100.00
      .00     .08     .17     .28     .44     .66     .99    1.46    2.16
     3.20    4.76    7.10   10.73   16.64   27.10   48.34  100.00
      .00     .00     .00     .00     .00     .00     .00     .00     .00
      .00     .00     .00     .00     .00     .00     .00  100.00

   ITERATIONS WITH OVER-RELAXATION FACTOR OF  1.50

   AFTER ITERATION NO.  26 MAX CHANGE IN U =     .0008 U MATRIX IS

      .00     .00     .00     .00     .00     .00     .00     .00     .00
      .00     .00     .00     .00     .00     .00     .00  100.00
      .00     .08     .17     .28     .44     .66     .99    1.46    2.16
     3.20    4.76    7.10   10.73   16.64   27.10   48.34  100.00
      .00     .14     .31     .52     .81    1.22    1.82    2.70    3.99
     5.90    8.72   12.91   19.19   28.72   43.41   66.25  100.00
      .00     .19     .40     .68    1.06    1.60    2.38    3.52    5.21
     7.68   11.31   16.63   24.40   35.64   51.59   73.24  100.00
      .00     .20     .43     .73    1.14    1.73    2.58    3.81    5.63
     8.30   12.20   17.90   26.14   37.87   54.04   75.13  100.00
      .00     .19     .40     .68    1.06    1.60    2.38    3.53    5.21
     7.68   11.31   16.63   24.40   35.64   51.59   73.24  100.00
      .00     .14     .31     .52     .81    1.22    1.82    2.70    3.99
     5.90    8.72   12.91   19.19   28.72   43.41   66.25  100.00
      .00     .08     .17     .28     .44     .66     .99    1.46    2.17
     3.20    4.76    7.10   10.73   16.64   27.10   48.34  100.00
      .00     .00     .00     .00     .00     .00     .00     .00     .00
      .00     .00     .00     .00     .00     .00     .00  100.00
```

values of the x- and y-coordinates of each point will need to be defined within the DO loops that end on statements 35 and 40, so that the arithmetic-statement function will be correctly evaluated.

If one does not wish to investigate the effect of ω on the number of iterations required, the program would be run with only one value for ω, presumably that calculated from Eq. (7.15).

The A.D.I. method is the basis for Program 3 (see Fig. 7.25). It solves for the steady-state potential from Laplace's equation based on a rectangle. The size of the region (as specified by M and N, the number of intervals in the x- and y-directions), the acceleration factor ρ, and constant boundary conditions are defined by DATA statements; DATA statements are also used to insert ones into the off-diagonal elements of the u- and v-coefficient matrices. Program 3 solves a problem with 105 interior points.

The u- and v-coefficient matrices are established, as are the boundary-value terms for computing the right-hand sides for equations analogous to those of Section 7.11. The LU decompositions of the coefficient matrices are then found, in preparation for solving the tridiagonal systems.

After this preliminary work, the right-hand-side vector for the first traverse is computed using initial estimates of the potential; then the equations are solved for the first iterate of the values of the potential at the interior points in the rectangular region.

These first calculated values are then used to compute the right-hand-side vector for the vertical traverse; the alternate set of equations is solved for the second iteration.

Iterations are continued, alternating the direction of traversing the points, until the final solution is obtained. A better way to terminate the iterations would be to compare the change in components of the solution vector against a tolerance criterion, rather than to perform a fixed number of iterations as done here.

```
      PROGRAM ADIELL(INPUT,OUTPUT)
C
C     -----------------------------------------------------------------
C
C     THIS PROGRAM SOLVES A STEADY STATE HEAT FLOW PROBLEM BY THE
C     ALTERNATING DIRECTION IMPLICIT METHOD.
C
C     -----------------------------------------------------------------
C
C     PARAMETERS ARE :
C
C     U      - VECTOR OF POTENTIAL VALUES FOR ODD TRAVERSES
C     V      - SAME FOR EVEN TRAVERSES
C     UCOEF  - MATRIX OF COEFFICIENTS FOR ODD TRAVERSES
C     VCOEF  - MATRIX OF COEFFICIENTS FOR EVEN TRAVERSES
C     BCNDU  - VECTOR OF BOUNDARY VALUES FOR UCOEF VALUES
C     BCNDV  - SAME FOR VCOEF
C     TOP    - VECTOR OF BOUNDARY VALUES ACROSS TOP
C     BOT    - SAME FOR BOTTOM OF REGION
C     LFT    - SAME FOR LEFT HAND EDGE
C     RT     - SAME FOR RIGHT HAND EDGE
C     M      - NUMBER OF ROWS OF NODE POINTS
C     N      - NUMBER OF COLUMNS OF NODE POINTS
C     RHO    - ACCELERATION FACTOR
```

Figure 7.25   Program 3.

Figure 7.25 (*continued*)

```
C
C     ------------------------------------------------------------------
C
C     SET UP THE VECTORS AND MATRICES
C
      REAL UCOEF(500,3),VCOEF(500,3),BCNDU(500),BCNDV(500)
      REAL TOP(100),BOT(100),LFT(100),RT(100),U(500),V(500)
      INTEGER M,N,MSIZE,ML1,MSL1,NL1,I,J,K,ITMAX,KNT,L,JROW
C
C     ------------------------------------------------------------------
C
C     INITIALIZE SOME VALUES WITH DATA STATEMENTS
C
      DATA UCOEF,VCOEF / 3000 * -1.0 /
      DATA M,N,RHO / 7,15,1.0 /
      DATA V / 500 * 0.0 /
      DATA RT,LFT,TOP,BOT / 100 * 100.0, 300 * 0.0 /
      DATA BCNDU,BCNDV / 1000 * 0.0 /
C
C     ------------------------------------------------------------------
C
C     ESTABLISH THE COEFFICIENT MATRICES BY OVER-WRITING ON THE
C     DIAGONAL AND CERTAIN OFF DIAGONAL TERMS.
C
      MSIZE = M * N
      DO 10 I = 1,MSIZE
         UCOEF(I,2) = 1.0/RHO + 2.0
         VCOEF(I,2) = 1.0/RHO + 2.0
   10 CONTINUE
      ML1 = M - 1
      NL1 = N - 1
      MSL1 = MSIZE - 1
      DO 20 I = N,MSL1,N
         UCOEF(I,3) = 0.0
         UCOEF(I+1,1) = 0.0
   20 CONTINUE
      DO 30 I = M,MSL1,M
         VCOEF(I,3) = 0.0
         VCOEF(I+1,1) = 0.0
   30 CONTINUE
C
C     ------------------------------------------------------------------
C
C     NOW GET VALUES INTO THE BCOND VECTORS
C
      DO 40 I = 1,N
         BCNDU(I) = TOP(I)
         J = MSIZE - N + I
         BCNDU(J) = BOT(I)
   40 CONTINUE
      DO 45 I = 1,M
         J = (I-1)*N + 1
         BCNDU(J) = BCNDU(J) + LFT(I)
         J = I * N
         BCNDU(J) = BCNDU(J) + RT(I)
   45 CONTINUE
      DO 50 I = 1,M
         BCNDV(I) = LFT(I)
         J = MSIZE - M + I
         BCNDV(J) = RT(I)
   50 CONTINUE
      DO 55 I = 1,N
```

Figure 7.25   (*continued*)

```
            J = (I-1)*M + 1
            BCNDV(J) = BCNDV(J) + TOP(I)
            J = I * M
            BCNDV(J) = BCNDV(J) + BOT(I)
   55 CONTINUE
C
C     ------------------------------------------------------------------
C
C  NOW WE GET THE LU DECOMPOSITIONS OF THE TWO COEFFICIENT MATRICES.
C  THESE ARE VERY EASY TO COMPUTE. WE STORE THEM BACK IN THE
C  ORIGINAL VECTOR SPACE.
C
      DO 60 I = 2,MSIZE
         UCOEF(I-1,3) = UCOEF(I-1,3)/UCOEF(I-1,2)
         UCOEF(I,2) = UCOEF(I,2) - UCOEF(I,1)*UCOEF(I-1,3)
         VCOEF(I-1,3) = VCOEF(I-1,3)/VCOEF(I-1,2)
         VCOEF(I,2) = VCOEF(I,2) - VCOEF(I,1)*VCOEF(I-1,3)
   60 CONTINUE
C
C     ------------------------------------------------------------------
C
C  NOW WE BEGIN THE ITERATIONS, LIMIT THEM TO ITMAX IN NUMBER.
C
      ITMAX = 30
      DO 190 KNT = 2,ITMAX,2
C
C     ------------------------------------------------------------------
C
C  COMPUTE R.H.S. FOR THE U EQUATIONS AND STORE THESE IN THE U
C  VECTOR. FIRST DO THE TOP AND BOTTOM SETS OF TERMS.
C
          DO 65 I = 1,N
             J = (I-1)*M + 1
             U(I) = (1.0/RHO-2.0)*V(J) + V(J+1) + BCNDU(I)
             K = MSIZE - N + I
             J = I * M
             U(K) = V(J-1) + (1.0/RHO-2.0)*V(J) + BCNDU(K)
   65     CONTINUE
C
C     ------------------------------------------------------------------
C
C  NOW DO THE INTERMEDIATE ONES.
C
          DO 75 I = 2,ML1
             DO 70 J = 1,N
                K = (I-1)*N + J
                L = I + (J-1)*M
                U(K) = V(L-1) + (1.0/RHO-2.0)*V(L) + V(L+1) + BCNDU(K)
   70        CONTINUE
   75     CONTINUE
C
C     ------------------------------------------------------------------
C
C  NOW GET THE SOLUTION FOR THE HORIZONTAL TRAVERSE. FIRST Y=L(-1)*B
C
          U(1) = U(1) / UCOEF(1,2)
          DO 80 I = 2,MSIZE
             U(I) = ( U(I) - UCOEF(I,1)*U(I-1) ) / UCOEF(I,2)
   80     CONTINUE
C
C     ------------------------------------------------------------------
C
```

Figure 7.25 (*continued*)

```
C  READY NOW TO GET X = U(-1)*Y.
C
       DO 90 JROW = MSL1,1,-1
          U(JROW) = U(JROW) - UCOEF(JROW,3)*U(JROW+1)
    90    CONTINUE
C
C  ----------------------------------------------------------------------
C
C  WE DO THE SAME FOR THE VERTICAL TRAVERSE - U AND V EXCHANGE
C  ROLES. COMPUTE R.H.S. FOR THE V EQUATIONS, STORE IN V. DO
C  THE TOP AND BOTTOM SETS.
C
       DO 95 I = 1,M
          J = (I-1)*N +1
          V(I) = (1.0/RHO-2.0)*U(J) + U(J+1) + BCNDV(I)
          K = MSIZE - M + I
          J = I * N
          V(K) = U(J-1) + (1.0/RHO-2.0)*U(J) + BCNDV(K)
    95    CONTINUE
C
C  ----------------------------------------------------------------------
C
C  DO THE INTERMEDIATE ROWS
C
       DO 105 I = 2,NL1
          DO 100 J = 1,M
             K = (I-1)*M + J
             L = I + (J-1)*N
             V(K) = U(L-1) + (1.0/RHO-2.0)*U(L) + U(L+1) +BCNDV(K)
   100       CONTINUE
   105    CONTINUE
C
C  ----------------------------------------------------------------------
C
C  GET THE SOLUTION - FIRST Y = L(-1)*B
C
       V(1) = V(1) / VCOEF(1,2)
       DO 110 I = 2,MSIZE
          V(I) = ( V(I) - VCOEF(I,1)*V(I-1) ) / VCOEF(I,2)
   110    CONTINUE
C
C  ----------------------------------------------------------------------
C
C  THEN X = U(-1) * Y
C
       DO 120 JROW = MSL1,1,-1
          V(JROW) = V(JROW) - VCOEF(JROW,3)*V(JROW+1)
   120    CONTINUE
C
C  ----------------------------------------------------------------------
C
C  PRINT OUT THE LATEST RESULT.
C
       PRINT 202, KNT, (( V(I),I=J,MSIZE,M ), J=1,M)
   202 FORMAT(///1X,' AFTER ITERATION NUMBER ',I3/(1X,15F8.4) )
C
C  END OF THE ITERATION LOOP
C
   190 CONTINUE
       STOP
       END
```

EXERCISES

Section 7.2

1. In Section 7.2, it is shown that flow of heat is a phenomenon for which the rate of flow is equal to

$$-kA\frac{\partial u}{\partial x},$$

    where k is a constant, A is the cross-sectional area for heat flow, and $\partial u/\partial x$ is the temperature gradient. Suppose that k, instead of being constant, is a function of the position (x, y). Derive the equation for net flow of heat into the element $dx\,dy$. Does this reduce to Laplace's equation?

2. Repeat Exercise 1, but for k varying with temperature, such as $k = a + bu + cu^2$. Does the equation for net heat flow now reduce to Laplace's equation? How does the equation differ from that in Exercise 1?

Section 7.3

▶ 3. The mixed second derivative $\partial^2 u/(\partial x\,\partial y)$ can be considered as

$$\frac{\partial}{\partial x}\left(\frac{\partial u}{\partial y}\right) = \frac{\partial^2 u}{\partial x\,\partial y} = \frac{\partial}{\partial y}\left(\frac{\partial u}{\partial x}\right).$$

   Show that, in terms of finite-difference quotients with $\Delta x = \Delta y$, this derivative can be approximated by the pictorial operator

$$\frac{1}{4h^2}\begin{Bmatrix} -1 & 1 \\ 1 & -1 \end{Bmatrix}u_{i,j} + O(h^2).$$

4. If

$$\frac{d^2 u}{dx^2} = \frac{-u_{i+2} + 16u_{i+1} - 30u_i + 16u_{i-1} - u_{i-2}}{12h^2} + O(h^4),$$

   find the fourth-order operator for the Laplacian. Assume that the function u has a continuous sixth derivative.

5. In some cases it is necessary to approximate a differential equation that contains the third derivative, $\partial^3 u/\partial x^3$, by a difference equation. Set up an expression to approximate this third derivative in terms of finite differences. Can you find an expression that uses function values symmetrical to the point at which the derivative is being evaluated? (In finding the expressions, you may wish to compare the Taylor-series method of Section 7.3 with the methods of Chapter 3 and the method of undetermined coefficients in the Appendix.)

6. Suppose we have a rectangular plate with top and bottom edges that are perfectly insulated ($\partial u/\partial y = 0$) while the right-hand edge is held at 100°, and the left edge at 0°. If heat flows in only two directions, it is obvious that the temperatures vary linearly in the x-direction, and are constant along vertical lines.

    a) Show that such a temperature distribution satisfies Eq. (7.6).
    b) Show that this temperature distribution also obeys the relationship derived in Exercise 4. What about points adjacent to the edges of the plate?

Section 7.4

▶ 7. Set up equations analogous to Eqs. (7.8) for the example problem of Section 7.4, but with a grid spacing of $3\frac{1}{3}$ cm. Solve the set of equations by elimination.

8. Solve for the steady-state temperatures in a rectangular plate, 12 by 15 in, if one 15-in edge is held at 100° while the other edges are all held at 20°. The material is aluminum. Take $\Delta x = \Delta y = 3$ in, and consider heat to flow only in the lateral directions. Sketch in the approximate location of the 50° isothermal curve.

9. Solve for the temperatures in the plate of Figure 7.26 when the edge temperatures are held as shown. The plate is 10×8 cm, and $\Delta x = \Delta y = 2$ cm.

10. Solve Exercise 7 using the nine-point formula of Eq. (7.6c). Assume that the temperatures at the corner points (20, 10) and (20, 0) are equal to 50°.

▶ 11. Solve Exercise 9 using the nine-point formula.

12. Suppose the differential equation for the plate were

$$u_{xx} + 2u_{yy} = 0.$$

What would Equation (7.6a) and the pictorial operator (7.6b) look like?

▶ 13. The region for which we can solve Laplace's equation does not have to be rectangular. We can apply the methods of Section 7.4 for any region so long as the meshes of our network coincide with the boundary. Solve for the steady-state potentials at the eight interior points shown in Fig. 7.27.

Figure 7.26

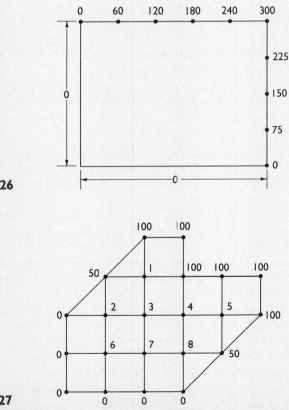

Figure 7.27

14. What fraction of the elements in the coefficient matrix of Exercise 8 are nonzero? What fraction if $h = k = 1$ in.? What fraction if $h = k = 0.1$ in.? How many equations must be solved in the latter case? Will your computer system handle that many coefficients in memory at the same time?

Section 7.5

▶15. Repeat Exercise 7, but use Liebmann's method.

16. Solve Exercise 9 by Liebmann's method with all the elements of the initial u vector equal to zero. Repeat with all the elements of the initial vector equal to 300, the upper bound to the steady-state temperatures. Repeat once more with the elements of the initial vector all equal to the arithmetic average of the boundary temperatures. Compare the number of iterations needed to reach a given tolerance for convergence in each case.

17. Solve Exercise 8 by Liebmann's method. Before you begin, make some quick hand computations to allow good estimates of starting values for the initial u vector.

▶18. Repeat Exercise 7, employing successive overrelaxation. Vary the overrelaxation factor to determine the optimum value. How does this compare to that predicted by Eq. (7.15)?

19. Repeat Exercise 16, but use S.O.R. with the value of ω_{opt} as given by Eq. (7.15).

20. Repeat Exercise 17, using S.O.R. with the value of ω_{opt} as given by Eq. (7.15).

Section 7.6

▶21. Find the torsion function ϕ for a 2×2 in square bar.

   a) Subdivide the cross section into nine equal squares, so that there are four interior points. Because of symmetry, all four values of ϕ are the same.

   b) Then subdivide the cross section into 36 equal squares so that there are 25 interior points. Use the results of (a) to estimate starting values for iteration.

22. Solve for the torsion in a hollow square bar, 5 in in outside dimension, and with walls 2 in thick (so that the inside hole is 1 in on a side). On the inner surface as well as on the outer, $\phi = 0$.

23. Solve for the torsion in a prismatic bar of a cross section similar to the region of Exercise 13. Assume the eight points are 0.5 cm apart.

24. Solve $\nabla^2 u = f(x, y)$ on the square region bounded by

$$x = 0, \quad x = 1, \quad y = 0, \quad y = 1,$$

with $f(x, y) = xy$. Use $h = \frac{1}{3}$. Take $u = 0$ on the boundary.

▶25. Suppose the function ϕ defined on the rectangular plate, Fig. 7.7, were given as

$$3\phi_{xx} + \phi_{yy} + 2 = 0, \quad \phi = 0 \text{ on the boundary.}$$

What would the pictorial diagram and the set of equations in (7.17) and (7.18) look like if you were using the five-point method?

▶26. Repeat Exercise 21(b), but use S.O.R. Vary the overrelaxation factor to find the optimum. How does this compare to the value from Eq. (7.15)?

27. Repeat Exercise 22, using S.O.R. Vary ω to find the optimum value.

28. Repeat Exercise 24, using $f(x, y) = (1 - x)(1 - y)$. Compare your results to those of Exercise 24.

Section 7.7

29. Given the pair of equations

$$2x_1 + x_2 = 24$$
$$x_1 + 2x_2 = -12,$$

a) Find B where

$$B = (D + \omega L)^{-1} (D - \omega D - \omega U) = \begin{bmatrix} 2 & 0 \\ \omega & 2 \end{bmatrix}^{-1} \begin{bmatrix} 2(1 - \omega) & -\omega \\ 0 & 2(1 - \omega) \end{bmatrix}.$$

b) Using the arguments of Section 7.7 that $\lambda_1 = \lambda_2 = \omega - 1$ and the trace argument, show that $\omega_{opt} = 1.0718$.

c) What is the largest-magnitude eigenvalue for the Gauss–Seidel matrix (for $\omega = 1$ in part (a))?

d) Evaluate $x^{(1)}$, $x^{(2)}$, $x^{(3)}$, using both Gauss–Seidel and with the optimal $\omega = 1.0718$. In both cases, use the zero vector as the starting vector. The exact answer is $\{20 \quad -16\}$.

30. Show that the optimal value for ω is less than 1 for the example used in Sections 2.11 and 6.7. Do this by writing a program that implements Eq. (7.21), using various values for ω between 0 and 2. (You should find the optimal value is between 0.9 and 1.0.)

31. What are the eigenvalues for the iteration matrix in Problem 30 when ω has the optimal value?

Section 7.8

32. Solve the set of equations in Eqs. (7.27), using (7.28) to eliminate u_a, u_b, u_c, and u_d.

a) Use elimination.

b) Use Liebmann's method.

c) Use S.O.R.

▶33. Solve the example problem of Section 7.8, except that the 8-cm edges are held at 20° and the 4-cm edges have an outward gradient of $-15°C/cm$.

34. Suppose the outward normal gradient on all the edges of the example problem of Section 7.8 is $-15°C/cm$. Can the problem be solved?

35. Solve for the steady-state temperatures of the region in Exercise 13, except that the plate is insulated at each exterior point marked zero. The edge temperatures are maintained at the temperatures given at the other exterior points.

36. Consider a region that is obtained from Exercise 13 by reflecting the figure across both of the edges marked zero (see Fig. 7.28). Show that the steady-state potentials at points in the original region are the same as for Exercise 35.

Section 7.9

▶37. Find the potential distribution in a region whose shape is a 3–4–5 right triangle. The potential is maintained at zero except on the hypotenuse, where it is 50. Use a square mesh with $h = 1$.

38. A coaxial cable has a circular outer conductor 10 cm in diameter, and a square inner conductor 3 cm on a side. (The square conductor is concentric.) The outer conductor is at zero volts while the inner is at 100 volts. Find the potential between the conductors. (Note that only one octant needs to be calculated, because of symmetry.) Use a grid of 1-cm squares.

39. If a hollow shaft has an outer circular cross section of diameter 10 in and a concentric square inner cross section which is 3 in on a side, what is the value of the torsion function at the nodes of a 1-in grid?

40. Find the solution to the Poisson equation over an equilateral triangle 5 in on a side if

$$\nabla^2 u = k, \qquad u = 0 \text{ on boundary,}$$

and values of k are as shown in Fig. 7.29.

▶41. Use the Laplacian in polar coordinates to set up the set of difference equations to solve

$$\nabla^2 u = 0.2,$$

on a semicircular region whose radius is 4. Take $\Delta r = 1$ and $\Delta \theta = \pi/6$. Boundary conditions are $u = 10$ on the straight edge and $u = 0$ on the curved boundary.

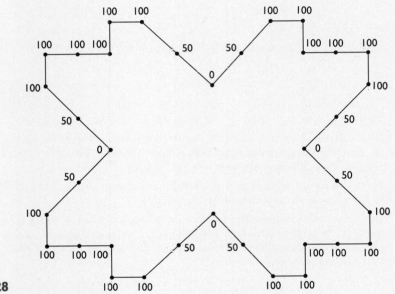

Figure 7.28

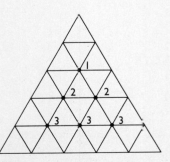

Figure 7.29

Section 7.10

▶42. A cube is 3 cm along its edge. Two opposite faces are held at 100°, the other four faces at 0°. Find the internal temperatures at the nodes of a 1-cm network.

43. Repeat Exercise 42 except that one of the 0° faces has been insulated.

Section 7.11

▶44. Solve Exercise 9 by the A.D.I. method, using $\rho = 1.0$. Begin with the initial u vector having all elements equal to the arithmetic average of the boundary values. Compare the number of iterations to those required with Liebmann's method (Exercise 16) and S.O.R. using ω_{opt} (Exercise 19).

45. Solve the sets of equations in (7.27) and (7.28) by A.D.I. with $\rho = 1.0$. Compare the number of iterations required with the number needed by S.O.R. as performed in Exercise 32(c).

46. Repeat Exercise 21, but use the A.D.I. method. Vary the value of ρ to find the optimum value.

Section 7.12

47. Find M^{-1}, $2A$, c, N, and $u(x, y)$ for these triangular elements:

	x_1	y_1	x_2	y_2	x_3	y_3	u_1	u_2	u_3	x	y
a)	1.25	3.3	−0.25	4.4	−2	−2	10	20	5	0	3
b)	30	40	10	50	20	10	12.3	1.5	22.2	23	34
c)	12.1	11.3	8.6	13.2	9.3	7.9	121	215	67	7.7	9.8

48. Solve Exercise 9 by finite elements. Locate four internal nodes at (4, 2), (8, 4), (2, 4), and (6, 6) relative to the lower left corner. Compare answers from this solution to the values obtained through finite differences with grid spacing of 2 cm.

49. Solve Exercise 13 by finite elements. Locate internal nodes at the points numbered 1, 4, and 7. Compare the answers to those of Exercise 13.

50. Use the finite-element method to solve for the torsion function within a bar whose cross section is an equilateral triangle. Each face of the bar is 3 in wide. If the internal nodes are placed symmetrically, symmetry in the problem can reduce the number of unknowns. Compare your solution to the analytical answer:

$$\phi = \frac{(1/2)(x^2 + y^2) + (1/2a)(x^3 - 3xy^2) - (2a^2)}{27}$$

where a is the distance from one face to the opposite vertex; x and y in the equation are relative to the centroid of the cross section.

APPLIED PROBLEMS AND PROJECTS

51. Modify and use the program for Poisson's equation in Section 7.14 to solve

$$\nabla^2 u = xy(x - 2)(y - 2),$$

on the region $0 \leqslant x \leqslant 2, 0 \leqslant y \leqslant 2$, with $u = 0$ on all boundaries except for $y = 0$, where $u = 1.0$.

52. Modify Program 2 so that it permits boundary conditions of the form

$$au + b\frac{\partial u}{\partial x} = c \quad \text{and} \quad a'u + b'\frac{\partial u}{\partial y} = c'.$$

Test it with the equation in Problem 51, modified in that the boundary where $y = 0$ has $\partial u/\partial y = 1$.

53. Repeat Problem 52, except make the modifications to Program 3. You will need to make changes to permit a Poisson-equation problem as well as derivative boundary conditions.

54. A classic problem in elliptic partial-differential equations is to solve $\nabla^2 u = 0$ on a region defined by $0 \leq x \leq \pi$, $0 \leq y \leq \infty$, with boundary condition of $u = 0$ at $x = 0$, at $x = \pi$, and at $y = \infty$. The boundary at $y = 0$ is held at $u = F(x)$. This can be quite readily solved by the method of separation of variables, to give the series solution

$$u = \sum_{n=1}^{\infty} B_n e^{-ny} \sin nx,$$

with

$$B_n = 2 \int_0^{\pi} F(x) \sin nx \, dx.$$

Solve this equation numerically for various definitions of $F(x)$. (You will need to redefine the region so that $0 \leq y \leq M$, where M is large enough that changes in u with y at $y = M$ are negligible.) Compare your results to the series solution. You might try

$$F(x) = 100; \quad F(x) = 100 \sin x; \quad F(x) = 400x(1 - x); \quad F(x) = 100(1 - |2x - 1|).$$

55. The equation

$$2\frac{\partial^2 u}{\partial x^2} + \frac{\partial^2 u}{\partial y^2} - \frac{\partial u}{\partial x} = 2$$

is an elliptic equation. Solve it on the unit square, subject to $u = 0$ on the boundaries. Approximate the first derivative by a central-difference approximation. Investigate the effect of size of Δx on the results, to determine at what size reducing it does not have further effect.

56. Solve the steady-state heat-flow problem that was discussed at the beginning of the chapter. Assume that the metal is aluminum and that the insulation is mineral wool. Look up the thermal properties in handbooks.

Solve it first as a three-dimensional problem. Then, considering that the plate is not very thick relative to its length and that aluminum is a good conductor so that there is little temperature variation across the thickness, solve it as a two-dimensional problem. Do you reach the same conclusion about the feasibility of the scheme?

Consider carefully how you will handle the problem of dissimilar materials—metal and insulation. Can you approximate the effect of the insulation by some proper value of the gradient at the surface of the aluminum?

8

Parabolic Partial-Differential Equations

8.0 CONTENTS OF THIS CHAPTER

The previous chapter discussed the solution of partial-differential equations that were time-independent. Such steady-state problems were described by elliptic equations. Unsteady-state problems in which the function is dependent on time are of great importance. We study these in this chapter.

The method of finite-difference equations will be further developed in this and the following chapter, since it is very important in solving parabolic and hyperbolic partial-differential equations. In the realm of available software, one can use PDECOL, distributed by IMSL, for solving a system of parabolic or hyperbolic partial-differential equations as well as boundary-value problems. Finite-element programs can also solve these.

The problem of unsteady-state flow of heat is a physical situation that can be represented by a parabolic partial-differential equation. The simplest situation is for flow of heat in one direction. Imagine a rod that is uniform in cross section and insulated around its perimeter so that heat flows only longitudinally. Consider a differential portion of the rod, dx in length with cross-sectional area A (see Fig. 8.1).

We let u represent the temperature at any point in the rod, whose distance from the left end is x. Heat is flowing from left to right under the influence of the temperature gradient $\partial u/\partial x$. Make a balance of the rate of heat flow into and

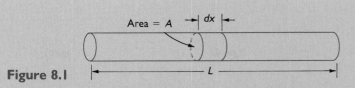

Figure 8.1

out of the element. Use k for thermal conductivity, cal/g $\cdot$ cm^2 $\cdot$ °C/cm, which we assume is constant.

Rate of flow of heat in: $-kA\dfrac{\partial u}{\partial x}$;

Rate of flow of heat out: $-kA\left(\dfrac{\partial u}{\partial x} + \dfrac{\partial}{\partial x}\left(\dfrac{\partial u}{\partial x}\right)dx\right)$.

The difference between the rate of flow in and the rate of flow out is the rate at which heat is being stored in the element. If c is the heat capacity, cal/g $\cdot$ °C, and ρ is the density, g/cm^3, we have, with t for time,

$$-kA\frac{\partial u}{\partial x} - \left(-kA\left(\frac{\partial u}{\partial x} + \frac{\partial}{\partial x}\left(\frac{\partial u}{\partial x}\right)dx\right)\right) = c\rho(A\,dx)\frac{\partial u}{\partial t}.$$

Simplifying, we have

$$k\frac{\partial^2 u}{\partial x^2} = c\rho\frac{\partial u}{\partial t}.$$

This is the basic mathematical model for unsteady-state flow. We have derived it for heat flow, but it applies equally to diffusion of material, flow of fluids (under conditions of laminar flow), flow of electricity in cables (the telegraph equations), and so on.

The differential equation is classed as parabolic because comparing to the standard form of an equation in two independent variables, x and t,

$$A\frac{\partial^2 u}{\partial x^2} + B\frac{\partial^2 u}{\partial x\,\partial t} + C\frac{\partial^2 u}{\partial t^2} + D\left(x,\,t,\,u,\,\frac{\partial u}{\partial x},\,\frac{\partial u}{\partial t}\right) = 0,$$

we find that $B^2 - 4AC = 0$.

In two or three space dimensions, analogous equations apply:

$$k\left(\frac{\partial^2 u}{\partial x^2} + \frac{\partial^2 u}{\partial y^2}\right) = c\rho\frac{\partial u}{\partial t},$$

$$k\left(\frac{\partial^2 u}{\partial x^2} + \frac{\partial^2 u}{\partial y^2} + \frac{\partial^2 u}{\partial z^2}\right) = c\rho\frac{\partial u}{\partial t}.$$

The function that we call the solution to the problem not only must obey the differential equation given above, but also must satisfy an initial condition and a set of boundary conditions. For the one-dimensional heat-flow problem that we first consider, the initial condition will be the initial temperatures at all points along the rod,

$$u(x, t)|_{t=0} = u(x, 0) = f(x).$$

The boundary conditions will describe the temperature at each end of the rod as functions of time. Our first examples will consider the case where these temperatures are held constant:

$$u(0, t) = c_1,$$

$$u(L, t) = c_2.$$

More general (and more practical) boundary conditions will involve not only the temperature but the temperature gradients, and these may vary with time:

$$A_1 u(0, t) + B_1 \frac{\partial u(0, t)}{\partial x} = F_1(t),$$

$$A_2 u(L, t) + B_2 \frac{\partial u(L, t)}{\partial x} = F_2(t).$$

In Chapter 8, you will study the application of finite differences to solve parabolic partial-differential equations. We begin with problems of one space dimension.

8.1 THE EXPLICIT METHOD

Divices space and time into discrete uniform subintervals and replaces both time and space derivatives by finite-difference approximations, permitting one to easily compute values of the function at a time Δt after the initial time. These values are then used to compute a second set of values and the process is repeated.

8.2 CRANK–NICOLSON METHOD

Overcomes the limitations that the explicit method places on the size of Δt and gives improved accuracy at the expense of having to solve a set of equations at each time step. Fortunately, the system is tridiagonal for a one-space-dimension problem.

8.3 DERIVATIVE BOUNDARY CONDITIONS

Really impose no new difficulties because we can extend the region artificially and replace the derivatives at the boundaries also by finite differences.

8.4 STABILITY AND CONVERGENCE CRITERIA

Shows you why there were limitations on the size of Δt in the explicit method and none with implicit methods.

8.1   THE EXPLICIT METHOD

Our approach to solving parabolic partial-differential equations by a numerical method is to replace the partial derivatives by finite-difference approximations. For the one-dimensional heat-flow equation,

$$\frac{\partial^2 u}{\partial x^2} = \frac{c\rho}{k}\frac{\partial u}{\partial t}, \tag{8.1}$$

we can use the relations

$$\left.\frac{\partial^2 u}{\partial x^2}\right|_{\substack{x=x_i \\ t=t_j}} = \frac{u^j_{i+1} - 2u^j_i + u^j_{i-1}}{(\Delta x)^2} + O(\Delta x)^2 \tag{8.2}$$

and

$$\left.\frac{\partial u}{\partial t}\right|_{\substack{x=x_i \\ t=t_j}} = \frac{u^{j+1}_i - u^j_i}{\Delta t} + O(\Delta t). \tag{8.3}$$

We use subscripts to denote position and superscripts for time. Note that the error terms are of different orders since a forward difference is used in Eq. (8.3). This introduces some special limitations, but when this is done, the procedure is simplified.

Substituting Eqs. (8.2) and (8.3) into (8.1) and solving for u_i^{j+1} gives the equation for the forward-difference method:

$$u_i^{j+1} = \frac{k\,\Delta t}{c\rho(\Delta x)^2}(u_{i+1}^j + u_{i-1}^j) + \left(1 - \frac{2k\,\Delta t}{c\rho(\Delta x)^2}\right)u_i^j. \qquad (8.4)$$

We have solved for u_i^{j+1} in terms of the temperatures at time t_j in Eq. (8.4) in view of the normally known conditions for a parabolic partial-differential equation. We subdivide the length into uniform subintervals and apply our finite-difference approximation to Eq. (8.1) at each point where u is not known. Equation (8.4) then gives the values of u at each interior point at $t = t_1$ since the values at $t = t_0$ are given by the initial conditions. It can then be used to get values at t_2 using the values at t_1 as initial conditions, so we can step the solution forward in time. At the endpoints, the boundary conditions will determine u.

The relative size of the time and distance steps, Δt and Δx, affects Eq. (8.4). If the ratio of $\Delta t/(\Delta x)^2$ is chosen so that $k\,\Delta t/c\rho(\Delta x)^2 = \frac{1}{2}$, the equation is simplified in that the last term vanishes and we have

$$u_i^{j+1} = \frac{1}{2}(u_{i+1}^j + u_{i-1}^j). \qquad (8.5)$$

If the value of $k\,\Delta t/c\rho(\Delta x)^2$ is chosen as less than one-half, there will be improved accuracy* (limited, of course, by the errors dependent on the size of Δx). If the value is chosen greater than one-half, which would reduce the number of calculations required to advance the solution through a given interval of time, the phenomenon of *instability* sets in. (We show later in this chapter why this ratio affects stability and convergence.)

We illustrate the method with a simple example, varying the value of $k\,\Delta t/c\rho(\Delta x)^2$ to demonstrate its effect.

EXAMPLE   A large flat steel plate is 2 cm thick. If the initial temperatures (°C) within the plate are given, as a function of the distance from one face, by the equations

$$u = 100x \qquad \text{for } 0 \le x \le 1,$$
$$u = 100(2 - x) \qquad \text{for } 1 \le x \le 2,$$

find the temperatures as a function of x and t if both faces are maintained at 0°C.

---

*It can be shown that choosing the ratio equal to $\frac{1}{6}$ is especially advantageous in minimizing the truncation error.

Since the plate is large, we can neglect lateral flow of heat relative to the flow perpendicular to the faces and hence use Eq. (8.1) for heat flow in one direction. For steel, $k = 0.13$ cal/sec $\cdot$ cm $\cdot$ °C, $c = 0.11$ cal/g $\cdot$ °C and $\rho = 7.8$ g/cm^3. In order to use Eq. (8.5) as an approximation to the physical problem, we subdivide the total thickness into an integral number of spaces. Let us use $\Delta x = 0.25$, giving eight subdivisions. If we wish to use Eq. (8.5), we then fix Δt by the relation

$$\frac{k\,\Delta t}{c\rho(\Delta x)^2} = \frac{1}{2}, \qquad \Delta t = \frac{(0.11)(7.8)(0.25)^2}{(2)(0.13)} = 0.206 \text{ sec.}$$

The boundary conditions are

$$u(0, t) = 0, \qquad u(2, t) = 0.$$

The initial condition is

$$u(x, 0) = 100x \qquad \text{for } 0 \le x \le 1,$$
$$u(x, 0) = 100(2 - x) \qquad \text{for } 1 \le x \le 2.$$

Our calculations are conveniently recorded in a table, as in Table 8.1, where each row of figures is at a particular time. We begin by filling in the initial conditions along the first row, at $t = 0.0$. The simple algorithm of Eq. (8.5) tells us that, at each interior point, the temperature at any point at the end of a time step is just the arithmetic average of the temperatures at the adjacent points at the beginning of that time step. The end temperatures are given by the boundary conditions. Because the temperatures are symmetrical on either side of the center line, we compute only for $x \le 1.0$. The temperature at $x = 1.25$ is the same as at $x = 0.75$.

Table 8.1 Numerical solution to heat-flow example

t	$x = 0$	$x = 0.25$	$x = 0.50$	$x = 0.75$	$x = 1.0$	$x = 1.25$
			Calculated temperatures at			
0.0	0	25.0	50.0	75.0	100.0	75.0
0.206	0	25.0	50.0	75.0	75.0	75.0
0.412	0	25.0	50.0	62.5	75.0	62.5
0.619	0	25.0	43.75	62.5	62.5	62.5
0.825	0	21.88	43.75	53.12	62.5	53.12
1.031	0	21.88	37.5	53.12	53.12	53.12
1.238	0	18.75	37.5	45.31	53.12	45.31
1.444	0	18.75	32.03	45.31	45.31	45.31
1.650	0	16.16	32.03	38.67	45.31	38.67
1.856	0	16.16	27.34	38.67	38.67	38.67
2.062	0	13.67	27.34	33.01	38.67	33.01

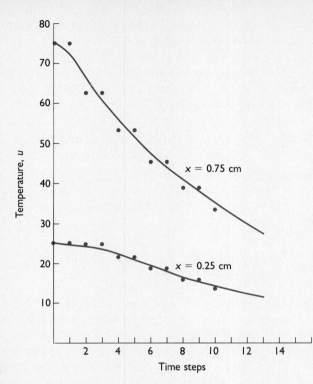

Figure 8.2

In Fig. 8.2, we compare some of the numerical results with the analytical solution, which is

$$u = 800 \sum_{n=0}^{\infty} \frac{1}{\pi^2(2n + 1)^2} \cos \frac{(2n + 1)\pi(x - 1)}{2} e^{-0.3738(2n+1)^2 t}.$$

Note that the numerical values are close to and oscillate about the curves that are drawn to represent the analytical solution. In general the errors of Table 8.1 are less than 4%. If the size of Δx were less, the errors would be smaller. Unfortunately, this easy method is not always so accurate.

When the value for $k \, \Delta t / c\rho(\Delta x)^2$ in Eq. (8.4) is other than 0.5, the resulting equation is slightly more complicated, but the related computer program is not significantly slower. The major effect is the size of the time steps; reducing the value requires more successive calculations to reach a given time after the start of heat flow. Increasing the ratio to a value greater than 0.5 would decrease the number of successive calculations and hence reduce the computer time needed to solve a given problem. An extremely important phenomenon occurs, however, when the ratio is >0.5; as illustrated in Fig. 8.3 where temperature profiles are shown at two different times, $t = 0.99$ sec and $t = 1.98$ sec, very inaccurate results may occur. The curves show the solution from the infinite series.

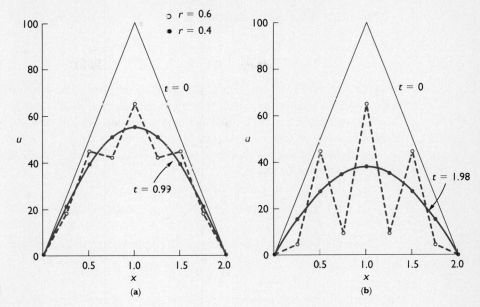

Figure 8.3

(a) (b)

The open circles, with $r = 0.6$, show extreme oscillation: The calculated values are very imprecise, and completely impossible behavior is represented. The data for $r = 0.4$ are smooth and almost exactly match the series solution.

The reason for this great difference in behavior is that the choice of the ratio $k \, \Delta t / c\rho(\Delta x)^2 = 0.6$ introduces instability in which errors grow at an accelerating rate as time increases. (At values of $t > 2$, negative values of u are computed, a patently impossible situation.) As indicated in Fig. 8.3, when this ratio is taken at 0.4, the results are excellent.

In Fig. 8.2 with $k \, \Delta t / c\rho(\Delta x)^2 = 0.5$, the limiting value to avoid instability was used. Even here, the oscillation of points around the curves indicates its borderline value, and the accuracy is hardly acceptable. The phenomena of stability and convergence, more fully discussed in a later section of this chapter, set a maximum value of 0.5 to the ratio $k \, \Delta t / c\rho(\Delta x)^2$. When derivative boundary conditions are involved, the ratio must be less than 0.5. Discontinuities in the initial conditions—even a discontinuity in the gradient, $\partial u / \partial x$, at $t = 0$—cause the accuracy to be poor at the maximum value. The next example will demonstrate this.

We will now solve a problem in diffusion, which is governed by the same mathematical equation as is heat conduction in a solid.

E X A M P L E A hollow tube 20 cm long is initially filled with air containing 2% of ethyl alcohol vapors. At the bottom of the tube is a pool of alcohol that evaporates into the stagnant gas above. (Heat transfers to the alcohol from the surroundings to maintain a constant temperature of 30°C, at which temperature the vapor pressure is 0.1 atm.) At the upper end of the tube, the alcohol vapors dissipate to the outside air, so the concentration is essentially zero. Considering only the effects of molecular diffusion, determine the concentration of alcohol as a function of time and the distance x measured from the top of the tube.

Molecular diffusion follows the law

$$\frac{\partial c}{\partial t} = D\frac{\partial^2 c}{\partial x^2},$$

where D is the diffusion coefficient, with units of cm^2/sec in the cgs system. (This is the same as for the ratio $k/c\rho$, which is often termed *thermal diffusivity*.) For ethyl alcohol, $D = 0.119$ cm^2/sec at 30°C, and the vapor pressure is such that 10 volume percent alcohol in air is present at the surface.

Subdivide the length of the tube into five intervals, so $\Delta x = 4$ cm. Using the maximum value permitted for Δt yields

$$D\frac{\Delta t}{(\Delta x)^2} = \frac{1}{2} = 0.119\frac{\Delta t}{(4)^2}, \qquad \Delta t = 67.2 \text{ sec.}$$

Our initial condition is $c(x, 0) = 2.0$. The boundary conditions are

$$c(0, t) = 0.0, \qquad c(20, t) = 10.0.$$

The concentrations are measured by the percent of alcohol vapor in the air.

The computations are shown in Table 8.2. Again, using Eq. (8.5), each interior value of c is given by the arithmetic average of concentrations on either side in the line above. A little reflection is required to determine the proper concentrations to be used for $x = 0$ and $x = 20$ at $t = 0$. While these are initially at 2%, we assume they are changed instantaneously to 0% and 10% because the effective concentrations acting during the first time interval are not the initial values but the changed values. We have accordingly rewritten these values as shown in the first row of Table 8.2.

Table 8.2   Diffusion of alcohol vapors in a tube—solution of the equation

$$u_i^{j+1} = \frac{1}{2}(u_{i-1}^j + u_{i+1}^j)$$

Time, sec	Concentration of alcohol at					
	$x = 0$	$x = 4$	$x = 8$	$x = 12$	$x = 16$	$x = 20$
0	0.0	2.0	2.0	2.0	2.0	10.0
67.2	0.0	1.00	2.00	2.00	6.00	10.0
134.4	0.0	1.00	1.50	4.00	6.00	10.0
201.6	0.0	0.75	2.50	3.75	7.00	10.0
268.8	0.0	1.25	2.25	4.75	6.875	10.0
336.0	0.0	1.125	3.00	4.562	7.375	10.0
403.2	0.0	1.500	2.844	5.188	7.281	10.0
470.4	0.0	1.422	3.344	5.062	7.594	10.0
537.6	0.0	1.672	3.242	5.469	7.531	10.0
604.8	0.0	1.621	3.570	5.386	7.734	10.0
Steady state	0.0	2.0	4.0	6.0	8.0	10.0

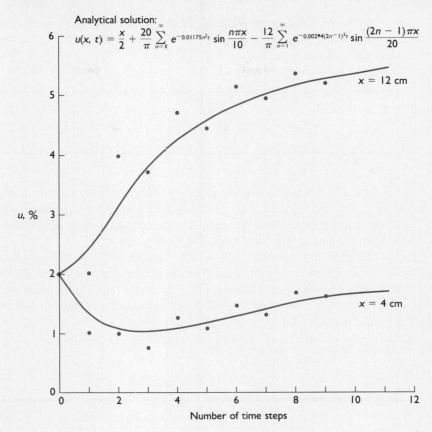

Analytical solution:
$$u(x, t) = \frac{x}{2} + \frac{20}{\pi} \sum_{n=1}^{\infty} e^{-0.01175n^2 t} \sin \frac{n\pi x}{10} - \frac{12}{\pi} \sum_{n=1}^{\infty} e^{-0.00294(2n-1)^2 t} \sin \frac{(2n-1)\pi x}{20}$$

$x = 12$ cm

$x = 4$ cm

u, %

Number of time steps

Figure 8.4

As time passes, the concentration will become a linear function of x, from 10% at $x = 20$ to 0% at $x = 0$. The calculated values approach the steady-state concentrations as time passes. However, as Fig. 8.4 shows, the successive calculated values oscillate about the values calculated from the analytical solution.* ∎

While the algorithm of Eq. (8.5) is simple to use, the results leave something to be desired in the way of accuracy. The oscillatory nature of the results, when a smooth trend of the concentration is expected, is also of concern. We can improve accuracy by smaller steps, but if we decrease Δx, the time steps must also decrease, because the ratio $D \Delta t/(\Delta x)^2$ cannot exceed $\frac{1}{2}$. Cutting Δx in half will require making Δt one-fourth of its previous value, giving a total of eight times as many calculations. With the mixed order of error terms, it is not obvious what reduction of error this would give, but they should be reduced about fourfold.

---

*In this instance the analytical solution is given by the infinite series shown in Fig. 8.4.

We can reduce Δt without change in Δx, however, because the method is stable for any value of the ratio less than $\frac{1}{2}$. Suppose we make $D \, \Delta t/(\Delta x)^2 = \frac{1}{4}$. The basic difference equation, (8.4), now becomes

$$u_i^{j+1} = \frac{1}{4}(u_{i+1}^j + u_{i-1}^j) + \frac{1}{2}u_i^j. \qquad (8.6)$$

Table 8.3 summarizes the use of Eq. (8.6) on the same example as before. As shown more clearly by Fig. 8.5, this last calculation causes the calculated concentrations to follow smooth curves, and the error is reduced considerably. The poorest accuracy is near the beginning. The later values are close to the curve.

This initial poor accuracy is a result of the discontinuities in the boundary conditions. The abrupt change of concentrations at the ends of the tube makes the explicit method inaccurate. In a situation with continuity of the initial values of the potential plus continuity of its derivatives and the absence of discontinuities in the boundary conditions, one would get better accuracy throughout.

Table 8.3 Diffusion of alcohol vapors in a tube:

$$u_i^{j+1} = \frac{1}{4}(u_{i-1}^j + u_{i+1}^j) + \frac{1}{2}u_i^j$$

Time, sec	Concentration of alcohol at					
	$x = 0$	$x = 4$	$x = 8$	$x = 12$	$x = 16$	$x = 20$
0.0	0.0	2.0	2.0	2.0	2.0	10.0
33.6	0.0	1.50	2.00	2.00	4.00	10.0
67.2	0.0	1.25	1.875	2.50	5.00	10.0
100.8	0.0	1.094	1.875	2.969	5.625	10.0
134.4	0.0	1.015	1.953	3.360	6.055	10.0
168.0	0.0	0.996	2.070	3.682	6.368	10.0
201.6	0.0	1.015	2.204	3.950	6.604	10.0
235.2	0.0	1.058	2.343	4.177	6.790	10.0
268.8	0.0	1.115	2.480	4.372	6.939	10.0
302.4	0.0	1.177	2.612	4.541	7.062	10.0
336.0	0.0	1.241	2.736	4.689	7.166	10.0
369.6	0.0	1.304	2.850	4.820	7.255	10.0
403.2	0.0	1.364	2.956	4.936	7.332	10.0
436.8	0.0	1.421	3.053	5.040	7.400	10.0
470.4	0.0	1.474	3.142	5.133	7.460	10.0
504.0	0.0	1.522	3.223	5.217	7.513	10.0
537.6	0.0	1.567	3.296	5.292	7.561	10.0
Steady state	0.0	2.0	4.0	6.0	8.0	10.0

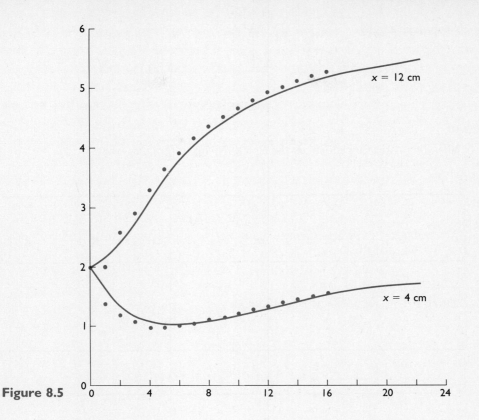

Figure 8.5

The method presented in this section is called the *explicit method* because each new value of the potential can be immediately calculated from quantities that are already known. It is simple and economical of calculation effort but has a severely limited upper value for the ratio $k\,\Delta t/c\rho(\Delta x)^2$. We can remove this limitation by going to an *implicit method*.

8.2 CRANK–NICOLSON METHOD

When the difference equation, Eq. (8.4), was derived, we noted that a mixed order of error was involved because a forward difference was used to replace the time derivative while a central difference was used for the distance derivative. However, the difference quotient, $(u_i^{j+1} - u_i^{j})/\Delta t$, can be considered a central difference if we take it as corresponding to the midpoint of the time interval. Suppose we do consider $(u_i^{j+1} - u_i^{j})/\Delta t$ as a central-difference approximation to $\partial u/\partial t$, and equate it to a central-difference quotient for the second derivative with distance, also corresponding to the midpoint in time, by averaging difference quotients at the beginning and end of the time step. Then

$$\frac{\partial^2 u}{\partial x^2} = \frac{c\rho}{k}\frac{\partial u}{\partial t}$$

is approximated by

$$\frac{1}{2}\left(\frac{u_{i+1}^j - 2u_i^j + u_{i-1}^j}{(\Delta x)^2} + \frac{u_{i+1}^{j+1} - 2u_i^{j+1} + u_{i-1}^{j+1}}{(\Delta x)^2}\right) = \frac{c\rho}{k}\left(\frac{u_i^{j+1} - u_i^j}{\Delta t}\right).$$

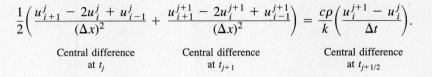

Central difference   Central difference   Central difference
at t_j at t_{j+1} at $t_{j+1/2}$

When this is rearranged we get the Crank–Nicolson formula, where

$$r = k\,\Delta t/c\rho(\Delta x)^2,$$
$$-ru_{i-1}^{j+1} + (2 + 2r)u_i^{j+1} - ru_{i+1}^{j+1} = ru_{i-1}^j + (2 - 2r)u_i^j + ru_{i+1}^j. \quad \textbf{(8.7)}$$

Letting $r = 1$, we get some simplification:

$$-u_{i-1}^{j+1} + 4u_i^{j+1} - u_{i+1}^{j+1} = u_{i-1}^j + u_{i+1}^j.$$

As we will discuss in the next section, one advantage of the Crank–Nicolson method is that it is stable for any value of r, although small values are more accurate.

Equation (8.7) is the usual formula for using the Crank–Nicolson method. Note that the new temperture u_i^{j+1} is not given directly in terms of known temperatures one time step earlier, but is a function of unknown temperatures at adjacent positions as well. It is therefore termed an *implicit method* in contrast to the explicit method of the previous section. With implicit methods, the values of u at $t = t_1$ are not just a function of values at $t = t_0$, but also involve the other u values at the same time step. This requires us to solve a set of simultaneous equations at each time step.

EXAMPLE   We illustrate the method by solving the problem of diffusion of alcohol vapors, the same as that previously attacked by the explicit method. Again, let us take $\Delta x = 4$ cm and take $r = 1$. Restating the problem, we have

$$\frac{\partial^2 u}{\partial x^2} = \frac{1}{D}\frac{\partial u}{\partial t}, \quad u(x, 0) = 2.0, \quad \begin{cases} u(0, t) = 0, \\ u(20, t) = 10.0, \end{cases}$$

$$D = 0.119 \text{ cm}^2/\text{sec}.$$

If $D\,\Delta t/(\Delta x)^2 = r = 1$ and $\Delta x = 4$ cm, then $\Delta t = 134.4$ sec. For $t = 134.4$, at the end of the first time step, we write the equations at each point whose concentration is unknown. The effective concentrations at the two ends, 0% and 10%, respectively, are again used:

$$-0.0 + 4u_1 - u_2 \quad = 0.0 + \quad 2.0,$$
$$-u_1 + 4u_2 - u_3 \quad = 2.0 + \quad 2.0,$$
$$-u_2 + 4u_3 - u_4 \quad = 2.0 + \quad 2.0,$$
$$-u_3 + 4u_4 - 10.0 = 2.0 + 10.0.$$

This method requires more work because, at each time step, we must solve a set of equations similar to the above. Fortunately, the system is tridiagonal. Since we will need to repeatedly solve the system with the same coefficient matrix (we transpose the constant values to the right side in the first and last equations), it is desirable to use an LU method. We then solve by back-substitutions, using the elements of L and U:

$$L = \begin{bmatrix} 1 & 0 & 0 & 0 \\ -1 & 3.75 & 0 & 0 \\ 0 & -1 & 3.733 & 0 \\ 0 & 0 & -1 & 3.732 \end{bmatrix}, \quad U = \begin{bmatrix} 1 & -0.25 & 0 & 0 \\ 0 & 1 & -0.267 & 0 \\ 0 & 0 & 1 & -0.268 \\ 0 & 0 & 0 & 1 \end{bmatrix}.$$

Remember that, to solve $Ax = b = LUx$, we first solve $Ly = b$ (which requires no reduction of L because it is already triangular) and then we solve $Ux = y$ by back-substitution. Observe that when A is tridiagonal, its zeros are preserved in L and U, and, in fact we need to compute only $2(n - 1)$ new values to get the LU equivalent of the $n \times n$ tridiagonal matrix A. Not only is the computational effort minimized, but we can also minimize computer memory space by storing only the three nonzero values in each row. In fact, as disussed in Chapter 2, we can store the elements of both L and U in place of the original values in A.

Table 8.4   Diffusion of alcohol vapors in a tube—Crank–Nicolson method

Time, sec	Calculated concentration values at					
	$x = 0$	$x = 4$	$x = 8$	$x = 12$	$x = 16$	$x = 20$
0.0	0.0	2.0	2.0	2.0	2.0	10.0
134.4	0.0	0.980	2.019	3.072	5.992	10.0
268.8	0.0	1.070	2.363	4.305	6.555	10.0
403.2	0.0	1.276	2.861	4.762	6.962	10.0
537.6	0.0	1.471	3.165	5.115	7.159	10.0

Time, sec	Analytical values	
	$x = 4$	$x = 12$
134.4	1.078	3.191
268.8	1.108	4.272
403.2	1.340	4.873
537.6	1.543	5.248

In Table 8.4 the results of calculations by the Crank–Nicolson method are listed. For the first line of calculations the b-vector has components 2.0, 4.0, 4.0, and 22.0, and the calculated values are the components of $A^{-1}b$. In computing the second line, the components of b are 2.019, 4.052, 8.011, and 23.072. Succeeding lines use the proper sums of values from the line above to determine the b-vector.

When the calculated values are compared to the analytical results, it is observed that the errors of this method, in this example, are about the same as those made in the explicit method with $D\,\Delta t/(\Delta x)^2 = \frac{1}{4}$. We require only one-fourth as many time steps, however, to reach 537.6 sec. With smaller values of Δx, this freedom of choice in the value of $D\,\Delta t/(\Delta x)^2$ is especially valuable. ∎

8.3 DERIVATIVE BOUNDARY CONDITIONS

In heat-conduction problems, the most usual situation at the endpoints is not that they are held at a constant temperature, but that heat is lost by conduction or radiation at a rate proportional to some power of the temperature difference between the surface of the body and its surroundings. This leads to a relationship involving the space derivative of temperature at the surface. In the analytical solution of heat-flow problems through Fourier series, this adds considerable complication in determining the coefficients, but our numerical technique requires only minor modifications.

The rate of heat loss from the surface of a solid is generally expressed as

$$\text{Rate of heat loss} = hA(u - u_0), \qquad (8.8)$$

where A is the surfce area, u is the surface temperature, u_0 is the temperature of the surrounding medium, and h is a coefficient of heat transfer; h is increased by motion of the surrounding medium. To facilitate heat flow, the surrounding medium is often caused to flow rapidly by mechanical means (as in some salt evaporators) or by proper design (as in vertical-tube heat exchangers in distillation columns). This situation is called *forced convection*.

In other situations, h is also a function of the surface temperature, as in *natural convection*, in which the motion of surrounding fluid is caused by thermal currents. Another important situation is heat loss by radiation, in which the rate of heat loss is proportional to the fourth power of the temperature difference between the surface and the surrounding surfaces to which heat is being radiated. In both these situations, the rate of heat loss is proportional to some power of the surface temperature. This gives rise to nonlinear equations that are not as readily solved. One usually prefers to force the mathematical model to a linear form even though this is not exactly true. One way to approximate this is to use Eq. (8.8), absorbing the nonlinear aspects into the coefficient h, and to change h appropriately through the progress of the calculations so that it takes on a reasonably correct average value. Our examples will treat only the simpler situations where Eq. (8.8) applies directly.

As we will see, heat loss from the surface by conduction leads to a derivative boundary condition. Our procedure will be to replace the derivatives in both the differential equation and in the boundary conditions by difference quotients. We illustrate with a simple example.

EXAMPLE   An aluminum cube is $4 \times 4 \times 4$ in. ($k = 0.52$, $c = 0.226$, $\rho = 2.70$ in cgs units). All but one face is perfectly insulated, and the cube is initially at 1000°F. Heat is lost from the uninsulated face to a fluid flowing past it according to the equation

$$\text{Rate of heat loss (in Btu per sec)} = hA(u - u_0),$$

where $h = 0.15$ Btu/sec $\cdot$ ft^2 $\cdot$ °F,
  A = surface area in ft^2,
  u = surface temperature, °F,
  u_0 = temperature of fluid, °F.

If u_0, the temperature of fluid flowing past the aluminum cube, is constant at 70°F, find the temperatures inside the cube as a function of time. While we could work the problem in cgs units, we elect to use English units, making suitable changes in k, c, and ρ.

Because of the insulation on the lateral faces, again the only direction in which heat will flow is perpendicular to the uninsulated face, and the equation is

$$\frac{\partial u}{\partial t} = \frac{k}{c\rho} \frac{\partial^2 u}{\partial x^2}.$$

The intitial condition, with x representing distance from the uninsulated face, is

$$u(x, 0) = 1000.$$

For boundary conditions, at the open surface we have

$$-kA \frac{\partial u}{\partial x}\bigg|_{x=0} = -hA(u - 70),$$

because the rate at which heat leaves the surface must be equal to the rate at which heat flows to the surface. The negative sign on the left is required because heat flows in a direction opposite to a positive gradient; on the right it occurs because heat is being lost in the direction of negative x. At the other side of the cube, the rate of heat flow is zero because of insulation:

$$\frac{\partial u}{\partial x}\bigg|_{x=4} = 0.$$

We plan to use the explicit method with

$$r = k \, \Delta t / c\rho(\Delta x)^2 = \frac{1}{4}.*$$

---

*The ratio must be smaller than $\frac{1}{2}$ to give stability with the derivative end conditions.

Suppose we let $\Delta x = 1$ in. To calculate Δt, we need k, c, and ρ expressed in the units of inches, pounds, seconds, and °F. We first convert units:

$$k = \left(0.52 \frac{\text{cal}}{\text{sec} \cdot \text{cm} \cdot °\text{C}}\right)\left(\frac{1 \text{ Btu}}{252 \text{ cal}}\right)\left(\frac{2.54 \text{ cm}}{1 \text{ in}}\right)\left(\frac{1°\text{C}}{1.8°\text{F}}\right)$$

$$= 0.00291 \frac{\text{Btu}}{\text{sec} \cdot \text{in} \cdot °\text{F}};$$

$$c = \left(0.226 \frac{\text{cal}}{\text{g} \cdot °\text{C}}\right)\left(\frac{1 \text{ Btu}}{252 \text{ cal}}\right)\left(\frac{454 \text{ g}}{1 \text{ lb}}\right)\left(\frac{1°\text{C}}{1.8°\text{F}}\right)$$

$$= 0.226 \frac{\text{Btu}}{\text{lb} \cdot °\text{F}};$$

$$\rho = \left(2.70 \frac{\text{g}}{\text{cm}^3}\right)\left(\frac{1 \text{ lb}}{454 \text{ g}}\right)\left(\frac{2.54 \text{ cm}}{1 \text{ in}}\right)^3$$

$$= 0.0975 \frac{\text{lb}}{\text{in}^3};$$

$$\Delta t = \frac{c\rho(\Delta x)^2}{4k} = \frac{(0.226)(0.0975)(1)^2}{(4)(0.00291)} = 1.89 \text{ sec.}$$

For the ratio of values of Δt and Δx that we have chosen, our differential equation becomes

$$u_i^{j+1} = \frac{1}{4}(u_{i+1}^j + u_{i-1}^j) + \frac{1}{2}u_i^j, \tag{8.9}$$

which is to be applied at every point where u is unknown. In this example, this includes the points at $x = 0$ and $x = 4$ as well as the interior points. To enable us to write Eq. (8.9), we extend the domain of u one step on either side of the boundary. Let x_R be a fictitious point to the right of $x = 4$, and let x_L be a fictitious point to the left of $x = 0$. If x_1 signifies $x = 0$, and if x_5 signifies $x = 4$, we have the relations, from Eq. (8.9)

$$u_1^{j+1} = \frac{1}{4}(u_2^j + u_L^j) + \frac{1}{2}u_1^j,$$

$$u_5^{j+1} = \frac{1}{4}(u_R^j + u_4^j) + \frac{1}{2}u_5^j.$$

Now we use the boundary conditions to eliminate the fictitious points, writing them as central-difference quotients:

$$-k\frac{\partial u}{\partial x}\bigg|_{x=0} \doteq -(0.00291)\left(\frac{u_2^j - u_L^j}{2(1)}\right) = -\frac{0.15}{144}(u_1^j - 70);$$

$$-k\frac{\partial u}{\partial x}\bigg|_{x=4} \doteq -(0.00291)\left(\frac{u_R^j - u_4^j}{2(1)}\right) = 0.$$

The 144 factor in the first equation changes h to a basis of in^2. Solving for u_L and u_R, we have

$$u_L = u_2 - 0.716u_1 + 50.1,$$

$$u_R = u_4,$$

and the set of equations that give the temperatures becomes

$$u_1^{j+1} = 0.32u_1^j + \frac{1}{2}u_2^j + 12.525,$$

$$u_2^{j+1} = \frac{1}{4}(u_1^j + u_3^j) + \frac{1}{2}u_2^j,$$

$$u_3^{j+1} = \frac{1}{4}(u_2^j + u_4^j) + \frac{1}{2}u_3^j,$$

$$u_4^{j+1} = \frac{1}{4}(u_3^j + u_5^j) + \frac{1}{2}u_4^j,$$

$$u_5^{j+1} = \frac{1}{2}u_4^j + \frac{1}{2}u_5^j. \qquad (8.10)$$

A similar treatment of the boundary conditions using the Crank–Nicolson method with $r = k\,\Delta t/c\rho(\Delta x)^2 = 1$ leads to the set of simultaneous equations:

$$4.716u_1^{j+1} - 2u_2^{j+1} = -0.716u_1^j + 2u_2^j + 100.2,$$

$$-u_1^{j+1} + 4u_2^{j+1} - u_3^{j+1} = u_1^j + u_3^j,$$

$$-u_2^{j+1} + 4u_3^{j+1} - u_4^{j+1} = u_2^j + u_4^j,$$

$$-u_3^{j+1} + 4u_4^{j+1} - u_5^{j+1} = u_3^j + u_5^j,$$

$$-2u_4^{j+1} + {}_4 4u_5^{j+1} = 2u_4^j. \qquad (8.11)$$

For the set of equations in (8.11), Δt will be 7.56 sec. We advance the solution one time step at a time by repeatedly solving either (8.10) or (8.11), using the proper values of u_i^j. ∎

8.4   STABILITY AND CONVERGENCE CRITERIA

We have previously stated that in order to ensure stability and convergence in the explicit method, the ratio $r = k \, \Delta t / c\rho(\Delta x)^2$ must be $\frac{1}{2}$ or less. The implicit Crank–Nicolson method has no such limitation. In this section, these phenomena and criteria will be studied in more detail.

By *convergence*, we mean that the results of the method approach the analytical values as Δt and Δx both approach zero. By *stability*, we mean that errors made at one stage of the calculations do not cause increasingly large errors as the computations are continued, but rather will eventually damp out.

We will first discuss convergence, limiting ourselves to the simple case of the unsteady-state heat-flow equation in one dimension:*

$$\frac{\partial U}{\partial t} = \frac{k}{c\rho} \frac{\partial^2 U}{\partial x^2}. \tag{8.12}$$

Let us use the symbol U to represent the exact solution to Eq. (8.12), and u to represent the numerical solution. At the moment we assume that u is free of round-off errors, so the only difference between U and u is the error made by replacing Eq. (8.12) by the difference equation. Let $e_i^j = U_i^j - u_i^j$, at the point $x = x_i$, $t = t_j$. By the explicit method, Eq. (8.12) becomes

$$u_i^{j+1} = r(u_{i+1}^j + u_{i-1}^j) + (1 - 2r)u_i^j, \tag{8.13}$$

where $r = k \, \Delta t / c\rho(\Delta x)^2$. Substituting $u = U - e$ into Eq. (8.13), we get

$$e_i^{j+1} = r(e_{i+1}^j + e_{i-1}^j) + (1 - 2r)e_i^j - r(U_{i+1}^j + U_{i-1}^j) - (1 - 2r)U_i^j + U_i^{j+1}. \tag{8.14}$$

By using Taylor-series expansions, we have

$$U_{i+1}^j = U_i^j + \left(\frac{\partial U}{\partial x}\right)_{i,j} \Delta x + \frac{(\Delta x)^2}{2} \frac{\partial^2 U(\xi_1, t_j)}{\partial x^2}, \qquad x_i < \xi_1 < x_{i+1},$$

$$U_{i-1}^j = U_i^j - \left(\frac{\partial U}{\partial x}\right)_{i,j} \Delta x + \frac{(\Delta x)^2}{2} \frac{\partial^2 U(\xi_2, t_j)}{\partial x^2}, \qquad x_{i-1} < \xi_2 < x_i,$$

$$U_i^{j+1} = U_i^j + \Delta t \frac{\partial U(x_i, \eta)}{\partial t}, \qquad t_j < \eta < t_{j+1}.$$

---

*We could have treated the simpler equation $\partial U / \partial T = \partial^2 U / \partial X^2$ without loss of generality, since with the change of variables $X = \sqrt{c\rho} \, x$, $T = kt$, the two equations are seen to be identical.

Substituting these into Eq. (8.14) and simplifying, remembering that $r(\Delta x)^2 = k\,\Delta t/c\rho$, we get

$$
e_i^{j+1} = r(e_{i+1}^j + e_{i-1}^j) + (1 - 2r)e_i^j + \Delta t\left[\frac{\partial U(x_i, \eta)}{\partial t} - \frac{k}{c\rho}\frac{\partial^2 U(\xi, t_j)}{\partial x^2}\right],
$$

$$
t_j \le \eta \le t_{j+1}, \qquad x_{i-1} \le \xi \le x_{i+1}. \tag{8.15}
$$

Let E^j be the magnitude of the maximum error in the row of calculations for $t = t_j$, and let $M > 0$ be an upper bound for the magnitude of the expression in brackets in Eq. (8.15). If $r \le \frac{1}{2}$, all the coefficients in Eq. (8.15) are positive (or zero) and we may write the inequality

$$
\left|e_i^{j+1}\right| \le 2rE^j + (1 - 2r)E^j + M\,\Delta t = E^j + M\,\Delta t.
$$

This is true for all the e_i^{j+1} at $t = t_{j+1}$, so

$$
E^{j+1} \le E^j + M\,\Delta t.
$$

Since this is true at each time step,

$$
E^{j+1} \le E^j + M\,\Delta t \le E^{j-1} + 2M\,\Delta t \le \cdots \le E^0 + (j+1)M\,\Delta t = E^0 + Mt_{j+1}
$$
$$
= Mt_{j+1},
$$

because E^0, the errors at $t = 0$, are zero, since U is given by the initial conditions.

Now, as $\Delta x \to 0$, $\Delta t \to 0$ if $k\,\Delta t/c\rho(\Delta x)^2 \le \frac{1}{2}$, and $M \to 0$, because, as both Δx and Δt get smaller,

$$
\left[\frac{\partial U(x_i, \eta)}{\partial t} - \frac{k}{c\rho}\frac{\partial^2 U(\xi, t_j)}{\partial x^2}\right] \to \left(\frac{\partial U}{\partial t} - \frac{k}{c\rho}\frac{\partial^2 U}{\partial x^2}\right)_{i,j} = 0.
$$

This last is by virtue of Eq. (8.12), of course. Consequently, we have shown that the explicit method is convergent for $r \le \frac{1}{2}$, because the errors approach zero as Δt and Δx are made smaller.

For the solution to the heat-flow equation by the Crank–Nicolson method, the analysis of convergence may be made by similar methods. The treatment is more complicated, but it can be shown that each E^{j+1} is no greater than a finite multiple of E^j plus a term that vanishes as both Δx and Δt become small, and this is independent of r. Hence, since the initial errors are zero, the finite-difference solution approaches the analytical solution as $\Delta t \to 0$ and $\Delta x \to 0$, requiring only that r stay finite.

Let us begin our discussion of stability with a numerical example. Since the heat-flow equation is linear, if two solutions are known, their sum is also a solution. We are interested in what happens to errors made in one line of the computations as the calculations are continued, and because of the additivity feature, the effect of a succession of errors is just the sum of the effects of the individual errors. We follow, then, a single error,* which most likely occurred due to round-off. If this single error does not grow in magnitude, we will call the method *stable*, since then the cumulative effect of all errors affects the later calculations no more than a linear combination of the previous errors would.

Table 8.5 illustrates the principle. We have calculated for the simple case where the boundary conditions are fixed, so that the errors at the endpoints are zero. We assume that a single error of size e occurs at $t = t_1$ and $x = x_2$. The explicit method, $k \Delta t/c\rho(\Delta x)^2 = \frac{1}{2}$, was used. The original error quite obviously dies out. As an exercise, it is left to the student to show that with $r > 0.5$, errors have an increasingly large effect on later computations. Table 8.6 shows that errors damp out for the Crank–Nicolson method with $r = 1$ even more rapidly than in the explicit method with $r = 0.5$.

Table 8.5 Propagation of errors—explicit method

t	Endpoint x_1	x_2	x_3	x_4	Endpoint x_5
t_0	0	0	0	0	0
t_1	0	e	0	0	0
t_2	0	0	$0.50e$	0	0
t_3	0	$0.25e$	0	$0.25e$	0
t_4	0	0	$0.25e$	0	0
t_5	0	$0.125e$	0	$0.125e$	0
t_6	0	0	$0.125e$	0	0
t_7	0	$0.062e$	0	$0.062e$	0
t_8	0	0	$0.062e$	0	0

Table 8.6 Propagation of errors—Crank–Nicolson method

t	x_1	x_2	x_3	x_4	x_5
t_0	0	0	0	0	0
t_1	0	e	0	0	0
t_2	0	$0.071e$	$0.286e$	$0.071e$	0
t_3	0	$0.107e$	$0.143e$	$0.107e$	0
t_4	0	$0.049e$	$0.092e$	$0.049e$	0
t_5	0	$0.036e$	$0.053e$	$0.036e$	0
t_6	0	$0.022e$	$0.033e$	$0.022e$	0
t_7	0	$0.013e$	$0.020e$	$0.013e$	0
t_8	0	$0.008e$	$0.013e$	$0.008e$	0

*A computation made assuming that each of the interior points has an error equal to e at $t = t_1$ demonstrates the effect more rapidly.

In order to discuss stability in a more analytical sense, we need some material from linear algebra. In Chapter 6 we discussed eigenvalues and eigenvectors of a matrix. We recall that for the matrix A and vector x, if

$$Ax = \lambda x,$$

then the scalar λ is an eigenvalue of A and x is the corresponding eigenvector. If the N eigenvalues of the $N \times N$ matrix A are all different, then the corresponding N eigenvectors are linearly independent, and any N-component vector can be written uniquely in terms of them.

Consider the unsteady-state heat-flow problem with fixed boundary conditions. Suppose we subdivide into $N + 1$ subintervals so there are N unknown values of the temperature being calculated at each time step. Think of these N values as the components of a vector. Our algorithm for the explicit method (Eq. (8.4)) can be written as the matrix equation*

$$\begin{bmatrix} u_1^{j+1} \\ u_2^{j+1} \\ \vdots \\ u_N^{j+1} \end{bmatrix} = \begin{bmatrix} (1 - 2r) & r & & & \\ r & (1 - 2r) & r & & \\ & \cdot & & \cdot & \\ & & r & (1 - 2r) \end{bmatrix} \begin{bmatrix} u_1^j \\ u_2^j \\ \vdots \\ u_N^j \end{bmatrix}$$

(8.16)

or

$$u^{j+1} = Au^j,$$

where A represents the coefficient matrix and u^j and u^{j+1} are the vectors whose N components are the successive calculated values of temperature. The components of u^0 are the initial values from which we begin our solution. The successive rows of our calculations are

$$u^1 = Au^0,$$
$$u^2 = Au^1 = A^2u^0,$$
$$\vdots$$
$$u^j = Au^{j-1} = A^2u^{j-2} = \cdots = A^ju^0.$$

(The superscripts on the A are here exponents; on the vectors they indicate time.)

---

*A change of variable is required to give boundary conditions of $u = 0$ at each end. This can always be done for fixed end conditions.

Suppose that errors are introduced into u^0, so that it becomes $\bar{u}^0$. We will follow the effects of this error through the calculations. The successive lines of calculation are now

$$\bar{u}^j = A\bar{u}^{j-1} = \cdots = A^j\bar{u}^0.$$

Let us define the vector e^j as $u^j - \bar{u}^j$ so the e^j represents the errors in u^j caused by the errors in $\bar{u}^0$. We have

$$e^j = u^j - \bar{u}^j = A^ju^0 - A^j\bar{u}^0 = A^je^0. \tag{8.17}$$

This shows that errors are propagated by using the same algorithm as that by which the temperatures are calculated, as was implicitly assumed earlier in this section.

Now the N eigenvalues of A are distinct (see below) so that its N eigenvectors x_1, $x_2, \ldots, x_N$ are independent, and

$$Ax_1 = \lambda_1 x_1,$$
$$Ax_2 = \lambda_2 x_2,$$
$$\vdots$$
$$Ax_N = \lambda_N x_N.$$

We now write the error vector e^0 as a linear combination of the x_i:

$$e^0 = c_1 x_1 + c_2 x_2 + \cdots + c_N x_N,$$

where the c's are constants. Then e^1 is, in terms of the x_i,

$$e^1 = Ae^0 = \sum_{i=1}^{N} Ac_i x_i = \sum_{i=1}^{N} c_i Ax_i = \sum_{i=1}^{N} c_i \lambda_i x_i,$$

and for e^2,

$$e^2 = Ae^1 = \sum_{i=1}^{N} Ac_i \lambda_i x_i = \sum_{i=1}^{N} c_i \lambda_i^2 x_i.$$

(Again, the superscripts on vectors indicate time; on λ they are exponents.) After j steps, Eq. (8.17) can be written

$$e^j = \sum_{i=1}^{N} c_i \lambda_i^j x_i.$$

If the magnitudes of all of the eigenvalues are less than or equal to unity, errors will not grow as the computations proceed; that is, the computational scheme is stable. This then is the analytical condition for stability: that the largest eigenvalue of the coefficient matrix for the algorithm be one or less in magnitude.

The eigenvalues of matrix A (Eq. (8.16)) can be shown to be (note they are all distinct):

$$1 - 4r \sin^2 \frac{n\pi}{2(N + 1)}, \qquad n = 1, 2, \ldots, N.$$

We will have stability for the explicit scheme if

$$-1 \leq 1 - 4r \sin^2 \frac{n\pi}{2(N+1)} \leq 1.$$

The limiting value of r is given by

$$-1 \leq 1 - 4r \sin^2 \frac{n\pi}{2(N+1)}$$

$$r \leq \frac{1/2}{\sin^2 (n\pi/2(N+1))}.$$

Hence, if $r \leq \frac{1}{2}$, the explicit scheme is stable.

The Crank–Nicolson scheme, in matrix form, is

$$\begin{bmatrix} (2+2r) & -r & & \\ -r & (2+2r) & -r & \\ & \cdot & \cdot & \\ & & \cdot & \\ & & -r & (2+2r) \end{bmatrix} \begin{bmatrix} u_1^{j+1} \\ u_2^{j+1} \\ \cdot \\ \cdot \\ u_N^{j+1} \end{bmatrix} = \begin{bmatrix} (2-2r) & r & & \\ r & (2-2r)r & & \\ & \cdot & \cdot & \\ & & \cdot & \\ & & r & (2-2r) \end{bmatrix} \begin{bmatrix} u_1^j \\ u_2^j \\ \cdot \\ \cdot \\ u_N^j \end{bmatrix},$$

or

$$Au^{j+1} = Bu^j.$$

We can write

$$u^{j+1} = (A^{-1}B)u^j, \tag{8.18}$$

so that stability is given by the magnitudes of the eigenvalues of $A^{-1}B$. These are

$$\frac{2 - 4r \sin^2 (n\pi/2(N-1))}{2 + 4r \sin^2 (n\pi/2(N-1))}, \quad n = 1, 2, \ldots, N.$$

Clearly all the eigenvalues are no greater than one in magnitude for any positive value of r.

With derivative boundary conditions, a similar analysis shows that the Crank–Nicolson method is stable for any positive value of r. For the explicit scheme, $r = \frac{1}{2}$ leads to instability with a finite surface coefficient. Smith (1978) shows that the limitation on r for stability is

$$r \le \frac{1}{2 + P\,\Delta x},$$

where P is the ratio of surface coefficient to conductivity, h/k.

8.5 PARABOLIC EQUATIONS IN TWO OR MORE DIMENSIONS

In principle, we can readily extend the preceding methods to higher space dimensions, especially when the region is rectangular. The heat-flow equation in two directions is

$$\frac{\partial u}{\partial t} = \frac{k}{c\rho}\left(\frac{\partial^2 u}{\partial x^2} + \frac{\partial^2 u}{\partial y^2}\right).$$

Taking $\Delta x = \Delta y$, and letting $r = k\,\Delta t/c\rho(\Delta x)^2$, we find that the explicit scheme becomes

$$u_{i,j}^{k+1} - u_{i,j}^{k} = r(u_{i+1,j}^{k} - 2u_{i,j}^{k} + u_{i-1,j}^{k} + u_{i,j+1}^{k} - 2u_{i,j}^{k} + u_{i,j-1}^{k})$$

or

$$u_{i,j}^{k+1} = r(u_{i+1,j}^{k} + u_{i-1,j}^{k} + u_{i,j+1}^{k} + u_{i,j-1}^{k}) + (1 - 4r)u_{i,j}^{k}.$$

In this scheme, the maximum value permissible for r in the simple case of constant end conditions is $\frac{1}{4}$. (Note that this corresponds again to the numerical value that gives a particularly simple formula.) In the more general case with $\Delta x \neq \Delta y$, the criterion is

$$\frac{k\,\Delta t}{c\rho[(\Delta x)^2 + (\Delta y)^2]} \le \frac{1}{8}.$$

The analogous equation in three dimensions, with equal grid spacing each way, has the coefficient $(1 - 6r)$, and $r \le \frac{1}{6}$ is required for convergence and stability.

The difficulty with the use of the explicit scheme is that the restrictions on Δt require inordinately many rows of calculations. One then looks for a method in which Δt can be made larger without loss of stability. In one dimension, the Crank–Nicolson method was such a method. In the two-dimensional case, using averages of central-difference approximations to give $\partial^2 u/\partial x^2$ and $\partial^2 u/\partial y^2$ at the midvalue of time, we get

$$
\begin{aligned}
u_{i,j}^{k+1} - u_{i,j}^k = \frac{r}{2}[&u_{i+1,j}^{k+1} - 2u_{i,j}^{k+1} + u_{i-1,j}^{k+1} + u_{i+1,j}^k - 2u_{i,j}^k + u_{i-1,j}^k \\
&+ u_{i,j+1}^{k+1} - 2u_{i,j}^{k+1} + u_{i,j-1}^{k+1} + u_{i,j+1}^k - 2u_{i,j}^k + u_{i,j-1}^k].
\end{aligned}
$$

The problem now is that a set of $(M)(N)$ simultaneous equations must be solved at each time step, where M is the number of unknown values in the x-direction and N in the y-direction. Furthermore, the coefficient matrix is no longer tridiagonal, so the solution to each set of equations is slower and memory space to store the elements of the matrix becomes exorbitant.

The advantage of a tridiagonal matrix is retained in the alternating-direction-implicit scheme (A.D.I.) proposed by Peaceman and Rachford (1955). It is widely used in modern computer programs for the solution of parabolic partial-differential equations. In this method, we approximate $\nabla^2 u$ by adding a central-difference approximation to $\partial^2 u/\partial x^2$ written at the beginning of the interval to a similar expression for $\partial^2 u/\partial y^2$ written at the end:

$$
u_{i,j}^{k+1} - u_{i,j}^k = r[\underbrace{u_{i+1,j}^k - 2u_{i,j}^k + u_{i-1,j}^k}_{\text{From } \partial^2 u/\partial x^2 \text{ at start}} + \underbrace{u_{i,j+1}^{k+1} - 2u_{i,j}^{k+1} + u_{i,j-1}^{k+1}}_{\text{From } \partial^2 u/\partial y^2 \text{ at end}}]. \tag{8.19}
$$

The obvious bias in this formula is balanced by reversing the order for the second-derivative approximations in the next time span:

$$
u_{i,j}^{k+2} - u_{i,j}^{k+1} = r[\underbrace{u_{i+1,j}^{k+2} - 2u_{i,j}^{k+2} + u_{i-1,j}^{k+2}}_{\text{From } \partial^2 u/\partial x^2 \text{ at end}} + \underbrace{u_{i,j+1}^{k+1} - 2u_{i,j}^{k+1} + u_{i,j-1}^{k+1}}_{\text{From } \partial^2 u/\partial y^2 \text{ at start}}]. \tag{8.20}
$$

We illustrate the A.D.I. method with a very simple example.

EXAMPLE   A square plate of steel is 15 cm on a side. Initially, all points are at 0°C. Follow the interior temperatures if two adjacent sides are suddenly brought to 100°C and held at that temperature. The plate is insulated on its flat surfaces, so that heat flows only in the x- and y-directions. We are given that $k = 0.13$, $\rho = 7.8$, $c = 0.11$ in cgs units. Take $\Delta x = \Delta y = 5$ cm, so that there are four interior points. Label the points as shown in Fig. 8.6, with u being used for horizontal traverses, v for vertical traverses.

Equations (8.19) and (8.20) become, with $r = k\,\Delta t/c\rho(\Delta x)^2$,

$$\begin{bmatrix} (1/r)+2 & -1 & 0 & 0 \\ -1 & (1/r)+2 & 0 & 0 \\ 0 & 0 & (1/r)+2 & -1 \\ 0 & 0 & -1 & (1/r)+2 \end{bmatrix} \begin{bmatrix} u_1 \\ u_2 \\ u_3 \\ u_4 \end{bmatrix} = \begin{bmatrix} 0 + 100 + [(1/r)-2]v_1 + v_2 \\ 100 + 100 + [(1/r)-2]v_3 + v_4 \\ 0 + v_1 + [(1/r)-2]v_2 + 0 \\ 100 + v_3 + [(1/r)-2]v_4 + 0 \end{bmatrix},$$

$$\begin{bmatrix} (1/r)+2 & -1 & 0 & 0 \\ -1 & (1/r)+2 & 0 & 0 \\ 0 & 0 & (1/r)+2 & -1 \\ 0 & 0 & -1 & (1/r)+2 \end{bmatrix} \begin{bmatrix} v_1 \\ v_2 \\ v_3 \\ v_4 \end{bmatrix} = \begin{bmatrix} 100 + 0 + [(1/r)-2]u_1 + u_2 \\ 0 + 0 + [(1/r)-2]u_3 + u_4 \\ 100 + u_1 + [(1/r)-2]u_2 + 100 \\ 0 + u_3 + [(1/r)-2]u_4 + 100 \end{bmatrix}.$$

Table 8.7 Temperature changes in a 15 × 15 cm steel plate, computed by the A.D.I. method

	Temperatures at			
t, sec	u_1 or v_1	u_2 or v_3	u_3 or v_2	u_4 or v_4
0.0	0.000	0.000	0.000	0.000
16.5*	8.392	17.482	0.699	9.790
33.0	16.529	30.515	2.543	16.529
49.5*	22.306	40.298	4.932	22.923
66.0	27.594	47.741	7.466	27.594
82.5*	31.532	53.444	9.891	31.803
99.0	35.000	57.863	12.138	35.001
115.5*	37.668	61.308	14.148	37.788
132.0	39.959	64.017	15.901	39.959
148.5*	41.758	66.159	17.410	41.811
165.0	43.278	67.864	18.693	43.278
181.5*	44.489	69.225	19.776	44.512
198.0	45.500	70.318	20.683	45.500
214.5*	46.314	71.197	21.440	46.324
231.0	46.988	71.907	22.068	46.988
247.5*	46.533	72.482	22.589	47.538
264.0	46.984	72.948	23.019	47.984
⋮	⋮	⋮	⋮	⋮
∞	50.00	75.00	25.00	50.00

*Calculations are less accurate at these times.

(Note the similarity to the equations for the A.D.I. method for steady-state temperatures in Section 7.11. This is hardly surprising, since the temperatures for the unsteady-state problem will eventually reach the steady state. One can think of this as the rationale behind the A.D.I. method for elliptic equations.)

We solve for the temperature history by using the above equations in succession. One normally discards the alternate computations, beginning with the first, because they tend to be inaccurate.

Table 8.7 and Fig. 8.7 show the results, with $\Delta t = 16.5$ sec, corresponding to $r = 0.1$.

Figure 8.6

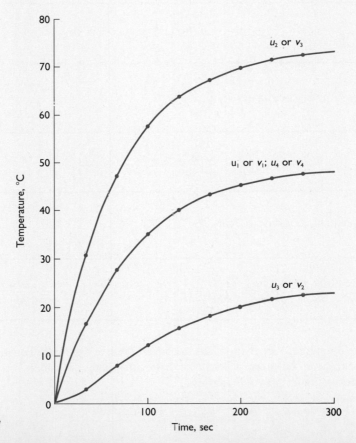

Figure 8.7

The compensation of errors produced by this alternation of direction gives a scheme that is convergent and stable for all values of r, although accuracy requires that r not be too large. The three-dimensional analog alternates three ways, returning to each of the three formulas after every third step. (Unfortunately the three-dimensional case is not stable for all fixed values of $r > 0$. A variant due to Douglas (1962) is unconditionally stable, however.) When the formulas are rearranged, in each case tridiagonal coefficient matrices result.

Note that the equations can be broken up into two independent subsets, each containing only two equations. This is always true in the A.D.I. method; each row gives a set independent of the equations from the other rows. For columns, the same thing occurs. For very large problems, this is important, because it permits the ready overlay of main memory in solving the independent sets.

When the region in which the heat-flow equation is to be satisfied is not rectangular, one may perturb the boundary to make it agree with a square mesh or interpolate from boundary points to estimate u at adjacent mesh points as discussed in Chapter 7 for elliptic equations.

The frequency with which circular or spherical regions occur makes it worthwhile to mention the heat equation in polar and spherical coordinates. The basic equation

$$\frac{\partial u}{\partial t} = \frac{k}{c\rho} \nabla^2 u$$

becomes, in polar coordinates (r, θ),

$$\frac{\partial u}{\partial t} = \frac{k}{c\rho} \left(\frac{\partial^2 u}{\partial r^2} + \frac{1}{r} \frac{\partial u}{\partial r} + \frac{1}{r^2} \frac{\partial^2 u}{\partial \theta^2} \right),$$

and in spherical coordinates (r, ϕ, θ),

$$\frac{\partial u}{\partial t} = \frac{k}{c\rho} \left(\frac{\partial^2 u}{\partial r^2} + \frac{2}{r} \frac{\partial u}{\partial r} + \frac{1}{r^2} \frac{\partial^2 u}{\partial \theta^2} + \frac{\cot \phi}{r} \frac{\partial u}{\partial \theta} + \frac{1}{r^2 \sin^2 \theta} \frac{\partial^2 u}{\partial \phi^2} \right).$$

Using finite-difference approximations to convert these to difference equations is straightforward except at the origin where $r = 0$. For this point, consider $\nabla^2 u$ in rectangular coordinates, so that, in two dimensions,

$$\partial^2 u = \frac{u_{i+1,j} + u_{i-1,j} + u_{i,j+1} + u_{i,j-1} - 4u_{i,j}}{(\Delta r)^2}, \qquad u_{ij} \text{ at } r = 0.$$

This is exactly the same as the expression for the Laplacian in Chapter 7, Eq. (7.6a). This expression for $\nabla^2 u$ is obviously independent of the orientation of the axes. We get the best value by using the average value of all points that are a distance Δr from the origin, so that for $r = 0$,

$$\nabla^2 u = \frac{4(u_{av} - u_0)}{(\Delta r)^2} \quad \text{at } r = 0.$$

The corresponding relation for spherical coordinates is

$$\nabla^2 u = \frac{6(u_{av} - u_0)}{(\Delta r)^2} \quad \text{at } r = 0.$$

8.6 FINITE ELEMENTS FOR HEAT FLOW

In Chapters 6 and 7 we observed that the finite-element method often is preferred for obtaining approximate solutions to differential equations whose conditions are specified on the boundaries of a region. They are also an important alternative to finite differences in solving parabolic equations. You should have some exposure to this application of finite elements, but we do not have space enough to give a full treatment.

Consider first the one-dimensional problem that we discussed in Sections 8.1–8.3 The equation for this is

$$\frac{\partial u}{\partial t} = \frac{k}{c\rho} \frac{\partial^2 u}{\partial x^2} \quad \text{over } [x_a, x_b] \tag{8.21}$$

subject to initial conditions that give $u(x, 0)$ and boundary conditions that give $u(x_a, t)$ and $u(x_b, t)$. It is possible to consider this a two-dimensional problem (in x and t) and treat it as in Section 7.12. However, it is customary to apply finite elements only to the x-domain and approximate the time derivative by finite differences. We will adopt this approach, using a forward-difference approximation, so Eq. (8.21) becomes

$$\frac{u^{m+1} - u^m}{\Delta t} - \frac{k}{c\rho} \frac{\partial^2 u}{\partial x^2} = 0. \tag{8.22}$$

Superscripts on u indicate time; subscripts on u in the following indicate position.

We break the region of the problem into discrete contiguous subintervals and approximate u within element (e), whose endpoints are x_k and x_{k+1}, as a linear interpolation of u_k and u_{k+1}. In other words,

$$u^{(e)} = N_k u_k + N_{k+1} u_{k+1} \quad \text{within } (e).$$

This is exactly what we did in Section 6.4. Of course in our present application, all of the u-values are functions of t. We assume that t has some specified value, say t_m, for the present. As explained in Chapter 6,

$$N_k = \frac{x_{k+1} - x}{x_{k+1} - x_k}, \qquad N_{k+1} = \frac{x - x_k}{x_{k+1} - x_k}.$$

These shape functions are identically zero outside element (e). Clearly $N_k(x_k) = 1$, $N_{k+1}(x_{k+1}) = 1$, $N_k(x_{k+1}) = 0$, and $N_{k+1}(x_k) = 0$. Also, $dN_k/dx = -1/h$ and $dN_{k+1}/dx = 1/h$ (where $h = x_{k+1} - x_k$) within (e) and both derivatives are zero outside of (e).

To find an approximation to u at the nodes, the finite-element method evaluates a set of integrals to get the coefficients of a set of linear equations whose variables are the unknown nodal values. For Eq. (8.22), the integral for node i in element (e) at time t_m is (this is usually obtained by the Galerkin procedure):

$$\int^{(e)} \frac{u_i^{m+1} - u_i^m}{\Delta t} N_i \, dx + \frac{k}{c\rho} \int^{(e)} \left(u_i^m \frac{dN_i}{dx} \frac{dN_i}{dx} + u_j^m \frac{dN_j}{dx} \frac{dN_i}{dx} \right) dx. \qquad \textbf{(8.23)}$$

In Eq. (8.23), j refers to the other node of element (e). (If there is a derivative boundary condition or if there are heat sources or sinks within (e), this nodal integral is more complicated.)

Since our shape functions are linear, the integrals are easy to evaluate:

$$\int^{(e)} N_i \, dx = \frac{h^{(e)}}{2}, \qquad h^{(e)} = \text{width of element},$$

$$\int^{(e)} \left(\frac{dN_i}{dx} \right) \left(\frac{dN_j}{dx} \right) dx = \begin{cases} 1/h^{(e)} & \text{if } i = j, \\ -1/h^{(e)} & \text{if } i \neq j. \end{cases}$$

Consider two adjoining elements, (e) and $(e + 1)$, that are joined at node n. The coefficients of the system are found by adding the nodal integrals for node n from both elements.

$$\frac{u_n^{m+1} - u_n^m}{\Delta t} \int^{(e)} N_n \, dx + u_{n-1}^m \frac{k}{c\rho} \int^{(e)} \frac{dN_{n-1}}{dx} \frac{dN_n}{dx} \, dx + u_n^m \frac{k}{c\rho} \int^{(e)} \frac{dN_n}{dx} \frac{dN_n}{dx} \, dx$$

$$+ \frac{u_n^{m+1} - u_n^m}{\Delta t} \int^{(e+1)} N_n \, dx + u_n^m \frac{k}{c\rho} \int^{(e+1)} \frac{dN_n}{dx} \frac{dN_n}{dx} \, dx$$

$$+ u_{n+1}^m \frac{k}{c\rho} \int^{(e+1)} \frac{dN_{n+1}}{dx} \frac{dN_n}{dx} \, dx = 0.$$

Evaluating the integrals and assuming the width of both elements is h, we get

$$\frac{u_n^{m+1} - u_n^m}{\Delta t}\left(\frac{h}{2}\right) + u_{n-1}^m\frac{k}{c\rho}\left(\frac{-1}{h}\right) + u_n^m\frac{k}{c\rho}\left(\frac{1}{h}\right) + \frac{u_n^{m+1} - u_n^m}{\Delta t}\left(\frac{h}{2}\right)$$

$$+ u_n^m\frac{k}{c\rho}\left(\frac{1}{h}\right) + u_{n+1}^m\frac{k}{c\rho}\left(\frac{-1}{h}\right) = 0.$$

Collecting terms, putting those at the new time t_{m+1} on the left and those at t_m on the right, gives

$$u_n^{m+1} = u_n^m + \frac{k}{c\rho}\frac{\Delta t}{h^2}(u_{n-1}^m - 2u_n^m + u_{n+1}^m). \tag{8.24}$$

This is exactly the same as Eq. (8.4), so we see that, if equal-sized elements are used and if the time derivative is approximated by a forward difference, the one-dimensional finite-element procedure reduces to the explicit finite-difference method. If different approximations are used for the time derivative, we can get implicit methods such as Crank–Nicolson. Of course, one often varies the size of elements to get greater precision in some subintervals. This is where finite-element formulations shine.

We have applied the finite-element method to only two contiguous elements. Extending to many simplex elements is easy. When we reach a boundary of the region, the value of u will be known, or is specified in terms of space derivatives. (Remember that derivative conditions may add terms to the element equations.) Of course, we start off in the usual manner, employing the known values of u given by the initial conditions. There is a limit to the size of Δt that can be used without instability, exactly as with finite differences.

For heat flow in two space dimensions, we can parallel the above development. Eq. (8.23) becomes, for node i in triangular element (e) at time t_m, and using a forward-difference approximation for $\partial u/\partial t$:

$$\iint^{(e)} \frac{u_i^{m+1} - u_i^m}{\Delta t} N_i \, dx \, dy$$

$$+ \frac{k}{c\rho} \iint^{(e)} \left(u_i^m\frac{\partial N_i}{\partial x}\frac{\partial N_i}{\partial x} + u_j^m\frac{\partial N_j}{\partial x}\frac{\partial N_i}{\partial x} + u_k^m\frac{\partial N_k}{\partial x}\frac{\partial N_i}{\partial x}\right) dx \, dy$$

$$+ \frac{k}{c\rho} \iint^{(e)} \left(u_i^m\frac{\partial N_i}{\partial y}\frac{\partial N_i}{\partial y} + u_j^m\frac{\partial N_j}{\partial y}\frac{\partial N_i}{\partial y} + u_k^m\frac{\partial N_k}{\partial y}\frac{\partial N_i}{\partial y}\right) dx \, dy. \tag{8.25}$$

In Eq. (8.25), we assume that element (e) has nodes i, j, k. There will be additional contributions from integrals of similar form to the coefficients of u_i and the right-hand side of row i from all other elements that share node i. Row i will also have terms for the columns that correspond to the other vertices of elements that join at node i.

Again, because we have used shape functions that are linear in x and y, the integrals are easy to evaluate:

$$\iint^{(e)} N_i \, dx \, dy = A^{(e)}/3, \text{ where } A^{(e)} \text{ is the area of } (e),$$

$$\iint^{(e)} \frac{\partial N_i}{\partial x} \frac{\partial N_j}{\partial x} \, dx \, dy = \frac{b_i b_j}{4A^{(e)}},$$

$$\iint^{(e)} \frac{\partial N_i}{\partial y} \frac{\partial N_j}{\partial y} \, dx \, dy = \frac{c_i c_j}{4A^{(e)}},$$

where the b_i and c_i are elements in the second and third row, respectively, of the inverse of

$$M = \begin{bmatrix} 1 & x_i & y_i \\ 1 & x_j & y_j \\ 1 & x_k & y_k \end{bmatrix}, \qquad M^{-1} = \begin{bmatrix} a_i & a_j & a_k \\ b_i & b_j & b_k \\ c_i & c_j & c_k \end{bmatrix}.$$

The extension of the finite-element method to time-dependent heat flow in a two-dimensional region using triangular elements, while straightforward, is computationally tedious. Setting up the matrices when there are many elements and many nodes really requires a computer.

Equation (8.25), as written, gives an explicit formula. Modifying the way the time derivative is approximated may lead to more accurate implicit methods that are less stringent on the size of Δt.

8.7 CHAPTER SUMMARY

Test your understanding of this chapter by asking yourself if you can

1. Solve a parabolic partial-differential equation in one space dimension by both the explicit and implicit (Crank–Nicolson) methods with boundary conditions that may involve derivatives.

2. Explain the differences between explicit and implicit methods, giving the advantages of each.

3. Outline the arguments that show if a method is stable and convergent.

4. Discuss how finite-difference methods can be applied to parabolic equations in two and three space dimensions, describing the problems that are involved.

5. Set up the equations for the A.D.I. method in a two-dimensional problem, and employ a computer program for this on a rectangular region. You can outline how the method might be applied to an irregular region.

6. Explain how finite elements can be applied to parabolic equations and solve simpler situations with this method. You should be able to tell when this method reduces to the same equations as with a finite-difference method.

7. Use, modify, and critique computer programs that solve parabolic equations.

8.8 COMPUTER PROGRAMS

Three computer programs are presented as examples of how parabolic partial-differential equations in one or two space dimensions can be solved on a computer. Program 1 (Fig. 8.8) uses the explicit method of Section 8.1; Program 2 (Fig. 8.9) employs the implicit Crank–Nicolson method of Section 8.2; Program 3 (Fig. 8.10) solves the unsteady-state heat-flow equation in two space dimensions by the A.D.I. method. In the first program, both ends are held at a known temperature, and these end temperatures are allowed to vary with time. In the second, however, derivative end conditions are permitted.

For Program 1, the fundamental difference equation that approximates the differential equation is Eq. (8.4):

$$u_i^{j+1} = r(u_{i+1}^j + u_{i-1}^j) + (1 - 2r)u_i^j,$$

where

$$r = \frac{k\,\Delta t}{c\rho(\Delta x)^2}.$$

This relation is applied at each interior point. The end conditions, $u(0, t)$ and $u(L, t)$, are defined by arithmetic statement functions. At the end of each time step, the interior temperatures are computed and a line of temperatures is printed.

The program is tested by computing the temperatures in a 10-cm-long bar with $k = 0.53$, $c = 0.226$, $\rho = 2.70$, $\Delta x = 1.0$, $r = 0.25$ (all in cgs units). The bar is initially at 20°C at all points, and temperatures are caused to change by suddenly cooling one end to 0°C and heating the other end to 100°C. The program computes values until $T > 20$, but only part of the output is shown.

Program 2, using the Crank–Nicolson method, is more elaborate in that derivative end conditions are permitted, in the form

$$au + b\frac{\partial u}{\partial x} = c,$$

where u is the temperature at the end of the bar. This equation can simulate the loss of heat by convection or by imperfect insulation at each end. The program generates the coefficients of the set of simultaneous equations that are to be solved, which are, when $r = 1$,

At left end: $(4 - 2\alpha\ \Delta x)u_1^{j+1} - 2u_2^{j+1} = 2\alpha\ \Delta x u_1^j + 2u_2^j - 4v\ \Delta x;$

Interior points: $-u_{i-1}^{j+1} + 4u_i^{j+1} - u_{i+1}^{j+1} = u_{i-1}^j + u_{i+1}^j;$

At right end: $-2u_n^{j+1} + (4 + 2\alpha\ \Delta x)u_{n+1}^{j+1} = 2u_n^j - 2\alpha\ \Delta x u_{n+1}^j + 4v\ \Delta x.$

In these relations, $\alpha = -a/b$, $v = c/b$. The coefficients are compressed into an $n \times 4$ array and the tridiagonal system is solved by first finding the LU decomposition. Temperatures are printed after each time step. The program calculates temperatures until $T = 1000$ for a bar of length 20 cm whose initial temperatures are given by $u = 100 - 10|x - 10|$, with the left end condition of $u_x = 0.2(u - 15)$ and right end condition of $u = 100$. Partial output is shown.

Program 3 illustrates how the A.D.I. method can be implemented with a computer program. Two vectors are used, U and V, to hold the function values as calculated by the alternating horizontal and vertical traverses. The program sets up the coefficient matrices for each traverse, computes the right-hand sides from the last computed values of U and V, and solves the system using LU decomposition. Function values are output only after each second set of computations because the odd traverses are less accurate. The output for this program is for a rectangular region with 105 interior nodes when the boundary is held at 0 on three sides and at 100 on the fourth side. Initially, the potential function is 0 at all points.

```
      PROGRAM EXPLPA(INPUT,OUTPUT)
C
C     ------------------------------------------------------------------
C
C     THIS PROGRAM COMPUTES ONE DIMENSIONAL UNSTEADY STATE POTENTIAL
C     FLOW PROBLEMS BY THE EXPLICIT METHOD. THE POTENTIAL AT EACH END
C     IS SPECIFIED AS A FUNCTION OF TIME THROUGH ARITHMETIC STATEMENT
C     FUNCTIONS.
C
C     ------------------------------------------------------------------
C
C     PARAMETERS ARE :
C
C     U, V    - VALUES OF THE POTENTIAL AT NODES
C     T       - TIME
C     TF      - FINAL TIME FOR WHICH VALUES ARE COMPUTED
C     DT      - DELTA T
C     LEN     - LENGTH
C     DX      - DELTA X
C     N       - NUMBER OF X INTERVALS
C     K       - CONDUCTIVITY
C     C       - HEAT CAPACITY
C     RHO     - DENSITY
C     RATIO   - RATIO OF K(DT)/C(RHO)(DX)**2
C
```

Figure 8.8 Program 1.

Figure 8.8 (*continued*)

```
C
C      FLFT    - BOUNDARY CONDITION ON LEFT END
C      FRT     - BOUNDARY CONDITION ON RIGHT END
C -----------------------------------------------------------------
C
       REAL U(500),V(500),LEN,K,T,TF,DT,DX,C,RHO,THALF,RATIO
       INTEGER N,NP1,I
C
C -----------------------------------------------------------------
C
C WE DEFINE SOME CONSTANTS WITH A DATA STATEMENT
C
       DATA T,TF,LEN,N,K,C,RHO,RATIO/0.0,20.0,10.0,10,0.53,0.226,
      +     2.70,0.25/
C
C -----------------------------------------------------------------
C
C READ IN THE INITIAL TEMPERATURES AND WRITE THEM OUT
C
       NP1 = N + 1
       DX = LEN / N
       DT = RATIO*C*RHO*DX*DX / K
       READ *, ( U(I), I = 1,NP1 )
       PRINT 201, T,LEN,DX,T,( U(I), I = 1,NP1 )
C
C -----------------------------------------------------------------
C
C WE GET THE POTENTIAL FROM THE EXPLICIT RELATION, PRINTING EACH
C SET OF VALUES AS THEY ARE COMPUTED. TWO ARRAYS ARE USED,
C HOLDING ALTERNATE SETS OF VALUES.
C
10 THALF = T + DT/2.0
       U(1) = FLFT(THALF)
       U(N+1) = FRT(THALF)
       DO 20 I = 2,N
         V(I) = RATIO * ( U(I+1) + U(I-1)) + ( 1.0 - 2.0*RATIO )*U(I)
20 CONTINUE
       T = T + DT
       V(1) = FLFT(T)
       V(NP1) = FRT(T)
       PRINT 202, T,( V(I), I = 1,NP1 )
       IF ( T .GT. TF ) STOP
C
C -----------------------------------------------------------------
C
C NOW DO SECOND SET OF VALUES
C
       THALF = T + DT/2.0
       V(1) = FLFT(THALF)
       V(N+1) = FRT(THALF)
       DO 30 I = 2,N
         U(I) = RATIO * ( V(I+1) + V(I-1)) + ( 1.0 -2.0*RATIO )*V(I)
30 CONTINUE
       T = T + DT
       U(1) = FLFT(T)
       U(NP1) = FRT(T)
       PRINT 202, T,( U(I), I = 1,NP1 )
       IF ( T .GT. TF ) STOP
       GO TO 10
C
C -----------------------------------------------------------------
C
```

Figure 8.8 (*continued*)

```
    201 FORMAT(/' POTENTIAL VALUES IN ONE DIMENSION BY ',
       +       'EXPLICIT METHOD '/1X,'  FOR X = ',F4.1,' TO X = ',
       +       F4.1,' WITH DELTA X OF ',F6.3//1X,' AT T = ',F6.3,
       +       /(1X,6F9.3) )
    202 FORMAT(/1X,' VALUES AT T = ',F8.3/(1X,6F9.3) )
        END
C
C ------------------------------------------------------------------
C
C          FUNCTIONS DEFINED FOR THE LEFT AND
C          RIGHT HAND BOUNDARIES.
C
C ------------------------------------------------------------------
C
        REAL FUNCTION FLFT(X)
        REAL X
        FLFT = 0.0
        RETURN
        END
C
        REAL FUNCTION FRT(X)
        REAL X
        FRT = 100.0
        RETURN
        END


            OUTPUT FOR PROGRAM 1


POTENTIAL VALUES IN ONE DIMENSION BY EXPLICIT METHOD
  FOR X =  0.0 TO X = 10.0 WITH DELTA X OF  1.000

 AT T =  0.000
    20.000   20.000   20.000   20.000   20.000   20.000
    20.000   20.000   20.000   20.000   20.000

 VALUES AT T =      .288
     0.000   15.000   20.000   20.000   20.000   20.000
    20.000   20.000   20.000   40.000  100.000

 VALUES AT T =      .576
     0.000   12.500   18.750   20.000   20.000   20.000
    20.000   20.000   25.000   50.000  100.000

 VALUES AT T =      .863
     0.000   10.938   17.500   19.688   20.000   20.000
    20.000   21.250   30.000   56.250  100.000

 VALUES AT T =     1.151
     0.000    9.844   16.406   19.219   19.922   20.000
    20.313   23.125   34.375   60.625  100.000

 VALUES AT T =     1.439
     0.000    9.023   15.469   18.691   19.766   20.059
    20.938   25.234   38.125   63.906  100.000

        *****   OUTPUT CONTINUED *****


 VALUES AT T =    18.997
     0.000    7.742   15.699   24.065   33.002   42.616
    52.954   63.988   75.621   87.694  100.000
```

Figure 8.8 *(continued)*

```
VALUES AT T =    19.285
     0.000      7.796     15.801     24.208     33.171     42.797
    53.128     64.138     75.731     87.752    100.000

VALUES AT T =    19.572
     0.000      7.848     15.902     24.347     33.337     42.973
    53.298     64.284     75.838     87.809    100.000

VALUES AT T =    19.860
     0.000      7.900     16.000     24.483     33.499     43.145
    53.463     64.426     75.942     87.864    100.000

VALUES AT T =    20.148
     0.000      7.950     16.096     24.616     33.656     43.313
    53.624     64.564     76.043     87.918    100.000
```

```
          PROGRAM CNIMPL(INPUT,OUTPUT)
C
C     THIS PROGRAM COMPUTES UNSTEADY STATE POTENTIALS IN ONE
C DIMENSION USING CRANK-NICHOLSON IMPLICIT METHOD. THE POTENTIAL
C AT EACH END IS DETERMINED BY A RELATION OF THE FORM :
C
C               AU + BU' = C
C
C
C     ----------------------------------------------------------------
C
C     PARAMETERS ARE :
C
C     U       - VALUES OF POTENTIAL AT NODES
C     T       - TIME
C     TF      - FINAL TIME VALUE FOR WHICH SOLUTION IS DESIRED
C     DT      - DELTA T
C     LEN     - LENGTH
C     DX      - DELTA X
C     N       - NUMBER OF X INTERVALS
C     DIF     - DIFFUSIVITY, K/C*RHO FOR HEAT FLOW
C     RATIO   - RATIO OF DT(DIF)/(DX)*2
C     COEF    - COEFFICIENT MATRIX FOR IMPLICIT RELATIONS
C
C     ----------------------------------------------------------------
C
      REAL U(500),COEF(500,3),RHS(500),LEN,T,TF,DIF,RATIO
     +      ,AL,BL,CL,AR,BR,CR,DX,DT
      INTEGER N,NP1,I
C
C     ----------------------------------------------------------------
C
C DEFINE SOME CONSTANTS
C
      DATA T,TF,LEN,N,DIF,RATIO/0.0,1000.0,20.0,20,0.119,1.0/
      DATA AL,BL,CL/-0.2,1.0,-3.0/
      DATA AR,BR,CR/1.0,0.0,100.0/
C
C     ----------------------------------------------------------------
```

Figure 8.9 Program 2.

Figure 8.9 (*continued*)

```
C
C  READ IN INITIAL VALUES
C
      NP1 = N + 1
      READ*, ( U(I), I = 1,NP1 )
      DX = LEN / N
      DT = RATIO*DX*DX / DIF
      PRINT 201, T,LEN,DX,T,( U(I), I = 1,NP1 )
C
C  ------------------------------------------------------------------------
C
C  ESTABLISH COEFFICIENT MATRIX
C
      IF ( BL .NE. 0.0 ) THEN
        COEF(1,2) = 2.0/RATIO + 2.0 - 2.0*AL*DX/BL
        COEF(1,3) = -2.0
      ELSE
        COEF(1,2) = 1.0
        COEF(1,3) = 0.0
      END IF
   20 DO 25 I = 2,N
        COEF(I,1) = -1.0
        COEF(I,2) = 2.0/RATIO + 2.0
        COEF(I,3) = -1.0
   25 CONTINUE
      IF ( BR .NE. 0.0 ) THEN
        COEF(N+1,1) = -2.0
        COEF(N+1,2) = 2.0/RATIO + 2.0 + 2.0*AR*DX/BR
      ELSE
        COEF(N+1,1) = 0.0
        COEF(N+1,2) = 1.0
      END IF


C
C  ------------------------------------------------------------------------
C
C  GET THE LU DECOMPOSITION TO PREPARE FOR SOLVING EQUATIONS
C
   40 DO 50 I = 2,NP1
        COEF(I-1,3) = COEF(I-1,3) / COEF(I-1,2)
        COEF(I,2) = COEF(I,2) - COEF(I,1)*COEF(I-1,3)
   50 CONTINUE
C
C  ------------------------------------------------------------------------
C
C  ESTABLISH THE R.H.S. VECTOR - FIRST THE TOP AND BOTTOM ROWS
C
   55 IF ( BL .NE. 0.0 ) THEN
        RHS(1) = ( 2.0/RATIO - 2.0 + 2.0*AL*DX/BL ) * U(1) +
     +              2.0*U(2) - 4.0*CL*DX/BL
      ELSE
        RHS(1) = CL / AL
      END IF
   60 IF ( BR .NE. 0.0 ) THEN
        RHS(N+1) = 2.0*U(N) + ( 2.0/RATIO - 2.0 - 2.0*AR*DX/BR ) *
     +              U(N+1) + 4.0*CR*DX/BR
      ELSE
        RHS(N+1) = CR / AR
      END IF
C
C  ------------------------------------------------------------------------
```

Figure 8.9 (*continued*)

```
C
C  NOW FOR THE OTHER ROWS OF THE RHS VECTOR
C
      DO 100 I = 2,N
         RHS(I) = U(I-1) + ( 2.0/RATIO - 2.0 ) * U(I) + U(I+1)
  100 CONTINUE
C
C  ------------------------------------------------------------------
C
C  WE GET THE SOLUTION FOR THE CURRENT TIME STEP
C
      U(1) = RHS(1) / COEF(1,2)
      DO 110 I = 2,NP1
         U(I) = ( RHS(I) - COEF(I,1)*U(I-1) ) / COEF(I,2)
  110 CONTINUE
      DO 120 I = 1,N
         JROW = N - I + 1
         U(JROW) = U(JROW) - COEF(JROW,3)*U(JROW+1)
  120 CONTINUE
C
C  ------------------------------------------------------------------
C
C  WRITE OUT THE SOLUTION JUST FOUND
C
      T = T + DT
      PRINT 202, T, ( U(I), I = 1,NP1 )
      IF ( T .LT. TF ) GO TO 55
      STOP
C
C  ------------------------------------------------------------------
C
  201 FORMAT('  POTENTIAL VALUES IN ONE DIMENSION BY THE ',
     +        'IMPLICIT METHOD ',/1X,'   FOR X = ',F4.1,' TO X = ',
     +        F4.1,' WITH DELTA X OF ',F6.3//1X,' AT T = ',F7.3,
     +        /(1X,10F8.2))
  202 FORMAT(1X,' VALUES AT T = ',F3.3,/(1X,10F8.2))
      END


                 OUTPUT FOR PROGRAM 2

POTENTIAL VALUES IN ONE DIMENSION BY THE IMPLICIT METHOD
  FOR X =  0.0 TO X = 20.0 WITH DELTA X OF  1.000

AT T =   0.000
   0.00   10.00   20.00   30.00   40.00   50.00   60.00   70.00   80.00   90.00
 100.00   90.00   80.00   70.00   60.00   50.00   40.00   30.00   20.00   10.00
   0.00
VALUES AT T =    8.403
  13.46   13.61   20.97   30.26   40.07   50.00   59.95   69.78   79.17   86.91
  88.45   86.91   79.17   69.79   59.98   50.12   40.51   31.92   27.18   36.79
 100.00
VALUES AT T =   16.807
  16.09   18.47   23.38   31.20   40.37   50.02   59.69   68.95   76.98   82.30
  84.61   82.31   77.01   69.02   59.94   50.82   42.86   38.58   43.76   67.73
 100.00
VALUES AT T =   25.210
  19.28   21.15   25.87   32.65   40.98   50.04   59.15   67.58   74.51   79.22
  80.77   79.25   74.63   67.91   60.09   52.59   47.46   47.86   57.36   75.28
 100.00
```

Figure 8.9 (*continued*)

```
                     **** OUTPUT CONTINUED ****
VALUES AT T =   983.193
    31.70    35.04    38.39    41.74    45.11    48.48    51.85    55.24    58.64    62.04
    65.46    68.88    72.32    75.76    79.21    82.66    86.12    89.59    93.06    96.53
   100.00
VALUES AT T =   991.597
    31.70    35.05    38.40    41.75    45.11    48.48    51.86    55.25    58.65    62.05
    65.47    68.89    72.32    75.77    79.21    82.67    86.13    89.59    93.06    96.53
   100.00
VALUES AT T = 1000.000
    31.71    35.05    38.40    41.76    45.12    48.49    51.87    55.26    58.66    62.06
    65.48    68.90    72.33    75.77    79.22    82.67    86.13    89.59    93.06    96.53
   100.00
```

```
      PROGRAM ADIUN(INPUT,OUTPUT)
C
C     ----------------------------------------------------------------
C
C     THIS PROGRAM SOLVES THE UNSTEADY STATE HEAT FLOW EQUATION IN TWO
C     SPACE DIMENSIONS USING THE A.D.I. METHOD.
C
C     ----------------------------------------------------------------
C
C     PARAMETERS ARE :
C
C     U     - VECTOR OF TEMPERATURES AFTER ODD TRAVERSES
C     V     - TEMPERATURES AFTER EVEN TRAVERSES
C     UCOEF - MATRIX OF COEFFICIENTS FOR THE U'S
C     VCOEF - MATRIX FOR THE V'S
C     BCNDU - VECTOR OF BOUNDARY VALUES FOR THE U EQUATIONS
C     BCNDV - SAME FOR THE V'S
C     TOP   - VECTOR TO HOLD BOUNDARY VALUES ACROSS TOP OF REGION
C     BOT   - SAME FOR THE VALUES ACROSS THE BOTTOM
C     RT    - HOLD THE RIGHT HAND SIDE VALUES
C     LFT   - HOLD VALUES FOR THE LEFT HAND EDGE
C     M     - NUMBER OF ROWS OF NODES IN THE GRID
C     N     - NUMBER OF COLUMNS OF NODES
C     DIFF  - THERMAL DIFFUSIVITY = K/C*DENSITY
C     H     - THE VALUE OF DELTA X = DELTA Y
C     TIME  - THE TIME VARIABLE
C     TMAX  - MAXIMUM VALUE OF TIME FOR WHICH COMPUTATIONS ARE
C             DESIRED
C     DT    - TIME STEP SIZE, RELATED TO DELTA X THROUGH R
C
C     ----------------------------------------------------------------
C
C     ESTABLISH THE MATRICES AND VECTORS, AND PUT IN SOME VALUES WITH
C     DATA STATEMENTS.
C
      REAL U(500),V(500),UCOEF(500,3),VCOEF(500,3)
      REAL BCNDU(500),BCNDV(500)
      REAL TOP(100),BOT(100),LFT(100),RT(100)
      DATA UCOEF,VCOEF / 3000 * -1.0 /
      DATA BCNDU,BCNDV / 1000 * 0.0 /
      DATA M,N,R / 7,15,0.5 /
```

Figure 8.10 Program 3.

Figure 8.10   (*continued*)

```
          DATA H,DIFF / 0.125,0.152 /
          DATA TMAX / 20.0 /
          DATA RT,LFT,TOP,BOT / 100 * 100.0, 300 * 0.0 /
C
C     -----------------------------------------------------------------
C
C     FOR THE TEST CASE, WE BEGIN WITH TEMPERATURES EVERYWHERE EQUAL TO
C     ZERO. ESTABLISH THESE BY A DATA STATEMENT.
C
          DATA V / 500 * 0.0 /
C
C     -----------------------------------------------------------------
C
C     SET UP THE COEFFICIENT MATRICES BY OVER-WRITING ON THE DIAGONAL
C     AND CERTAIN OFF DIAGONAL TERMS.
C
          MSIZE = M * N
          DO 10 I = 1,MSIZE
            UCOEF(I,2) = 1.0/R + 2.0
            VCOEF(I,2) = 1.0/R + 2.0
       10 CONTINUE
          ML1 = M - 1
          NL1 = N - 1
          MSL1 = MSIZE - 1
          DO 20 I = N,MSL1,N
            UCOEF(I,3) = 0.0
            UCOEF(I+1,1) = 0.0
       20 CONTINUE
          DO 30 I = M,MSL1,M
            VCOEF(I,3) =0.0
            VCOEF(I+1,1) = 0.0
       30 CONTINUE
C
C     -----------------------------------------------------------------
C
C     NOW GET VALUES INTO THE BCND VECTORS
C
          DO 40 I = 1,N
            BCNDU(I) = TOP(I)
            J = MSIZE - N + I
            BCNDU(J) = BOT(I)
       40 CONTINUE
          DO 45 I =1,M
            J = (I-1)*N + 1
            BCNDU(J) = BCNDU(J) + LFT(I)
            J = I * N
            BCNDU(J) = BCNDU(J) + RT(I)
       45 CONTINUE
          DO 50 I = 1,M
            BCNDV(I) = LFT(I)
            J = MSIZE - M + I
            BCNDV(J) = RT(I)
       50 CONTINUE
          DO 55 I = 1,N
            J = (I-1)*M + 1
            BCNDV(J) = BCNDV(J) + TOP(I)
            J = I * M
            BCNDV(J) = BCNDV(J) + BOT(I)
       55 CONTINUE
C
C     -----------------------------------------------------------------
C
C     WE NOW GET THE LU DECOMPOSITIONS OF UCOEF AND VCOEF
```

Figure 8.10 *(continued)*

```
C
      DO 60 I = 2,MSIZE
         UCOEF(I-1,3) = UCOEF(I-1,3)/UCOEF(I-1,2)
         UCOEF(I,2) = UCOEF(I,2) - UCOEF(I,1)*UCOEF(I-1,3)
         VCOEF(I-1,3) = VCOEF(I-1,3)/VCOEF(I-1,2)
         VCOEF(I,2) = VCOEF(I,2) - VCOEF(I,1)*VCOEF(I-1,3)
   60 CONTINUE
C
C -----------------------------------------------------------------
C
C NOW WE DO THE ITERATIONS UNTIL TIME EQUALS TMAX.
C
      TIME = 0.0
      DT = R / DIFF*H*H
   62 IF ( TIME .GT. TMAX ) STOP
C
C -----------------------------------------------------------------
C
C COMPUTE THE R.H.S. FOR THE U EQUATIONS AND STORE IN THE U VECTOR.
C FIRST DO THE TOP AND BOTTOM ROWS.
C
      DO 65 I = 1,N
         J = (I-1)*M + 1
         U(I) = ( 1.0/R - 2.0 )*V(J) + V(J+1) + BCNDU(I)
         K = MSIZE - N + I
         J = I * M
         U(K) = V(J-1) + ( 1.0/R - 2.0 )*V(J) + BCNDU(K)
   65 CONTINUE
C
C -----------------------------------------------------------------
C
C NOW FOR THE OTHER ONES
C
      DO 75 I = 2,ML1
         DO 70 J = 1,N
            K = (I-1)*N + J
            L = I + (J-1)*M
            U(K) = V(L-1) + ( 1.0/R - 2.0 )*V(L) + V(L+1) + BCNDU(K)
   70    CONTINUE
   75 CONTINUE
C
C -----------------------------------------------------------------
C
C NOW GET THE SOLUTION FOR THE ODD TRAVERSE
C
      U(1) = U(1) / UCOEF(1,2)
      DO 80 I = 2,MSIZE
         U(I) = ( U(I) - UCOEF(I,1)*U(I-1)) / UCOEF(I,2)
   80 CONTINUE
      DO 90 I = 1,MSL1
         JROW = MSIZE - I
         U(JROW) = U(JROW) - UCOEF(JROW,3)*U(JROW+1)
   90 CONTINUE
C
C -----------------------------------------------------------------
C
C COMPUTE THE R.H.S. FOR THE EVEN TRAVERSE, STORE IN V. DO THE
C TOP AND BOTTOM ONES.
C
      DO 95 I = 1,M
         J = (I-1)*N + 1
         V(I) = ( 1.0/R - 2.0 )*U(J) + U(J+1) + BCNDV(I)
```

Figure 8.10   (*continued*)

```
            K = MSIZE - M + I
            J = I * N
            V(K) = U(J-1) + ( 1.0/R - 2.0 )*U(J) + BCNDV(K)
    95 CONTINUE
C
C    ------------------------------------------------------------
C
C  NOW THE REST OF THE ROWS
C
        DO 105 I = 2,NL1
          DO 100 J = 1,M
            K = (I-1)*M + J
            L = I + (J-1)*N
            V(K) = U(L-1) + (1.0/R-2.0)*U(L) + U(L+1) + BCNDV(K)
   100    CONTINUE
   105 CONTINUE
C
C    ------------------------------------------------------------
C
C  GET THE SOLUTION FOR THE EVEN TRAVERSE
C
        V(1) = V(1) / VCOEF(1,2)
        DO 110 I = 2,MSIZE
          V(I) = ( V(I) - VCOEF(I,1)*V(I-1)) / VCOEF(I,2)
   110 CONTINUE
        DO 120 I = 1,MSL1
          JROW = MSIZE - I
          V(JROW) = V(JROW) - VCOEF(JROW,3)*V(JROW+1)
   120 CONTINUE
        TIME = TIME + 2.0*DT
C
C    ------------------------------------------------------------
C
C  PRINT OUT THE LAST RESULT.
C
        PRINT 202, TIME, ( V(I), I = 1,MSIZE )
   202 FORMAT(//1X,' WHEN T = ',F6.3,' V VALUES ARE' /(1X,15F8.4))
        GO TO 62
        END
```

EXERCISES

Section 8.1

1. The parameters of the basic equation for unsteady-state heat transfer are dimensional. If it is desired to measure u in °F and x in inches, how must the units of k, c, and ρ be chosen in

$$\frac{\partial^2 u}{\partial x^2} = \frac{c\rho}{k}\frac{\partial u}{\partial t}?$$

▶ 2. Solve for the temperatures at $t = 2.062$ sec in the 2-cm-thick steel slab of Section 8.1 if the initial temperatures are given by the relation

$$u(x, 0) = 100 \sin \frac{\pi x}{2}.$$

Use the explicit method with $\Delta x = 0.25$ cm. Compare to the analytical solution: $100e^{-0.3738t} \sin (\pi x/2)$.

3. Solve for the temperatures in a copper rod 10 in long, with the outer curved surface insulated so that heat flows in only one direction. The initial temperature is linear from 0°C at one end to 100°C at the other, when suddenly the hot end is brought to 0°C, and the cold end is held at 0°C. Use $\Delta x = 1$ in and an appropriate value of Δt so that $k \, \Delta t / c\rho(\Delta x)^2 = \frac{1}{2}$. Look up values of k, c, and ρ in a handbook. Carry out the solution for 10 time steps.

4. Repeat Exercise 3 with $\Delta x = 0.5$ in, and compare the temperatures at points 1 in, 3 in, and 6 in from the cold end in the two calculations.

5. Repeat Exercise 3 with $\Delta x = 1$ in and Δt such that $k \, \Delta t / c\rho(\Delta x)^2 = \frac{1}{4}$. Compare results with Exercises 3 and 4.

▶ 6. Repeat computations with Δx as given below for the diffusion example of Section 8.1, and compare to the analytical solution at $x = 4$ and $x = 12$ cm. Carry them out until $t = 268.8$ sec.

   a) With $\Delta x = 4$ cm, $D \, \Delta t / (\Delta x)^2 = 0.125$;
   b) With $\Delta x = 2$ cm, $D \, \Delta t / (\Delta x)^2 = \frac{1}{2}$;
   c) With $\Delta x = 1$ cm, $D \, \Delta t / (\Delta x)^2 = \frac{1}{2}$.

   If you use the results shown in Section 8.1 of the text, you now have data that illustrate the effect of size of Δx and of r on accuracy. Considering the amount of calculations required, which is the more effective way to improve accuracy?

Section 8.2

7. Solve Exercise 2 by the Crank–Nicolson method, $r = 1$.

8. Solve Exercise 3 by the Crank–Nicolson method, $r = 1$. Compare results with Exercises 3, 4, and 5.

▶ 9. The methods of Sections 8.1 and 8.2 can be applied readily to more complicated situations. For example, if heat is being generated at various points along a bar at a rate that is a function of x, the unsteady-state heat equation becomes

$$k\frac{\partial^2 u}{\partial x^2} - c\rho\frac{\partial u}{\partial t} = f(x).$$

Solve this equation where $f(x) = x$ cal/cm$^3 \cdot$ sec, subject to conditions

$$u(0, t) = 0, \qquad u(1, t) = 0, \qquad u(x, 0) = 0.$$

Take $\Delta x = 0.2$, $k = 0.37$, $c\rho = 0.433$ in cgs units. (These are properties of magnesium.) Use the Crank–Nicolson method, $r = 1$, and solve for five time steps.

Section 8.3

10. Use the set of equations in Eq. (8.10) to find the solution to the example in Section 8.3 through eight time steps.

11. Solve the set of equations in Eq. (8.11) to find the solution to the example in Section 8.3 by the Crank–Nicolson method. Compare results with those from Exercise 10.

▶ 12. Solve the example of Section 8.3 but with two opposite faces of the cube losing heat at a rate equal to $0.15A(u - 70)$, where u is the surface temperature in °F, and A is the area. Use $\Delta x = 1$ in, and employ the explicit method with $k \, \Delta t / c\rho(\Delta x)^2 = \frac{1}{4}$.

13. Solve Exercise 12 using the Crank–Nicolson method with $r = 1$. Compare results by the two methods of solution.

▶ 14. Heat is added to one end of a 2-ft-long bar of copper, at a rate given by

$$-hA(u - u_0),$$

and lost at the other end through a similar rate equation. Temperatures are $u_0 = 500°F$ at the hot end, and $u_0 = 60°F$ at the cold end, with resistance to heat flow being such that $h = 0.3$ Btu/sec $\cdot$ ft$^2 \cdot$ °F at both ends. The circumference of the bar is carefully insulated so that heat flows in only one dimension. Find the time required for the midpoint of the bar to reach 200°F. The bar is initially at 60°F at all points.

Section 8.4

►15. Demonstrate, by performing calculations similar to those given in Table 8.5, that the explicit method is unstable with $k \, \Delta t / c \rho (\Delta x)^2 = 0.6$.

16. Demonstrate, by performing calculations similar to those given in Table 8.5, that the explicit method with $k \, \Delta t / c \rho (\Delta x)^2 = \frac{1}{4}$ has errors that damp out less rapidly than those in Table 8.5, but that the method is still stable.

17. Demonstrate, by performing calculations similar to those in Table 8.5, except that at both $x = x_1$ and $x = x_5$ the gradient is zero ($\partial u / \partial x = 0$ instead of $u = $ constant), that the explicit method is still stable with $k \, \Delta t / c \rho (\Delta x)^2 = \frac{1}{2}$. Note, however, how much more slowly an error damps out, and that the error at a later step becomes a linear combination of earlier errors.

18. Demonstrate, by performing calculations similar to those in Table 8.6, that the Crank–Nicolson method is still stable even though the value of $k \, \Delta t / c \rho (\Delta x)^2$ is taken as 10. You will probably wish to use the LU method to solve the system.

►19. Compute the largest eigenvalue of the coefficient matrix of Eq. (8.16) for $r = 0.5$, then for $r = 0.6$, when $N = 10$. (You may wish to use the power-method program of Chapter 6.) Do you find that the statements in the text about the value of the eigenvalues are confirmed?

20. Repeat Exercise 19 but for Eq. (8.18) using matrix $A^{-1}B$. Use $r = 1.0$ and $r = 2.0$.

Section 8.5

21. A rectangular plate 2×3 in. is initially at 50°. At $t = 0$, one 2-in. edge is suddenly raised to 100°, and one 3-in. edge is suddenly cooled to 0°. The temperature on these two edges is then held constant at these temperatures. The other two faces are perfectly insulated. Use a 1-in. grid to subdivide the plate and write the A.D.I. equations for each of the six points where unknown temperatures are involved. Use $r = 2$, and solve the equations for four time intervals.

►22. Suppose the cube used for the example problem in Section 8.3 ($4 \times 4 \times 4$ in.) had heat flowing in all three directions, such as by having three adjacent faces lose heat by conduction to the flowing liquid. Set up the equations to solve for the temperature at any point. Do this in terms of the surrounding temperatures one Δt previously, using the explicit method with $\Delta x = \Delta y = \Delta z = 1$. How many time steps are needed to reach $t = 15.12$ sec? How many equations are involved in each stage of the calculation?

23. Repeat Exercise 22 for the Crank–Nicolson method, $r = 1$.

►24. Repeat Exercise 22 using the A.D.I. method.

Section 8.6

25. In the finite-element method, one usually spaces the nodes more closely where u is expected to vary more rapidly. Suppose, in two adjoining elements, that the widths are h_1 and h_2. Rederive the equivalent of Eq. (8.24) for this case. Can you get the same result by applying finite-difference approximations to $\partial^2 u / \partial x^2$ over three points, x_1, x_2, x_3, that are unevenly spaced?

26. Show that Eq. (8.25) does give the explicit formula of Section 8.5 for equispaced nodes that are connected to give four adjacent triangular elements that share the common central node.

27. Solve Exercise 21 by the finite-element method. While we have not given the element integrals for a derivative boundary, in this case you can avoid having derivatives on the boundary by reflecting across the boundaries where these occur.

APPLIED PROBLEMS AND PROJECTS

28. When steel is forged, billets are heated in a furnace until the metal is of the proper temperature, between 2000°F and 2300°F. It can then be formed by the forging press into rough shapes that are later given their final finishing operations. In order to produce a certain machine part, a billet of size $4 \times 4 \times 20$ in is heated in a furnace whose temperature is maintained at 2350°F. You have been requested to estimate how long it will take all parts of the billet to reach a temperature above 2000°F. Heat transfers to the surface of the billet at a very high rate, principally through radiation. It has been suggested that you can solve the problem by assuming that the surface temperature becomes 2250°F instantaneously and remains at that temperature. Using this assumption, find the required heating time.

   Since the steel piece is relatively long compared to its width and thickness, it may not introduce significant error to calculate as if it were infinitely long. This will simplify the problem, permitting a two-dimensional treatment rather than three-dimensional. Such a calculation should also give a more conservative estimate of heating time. Compare the estimates from two- and three-dimensional approaches.

29. After you have calculated the answers to Problem 28, your results have been challenged on the basis of assuming constant surface temperature of the steel. Radiation of heat flows according to the equation

$$q = E\sigma(u_F^4 - u_S^4) \text{ Btu/hr} \cdot \text{ft}^2,$$

   where E = emissivity (use 0.80), σ is the Stefan–Boltzmann constant (0.171×10^{-8} Btu/hr $\cdot$ ft^2 $\cdot$ °F^4), u_F and u_S are the furnace and surface absolute temperatures, respectively (°F + 460°).

   The heat radiating to the surface must also flow into the interior of the billet by conduction, so

$$q = -k\frac{\partial u}{\partial x},$$

   where k is the thermal conductivity of steel (use 26.2 Btu/hr $\cdot$ ft^2 $\cdot$ °F/ft) and ($\partial u/\partial x$) is the temperature gradient at the surface in a direction normal to the surface. Solve the problem with this boundary conditon and compare your solution to that of Problem 28. (Observe that this is now a nonlinear problem. Think carefully how your solution can cope with it.)

30. Shipment of liquefied natural gas by refrigerated tankers to industrial nations is becoming an important means of supplying the world's energy needs. It must be stored at the receiving port, however. (A. R. Duffy and his coworkers (1967) discuss the storage of liquefied natural gas in underground tanks.) A commercial design, based on experimental verifiction of its feasibility, contemplated a prestressed concrete tank 270 ft in diameter and 61 ft deep, holding some 600,000 bbl of liquefied gas at −258°F. Convection currents in the liquid were shown to keep the temperature uniform at this value, the boiling point of the liquid.

   Important considerations of the design are the rate of heat gained from the surroundings (causing evaporation of the liquid gas) and variation of temperatures in the earth below the tank (relating to the safety of the tank, which could be affected by possible settling or frost-heaving.)

The tank itself is to be made of concrete 6 in thick, covered with 8 in of insulation (on the liquid side). (A sealing barrier keeps the insulation free of liquid; otherwise its insulating capacity would be impaired.) The experimental tests showed that there is a very small temperature drop through the concrete: 12°F. This observed 12°F temperature difference seems reasonable in light of the relatively high thermal conductivity of concrete. We expect then that most of the temperature drop occurs in the insulation or in the earth below the tank.

Since the commercial-design tank is very large, if we are interested in ground temperatures near the center of the tank (where penetration of cold will be a maximum) it should be satisfactory to consider heat flowing in only one dimension, in a direction directly downward from the base of the tank. Making this simplifying assumption, compute how long it will take for the temperature to decrease to 32°F (freezing point of water) at a point 8 ft away from the tank wall. The necessary thermal data are

	Insulation	Concrete	Earth
Thermal conductivity (Btu/hr · ft · °F)	0.013	0.90	2.6
Density (lb/ft³)	2.0	150	132
Specific heat (Btu/lb · °F)	0.195	0.200	0.200

Assume that the initial conditions are: temperature of liquid: −258°F; temperature of insulation: −258°F to 72°F (inner surface to outer); temperature of concrete: 72°F to 60°F; temperature of earth: 60°F.

31. Modify Program 1 so that end conditions of the form

$$au + b\frac{\partial u}{\partial x} = c$$

are permitted, similar to Program 2. Then use your program to solve Exercise 14. Vary the size of Δx and/or $r = k \Delta t/c\rho(\Delta x)^2$ until you are sure that the time for the midpoint temperature to become 200°F is known to within 0.5% relative error.

32. Solve Exercise 14 by Program 2. Critically compare the efficiency of Program 2 in obtaining the solution with the efficiency of the program you wrote in Problem 31.

33. Write a program similar to Program 3 but employing the A.D.I. method in three space dimensions with a parallelopiped-shaped region. Provide for derivative boundary conditions as an alternative to fixed boundary conditions by allowing for conditions of the form

$$au + b\frac{\partial u}{\partial x} = c.$$

9

Hyperbolic Partial-Differential Equations

9.0 CONTENTS OF THIS CHAPTER

The third classification of partial-differential equations, hyperbolic differential equations, includes the "wave equation" that is fundamental to the study of vibrating systems. They are also involved in transport problems (diffusion of matter, neutron diffusion, and radiation transfer), wave mechanics, gas dynamics, supersonic flow, and other important areas. The technology in most of these applications is so complex that their study is beyond our scope. We will settle for simple situations within the present experience of our readers. One such case is the vibrations of a string held taut by fixed ends, as in a violin string. We outline the derivation of the simple wave equation in one dimension. Imagine an elastic string stretched between two fixed endpoints and set to vibrating by plucking it with a finger.

We make a number of simplifying assumptions: the string is perfectly elastic, and we neglect gravitational forces, so that the only force is the tension force in the direction of the string; the string is uniform in density and thickness and has a weight per unit length of w lb/ft; the lateral displacement of the string is so small that the tension can be considered to be a constant value of T lb; and the slope of the string is hence small enough that $\sin \alpha \doteq \tan \alpha$, where α is the angle of inclination. Figure 9.1 illustrates the problem with the lateral displacement greatly exaggerated.

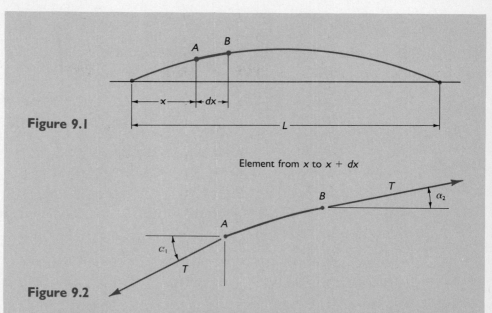

Figure 9.1

Element from x to $x + dx$

Figure 9.2

We take $x = 0$ at the left end. L is the total length of the string. Focus attention on the element of length dx between points A and B. The uniform tension T acts at each end. We are interested in how y, the lateral displacement, varies with time t and with distance x along the string.

The forces acting in the y-direction are the vertical components of the two tensions (see Fig. 9.2). We take the upward direction as positive:

Upward force at left end: $-T \sin \alpha_1 \doteq -T \tan \alpha_1 = -T\left(\dfrac{\partial y}{\partial x}\right)_A$;

Upward force at right end: $+T \sin \alpha_2 \doteq +T \tan \alpha_2 = T\left(\dfrac{\partial y}{\partial x}\right)_B$

$$= T\left(\left(\frac{\partial y}{\partial x}\right)_A + \frac{\partial}{\partial x}\left(\frac{\partial y}{\partial x}\right) dx\right);$$

Net force: $T\dfrac{\partial^2 y}{\partial x^2}\, dx.$

Partials are used to express the slope because y is a function of both t (time) and x (horizontal distance). We use Newton's law, and equate the force to mass times acceleration in the vertical direction. Our simplifying assumptions permit us to use $w\, dx$ as the weight:

$$\left(\frac{w\, dx}{g}\right)\frac{\partial^2 y}{\partial t^2} = T\frac{\partial^2 y}{\partial x^2}\, dx, \qquad \frac{\partial^2 y}{\partial t^2} = \frac{Tg}{w}\frac{\partial^2 y}{\partial x^2}. \qquad \textbf{(9.1)}$$

This is the wave equation in one dimension. The conditions imposed on the solution are the end conditions $[ay + b(\partial y/\partial x) = f(t)]$ at $x = 0$ and $x = L$, and the initial conditions at $t = 0$. Initial conditions specifying both $y = f(x)$ and the velocity $\partial y/\partial t = g(x)$ are usual for this problem.

Comparing Eq. (9.1) to the general form of a second-order partial-differential equation,

$$A\frac{\partial^2 y}{\partial t^2} + B\frac{\partial^2 y}{\partial t\, \partial x} + C\frac{\partial^2 y}{\partial x^2} + D\left(t, x, y, \frac{\partial y}{\partial t}, \frac{\partial y}{\partial x}\right) = 0,$$

shows $B^2 - 4AC > 0$, since $B = 0$, $A = 1$, and $C = -Tg/w$. We see that Eq. (9.1) falls in the class of hyperbolic equations.

We will solve some typical hyperbolic partial-differential equations by two methods in this chapter.

9.1 SOLVING THE WAVE EQUATION BY FINITE DIFFERENCES

Uses finite-difference approximations of the derivatives similar to the technique of Chapters 6, 7, and 8, but we discover some unexpected situations.

9.2 COMPARISON TO THE D'ALEMBERT SOLUTION

Shows that the finite-difference solution can exactly match the analytical solution!

9.3 STABILITY OF THE NUMERICAL METHOD

Demonstrates that the finite-difference solution is stable when we use restricted values for Δt and Δx.

9.4 METHOD OF CHARACTERISTICS

Leads you into the intricacies of a method that is tedious but that can handle discontinuities. It is therefore of great value in many practical applications.

9.5 THE WAVE EQUATION IN TWO SPACE DIMENSIONS

Shows that finite-difference approximations permit the solution when there are two or even three spatial variables.

9.6 CHAPTER SUMMARY

Reviews the chapter in our accustomed style.

9.7 COMPUTER PROGRAM

Illustrates how the computer solves problems similar to those of Section 9.1.

9.1 SOLVING THE WAVE EQUATION BY FINITE DIFFERENCES

We attack the problem of solving the one-dimensional wave equation in the usual way by replacing derivatives by difference quotients. We use superscripts to denote time, and subscripts for position:

$$\frac{\partial^2 y}{\partial t^2} = \frac{Tg}{w}\frac{\partial^2 y}{\partial x^2},$$

$$\frac{y_i^{j+1} - 2y_i^j + y_i^{j-1}}{(\Delta t)^2} = \frac{Tg}{w}\left(\frac{y_{i+1}^j - 2y_i^j + y_{i-1}^j}{(\Delta x)^2}\right).$$

Solving for the displacement at the end of the current interval, at $t = t_{j+1}$, we get

$$y_i^{j+1} = \frac{Tg(\Delta t)^2}{w(\Delta x)^2}(y_{i+1}^j + y_{i-1}^j) - y_i^{j-1} + 2\left(1 - \frac{Tg(\Delta t)^2}{w(\Delta x)^2}\right)y_i^j.$$

Note that selecting the ratio $Tg(\Delta t)^2/w(\Delta x)^2 = 1$ gives some simplification, though y_i^{j+1} is still a function of conditions at both t_j and t_{j-1}:

$$y_i^{j+1} = y_{i+1}^j + y_{i-1}^j - y_i^{j-1}, \qquad \Delta t = \frac{\Delta x}{\sqrt{Tg/w}}. \qquad (9.2)$$

Equation (9.2) is the usual way that the one-dimensional wave equation is solved numerically. We are, of course, interested in the validity of choosing the ratio at unity. Our finite-difference replacements to the derivatives do not have mixed error terms here, as they did in the explicit method for the heat equation; but, as we will later show, after discussing the method of characteristics, if the ratio is greater than one, we cannot be sure of convergence. Stability also sets a limit of unity to the ratio. It is surprising to find that, if one uses a value of $w(\Delta t)^2/Tg(\Delta x)^2$ of less than one, the results are less accurate, while with the ratio equal to one we can get exact analytical answers!*

There is still a problem in applying Eq. (9.2). We know y at $t = t_0 = 0$ from the initial condition, but to compute y at $t = \Delta t = t_1$, we need values at $t = t_{-1}$. We may at first be bothered by a need to know displacements *before* the start of the problem; but if we imagine the function $y = y(x, t)$ to be extended backward in time, the term t_{-1} makes good sense. Since we will ordinarily get periodic functions for y versus t at a given point, we can consider zero time as an arbitrary point at which we know the displacement and velocity, within the duration of an ongoing process.

One commonly used way to get values for the fictitious point at $t = t_{-1}$ is by employing the specification for the initial velocity $\partial y/\partial t = g(x)$ at $t = 0$. Using a central-difference approximation, we have

$$\frac{\partial y}{\partial t}(x_i, 0) \doteq \frac{y_i^1 - y_i^{-1}}{2\Delta t} = g(x_i),$$

$$y_i^{-1} = y_i^1 - 2g(x_i)\Delta t \qquad \text{at } t = 0 \text{ only.} \qquad (9.3)$$

Equation (9.3) is valid only at $t = 0$; substituting into Eq. (9.2) gives us, for $t = t_1$,

$$y_i^1 = \frac{1}{2}(y_{i+1}^0 + y_{i-1}^0) + g(x_i)\Delta t. \qquad (9.4)$$

---

*What this means is that, with $w(\Delta t)^2/Tg(\Delta x)^2$ less than one, errors will decrease as $\Delta x \to 0$; that is, the method is stable. However, with finite values for Δx, the computed values are not exact unless the ratio is equal to one. Our discussion of characteristics later in this chapter will clarify this unusual situation.

After computing the first line with Eq. (9.4), we use Eq. (9.2) thereafter.* If the boundary conditions involve derivatives, we rewrite them in terms of difference quotients to incorporate them in our difference equation.

EXAMPLE   A banjo string is 80 cm long and weighs 1.0 g. It is stretched with a tension of 40,000 g. At a point 20 cm from one end it is pulled 0.6 cm from the equilibrium position and then released. Find the displacement of points along the string as a function of time. How long does it take for one complete cycle of motion? From this, compute the frequency with which it vibrates.

Let $\Delta x = 10$ cm. In Table 9.1 we show the results of using Eqs. (9.4) and (9.2). Because the string is just released from its initially displaced position, the initial velocity at all points is zero, and Eq. (9.4) becomes simply

$$y_i^1 = \frac{1}{2}(y_{i+1}^0 + y_{i-1}^0).$$

The initial conditions imply that y is linear in x from $y = 0$ at $x = 0$ to $y = 0.6$ at $x = 20$ and also linear to $x = 80$. The size of time steps is given by

$$\frac{Tg(\Delta t)^2}{w(\Delta x)^2} = 1 = \frac{(40,000)(980)(\Delta t)^2}{(1.0/80)(10)^2}, \qquad \Delta t = 0.000179 \text{ sec.}$$

After we have completed calculations for eight time steps, we observe that the y-values are reproducing the original steps but with negative signs and the end of the string reversed; that is, half a cycle has been completed in eight steps. For a complete cycle, $16\ \Delta t$'s are needed. The frequency of vibration is then

$$f = \frac{1}{(16)(0.000179)} = 350 \text{ cycles/sec.}$$

(This value is exactly the same as that given by the standard formula,

$$f = \frac{1}{2L}\sqrt{Tg/w}.\Big)\quad\blacksquare$$

The solution to our example as given in Table 9.1 is exactly the analytical solution, as we will now demonstrate. We will compare our finite-difference equation (Eq. (9.2)) with the d'Alembert solution to the wave equation.

9.2 COMPARISON TO THE D'ALEMBERT SOLUTION

The method of d'Alembert lets us find the analytical solution to the wave equation in one dimension. The one-dimensional wave equation, with $\sqrt{Tg/w} = c$, is

---

*We will later discuss a more accurate way to begin the solution. Equation (9.4) is satisfactory when the initial velocity is zero.

Table 9.1   Solution to banjo string example

t	0	10	20	30	40	50	60	70	80
0	0.0	0.3	0.6	0.5	0.4	0.3	0.2	0.1	0.0
Δt	0.0	0.3	0.4	0.5	0.4	0.3	0.2	0.1	0.0
$2\Delta t$	0.0	0.1	0.2	0.3	0.4	0.3	0.2	0.1	0.0
$3\Delta t$	0.0	−0.1	0.0	0.1	0.2	0.3	0.2	0.1	0.0
$4\Delta t$	0.0	−0.1	−0.2	−0.1	0.0	0.1	0.2	0.1	0.0
$5\Delta t$	0.0	−0.1	−0.2	−0.3	−0.2	−0.1	0.0	0.1	0.0
$6\Delta t$	0.0	−0.1	−0.2	−0.3	−0.4	−0.3	−0.2	−0.1	0.0
$7\Delta t$	0.0	−0.1	−0.2	−0.3	−0.4	−0.5	−0.4	−0.3	0.0
$8\Delta t$	0.0	−0.1	−0.2	−0.3	−0.4	−0.5	−0.6	−0.3	0.0
$9\Delta t$	0.0	−0.1	−0.2	−0.3	−0.4	−0.5	−0.4	−0.3	0.0
$10\Delta t$	0.0	−0.1	−0.2	−0.3	−0.4	−0.3	−0.2	−0.1	0.0
$11\Delta t$	0.0	−0.1	−0.2	−0.3	−0.2	−0.1	0.0	0.1	0.0
$12\Delta t$	0.0	−0.1	−0.2	−0.1	0.0	0.1	0.2	0.1	0.0
$13\Delta t$	0.0	−0.1	0.0	0.1	0.2	0.3	0.2	0.1	0.0
$14\Delta t$	0.0	0.1	0.2	0.3	0.4	0.3	0.2	0.1	0.0
$15\Delta t$	0.0	0.3	0.4	0.5	0.4	0.3	0.2	0.1	0.0
$16\Delta t$	0.0	0.3	0.6	0.5	0.4	0.3	0.2	0.1	0.0

$$\frac{\partial^2 y}{\partial t^2} = c^2 \frac{\partial^2 y}{\partial x^2}. \tag{9.5}$$

By direct substitution, it is readily seen that, for any arbitrary functions F and G, Eq. (9.5) is solved by

$$y(z, t) = F(x + ct) + G(x - ct). \tag{9.6}$$

The demonstration is easy, since

$$\frac{\partial y}{\partial t} = F' \frac{\partial(x + ct)}{\partial t} + G' \frac{\partial(x - ct)}{\partial t} = cF' - cG',$$

$$\frac{\partial^2 y}{\partial t^2} = c^2 F'' - c(-c)G'' = c^2 F'' + c^2 G''; \tag{9.7}$$

$$\frac{\partial y}{\partial x} = F' \frac{\partial(x + ct)}{\partial x} + G' \frac{\partial(x - ct)}{\partial x} = F' + G',$$

$$\frac{\partial^2 y}{\partial x^2} = F'' + G''. \tag{9.8}$$

Equation (9.5) results immediately from Eqs. (9.7) and (9.8).

The solution to the problem is found, then, if we can find a pair of functions of the form of (9.6) whose sum matches the initial conditions and the boundary conditions for the problem. If the initial conditions are

$$y(x, 0) = f(x),$$

$$\frac{\partial y}{\partial t}(x, 0) = g(x),$$

it is again readily seen that $y(x, t)$ is given by

$$y(x, t) = \frac{1}{2}[f(x + ct) + f(x - ct)] + \frac{1}{2c}\int_{x-ct}^{x+ct} g(v)\, dv. \tag{9.9}$$

Note that we can rewrite Eq. (9.9) in the form of Eq. (9.6). In Eq. (9.9) we have changed to the dummy variable v under the integral sign. The demonstration parallels that above when we recall how to differentiate an integral:

$$\frac{\partial}{\partial t}\left[\frac{1}{2c}\int_{x-ct}^{x+ct} g(v)\, dv\right] = \frac{1}{2c}[cg(x + ct) - (-c)g(x - ct)],$$

$$\frac{\partial^2}{\partial t^2}\left[\frac{1}{2c}\int_{x-ct}^{x+ct} g(v)\, dv\right] = \frac{\partial}{\partial t}\left[\frac{1}{2}g(x + ct) + \frac{1}{2}g(x - ct)\right] = \frac{1}{2}cg' - \frac{1}{2}cg' = 0,$$

$$\frac{\partial}{\partial x}\left[\frac{1}{2c}\int_{x-ct}^{x+ct} g(v)\, dv\right] = \frac{1}{2c}[g(x + ct) - g(x - ct)],$$

$$\frac{\partial^2}{\partial x^2}\left[\frac{1}{2c}\int_{x-ct}^{x+ct} g(v)\, dv\right] = \frac{\partial}{\partial x}\left[\frac{1}{2c}g(x + ct) - \frac{1}{2c}g(x - ct)\right] = \frac{1}{2c}g' - \frac{1}{2c}g' = 0.$$

Equation (9.9) gives the value of y at any time t provided that y is known at points ct to the right and left of the point at $t = 0$, and in terms of the integral of the initial velocity between the lateral points. Hence it is useful to find the displacement of points in the interior portions of a vibrating string.

We are now in a position to verify that our numerical procedure, given by Eq. (9.2), gives the exact solution (except for round-off), provided that two lines of correct y-values are known. The algorithm we use for the numerical procedure is

$$y_i^{j+1} = y_{i-1}^j + y_{i+1}^j - y_i^{j-1}. \tag{9.10}$$

We now show that the solution to this difference equation is a solution to the differential equation. Consider the function

$$y_i^j = F(x_i + ct_j) + G(x_i - ct_j). \tag{9.11}$$

Because we use

$$\frac{Tg(\Delta t)^2}{w(\Delta x)^2} = 1 = \frac{c^2(\Delta t)^2}{(\Delta x)^2},$$

then

$$\Delta x = c \, \Delta t.$$

Write x_i and t_j in terms of starting values, x_0, and $t_0 = 0$:

$$x_i = x_0 + i \, \Delta x,$$

$$ct_j = c(t_0 + j \, \Delta t) = cj \, \Delta t = j \, \Delta x.$$

Substituting into Eq. (9.11), we get

$$y_i^j = F(x_0 + i \, \Delta x + j \, \Delta x) + G(x_0 + i \, \Delta x - j \, \Delta x)$$
$$= F[x_0 + (i + j)\Delta x] + G[x_0 + (i - j)\Delta x].$$

Use this relation to rewrite the right-hand side of the difference equation, Eq. (9.10):

$$
\begin{aligned}
y_{i-1}^j + y_{i+1}^j - y_i^{j-1} = {} & F[x_0 + (i - 1 + j)\Delta x] \\
& + G[x_0 + (i - 1 - j)\Delta x] \\
& + F[x_0 + (i + 1 + j)\Delta x] \\
& + G[x_0 + (i + 1 - j)\Delta x] \\
& - F[x_0 + (i + j - 1)\Delta x] \\
& - G[x_0 + (i - j + 1)\Delta x] \\
= {} & F[x_0 + (i + j + 1)\Delta x] \\
& + G[x_0 + (i - j - 1)\Delta x] = y_i^{j+1}.
\end{aligned}
\tag{9.12}
$$

Equation (9.12) shows that, if the previous two lines of the numerical solution are of the form of Eq. (9.11) (and hence must be exact solutions of the wave equation in view of Eq. (9.6)), it follows that the values on the next line are exact solutions also, because they also are of the form of Eq. (9.11).

   In order that our simple algorithm give the analytical solution, it is then necessary only that two lines of the computation be correct. The first line is correct because it is given by the initial conditions. While Eq. (9.4) is frequently recommended for giving the second line in terms of the initial velocity, it is sometimes inaccurate because it assumes that the initial velocity and the average velocity from t_{-1} to t_1 are the same, and this may not be true.*

---

*The relationship of Eq. (9.4) is correct whenever $g(x)$ = constant. It is therefore correct for the case of $g(x) = 0$.

Equation (9.9) offers a more exact way to get the second line of y-values. If we employ that relationship at $t = \Delta t$, we have

$$
y(x_i, \Delta t) = y_i^1 = \frac{1}{2}[f(x_i + \Delta x) + f(x_i - \Delta x)] + \frac{1}{2c}\int_{x_i - \Delta x}^{x_i + \Delta x} g(v)\, dv
$$

$$
= \frac{1}{2}[y_{i+1}^0 + y_{i-1}^0] + \frac{1}{2c}\int_{x_{i-1}}^{x_{i+1}} g(v)\, dv. \tag{9.13}
$$

Equation (9.13) differs from Eq. (9.4) only in the last term. We now see why the banjo string example of Section 9.1 gave correct values in spite of using Eq. (9.4): In the example, $g(x)$ was everywhere zero, so the form of the last term was unimportant. In the next example, we illustrate the difference between the two different ways to begin the solution.

E X A M P L E A string whose length is 9 units is initially in its equilibrium position. It is set into motion by striking it so that its initial velocity is given by $\partial y/\partial t = 3\sin(\pi x/L)$. Take $\Delta x = 1$ unit and assume that $c = \sqrt{Tg/w} = 2$. When the ratio of $c^2(\Delta t)^2/(\Delta x)^2$ is unity, $\Delta t = 0.5$ time unit.

If the ends are fixed, find the displacements at $t = 0.5$ time unit later. The length is subdivided into nine intervals because $\Delta x = 1$.

The displacements we require are after one time step. Table 9.2 summarizes the computations. We first compute them using Eq. (9.4). The values disagree by several percent from the analytical values, which are given by

$$
y(x, t) = \frac{1}{2c}\int_{x-ct}^{x+ct} 3\sin\frac{\pi v}{L}\, dv = \frac{3L}{2c\pi}\left\{\cos\left(\frac{\pi x}{L} - \frac{\pi ct}{L}\right) - \cos\left(\frac{\pi x}{L} + \frac{\pi ct}{L}\right)\right\}.
$$

When Δx is cut in half, Eq. (9.4) gives improved values; it now requires two time steps to reach $t = 0.5$, of course.

Using Eq. (9.13), and evaluating the integral with Simpson's rule, we obtain results with $\Delta x = 1$ that are accurate to within one in the fourth decimal place.

9.3   STABILITY OF THE NUMERICAL METHOD

Since we will ordinarily solve the one-dimensional wave equation numerically only with

$$
\frac{Tg(\Delta t)^2}{w(\Delta x)^2} = 1,
$$

it is sufficient to demonstrate stability for that scheme. We assume that a set of errorless computations have been made when a single error of size 1 occurs; we trace the effects

Table 9.2   Comparison of ways to begin the solution of the wave equation in one dimension:

$$\frac{\partial^2 y}{\partial t^2} = c^2 \frac{\partial^2 y}{\partial x^2} \quad \text{over } x = 0 \text{ to } x = 9, \text{ with } c = 2;$$

$$y(x, 0) = 0, \quad \frac{\partial y}{\partial t}(x, 0) = 3 \sin\left(\frac{\pi x}{9}\right)$$

	Value of displacements at $t = 0.5$			
Using Eq. (9.4): $\Delta x = 1$				
x	1	2	3	4
y	0.5130	0.9642	1.2990	1.4772
Using Eq. (9.4): $\Delta x = 0.5$				
x	1	2	3	4
y	0.5052	0.9495	1.2793	1.4548
Using Eq. (9.13): $\Delta x = 1$ Simpson's-rule integration				
x	1	2	3	4
y	0.5027	0.9448	1.2729	1.4475
Analytical solution:				
x	1	2	3	4
y	0.50267	0.94472	1.27282	1.44740

■

Table 9.3   Propagation of single error in numerical solution to wave equation

Initially error-free values	0	0	0	0	0	0	0
	0	0	0	0	0	0	0
Error made here	0	0	1	0	0	0	0
	0	1	0	1	0	0	0
	0	0	0	0	1	0	0
	0	−1	0	0	0	1	0
	0	0	−1	0	0	0	0
	0	0	0	−1	0	−1	0
	0	0	0	0	−2	0	0
	0	0	0	−1	0	−1	0
	0	0	−1	0	0	0	0
	0	−1	0	0	0	1	0

of the single error. If the error does not have increasingly great effect on subsequent calculations, we call the method *stable*.

This simple procedure is adequate because, since the problem is linear, the principle of superposition lets us add the effects of all errors together and lets us add these errors to the true solution to obtain the actual results. Table 9.3 demonstrates the principle, assuming that the displacement of the endpoints is specified so that these are always free of error.

As the arrows indicate, the wave equation propagates disturbances in opposite directions, with reflections occurring at fixed ends, with a reversal of sign on reflection. Stability is demonstrated because the original error does not grow in size.

9.4   METHOD OF CHARACTERISTICS

The properties of the solution to the wave equation are further elucidated by considering the "characteristic curves" of the equation. This will also permit us to extend our numerical method to more general hyperbolic equations.

Consider the second-order partial-differential equation in two variables x and t:

$$au_{xx} + bu_{xt} + cu_{tt} + e = 0. \qquad (9.14)$$

Here we have used the subscript notation to represent partial derivatives. The coefficients a, b, c, and e may be functions of x, t, u_x, u_t, and u, so the equation is very general.* We take $u_{xt} = u_{tx}$. To facilitate manipulations, let

$$p = \frac{\partial u}{\partial x} = u_x, \qquad q = \frac{\partial u}{\partial t} = u_t.$$

Write out the differentials of p and q:

$$dp = \frac{\partial p}{\partial x} dx + \frac{\partial p}{\partial t} dt = u_{xx} dx + u_{xt} dt,$$

$$dq = \frac{\partial q}{\partial x} dx + \frac{\partial q}{\partial t} dt = u_{xt} dx + u_{tt} dt.$$

Solving these last equations for u_{xx} and u_{tt}, respectively, we have

$$u_{xx} = \frac{dp - u_{xt}\, dt}{dx} = \frac{dp}{dx} - u_{xt}\frac{dt}{dx},$$

$$u_{tt} = \frac{dq - u_{xt}\, dx}{dt} = \frac{dq}{dt} - u_{xt}\frac{dx}{dt}.$$

Substituting in Eq. (9.14) and rearranging, we obtain

$$-au_{xt}\frac{dt}{dx} + bu_{xt} - cu_{xt}\frac{dx}{dt} + a\frac{dp}{dx} + c\frac{dq}{dt} + e = 0.$$

---

*When the coefficients are independent of the function u or its derivatives, it is linear. If they are functions of u, u_x, or u_t (but not u_{xx} or u_{tt}), it is called quasilinear.

Now multiplying by $-dt/dx$, we get

$$u_{xt}\left[a\left(\frac{dt}{dx}\right)^2 - b\left(\frac{dt}{dx}\right) + c\right] - \left[a\frac{dp}{dx}\frac{dt}{dx} + c\frac{dq}{dx} + e\frac{dt}{dx}\right] = 0.$$

Suppose, in the xt-plane, we define curves such that the first bracketed expression is zero. On such curves, the original differential equation is equivalent to setting the second bracketed expression equal to zero. That is, if

$$am^2 - bm + c = 0, \qquad (9.15)$$

where $m = dt/dx$, then the solution to the original equation (Eq. (9.14)) can be found by solving

$$am\, dp + c\, dq + e\, dt = 0. \qquad (9.16)$$

We have elected to write Eq. (9.16) in the form of differentials. It will be seen that we reduce the original problem, which is a second-order partial-differential equation, to solving a pair of first-order equations of the form of Eq. (9.16).

The curves whose slope m is given by Eq. (9.15) are called the *characteristics* of the differential equation. Since the equation is a quadratic, it may have one, two, or no real solutions, depending on the value of $b^2 - 4ac$. The value of this discriminant is the usual basis for classifying partial-differential equations. If

$$b^2 - 4ac < 0,$$

the equation is called *elliptic*, and there are no (real) characteristics. If $b^2 - 4ac = 0$, there is a single characteristic at any point, and the equation is termed *parabolic*. When $b^2 - 4ac > 0$, at every point there will be a pair of characteristic curves whose slopes are given by the two distinct, real roots of Eq. (9.15). Such equations are called *hyperbolic*, and our present discussion considers only this type.

Along the characteristics, the solution has special and desirable properties. For example, discontinuities in the initial conditions are propagated along them. On the characteristic curves, the numerical solution can be developed in the general case also.

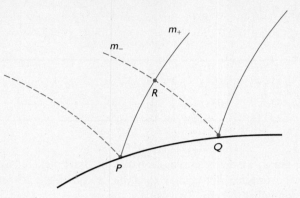

Figure 9.3

We will now outline a method of solving equations of the form of Eq. (9.14) by numerical integration along the characteristics.* We visualize the initial conditions as specifying the function u on some curve in the tx-plane,[†] as well as its normal derivative. Consider two points, P and Q on this initial curve (Fig. 9.3). When Eq. (9.14) is hyperbolic, there are two characteristic curves through each point. The rightmost curve through P intersects the leftmost curve through Q, and these curves are such that their slopes are given by the appropriate roots of Eq. (9.15). Call m_+ the values of the slope on curve PR, and m_- the values on curve QR.

Since these curves are characteristics, the solution to the problem can be found by solving Eq. (9.16) along them.

Our procedure will be to first find point R (perhaps only as a first approximation if a, b, or c involve the unknown function u). This is done by solving the equation

$$\Delta t = \left(\frac{dt}{dx}\right)_{av} \Delta x = m_{av}\, \Delta x, \qquad (9.17)$$

applied over the arcs PR and QR simultaneously. When dt/dx is a function of x and/or t, it may be possible to integrate Eq. (9.17) analytically. When dt/dx varies with u, we will use a procedure resembling the Euler predictor–corrector method, by predicting with m_{av} taken as equal to m_+ at P or m_- at Q to start the solution. We correct by using the arithmetic average of m at the endpoints of each arc as soon as the value of m at R can be evaluated.

We then integrate Eq. (9.16) in the form

$$a_{av}\, m_{av}\, \Delta p + c_{av}\, \Delta q + e_{av}\, \Delta t = 0,$$

---

*We discuss only the solution of hyperbolic differential equations by the method of characteristics, but the technique can also be applied to parabolic equations.

[†]This curve must not itself be one of the characteristics, or advancing the solution is impossible.

starting first from P and then from Q, using the appropriate values of m for each. This will estimate p and q at point R. Finally we evaluate the function u at R from

$$du = \frac{\partial u}{\partial x} \, dx + \frac{\partial u}{\partial t} \, dt,$$

used in the form

$$\Delta u = p_{av} \, \Delta x + q_{av} \, \Delta t.$$

The equation for Δu can be applied to either the change along

$$P \rightarrow R \quad \text{or} \quad Q \rightarrow R;$$

when these do not give the same result, we will *average* the two values. It may be necessary to iterate this procedure to get improved values at R. Average values are used for all varying quantities. The calculations are repeated for a second point S that is the intersection of chararacteristics through Q and another initial point W, and then continued in a like manner to evaluate u throughout the region in the xt-plane as desired. We illustrate with three examples, first with dt/dx equal to a constant, then with dt/dx varying with x, and finally a more complex example with dt/dx varying with u.

For the simple wave equation,

$$u_{tt} = c^2 u_{xx},$$

the slopes of the characteristics are

$$m = \pm \frac{1}{c},$$

and the characteristics are the lines

$$t = \pm \frac{1}{c}(x - x_i).$$

Consider the curves from points P and Q, taken on the line for $t = 0$ (Fig. 9.4). The network of points used in the explicit finite-difference method of Section 9.1 are seen to be the intersections of characteristics through pairs of points spaced $2\Delta x$ apart. The finite-difference method, with

$$\frac{Tg(\Delta x)^2}{w(\Delta t)^2} = 1,$$

will be found to be the equivalent of integration along the characteristics, lending further support to the likelihood that it will give exact answers.

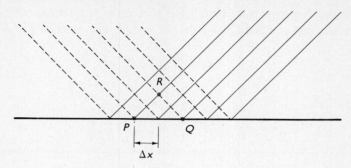

Figure 9.4

EXAMPLE I   Solve

$$\frac{\partial^2 u}{\partial t^2} = 2\frac{\partial^2 u}{\partial x^2} - 4,$$

with initial conditions

$$u = 12x \quad \text{for} \quad 0 \le x \le 0.25,$$
$$u = 4 - 4x \quad \text{for} \quad 0.25 \le x \le 1.0,$$
$$\frac{\partial u}{\partial t} = 0 \quad \text{for} \quad 0 \le x \le 1.0;$$

boundary conditions are $u = 0$ at $x = 0$ and at $x = 1.0$.
Putting the equation into the standard form,

$$a\frac{\partial^2 u}{\partial x^2} + b\frac{\partial^2 u}{\partial x\,\partial t} + c\frac{\partial^2 u}{\partial t^2} + e = 0,$$

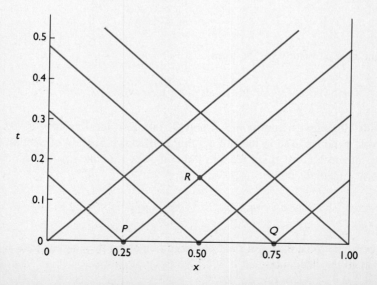

Figure 9.5

gives $a = -2$, $b = 0$, $c = 1$, and $e = 4$. (The equation is linear since a, b, c, and e are independent of u, u_x, and u_t.)

The slopes of the characteristics are the roots of

$$-2m^2 + 1 = 0,$$

$$m = \pm\frac{\sqrt{2}}{2},$$

so the characteristic curves are straight lines in the xt-plane, as shown in Fig. 9.5. Consider points P, Q, and R—(0.25, 0), (0.75, 0), and (0.5, 0.1768)—and solve Eq. (9.16), which is

$$am_{av}\,\Delta p + c\,\Delta q + e\,\Delta t = 0.$$

Along $P \to R$: $-2\left(\dfrac{\sqrt{2}}{2}\right)\Delta p + \Delta q + 4\Delta t = 0,$

$-\sqrt{2}(p_R - p_P) + (q_R - q_P) + 4(0.1768) = 0;$

Along $Q \to R$: $-2\left(\dfrac{-\sqrt{2}}{2}\right)\Delta p + \Delta q + 4\Delta t = 0,$

$\sqrt{2}(p_R - p_Q) + (q_R - q_Q) + 4(0.1768) = 0.$

Using

$$p_P = \left(\frac{\partial u}{\partial x}\right)_P = -4,^* \qquad q_P = \left(\frac{\partial u}{\partial t}\right)_P = 0, \qquad p_Q = \left(\frac{\partial u}{\partial x}\right)_Q = -4,$$

$$q_Q = \left(\frac{\partial u}{\partial t}\right)_Q = 0,$$

we find $p_R = -4$, $q_R = -\sqrt{2}/2$ by solving the equations $P \to R$ and $Q \to R$ simultaneously.

Now we evaluate u at point R through its change along $P \to R$:

$$\Delta u = p_{av}\,\Delta x + q_{av}\,\Delta t = -4(0.25) + \left(\frac{0 - \sqrt{2}/2}{2}\right)(0.1768)$$

$$= -1.0625;$$

$$u_R = 3 + (-1.0625) = 1.9375.$$

(If we compute through evaluating Δu along $Q \to R$, we get the same result.)

---

*The gradient has a discontinuity at $x = 0.25$. The value of $\partial u/\partial x$ for points to the right of P applies for the region PRQ.

Table 9.4

x	0	0.25	0.5	0.75	1.0
$u(t = 0)$	0.0	3.0	2.0	1.0	0.0
$u(t = 0.1768)$	0.0	0.9375	(1.9375)	0.9375	0.0
$u(t = 0.3535)$	0.0	−1.1875	−0.2500	0.8125	0.0
$u(t = 0.5303)$	0.0	−1.3125	−2.4375	−1.3125	0.0

For this simple problem, the finite-difference method is much simpler, and we expect it to give the same results. Following the procedure of Section 9.1* we compute with $\Delta x = 0.25$, $\Delta t = \Delta x / \sqrt{c} = 0.1768$, and obtain Table 9.4. The circled value agrees exactly with that calculated by the method of characteristics. ∎

EXAMPLE 2   Solve

$$\frac{\partial^2 u}{\partial t^2} = (1 + 2x)\frac{\partial^2 u}{\partial x^2}$$

over $(0, 1)$ with fixed boundaries and the initial conditions:

$$u(x, 0) = 0, \qquad \frac{\partial u}{\partial t}(x, 0) = x(1 - x).$$

For this problem, $a = -(1 + 2x)$, $b = 0$, $c = 1$, $e = 0$. Then $am^2 + bm + c = 0$ gives

$$m = \pm \sqrt{\frac{1}{(1 + 2x)}}.$$

The characteristic curves are found by solving the differential equations $dt/dx = \sqrt{1/(1 + 2x)}$ and $dt/dx = -\sqrt{1/(1 + 2x)}$. Integrating[†] from the initial point x_0 and t_0, we have

$$t = t_0 + \sqrt{1 + 2x} - \sqrt{1 + 2x_0} \quad \text{from } m_+,$$
$$t = t_0 - \sqrt{1 + 2x} + \sqrt{1 + 2x_0} \quad \text{from } m_-.$$

Figure 9.6 shows several of the characteristic curves. We select two points on the initial curve for $t = 0$, at $P = (0.25, 0)$ and $Q = (0.75, 0)$, whose characteristics intersect at point R. Solving for the intersection, we find $R = (0.4841, 0.1782)$.

---

*The algorithm is $u_i^{j+1} = (u_{i+1}^j + u_{i-1}^j) - u_i^{j-1} - 4(\Delta t)^2$ with $\Delta t = \Delta x/\sqrt{2}$. For the first time step, $u_i^1 = \frac{1}{2}(u_{i+1}^0 + u_{i-1}^0) - \frac{1}{2}(4)(\Delta t)^2$.

[†]In this example, the integration methods of calculus are easy to use. We could use a numerical method if they were not.

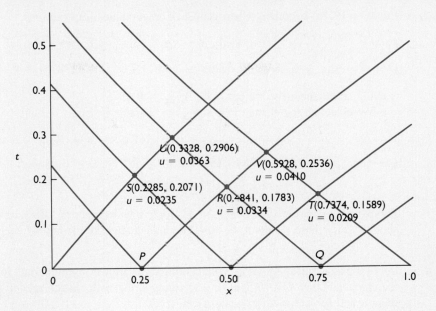

Figure 9.6

We now solve Eq. (9.16) to obtain $p = \partial u/\partial x$ and $q = \partial u/\partial t$ at R:

At point P: $\quad x = 0.25, \quad t = 0, \quad u = 0, \quad p = \left(\dfrac{\partial u}{\partial x}\right)_P = 0,$

$$q = \left(\frac{\partial u}{\partial t}\right)_P = x - x^2 = 0.1875,$$

$$m = \sqrt{1/(1 + 2x)} = 0.8165,$$

$$a = -(1 + 2x) = -1.5, \quad b = 0, \quad c = 1, \quad e = 0.$$

At point Q: $\quad x = 0.75, \quad t = 0, \quad u = 0, \quad p = 0, \quad q = x - x^2 = 0.1875,$

$$m = -\sqrt{1/(1 + 2x)} = -0.6325,$$

$$a = -2.5, \quad b = 0, \quad c = 1, \quad e = 0.$$

At point R: $\quad x = 0.4841, \quad t = 0.1783,$

$$m_+ = \sqrt{1/(1 + 2x)} = 0.7128,$$

$$m_- = -\sqrt{1/(1 + 2x)} = -0.7128,$$

$$a = -1.9682, \quad b = 0, \quad c = 1, \quad e = 0.$$

Equation (9.16) becomes, when we use average values for a and m,

$P \rightarrow R$: $\qquad -1.7341(0.7646)(p_R - 0) + (1)(q_R - 0.1875) = 0;$

$Q \rightarrow R$: $\qquad -2.2341(-0.6726)(p_R - 0) + (1)(q_R - 0.1875) = 0.$

Solving simultaneously, we get $p_R = 0$, $q_R = 0.1875$.
We calculate the change in u along the characteristics:

$P \rightarrow R$: $\qquad \Delta u = 0(0.2341) + 0.1875(0.1783) = 0.0334,$

$Q \rightarrow R$: $\qquad \Delta u = 0(-0.2659) + 0.1875(0.1783) = 0.0334,$

$\qquad\qquad\qquad u_R = 0 + 0.0334 = 0.0334.$

Figure 9.6 gives the results at several other intersections of characteristics. Students should verify these results to be sure they understand the method of characteristics. ∎

EXAMPLE 3 Solve the quasilinear equation, with conditions as shown, by numerical integration along the characteristics. (This might be a vibrating string with tension related to the displacement u and subject to an external lateral force.)

$$\frac{\partial^2 u}{\partial x^2} - u\frac{\partial^2 u}{\partial t^2} + (1 - x^2) = 0, \qquad u(x, 0) = x(1 - x), \qquad u_t(x, 0) = 0,$$

$$u(0, t) = 0, \qquad u(1, t) = 0. \tag{9.18}$$

We will advance the solution beyond the start from P, at $x = 0.2$, $t = 0$, and Q, at $x = 0.4$, $t = 0$, to one new point R. Comparing Eq. (9.18) to the standard form,

$$au_{xx} + bu_{xt} + cu_{tt} + e = 0,$$

we have $a = 1$, $b = 0$, $c = -u$, $e = 1 - x^2$. We first compute u, p, and q at points P and Q

$$u = x(1 - x)$$

(from the initial conditions), so

$$u_P = 0.2(1 - 0.2) = 0.16,$$

$$u_Q = 0.4(1 - 0.4) = 0.24.$$

Also,

$$p = \frac{\partial u}{\partial x} = 1 - 2x$$

(by differentiating the initial conditions), so

$$p_P = 1 - 2(0.2) = 0.6,$$

$$p_Q = 1 - 2(0.4) = 0.2;$$

and

$$q = \frac{\partial u}{\partial t} = 0$$

(from the initial conditions), so

$$q_P = 0,$$
$$q_Q = 0.$$

To locate point R, we need the slope m of the characteristic. Using $am^2 - bm + c = 0$, we get

$$m = \frac{b \pm \sqrt{b^2 - 4ac}}{2a},$$

$$m = \frac{\pm\sqrt{4u}}{2} = \pm\sqrt{u}.$$

Since m depends on the solution u, we will need to find point R through the predictor–corrector approach. In the first trial, use the initial values over the whole arc; that is, take $m_+ = +m_P$ and $m_- = -m_Q$:

$$m_+ = \sqrt{u_P} = \sqrt{0.16} = 0.4,$$
$$m_- = \sqrt{u_Q} = -\sqrt{0.24} = -0.490.$$

We now estimate the coordinates of R by solving simultaneously

$$t_R = m_+(x_R - x_P) = 0.4(x_R - 0.2),$$
$$t_R = m_-(x_R - x_Q) = -0.490(x_R - 0.4).$$

These give

$$x_R = 0.310, \qquad t_R = 0.044.$$

We write Eq. (9.16) along each characteristic, again using the initial values of m, since m at R is still unknown:

$$am\,\Delta p + c\,\Delta q + e\,\Delta t = 0,$$

$$(1)(0.4)(p_R - 0.6) + (-0.16)(q_R - 0) + \left(1 - \frac{0.04 + 0.096}{2}\right)(0.044) = 0,$$

$$(1)(-0.490(p_R - 0.2) + (-0.24)(q_R - 0) + \left(1 - \frac{0.16 + 0.096}{2}\right)(0.044) = 0.$$

In these equations we used the arithmetic average of x^2 in the last terms. Solving simultaneously, we get

$$p_R = 0.399, \qquad q_R = -0.246.$$

As a first approximation for u at R, then,

$$\Delta u = p \, \Delta x + q \, \Delta t,$$

$$u_R - 0.16 = \frac{0.6 + 0.399}{2}(0.310 - 0.2) + \frac{0 - 0.246}{2}(0.044 - 0),$$

$$u_R = 0.2095.$$

The last computation was along PR, using average values of p and q. We could have alternatively proceeded along QR. If this is done,

$$u_R - 0.24 = \frac{0.2 + 0.399}{2}(0.310 - 0.4) + \frac{0 - 0.246}{2}(0.044 - 0),$$

$$u_R = 0.2076.$$

The two values should be close to each other. Let us use the average value, 0.2086, as our initial estimate of u_R. We now repeat the work. In getting the coordinates of R, we now use average values of the slopes,

$$t_R = \frac{0.4 + \sqrt{0.2086}}{2}(x_R - 0.2),$$

$$t_R = \frac{-0.490 - \sqrt{0.2086}}{2}(x_R - 0.4),$$

$$x_R = 0.305, \qquad t_R = 0.045;$$

$$(1)\left(\frac{0.4 + \sqrt{0.2086}}{2}\right)(p_R - 0.6) - \left(\frac{0.16 + 0.2086}{2}\right)(q_R - 0)$$

$$+ \left(1 - \frac{0.04 + 0.0930}{2}\right)(0.045) = 0,$$

$$(1)\left(\frac{-0.490 - \sqrt{0.2086}}{2}\right)(p_R - 0.2) - \left(\frac{0.24 + 0.2086}{2}\right)(q_R - 0)$$

$$+ \left(1 - \frac{0.16 + 0.0930}{2}\right)(0.045) = 0,$$

$$p_R = 0.398, \qquad q_R = -0.242;$$

$$u_R = 0.16 + \frac{0.6 + 0.398}{2}(0.305 - 0.2) + \frac{0 - 0.242}{2}(0.045 - 0),$$

$$u_R = 0.2071 \qquad \text{(along } PR\text{)};$$

$$u_R = 0.24 + \frac{0.2 + 0.398}{2}(0.305 - 0.4) + \frac{0 - 0.242}{2}(0.045 - 0),$$

$$u_R = 0.2063 \qquad \text{(along } QR\text{)}.$$

The average value is 0.2067.

Another round of calculations gives $u_R = 0.2066$, which checks the previous value sufficiently. This method is, of course, very tedious by hand. ∎

9.5 THE WAVE EQUATION IN TWO SPACE DIMENSIONS

The finite-difference method can be applied to hyperbolic partial-differential equations in two or more space dimensions. A typical problem is the vibrating membrane. Consider a thin flexible membrane stretched over a rectangular frame and set to vibrating. A development analogous to that for the vibrating string gives

$$\frac{\partial^2 u}{\partial t^2} = \frac{Tg}{w}\left(\frac{\partial^2 u}{\partial x^2} + \frac{\partial^2 u}{\partial y^2}\right),$$

in which u is the displacement, t is the time, x and y are the space coordinates, T is the uniform tension per unit length, g is the acceleration of gravity, and w is the weight per unit area. For simplification, let $Tg/w = \alpha^2$. Replacing each derivative by its central-difference approximation, and using $h = \Delta x = \Delta y$, gives (we recognize the Laplacian on the right-hand side)

$$\frac{u_{i,j}^{k+1} - 2u_{i,j}^k + u_{i,j}^{k-1}}{(\Delta t)^2} = \alpha^2 \frac{u_{i+1,j}^k + u_{i-1,j}^k + u_{i,j+1}^k + u_{i,j-1}^k - 4u_{i,j}^k}{h^2}. \quad \textbf{(9.19)}$$

Solving for the displacement at time t_{k+1}, we obtain

$$u_{i,j}^{k+1} = \frac{\alpha^2(\Delta t)^2}{h^2}\left\{\begin{matrix} & 1 & \\ 1 & 0 & 1 \\ & 1 & \end{matrix}\right\}u_{i,j}^k - u_{i,j}^{k-1} + \left(2 - 4\frac{\alpha^2(\Delta t)^2}{h^2}\right)u_{i,j}^k. \quad \textbf{(9.20)}$$

In Eqs. (9.19) and (9.20), we use superscripts to denote the time. If we let $\alpha^2(\Delta t)^2/h^2 = \frac{1}{2}$, the last term vanishes and we get

$$u_{i,j}^{k+1} = \frac{1}{2}\left\{\begin{matrix} & 1 & \\ 1 & 0 & 1 \\ & 1 & \end{matrix}\right\}u_{i,j}^k - u_{i,j}^{k-1}. \quad \textbf{(9.21)}$$

For the first time step, we get displacements from Eq. (9.22), which is obtained by approximating $\partial u/\partial t$ at $t = 0$ by a central-difference approximation involving $u_{i,j}^1$ and $u_{i,j}^{-1}$:

$$u_{i,j}^1 = \frac{1}{4}\left\{\begin{matrix} & 1 & \\ 1 & 0 & 1 \\ & 1 & \end{matrix}\right\}u_{i,j}^0 + (\Delta t)g(x_i, y_j). \quad \textbf{(9.22)}$$

In Eq. (9.22), $g(x, y)$ is the initial velocity.

It should not surprise us to learn that this ratio $\alpha^2(\Delta t)^2/h^2 = \frac{1}{2}$ is the maximum value for stability, in view of our previous experience with explicit methods. However, in contrast with the wave equation in one space dimension, we do not get exact answers from the numerical procedure of Eq. (9.21), and we further observe that we must use smaller time steps in relation to the size of the space interval. Therefore we advance in time more slowly. However, the numerical method is straightforward, as the following example will show.

EXAMPLE   A membrane for which $\alpha^2 = Tg/w = 3$ is stretched over a square frame that occupies the region $0 \leqslant x \leqslant 2$, $0 \leqslant y \leqslant 2$, in the xy-plane. It is given an initial displacement described by

$$u = x(2 - x)y(2 - y),$$

and has an initial velocity of zero. Find how the displacement varies with time.

We divide the region with $h = \Delta x = \Delta y = \frac{1}{2}$, obtaining nine intermediate points. Initial displacements are calculated from the initial conditions: $u^0(x, y) = x(2 - x)y(2 - y)$; Δt is taken at its maximum value for stability, $h/(\sqrt{2}\alpha) = 0.0204$. The values at the end of one time step are given by

$$u^1_{i,j} = \frac{1}{4}\left\{\begin{matrix} & 1 & \\ 1 & 0 & 1 \\ & 1 & \end{matrix}\right\}u^0_{i,j}$$

because $g(x, y)$ in Eq. (9.22) is everywhere zero. For succeeding time steps, Eq. (9.21) is used. Table 9.5 gives the results of our calculations. Also shown in Table 9.5 (in parentheses) are analytical values, computed from the double infinite series:

$$u(x, y, t) = \sum_{m=1}^{\infty} \sum_{n=1}^{\infty} B_{mn} \sin\frac{m\pi x}{a} \sin\frac{n\pi y}{b} \cos\left(\alpha\pi t\sqrt{\frac{m^2}{a^2} - \frac{n^2}{b^2}}\right),$$

$$B_{mn} = \frac{16a^2b^2A}{\pi^6 m^3 n^3}(1 - \cos m\pi)(1 - \cos n\pi),$$

which gives the displacement of a membrane fastened to a rectangular framework, $0 \leqslant x \leqslant a$, $0 \leqslant y \leqslant b$, with initial displacements of $Ax(a - x)y(b - y)$.

We observe that the finite-difference results do not agree exactly with the analytical calculations. The finite-difference values are symmetrical with respect to position and repeat themselves with a regular frequency. The very regularity of the values itself indicates that the finite-difference computations are in error, since they predict that the membrane could emit a musical note. We know from experience that a drum does not give a musical tone when struck; therefore the vibrations do not have a cyclic pattern of constant frequency, as exhibited by our numerical results.

Decreasing the ratio of $\alpha^2(\Delta t)^2/h^2$ and using Eq. (9.20) gives little or no improvement in the average accuracy; to approach closely to the analytical results, $h = \Delta x = \Delta y$ must be made smaller. When this is done, Δt will need to decrease in proportion, requiring many time steps and leading to many repetitions of the algorithm and extravagant use

Table 9.5 Displacements of a vibrating membrane—finite-difference method: $\Delta t = h/(\sqrt{2}\alpha)$

t	(0.5, 0.5)	(1.0, 0.5)	(1.5, 0.5)	(0.5, 1.0)	(1.0, 1.0)	(1.5, 1.0)	(0.5, 1.5)	(1.0, 1.5)	(1.5, 1.5)
			←—————————————— Grid location ——————————————→						
0	0.5625	0.750	0.5625	0.750	1.000	0.750	0.5625	0.750	0.5625
	(0.5625)	(0.750)			(1.000)				
0.204	0.375	0.531	0.375	0.531	0.750	0.531	0.375	0.531	0.375
	(0.380)	(0.536)			(0.755)				
0.408	−0.031	0.000	−0.031	0.000	0.062	0.000	−0.031	0.000	−0.031
	(−0.044)	(−0.009)			(0.083)				
0.612	−0.375	−0.531	−0.375	−0.531	−0.750	−0.531	−0.375	−0.531	−0.375
	(−0.352)	(−0.539)			(−0.813)				
0.816	−0.500	−0.750	−0.500	−0.750	−1.125	−0.750	−0.500	−0.750	−0.500
	(−0.502)	(−0.746)			(−1.114)				
1.021	−0.375	−0.531	−0.375	−0.531	−0.750	−0.531	−0.375	−0.531	−0.375
	(−0.407)	(−0.535)			(−0.691)				
1.225	−0.031	0.000	−0.031	0.000	0.062	0.000	−0.031	0.000	−0.031
	(−0.015)	(0.008)			(0.030)				
1.429	0.375	0.531	0.375	0.531	0.750	0.531	0.375	0.531	0.375
	(0.410)	(0.534)			(0.688)				

of computer time. One remedy is the use of implicit methods, which allow the use of larger ratios of $\alpha^2(\Delta t)^2/h^2$. However, with many nodes in the xy-grid, this requires large, sparse matrices similar to the Crank–Nicolson method for parabolic equations in two space dimensions. A.D.I. methods have been used for hyperbolic equations—tridiagonal systems result. We do not discuss these methods. ∎

9.6 CHAPTER SUMMARY

You can test your understanding of Chapter 9 by seeing if you can

1. Determine if a partial-differential equation falls in the hyperbolic class.
2. Set up equations to find the displacements of a vibrating string after one time step, and, from these, determine the displacements after n time steps.
3. Explain the D'Alembert method for the wave equation and how this shows that the finite-difference method can give exact answers.
4. Outline the argument that demonstrates stability for the finite-difference procedure.
5. Explain what characteristic curves are, find them for a typical problem, and compute a few points on the characteristics.
6. Apply the finite-difference method to a simple hyperbolic partial-differential equation with two space dimensions.

SELECTED READING FOR CHAPTER 9

Smith (1978).

9.7 COMPUTER PROGRAM

Program 1 (Fig. 9.7) uses the easy algorithm of Section 9.1, but starts the solution by use of Eq. (9.13). The integral of the initial velocity is computed by Simpson's $\frac{1}{3}$ rule with the interval from x_{i-1} to x_{i+1} subdivided into 20 panels. The initial displacement $f(x)$ and the initial velocity $g(x)$ are each defined by function subprograms. The program is tested by calculating the example problem of Section 9.2, in which a string initially at rest is set vibrating by giving it an initial velocity of $g(x) = 3 \sin (\pi x/L)$. Input values are $L = 9$, $\Delta x = 1$, $\alpha = \sqrt{Tg/w} = 2$, so that nine time steps are required here to reach the time $t = \alpha L/4$.

```
      PROGRAM WAVE(INPUT,OUTPUT)
C
C     THIS PROGRAM COMPUTES SOLUTIONS FOR THE ONE DIMENSIONAL
C  WAVE EQUATION. THE INITIAL DISPLACEMENTS OF THE VIBRATING
C  STRING, FROM X=0 TO X=XLEN ARE GIVEN BY F(X), THE INITIAL
C  VELOCITIES ARE GIVEN BY G(X). F(X) AND G(X) ARE DEFINED BY
C  FUNCTION SUBPROGRAMS. THE END POINTS ARE ASSUMED FIXED AT
C  ZERO DISPLACEMENT.
C
C  ----------------------------------------------------------------------
C
C  THE RELATION USED IS U(I,J+1) = U(I+1,J) + U(I-1,J) - U(I,J-1)
C  EXCEPT FOR THE FIRST TIME STEP, WHEN THE VALUE IS GIVEN BY
C  U(I,1) = 0.5*( U(I-1,0) + U(I+1,0) ) + 0.5/C*INTEGRAL OF INIT VEL.
C
C  ----------------------------------------------------------------------
C
C     PARAMETERS ARE :
C
C     X       - DISTANCE ALONG THE STRING
C     DX      - INCREMENTS OF DISTANCE
C     XLEN    - TOTAL LENGTH OF STRING
C     N       - NUMBER OF SUBDIVISIONS
C     T       - TIME
C     TLAST   - FINAL VALUE OF TIME FOR WHICH SOLUTION IS DESIRED
C     F(X)    - INITIAL DISPLACEMENTS
C     G(X)    - INITIAL VELOCITIES
C     TDM     - VALUE OF TENSION / MASS = C SQUARED
C     U       - DISPLACEMENTS AT EVEN TIME INTERVALS
C     V       - DISPLACEMENTS AT ODD TIME INTERVALS
C
C  ----------------------------------------------------------------------
C
      REAL U(100),V(100),X,DX,XLEN,T,TLAST,F,G,TDM,SUBDX,XSUB,PI
      INTEGER N,NP1,I,J
      COMMON XLEN,PI
```

Figure 9.7 Program 1.

Figure 9.7 (*continued*)

```
C
C     ------------------------------------------------------------------
C
C     DEFINE SOME DATA STATEMENTS FOR INITIAL VALUES
C
      DATA X,N,T/0.0,9,0.0/
      DATA TDM,TLAST/4.0,4.5/
      PI = 4.0*ATAN(1.0)
      XLEN = 9.0
C
C     ------------------------------------------------------------------
C
C     GET INITIAL DISPLACEMENTS
C
      NP1 = N + 1
      DX = XLEN / FLOAT(N)
      U(1) = 0.0
      U(NP1) = 0.0
      V(1) = 0.0
      V(NP1) = 0.0
      DO 10 I = 2,N
         X = X + DX
         U(I) = F(X)
   10 CONTINUE
C
C     ------------------------------------------------------------------
C
C     WRITE OUT THE INITIAL DISPLACEMENTS
C
      PRINT 200, ( U(I), I = 1,NP1/2)
C
C     ------------------------------------------------------------------
C
C     NOW GET DISPLACEMENTS AFTER FIRST TIME STEP. USE SIMPSON'S RULE
C     TO PERFORM INTEGRATIONS, USING 20 INTERVALS BETWEEN X(N-1) AND
C     X(N+1). WE DO THIS FOR EACH INTERMEDIATE X VALUE.
C
      SUBDX = DX / 10.0
      XSUB = DX
      DO 30 I = 2,N
         SUM = 0.0
         XSUB = XSUB - DX
         DO 20 J = 1,19,2
            SUM = SUM + G(XSUB) + 4.0*G(XSUB+SUBDX) + G(XSUB+2.0*SUBDX)
            XSUB = XSUB + 2.0*SUBDX
   20    CONTINUE
         V(I) = 0.5*( U(I-1) + U(I+1) ) + 0.5/SQRT(TDM)*SUBDX/3.0*SUM
   30 CONTINUE
C
C     ------------------------------------------------------------------
C
C     WRITE OUT DISPLACEMENTS AFTER FIRST TIME STEP
C
      T = DX / SQRT(TDM)
      PRINT 201, T,( V(I), I = 1,NP1/2)
C
C     ------------------------------------------------------------------
C
C     NOW COMPUTE UNTIL TLAST IS REACHED
C
   35 IF ( T .GE. TLAST ) STOP
      DO 40 I = 2,N
         U(I) = V(I-1) + V(I+1) -U(I)
```

Figure 9.7 (*continued*)

```
   40 CONTINUE
      T = T + DX/SQRT(TDM)
      PRINT 201, T,( U(I), I = 1,NP1/2)
      DO 50 I = 2,N
         V(I) = U(I-1) + U(I+1) - V(I)
   50 CONTINUE
      T = T + DX/SQRT(TDM)
      PRINT 201, T,( V(I), I = 1,NP1/2)
      GO TO 35
C
C -------------------------------------------------------------------
C
  200 FORMAT(//' SOLUTION TO VIBRATING STRING PROBLEM ',///,
     +        ' INITIAL DISPLACEMENTS ARE '//(1X,11F9.4) )
  201 FORMAT(/' AT T = ',F5.2/(1X,11F9.4) )
      END
C
C -------------------------------------------------------------------
C
      REAL FUNCTION F(X)
C
      REAL X
         F = 0.0
      RETURN
      END
C
C -------------------------------------------------------------------
C
      REAL FUNCTION G(X)
C
      REAL X
      COMMON XLEN,PI
      G = 3.0*SIN(PI*X/XLEN)
      RETURN
      END



            OUTPUT FOR PROGRAM 1


  SOLUTION TO VIBRATING STRING PROBLEM


  INITIAL DISPLACEMENTS ARE

     .0000    .0000    .0000    .0000    .0000

  AT T =   .50
     .0000    .5027    .9447   1.2728   1.4474

  AT T =  1.00
     .0000    .9447   1.7755   2.3921   2.7202

  AT T =  1.50
     .0000   1.2728   2.3921   3.2229   3.6649

  AT T =  2.00
     .0000   1.4474   2.7202   3.6649   4.1676

  AT T =  2.50
     .0000   1.4474   2.7202   3.6649   4.1676
```

Figure 9.7 (*continued*)

```
AT T =  3.00
    .0000   1.2728   2.3921   3.2229   3.6649

AT T =  3.50
    .0000    .9447   1.7755   2.3921   2.7202

AT T =  4.00
    .0000    .5027    .9447   1.2728   1.4474

AT T =  4.50
    .0000    .0000    .0000    .0000    .0000
```

EXERCISES

Section 9.1

▶ 1. If the banjo string in the example of Section 9.1 is tightened, its frequency of vibration is increased. Likewise, if the length of the vibrating portion of the string is shortened, as by holding it against one of the frets with the finger, the pitch is raised. What would the frequency be if the tension is made 42,000 g and the effective length is 70 cm? Determine the answer by finding the number of Δt steps for the original displacement to be duplicated and compare to

$$f = \frac{1}{2L}\sqrt{\frac{Tg}{w}}.$$

2. A vibrating string system has $Tg/w = 4 \ cm^2/sec^2$. Divide the length L ($L = 80$ cm) into intervals so that $\Delta x = L/8$ cm. Find the displacement for $t = 0$ up to $t = L$ if both ends are fixed, and the initial conditions are

a) $y = \dfrac{x(L - x)}{L^2}, \qquad \dfrac{\partial y}{\partial t} = 0.$

b) The string is displaced $+1$ units at $L/4$ and -1 units at $5L/8$, $\partial y/\partial t = 0$.

c) $y = 0, \qquad \dfrac{\partial y}{\partial t} = \dfrac{-x(L - x)}{L^2}.$ (Use Eq. (9.3) to begin the solution.)

d) The string is displaced 1 unit at $L/2$, $\partial y/\partial t = -y$.

In part (a), compare to the analytical solution:

$$y = \frac{8}{\pi^3} \sum_{r=1}^{\infty} \frac{1}{(2n - 1)^3} \sin [(2n - 1)\pi x] \cos [(4n - 2)\pi t].$$

▶ 3. A function u satisfies the equation

$$\frac{\partial^2 u}{\partial x^2} = \frac{\partial^2 u}{\partial t^2},$$

with boundary conditions $u = 0$ at $x = 0$, $u = 0$ at $x = 1$, and with initial conditions

$$u = \sin \pi x, \qquad \partial u/\partial t = 0 \qquad \text{for} \qquad 0 \leq x \leq 1.$$

Solve by the finite-difference method and show that the results are the same as the analytical solution:

$$U(x, t) = \sin \pi x \cos \pi t.$$

4. The ends of the vibrating string do not need to be fixed. Solve $\partial^2 y / \partial t^2 = \partial^2 y / \partial x^2$ with initial conditions $y = 0$, $\partial y / \partial t = 0$, for $0 \leqslant x \leqslant 1$, and end conditions

$$y = 0 \quad \text{at} \quad x = 0, \qquad y = \sin \pi t / 4, \quad \partial y / \partial x = 0 \quad \text{at} \quad x = 1.$$

Section 9.2

5. Why can't we use Eq. (9.9) to solve the example in Section 9.1?

▶ 6. Since Eq. (9.4) is sometimes inaccurate when the initial velocity is not zero, the solutions to parts (c) and (d) of Exercise 2 are not exact. Repeat these problems, getting more accurate y-values by using Eq. (9.13) and evaluating the integral by Simpson's $\frac{1}{3}$ rule.

7. A string that weighs w lb/ft is tightly stretched from $x = 0$ to $x = L$ and is initially at rest. Each point of the string is given initially a velocity of

$$\left. \frac{\partial y}{\partial t} \right|_{t=0} = v_0 \sin^3 \frac{\pi x}{L}.$$

The analytical solution to this problem is

$$y(x, t) = \frac{v_0 L}{12 a \pi} \left(9 \sin \frac{\pi x}{L} \sin \frac{a \pi t}{L} - \sin \frac{3 \pi x}{L} \sin \frac{3 a \pi t}{L} \right),$$

where $a = \sqrt{Tg/w}$, T being the tension and g the acceleration of gravity. When $L = 3$ ft, $w = 0.02$ lb/ft, and $T = 5$ lb, with $v_0 = 1$ ft/sec, the analytical equation predicts $y = 0.081$ in at the midpoint when $t = 0.01$ sec. Solve the problem numerically to confirm this prediction. Does your solution conform to the analytical solution at other values of x and t?

8. Demonstrate that if $Tg(\Delta t)^2 / w(\Delta x)^2 > 1$, the finite-difference method for the wave equation will propagate an error with increasing effect by tracing the growth of the error similar to that in Table 9.3. Take the simple case with ends fixed. Repeat for the ratio < 1.

▶ 9. Trace the magnitude of an isolated error in solving the wave equation with $Tg(\Delta t)^2 / w(\Delta x)^2 = 1$, similar to Table 9.3, but for the case when the left-hand end is fixed and the right-hand end moves according to

$$y(1, t) = \sin \frac{\pi t}{4}, \qquad \frac{\partial y}{\partial x} = 0 \qquad \text{at} \qquad x = 1.$$

Section 9.4

10. For the partial-differential equation

$$a u_{xx} + b u_{xt} + c u_{tt} + e = 0,$$

sketch the characteristic curves through the point $x = 0.5$, $t = 0$ when:

a) $a = 1$, $b = 2$, $c = 1$
b) $a = 1$, $b = 2$, $c = -3$
c) $a = 1$, $b = 2$, $c = 3$
d) $a = 1$, $b = 2$, $c = -1$
e) $a = t^2$, $b = xt$, $c = -2x^2$

11. Verify the values at points S, T, U, and V in Fig. 9.6 for Example 2 of Section 9.4 by the method of characteristics.

▶12. The equation

$$\frac{\partial^2 u}{\partial t^2} = (1 + 2x) \frac{\partial^2 u}{\partial x^2}$$

can be solved by the difference-equation method discussed in Sections 9.1 and 9.2. Solve it by that method with initial conditions of

$$u(x, 0) = 0, \qquad \frac{\partial u}{\partial t}(x, 0) = x(1 - x).$$

Compare these results with those from the method of characteristics in Fig. 9.6.

13. Get part of the solution by integrating along characteristics for the equation

$$\frac{\partial^2 u}{\partial x^2} + \frac{\partial^2 u}{\partial x\, \partial t} - \frac{\partial^2 u}{\partial t^2} = 1,$$

subject to initial conditions

$$u = 0, \qquad \frac{\partial u}{\partial t} = x(1 - x).$$

Find the solution at three points in the xt-plane, where the characteristics through $(0.4, 0)$, $(0.5, 0)$, and $(0.6, 0)$ intersect.

▶14. Solve the equation

$$u_{xx} + xtu_{xt} - u_{tt} = 0,$$

with initial conditions

$$u = 2x, \qquad \frac{\partial u}{\partial t} = 0.$$

Boundary conditions are $u(0, t) = 0$, $u(1, t) = 2$.

Find the solution at several points in the region of the xt-plane bounded by the lines $x = 0.4$, $x = 0.6$.

15. Continue the solution of Example 3 of Section 9.4,

$$u_{xx} - uu_{tt} + 1 - x^2 = 0, \qquad u(x, 0) = x(1 - x), \qquad u_t(x, 0) = 0,$$
$$u(0, t) = 0, \qquad u(1, t) = 0,$$

by finding the solution at the intersection of characteristics through points at $(0.4, 0)$ and $(0.6, 0)$. Then find the solution at the intersection of characteristics through $(0.2, 0)$ and $(0.6, 0)$ as an example of how the solution can be advanced in time.

Section 9.5

16. Solve the vibrating membrane example of Section 9.5 but with initial conditions of

$$u(x, y)\big|_{t=0} = 0, \qquad \frac{\partial u}{\partial t}(x, y)\Big|_{t=0} = x(2 - x)y(2 - y).$$

▶17. A membrane is stretched over a frame that occupies the region in the xy-plane bounded by

$$x = 0, \qquad x = 2, \qquad y = 0, \qquad y = 4.$$

Initially the point at $(1, 3)$ is lifted 1 in above the xy-plane, and then released. If $T = 7$ lb/in and $w = 0.07$ lb/in^2, find the displacement at the point $(1, 2)$ as a function of time.

18. How do the vibrations in Exercise 17 change if $w = 0.7$, the other parameters being unchanged?

19. The frame holding the membrane in Exercise 17 is distorted by lifting the corner at $(2, 4)$ so it is 1 in above the xy-plane. (The frame members stretch at the same time, so the corner

point moves only vertically.) The membrane is set to vibrating in the same way as in Exercise 17. Follow its vibrations through time. (Assume that the initial positions of points on the membrane lie on intersecting planes.)

APPLIED PROBLEMS AND PROJECTS

20. A vibrating string that has a damping force opposing its motion that is proportional to the velocity, follows the equation

$$\frac{\partial^2 y}{\partial t^2} = \frac{Tg}{w}\frac{\partial^2 y}{\partial x^2} - B\frac{\partial y}{\partial t},$$

where B is the magnitude of the damping force. Solve the problem if the length of the string is 5 ft with $T = 24$ lb, $w = 0.1$ lb/ft, and $B = 2.0$. Initial conditions are

$$y(x)\big|_{t=0} = \frac{x}{3}, \qquad 0 \leq x < 3,$$

$$y(x)\big|_{t=0} = \frac{5}{2} - \frac{x}{2}, \qquad 3 \leq x \leq 5,$$

$$\frac{\partial y}{\partial t}\bigg|_{t=0} = x(x - 5).$$

Compute a few points of the solution by both difference equations and the method of characteristics. Compare the effort involved in the two methods.

21. A horizontal elastic rod is initially undeformed and is at rest. One end, at $x = 0$, is fixed, and the other end, at $x = L$ (when $t = 0$), is pulled with a steady force of F lb/ft^2. It can be shown that the displacements $y(x, t)$ of points originally at the point x are given by

$$\frac{\partial^2 y}{\partial t^2} = a^2\frac{\partial^2 y}{\partial x^2}, \qquad y(0, t) = 0, \qquad \frac{\partial y}{\partial t}\bigg|_{x=L} = \frac{F}{E},$$

$$y(x, 0) = 0, \qquad \frac{\partial y}{\partial t}\bigg|_{t=0} = 0$$

where $a^2 = Eg/\rho$; $E =$ Young's modulus (lb/ft^2); $g =$ acceleration of gravity; $\rho =$ density (lb/ft^3). Find y versus t for the midpoint of a 2-ft-long piece of rubber for which $E = 1.8 \times 10^6$ and $\rho = 70$ if $F/E = 0.7$.

22. A circular membrane, when set to vibrating, obeys the equation (in polar coordinates)

$$\frac{1}{r}\frac{\partial}{\partial r}\left(r\frac{\partial u}{\partial r}\right) + \frac{1}{r^2}\frac{\partial^2 u}{\partial \theta^2} = \frac{w}{Tg}\frac{\partial^2 y}{\partial t^2}.$$

A 3-ft-diameter kettledrum is started to vibrating by depressing the center $\frac{1}{2}$ in. If $w = 0.072$ lb/ft^2 and $T = 80$ lb/ft, find how the displacements at 6 in and 12 in from the center vary with time. The problem can be solved in polar coordinates, or it can be solved in rectangular coordinates using the method of Section 7.8 to approximate $\nabla^2 u$ near the boundaries.

23. A flexible chain hangs freely, as shown in Fig. 9.8. For small disturbances from its equilibrium position (hanging vertically), the equation of motion is

$$x\frac{\partial^2 y}{\partial x^2} + \frac{\partial y}{\partial x} = \frac{1}{g}\frac{\partial^2 y}{\partial t^2}.$$

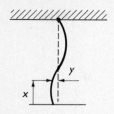

Figure 9.8

In this equation, x is the distance from the end of the chain, y is the displacement from the equilibrium position, t is the time, and g is the acceleration of gravity. A 10-ft-long chain is originally hanging freely. It is set into motion by striking it sharply at its midpoint, imparting a velocity there of 1 ft/sec. Find how the chain moves as a result of the blow. If you find you need additional information at $t = 0$, make reasonable assumptions.

24. Write a computer program that solves hyperbolic partial-differential equations by the method of characteristics, given the values of u and $\partial u / \partial t$ along a curve that is not one of the characteristics. Your program should be a subroutine that accepts the coordinates of two points on the initial-condition curve and computes the value of u and $\partial u / \partial t$ at the intersection of the characteristic curves through the two points.

10

Curve-Fitting and Approximation of Functions

10.0 CONTENTS OF THIS CHAPTER

In an early chapter, when we studied how to fit a polynomial to a set of data, we made the implicit assumption that the data were free of error (except round-off) so that it was appropriate to match the interpolating polynomial exactly at the data points. In the case of experimental data, this assumption as to accuracy is often not true. Each data point is subject to experimental errors that, in the case of complicated measurements, may be relatively large in magnitude. Again, in the previous chapter, the true function that relates the data was generally unknown, while, in the case of experimental results, the form of the function is frequently known from the physical laws that apply.

We wish to consider the problem of finding the "best" curve that represents data that are subject to error. "Best" is in quotation marks because the criterion of goodness of fit is to some degree arbitrary, although the least-squares criterion is commonly applied.

We will also consider in this chapter the most efficient way in which functions can be represented by a polynomial or a ratio of polynomials over a given range of values of the argument. This is of special importance in connection with the library function subroutines for digital computers.

The final topic of this chapter is a discussion of the approximation of functions by trigonometric series. This leads us right away to a study of the Fourier transform and, in particular, the consideration of FFTs (fast Fourier transforms).

10.1   LEAST-SQUARES APPROXIMATIONS

Suppose we wish to fit a curve to an approximate set of data, such as from the determination of the effects of temperature on a resistance by students in their physics laboratory. They have recorded the temperature and resistance measurements as shown in Fig. 10.1, where the graph suggests a linear relationship. We want to suitably determine the constants a and b in the equation relating resistance R and temperature T,

$$R = aT + b, \tag{10.1}$$

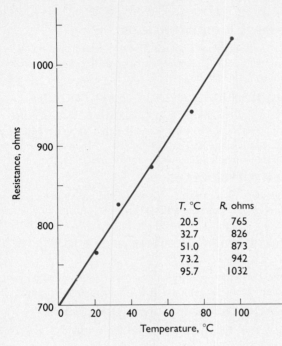

T, °C	R, ohms
20.5	765
32.7	826
51.0	873
73.2	942
95.7	1032

Figure 10.1

so that in subsequent use the resistance can be predicted at any temperatures. The line as sketched by eye represents the data fairly well, but if we replotted the data and asked someone else to draw a line, exactly the same line would rarely be obtained. One of our requirements for fitting a curve to data is that the process be *unambiguous*. We would also like, in some sense, to minimize the deviations of the points from the line. The deviations are measured by the distances from the points to the line—how these distances are measured depends on whether or not both variables are subject to error. We will assume that the error of reading the temperatures in Fig. 10.1 is negligible, so that all the errors are in the resistance measurements, and use vertical distances. (If both were subject to error, we might use perpendicular distances, and would modify the following. In this way the problem also becomes considerably more complicated. We will treat only the simpler case.)

We might first suppose we could minimize the deviations by making their sum a minimum, but this is not an adequate criterion. Consider the case of only two points (Fig. 10.2). Obviously, the best line passes through each, but any line that passes through the midpoint of the segment connecting them has a sum of errors equal to zero.

Then what about making the sum of the magnitudes of the errors a minimum? This also is inadequate, as the case of three points shows (Fig. 10.3). Assume that two of the points are at the same x-value (which is not an abnormal situation since frequently experiments are duplicated). The best line will obviously pass through the average of the duplicated tests. However, any line that falls between the dotted lines shown will have the same sum of the magnitudes of the vertical distances. Since we wish an unambiguous result, we cannot use this as a basis for our work.

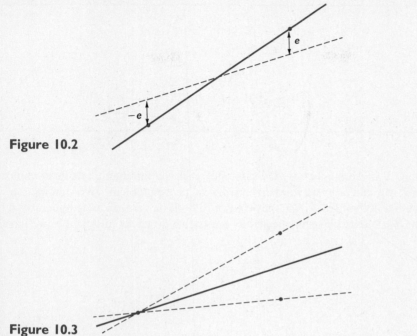

Figure 10.2

Figure 10.3

We might accept the criterion that we make the magnitude of the maximum error a minimum (the so-called *minimax* criterion), but for the problem at hand this is rarely done.* This criterion is awkward because the absolute-value function has no derivative at the origin, and it also is felt to give undue importance to a single large error. The usual criterion is to minimize the sum of the *squares* of the errors, the "least-squares" principle.[†]

In addition to giving a unique result for a given set of data, the least-squares method is also in accord with the *maximum-likelihood* principle of statistics. If the measurement errors have a so-called normal distribution and if the standard deviation is constant for all the data, the line determined by minimizing the sum of squares can be shown to have values of slope and intercept that have maximum likelihood of occurrence.

Let Y_i represent an experimental value, and let y_i be a value from the equation

$$y_i = ax_i + b,$$

where x_i is a particular value of the variable assumed free of error. We wish to determine the best values for a and b so that the y's predict the function values that correspond to x-values. Let $e_i = Y_i - y_i$. The least-squares criterion requires that

---

*We will use this criterion later in this chapter, however.

†The various criteria for a "best fit" can be described by minimizing a norm of the error vector. Relate each criterion to its corresponding vector norm to review the definition of such norms.

$$S = e_1^2 + e_2^2 + \cdots + e_N^2$$

$$= \sum_{i=1}^{N} e_i^2$$

$$= \sum_{i=1}^{N} (Y_i - ax_i - b)^2$$

be a minimum. N is the number of x,Y-pairs. We reach the minimum by proper choice of the parameters a and b, so they are the "variables" of the problem. At a minimum for S, the two partial derivatives $\partial S/\partial a$ and $\partial S/\partial b$ will both be zero. Hence, remembering that the x_i and Y_i are data points unaffected by our choice of values for a and b, we have

$$\frac{\partial S}{\partial a} = 0 = \sum_{i=1}^{N} 2(Y_i - ax_i - b)(-x_i),$$

$$\frac{\partial S}{\partial b} = 0 = \sum_{i=1}^{N} 2(Y_i - ax_i - b)(-1).$$

Dividing each of these equations by -2 and expanding the summation, we get the so-called *normal equations*

$$a \sum x_i^2 + b \sum x_i = \sum x_i Y_i,$$

$$a \sum x_i + bN = \sum Y_i. \tag{10.2}$$

All the summations in Eq. (10.2) are from $i = 1$ to $i = N$. Solving these equations simultaneously gives the values for slope and intercept a and b.

For the data in Fig. 10.1 we find that

$$N = 5, \quad \sum T_i = 273.1, \quad \sum T_i^2 = 18{,}607.27, \quad \sum R_i = 4438,$$

$$\sum T_i R_i = 254{,}932.5.$$

Our normal equations are then

$$18{,}607.27a + 273.1b = 254{,}932.5,$$

$$273.1a + 5b = 4438.$$

From these we find $a = 3.395$, $b = 702.2$, and hence write Eq. (10.1) as

$$R = 702 + 3.39T.$$

10.2   FITTING NONLINEAR CURVES BY LEAST SQUARES

In many cases, of course, data from experimental tests are not linear, so we need to fit some other function than a first-degree polynomial to them. Popular forms that are tried are the exponential forms

$$y = ax^b$$

or

$$y = ae^{bx}.$$

We can develop normal equations for these analogously to the preceding development for a least-squares line by setting the partial derivatives equal to zero. Such nonlinear simultaneous equations are much more difficult to solve* than linear equations. Because of this, the exponential forms are usually linearized by taking logarithms before determining the parameters:

$$\ln y = \ln a + b \ln x$$

or

$$\ln y = \ln a + bx.$$

We now fit the new variable $z = \ln y$ as a linear function of $\ln x$ or x as described earlier. Here we do not minimize the sum of squares of the deviations of Y from the curve, but rather the deviations of $\ln Y$. In effect, this amounts to minimizing the squares of the percentage errors, which itself may be a desirable feature. An added advantage of the linearized forms is that plots of the data on either log-log or semilog graph paper show at a glance whether these forms are suitable by whether a straight line represents the data when so plotted.

In cases when such linearization of the function is not desirable, or when no method of linearization can be discovered, graphical methods are frequently used; one merely plots the experimental values and sketches in a curve that seems to fit well. Special forms of graph paper, in addition to log-log and semilog, may be useful (probability, log-probability, and so on). Transformation of the variables to give near linearity, such as by plotting against $1/x$, $1/(ax + b)$, $1/x^2$, and other nonpolynomial forms of the argument may give curves with gentle enough changes in slope to allow a smooth curve to be drawn. S-shaped curves are not easy to linearize; the Gompertz relation

$$y = ab^{c^x}$$

is sometimes employed. The constants a, b, and c are determined by special procedures. Another relation that fits data to an S-shaped curve is

$$\frac{1}{y} = a + be^{-x}.$$

---

*They are treated briefly in Chapter 2.

In awkward cases, subdividing the region of interest into subregions with a piecewise fit in the subregions can be used.

The objection to the last-mentioned methods is the *lack of uniqueness*. Two individuals will usually not draw the same curve through the points. One's judgment is frequently distorted by one or two points that deviate widely from the remaining data. Often one tends to pay too much attention to the extremities in comparison to the points in the central parts of the region of interest.

Further problems are caused if we wish to integrate or differentiate the function. Our discussion of least-squares polynomials is one solution to these difficulties.

Because polynomials can be readily manipulated, fitting such functions to data that do not plot linearly is common. We now consider this case. It will turn out that the normal equations are linear for this situation, which is an added advantage. In the development, we use n as the degree of the polynomial and N as the number of data pairs. Obviously if $N = n + 1$, the polynomial passes exactly through each point and the methods of Chapter 3 apply, so we will always have $N > n + 1$ in the following.

We assume the functional relationship

$$y = a_0 + a_1x + a_2x^2 + \cdots + a_nx^n, \tag{10.3}$$

with errors defined by

$$e_i = Y_i - y_i = Y_i - a_0 - a_1x_i - a_2x_i^2 - \cdots - a_nx_i^n.$$

We again use Y_i to represent the observed or experimental value corresponding to x_i, with x_i free of error. We minimize the sum of squares,

$$S = \sum_{i=1}^{N} e_i^2 = \sum_{i=1}^{N} (Y_i - a_0 - a_1x_i - a_2x_i^2 - \cdots - a_nx_i^n)^2.$$

At the minimum, all the partial derivatives $\partial S/\partial a_0, \partial S/\partial a_1, \ldots, \partial S/\partial a_n$ vanish. Writing the equations for these gives $n + 1$ equations:

$$\frac{\partial S}{\partial a_0} = 0 = \sum_{i=1}^{N} 2(Y_i - a_0 - a_1x_i - \cdots - a_ix_i^n)(-1),$$

$$\frac{\partial S}{\partial a_1} = 0 = \sum_{i=1}^{N} 2(Y_i - a_0 - a_1x_i - \cdots - a_ix_i^n)(-x_i),$$

$$\vdots \qquad \vdots$$

$$\frac{\partial S}{\partial a_n} = 0 = \sum_{i=1}^{N} 2(Y_i - a_0 - a_1x_i - \cdots - a_nx_i^n)(-x_i^n).$$

Dividing each by -2 and rearranging gives the $n + 1$ normal equations to be solved simultaneously:

$$a_0N + a_1 \sum x_i + a_2 \sum x_i^2 + \cdots + a_n \sum x_i^n = \sum Y_i,$$

$$a_0 \sum x_i + a_1 \sum x_i^2 + a_2 \sum x_i^3 + \cdots + a_n \sum x_i^{n+1} = \sum x_iY_i,$$

$$a_0 \sum x_i^2 + a_1 \sum x_i^3 + a_2 \sum x_i^4 + \cdots + a_n \sum x_i^{n+2} = \sum x_i^2 Y_i,$$

$$\vdots \qquad\qquad\qquad\qquad\qquad\qquad\qquad\qquad \vdots$$

$$a_0 \sum x_i^n + a_1 \sum x_i^{n+1} + a_2 \sum x_i^{n+2} + \cdots + a_n \sum x_i^{2n} = \sum x_i^n Y_i. \quad \textbf{(10.4)}$$

Putting these equations in matrix form shows an interesting pattern in the coefficient matrix.

$$
\begin{bmatrix}
N & \sum x_i & \sum x_i^2 & \sum x_i^3 & \dots & \sum x_i^n \\
\sum x_i & \sum x_i^2 & \sum x_i^3 & \sum x_i^4 & \dots & \sum x_i^{n+1} \\
\sum x_i^2 & \sum x_i^3 & \sum x_i^4 & \sum x_i^5 & \dots & \sum x_i^{n+2} \\
\vdots & & & & & \vdots \\
\sum x_i^n & \sum x_i^{n+1} & \sum x_i^{n+2} & \sum x_i^{n+3} & \dots & \sum x_i^{2n}
\end{bmatrix}
a =
\begin{bmatrix}
\sum Y_i \\
\sum x_i Y_i \\
\sum x_i^2 Y_i \\
\vdots \\
\sum x_i^n Y_i
\end{bmatrix}.
$$

$$\textbf{(10.5)}$$

All the summations in Eqs. (10.4) and (10.5) run from 1 to N.

Solving large sets of linear equations is not a simple task. Methods for this are the subject of Chapter 2. These particular equations have an added difficulty in that they have the undesirable property known as *ill-conditioning*. The result of this is that round-off errors in solving them cause unusually large errors in the solutions, which of course are the desired values of the coefficients a_i in Eq. (10.3). Up to $n = 4$ or 5, the problem is not too great (that is, double-precision arithmetic in computer solutions is only desirable and not essential), but beyond this point special methods are needed. Such special methods use orthogonal polynomials in an equivalent form of Eq. (10.3). We will not pursue this matter further,* although we will treat one form of orthogonal polynomials later in this chapter in connection with representation of functions. From the point of view of the experimentalist, functions more complex than fourth-degree polynomials are rarely needed, and when they are, the problem can often be handled by fitting a series of polynomials to subsets of the data.

The matrix of Eq. (10.5) is called the *normal matrix* for the least-squares problem. There is another matrix that corresponds to this, called the *design matrix*. It is of the form

$$
A =
\begin{bmatrix}
1 & 1 & 1 & \dots & 1 \\
x_1 & x_2 & x_3 & \dots & x_N \\
x_1^2 & x_2^2 & x_3^2 & \dots & x_N^2 \\
\vdots & & & & \vdots \\
x_1^n & x_2^n & x_3^n & \dots & x_N^n
\end{bmatrix}
$$

---

*Ralston (1965) is a good source of further information. The ill-conditioning problem, though very real, is often academic, since it is seldom that a degree above 4 or 5 is needed to give a curve that fits the data with adequate precision.

It is easy to show that AA^T is just the coefficient matrix of Eq. (10.5). It is also easy to see that Ay, where y is the column vector of Y-values, gives the right-hand side of Eq. (10.5). (You ought to try this for, say, a 3×3 case to reassure yourself.) This means that we can rewrite Eq. (10.5), in matrix form, as

$$AA^T a = Ba = Ay.$$

Usually we would use Gaussian elimination to solve the system, but because B has special properties, we can use other methods that avoid the problem of ill-conditioning that was pointed out above.

1. The matrix $B = AA^T$, is symmetric and positive semidefinite. A n $\times$ n matrix is said to be positive semidefinite if, for every n-component vector, $x^T Mx \geq 0$. If we add the condition that $x^T Mx = 0$ only if x is the zero vector, M is said to be positive definite. (You should show that B is positive semidefinite and symmetric.)

2. In linear algebra, it is shown that B can be diagonalized by an orthogonal matrix P:

$$PBP^T = PAA^T P^T = D,$$

where the diagonal elements of D are the eigenvalues of B. Note that orthogonality implies that $PP^T = I$, the identity matrix.

3. Since B is positive semidefinite, all of its eigenvalues are nonnegative. This means that we can define a matrix S as

$$S = \sqrt{D}, \quad \text{or} \quad S^2 = D.$$

The diagonal elements of S are called the singular values of A.

4. We can rewrite Eq. (10.5) and its solution as follows:

$$AA^T a = P^T DPa = PS(PS)^T a = Ay,$$
$$a = PD^{-1}P^T Ay.$$

This last eliminates having to multiply out AA^T and, by extending this approach, leads to an important method for solving Eq. (10.5) called *singular-value decomposition*.

Table 10.1   Data to illustrate curve-fitting

x_i	0.05	0.11	0.15	0.31	0.46	0.52	0.70	0.74	0.82	0.98	1.17
Y_i	0.956	0.890	0.832	0.717	0.571	0.539	0.378	0.370	0.306	0.242	0.104

$$\sum x_i = 6.01 \qquad\qquad N = 11$$
$$\sum x_i^2 = 4.6545 \qquad\qquad \sum Y_i = 5.905$$
$$\sum x_i^3 = 4.1150 \qquad\qquad \sum x_i Y_i = 2.1839$$
$$\sum x_i^4 = 3.9161 \qquad\qquad \sum x_i^2 Y_i = 1.3357$$

We illustrate the use of Eqs. (10.4) to fit a quadratic to the data of Table 10.1. Figure 10.4 shows a plot of the data. (The data are actually a perturbation of the relation $y = 1 - x + 0.2x^2$. It will be of interest to see how well we approximate this function.) To set up the normal equations, we need the sums tabulated in Table 10.1. A calculator can give these as accumulated totals directly. A computer program at the end of this chapter is designed to set up the matrix and the right-hand-side vector. We need to solve the set of equations

$$11a_0 + 6.01a_1 + 4.6545a_2 = 5.905,$$

$$6.01a_0 + 4.6545a_1 + 4.1150a_2 = 2.1839,$$

$$4.6545a_0 + 4.1150a_1 + 3.9161a_2 = 1.3357.$$

The result is $a_0 = 0.998$, $a_1 = -1.018$, $a_2 = 0.225$, so the least-squares method gives

$$y = 0.998 - 1.018x + 0.225x^2.$$

Compare this to $y = 1 - x + 0.2x^2$. We do not expect to reproduce the coefficients exactly because of the errors in the data.

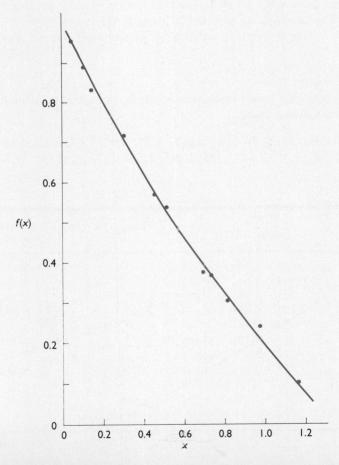

Figure 10.4

In the general case, we may wonder what degree of polynomial should be used. As we use higher-degree polynomials, we of course will reduce the deviations of the points from the curve until, when the degree of the polynomial, n, equals $N - 1$, there is an exact match (assuming no duplicate data at the same x-value) and we have the interpolating polynomials of Chapter 3. The answer to this problem is found in statistics. One increases the degree of approximating polynomial so long as there is a statistically significant decrease in the variance, σ^2, which is computed by

$$\sigma^2 = \frac{\Sigma\, e_i^2}{N - n - 1}. \tag{10.6}$$

For the above example, when the degree of the polynomial made to fit the points is varied from 1 to 7, we obtain the results shown in Table 10.2.

The criterion of Eq. (10.6) chooses the optimum degree as 2. This is no surprise, in view of how the data were constructed. It is important to realize that the numerator of Eq. (10.6), the *sum of the deviations squared* of the points from the curve, should continually decrease as the degree of the polynomial is raised. It is the denominator of Eq. (10.6) that makes σ^2 increase as we go above the optimum degree. In this example, this behavior is observed for $n = 3$. Above $n = 3$, a second effect sets in. Due to ill-conditioning, the coefficients of the least-squares polynomials are determined with poor precision. This modifies the expected increases of the values of σ^2.

Before leaving this section, we illustrate how to apply these methods to more complicated functions.

EXAMPLE The results of a wind tunnel experiment on the flow of air on the wing tip of an airplane provide the following data:

R/C: 0.73, 0.78, 0.81, 0.86, 0.875, 0.89, 0.95, 1.02, 1.03, 1.055, 1.135, 1.14, 1.245, 1.32, 1.385, 1.43, 1.445, 1.535, 1.57, 1.63, 1.755;

Table 10.2

Degree	Equation	σ^2 (Eq. (10.6))	$\Sigma\, e^2$
1	$y = 0.952 - 0.760x$	0.0010	0.0092
2	$y = 0.998 - 1.018x + 0.225x^2$	0.0002	0.0018
3	$y = 1.004 - 1.079x + 0.351x^2 - 0.069x^3$	0.0003	0.0018
4	$y = 0.998 - 0.838x - 0.522x^2 + 1.040x^3$ $- 0.454x^4$	0.0003	0.0016
5	$y = 1.031 - 1.704x + 4.278x^2 - 9.477x^3$ $+ 9.394x^4 - 3.290x^5$	0.0001	0.0007
6	$y = 1.038 - 1.910x + 5.952x^2 - 15.078x^3$ $+ 18.277x^4 - 9.835x^5 + 1.836x^6$	0.0002	0.0007
7	$y = 1.032 - 1.742x + 4.694x^2 - 11.898x^3$ $+ 16.645x^4 - 14.346x^5 + 8.141x^6 - 2.293x^7$	0.0002	0.0007

V_θ/V_∞:   0.0788, 0.0788, 0.064, 0.0788, 0.0681, 0.0703, 0.0703, 0.0681, 0.0681, 0.079, 0.0575, 0.0681, 0.0575, 0.0511, 0.0575, 0.049, 0.0532, 0.0511, 0.049, 0.0532, 0.0426;

where R is the distance from the vortex core, C is the aircraft wing chord, V_θ is the vortex tangential velocity, and V_∞ is the aircraft free-stream velocity. Let $x = R/C$ and $y = V_\theta/V_\infty$. We would like our curve to be of the form

$$g(x) = \frac{A}{x}(1 - e^{-\lambda x^2}),$$

and our least-squares equations become

$$S = \sum_{i=1}^{21} (y_i - g(x_i))^2$$

$$= \sum_{i=1}^{21} \left(y_i - \frac{A}{x_i}(1 - e^{-\lambda x_i^2}) \right)^2.$$

Setting $S_A = S_\lambda = 0$ gives the following equations:

$$\sum_{i=1}^{21} \left(\frac{1}{x_i}\right)(1 - e^{-\lambda x_i^2})\left(y_i - \frac{A}{x_i}(1 - e^{-\lambda x_i^2}) \right) = 0,$$

$$\sum_{i=1}^{21} x_i(e^{-\lambda x_i^2})\left(y_i - \frac{A}{x_i}(1 - e^{-\lambda x_i^2}) \right) = 0.$$

Program 3 (Fig. 10.10) solves this system of nonlinear equations using NLSYST, to give us

$$g(x) = \frac{0.07618}{x}(1 - e^{-2.30574x^2}).$$

For these values of A and λ, $S = 0.000016$. The graph of this function is presented in Fig. 10.5. ■

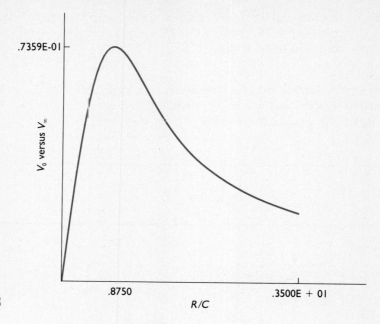

Figure 10.5

10.3 CHEBYSHEV POLYNOMIALS

We turn now to the problem of representing a function with a minimum error. This is a central problem in the software development of digital computers because it is more economical to compute the values of the common functions using an efficient approximation than to store a table of values and employ interpolation techniques. Since digital computers are essentially only arithmetic devices, the most elaborate function they can compute is a rational function, a ratio of polynomials. We will hence restrict our discussion to representation of functions by polynomials or rational functions.

One way to approximate a function by a polynomial is to use a truncated Taylor series. This is not the best way, in most cases. In order to study better ways, we need to introduce the Chebyshev polynomials.

The familiar Taylor-series expansion represents the function with very small error near the point of the expansion, but the error increases rapidly (proportional to a power) as we employ it at points farther away. In a digital computer, we have no control over where in an interval the approximation will be used, so the Taylor series is not usually appropriate. We would prefer to trade some of its excessive precision at the center of the interval to reduce the errors at the ends.

We can do this while still expressing functions as polynomials by the use of Chebyshev polynomials. The first few of these are*

---

*The commonly accepted symbol $T(x)$ comes from the older spelling, Tschebycheff.

$$T_0(x) = 1,$$
$$T_1(x) = x,$$
$$T_2(x) = 2x^2 - 1,$$
$$T_3(x) = 4x^3 - 3x,$$
$$T_4(x) = 8x^4 - 8x^2 + 1,$$
$$T_5(x) = 16x^5 - 20x^3 + 5x,$$
$$T_6(x) = 32x^6 - 48x^4 + 18x^2 - 1,$$
$$T_7(x) = 64x^7 - 112x^5 + 56x^3 - 7x,$$
$$T_8(x) = 128x^8 - 256x^6 + 160x^4 - 32x^2 + 1,$$
$$T_9(x) = 256x^9 - 576x^7 + 432x^5 - 120x^3 + 9x,$$
$$T_{10}(x) = 512x^{10} - 1280x^8 + 1120x^6 - 400x^4 + 50x^2 - 1. \quad \textbf{(10.7)}$$

The members of this series of polynomials can be generated from the two-term recursion formula

$$T_{n-1}(x) = 2xT_n(x) - T_{n-1}(x), \qquad T_0(x) = 1, \qquad T_1(x) = x. \quad \textbf{(10.8)}$$

Note that the coefficient of x^n in $T_n(x)$ is always 2^{n-1}. In Fig. 10.6 we plot the first four polynomials of Eq. (10.7).

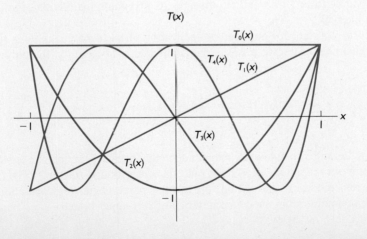

Figure 10.6

These polynomials have some unusual properties. They form an orthogonal set, in that

$$\int_{-1}^{1} \frac{T_n(x)T_m(x)}{\sqrt{1-x^2}} \, dx = \begin{cases} 0, & n \neq m, \\ \pi, & n = m = 0, \\ \pi/2, & n = m \neq 0. \end{cases} \tag{10.9}$$

The orthogonality of these functions will not be of immediate concern to us.

The Chebyshev polynomials are also terms of Fourier series, since

$$T_n(x) = \cos n\theta, \tag{10.10}$$

where $\theta = \arccos x$. Observe that $\cos 0 = 1$, $\cos \theta = \cos (\arccos x) = x$.

In order to demonstrate the equivalence of Eq. (10.10) to Eqs. (10.7) and (10.8), we recall some trigonometric identities, such as

$$\cos 2\theta = 2 \cos^2 \theta - 1,$$
$$T_2(x) = 2x^2 - 1;$$

$$\cos 3\theta = 4 \cos^3 \theta - 3 \cos \theta,$$
$$T_3(x) = 4x^3 - 3x;$$

$$\cos (n + 1)\theta + \cos (n - 1)\theta = 2 \cos \theta \cos n\theta,$$
$$T_{n+1}(x) + T_{n-1}(x) = 2xT_n(x).$$

Because of the relation $T_n(x) = \cos n\theta$, it is apparent that the Chebyshev polynomials have a succession of maximums and minimums of alternating signs, each of magnitude one. Further, since $|\cos n\theta| = 1$ for $n\theta = 0$, π, 2π, . . . , and since θ varies from 0 to π as x varies from 1 to -1, $T_n(x)$ assumes its maximum magnitude of unity $n + 1$ times on the interval $[-1, 1]$.

Most important for our present application of these polynomials is the fact that, of all polynomials of degree n where the coefficient of x^n is unity, the polynomial

$$\frac{1}{2^{n-1}} T_n(x)$$

has a smaller upper bound to its magnitude in the interval $[-1, 1]$ than any other. Because the maximum magnitude of $T_n(x)$ is one, the upper bound referred to is $1/2^{n-1}$. This is of importance because we will be able to write power-series representations of functions whose maximum errors are given in terms of this upper bound.

We first prove this assertion about bounds on the magnitude of polynomials. The proof is by contradiction. Let $P_n(x)$ be a polynomial whose leading term* is x^n and suppose that its maximum magnitude on $[-1, 1]$ is less than that of $T_n(x)/2^{n-1}$. Write

$$T_n(x)/2^{n-1} - P_n(x) = P_{n-1}(x),$$

where $P_{n-1}(x)$ is a polynomial of degree $n - 1$ or less, since the x^n terms cancel. The polynomial $T_n(x)$ has $n + 1$ extremes (counting endpoints), each of magnitude one, so $T_n(x)/2^{n-1}$ has $n + 1$ extremes each of magnitude $1/2^{n-1}$, and these successive extremes alternate in sign. By our supposition about $P_n(x)$, at each of these maximums or minimums, the magnitude of $P_n(x)$ is less than $1/2^{n-1}$; hence $P_{n-1}(x)$ must change its sign at least for every extreme of $T_n(x)$, which is then at least $n + 1$ times. $P_{n-1}(x)$ hence crosses the axis at least n times and would have n zeros. But this is impossible if $P_{n-1}(x)$ is only of degree $n - 1$, unless it is identically zero. The premise must then be false and $P_n(x)$ has a larger magnitude than the polynomial we are testing or, alternatively, $P_n(x)$ is exactly the same polynomial.

10.4   APPROXIMATION OF FUNCTIONS WITH ECONOMIZED POWER SERIES

We are now ready to use Chebyshev polynomials to "economize" a power series. Consider the Maclaurin series for e^x:

$$e^x = 1 + x + \frac{x^2}{2} + \frac{x^3}{6} + \frac{x^4}{24} + \frac{x^5}{120} + \frac{x^6}{720} + \cdots .$$

If we would like to use a truncated series to approximate e^x on the interval $[0, 1]$ with precision of 0.001, we will have to retain terms through that in x^6, since the error after the term in x^5 will be more than $1/720$. Suppose we subtract

$$\left(\frac{1}{720}\right)\left(\frac{T_6}{32}\right)$$

from the truncated series. We note from Eq. (10.7) that this will exactly cancel the x^6 term and at the same time make adjustments in other coefficients of the Maclaurin series. Since the maximum value of T_5 on the interval $[0, 1]$ is unity, this will change the sum of the truncated series by only

$$\frac{1}{720} \cdot \frac{1}{32} < 0.00005,$$

---

*We restrict the polynomials to those whose leading term is x^n so that all are scaled alike.

which is small with respect to our required precision of 0.001. Performing the calculations, we have

$$e^x \doteq 1 + x + \frac{x^2}{2} + \frac{x^3}{6} + \frac{x^4}{24} + \frac{x^5}{120} + \frac{x^6}{720}$$

$$- \frac{1}{720}\left(\frac{1}{32}\right)(32x^6 - 48x^4 + 18x^2 - 1),$$

$$e^x \doteq 1.000043 + x + 0.499219x^2 + \frac{x^3}{6} + 0.043750x^4$$

$$+ \frac{x^5}{120}. \tag{10.11}$$

This gives a fifth-degree polynomial that approximates e^x on [0, 1] almost as well as the sixth-degree one derived from the Maclaurin series. (The actual maximum error of the fifth-degree expression is 0.000270; for the sixth-degree expression it is 0.000226.) We hence have "economized" the power series in that we get nearly the same precision with fewer terms.

By subtracting $\frac{1}{120}(T_5/16)$ we can economize further, getting a fourth-degree polynomial that is almost as good as the economized fifth-degree one. It is left as an exercise to do this and to show that the maximum error is now 0.000781, so that we have found a fourth-degree power series that meets an error criterion that requires us to use two additional terms of the original Maclaurin series. Because of the relative ease with which they can be developed, such economized power series are frequently used for approximations to functions and are much more efficient than power series of the same degree obtained by merely truncating a Taylor or Maclaurin series. Table 10.3 compares the errors of these power series.

Table 10.3 Comparison of errors of economized power series and a Maclaurin series for e^x

x	e^x	Maclaurin, sixth-degree	Economized, fifth-degree	Economized, fourth-degree	Maclaurin, fourth-degree
0	1.00000	1.00000	1.00004	1.00004	1.00000
0.2	1.22140	1.22140	1.22142	1.22098	1.22140
0.4	1.49182	1.49182	1.49179	1.49133	1.49173
0.6	1.82212	1.82211	1.82208	1.82212	1.82140
0.8	2.22554	2.22549	2.22553	2.22605	2.22240
1.0	2.71828	2.71806	2.71801	2.71749	2.70833
Maximum error		0.00023	0.00027	0.00078	0.00995

The maximum error in the economized fifth-degree polynomial is only slightly greater than the sixth-degree Maclaurin series. The economized fourth-degree polynomial incurs a maximum error about three and one-half times as much, but still within the 0.001 limit that was initially imposed, and will require significantly reduced computational effort. In addition, there is a proportionately reduced memory-space requirement to store the constants of the polynomial. In contrast, a fourth-degree Maclaurin series has an error nearly ten times greater than the 0.001 tolerance, and its error is over twelve times that of the fourth-degree economized form.

By rearranging the Chebyshev polynomials, we can express powers of x in terms of them:

$$1 = T_0,$$

$$x = T_1,$$

$$x^2 = \frac{1}{2}(T_0 + T_2),$$

$$x^3 = \frac{1}{4}(3T_1 + T_3),$$

$$x^4 = \frac{1}{8}(3T_0 + 4T_2 + T_4),$$

$$x^5 = \frac{1}{16}(10T_1 + 5T_3 + T_5),$$

$$x^6 = \frac{1}{32}(10T_0 + 15T_2 + 6T_4 + T_6),$$

$$x^7 = \frac{1}{64}(35T_1 + 21T_3 + 7T_5 + T_7),$$

$$x^8 = \frac{1}{128}(35T_0 + 56T_2 + 28T_4 + 8T_6 + T_8),$$

$$x^9 = \frac{1}{256}(126T_1 + 84T_3 + 36T_5 + 9T_7 + T_9). \tag{10.12}$$

By substituting these identities into an infinite Taylor series and collecting terms in $T_i(x)$, we create a Chebyshev series. For example, we can get the first four terms of a Chebyshev series by starting with the Maclaurin expansion for e^x. Such a series converges more rapidly than does a Taylor series on $[-1, 1]$:

$$e^x = 1 + x + \frac{x^2}{2} + \frac{x^3}{6} + \frac{x^4}{24} + \cdots .$$

Replacing terms by Eq. (10.12), but omitting polynomials beyond $T_3(x)$, since we want only four terms,* we have

---

*The number of terms that are employed determines the accuracy of the computed values, of course.

$$e^x = T_0 + T_1 + \frac{1}{4}(T_0 + T_2) + \frac{1}{24}(3T_1 + T_3) + \frac{1}{192}(3T_0 + 4T_2 + \cdots)$$

$$+ \frac{1}{1920}(10T_1 + 5T_3 + \cdots) + \frac{1}{23,040}(10T_0 + 15T_2 + \cdots) + \cdots$$

$$= 1.2661T_0 + 1.1303T_1 + 0.2715T_2 + 0.0444T_3 + \cdots.$$

In order to compare the Chebyshev expansion with the Maclaurin series, we convert back to powers of x, using Eq. (10.7):

$$e^x = 1.2661 + 1.1303(x) + 0.2715(2x^2 - 1) + 0.0444(4x^3 - 3x) + \cdots.$$

$$e^x = 0.9946 + 0.9971x + 0.5430x^2 + 0.1776x^3 + \cdots. \tag{10.13}$$

Table 10.4 and Fig. 10.7 compare the error of the Chebyshev expansion, Eq. (10.13), with the Maclaurin series, using terms through x^3 in each case. The figure shows how the Chebyshev expansion attains a smaller maximum error by permitting the error at the origin to increase. The errors can be considered to be distributed more or less uniformly throughout the interval. In contrast to this, the Maclaurin expansion, which gives very small errors near the origin, allows the error to bunch up at the ends of the interval.

If the function is to be expressed directly as an expansion in Chebyshev polynomials, the coefficients can be obtained by integration. Based on the orthogonality property, the coefficients are computed from

$$a_i = \frac{2}{\pi} \int_{-1}^{1} \frac{f(x)T_i(x)}{\sqrt{1 - x^2}} \, dx,$$

Table 10.4   Comparison of Chebyshev series for e^x with Maclaurin series:

$$e^x = 0.9946 + 0.9971x + 0.5430x^2 + 0.1776x^3;$$

$$e^x = 1 + x + 0.5x^2 + 0.1667x^3$$

x	e^x	Chebyshev	Error	Maclaurin	Error
-1.0	0.3679	0.3629	0.0050	0.3333	0.0346
-0.8	0.4493	0.4535	-0.0042	0.4347	0.0146
-0.6	0.5488	0.5535	-0.0047	0.5440	0.0048
-0.4	0.6703	0.6713	-0.0010	0.6693	0.0010
-0.2	0.8187	0.8155	0.0032	0.8187	0.0000
0	1.0000	0.9946	0.0054	1.0000	0.0000
0.2	1.2214	1.2172	0.0042	1.2213	0.0001
0.4	1.4918	1.4917	0.0001	1.4907	0.0011
0.6	1.8221	1.8267	-0.0046	1.8160	0.0061
0.8	2.2255	2.2307	-0.0052	2.2053	0.0202
1.0	2.7183	2.7123	0.0060	2.6667	0.0516

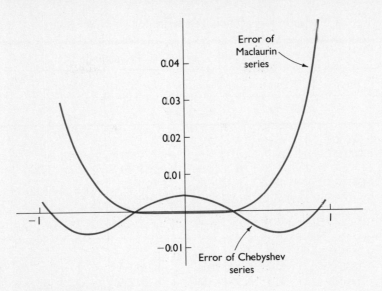

Figure 10.7

and the series is expressed as

$$f(x) = \frac{a_0}{2} + \sum_{i=1}^{\infty} a_i T_i(x).$$

A change of variable will be required if the desired interval is other than $(-1, 1)$. In some cases, the definite integral which defines the coefficients can be profitably evaluated by numerical methods as described in Chapter 4.

Since the coefficients of the terms of a Chebyshev expansion usually decrease even more rapidly than the terms of a Maclaurin expansion, one can get an estimate of the magnitude of the error from the next nonzero term after those that were retained. For the truncated Chebyshev series given by Eq. (10.13), the $T_4(x)$ term would be

$$\frac{1}{192}(T_4) + \frac{1}{23{,}040}(6T_4) + \cdots = 0.00525T_4.$$

Since the maximum value of $T_4(x)$ on $(-1, 1)$ is 1.0, we estimate the maximum errors of Eq. (10.13) to be 0.00525. The maximum error in Table 10.4 is 0.0060. This good agreement is caused by the very rapid decrease in coefficients in this example.

The computational economy to be gained by economizing a Maclaurin series, or by using a Chebyshev series, is even more dramatic when the Maclaurin series is slowly convergent. The previous example for $f(x) = e^x$ is a case in which the Maclaurin series converges rapidly. The power of the methods of this section is better demonstrated in the following example.

EXAMPLE   A Maclaurin series for $(1 + x)^{-1}$ is

$$(1 + x)^{-1} = 1 - x + x^2 - x^3 + x^4 - \cdots \quad (-1 < x < 1).$$

Table 10.5 compares the accuracy of truncated Maclaurin series with the economized series derived from them.

Table 10.5 Comparison of Maclaurin and economized series for $(1 + x)^{-1}$

Maclaurin			Economized					
Degree	Value	Error	Degree*	Value	Error	Degree*	Value	Error*
				$x = 0.2$				
2	0.840000	0.006667						
4	0.833600	0.000267	2	0.758600	−0.0747333			
6	0.833344	$11 * 10^{-6}$	4	0.764594	−0.068739			
8	0.833334	$1 * 10^{-6}$	6	0.803646	−0.029687			
10	0.833333	0	8	0.822786	−0.010547			
				$x = 0.8$				
2	0.840000	0.284445						
4	0.737600	0.182045	2	0.812600	0.257045			
6	0.672064	0.116509	4	0.678314	0.122759			
8	0.630121	0.074566	6	0.628558	0.073003	4	0.658246	0.102691
10	0.603277	0.047722	8	0.602106	0.046551	6	0.598199	0.042644

*Economized series were derived from Maclaurin series of corresponding degree.

In Table 10.5, we see that the error of the Maclaurin series is small for $x = 0.2$, and this also would be true for other values near $x = 0$, while the economized polynomial has less accuracy. At $x = 0.8$, the situation is reversed, however. Economized polynomials of degrees 8 and 6, derived from truncated Maclaurin series of degrees 10 and 8, actually have smaller errors than their precursors. Further economization, giving polynomials of degrees 6 and 4, have lesser or only slightly greater errors than their precursors, at significant savings of computational effort and with smaller storage requirements in a computer's memory for the coefficients of the polynomials. ■

10.5 APPROXIMATION WITH RATIONAL FUNCTIONS

We have seen that expansion of a function in terms of Chebyshev polynomials gives a power-series expansion that is much more efficient on the interval $(-1, 1)$ than the Maclaurin expansion, in that it has a smaller maximum error with a given number of terms. These are not the best approximations for use in most digital computers, however. In this application, we measure efficiency by the computer time required to evaluate the function, plus some consideration of storage requirements for the constants. Since the arithmetic operations of a computer can directly evaluate only polynomials, we limit our discussion of more efficient approximations to rational functions, which are the ratios of two polynomials.

Our discussion of methods of finding efficient rational approximations will be elementary and introductory only. Obtaining truly best approximations is a difficult subject. In its present stage of development it is as much art as science, and requires successive approximations from a "suitably close" initial approximation. Our study will serve to introduce the student to some of the ideas and procedures used. The topic is of great importance, however, since the saving of just 1 msec of time in the generation of a frequently used elementary function may save hundreds of dollars' worth of machine time each year.

We start with a discussion of Padé approximations. Suppose we wish to represent a function as the quotient of two polynomials:

$$f(x) \doteq R_N(x) = \frac{a_0 + a_1 x + a_2 x^2 + \cdots + a_n x^n}{1 + b_1 x + b_2 x^2 + \cdots + b_m x^m}, \qquad N = n + m.$$

The constant term in the denominator can be taken as unity without loss of generality, since we can always convert to this form by dividing numerator and denominator by b_0. The constant b_0 will generally not be zero, for, in that case, the fraction would be undefined at $x = 0$. The most useful of the Padé approximations are those with the degree of the numerator equal to, or one greater than, the degree of the denominator. Note that the number of constants in $R_N(x)$ is $N + 1 = n + m + 1$.

The Padé approximations are related to Maclaurin expansions in that the coefficients are determined in a similar fashion to make $f(x)$ and $R_N(x)$ agree at $x = 0$ and also to make the first N derivatives agree at $x = 0$.*

We begin with the Maclaurin series for $f(x)$ (we use only terms through x^N) and write

$$f(x) - R_N(x)$$

$$\doteq (c_0 + c_1 x + c_2 x^2 + \cdots + c_N x^N) - \frac{a_0 + a_1 x + \cdots + a_n x^n}{1 + b_1 x + \cdots + b_m x^m}$$

$$= \frac{(c_0 + c_1 x + \cdots + c_N x^N)(1 + b_1 x + \cdots + b_m x^m) - (a_0 + a_1 x + \cdots + a_n x^n)}{1 + b_1 x + \cdots + b_m x^m}.$$

$$(10.14)$$

The coefficients c_i are $f^{(i)}(0)/(i!)$ of the Maclaurin expansion. Now if $f(x) = R_N(x)$ at $x = 0$, the numerator of Eq. (10.14) must have no constant term. Hence

$$c_0 - a_0 = 0. \qquad (10.15)$$

---

*A similar development can be derived for the expansion about a nonzero value of x, but the manipulations are not as easy. By a change of variable we can always make the region of interest contain the origin.

In order for the first N derivatives of $f(x)$ and $R_N(x)$ to be equal at $x = 0$, the coefficients of the powers of x up to and including x^N in the numerator must all be zero also. This gives N additional equations for the a's and b's. The first n of these involve a's, the rest only b's and c's:

$$b_1 c_0 + c_1 - a_1 = 0,$$
$$b_2 c_0 + b_1 c_1 + c_2 - a_2 = 0,$$
$$b_3 c_0 + b_2 c_1 + b_1 c_2 + c_3 - a_3 = 0,$$
$$\vdots$$
$$b_m c_{n-m} + b_{m-1} c_{n-m+1} + \cdots + c_n - a_n = 0,$$
$$b_m c_{n-m+1} + b_{m-1} c_{n-m+2} + \cdots + c_{n+1} = 0,$$
$$b_m c_{n-m+2} + b_{m-1} c_{n-m+3} + \cdots + c_{n+2} = 0,$$
$$\vdots$$
$$b_m c_{N-m} + b_{m-1} c_{N-m+1} + \cdots + c_N = 0. \tag{10.16}$$

Note that, in each equation, the sum of the subscripts on the factors of each product is the same, and is equal to the exponent of the x-term in the numerator. The $N + 1$ equations of Eqs. (10.15) and (10.16) give the required coefficients of the Padé approximation. We illustrate by an example.

E X A M P L E   Find $\arctan x \doteq R_9(x)$. Use in the numerator a polynomial of degree five.
The Maclaurin series through x^9 is

$$\arctan x \doteq x - \frac{1}{3}x^3 + \frac{1}{5}x^5 - \frac{1}{7}x^7 + \frac{1}{9}x^9. \tag{10.17}$$

We form, analogously to Eq. (10.14),

$$f(x) - R_9(x)$$
$$= \frac{\left(x - \frac{1}{3}x^3 + \frac{1}{5}x^5 - \frac{1}{7}x^7 + \frac{1}{9}x^9\right)(1 + b_1 x + b_2 x^2 + b_3 x^3 + b_4 x^4) - (a_0 + a_1 x + \cdots + a_5 x^5)}{(1 + b_1 x + b_2 x^2 + b_3 x^3 + b_4 x^4)}.$$

$$\tag{10.18}$$

Making coefficients through that of x^9 in the numerator equal to zero, we get

$$a_0 = 0,$$
$$a_1 = 1,$$
$$a_2 = b_1,$$
$$a_3 = -\frac{1}{3} + b_2,$$

$$a_4 = -\frac{1}{3}b_1 + b_3,$$

$$a_5 = \frac{1}{5} - \frac{1}{3}b_2 + b_4,$$

$$\frac{1}{5}b_1 - \frac{1}{3}b_3 = 0,$$

$$-\frac{1}{7} + \frac{1}{5}b_2 - \frac{1}{3}b_4 = 0,$$

$$-\frac{1}{7}b_1 + \frac{1}{5}b_3 = 0,$$

$$\frac{1}{9} - \frac{1}{7}b_2 + \frac{1}{5}b_4 = 0.$$

Solving first the last four equations for the b's, and then getting the a's, we have

$$a_0 = 0, \quad a_1 = 1, \quad a_2 = 0, \quad a_3 = \frac{7}{9}, \quad a_4 = 0, \quad a_5 = \frac{64}{945},$$

$$b_1 = 0, \quad b_2 = \frac{10}{9}, \quad b_3 = 0, \quad b_4 = \frac{5}{21}.$$

A rational function that approximates arctan x is then

$$\arctan x \doteq \frac{x + \dfrac{7}{9}x^3 + \dfrac{64}{945}x^5}{1 + \dfrac{10}{9}x^2 + \dfrac{5}{21}x^4}. \tag{10.19}$$

In Table 10.6 we compare the errors for Padé approximation (Eq. 10.19) to the Maclaurin series expansion (Eq. 10.17). Enough terms are available in the Maclaurin series to give five-decimal precision at $x = 0.2$ and 0.4, but at $x = 1$ (the limit for convergence of the

Table 10.6   Comparison of Padé approximation to Maclaurin series for arctan x

x	True Value	Padé Eq. (10.19)	Error	Maclaurin Eq. (10.17)	Error
0.2	0.19740	0.19740	0.00000	0.19740	0.00000
0.4	0.38051	0.38051	0.00000	0.38051	0.00000
0.6	0.54042	0.54042	0.00000	0.54067	−0.00025
0.8	0.67474	0.67477	−0.00003	0.67982	−0.00508
1.0	0.78540	0.78558	−0.00018	0.83492	−0.04952

series) the error is sizable. Even though we used no more information in establishing it, the Padé formula is surprisingly accurate, having an error only $1/275$ as large at $x = 1$. It is then particularly astonishing to realize that the Padé approximation is still not the best one of its form, for it violates the minimax principle. If the extreme precision near $x = 0$ is relaxed, we can make the maximum error smaller in the interval. ∎

Before we discuss such better approximations in the form of rational functions, remarks on the amount of effort required for the computation using Eq. (10.19) are in order. If we implement the equation in a computer as it stands, we would, of course, use the constants in decimal form, and we would evaluate the polynomials in nested form:

$$\text{Numerator} = [(0.0677x^2 + 0.7778)x^2 + 1]x,$$

$$\text{Denominator} = (0.2381x^2 + 1.1111)x^2 + 1.$$

Since additions and subtractions are generally much faster than multiplications or divisions, we generally neglect them in a count of operations. We have then three multiplications for the numerator, two for the denominator, plus one to get x^2, and one division, for a total of seven operations. The Maclaurin series is evaluated with six multiplications, using the nested form. If division and multiplication consume about the same time, there is about a standoff in effort, but greater precision for Eq. (10.19).*

Since small differences in effort accumulate for a frequently used function, it is of interest to see if we can further decrease the number of operations to evaluate Eq. (10.19). By means of a succession of divisions we can re-express it in continued-fraction form:

$$\frac{0.0677x^5 + 0.7778x^3 + x}{0.2381x^4 + 1.1111x^2 + 1} = \frac{0.2844x^5 + 3.2667x^3 + 4.2x}{x^4 + 4.6667x^2 + 4.2}$$

$$= \frac{0.2844x(x^4 + 11.4846x^2 + 14.7659)}{x^4 + 4.6667x^2 + 4.2}$$

$$= \frac{0.2844x}{(x^4 + 4.6667x^2 + 4.2)/(x^4 + 11.4846x^2 + 14.7659)}$$

$$= \frac{0.2844x}{1 - (6.8179x^2 + 10.5659)/(x^4 + 11.4846x^2 + 14.7659)}$$

$$= \frac{0.2844x}{1 - 6.8179(x^2 + 1.5497)/(x^4 + 11.4846x^2 + 14.7659)}$$

$$= \frac{0.2844x}{1 - 6.8179/[(x^4 + 11.4846x^2 + 14.7659)/(x^2 + 1.5497)]}$$

$$= \frac{0.2844x}{1 - 6.8179/[x^2 + 9.9348 - 0.6304/(x^2 + 1.5497)]}.$$

In this last form, we see that three divisions and two multiplications are needed (one multiplication by x and one to get x^2), for a total of five operations. We have saved two steps. In most cases there is an even greater advantage to the continued-fraction form; in this example the missing powers of x favored the evaluation as polynomials.

---

*On many computers, division is slower than multiplication. This will modify the conclusion reached here.

The error of a Padé approximation can often be roughly estimated by computing the next nonzero term in the numerator of Eq. (10.19). For the above example, the coefficient of x^{10} is zero, and the next term is

$$\left(-\frac{1}{7}b_4 + \frac{1}{9}b_2 - \frac{1}{11}\right)x^{11} = \left[-\frac{1}{7}\left(\frac{5}{21}\right) + \frac{1}{9}\left(\frac{10}{9}\right) - \frac{1}{11}\right]x^{11}$$
$$= -0.0014x^{11}.$$

Dividing by the denominator, we have

$$\text{Error} \doteq \frac{-0.0014x^{11}}{1 + 1.1111x^2 + 0.2381x^4}.$$

At $x = 1$ this estimate gives -0.00060, which is about three times too large, but still of the correct order of magnitude. It is not unusual that such estimates be rough; analogous estimates of error by using the next term in a Maclaurin series behave similarly. The validity of the rule of thumb that "next term approximates the error" is poor when the coefficients do not decrease rapidly.

The preference for Padé approximations with the degree of the numerator the same as or one more than the degree of the denominator rests on the empirical fact that the errors are usually less for these. Ralston (1965) gives examples demonstrating this.

One can get somewhat improved rational-function approximations by starting with the Chebyshev expansion and operating analogously to the method for Padé approximations. We illustrate with an approximation for e^x. The Chebyshev series was derived in Section 10.4, Eq. (10.13):

$$e^x = 1.2661T_0 + 1.1303T_1 + 0.2715T_2 + 0.0444T_3.$$

Using this approximation, we form the difference

$$f(x) - \frac{P_n(x)}{Q_m(x)} = \frac{(1.2661 + 1.303T_1 + 0.2715T_2 + 0.0444T_3)(1 + b_1T_1) - (a_0 + a_1T_1 + a_2T_2)}{1 + b_1T_1}$$

Here we have chosen the numerator as a second-degree Chebyshev polynomial and the denominator as first degree. We again make the first $N = n + m$ powers of x in the numerator vanish. Expanding the numerator, we get

$$\begin{aligned}
\text{Numerator} = {} & 1.2661 + 1.1303T_1 + 0.2715T_2 + 0.0444T_3 + 1.2661b_1T_1 \\
& + 1.1303b_1T_1^2 + 0.2715b_1T_1T_2 + 0.0444b_1T_1T_3 - a_0 \\
& - a_1T_1 - a_2T_2.
\end{aligned}$$

Before we can equate coefficients to zero, we need to resolve the products of Chebyshev polynomials that occur. Recalling that $T_n(x) = \cos n\theta$, we can use the trigonometric identity

$$\cos n\theta \cos m\theta = \frac{1}{2}[\cos(n + m)\theta + \cos(n - m)\theta],$$

$$T_n(x)T_m(x) = \frac{1}{2}[T_{n+m}(x) + T_{|n-m|}(x)].$$

The absolute value of the difference $n - m$ occurs because $\cos(z) = \cos(-z)$. Using this relation we can write the equations

$$a_0 = 1.2661 + \frac{1.1303}{2}b_1,$$

$$a_1 = 1.1303 + \left(\frac{0.2715}{2} + 1.2661\right)b_1,$$

$$a_2 = 0.2715 + \left(\frac{1.1303}{2} + \frac{0.0444}{2}\right)b_1,$$

$$0 = 0.0444 + \frac{0.2715}{2}b_1.$$

Solving, we get $b_1 = -0.3266$, $a_0 = 1.0815$, $a_1 = 0.6724$, $a_2 = 0.07966$, and

$$e^x \doteq \frac{1.0815 + 0.6724T_1 + 0.07966T_2}{1 - 0.3266T_1},$$

$$e^x \doteq \frac{1.0018 + 0.6724x + 0.1593x^2}{1 - 0.3266x}. \tag{10.20}$$

The last expression results when the Chebyshev polynomials are written in terms of powers of x. In Table 10.7 the error of this rational approximation is compared to the Chebyshev expansion. We see that the maximum error is reduced by 22%. Note that we do not, nevertheless, yet have a "best approximation." The error should reach equal maximums at five points in the interval—instead the error is large near $x = 1$ and too small elsewhere.

The basis for this last statement is the minimax theorem. Based on a theorem due to Chebyshev, we may state a principle whereby we may determine whether the approximation represented by a given polynomial or rational function is optimum, in the sense that it gives the least maximum error of any rational function of the same degree of numerator and denominator on a given interval. An expression is "minimax" if and only if there are at least $n + 2$ maxima in the deviations, and these are all equal in magnitude and of alternating sign, on the interval of approximation. (Here n is the sum of the degrees of numerator and denominator of the rational function.) In the discussion that follows,

Table 10.7   Comparison of rational approximations (Eq. (10.20)) with Chebyshev series for e^x

x	e^x	Chebyshev	Error	Rational function	Error
−1.0	0.3679	0.3629	0.0050	0.3684	−0.0005
−0.8	0.4493	0.4535	−0.0042	0.4486	0.0007
−0.6	0.5488	0.5535	−0.0047	0.5482	0.0006
−0.4	0.6703	0.6713	−0.0010	0.6707	−0.0004
−0.2	0.8187	0.8155	0.0032	0.8201	−0.0014
0	1.0000	0.9946	0.0054	1.0018	−0.0018
0.2	1.2214	1.2172	0.0042	1.2225	−0.0011
0.4	1.4918	1.4917	0.0001	1.4911	0.0007
0.6	1.8221	1.8267	−0.0046	1.8191	0.0030
0.8	2.2255	2.2307	−0.0052	2.2225	0.0031
1.0	2.7183	2.7123	0.0060	2.7230	−0.0047

we shall be referring to the magnitude of the errors and will not be concerned about their sign.

A second important consequence of this principle is that we can put bounds on the error of the minimax expression from the range of the errors of a function that is not minimax. Suppose we have an expression like Eq. (10.20). It has five maxima of alternating sign on the interval $[−1, 1]$, as shown by Table 10.7. This is the correct number, but we know the function isn't minimax because the maxima aren't equal in magnitude. While some of the maxima are not given precisely by the table, we can see that the smallest is 0.0005 (at $x = −1$) and the largest appears to be 0.0047 (at $x = 1$). From this range of maximum error values, we can bound the maximum error of the minimax rational function [of degree $(2, 1)$]: the minimax expression will have a maximum error on $[−1, 1]$ no less than 0.0005 and no greater than 0.0047.

Similarly, the truncated Chebyshev series of degree 3 whose errors are tabulated in Table 10.4 is not truly minimax; it has five alternating maxima to its errors but they are not quite equal in magnitude. From an examination of Table 10.4 we can say, however, that the minimax polynomial of degree 3 will have a maximum error bounded by 0.0047 and 0.0060. (A tighter bound might result from a more careful computation of the errors of Eq. (10.13).) Such a prediction about the amount of improvement that will be provided by a minimax expression can help one decide whether the additional effort to find it is worthwhile.

To obtain the optimum rational function that approximates the function with equal-magnitude errors distributed through the interval is beyond the scope of this text. The approach that is used is to improve an initial estimate of a function, such as Eq. (10.20), by successive trials, often modifying the constants on the basis of experience until eventually one has a satisfactory formula. Systematic methods of determining the constants in such minimax rational approximations have also been determined. They are iteration methods beginning from an initial "sufficiently good" approximation. They are expensive to compute because the iterations involve solving a set of nonlinear equations. Ralston (1965) describes one such method. Prenter (1975) discusses the approximation of functions of several variables.

10.6   APPROXIMATION OF FUNCTIONS WITH TRIGONOMETRIC SERIES: FAST FOURIER TRANSFORMS

In nearly all of our discussions until now we have approximated functions by polynomials, but for many functions it is better to use trigonometric functions as the basis. This is particularly true when $f(x)$ is periodic or when it has discontinuities. A trigonometric series, also called a *Fourier series*, is another means for solving some partial-differential equations. Getting the coefficients of a trigonometric or Fourier series can be computationally expensive, but the fast Fourier transform (FFT) is a way to minimize the effort. In this section, we will first describe the Fourier series and then develop the FFT procedure.

Recall what is meant by a periodic function; $f(x)$ is periodic with the period T if $f(x) = f(x + T)$.

EXAMPLES

1. $\sin (x)$ and $\cos (x)$ are periodic with period 2π, since $\sin (x) = \sin (x + 2\pi)$ and $\cos (x) = \cos (x + 2\pi)$.

2. $\sin (4x)$ is periodic with period $\pi/2$.

3.   This function has a period of 2:

4.   This function has a period of 10:

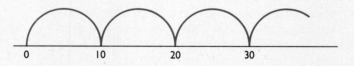

It will be convenient to assume that the period T is always 2π in the following. There is no loss of generality in this because, if $f(x) = f(x + T)$, we can define $g(x) = f(xT/2\pi)$, where $g(x)$ has period 2π.

If $f(x)$ is periodic of period 2π and obeys certain assumptions, we can represent it as a Fourier series:

$$f(x) = a_0/2 + \sum_{k=1}^{\infty} [a_k \cos (kx) + b_k \sin (kx)]. \qquad (10.21)$$

An alternative form of this series uses complex exponentials. Recall that a complex number z can be written as $a + bi$, where a and b are real and $i = \sqrt{-1}$, so $i^2 = -1$. Euler's identity relates sines and cosines to the exponential form:

$$e^{ijx} = \cos(jx) + i\sin(jx),$$

and this permits us to write Eq. (10.21) as

$$f(x) = \sum_{j=0}^{\infty} (c_j e^{ijx} + c_{-j} e^{-ijx})$$

$$= 2c_0 + \sum_{j=1}^{\infty} [(c_j + c_{-j})\cos(jx) + i(c_j - c_{-j})\sin(jx)]$$

$$= \sum_{j=-\infty}^{\infty} c_j e^{ijx}. \tag{10.22}$$

We can match up the a's and b's of Eq. (10.21) to the c's of (10.22):

$$a_j = c_j + c_{-j}, \qquad b_j = i(c_j - c_{-j}),$$

$$c_j = \frac{a_j - ib_j}{2}, \qquad c_{-j} = \frac{a_j + ib_j}{2}. \tag{10.23}$$

For $f(x)$ real, it is easy to show that $c_0 = \bar{c}_0$ and $c_j = \bar{c}_{-j}$, where the bars represent complex conjugates.

For integers j and k, it is true that

$$\int_0^{2\pi} (e^{ikx})(e^{ijx}) \, dx = \int_0^{2\pi} e^{i(k+j)x} \, dx = \begin{cases} 0 & \text{for } k \neq -j, \\ 2\pi & \text{for } k = -j. \end{cases}$$

(You can verify the first of these through Euler's identity.) This allows us to evaluate the c's of Eq. (10.22) by the following.

For each fixed k, we get

$$f(x)\, e^{-ikx} = \sum_{j=-\infty}^{\infty} c_k e^{i(j-k)x},$$

$$\int_0^{2\pi} f(x)\, e^{-ikx} \, dx = 2\pi c_k, \text{ or}$$

$$c_k = \frac{1}{2\pi} \int_0^{2\pi} f(x)\, e^{-ikx} \, dx, \; k = 0, \pm 1, \pm 2, \ldots. \tag{10.24}$$

EXAMPLES   (You should verify each of these.)

1. Let $f(x) = x$; then

$$c_k = \frac{1}{2\pi} \int_0^{2\pi} x e^{-ikx} \, dx = -\frac{i}{k}, \qquad k \neq 0.$$

2. Let $f(x) = x(2\pi - x)$; then

$$c_k = \frac{1}{2\pi} \int_0^{2\pi} x(2\pi - x) e^{-ikx} \, dx = \frac{2}{k^2}, \qquad k \neq 0.$$

3.   Let $f(x) = \cos(x)$; then

$$c_k = \frac{1}{2\pi} \int_0^{2\pi} \cos(x) e^{-ikx} \, dx = \begin{cases} \frac{1}{2} & \text{for } k = 1 \text{ or } -1, \\ 0 & \text{for all other } k. \end{cases}$$

Note that for Eq. (10.21) this makes $a_1 = 1$ and all the other a_j's $= 0$. Thus, for a given $f(x)$ that satisfies continuity conditions, we have

$$c_j = \frac{1}{2\pi} \int_0^{2\pi} f(x) e^{-ijx} \, dx, \quad j = 0, \pm 1, \pm 2, \ldots . \quad \blacksquare$$

The magnitudes of the Fourier series coefficients $|c_j|$ are the *power spectrum* of f; these show the frequencies that are represented in $f(x)$. If we know $f(x)$ in the time domain, we can identify f by computing the c_j's. In getting the Fourier series, we have transformed from the time domain to the frequency domain, an important aspect of wave analysis.

Suppose we have N values for $f(x)$ on the interval $[0, 2\pi]$ at equispaced points, $x_k = 2\pi k/N$, $k = 0, 1, \ldots, N - 1$. Since $f(x)$ is periodic, $f_N = f_0, f_{N+1} = f_1$, and so on. Instead of formal analytical integration, we would use a numerical integration method to get the coefficients. Even if $f(x)$ is known at all points in $[0, 2\pi]$, we might prefer to use numerical integration. This would use only certain values of $f(x)$, often those evaluated at uniform intervals. It is also often true that we do not know $f(x)$ everywhere, because we have sampled a continuous signal. In that case, however, it is better to use the discrete Fourier transform, which can be defined as

$$X(n) = \sum_{k=0}^{N-1} x_0(k) \, e^{-i2\pi nk/N}, \, n = 0, 1, 2, \ldots, N - 1. \quad \textbf{(10.25)}$$

In (10.25), we have changed notation to conform more closely to the literature on FFT. $X(n)$ corresponds to the coefficients of N frequency terms, and the $x_0(k)$ are the N values of the signal samples in the time domain. You can think of n as indexing the X-terms and k as indexing the x_0-terms. Equation (10.25) corresponds to a set of N linear equations that we can solve for the unknown $X(n)$. Since the unknowns appear on the left-hand side of (10.25), this requires only the multiplication of an N component vector by an $N \times N$ matrix.

It will simplify the notation if we let $W = e^{-i2\pi/N}$, making the right-hand-side terms of Eq. (10.25) become $x_0(k)W^{nk}$. To develop the FFT algorithm, suppose that $N = 4$. We write the four equations for this case:

$$X(0) = W^0 x_0(0) + W^0 x_0(1) + W^0 x_0(2) + W^0 x_0(3),$$

$$X(1) = W^0 x_0(0) + W^1 x_0(1) + W^2 x_0(2) + W^3 x_0(3),$$

$$X(2) = W^0 x_0(0) + W^2 x_0(1) + W^4 x_0(2) + W^6 x_0(3),$$

$$X(3) = W^0 x_0(0) + W^3 x_0(1) + W^6 x_0(2) + W^9 x_0(3).$$

In matrix form:

$$
\begin{bmatrix} X_0 \\ X_1 \\ X_2 \\ X_3 \end{bmatrix} = \begin{bmatrix} W^0 & W^0 & W^0 & W^0 \\ W^0 & W^1 & W^2 & W^3 \\ W^0 & W^2 & W^4 & W^6 \\ W^0 & W^3 & W^6 & W^9 \end{bmatrix} x_0. \tag{10.26}
$$

In solving the set of N equations in the form of Eq. (10.26), we will have to make N^2 complex multiplications plus $N(N-1)$ complex additions. The value of the FFT is that the number of such operations is greatly reduced. While there are several variations on the algorithm, we will concentrate on the Cooley–Tukey formulation. The matrix of Eq. (10.26) can be factored to give an equivalent form for the set of equations. At the same time we will use the fact that $W^0 = 1$ and $W^k = W^{k \bmod(N)}$:

$$
\begin{bmatrix} X(0) \\ X(2) \\ X(1) \\ X(3) \end{bmatrix} = \begin{bmatrix} 1 & W^0 & 0 & 0 \\ 1 & W^2 & 0 & 0 \\ 0 & 0 & 1 & W^1 \\ 0 & 0 & 1 & W^3 \end{bmatrix} \begin{bmatrix} 1 & 0 & W^0 & 0 \\ 0 & 1 & 0 & W^0 \\ 1 & 0 & W^2 & 0 \\ 0 & 1 & 0 & W^2 \end{bmatrix} x_0. \tag{10.27}
$$

You should verify that the factored form (10.27) is exactly equivalent to Eq. (10.26) by multiplying out. Note carefully that the elements of the X vector are scrambled. (The development can be done formally and more generally by representing n and k as binary values, but it will suffice to show the basis for the FFT algorithm by expanding on this simple $N = 4$ case.)

By using the factored form, we now get the values of $X(n)$ by two steps (stages), in each of which we multiply a matrix times a vector. In the first stage, we transform x_0 into x_1 by multiplying the right matrix of (10.27) and x_0. In the second stage, we multiply the left matrix and x_1, getting x_2. We get X by scrambling the components of x_2. By doing the operation in stages, the number of complex multiplications is reduced to $N(\log_2 N)$. For $N = 4$, this is a reduction by one-half, but for large N it is very significant; if $N = 1024$, there are 10 stages and the reduction in complex multiplies is a hundredfold!

It is convenient to represent the sequence of multiplications of the factored form (Eq. (10.27) or its equivalent for larger N) by flow diagrams. Figure 10.8 is for $N = 4$ and Fig. 10.9 is for $N = 16$. Each column holds values of x_{ST}, where the subscript tells which stage is being computed; ST ranges from 1 to 2 for $N = 4$ and from 1 to 4 for $N = 16$. (The number of stages, for N a power of 2, is $\log_2(N)$.) In each stage, we get x-values of the next stage from those of the present stage. Every new x-value is the sum of the two x-values from the previous stage that connect to it, with one of these multiplied by a power of W. The diagram tells which x_{ST} terms are combined to give an x_{ST+1} term, and the numbers shown within the lines are the powers of W that are used. For example, looking at Fig. 10.9, we see that

$$
x_2(6) = x_1(2) + W^8 \, x_1(6),
$$

$$
x_3(13) = x_2(13) + W^6 \, x_2(15),
$$

$$
x_4(9) = x_3(8) + W^9 \, x_3(9),
$$

and so on.

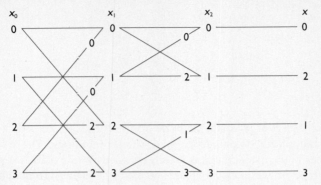

Figure 10.8

The last columns in Figs. 10.8 and 10.9 indicate how the final x-values are unscrambled to give the X-values. This relationship can be found by expressing the index k of x in the last stage as a binary number and reversing the bits; this gives n in $X(n)$. For example, in Fig. 10.9, we see that $x_4(3) = X(12)$ and $x_4(11) = X(13)$. From the bit-reversing rule, we get

$$3 = 0011_2 \rightarrow 1100_2 = 12, \quad 11 = 1011_2 \rightarrow 1101_2 = 13.$$

Observe also that the bit-reversing rule can give the powers of W that are involved in computing the next stage. For the last stage, the powers are identical to the numbers obtained by bit reversal. At each previous stage, however, only the first half of the powers are employed, but each power is used twice as often. It is of interest to see how we can generate these values. Computer languages that facilitate bit manipulations make this an easy job, but there is a good alternative. Observe how the powers in Fig. 10.8 differ from those in Fig. 10.9 and how they progress from stage to stage. The following table pinpoints this.

Stage	N = 4				N = 16															
1:	0	0	2	2	0	0	0	0	0	0	0	0	8	8	8	8	8	8	8	8
2:	0	2	1	3	0	0	0	0	8	8	8	8	4	4	4	4	12	12	12	12
3:					0	0	8	8	4	4	12	12	2	2	10	10	6	6	14	14
4:					0	8	4	12	2	10	6	14	1	9	5	13	3	11	7	15

Can you see what a similar table for $N = 2$ would look like? Its single row would be 0 1. Now we see that the row of powers for the last stage can be divided into two halves with the numbers in the second half always 1 greater than the corresponding entry in the first half. The row above is the left half of the current row with each value repeated. This observation leads to the following algorithm.

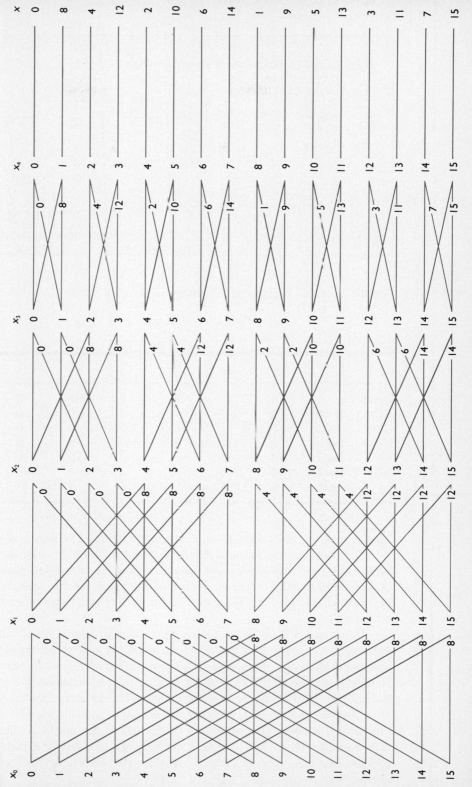

Figure 10.9

Algorithm to Generate Powers of W in FFT

For N a power of 2, let $Q = \log_2(N)$.

 Initialize an array P of length N to all zeros.
 Let ST = 1.
 Repeat
   Double the values of $P(K)$ for $K = 1 \mathinner{.\,.} 2^{ST-1}$,
   Let each $P(K + 2^{ST-1}) = P(K) + 1$ for $K = 0 \mathinner{.\,.} 2^{ST-1} - 1$,
   Increment ST,
 Until ST $> Q$.

The successive new values for powers of W are now in array P.

E X A M P L E Use the algorithm to generate the powers of W for $N = 8$:

$$Q = \log_2(8) = 3.$$

K:	0	1	2	3	4	5	6	7
Initial P array:	0	0	0	0	0	0	0	0
ST = 1, doubled:	0	0	0	0	0	0	0	0
add 1:	0	1	0	0	0	0	0	0
ST = 2, doubled:	0	2	0	0	0	0	0	0
add 1:	0	2	1	3	0	0	0	0
ST = 3, doubled:	0	4	2	6	0	0	0	0
add 1:	0	4	2	6	1	5	3	7

The last row of values corresponds to the bits of 000 to 111 after reversal.

 Our discussion has assumed that N is a power of 2; for this case the economy of the FFT is a maximum. When N is not a power of 2 but can be factored, there are adaptations of the general idea that reduce the number of operations, but they are more than $N\log_2(N)$. See Brigham (1974) for a discussion of this as well as a fuller treatment of the theory behind FFT.

 More recently there has been interest in another transform, called the discrete Hartley transform. A discussion of this transform would parallel our discussion of the Fourier transform. Moreover, it has been shown that this transform can be converted into a Fast Hartley Transform (FHT) that reduces to $N\log_2(N)$ computations. For a full coverage of the FHT, one should consult Bracewell (1986). The advantages of the FHT are that it is usually faster than the FFT. Moreover, it is easy to compute the FFT from the Hartley transform. However, the main power of the FHT is that all the computations are done in real arithmetic, so that one can use a language like PASCAL that does not have a complex data type. An interesting and easy introduction into the FHT is found in O'Neill (1988). ∎

10.7 CHAPTER SUMMARY

Here's what you are able to do if you understand the topics of the chapter:

1. Explain what is meant by the *least-squares criterion* and use it to derive the normal equations for the coefficients of a least-squares line.

2. Apply the least-squares criterion to find the coefficients of a polynomial that fits a set of data points. You should be able to show how the design matrix is related to the normal equations.

3. Get the coefficients of other functions than polynomials by the least-squares procedure and recognize why this is more difficult.

4. Explain how to get a Chebyshev polynomial of degree n, what is meant by a set of orthogonal functions, and what is the most important property of the Chebyshev polynomials.

5. Use Chebyshev polynomials to economize a power series and explain why such approximations are preferred in computers.

6. Obtain the coefficients of a Padé approximation and describe how one can get a further improvement in efficiency. You can explain the minimax principle.

7. Tell how the coefficients of a trigonometric series can be evaluated and outline the FFT method for fitting to a set of equispaced data.

8. Use computer programs like those of this chapter.

SELECTED READINGS FOR CHAPTER 10

Conte and de Boor (1980); Fike (1968); Brigham (1974); Ramirez (1985).

10.8 COMPUTER PROGRAMS

Three computer programs are presented. Program 1 (Fig. 10.10) finds least-squares polynomials that fit a set of x,y-pairs. The program reads in the coordinates of the points and computes the coefficients of the normal equations for polynomials whose degree ranges from MS to MF. Values of these parameters are also read in. To get these normal equations, the terms of the largest augmented matrix (Eq. (10.5)) are computed. (This matrix corresponds to degree MF). In the process, the powers of x are stored in a vector; this expedites forming the sums. The program recognizes the symmetry of the coefficient matrix, which lessens the need to recompute terms.

After the matrix that represents the normal equations has been formed, a subroutine is called to compute the LU equivalent. This subroutine is very similar to one in Chapter 2, except that pivoting is not done. The solution for the coefficients of each of the polynomials is obtained by calling a second subroutine. Using the LU decomposition process avoids triangularizing the normal equations repeatedly for the various degrees of polynomial. The values of Eq. (10.6), here called BETA, are computed and printed to

```
          PROGRAM LEASQR(INPUT,OUTPUT)
C
C     ------------------------------------------------------------------
C
C     THIS PROGRAM IS USED IN FITTING A POLYNOMIAL TO A SET OF DATA.
C     THE PROGRAM READS IN N PAIRS OF X AND Y VALUES AND COMPUTES THE
C     COEFFICIENTS OF THE NORMAL EQUATIONS FOR THE LEAST SQUARES
C     METHOD.
C
C     ------------------------------------------------------------------
C
C
C         PARAMETERS ARE :
C
C         X, Y   - ARRAY OF X AND Y VALUES
C         N      - NUMBER OF DATA PAIRS
C         MS,MF  - THE RANGE OF DEGREE OF POLYNOMIALS TO BE COMPUTED
C                  THE MAXIMUM DEGREE IS 9.
C         A      - AUGMENTED ARRAY OF THE COEFFICIENTS OF THE NORMAL
C                  EQUATIONS.
C         C      - ARRAY OF COEFFICIENTS OF THE LEAST SQUARES
C                  POLYNOMIALS.
C
C     ------------------------------------------------------------------
C
          REAL X(100),Y(100),C(100),A(10,11),XN(100),SUM,BETA
          INTEGER N,MS,MF,MFP1,MFP2,I,J,IM1,IPT,ICOEF,JCOEF
C
C     ------------------------------------------------------------------
C
C     READ IN N, THEN THE X AND Y VALUES.
C
C         READ *, N, ( X(I),Y(I), I = 1,N )
C
          DATA N/11/
          DATA X/0.05,0.11,0.15,0.31,0.46,0.52,0.7,0.74,0.82,
         *      0.98,1.17,89*0.0/
          DATA Y/0.956, 0.89, 0.832, 0.717, 0.571, 0.539, 0.378,
         *      0.37, 0.306, 0.242, 0.104, 89*0.0/
C
C     READ IN MS,MF. THE PROGRAM WILL FIND COEFFICIENTS FOR EACH
C     DEGREE OF POLYNOMIAL FROM DEGREE MS TO DEGREE MF.
C
C         READ *, MS,MF
C
          DATA MS,MF/1,7/
C     ------------------------------------------------------------------
C
C     COMPUTE MATRIX OF COEFFICIENTS AND R.H.S. FOR MF DEGREE.
C     HOWEVER, FIRST CHECK TO SEE IF MAX DEGREE REQUESTED IS TOO
C     LARGE. IT CANNOT EXCEED N-1. IF IT DOES, REDUCE TO EQUAL N-1
C     AND PRINT MESSAGE.
C
          IF ( MF .GT. (N-1) ) THEN
            MF = N - 1
            PRINT 200, MF
          END IF
        5 MFP1 = MF + 1
          MFP2 = MF + 2
C
C     ------------------------------------------------------------------
C
```

Figure 10.10   Program 1.

Figure 10.10   (*continued*)

```
C  PUT ONES INTO A NEW ARRAY. THIS WILL HOLD THE POWERS OF
C  THE X VALUES AS WE PROCEED.
C
      DO 10 I = 1,N
        XN(I) = 1.0
   10 CONTINUE
C
C  ----------------------------------------------------------------
C
C  COMPUTE FIRST COLUMN AND N+1 ST COLUMN OF A. I MOVES DOWN THE
C  ROWS, J SUMS OVER THE N VALUES.
C
      DO 30 I = 1,MFP1
        A(I,1) = 0.0
        A(I,MFP2) = 0.0
        DO 20 J = 1,N
          A(I,1) = A(I,1) + XN(J)
          A(I,MFP2) = A(I,MFP2) + Y(J)*XN(J)
          XN(J) = XN(J) * X(J)
   20   CONTINUE
   30 CONTINUE
C
C  ----------------------------------------------------------------
C
C  COMPUTE THE LAST ROW OF A. I MOVES ACROSS THE COLUMNS, J
C  SUMS OVER THE N VALUES.
C
      DO 50 I = 2,MFP1
        A(MFP1,I) = 0.0
        DO 40 J = 1,N
          A(MFP1,I) = A(MFP1,I) + XN(J)
          XN(J) = XN(J) * X(J)
   40   CONTINUE
   50 CONTINUE
C
C  ----------------------------------------------------------------
C
C  NOW FILL IN THE REST OF THE A MATRIX. I MOVES DOWN THE ROWS,
C  J MOVES ACROSS THE COLUMNS.
C
      DO 70 J = 2,MFP1
        DO 60 I = 1,MF
          A(I,J) = A(I+1,J-1)
   60   CONTINUE
   70 CONTINUE
C
C  ----------------------------------------------------------------
C
C  WRITE OUT THE MATRIX OF NORMAL EQUATIONS.
C
      PRINT '(///)'
      PRINT *, '         THE NORMAL EQUATIONS ARE:
      PRINT '(/)'
      PRINT 201, ((A(I,J), J=1,MFP2), I=1,MFP1)
      PRINT '(//)'
C
C  NOW CALL A SUBROUTINE TO SOLVE THE SYSTEM. DO THIS FOR EACH
C  DEGREE FROM MS TO MF. GET THE LU DECOMPOSITION OF A.
C
      CALL LUDCMQ(A,MFP1,10)
C
C  ----------------------------------------------------------------
C
```

Figure 10.10 *(continued)*

```
C   RESET THE R.H.S. INTO C. WE NEED TO DO THIS FOR EACH DEGREE.
C
      MSP1 = MS + 1
      DO 95 I = MSP1,MFP1
        DO 90 J = 1,I
          C(J) = A(J,MFP2)
   90   CONTINUE
        CALL SOLNQ(A,C,I,10)
        IM1 = I - 1
C
C   -----------------------------------------------------------------------
C
C   NOW WRITE OUT THE COEFFICIENTS OF THE LEAST SQUARE POLYNOMIAL.
C
      PRINT 202, IM1, ( C(J); J=1,I)
C
C   -----------------------------------------------------------------------
C
C   COMPUTE AND PRINT THE VALUE OF BETA = SUM OF DEV SQUARED /
C   ( N - M - 1 ).
C
      BETA = 0.0
      DO 94 IPT = 1,N
        SUM = 0.0
        DO 93 ICOEF = 2,I
          JCOEF = I - ICOEF + 2
          SUM = ( SUM + C(JCOEF) ) * X(IPT)
   93   CONTINUE
        SUM = SUM + C(1)
        BETA = BETA + ( Y(IPT) - SUM )**2
   94 CONTINUE
      BETA = BETA / (N - I)
      PRINT 203, BETA
   95 CONTINUE
C
C   -----------------------------------------------------------------------
C
  200 FORMAT(//' DEGREE OF POLYNOMIAL CANNOT EXCEED N - 1.',/
     +       '   REQUESTED MAXIMUM DEGREE TOO LARGE - ',
     +       'REDUCED TO ',I3)
  201 FORMAT(1X,9F8.2)
  202 FORMAT(/' FOR DEGREE OF ',I2,' COEFFICIENTS ARE'//
     +       ' ',5X,11F9.3)
  203 FORMAT(9X,' BETA IS ',F10.5//)
      STOP
      END
      SUBROUTINE LUDCMQ(A,N,NDIM)
C
C   -----------------------------------------------------------------------
C
C   SUBROUTINE LUDCMQ :
C                    THIS SUBROUTINE FORMS THE LU EQUIVALENT OF
C   THE SQUARE COEFFICIENT MATRIX A. THE LU, IN COMPACT FORM, IS
C   RETURNED IN THE A MATRIX SPACE. THE UPPER TRIANGULAR MATRIX U
C   HAS ONES ON ITS DIAGONAL - THESE VALUES ARE NOT INCLUDED IN
C   THE RESULT.
C
C   -----------------------------------------------------------------------
C
      REAL A(NDIM,NDIM),SUM
      INTEGER N,NDIM,I,J,JM1,IM1,K
C
```

Figure 10.10   (*continued*)

```
C   ------------------------------------------------------------------
C
      DO 30 I = 1,N
      DO 30 J = 2,N
        SUM = 0.0
        IF ( J .LE. I ) THEN
          JM1 = J - 1
          DO 10 K = 1,JM1
            SUM = SUM + A(I,K)*A(K,J)
   10     CONTINUE
          A(I,J) = A(I,J) - SUM
        ELSE
          IM1 = I - 1
          IF ( IM1 .NE. 0 ) THEN
            DO 20 K = 1,IM1
              SUM = SUM  + A(I,K)*A(K,J)
   20       CONTINUE
          END IF
C
C   ------------------------------------------------------------------
C
C
C TEST FOR SMALL VALUE ON THE DIAGONAL
C
   25     IF ( ABS(A(I,I)) .LT. 1.0E-10 ) THEN
            PRINT 100, I
            RETURN
          ELSE
            A(I,J) = ( A(I,J) - SUM ) / A(I,I)
          END IF
        END IF
   30 CONTINUE
      RETURN
C
  100 FORMAT(' REDUCTION NOT COMPLETED BECAUSE SMALL VALUE',
     +        ' FOUND FOR DIVISOR IN ROW ',I3)
      END
      SUBROUTINE SOLNQ(A,B,N,NDIM)
C
C   ------------------------------------------------------------------
C
C
C   SUBROUTINE SOLNQ :
C                    THIS SUBROUTINE FINDS THE SOLUTION TO A SET
C   OF N LINEAR EQUATIONS THAT CORRESPONDS TO THE RIGHT HAND SIDE
C   VECTOR B. THE A MATRIX IS THE LU DECOMPOSITION EQUIVALENT TO THE
C   COEFFICIENT MATRIX OF THE ORIGINAL EQUATIONS, AS PRODUCED BY
C   LUDCMQ. THE SOLUTION VECTOR IS RETURNED IN THE B VECTOR.
C
C
C   ------------------------------------------------------------------
C
      REAL A(NDIM,NDIM),B(NDIM),SUM
      INTEGER N,NDIM,I,IM1,K,J,NMJP1,NMJP2
C
C   ------------------------------------------------------------------
C
C DO THE REDUCTION STEP
C
      B(1) = B(1) / A(1,1)
      DO 20 I = 2,N
        IM1 = I - 1
        SUM = 0.0
        DO 10 K = 1,IM1
          SUM = SUM + A(I,K)*B(K)
```

Figure 10.10   (*continued*)

```
      10    CONTINUE
            B(I) = ( B(I) - SUM ) / A(I,I)
      20 CONTINUE
C
C     ------------------------------------------------------------------------
C
C     NOW WE ARE READY FOR BACK SUBSTITUTION. REMEMBER THAT THE ELEMENTS
C     OF U ON THE DIAGONAL ARE ALL ONES.
C
         DO 40 J = 2,N
            NMJP2 = N - J + 2
            NMJP1 = N - J + 1
            SUM = 0.0
            DO 30 K = NMJP2,N
               SUM = SUM + A(NMJP1,K)*B(K)
      30    CONTINUE
            B(NMJP1) = B(NMJP1) - SUM
      40 CONTINUE
         RETURN
         END


               OUTPUT FOR PROGRAM 1


            THE NORMAL EQUATIONS ARE:


   11.00     6.01     4.65     4.11     3.92     3.92     4.07     4.34     5.91
    6.01     4.65     4.11     3.92     3.92     4.07     4.34     4.72     2.18
    4.65     4.11     3.92     3.92     4.07     4.34     4.72     5.22     1.34
    4.11     3.92     3.92     4.07     4.34     4.72     5.22     5.84     1.00
    3.92     3.92     4.07     4.34     4.72     5.22     5.84     6.59      .83
    3.92     4.07     4.34     4.72     5.22     5.84     6.59     7.50      .74
    4.07     4.34     4.72     5.22     5.84     6.59     7.50     8.57      .70
    4.34     4.72     5.22     5.84     6.59     7.50     8.57     9.84      .68


FOR DEGREE OF  1 COEFFICIENTS ARE

       .952     -.760
       BETA IS     .00102


FOR DEGREE OF  2 COEFFICIENTS ARE

       .998    -1.018     .225
       BETA IS     .00023


FOR DEGREE OF  3 COEFFICIENTS ARE

      1.004    -1.079     .351    -.069
       BETA IS     .00026


FOR DEGREE OF  4 COEFFICIENTS ARE

       .988     -.837    -.527    1.046    -.456
       BETA IS     .00027
```

Figure 10.10   (*continued*)

```
FOR DEGREE OF  5 COEFFICIENTS ARE

        1.037   -1.824    4.895  -10.753   10.537   -3.659
        BETA IS    .00013




FOR DEGREE OF  6 COEFFICIENTS ARE

        1.041   -1.946    5.886  -14.081   15.818   -7.599    1.112
        BETA IS    .00017




FOR DEGREE OF  7 COEFFICIENTS ARE

        1.060   -2.562   12.442  -44.820   88.981  -99.706   59.449  -14.608
        BETA IS    .00021
```

```
        PROGRAM FFTIMSL(INPUT,OUTPUT)
C
C    ------------------------------------------------------------------
C
C    PROGRAM FFTIMSL CALLS THE IMSL SUBROUTINE FFTCC TO COMPUTE
C    THE FOURIER TRANSFORM OF N DATA POINTS.
C
C    ------------------------------------------------------------------
C
C
C    PARAMETERS ARE:
C         CHAT   - A COMPLEX ARRAY OF LENGTH N. ON INPUT CHAT CONTAINS THE
C                  FUNCTION VALUES, F(K), FOR K = 0 TO N-1.
C                  ON OUTPUT CHAT CONTAINS THE VALUES, CHAT(J) = SUM FROM K
C                  EQUALS 0 TO N-1 F(K)*EXP(2*PI*I*K/N).
C         N      - A INTEGER THAT REPRSENTS THE NUMBER OF DATA POINTS.
C         IWK    - AN INTEGER VECTOR ARRAY OF LENGTH 6*N + 150.
C         WK     - A REAL WORK ARRAY OF LENGTH 6*N + 150.
C
C
C    ------------------------------------------------------------------
C
      INTEGER N,IWK(534)
      REAL MAGN,EN,WK(534)
      COMPLEX CHAT(64)
C
      N=64
      EN=FLOAT(N)
      TPIDIVN = 2.0*ACOS(-1.0)/EN
C
C    ------------------------------------------------------------------
C
C              COMPUTE THE F(I)'S, I EQUAL 0 TO N-1.
C
C    ------------------------------------------------------------------
C
      DO 5 I=0,N-1
    5    CHAT(I+1) = CMPLX(SIN(TPIDIVN*I) + 4.0*COS(8.0*TPIDIVN*I)
     +         + SIN(10.0*TPIDIVN*I)*COS(15.0*TPIDIVN*I))
```

Figure 10.11   Program 2.

Figure 10.11 (*continued*)

```
C
C      ------------------------------------------------------------------
C
       CALL FFTCC(CHAT,N,IWK,WK)
C
C      ------------------------------------------------------------------
C
       PRINT 99
   99 FORMAT(///'        I       REAL PART         IMAG. PART',
      +        '        MAGNITUDE' /
      +        '        **      ***********         ***********',
      +        '        ************'/)
       DO 15 I=1,N
          MAGN = CABS(CHAT(I))/EN
   15     PRINT 98,    I-1,CONJG(CHAT(I))/EN,MAGN
   98 FORMAT('      ',I5,5X,F13.6,5X,F13.6,5X,F13.6)
C
       STOP
       END


          OUTPUT FOR PROGRAM 2


       I        REAL PART        IMAG. PART        MAGNITUDE
       **      ************      ************      ************

        0       .000000           .000000          .000000
        1       .000000          -.500000          .500000
        2       .000000           .000000          .000000
        3       .000000           .000000          .000000
        4       .000000           .000000          .000000
        5       .000000           .250000          .250000
        6       .000000           .000000          .000000
        7       .000000           .000000          .000000
        8      2.000000           .000000         2.000000
        9       .000000           .000000          .000000
       10       .000000           .000000          .000000
       11       .000000           .000000          .000000
       12       .000000           .000000          .000000
       13       .000000           .000000          .000000
       14       .000000           .000000          .000000
       15       .000000           .000000          .000000
       16       .000000           .000000          .000000
       17       .000000           .000000          .000000
       18       .000000           .000000          .000000
       19       .000000           .000000          .000000
       20       .000000           .000000          .000000
       21       .000000           .000000          .000000
       22       .000000           .000000          .000000
       23       .000000           .000000          .000000
       24       .000000           .000000          .000000
       25       .000000          -.250000          .250000
       26       .000000           .000000          .000000
       27       .000000           .000000          .000000
       28       .000000           .000000          .000000
       29       .000000           .000000          .000000
       30       .000000           .000000          .000000
       31       .000000           .000000          .000000
       32       .000000           .000000          .000000
       33       .000000           .000000          .000000
```

assist the user in selecting the optimum degree of polynomial. The program was run with the input data shown in Fig. 10.4.

The second program (Fig. 10.11) uses the IMSL subroutine FFTCC to calculate the $\hat{c}_j$'s for $f(x) = \sin(x) + 4\cos(8x) + \sin(10x)\cos(15x)$. It is obvious that from Eq. (10.21) this example will have $a_8 = 4$, $b_1 = 1$, $-b_5 = b_{25} = \frac{1}{2}$, since $\sin(10x)\cos(15x) = \frac{1}{2}(\sin(25x) + \sin(5x))$. By Eq. (10.23) we get $c_1 = -i/2$, $-c_5 = c_{25} = -i/4$, and $c_8 = 2$. Recall that, for real functions $f(x)$, $c_{-j} = \bar{c}_j$, and $c_{-j} = c_{n-j}$, $j = 1, 2, \ldots, 31$. Since this subroutine calculates $\sum_{k=0}^{N-1} f_k e^{ijx_k}$, our input values are the conjugates $(\bar{f}_0, \bar{f}_1, \ldots, \bar{f}_{n-1})$, and on output we again take the conjugates $(\bar{c}_0, \bar{c}_1, \ldots, \bar{c}_{n-1})$ to get the correct results.

The third program (Fig. 10.12) solves the example at the end of Section 10.2. It uses subroutine NLSYST from Chapter 2.

```
      PROGRAM AER0(INPUT,OUTPUT)
C
C     ------------------------------------------------------------
C
C        THIS PROGRAM SOLVES THE NONLINEAR LEAST SQUARES FIT EXAMPLE
C        THAT WAS GIVEN AT THE BEGINNING OF THIS CHAPTER AND DERIVED
C        IN SECTION 10.2.
C
C     ------------------------------------------------------------
C
C     PARAMETERS ARE:
C             ROVERC:  THE VECTOR CONTAINING X = R/C.
C             VOVERV:  THE VECTOR CONTAINING Y = V/V.
C             NEQUS:   THE NUMBER OF UNKNOWNS. HERE THERE ARE TWO.
C             NDATA:   THE NUMBER OF DATA POINTS. HERE TWENTY ONE.
C
C     SUBROUTINES USED:
C             NLSYST:  SEE CHAPTER 2.
C             ELIM:    CHAPTER 2.
C
C     ------------------------------------------------------------
C
      REAL X(2),F(2),ROVERC(21),VOVERV(21),XTOL,FTOL,DELTA,VALUE
     +    ,TEMP3
      INTEGER NDATA,NEQUS,MAXIT
      EXTERNAL FCN
      COMMON VALUE
      COMMON /BLK1/NDATA,NEQUS,ROVERC,VOVERV
      DATA I,MAXIT/0,50/
      DATA DELTA,XTOL,FTOL,X/0.000125,0.0001,0.0005,0.5,2.0/
C
      CALL NLSYST(FCN,NEQUS,MAXIT,X,F,DELTA,XTOL,FTOL,I)
C
C        WE COMPUTE THE SUM OF THE SQUARE ERROR TERMS
C                    AND PRINT OUT ITS VALUE.
C
C
      PRINT 100, VALUE
100   FORMAT(/60('*')/T9,'THE SUM OF THE SQUARES ERROR TERM IS:'//
     +       T15,F10.6//)
      STOP
      END
```

Figure 10.12 Program 3.

Figure 10.12 (*continued*)

```
C
C       ---------------------------------------------------------------
C
C           HERE WE COMPUTE THE TWO FIRST PARTIALS THAT MUST BE SET
C           EQUAL TO ZER0.
C
C       ---------------------------------------------------------------
C
        SUBROUTINE FCN(X,F)
        REAL X(2),F(2),ROVERC(21),VOVERV(21),   TEMP1,TEMP2,TEMP3,SUMF1,SUMF2
        INTEGER NDATA,NEQUS,I
        COMMON VALUE
        COMMON /BLK1/NDATA,NEQUS,ROVERC,VOVERV
        SUMF1 = 0.0
        SUMF2 = 0.0
        VALUE = 0.0
        DO 10 I = 1 , NDATA
            TEMP1 = 1.0/EXP(X(2)*ROVERC(I)**2)
            TEMP2 = 1.0 - TEMP1
            TEMP3 = VOVERV(I) - X(1)/ROVERC(I)*TEMP2
            SUMF1 =   SUMF1 + TEMP3*TEMP2/ROVERC(I)
            SUMF2 = SUMF2 + X(1)*TEMP3*TEMP1*ROVERC(I)
            VALUE = VALUE + TEMP3**2
10          CONTINUE
        F(1) = SUMF1
        F(2) = SUMF2
        RETURN
        END
C
C       ---------------------------------------------------------------
C
C           WE USE BLOCK DATA TO INITIALIZE THE VARIABLES WHICH ALSO
C           HAVE TO BE IN COMMON /BLK1/.
C
C       ---------------------------------------------------------------
        BLOCK DATA
        INTEGER NDATA,NEQUS
        REAL ROVERC(21),VOVERV(21)
        COMMON /BLK1/NDATA,NEQUS,ROVERC,VOVERV
        DATA ROVERC/0.73,0.78,0.81,0.86,0.875,0.89,0.95,1.02,1.03,1.055,1.135
       +,1.14,1.245,1.32,1.385,1.43,1.445,1.535,1.57,1.63,1.755/
        DATA VOVERV /0.0788,0.0788,0.064,0.0788,0.0681,0.0703,0.0703,0.0681
       +,0.0618,0.079,0.0575,0.0681,0.0575,0.0511,0.0575,0.049,0.0532,0.0511
       +,0.049,0.0532,0.0425/
        DATA NDATA,NEQUS/21,2/
        END
        SUBROUTINE NLSYST(FCN,N,MAXIT,X,F,DELTA,XTOL,FTOL,I)
C
C       ---------------------------------------------------------------
C
C       SUBROUTINE NLSYST :
C                       THIS SUBROUTINE SOLVES A SYSTEM OF N NON-
C       LINEAR EQUATIONS BY NEWTON'S METHOD. THE PARTIAL DERIVATIVES OF
C       THE FUNCTIONS ARE ESTIMATED BY DIFFERENCE QUOTIENTS WHEN A
C       VARIABLE IS PERTURBED BY AN AMOUNT EQUAL TO DELTA ( DELTA IS
C       ADDED ). THIS IS DONE FOR EACH VARIABLE IN EACH FUNCTION.
C       INCREMENTS TO IMPROVE THE ESTIMATES FOR THE X-VALUES ARE COMPU-
C       TED FROM A SYSTEM OF EQUATIONS USING SUBROUTINE ELIM.
C
C       ---------------------------------------------------------------
C
C       PARAMETERS ARE :
C
```

Figure 10.12 (*continued*)

```
C     FCN    - SUBROUTINE THAT COMPUTES VALUES OF THE FUNCTIONS. MUST
C              BE DECLARED EXTERNAL IN THE CALLING PROGRAM.
C     N      - THE NUMBER OF EQUATIONS
C     MAXIT  - LIMIT TO THE NUMBER OF ITERATIONS THAT WILL BE USED
C     X      - ARRAY TO HOLD THE X VALUES. INITIALLY THIS ARRAY HOLDS
C              THE INITIAL GUESSES. IT RETURNS THE FINAL VALUES.
C     F      - AN ARRAY THAT HOLDS VALUES OF THE FUNCTIONS
C     DELTA  - A SMALL VALUE USED TO PERTURB THE X VALUES SO PARTIAL
C              DERIVATIVES CAN BE COMPUTED BY DIFFERENCE QUOTIENT.
C     XTOL   - TOLERANCE VALUE FOR CHANGE IN X VALUES TO STOP ITERA-
C              TIONS. WHEN THE LARGEST CHANGE IN ANY X MEETS XTOL,
C              THE SUBROUTINE TERMINATES.
C     FTOL   - TOLERANCE VALUE ON F TO TERMINATE. WHEN THE LARGEST F
C              VALUE IS LESS THAN FTOL, SUBROUTINE TERMINATES.
C     I      - RETURNS VALUES TO INDICATE HOW THE ROUTINE TERMINATED
C
C        I=1    XTOL WAS MET
C        I=2    FTOL WAS MET
C        I=-1   MAXIT EXCEEDED BUT TOLERANCES NOT MET
C        I=-2   VERY SMALL PIVOT ENCOUNTERED IN GAUSSIAN ELIMINATION
C               STEP - NO RESULTS OBTAINED
C        I=-3   INCORRECT VALUE OF N WAS SUPPLIED - N MUST BE BETWEEN
C               2 AND 10
C
C     ----------------------------------------------------------------
C
      REAL X(N),F(N),DELTA,XTOL,FTOL
      INTEGER N,MAXIT,I
      REAL A(10,11),XSAVE(10),FSAVE(10)
      INTEGER NP,IT,IVBL,ITEST,IFCN,IROW,JCOL
C
C     ----------------------------------------------------------------
C
C     CHECK VALIDITY OF VALUE OF N
C
      IF ( N .LT. 2 .OR. N .GT. 10 ) THEN
        I = -3
        PRINT 1004, N
        RETURN
      END IF
C
C     ----------------------------------------------------------------
C
C     BEGIN ITERATIONS - SAVE X VALUES, THEN GET F VALUES
C
      NP = N + 1
      DO 100 IT = 1,MAXIT
        DO 10 IVBL = 1,N
          XSAVE(IVBL) = X(IVBL)
   10   CONTINUE
        CALL FCN(X,F)
C
C     ----------------------------------------------------------------
C
C     TEST F VALUES AND SAVE THEM
C
        ITEST = 0
        DO 20 IFCN = 1,N
          IF ( ABS(F(IFCN)) .GT. FTOL ) ITEST = ITEST + 1
          FSAVE(IFCN) = F(IFCN)
   20   CONTINUE
        IF ( I .EQ. 0 ) THEN
```

Figure 10.12   (*continued*)

```
               PRINT 1000, IT,X
               PRINT 1001, F
            END IF
C
C     ----------------------------------------------------------------
C
C     SEE IF FTOL IS MET. IF NOT, CONTINUE. IF SO, SET I = 2 AND RETURN.
C
            IF ( ITEST .EQ. 0 ) THEN
               I = 2
               RETURN
            END IF
C
C     ----------------------------------------------------------------
C
C     THIS DOUBLE LOOP COMPUTES THE PARTIAL DERIVATIVES OF EACH FUNCTION
C     FOR EACH VARIABLE AND STORES THEM IN A COEFFICIENT ARRAY.
C
            DO 50 JCOL = 1,N
               X(JCOL) = XSAVE(JCOL) + DELTA
               CALL FCN(X,F)
               DO 40 IROW = 1,N
                  A(IROW,JCOL) = (F(IROW) - FSAVE(IROW)) / DELTA
    40         CONTINUE
C
C     RESET X VALUES FOR NEXT COLUMN OF PARTIALS
C
               X(JCOL) = XSAVE(JCOL)
    50      CONTINUE
C
C     ----------------------------------------------------------------
C
C     NOW WE PUT NEGATIVE OF F VALUES AS RIGHT HAND SIDES AND CALL ELIM
C
            DO 60 IROW = 1,N
               A(IROW,NP) = -FSAVE(IROW)
    60      CONTINUE
            CALL ELIM(A,N,NP,10)
C
C     ----------------------------------------------------------------
C
C     BE SURE THAT THE COEFFICIENT MATRIX IS NOT TOO ILL-CONDITIONED
C
            DO 70 IROW = 1,N
               IF ( ABS(A(IROW,IROW)) .LE. 1.0E-6 ) THEN
                  I = -2
                  PRINT 1003
                  RETURN
               END IF
    70      CONTINUE
C
C     ----------------------------------------------------------------
C
C     APPLY THE CORRECTIONS TO THE X VALUES, ALSO SEE IF XTOL IS MET.
C
            ITEST = 0
            DO 80 IVBL = 1,N
               X(IVBL) = XSAVE(IVBL) + A(IVBL,NP)
               IF ( ABS(A(IVBL,NP)) .GT. XTOL ) ITEST = ITEST + 1
    80      CONTINUE
C
C     ----------------------------------------------------------------
```

Figure 10.12 (*continued*)

```
C
C  IF XTOL IS MET, PRINT LAST VALUES AND RETURN, ELSE DO ANOTHER
C  ITERATION.
C
        IF ( ITEST .EQ. 0 ) THEN
          I = 1
          IF ( I .EQ. 0 ) PRINT 1002, IT,X
          RETURN
        END IF
  100 CONTINUE
C
C  ------------------------------------------------------------------
C
C  WHEN WE HAVE DONE MAXIT ITERATIONS, SET I = -1 AND RETURN
C
      I = -1
      RETURN
C
 1000 FORMAT(/' AFTER ITERATION NUMBER',I3,' X AND F VALUES ARE'
     +        //10F13.5)
 1001 FORMAT(/10F13.5)
 1002 FORMAT(/' AFTER ITERATION NUMBER',I3,' X VALUES (MEETING',
     +        ' XTOL) ARE '//10F13.5)
 1003 FORMAT(/' CANNOT SOLVE SYSTEM. MATRIX NEARLY SINGULAR.')
 1004 FORMAT(/' NUMBER OF EQUATIONS PASSED TO NLSYST IS INVALID.',
     +        ' MUST BE 1 < N < 11. VALUE WAS ',I3)
C
      END
      SUBROUTINE ELIM(AB,N,NP,NDIM)
C
C  ------------------------------------------------------------------
C
C  SUBROUTINE ELIM :
C                    THIS SUBROUTINE SOLVES A SET OF LINEAR EQUATIONS
C  AND GIVES AN LU DECOMPOSITION OF THE COEFFICIENT MATRIX. THE GAUSS
C  ELIMINATION METHOD IS USED, WITH PARTIAL PIVOTING. MULTIPLE RIGHT
C  HAND SIDES ARE PERMITTED, THEY SHOULD BE SUPPLIED AS COLUMNS THAT
C  AUGMENT THE COEFFICIENT MATRIX.
C
C  ------------------------------------------------------------------
C
C  PARAMETERS ARE :
C
C  AB    - COEFFICIENT MATRIX AUGMENTED WITH R.H.S. VECTORS
C  N     - NUMBER OF EQUATIONS
C  NP    - TOTAL NUMBER OF COLUMNS IN THE AUGMENTED MATRIX
C  NDIM  - FIRST DIMENSION OF MATRIX AB IN THE CALLING PROGRAM
C
C  THE SOLUTION VECTOR(S) ARE RETURNED IN THE AUGMENTATION COLUMNS
C  OF AB.
C
C  ------------------------------------------------------------------
C
      REAL AB(NDIM,NP)
      INTEGER N,NP,NDIM
      REAL SAVE,RATIO,VALUE
      INTEGER NM1,IPVT,IP1,J,NVBL,L,KCOL,JCOL,JROW
C
C  ------------------------------------------------------------------
C
C  BEGIN THE REDUCTION
C
```

Figure 10.12 (*continued*)

```
      NM1 = N - 1
      DO 35 I = 1,NM1
C
C  FIND THE ROW NUMBER OF THE PIVOT ROW. WE WILL THEN INTERCHANGE ROWS TO
C  PUT THE PIVOT ELEMENT ON THE DIAGONAL.
C
         IPVT = I
         IP1 = I + 1
         DO 10 J = IP1,N
            IF ( ABS(AB(IPVT,I)) .LT. ABS(AB(J,I)) ) IPVT=J
  10     CONTINUE
C
C  -----------------------------------------------------------------------
C
C  CHECK FOR A NEAR SINGULAR MATRIX
C
         IF ( ABS(AB(IPVT,I)) .LT. 1.0 E-6 ) THEN
            PRINT 100
            RETURN
         END IF
C
C  NOW INTERCHANGE, EXCEPT IF THE PIVOT ELEMENT IS ALREADY ON THE
C  DIAGONAL, DON'T NEED TO.
C
         IF ( IPVT .NE. I ) THEN
            DO 20 JCOL = 1,NP
               SAVE = AB(I,JCOL)
               AB(I,JCOL) = AB(IPVT,JCOL)
               AB(IPVT,JCOL) = SAVE
  20        CONTINUE
         END IF
C
C  -----------------------------------------------------------------------
C
C  NOW REDUCE ALL ELEMENTS BELOW THE DIAGONAL IN THE I-TH ROW. CHECK FIRST
C  TO SEE IF A ZERO ALREADY PRESENT. IF SO, CAN SKIP REDUCTION ON THAT ROW.
C
         DO 32 JROW = IP1,N
            IF ( AB(JROW,I) .EQ. 0 ) GO TO 32
               RATIO = AB(JROW,I) / AB(I,I)
               AB(JROW,I) = RATIO
               DO 30 KCOL = IP1,NP
                  AB(JROW,KCOL) = AB(JROW,KCOL) - RATIO*AB(I,KCOL)
  30           CONTINUE
  32     CONTINUE
  35 CONTINUE
C
C  -----------------------------------------------------------------------
C
C  WE STILL NEED TO CHECK AB(N,N) FOR SIZE
C
      IF ( ABS(AB(N,N)) .LT. 1.0 E-6 ) THEN
         PRINT 100
         RETURN
      END IF
C
C  -----------------------------------------------------------------------
C
C  NOW WE BACK SUBSTITUTE
C
      NP1 = N + 1
      DO 50 KCOL = NP1,NP
```

Figure 10.12   (*continued*)

```
          AB(N,KCOL) = AB(N,KCOL) / AB(N,N)
          DO 45 J = 2,N
            NVBL = NP1 - J
            L = NVBL + 1
            VALUE = AB(NVBL,KCOL)
            DO 40 K = L,N
              VALUE = VALUE - AB(NVBL,K) * AB(K,KCOL)
   40       CONTINUE
            AB(NVBL,KCOL) = VALUE / AB(NVBL,NVBL)
   45     CONTINUE
   50 CONTINUE
      RETURN
C
  100 FORMAT(/' SOLUTION NOT FEASIBLE. A NEAR ZERO PIVOT ',
     +        'WAS ENCOUNTERED.')
      END



          OUTPUT FOR PROGRAM 3

AFTER ITERATION NUMBER  1 X AND F VALUES ARE

     .50000      2.00000

   -5.58847       -.40348

AFTER ITERATION NUMBER  2 X AND F VALUES ARE

     .16954      1.32113

    -.78095       -.03873

AFTER ITERATION NUMBER  3 X AND F VALUES ARE

     .11395      1.01639

    -.09659       -.00289

AFTER ITERATION NUMBER  4 X AND F VALUES ARE

     .12644       .58295

     .04134       .00989

AFTER ITERATION NUMBER  5 X AND F VALUES ARE

     .11310       .77840

     .01645       .00448

AFTER ITERATION NUMBER  6 X AND F VALUES ARE

     .08925      1.13874

     .06555       .00379

AFTER ITERATION NUMBER  7 X AND F VALUES ARE

     .08899      1.39420

     .00211       .00098
```

Figure 10.12 (*continued*)

```
AFTER ITERATION NUMBER  8 X AND F VALUES ARE
       .07895      1.82196
       .03043       .00071
AFTER ITERATION NUMBER  9 X AND F VALUES ARE
       .07851      2.04292
       .00222       .00015
AFTER ITERATION NUMBER 10 X AND F VALUES ARE
       .07651      2.24423
       .00288       .00005
AFTER ITERATION NUMBER 11 X AND F VALUES ARE
       .07618      2.30574
       .00019       .00000
************************************************************
       THE SUM OF THE SQUARES ERROR TERM IS:
           .000631
```

EXERCISES

Section 10.1

1. Find the individual deviations of the data in Fig. 10.1 from those computed from the least-squares line, $R = 702.2 + 3.395T$. Compare these deviations with those from the line drawn by eye, $R = 700 + 3.500T$. Find the sum of squares of deviations in each case and compare. Note that even though the maximum deviations from each of the two lines are not too different, the sums of squares differ significantly.

▶ 2. Show that the point whose x-coordinate is the mean of all the x-values and whose y-coordinate is the mean of all the y-values satisfies the least-squares line. Often a change of variable is made to relocate the origin at this point, with a corresponding reduction in the magnitude of the numbers worked with, making them more readily handled by hand or on some desk calculators.

▶ 3. Find the least-squares line that fits to the following data, assuming the x-values are free of error:

x	y	x	y
1	2.04	4	7.18
2	4.12	5	9.20
3	5.64	6	12.04

(The data are tabulated from $y = 2x$, with perturbations from a table of random numbers.)

4. In the data for Exercise 3, consider that the y-values are free of error and that all the errors are in the x-values. By suitable modifications of the normal equations, now determine the least-squares line for $x = ay + b$. Note that this is not the same line as determined in Exercise 3.

5. In multivariate analysis, the least-squares technique is used to determine the hyperplane from which the sum of the squares of the deviations is a minimum. For z, a function of two independent variables, x and y, find the normal equations to determine the parameters a, b, and c in:

    $$z = ax + by + c.$$

    Use this relation to find the least-squares values for the constants if the data below are available.

x	0	1.2	2.1	3.4	4.0	4.2	5.6	5.8	6.9
y	0	0.5	6.0	0.5	5.1	3.2	1.3	7.4	10.2
z	1.2	3.4	−4.6	9.9	2.4	7.2	14.3	3.5	1.3

Section 10.2

6. Observe that the data in Table 10.8 seem to be fit by a curve $y = ae^{bx}$ by plotting on semilog paper and noting that the points then fall near a straight line. The data are for the solubilities of n-butane in anhydrous hydrofluoric acid at high pressures, and were needed in the design of petroleum refineries.

Table 10.8

Temperature, °F	Solubility, weight %
77	2.4
100	3.4
185	7.0
239	11.1
285	19.6

By plotting on rectilinear graph paper, observe that the relationship is nonlinear.

▶ 7. Determine the constants for $y = ae^{bx}$ for the data in Exercise 6 by the least-squares method by fitting to the relation $\ln y = \ln a + bx$.

8. It is suspected (from theoretical considerations) that the rate of flow from a fire hose is proportional to some power of the pressure at the nozzle. Determine whether the speculation seems to be true, and what the exponent is from the data in Table 10.9. (Assume that the pressure data are more accurate.)

Table 10.9

Flow, gallons per minute	Pressure, psi
94	10
118	16
147	25
180	40
230	60

9. Since the data of Exercise 8, when plotted on log-log paper (pressure as a function of flow), seem to have a slope of nearly 2, we should expect that fitting a quadratic to the data would be successful. Do this and compare the deviations with those from the power-function relation of Exercise 8.

▶10. The data given in Exercise 3, while perturbations from a linear relation, seem to plot better along a curve because of the accidental occurrence of three negative deviations in succession. Fit a quadratic to the data.

11. To compare the results of an exact polynomial fit according to Chapter 3 with the least-squares procedure, find y-values at $x = 1.5, 2.5, 3.5, 4.5$, and 5.5 for the data of Exercise 3, utilizing a fifth-degree interpolating polynomial. Sketch the interpolating polynomial and compare to the least-squares line of Exercise 3, and the least-squares quadratic of Exercise 10. The slope of the "true" function is 2.0. How do the maximum and minimum values of the slope as determined from the three approximations (use a graphical procedure) compare to the "true" value?

▶12. The data in Table 10.10 seem to fit a cubic equation, but determine by least squares the optimum degree of polynomial.

Table 10.10

x	$f(x)$	x	$f(x)$	x	$f(x)$
0.1	1.9	6.6	36.1	13.2	−48.6
1.1	7.9	7.1	23.7	14.1	−40.2
1.6	24.9	8.2	13.0	15.6	−51.6
2.4	24.9	9.1	20.5	16.1	−30.5
2.5	34.9	9.4	−3.1	17.6	−34.6
4.1	42.7	11.1	−13.0	17.9	−16.4
5.2	29.7	11.4	−28.7	19.1	−13.8
6.1	49.8	12.2	−39.5	20.0	−1.1

13. Repeat Exercise 12, but now using every other point ($x = 0.1$, $x = 1.6$, $x = 2.5$, and so on). Do you get the same result for the optimum degree of polynomial? Are the coefficients the same? Repeat again, with the other half of the points.

14. The data of Exercise 12 also suggest a function of the form $y = A + B \sin Cx$. How could the least-squares method be used to evaluate A, B, and C? In solving the normal equations, are any difficulties encountered? What if theory suggested that $C = \pi/10$? Would this make any significant difference in solving the normal equations?

Section 10.3

15. Compute $T_{11}(x)$ and $T_{12}(x)$.

▶16. Show that Eq. (10.9) is satisfied for these values of (m, n): $(0, 1)$, $(1, 1)$, $(0, 0)$, $(1, 2)$.

17. a) Graph $T_5(x)$ on $[-1, 1]$.
    b) Extend the graphs of $T_1(x)$, $T_2(x)$, $T_3(x)$ to the interval $[-2, 2]$. Observe that the maximum magnitude of the Chebyshev polynomials is not equal to one outside of $[-1, 1]$.

Section 10.4

18. Reduce Eq. (10.11) to a fourth-degree economized polynomial and show that its maximum error is 0.000791.

19. The error curve for a truncated Maclaurin-series approximation increases monotonically as x varies from 0 to the ends of the interval. This is not true for an economized power series. Exhibit the form of the error curve by plotting the error of Eq. (10.11) on the interval [0, 1].

▶20. Given

$$\arctan x = x - \frac{x^3}{3} + \frac{x^5}{5} - \frac{x^7}{7} + \frac{x^9}{9} - \cdots .$$

Plot the error over the interval $[-1, 1]$ when the above series is truncated after the term in x^7. Economize the ninth-degree truncated power series three times (giving a third-degree expression), and plot its error over the interval $[-1, 1]$.

21. Find the first few terms of the Chebyshev series for $\sin x$ by rewriting the Maclaurin series in terms of $T_i(x)$ and collecting terms. Express this as a power series. Compare the errors when both series are truncated to third-degree polynomials.

22. A power-series expansion of $(1 + x/5)^{1/2}$ is

$$1 + \frac{x}{10} - \frac{x^2}{200} + \frac{x^3}{2000} - \frac{x^5}{16,000} + \frac{x^7}{800,000} - \cdots .$$

Convert this to a Chebyshev series including terms to $T_2(x)$. What is the maximum error of the truncated Chebyshev series? Compare this to the error of the power series truncated after the x^2 term.

23. The Chebyshev series, and economized polynomials as well, require us to approximate the function on the interval $[-1, 1]$ only. Show that an appropriate change of variable will change $f(x)$ on $[a, b]$ to $f(y)$ on $[-1, 1]$. Find the linear relation between x and y that makes this transformation.

Section 10.5

▶24. Find Padé approximations to the following functions, with numerators and denominators each of degree three:

a) $\sin x$;   b) $\cos x$;   ▶c) e^x.

25. Compare the errors on the interval $[-1, 1]$ for the Padé approximations of Exercise 24 with the errors of the corresponding Maclaurin series.

26. Express these rational functions in continued-fraction form:

a) $\dfrac{x^2 - 2x + 2}{x^2 + 2x - 2}$ b) $\dfrac{2x^2 + x^2 + x + 3}{x^2 - x - 4}$

c) $\dfrac{2x^4 + 45x^3 + 381x^2 + 1353x + 1511}{x^3 + 21x^2 + 157x + 409}$.

In each case, compare the number of multiplication and division operations in continued fraction form with initial evaluation of the polynomials by nested multiplication.

27. Express the Padé approximations of Exercise 24 as continued fractions.

28. Estimate the errors of the Padé approximations of Exercise 24 by computing the coefficient of the next nonzero term in the numerator. Compare these estimates with the actual errors at $x = 1$, and at $x = -1$.

▶29. The Chebyshev series for $\cos(\pi x/4)$ is

$$0.851632 - 0.146437T_2 + 0.00192145T_4 - 9.965 \times 10^{-6}T_6.$$

Develop a Padé-like rational function from this by the method of Section 10.5, where the function is $R_{4,2}$.

30. Fike (1968) gives this example of a rational fraction approximation to $\Gamma(1 + x)$ on [0, 1]:

$$R_{3,4}(x) = \frac{0.999999 + 0.601781x + 0.186145x^2 + 0.0687440x^3}{1 + 1.17899x - 0.122321x^2 - 0.260996x^3 + 0.0609927x^4}.$$

How could you determine if this is a minimax approximation? Could one find bounds to the error of the equivalent minimax $R_{3,4}(x)$ if it is not?

Section 10.6

31. Verify the results of the first example of Section 10.6.

32. Show that for $N = 16$ equispaced data points, one can find the $\hat{c}_j$'s by 64 multiplications rather than the expected 256.

33. Find the c_j's for the "sawtooth" function, $f(x) = x$ on $[0, \pi]$ and $f(x) = 2\pi - x$ on $[\pi, 2\pi]$, Eq. (10.24).

34. Use FFTCC or some other subroutine to compute the results in Exercise 31.

35. Show that for N given points, (x_k, f_k), $k = 0, 1, \ldots, N - 1$ and $x_k = 2\pi k/N$, then $\hat{c}_j = \hat{c}_{j+N}$. In fact, show that for any multiple of N, say M, $\hat{c}_j = \hat{c}_{j+M}$.

36. For a periodic function $f(x)$ to be represented by a Fourier series, the following conditions are sometimes given:

a) The function has at most a finite number of discontinuities in any period.
b) The function must contain only a finite number of maxima and minima in any period.
c) The function must be absolutely integrable.

$$\int_0^L |f(x)|\, dx < \infty, \text{ where } L \text{ is the period of the function.}$$

Which of the following functions satisfy or do not satisfy these conditions?

$$i)\ f(x) = \begin{cases} \sin \dfrac{1}{x} & (0, \pi] \\ 0 & x = 0 \end{cases}$$

$$ii)\ f(x) = \begin{cases} e^{-1/t^2} & (0, 1] \\ 0 & x = 0 \end{cases}$$

$$iii)\ f(x) = \begin{cases} 1 & [0, 0.5) \\ -1 & [0.5, 1) \end{cases}$$

$$iv)\ f(x) = \begin{cases} x^2 \sin \dfrac{1}{x} & (0, \pi] \\ 0 & x = 0 \end{cases}$$

APPLIED PROBLEMS AND PROJECTS

37. In many situations, experimental data fit to an exponential relation: $f(x) = ae^{bx}$. To determine the constants through least squares, one frequently fits the function $\ln f = \ln a + bx$ to the data by least squares. In this manner one can find the proper values for the constants b and $\ln a$ quite readily, for the log relation is linear in these terms. However, we might wish to determine the constants a and b of the original function, without changing it to a logarithm. Develop a procedure for doing this, which will require solving a set of nonlinear normal equations. Use the data in Exercise 6 to test your procedure; then compare your values for a and b with those obtained in Exercise 7.

38. Penrod and Prasanna (1962) measured the total sun energy reaching their test site in Lexington, Ky., during the winter months. The insolation was recorded both for a vertical receiver and for one tilted at an angle of 25°45′. Fit the data by least-squares polynomials.

Month	Sept.	Oct.	Nov.	Dec.	Jan.	Feb.	Mar.	Apr.
I_{Tilt}	2177	1952	1545	1215	1140	1523	1615	1713
I_{Vert}	1484	1523	1290	1041	963	1208	1143	1043

39. Look up the algorithms used on your computer to compute the functions built into FORTRAN. Classify them into Taylor-series formulas, Chebyshev-polynomial formulas, rational functions, and other types. Which of them are minimax? If more than one computer system is accessible to you, compare the different systems. How does BASIC compute these functions?

40. Write a computer program to get the $\hat{c}_i$ for a discrete Fourier transform by FFT for N equal to a power of 2.

Appendixes

APPENDIX A

Some Basic Information from Calculus

Since a number of results and theorems from the calculus are frequently used in the text, we collect here a number of these items for ready reference, and to refresh the student's memory.

OPEN AND CLOSED INTERVALS

We use for the open interval $a < x < b$, the notation (a, b), and for the closed interval $a \leq x \leq b$, the notation $[a, b]$.

CONTINUOUS FUNCTIONS

If a real-valued function is defined on the interval (a, b), it is said to be continuous at a point x_0 in that interval if for every $\varepsilon > 0$ there exists a positive nonzero number δ such that $|f(x) - f(x_0)| < \varepsilon$ whenever $|x - x_0| < \delta$ and $a < x < b$. In simple terms, we can meet any criterion of matching the value of $f(x_0)$ (the criterion is the quantity ε) by choosing x near enough to x_0, without having to make x equal to x_0, when the function is continuous.

If a function is continuous for all x-values in an interval, it is said to be continuous on the interval. A function that is continuous on a closed interval $[a, b]$ will assume a maximum value and a minimum value at points in the interval (perhaps the endpoints). It will also assume any value between the maximum and the minimum at some point in the interval.

Similar statements can be made about a function of two or more variables. We then refer to a domain in the space of the several variables instead of to an interval.

SUMS OF VALUES OF CONTINUOUS FUNCTIONS

When x is in $[a, b]$, the value of a continuous function $f(x)$ must be no greater than the maximum and no less than the minimum value of $f(x)$ on $[a, b]$. The sum of n such values must be bounded by $(n)(m)$ and $(n)(M)$, where m and M are the minimum and maximum values. Consequently the sum is n times some intermediate value of the function. Hence

$$\sum_{i=1}^{n} f(\xi_i) = nf(\xi) \quad \text{if } a \leq \xi_i \leq b, \quad i = 1, 2, \ldots, n, \quad a \leq \xi \leq b.$$

Similarly, it is obvious that

$$c_1 f(\xi_1) + c_2 f(\xi_2) = (c_1 + c_2) f(\xi), \quad \xi_1, \xi_2, \xi \text{ in } [a, b],$$

for the continuous function f when c_1 and c_2 are both equal to or greater than one. If the coefficients are positive fractions, dividing by the smaller gives

$$c_1 f(\xi_1) + c_2 f(\xi_2) = c_1 \left[f(\xi_1) + \frac{c_2}{c_1} f(\xi_2) \right] = c_1 \left(1 + \frac{c_2}{c_1} \right) f(\xi) = (c_1 + c_2) f(\xi),$$

so the rule holds for fractions as well. If c_1 and c_2 are of unlike sign, this rule does not hold unless the values of $f(\xi_1)$ and $f(\xi_2)$ are narrowly restricted.

MEAN-VALUE THEOREM FOR DERIVATIVES

When $f(x)$ is continuous on the closed interval $[a, b]$, then at some point ξ in the interior of the interval

$$f'(\xi) = \frac{f(b) - f(a)}{b - a}, \quad a < \xi < b,$$

provided, of course, that $f'(x)$ exists at all interior points. Geometrically this means that the curve has at one or more interior points a tangent parallel to the secant line connecting the ends of the curve (Fig. A.1).

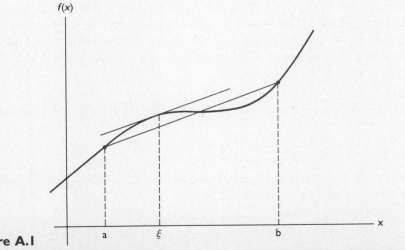

Figure A.1

MEAN-VALUE THEOREMS FOR INTEGRALS

If $f(x)$ is continuous and integrable on $[a, b]$, then

$$\int_a^b f(x)\ dx = (b - a)f(\xi), \qquad a < \xi < b.$$

This says, in effect, that the value of the integral is an average value of the function times the length of the interval. Since the average value lies between the maximum and minimum values, there is some point ξ at which $f(x)$ assumes this average value.

If $f(x)$ and $g(x)$ are continuous and integrable on $[a, b]$, and if $g(x)$ does not change sign on $[a, b]$, then

$$\int_a^b f(x)g(x)\ dx = f(\xi) \int_a^b g(x)\ dx, \qquad a < \xi < b.$$

Note that the previous statement is a special case $[g(x) = 1]$ of this last theorem, which is called the *second theorem of the mean for integrals.*

TAYLOR SERIES

If a function $f(x)$ can be represented by a power series on the interval $(-a, a)$, then the function has derivatives of all orders on that interval and the power series is

$$f(x) = f(0) + f'(0)x + \frac{f''(0)}{2!}x^2 + \frac{f'''(0)}{3!}x^3 + \cdots .$$

The preceding power-series expansion of $f(x)$ about the origin is called a *Maclaurin* series. Note that if the series exists, it is unique and any method of developing the coefficients gives this same series.

If the expansion is about the point $x = a$, we have the Taylor series

$$f(x) = f(a) + f'(a)(x - a) + \frac{f''(a)}{2!}(x - a)^2 + \frac{f'''(a)}{3!}(x - a)^3 + \cdots .$$

We frequently represent a function by a polynomial approximation, which we can regard as a truncated Taylor series. Usually we cannot represent a function exactly by this means, so we are interested in the error. Taylor's formula with a remainder gives us the error term. The remainder term is usually derived in elementary calculus texts in the form of an integral:

$$f(x) = f(a) + f'(a)(x - a) + \frac{f''(a)}{2!}(x - a)^2 + \cdots + \frac{f^{(n)}(a)}{n!}(x - a)^n$$

$$+ \int_a^x \frac{(x - t)^n}{n!} f^{(n+1)}(t)\ dt.$$

Since $(x - t)$ does not change sign as t varies from a to x, the second theorem of the mean allows us to write the remainder term as

$$\text{Remainder of Taylor series} = \frac{(x - a)^{n+1}}{(n + 1)!} f^{(n+1)}(\xi), \qquad \xi \text{ in } [a, x].$$

The derivative form is the more useful for our purposes. It is occasionally useful to express a Taylor series in a notation that shows how the function behaves at a distance h from a fixed point a. If we call $x = a + h$ in the above, so $x - a = h$, we get

$$f(a + h) = f(a) + f'(a)h + \frac{f''(a)}{2!}h^2 + \cdots + \frac{f^{(n)}(a)}{n!}h^n + \frac{f^{(n+1)}(\xi)}{(n+1)!}h^{n+1}.$$

TAYLOR SERIES FOR FUNCTIONS OF TWO VARIABLES

For a function of two variables, $f(x, y)$, the rate of change of the function can be due to changes in either x or y. The derivatives of f can be expressed in terms of the partial derivatives. For the expansion in the neighborhood of the point (a, b),

$$f(x, y) = f(a, b) + f_x(a, b)(x - a) + f_y(a, b)(y - b)$$
$$+ \frac{1}{2!}[f_{xx}(a, b)(x - a)^2 + 2f_{xy}(a, b)(x - a)(y - b) + f_{yy}(a, b)(y - b)^2]$$
$$+ \cdots .$$

DESCARTES' RULE OF SIGNS

Let $p(x)$ be a polynomial with real coefficients and consider the equation $p(x) = 0$. Then

1. The number of positive real roots is equal to the number of variations in the signs of the coefficients of $p(x)$ or is less than that number by an even integer.

2. The number of negative real roots is determined the same way, but for $p(-x)$. Here also the number of negative roots is equal to the number of variations in the signs of the coefficients of $p(-x)$ or is less than that number by an even integer.

For example, the polynomial $p(x) = 3x^5 - 2x^4 + 7x^2 - 12 = 0$ will have 3 or 1 positive and 2 or 0 negative real roots.

APPENDIX B

Deriving Formulas by the Method of Undetermined Coefficients

In Chapter 3 we developed formulas for integration and differentiation of functions by replacing the actual function with a polynomial that agrees at a number of points (a so-called *interpolating polynomial*), and then integrating or differentiating the polynomial. These formulas are valuable if one wishes to write computer programs for integration and differentiation, but the most important use is for solving differential equations numerically. Because computers calculate their functional values rather than interpolating in a table, there is less interest today in interpolating polynomials than in earlier times. Hence, there is reason to present an alternative method of deriving the formulas for derivatives and integrals that are needed to solve differential equations.

We will call the method that we employ in this appendix the *method of undetermined coefficients*. Basically, we impose certain conditions on a formula of desired form and use these conditions to determine values of the unknown coefficients in the formula. Hamming (1962) presents the method in considerable detail.

B.1 DERIVATIVE FORMULAS BY THE METHOD OF UNDETERMINED COEFFICIENTS

Since the derivative of a function is the rate of change of the function relative to changes in the independent variable, we should expect that formulas for the derivative would involve differences between function values in the neighborhood of the point where we wish to evaluate the derivative. It is, in fact, possible to approximate the derivative as a linear combination of such function values. While one can argue that formulas of greatest accuracy will use function values very near to the point in question, the formulas important in practice impose the restriction that only function values at equally spaced x-values are to be used.

For example, we can write a formula for the first derivative in terms of $n + 1$ equispaced points:

$$f'(x_0) = c_0 f(x_0) + c_1 f(x_1) + c_2 f(x_2) + \cdots + c_n f(x_n),$$
$$x_{i+1} - x_i = h = \text{constant}. \tag{B.1}$$

The more terms we employ, the greater accuracy we shall expect, since more information about the function is being fed in. We will evaluate the coefficients in the equation by requiring that the formula be exact whenever the function is a polynomial of degree n or less.* (We will find, throughout this chapter, that the method of undetermined coefficients uses this criterion to develop formulas. It has validity because any function that is continuous on an interval can be approximated to any specified precision by a polynomial of sufficiently high degree. Using polynomials to replace the function also greatly simplifies the work, in contrast to replacing with other functions.)

Let us illustrate the method of undetermined coefficients with a simple case. We shall simplify the notation by defining

$$f_i = f(x_i).$$

If we write the derivative in terms of only two function values, we would have

$$f_0' = c_0 f_0 + c_1 f_1. \tag{B.2}$$

We require that the formula be exact if $f(x)$ is a polynomial of degree 1 or less. (The maximum degree is always one less than the number of undetermined constants.) Hence, since the formula is to be exact if $f(x)$ is any first-degree polynomial, it must be exact if either $f(x) = x$ or $f(x) = 1$, for these definitions of $f(x)$ are just two special cases of the general first-degree function $ax + b$. We write Eq. (B.2) for both these cases:

$$f(x) = x, \quad f'(x) = 1, \quad 1 = c_0(x_0) + c_1(x_0 + h),$$
$$f(x) = 1, \quad f'(x) = 0, \quad 0 = c_0(1) + c_1(1). \tag{B.3}$$

We solve Eqs. (B.3) simultaneously to get $c_0 = -1/h$, $c_1 = 1/h$. Consequently,

$$f'(x_0) \doteq \frac{f_1 - f_0}{h}. \tag{B.4}$$

The dot over the equal sign in Eq. (B.4) is to remind us that the formula is only an approximation. It is exact if $f(x)$ is a polynomial of degree 1, but not exact if a polynomial of higher degree, or some transcendental function.

---

*Intuitively, it seems reasonable that we can satisfy this criterion, for a polynomial of degree n is determined uniquely by its $n + 1$ coefficients and our formula contains $n + 1$ constants.

Similarly, a three-term formula can be derived by replacing the function with x^2 and x and 1 in

$$f_0' = c_0 f_0 + c_1 f_1 + c_2 f_2.$$

The set of equations to solve is

$$2x_0 = c_0(x_0)^2 + c_1(x_0 + h)^2 + c_2(x_0 + 2h)^2,$$
$$1 = c_0(x_0) + c_1(x_0 + h) + c_2(x_0 + 2h),$$
$$0 = c_0(1) + c_1(1) + c_2(1). \tag{B.5}$$

The arithmetic is simplified by letting $x_0 = 0$. That this is valid is readily seen. Imagine the graph of $f(x)$ versus x. The derivative we desire is the slope of the curve at the point where $x = x_0$, which obviously is unchanged by a translation of the axes. Taking $x_0 = 0$ is the equivalent of a translation of axes so that the origin corresponds to x_0. Equations (B.5) become, with this change,

$$0 = c_0(0) + c_1(h)^2 + c_2(2h)^2,$$
$$1 = c_0(0) + c_1(h) + c_2(2h),$$
$$0 = c_0(1) + c_1(1) + c_2(1).$$

Solving, we get $c_0 = -3/2h$, $c_1 = 2/h$, $c_2 = -1/2h$; so a three-term formula for the derivative is

$$f_0' \doteq \frac{-3f_0 + 4f_1 - f_2}{2h}. \tag{B.6}$$

We could extend these formulas to include more and more terms, but after a little reflection we can conclude that the original form, Eq. (B.1), is not the best to use. It utilizes only functional values to one side of the point in question, while it would be better to utilize information from both sides of x. After all, the limit in the definition of the derivative is two-sided, and further, we utilize closer and hence more pertinent information by going to both the left and the right.

We expect an improvement over Eq. (B.6) if we begin with

$$f_0' = c_{-1} f_{-1} + c_0 f_0 + c_1 f_1, \tag{B.7}$$

where $f_{-1} = f(x_0 - h)$. Adopting the simplification of letting $x_0 = 0$ as before, and taking x^2, x, and 1 for $f(x)$, we have to solve

$$0 = c_{-1}(-h)^2 + c_0(0) + c_1(h)^2,$$
$$1 = c_{-1}(-h) + c_0(0) + c_1(h),$$
$$0 = c_{-1}(1) + c_0(1) + c_1(1). \tag{B.8}$$

Completing the algebra, we get $c_{-1} = -1/2h$, $c_0 = 0$, $c_1 = 1/2h$.

The following equation is particularly important:

$$f_0' \doteq \frac{f_1 - f_{-1}}{2h}. \tag{B.9}$$

In the next section we will compare its accuracy with Eqs. (B.4) and (B.6). Because the point where the derivative is evaluated is centered among the function values whose differences appear in the formula, it is called a *central-difference approximation*. Higher-order central difference approximations to the first derivative, utilizing five or seven or more points, can be derived by this same procedure.

We now apply the method of undetermined coefficients to higher derivatives. We will discuss only the central difference approximations since they are the more widely used. In terms of values both to the right and left of x_0,

$$f_0'' = c_{-1}f_{-1} + c_0 f_0 + c_1 f_1.$$

Taking $f(x) = x^2$, x, and 1, we get the relations ($x_0 = 0$),

$$2 = c_{-1}(-h)^2 + c_0(0) + c_1(h)^2,$$
$$0 = c_{-1}(-h) + c_0(0) + c_1(h),$$
$$0 = c_{-1}(1) + c_0(1) + c_1(1).$$

The resulting formula is

$$f_0'' \doteq \frac{f_{-1} - 2f_0 + f_1}{h^2}. \tag{B.10}$$

Equation (B.10), like Eq. (B.9), is particularly useful.

We are not confined to using only functional values in the method of undetermined coefficients.* Suppose we have values of the first derivative as well as functional values. A formula for the second derivative might then be written as

$$f_0'' = c_0 f_0 + c_1 f_1 + c_3 f_0'.$$

As before, we take x^2, x, and 1 for $f(x)$, with $x_0 = 0$:

$$2 = c_0(0)^2 + c_1(h)^2 + c_3(0),$$
$$0 = c_0(0) + c_1(h) + c_3(1),$$
$$0 = c_0(1) + c_1(1) + c_3(0).$$

Solving, we obtain

$$f_0'' = 2\frac{f_1 - f_0 - hf_0'}{h^2}.$$

In the same way, we can derive formulas for the third and fourth derivatives. Derivatives beyond these do not often appear in applied problems. We must use a minimum of four and five terms, however, since only polynomials of degree 3 and 4 have nonzero third or fourth derivatives. For the third-degree formula, complete symmetry with four

---

*We must be sure that the set of values is sufficient to determine a polynomial uniquely, however. For example, f_{-1}, f_1, and f_0' will not give a formula for f_0''.

points is impossible; five-term formulas for both derivatives are therefore given. We present the results only, leaving the derivations as an exercise:

$$f_0''' = \frac{-f_{-2} + 2f_{-1} - 2f_1 + f_2}{2h^3}, \tag{B.11}$$

$$f_0^{iv} = \frac{f_{-2} - 4f_{-1} + 6f_0 - 4f_1 + f_2}{h^4}. \tag{B.12}$$

B.2 ERROR TERMS FOR DERIVATIVE FORMULAS

In the previous section, we used the method of undetermined coefficients to derive several formulas for the first derivative of a function, utilizing function values at equispaced x-values. These were, using the notation that $f_i = f(x_i)$,

$$f_0' \doteq \frac{f_1 - f_0}{h}, \tag{B.13}$$

$$f_0' \doteq \frac{-3f_0 + 4f_1 - f_2}{2h}, \tag{B.14}$$

$$f_0' \doteq \frac{f_1 - f_{-1}}{2h}. \tag{B.15}$$

While each of these formulas is not exact, as suggested by the $\doteq$ symbol, we argued heuristically that the error should decrease in each succeeding one. We now wish to develop expressions for the errors.

Begin with the Taylor-series expansion of $f(x_1) = f(x_0 + h)$ in terms of $x_1 - x_0 = h$:

$$f(x_1) = f(x_0) + hf'(x_0) + \frac{1}{2}h^2f''(x_0) + \frac{1}{6}h^3f'''(x_0) + \cdots.$$

Changing to our subscript notation, and truncating after the term in h, with the usual error term, we have

$$f_1 = f_0 + hf_0' + \frac{1}{2}h^2f''(\xi), \qquad x_0 < \xi < x_0 + h. \tag{B.16}$$

Solving for f_0', we have

$$f_0' = \frac{f_1 - f_0}{h} - \frac{1}{2}hf''(\xi), \qquad x_0 < \xi < x_0 + h, \tag{B.17}$$

so that the error term is $-\frac{1}{2}hf''(\xi)$ for the derivative formula, Eq. (B.13).

In Eq. (B.17), the error term involves the second derivative of the function evaluated at a place which is known only within a certain interval. In the majority of applications for numerical differentiation, not only is the point of evaluation uncertain, but the function $f(x)$ is also unknown. If we do not know $f(x)$, we can hardly expect to know its derivatives.

We do know, however, that the error involves the first power of $h = x_{i+1} - x_i$, and, in fact, the only way we can change the error is to change h. Making h smaller will decrease h, and in the limit as $h \to 0$ the error will go to zero. Further, as $h \to 0$, $x_1 \to x_0$, and the value of ξ is squeezed into a smaller and smaller interval. In other words, $f''(\xi)$ approaches a fixed value, specifically $f''(x_0)$, as h goes to zero.

The special importance of h in the error term is denoted by a special notation in numerical analysis, the *order relation*. We say the error of Eq. (B.17) is "of order h" and write

$$\text{Error} = O(h), \qquad \text{when} \qquad \lim_{h \to 0} (\text{error}) = ch,$$

where c is a fixed value not equal to zero.

To develop the error term for Eq. (B.14), we proceed similarly, except that we write expansions for both f_1 and f_2. It is also necessary to carry terms through h^3 because the second derivatives cancel, resulting in an error term involving the third derivative:

$$f_1 = f_0 + hf_0' + \frac{1}{2}h^2 f_0'' + \frac{1}{6}h^3 f'''(\xi_1), \qquad x_0 < \xi_1 < x_0 + h;$$

$$f_2 = f_0 + 2hf_0' + \frac{1}{2}(2h)^2 f_0'' + \frac{1}{6}(2h)^3 f'''(\xi_2), \qquad x_0 < \xi_2 < x_0 + 2h.$$

Note that the values of ξ_1 and ξ_2 may not be identical. If we multiply the first equation by 4, the second by -1, and add $-3f_0$ to their sum, we get

$$-3f_0 + 4f_1 - f_2 = (-3 + 4 - 1)f_0 + 2hf_0' + (2 - 2)f_0'' + \frac{4h^3}{6}[f'''(\xi_1) - 2f'''(\xi_2)].$$

Solving for f_0', we get

$$f_0' = \frac{-3f_0 + 4f_1 - f_2}{2h} + \frac{1}{3}h^2[2f'''(\xi_2) - f'''(\xi_1)],$$

$$x_0 < \xi_1 < x_0 + h, \qquad x_0 < \xi_2 < x_0 + 2h. \tag{B.18}$$

The last term of Eq. (B.18) is the error term. As $h \to 0$, the two values of ξ approach the same values. Consequently the error term approaches $\frac{1}{3}h^2 f'''(x_0)$. We then conclude that the error of Eq. (B.14) is $O(h^2)$.

For the error of Eq. (B.15), we proceeded similarly. We leave the development as an exercise to show that

$$f_0' = \frac{f_1 - f_{-1}}{2h} - \frac{1}{6}h^2 f'''(\xi), \qquad x_0 - h < \xi < x_0 + h,$$

$$= \frac{f_1 - f_{-1}}{2h} - O(h^2). \tag{B.19}$$

The error terms of both Eqs. (B.18) and (B.19) are $O(h^2)$, but the coefficient in Eq. (B.19) is only half the magnitude of the coefficient in Eq. (B.18). The progressive increase in accuracy we anticipated is confirmed.

By similar arguments, we can show that the formulas for third and fourth derivatives, Eqs. (B.11) and (B.12), both have errors $O(h^2)$.

B.3   INTEGRATION FORMULAS BY THE METHOD OF UNDETERMINED COEFFICIENTS

A numerical integration formula will estimate the value of the integral of the function by a formula involving the values of the function at a number of points in or near the interval of integration. Figure B.1 illustrates the general situation. If we desire to evaluate $\int_a^b f(x)\,dx$, it is obvious that if we could find some average value of the function on the interval $[a, b]$, the integral would be:

$$\int_a^b f(x)\,dx = (b - a)f_{av}.$$

It seems reasonable to assume that f_{av} could be approximated by a linear combination of function values within or near the interval.* We therefore write, similarly to our procedure for derivatives,

$$\int_a^b f(x)\,dx = c_0 f(x_0) + c_1 f(x_1) + \cdots + c_n f(x_n),$$

where the coefficients c_i are to be determined. The points x_i at which the function is to be evaluated can also be left undetermined but, for the formulas which we need to derive, we will again impose the restriction that they are equally spaced with the value of $x_{i+1} - x_i = \Delta x = h = $ constant. We will impose the further restriction that the boundaries of the interval of integration, a and b, coincide with two of the x_i-values.

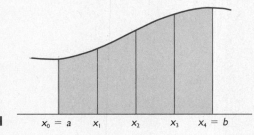

Figure B.1 $x_0 = a$ x_1 x_2 x_3 $x_4 = b$

We start with a simple case. Suppose we wished to express the integral in terms of just two functional values, specifically $f(a)$ and $f(b)$. Our formula takes the form

$$\int_a^b f(x)\,dx = c_0 f(a) + c_1 f(b). \tag{B.20}$$

Obviously, the values of c_0 and c_1 depend on $f(x)$, and probably on the values of a and b as well. The method of undetermined coefficients assumes that $f(x)$ can be approximated by a polynomial over the interval $[a, b]$ and determines c_0 and c_1 so that Eq. (B.20) is exact for all polynomials of a certain maximum degree or less.

Equation (B.20) has only two arbitrary constants, so we would not expect that it could be made exact for polynomials of degree higher than the first, since more than two parameters appear in second- or higher-degree polynomials. We therefore force Eq. (B.20)

---

*One could generalize the concept by extending this to include derivatives of the function as well, but we stay with the simpler case.

to be exact when $f(x)$ is replaced by either a first-degree polynomial or one of zero degree (a constant function). If it is to be exact for all first-degree polynomials, it must be exact for the very simple function $f(x) = x$. If it is to be exact when the function is any constant, it must hold if $f(x) = 1$. We express these two relations:

$$\int_a^b x \, dx = c_0(a) + c_1(b),$$

$$\int_a^b (1) \, dx = c_1(1) + c_1(1). \tag{B.21}$$

The integrals of Eqs. (B.21) are easily evaluated, so we have, as conditions that the constants must satisfy,

$$\int_a^b x \, dx = \frac{x^2}{2}\bigg|_a^b = \frac{b^2}{2} - \frac{a^2}{2} = c_0 a + c_1 b,$$

$$\int_a^b dx = x \bigg|_a^b = b - a = c_0 + c_1.$$

Solving these equations gives $c_0 = (b - a)/2$, $c_1 = (b - a)/2$.

Consequently,

$$\int_c^b f(x) \doteq \frac{b - a}{2}[f(a) + f(b)]. \tag{B.22}$$

This formula is the *trapezoidal rule*, and was derived in Chapter 3 as Eq. (3.10) through an entirely different approach. We have put a dot over the equality sign in Eq. (B.22) to remind ourselves that the relation is only approximately true, for the function $f(x)$ cannot ordinarily be replaced without error by a first-degree polynomial. It is intuitively obvious that unless the interval $[a, b]$ is very small, the error will be considerable.

We can reduce the error in the above formula by using a higher-degree polynomial to replace $f(x)$, but we would then need to take a linear combination of more than two function values. In fact, we will have to preserve a balance between the number of undetermined coefficients in the formula and the number of parameters in the polynomial. The number of coefficients hence must be one greater than the degree of the polynomial.

We now look at the next case, a three-term formula corresponding to replacing $f(x)$ by a quadratic. For three terms, using $x_0 = a$ and $x_2 = b$, with x_1 at the midpoint,

$$\int_a^b f(x) \, dx = c_0 f(a) + c_1 f\left(\frac{a + b}{2}\right) + c_2 f(b).$$

The formula must be exact if $f(x) = x^2$, or $f(x) = x$, or $f(x) = 1$, so

$$\int_a^b x^2 \, dx = \frac{b^3}{3} - \frac{a^3}{3} = c_0 a^2 + c_1\left(\frac{a + b}{2}\right)^2 + c_2 b^2,$$

$$\int_a^b x \, dx = \frac{b^2}{2} - \frac{a^2}{2} = c_0 a + c_1\left(\frac{a + b}{2}\right) + c_2 b,$$

$$\int_a^b dx = b - a = c_0 + c_1 + c_2.$$

The solution is $c_0 = c_2 = \frac{1}{3}(b - a)$, $c_1 = \frac{4}{3}(b - a)$. Hence

$$\int_a^b f(x)\, dx \doteq \frac{b - a}{6}\left(f(a) + 4f\left(\frac{a + b}{2}\right) + f(b)\right). \qquad \text{(B.23)}$$

A more common form of Eq. (B.23) is found by writing $b - a = 2h$ (the interval from a to b is subdivided into two panels) and substituting x_0 for a, x_2 for b, and x_1 for the midpoint. We then have

$$\int_{x_0}^{x_0 + 2h} f(x)\, dx \doteq \frac{h}{3}(f_0 + 4f_1 + f_2). \qquad \text{(B.24)}$$

Formula (B.24) is Simpson's $\frac{1}{3}$ rule, a particularly popular formula. Again, it is not exact—as indicated by the dot over the equality sign.

The application of Eqs. (B.22) and (B.24) over an extended interval is straightforward. It is intuitively apparent that the error in these formulas will be large if the interval is not small. (We discuss these errors quantitatively in Section B.5.) To apply these to a large interval of integration and still maintain control over the error, we break the interval into a large number of small subintervals and add together the formulas applied to the subintervals. When this is done we get the *extended trapezoidal rule*:

$$\int_a^b f(x)\, dx \doteq \frac{b - a}{2n}[f(x_0) + 2f(x_1) + 2f(x_2) + \cdots + 2f(x_{n-1}) + f(x_n)]; \qquad \text{(B.25)}$$

and the extended Simpson's $\frac{1}{3}$ rule:

$$\int_a^b f(x)\, dx \doteq \frac{b - a}{3(2n)}[f(x_0) + 4f(x_1) + 2f(x_2) + 4f(x_3) + \cdots$$
$$+ 2f(x_{2n-2}) + 4f(x_{2n-1}) + f(x_{2n})]. \qquad \text{(B.26)}$$

These forms of the trapezoid rule (Eq. (B.25)) and of Simpson's $\frac{1}{3}$ rule (Eq. (B.26)) are widely used in computer programs for integration. By applying them with n increasing, the error can be made arbitrarily small.* For Simpson's rule, observe that the number of subintervals must be even.

B.4 INTEGRATION FORMULAS USING POINTS OUTSIDE THE INTERVAL

In studying numerical methods to solve differential equations we shall have need for some specific integral formulas that involve function values computed at points outside the interval of integration. Figures B.2, B.3, and B.4 sketch three special cases of importance.

---

*Except for round-off error effects, which eventually will dominate since they are not decreased by small subdivision of the interval and, in fact, may increase as the number of computations increases.

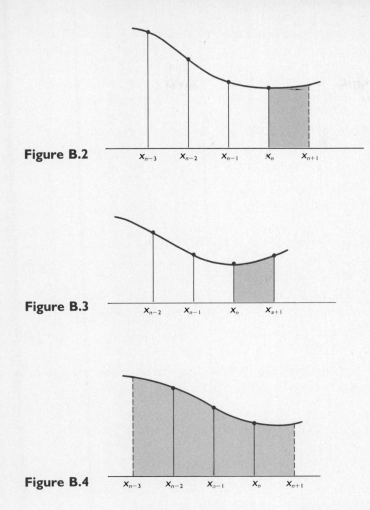

Figure B.2

Figure B.3

Figure B.4

In Fig. B.2 the curve that passes through the four points whose abscissas are x_{n-3}, x_{n-2}, x_{n-1}, and x_n is extrapolated to x_{n+1}, and we desire the integral only over the extrapolated interval. Figure B.3 presents the case where the curve passes through four points, and the integral is taken only over the last panel. In Fig. B.4 we consider a case where a curve that fits at three points is extrapolated in both directions, and integration is taken over four panels.

In the derivations of this section it will be convenient to adopt the notation that $f_i = f(x_i)$, so that subscripts on the function indicate the x-value at which it is evaluated.

For the first case (Fig. B.2) we desire a formula of the form

$$\int_{x_n}^{x_{n+1}} f(x) \, dx = c_0 f_{n-3} + c_1 f_{n-2} + c_2 f_{n-1} + c_3 f_n.$$

With four constants, we can make the formula exact when $f(x)$ is any polynomial of degree 3 or less. Accordingly, we replace $f(x)$ successively by x^3, x^2, x, and 1 to evaluate the coefficients.

It is apparent that the formula must be independent of the actual x-values. To simplify the equations, let us shift the origin to the point $x = x_n$; our integral is then taken over the interval from 0 to h, where $h = x_{n+1} - x_n$:

$$\int_0^h f(x)\, dx = c_0 f(-3h) + c_1 f(-2h) + c_2 f(-h) + c_3 f(0).$$

Carrying out the computations by replacing $f(x)$ with the particular polynomials, we have

$$\frac{h^4}{4} = c_0(-3h)^3 + c_1(-2h)^3 + c_2(-h)^3 + c_3(0),$$

$$\frac{h^3}{3} = c_0(-3h)^2 + c_1(-2h)^2 + c_2(-h)^2 + c_3(0),$$

$$\frac{h^2}{2} = c_0(-3h) + c_1(-2h) + c_2(-h) + c_3(0),$$

$$h = c_0(1) + c_1(1) + c_2(1) + c_3(1). \tag{B.27}$$

After completion of the algebra we get

$$\int_{x_n}^{x_{n+1}} f(x)\, dx \doteq \frac{h}{24}(-9f_{n-3} + 37f_{n-2} - 59f_{n-1} + 55f_n). \tag{B.28}$$

For the second case, illustrated by Fig. B.3, we again translate the origin to x_n to simplify. The set of equations analogous to Eqs. (B.27) is

$$\frac{h^4}{4} = c_0(-2h)^3 + c_1(-h)^3 + c_2(0) + c_3(h)^3,$$

$$\frac{h^3}{3} = c_0(-2h)^2 + c_1(-h)^2 + c_2(0) + c_3(h)^2,$$

$$\frac{h^2}{2} = c_0(-2h) + c_1(-h) + c_2(0) + c_3(h),$$

$$h = c_0(1) + c_1(1) + c_2(1) + c_3(1).$$

These give values of the constants in the integration formula:

$$\int_{x_n}^{x_{n+1}} f(x)\, dx \doteq \frac{h}{24}(f_{n-2} - 5f_{n-1} + 19f_n + 9f_{n+1}). \tag{B.29}$$

The third case, Fig. B.4, where extrapolation of a quadratic in both directions is involved, leads to the equation

$$\int_{x_{n-3}}^{x_{n+1}} f(x)\, dx \doteq \frac{4h}{3}(2f_{n-2} - f_{n-1} + 2f_n). \tag{B.30}$$

Setting up the equations for this case is left as an exercise for the student.

The particular methods for solving differential equations that use the formulas derived in this section are known as *Adams' method* and *Milne's method*.

B.5 ERROR OF INTEGRATION FORMULAS

In each of the formulas that we derived above, we used the symbol $\doteq$ to remind us that the formulas are not exact unless $f(x)$ is in fact a polynomial of a certain degree or less. It is important to derive expressions for the error.

We first determine the error term for the trapezoidal rule over one panel, Eq. (B.22). Begin with a Taylor expansion:

$$F(x_1) = F(x_0) + F'(x_0)h + \frac{1}{2}F''(x_0)h^2 + \frac{1}{6}F'''(\xi_1)h^3,$$

$$x_0 < \xi_1 < x_1 = x_0 + h. \qquad \text{(B.31)}$$

Let us define $F(x) = \int_a^x f(t)\, dt$ so that $F'(x) = f(x)$, $F''(x) = f'(x)$, $F'''(x) = f''(x)$. Equation (B.31) becomes, with this definition of $F(x)$,

$$\int_a^{x_1} f(x)\, dx = \int_a^{x_0} f(x)\, dx + f(x_0)h + \frac{1}{2}f'(x_0)h^2 + \frac{1}{6}f''(\xi_1)h^3.$$

On rearranging, and using subscript notation, we get

$$\int_a^{x_1} f(x)\, dx - \int_a^{x_0} f(x)\, dx = \int_{x_0}^{x_1} f(x)\, dx = f_0 h + \frac{1}{2}f_0' h^2 + \frac{1}{6}f''(\xi_1)h^3. \qquad \text{(B.32)}$$

We now replace the term f_0' by the forward-difference approximation with error term from Section B.2:

$$f_0' = \frac{f_1 - f_0}{h} - \frac{1}{2}hf''(\xi_2), \qquad x_0 < \xi_2 < x_1.$$

Equation (B.32) becomes

$$\int_{x_0}^{x_1} f(x)\, dx = f_0 h + \frac{1}{2}h(f_1 - f_0) - \frac{1}{4}h^3 f''(\xi_2) + \frac{1}{6}h^3 f''(\xi_1).$$

We can combine the last two terms into $-\frac{1}{12}h^3 f''(\xi)$, where ξ also lies in $[x_0, x_1]$. Hence,

$$\int_{x_0}^{x_1} f(x)\, dx = \frac{h}{2}(f_0 + f_1) - \frac{1}{12}h^3 f''(\xi).$$

The error of the trapezoidal rule over one panel is then $O(h^3)$. This is called the *local error*. Since we normally use a succession of applications of this rule over the interval $[a, b]$ to evaluate $\int_a^b f(x)\, dx$, by subdividing into $n = (b - a)/h$ panels, we need to determine the so-called global error for such intervals of integration. The global error will be the sum of the local errors in each of the n panels:

$$\text{Global error} = -\frac{1}{12}h^3 f''(\xi_1) - \frac{1}{12}h^3 f''(\xi_2) - \cdots - \frac{1}{12}h^3 f''(\xi_n)$$

$$= -\frac{1}{12}h^3 [f''(\xi_1) + f''(\xi_2) + \cdots + f''(\xi_n)].$$

Here the subscripts on ξ indicate the panel in whose interval the value lies. If $f''(x)$ is continuous throughout the interval of integration,

$$\sum_{i=1}^{n} f''(\xi_i) = nf''(\xi), \qquad a < \xi < b.$$

Using $n = (b - a)/h$, we get

$$\text{Global error} = -\frac{1}{12}h^3\left(\frac{b-a}{h}\right)f''(\xi) = -\frac{b-a}{12}h^2f''(\xi) = O(h^2).$$

Therefore the global error is $O(h^2)$, even though the local error is $O(h^3)$.

Similar treatment shows that the local error for Simpson's $\frac{1}{3}$ rule is

$$-\frac{1}{90}h^5f^{\text{iv}}(\xi) = O(h^5),$$

and the global error is

$$-\frac{1}{180}(b - a)h^4f^{\text{iv}}(\xi) = O(h^4).$$

One can also show that the local error of each of the integration formulas in Section B.4 is $O(h^5)$.

APPENDIX C

Software Libraries

Besides the programs and subroutines presented in this book, there are many other excellent programs and subroutines. Among these are:

IMSL (INTERNATIONAL MATHEMATICAL AND STATISTICAL LIBRARY)

This is a library of well over 400 subroutines available commercially and perhaps the most widely used for scientific programming. These subroutines cover both mathematical and statistical areas.

The library is divided into the following chapters, each of which consists of many subprograms.

A. Analysis of Variance.

B. Basic Statistics.

C. Categorized Data Analysis.

D. Differential Equations, Quadrature, and Differentiation.

E. Eigensystem Analysis.

F. Forecasting, Econometrics, Time Series, and Transforms.

G. Generation and Testing of Random Numbers.

I. Interpolation, Approximation, and Smoothing.

L. Linear Algebraic Equations.

M. Mathematical and Statistical Functions.

N. Non-Parametric Statistics.

O. Observation Structure and Multivariate Statistics.

R. Regression Analysis.

S. Sampling.

U. Utility Functions.

V. Vector and Matrix Arithmetic.

Z. Zeroes and Extrema, Linear Programming.

In order to use any IMSL subroutine, one must write a FORTRAN main program that calls the appropriate IMSL subroutine. More recently, IMSL, Inc. has introduced the PROTRAN system that uses the IMSL subroutines and allows access to them in a simple, easy language. Both IMSL and PROTRAN are products of IMSL, Inc.

Source: IMSL, 2500 ParkWest Tower One, 2500 CityWest Blvd., Houston, TX 77042.

TWODEPEP

This is a package of programs devoted to solving two-dimensional elliptic and parabolic problems in partial-differential equations by the finite-element method. This work is also a product of IMSL, Inc.

LINPACK

This is a collection of subroutines for solving linear systems of equations. The listing of the programs is published in a book by Dongarra et al. (1979). The programs themselves are distributed by IMSL, Inc.

ELLPACK

The ELLPACK system solves elliptic problems in two and three dimensions by a large number of methods. ELLPACK allows the user to state the problem in simple, mathematical language. ELLPACK consists of over 50,000 lines of FORTRAN code, but using a FORTRAN preprocessor allows for easy interaction with the large list of programs in the ELLPACK system.

Source: Professor John R. Rice, Math Science 428, Purdue University, West Lafayette, IN 47907.

PDECOL

These are a collection of subroutines for solving parabolic and hyperbolic partial equations. This program is listed as ACM Algorithm 540 and is distributed through IMSL, Inc.

ACM ALGORITHMS

ACM (The Association for Computing Machinery) provided a listing of the algorithms submitted to its journal, *Transaction on Mathematical Software*. The titles of many of them are listed in the back of this journal. For a listing and tapes one can write to: ACM Algorithms Distribution Service, IMSL, Inc., and so on.

For a detailed study of these and other software items, one should consult Rice (1983).

SOFTWARE FOR PERSONAL COMPUTERS

In the past few years many of the good software libraries have been ported to personal computers—most commonly, PC/XT/ATs. In addition, there are two new ones that were written especially for the microcomputer. Only a few of the many excellent ones are given here. The first three presented here are well known to scientific programmers.

IMSL

IMSL has subdivided its subroutines into three distinct libraries: MATH, STAT, and SFUN. This last stands for special functions. These libraries can be leased separately or together. The programs are in FORTRAN with the same excellent documentation IMSL provides for its original mainframe version.

 Source: IMSL, 2500 ParkWest Tower One, 2500 CityWest Blvd., Houston, TX 77042.

NAG

The Numerical Algorithms Group provides a library of 688 FORTRAN subprograms for mainframes. However, the NAG Fortran PC50 Library contains 50 of the most commonly used programs that have been implemented for personal computers.

 Source: Numerical Algorithms Group, Inc., 1101 31st St., Suite 100. Downers Grove, IL 60515-1263.

THE SCIENTIFIC DESK

This library contains more than 350 routines that cover both mathematics and statistics. The library also includes FORTRAN-callable graphics, plotting, and cursor control subroutines. It includes a set of interactive problem solvers that require no programming. The tutorials, problem solvers, and documentation can be accessed through a menu-driven interface. This package is also available for the mainframe. The work was begun by the founder and first president of IMSL.

 Source: C. Abaci, 208 St. Mary's St., Raleigh, NC 27605.

BORLAND

Turbo PASCAL Numerical Methods Toolbox contains a large selection of numerical programs written in Turbo PASCAL.

 Source: Borland International, 4585 Scotts Valley Dr., Scotts Valley, CA 95066.

NUMERICAL RECIPES

The FORTRAN subroutines and the PASCAL procedures as listed in the book *Numerical Recipes: The Art of Scientific Computing* are available on separate disks.

 Source: Customer Services Department, Cambridge University Press, Edinburgh Building, Shaftesbury Rd., Cambridge CB2 2RU, England.

References

Acton, F. S. (1970). *Numerical Methods That Work*. New York: Harper and Row.

Allaire, Paul E. (1985). *Basics of the Finite Element Method*. Dubuque, Iowa: Brown.

Allen, D. N. (1954). *Relaxation Methods*. New York: McGraw-Hill.

Atkinson, K. E. (1978). *An Introduction to Numerical Analysis*. New York: Wiley.

Bartels, Richard, J. Beatty, and B. Barsky (1987). *An Introduction to Splines for Use in Computer Graphics and Geometric Modeling*. Los Altos, Calif.: Morgan Kaufmann.

Birkhoff, Garrett, Richard Varga, and David Young (1962). Alternating direction implicit methods. *Advances in Computers*, 3:187–273.

Bracewell, Ronald N. (1986). *The Fast Hartley Transform*. New York: Oxford University Press.

Brigham, E. Oron (1974). *The Fast Fourier Transform*. Englewood Cliffs, N.J.: Prentice-Hall.

Burnett, David S. (1987). *Finite Element Analysis: From Concepts to Applications*. Reading, Mass.: Addison-Wesley.

Carnahan, Brice (1964). *Radiation Induced Cracking of Pentanes and Dimethylbutanes*. Ph.D. dissertation, University of Michigan.

Carnahan, Brice, et al. (1969). *Applied Numerical Methods*. New York: Wiley.

Carslaw, H. S., and J. C. Jaeger (1959). *Conduction of Heat in Solids*. 2nd ed. London: Oxford University Press.

Cheney, W., and D. Kincaid (1985). *Numerical Mathematics and Computing*. Monterey, Calif.: Brooks/Cole.

Condon, Edward, and Hugh Odishaw, eds. (1967). *Handbook of Physics*. New York: McGraw-Hill.

Conte, S. D., and C. de Boor (1980). *Elementary Numerical Analysis*. 3rd ed. New York: McGraw-Hill.

Cooley, J. W., and J. W. Tukey (1965). An algorithm for the machine calculations of complex Fourier series. *Mathematics of Computation*, 19:297–301.

Corliss, G., and Y. F. Chang (1982). Solving ordinary differential equations using Taylor series. *ACM Transactions on Mathematical Software*, 8:114–144.

Crow, Frank (1987). Origins of a teapot. *IEEE Computer Graphics and Applications*, 7(1):8–19.

Davis, Alan J. (1980). *The Finite Element Method*. Oxford: Clarendon Press.

Davis, Phillip J., and Philip Rabinowitz (1967). *Numerical Integration*. Waltham, Mass.: Blaisdell.

de Boor, C. (1978). *A Practical Guide to Splines*. New York: Springer-Verlag.

De Santis, R., F. Gironi, and L. Marelli (1976). Vector-liquid equilibrium from a hard-sphere equation of state. *Industrial and Engineering Chemistry Fundamentals*, 15(3):183–189.

Dongarra, J. J., J. R. Bunch, C. B. Moler, and G. W. Stewart (1979). *LINPACK User's Guide*. Philadelphia, Pa.: SIAM.

Douglas, J. (1962). Alternating direction methods for three space variables, *Numerical Mathematics*, 4:41–63.

Duffy, A. R., J. E. Sorenson, and R. E. Mesloh (1967). Heat transfer characteristics of belowground LNG storage. *Chemical Engineering Progress*, 63(6):55–61.

Fike, C. T. (1968). *Computer Evaluation of Mathematical Functions*. Englewood Cliffs, N.J.: Prentice-Hall.

Forsythe, G. E., M. A. Malcolm, and C. B. Moler (1977). *Computer Methods for Mathematical Computation*. Englewood Cliffs, N.J.: Prentice-Hall.

Forsythe, G. E., and C. B. Moler (1967). *Computer Solution of Linear Algebraic Systems*. Englewood Cliffs, N.J.: Prentice-Hall.

Gear, C. W. (1967). The numerical integration of ordinary differential equations. *Mathematics of Computation*, 21:146–156.

Gear, C. W. (1971). *Numerical Initial Value Problems in Ordinary Differential Equations*. Englewood Cliffs, N.J.: Prentice-Hall.

Hageman, L. A., and D. M. Young (1981). *Applied Iterative Methods*. New York: Academic Press.

Hamming, R. W. (1962): *Numerical Methods for Scientists and Engineers*. New York: McGraw-Hill.

Hamming, R. W. (1971). *Introduction to Applied Numerical Analysis*. New York: McGraw-Hill.

Harrington, Steven (1987). *Computer Graphics: A Programming Approach*. New York: McGraw-Hill.

Henrici, P. H. (1964). *Elements of Numerical Analysis*. New York: Wiley.

Hornbeck, R. W. (1975). *Numerical Methods*. New York: Quantum.

IEEE Standard for Binary Floating-Point Arithmetic (1985). Institute of Electrical and Electronics Engineers, Inc., New York.

Jones, B. (1982). A note on the T transformation. *Nonlinear Analysis, Theory, Methods and Applications*, 6:303–305.

Lee, Peter, and Geoffrey Duffy (1976). Relationships between velocity profiles and drag reduction in turbulent fiber suspension flow. *Journal of the American Institute of Chemical Engineering*, :750–753.

Love, Carl H. (1966). *Abscissas and Weights for Gaussian Quadrature*. National Bureau of Standards, Monograph 98.

Muller, D. E. (1956). A method of solving algebraic equations using an automatic computer. *Math Tables and Other Aids to Computation*, 10:208–215.

O'Neill, Mark A. (1988). Faster Than Fast Fourier. *BYTE*, 13(4):293–300.

Peaceman, D. W., and H. H. Rachford (1955). The numerical solution of parabolic and elliptic differential equations. *Journal of the Society for Industrial and Applied Mathematics*, 3:28–41.

Penrod, E. B., and K. V. Prasanna (1962). Design of a flat-plate collector for a solar earth heat pump. *Solar Energy*, 6(1):9–22.

Pokorny, C., and C. Gerald (1989). *Computer Graphics: The Principles Behind the Art and Science*. Irvine, Calif.: Franklin, Beedle, and Associates.

Prenter, P. M. (1975). *Splines and Variational Methods*. New York: Wiley.

Press, W., B. Flannery, S. Teukolsky, and W. Vetterling (1986). *Numerical Recipes: The Art of Scientific Computing*. New York: Cambridge University Press.

Ralston, Anthony (1965). *A First Course in Numerical Analysis*. New York: McGraw-Hill.

Ramirez, Robert W. (1985). *The FFT, Fundamentals and Concepts*. Englewood Cliffs, N.J.: Prentice-Hall.

Rice, John R. (1983). *Numerical Methods, Software, and Analysis*. New York: McGraw-Hill.

Richtmyer, R. D. (1957). *Difference Methods for Initial Value Problems*. New York: Wiley Interscience.

Scarborough, J. B. (1950): Numerical Mathematical Analysis, Johns Hopkins Press, Baltimore.

Shampine, L., and R. Allen (1973). *Numerical Computing*. Philadelphia: Saunders.

Smith, G. D. (1978). *Numerical Solution of Partial Differential Equations*. 2nd ed. London: Oxford University Press.

Stewart, G. W. (1973). *Introduction to Matrix Computations*. New York: Academic Press.

Stoer, J., and R. Bulirsch (1980). *Introduction to Numerical Analysis*. New York: Springer-Verlag.

Stubbs, D. and N. Webre (1987). *Data Structures with Abstract Data Types*. Monterey, Calif.: Brooks/Cole.

Traub, J. F. (1964). *Iterative Methods for the Solution of Equations*. Englewood Cliffs, N.J.: Prentice-Hall.

Varga, Richard (1959). *p*-Cyclic matrices: A generalization of the Young–Frankel successive overrelaxation scheme. *Pacific Journal of Mathematics*, 9:617–628.

Vichnevetsky, R. (1981). *Computer Methods for Partial Differential Equations. Vol. 1: Elliptic Equations and the Finite Element Method*. Englewood Cliffs, N.J.: Prentice-Hall.

Waser, S., and M. J. Flynn (1982). *Introduction to Arithmetic for Digital Systems Designers*. New York: Holt, Rinehart and Winston.

Wilkinson, J. H. (1963). *Rounding Errors in Algebraic Processes*. Englewood Cliffs, N.J.: Prentice-Hall.

Wilkinson, J. H. (1965). *The Algebraic Eigenvalue Problem*. London: Oxford University Press.

Answers to Selected Exercises

CHAPTER I

1. Successive iterates are 0.5, 0.75, 0.625, 0.5625, 0.59375, 0.60938. For 5 accurate digits, the error must be ≤ 0.00005, which requires 15 iterations. The error bound is then $\pm 0.5^{15} = \pm 0.00003$.

4. b) 0.4450   c) 0.9210.

7. 0.55496, 2.2470, -0.80194.

10. -0.458962.

21. Roots are $0 \pm i$, $-2 \pm 4i$. For the particular computer programs used, the execution times were 0.13 sec for the Bairstow method and 0.49 sec for the Muller method.

23. Ten iterations. Using Aitken acceleration, 4 iterations are required (2 extrapolations).

25. Suitable rearrangements are

$$x = \sqrt[3]{\frac{-4x^2 + 2x + 5}{2}}, \qquad x = \sqrt{\frac{-2x^3 + 2x + 5}{4}}, \qquad x = \sqrt{\frac{2x + 5}{2x + 4}}.$$

Some of these may converge slowly.

29. Iterates: 1.0, 1.5, 1.75, 1.875, 1.9375, . . . , 2.0. Errors: 1.0, 0.5, 0.25, 0.125, 0.0625, . . . , 0.0. This is linearly convergent; Aitken acceleration applies.

32. Roots found are 0.791288 and 1.61803, leaving a quadratic whose roots are -0.618034 and -3.79129.

34. $P(x) = (x + 1)(x^3 + 3.6x^2 + 3x - 14) = (x + 1)(x - 1.4)(x^2 + 5x + 10)$.

36. After three iterations $r = 4.1$, $s = -5.2$; factor is $x^2 - 4.1x + 5.2$. From the reduced polynomial, the other factor is $x^2 + x + 1$. Zeros of the polynomial are at $2.05 \pm 0.9987i$, $-0.5 \pm 0.8660i$.

40. a) 1.8019, -1.2470, 0.44504.
    d) Two e-values oscillate, indicating two pairs of complex roots. Factors are $x^2 - 4.1x + 5.2$ and $x^2 + x + 1$.

47. b) Chopped: 0.176×10^6, error 0.6×10^3.
   Rounded: 0.177×10^6, error -0.4×10^3.

49. c) 0.135×10^{-4}. Absolute error -0.24×10^{-7}, relative error -0.18×10^{-2}.

50. b) $(1.32 * 1.98) * 1.01 \neq 1.32 * (1.98 * 1.01)$ with rounding.

53. The results depend on the computer system. With IBM/360, single precision:

   a) 0.9999991   b) 0.9999878   c) 0.9995956
   d) 14.017 after about 1,050,000 terms. It converges because the computer evaluates $1/n$ as 0, for large enough n, in comparison to the current sum.
   e) The value is greater than the result in (d) because less precision is lost due to exponent alignment.

62. Absolute min at $x = 0$ and 1. A relative max $(f'(x) = 0)$ at $x = 0.9499$.

CHAPTER 2 1. b)
$$Ax = \begin{bmatrix} -3 \\ 18 \\ 38 \end{bmatrix}, \quad By = \begin{bmatrix} -17 \\ 76 \\ 55 \end{bmatrix}, \quad x^T y = 19.$$

3. c)
$$AC = \begin{bmatrix} -3 & -1 & -2 \\ -4 & 4 & 10 \\ -3 & 2 & 5 \end{bmatrix}, \quad CA = \begin{bmatrix} 6 & 3 & 1 \\ 9 & 1 & 4 \\ 1 & 2 & -1 \end{bmatrix}.$$

6. b) $x_1 = -2$, $x_2 = (-13 + 4)/-3 = 3$, $x_3 = (7 + 8 - 9)/-6 = -1$.

12. a) $x^T = (1.30, -1.35, -0.275)$.
   b) $x^T = (1.45, -1.59, -0.276)$.
   c) Calculated right-hand sides are

$$(0.02, 1.02, -0.21) \quad \text{and} \quad (0.04, 1.03, -0.54).$$

14. $x_1^T = (0.5834, 0.6370, 0.4019, -0.6439)$,
   $x_2^T = (-0.8333, 0.3636, -0.6515, 0.8939)$,
   $x_3^T = (2.419, -0.4401, 2.699, 0.007576)$.

16. $Az = b$ implies

$$(B + Ci)(x + yi) = (Bx - Cy) + (By + Cx)i$$
$$= p + qi.$$

   a) $\begin{bmatrix} B & -C \\ C & B \end{bmatrix}\begin{bmatrix} x \\ y \end{bmatrix} = \begin{bmatrix} p \\ q \end{bmatrix}.$
   b) $2n^2 + 2n$ versus $4n^2 + 2n$.

20. b) $x^T = (27.051, 8.2051, 5.7692, 14.872, 53.718)$.

23. b) $7c_1 - 12c_2 - 3c_3 = \bar{0}$.

25. a) $\det(H)$ is very small (about 1.65×10^{-5}); one also cannot avoid a very small divisor in solving the system.
   b) $x^T = (1.11, 0.228, 1.95, 0.797)$.
   c) $x^T = (0.988, 1.42, -0.428, 2.10)$.
   d) $x^T = (1.0000, 0.9995, 1.0017, 0.9990)$.

27. $\det = 51$.

32.
$$H^{-1} = \begin{bmatrix} 16.00 & -120.0 & 239.9 & -139.9 \\ -120.0 & 1199. & -2699. & 1679. \\ 239.9 & -2699. & 6477. & -4198. \\ -139.9 & 1679. & -4198. & 2799. \end{bmatrix}.$$

35. 10.337.

36. a) $x^T = (1592.71, -631.956, -493.653)$.

38. $x^T = (119.5, -47.14, -36.84)$. This is further evidence of ill-condition, in that small changes in the coefficients make large changes in the solution vector.

41. $\dfrac{1}{\text{C.N.}} \dfrac{\|r\|_2}{\|b\|_2} = 1.82 \times 10^{-5}$, $\quad$ C.N. $\dfrac{\|r\|_2}{\|b\|_2} = 55{,}642$.

$$\frac{\|e\|_2}{\|\bar{x}\|_2} = 25.5, \qquad \frac{\|e\|_2}{\|x\|_2} = 1.04.$$

The relation is verified no matter whether $\|\bar{x}\|$ or $\|x\|$ is used.

45. Using 3-digit arithmetic to compute $\bar{e}$:

$\bar{e} = \{-0.00271, 0.00126, 0.00103\}$, $\bar{x} = \{0.153, 0.144, -0.166\}$.

$(\bar{x} + \bar{e}) = \{0.15029, 0.14526, -0.16497\}$, to be compared with $x = \{0.15094, 0.14525, -0.16592\}$. The improvement is remarkable even in one iteration.

48. After six iterations, the Jacobi vector has diverged to $x^T = (3.16, 3.16)$ while with Gauss–Seidel, $x^T = (-154.5, 467.6)$. Gauss–Seidel is diverging much more rapidly.

50. $x^T = (1, -1, 2)$.

54. $(0.72595, 0.50295), (-1.6701, 0.34513)$.

57. $(1.64304, -2.34978), (-2.07929, -3.16173)$.

CHAPTER 3

2. $\dfrac{(x - 1)(x + 2)(x - 2)}{-12}(15) + \dfrac{(x - 1)(x + 2)(x + 1)}{12}(9) = -0.5x^3 + 4x^2 + 0.5x - 4$.

4. 1.2218. Actual error $= -0.0004$, estimates $= \begin{cases} -0.00033 & \text{min,} \\ -0.00045 & \text{max.} \end{cases}$

8. a) 1.0919 $\qquad$ b) 1.0973 $\qquad$ c) 1.0941 $\qquad$ d) 1.0951 $\qquad$ e) 1.0920

$$f(x) = \frac{1}{\sin(x + 1)} \quad \text{and} \quad f(0.15) = 1.0956.$$

12. A sixth-degree polynomial is required to fit exactly. Since the third differences are almost constant (in fact, their variation could be due to round-off errors in $f(x)$), a third-degree polynomial will almost fit all seven points.

14. $f(0.158) = 0.78801$, $f(0.636) = 0.65178$.

21. All results are identical $(f(0.385) = 0.74091)$ because all polynomials are the same though of different forms.

23.

Formula	y_0	s	$y(0.92)$
NGF	0.148	2.1	0.3836
NGB	0.518	-0.9	0.3836
GF	0.248	1.1	0.3836
GB	0.370	0.1	0.3836
Bessel	0.309	1.1	0.3836

(The precise value of each $y(0.92)$ was 0.383564.)

25. $\left(\dfrac{s}{2}\right)h^2 f''(\xi)$, $x_0 \le \xi \le x_1$ (based on NGF).

Set the derivative of $[s(s - 1)]/2$ equal to zero; this gives a maximum at $s = 0.5$.

32. a) $\Delta(f_0 g_0) = (f_1 g_1) - (f_0 g_0) = f_1 g_1 - f_0 g_1 + f_0 g_1 - f_0 g_0$
$= f_0(g_1 - g_0) + g_1(f_1 - f_0) = f_0 \Delta g_0 + g_1 \Delta f_0.$
b) $\Delta^n x^n = n!$ by the argument in Section 3.4.

$$\nabla^n E^n x^n = (1 - E^{-1})^n E^n x^n = (E - 1)^n x^n = \Delta^n x^n = n!$$
$$\nabla^n x^n = (E^{-1}\Delta)^n x^n = \Delta^n E^{-n} x^n = \Delta^n (E^{-1}x)^n = \Delta^n (x - 1)^n = n!$$

because $(x - 1)^n$ is an nth-degree polynomial whose leading term is x^n.

34. $\Delta^n y_s = (\Delta^n E^{-n})E^n y_s = \nabla^n y_{s+n}$ so $r = s + n$.

38. $x(25) = 0.6$. Estimated error $= 0.1613$ (min) to 504 (max).

42. $$\begin{bmatrix} 4.9523 & -2.7323 & & \\ 0.13 & 1.1568 & 0.18 & \\ & 0.18 & 1.6800 & 0.66 \\ & & 0.4412 & 2.6788 \end{bmatrix} \begin{bmatrix} S_1 \\ S_2 \\ S_3 \\ S_4 \end{bmatrix} = \begin{bmatrix} -0.46275 \\ -0.04359 \\ 0.21212 \\ 0.50507 \end{bmatrix}.$$
$S_0 = 5.6923 S_1 - 4.6923 S_2,$
$S_5 = 1.5758 S_4 - 0.57576 S_3.$

57. $f(1.6, 0.33) = 1.8330$. Compare to the true value of 1.8350.

58. $f(1.62, 0.31) = 1.7524$. (Analytical value $= 1.7515$.)

CHAPTER 4

1. $f'(0.15)$: a) 2.715, error 0.180   b) 2.862, error 0.033   c) 2.878, error 0.018.
$f'(0.19)$: a) 2.170, error 0.116   b) 2.268, error 0.018   c) 2.278, error 0.008.
$f'(0.23)$: a) 1.810, error 0.078   b) 1.878, error 0.011   c) 1.881, error 0.007.

2. $x = 0.15$: a) 0.150 to 0.193   b) 0.017 to 0.034   c) 0.006 to 0.010.
$x = 0.19$: a) 0.098 to 0.120   b) 0.010 to 0.017   c) 0.001 to 0.004.
$x = 0.23$: a) 0.069 to 0.082   b) 0.006 to 0.010   c) 7×10^{-4} to 0.002.

(Round-off errors cause the actual error to fall outside these bounds in some cases.)

9. Use $h^{-n}\Delta^n f$ as an approximation of $f^{(n)}(\xi)$. The best average value for the derivative uses a difference in the center of the range of interpolation. Estimates of errors are

$x = 0.15$:	a) 0.148	b) 0.015	c) -0.002;
$x = 0.19$:	a) 0.111	b) 0.009	c) -0.002;
$x = 0.23$:	a) 0.075	b) 0.006	c) 0.

Round-off may invalidate such estimates.

17. Successive computations and extrapolations are

$\Delta x = 0.08$:   1.9706

    1.8865

$\Delta x = 0.04$:   1.9075   1.8874

    1.8875

$\Delta x = 0.02$:   1.8925

21. a)

h:	0.0001	0.001	0.01	0.1
f':	∞	3.50	3.45	3.52
Error:	∞	-0.074	-0.024	-0.094

b)

h:	0.0001	0.001	0.01	0.1
f':	5.00	3.50	3.40	3.52
Error:	-1.574	-0.074	0.026	-0.094

26. The error terms are of one order of h greater than expected because

$$\int_0^{n-1} \binom{s}{n} ds = 0 \qquad \text{when } n \text{ is odd.}$$

This can be shown by a change of variable so that the range of integration is symmetrical about the origin; the integrand changes to an odd function for which the integral is zero. For example, with $n = 3$, take $s = t + 1$, giving

$$\int_0^2 \binom{s}{3} ds = \int_{-1}^1 \frac{(t+1)(t)(t-1)}{6} dt = \int_{-1}^1 \frac{t^3 - t}{6} dt = 0.$$

30. a) 1.7683   b) 1.7728   c) 1.7904.

33. $h \leq 1.1 \times 10^{-3}$, using the upper bound of the error estimate.

36.

h	Integral	Extrapolations	
1.0	0.32045		
		0.34156	
0.5	0.33628		0.34154
		0.34136	
0.25	0.34009		

40. Using the upper bound to the error, h must be ≤ 0.135. With $h = 0.1$, the integral is 1.718283 (exact value is 1.718282).

44. -0.516845.

46. $$\int_{x_0}^{x_3} f(x)\, dx = h\left(3f_0 + \frac{9}{2}\Delta f_0 + \frac{9}{4}\Delta^2 f_0 + \frac{9}{12}\Delta^3 f_0 - \frac{3}{80}\Delta^4 f_0 + \cdots\right).$$

51. 0.94608.

56. a) 2.8285.

58. $0.785353 + 0.000045 = 0.785398$.

62.

x:	1.5	2.0	2.5	
y':	-0.505	-0.234	-0.162	Spline,
y'':	0.966	0.118	0.168	condition 1.
y':	-0.476	-0.240	-0.164	Spline,
y'':	0.761	0.181	0.122	condition 2.
y':	-0.462	-0.245	-0.161	Spline,
y'':	0.664	0.203	0.132	condition 3.
y':	-0.500	-0.267	-0.167	Central
y'':	0.644	0.268	0.132	differences.
y':	-0.4444	-0.250	-0.1600	Analytical
y'':	0.5926	0.2500	0.1280	values.

The natural spline gives poor results, in comparison to the other techniques.

65. Integral $= 1.29919$. Analytical value $= 1.30176$. Simpson's $\frac{1}{3}$ rule $= 1.30160$. In this example, Simpson's rule has a considerably smaller error.

69. b) $\begin{Bmatrix} 1 & 4 & 2 & 4 & 1 \\ 4 & 16 & 8 & 16 & 4 \\ 2 & 8 & 4 & 8 & 2 \\ 4 & 16 & 8 & 16 & 4 \\ 1 & 4 & 2 & 4 & 1 \end{Bmatrix}$

72. a) 0.140586   b) 0.140587.

75. a) 0.1046.

77. Integrating with x constant, $\Delta x = 0.2$, and taking four intervals in the y-direction: Integral = 0.64307.

CHAPTER 5

1. $y(x) = 1 + x + \frac{3}{2}x^2 + \frac{5}{6}x^3 + \frac{7}{12}x^4 + \frac{17}{60}x^5$;
$y(0.1) = 1.11589,\quad y(0.5) = 2.0245.$

3. $y(0.2) = 1.2015,\quad y(0.4) = 1.4130,\quad y(0.6) = 1.6487.$

5. With $h = 0.01$, $y(0.1) = 1.11418$, error = 0.00171. To reduce the error to 0.00005 (34-fold), we must reduce h 34-fold, or to about 0.00029.

7. $y(0.1) = 1.11587$ (four decimals are correct). The simple Euler method will require about 340 steps and 340 function evaluations for similar accuracy, compared to four steps and eight function evaluations with the modified Euler method.

9. x: 0.1   0.2   0.3   0.4   0.5
y: 2.2150 2.4630 2.7473 3.0715 3.4394

13. t: 0.2   0.4   0.6
y: 2.0933 2.1755 2.2493

18.

x	y	Analytical
0.0	−1.000000	−1.000000
0.2	−0.781269	−0.781269
0.4	−0.529680	−0.529680
0.6	−0.251188	−0.251188
0.8	0.049329	0.049329
1.0	0.367880	0.367880

20. a) $y(0.5) = -0.28326$, error = −0.00077.
b) $y(0.5) = -0.28387$, error = −0.00016.
c) $y(0.5) = -0.28396$, error = −0.00007.

25. $y(4)$ predicted = 4.2229, $y(4)$ corrected = 4.1149. From these, the error in y_c is estimated as 0.0037; equivalently y_c has three correct digits. (The actual error is 0.0082.) Obviously the starting values must have at least three digits of accuracy also.

27.

x:	0.8	1.0	1.2	1.6	2.0
y:	2.3163	2.3780	2.4350	2.5380	2.6294
Error estimate:	0.0003	<0.00005	≈0	−0.00005	−0.00002

After $x = 1.2$, h was increased to 0.4.

33. By Runge–Kutta:
x: 0   0.2   0.4   0.6
y: 0   0.0004 0.0064 0.0325

By Adams–Moulton:

x:	0.8	0.9	1.0	1.1	1.2	1.25	1.3	1.35	1.4
y:	0.1035	0.1669	0.2574	0.3836	0.5581	0.6688	0.7994	0.9541	1.1387

The step size was halved after $x = 0.8$ and $x = 1.2$, to provide accuracy to four decimals.

35. a) $\partial f/\partial y = \sin x$ so $h_{max} = (24/9)/1 = 2.67$.
b) With $h = 0.267$, D cannot exceed 10×10^{-N} for N-decimal-place accuracy.
c) If $D = 14.2 \times 10^{-N}$, h cannot exceed $(1/14.2)h_{max}$.

41.

x	y	f	$1 + hf_y$	$\frac{1}{2}h^2y''$	Est'd. error	Actual error
1.0	1.000	1.000	1.200	0.015	0	0
1.1	1.100	1.331	1.242	0.022	0.019	0.017
1.2	1.233	1.825	1.296	0.035	0.052	0.049
1.3	1.416	2.605	1.368	0.058	0.115	0.111
1.4	1.676	3.933	1.437	0.087	0.234	0.247
1.5	2.069	6.424	1.560	0.173	0.481	0.597
1.6	2.712	11.766	1.782	0.403	1.091	1.834

In the above calculations, average values were used to propagate the errors.

44. $y' = z,$ $y(0) = 0;$

$$z' = \frac{M(x)}{EI}(1 - z^2)^{3/2}, \qquad z(0) = 0.$$

46.

t:	0	0.2	0.4	0.6
y:	1	0.982	0.933	0.864
x:	0	0.022	0.093	0.220

49.

x	y	y'
0	1	-1
0.1	0.8950	-1.0995
0.2	0.7802	-1.1956
0.3	0.6561	-1.2847
0.4	0.5236	-1.3629
0.5	0.3840	-1.4263
0.6	0.2389	-1.4715

52.

t	x	Analytical
0	0	0
0.1	0.0715	0.0707
0.2	0.1983	0.1999
0.3	0.2004	0.2026
0.4	-0.0237	-0.0233
0.5	-0.3747	-0.3784
0.6	-0.5912	-0.5977
0.7	-0.4377	-0.4419
0.8	0.0898	0.0924

CHAPTER 6 1. With $y'(0) = 1.0,$ $y(1) = 2.08536.$
With $y'(0) = 0.8,$ $y(1) = 1.81534.$

Interpolating gives $y'(0) = 0.86271$ for $y(1) = 1.9$;
Computations with $y'(0) = 0.86271$:

t:	0	0.25	0.50	0.75	1.00
y:	0	0.2388	0.5724	1.0948	1.9000
Analytical:	0	0.2406	0.5750	1.0969	1.9000

5. Using the modified Euler method with $h = 0.2$, initial slope is -2.4270.

t:	0	0.2	0.4	0.6	0.8	1.0
y:	1	0.4494	-0.0119	-0.4449	-0.7863	-1.0000

8. c)

θ:	0	$\pi/8$	$\pi/4$	$3\pi/8$	$\pi/2$
y:	0	0.3851	0.7108	0.9269	1.0

13. c)

θ:	0	$\pi/8$	$\pi/4$	$3\pi/8$	$\pi/2$
y:	-1.0199	-0.9413	-0.7175	-0.3830	0.0105

17. Using $h = 0.2$:

x:	0.2	0.4	0.6	0.8
y:	0.2607	0.7844	1.3592	1.8109

21. Analytical: $y = \dfrac{x^3 - x}{3}$; Approximating: $y = \dfrac{x^2 - x}{2}$.

22. b) Change variable. Let $u = y - 1 - 2x$; so $u'' = 2x$, $u(0) = 0$, $u(1) = 0$ gives $y = u + 1 + 2x$.

27. a) ± 2.82843 b) ± 3.08459 c) ± 3.17664 d) $3.29271, 3.29499$.

30. Analytical: $y = Ae^{-x} \sin 2\pi x$

From finite differences, using $h = \frac{1}{8}$ and $k = \pm 6.3623$:

x:	0.25	0.50	0.75
y:	0.7788	-0.0607	-0.4664
Analytical:	0.7788	0	-0.4724

31. b)   Let $X_0 = \begin{bmatrix} 1 \\ 1 \end{bmatrix}$, $AX_i = \mu_{i+1}X_{i+1}$.

Then $\mu_1 = 9$, $\mu_2 = \frac{17}{3}$, $\mu_3 = \frac{103}{17}$, $\to \lambda = 6$;

$$X_1 = \begin{bmatrix} \frac{1}{3} \\ 1 \end{bmatrix}, X_2 = \begin{bmatrix} \frac{7}{17} \\ 1 \end{bmatrix}, X_3 = \begin{bmatrix} \frac{41}{103} \\ 1 \end{bmatrix}, \to \begin{bmatrix} 0.4 \\ 1.0 \end{bmatrix}.$$

e) $\lambda = 3.49086$, $x^T = (1, 0.9362, 0.7733)$.

35. $\lambda = -9.9404, 7.0319, -4.0915$.

For the inverse, $\lambda = -0.10060, 0.14221, -0.24441$, which are reciprocals of the above values.

CHAPTER 7   3. $\dfrac{\partial u}{\partial x} = \dfrac{u_{i+1,j} - u_{i-1,j}}{2\Delta x}$;

$$\frac{\partial}{\partial y}\left(\frac{\partial u}{\partial x}\right) = \frac{(u_{i+1,j+1} - u_{i-1,j+1})/2\Delta x - (u_{i+1,j-1} - u_{i-1,j-1})/2\Delta x}{2\Delta y}$$

$$= \frac{1}{4h^2}\begin{Bmatrix} -1 & 1 \\ 1 & -1 \end{Bmatrix} u_{i,j}, \text{ when } \Delta x = \Delta y = h.$$

7. Temperatures are

0.69	2.08	5.56	14.58	38.19
0.69	2.08	5.56	14.58	38.19

9. The interior temperatures of the plate are

45.000	90.000	135.000	180.000
30.000	60.000	90.000	120.000
15.000	30.000	45.000	60.000

11. Starting at the upper left-hand corner, going left to right and down, we get the following 12 values for u:

44.9	89.3	132.1	168.8
30.5	59.7	88.2	117.3
18.2	30.4	44.6	60.2

13. For points 1 to 8:

74.46, 27.69, 47.86, 65.38, 78.85, 12.90, 23.89, 34.82.

15. Same answers as Exercise 7. Beginning with initial values of $0°$ at all interior points, 18 iterations are required to achieve a maximum change at any point less than $0.0001°$.

18. Same answers as Exercise 7. Beginning with all interior points at 0 and a tolerance for maximum change at any point of 0.0001, the number of iterations required is

Value of ω:	1.05	1.10	1.15	1.20	1.25	1.30
Number of iterations:	16	14	12	11	12	13

21. a) Each interior point = 0.444.
    b) Values at interior points:

0.211	0.312	0.342	0.312	0.211
0.312	0.472	0.521	0.472	0.312
0.342	0.521	0.577	0.521	0.342
0.312	0.472	0.521	0.472	0.311
0.211	0.312	0.342	0.312	0.211

25. $\dfrac{1}{h^2}\left\{3 \quad \begin{matrix} 1 \\ -8 \\ 1 \end{matrix} \quad 3\right\} + 2 = 0.$

$$0 + 0 + 3\phi_{12} + \phi_{21} - 8\phi_{11} + 8 = 0$$

$$3\phi_{11} + 0 + 0 + \phi_{22} - 8\phi_{12} + 8 = 0$$

$$0 + \phi_{11} + 3\phi_{22} + \phi_{31} - 8\phi_{21} + 8 = 0$$

$$3\phi_{21} + \phi_{12} + 0 + \phi_{32} - 8\phi_{22} + 8 = 0$$

$$0 + \phi_{21} + 3\phi_{32} + 0 - 8\phi_{31} + 8 = 0$$

$$3\phi_{31} + \phi_{22} + 0 + 0 - 8\phi_{32} + 8 = 0$$

26. Same values as in Exercise 21(b). The number of iterations to achieve a change in any value less than 0.0001 is

ω:	1.0	1.1	1.2	1.3	1.35	1.4	1.5
No. of iterations:	13	11	10	9	10	11	13

The calculated value for ω optimum is 1.33. The smallest number of iterations occurs with that value.

33. There is fourfold symmetry. Values in the upper left quadrant are

20.00	20.00	20.00	20.00	20.00
50.30	59.21	63.16	64.83	65.30
61.53	72.13	77.33	79.63	80.28

37. Interior points:

41.08

19.06   35.17

41. Utilizing symmetry, nine points are sufficient.

    At $r = 1$, clockwise from the center: 10.45, 10.41, 10.27.

    At $r = 2$:   10.79, 10.78, 10.70.

    At $r = 3$:   7.01, 7.06, 7.41.

42. There is sixfold symmetry, so each node is the same, 33.33°.

44. Same answers as Exercise 9. The number of iterations (one vertical and one horizontal traverse) is 8. Compare this to 13 iterations by S.O.R. using optimum ω. S.O.R. with $\omega = 1$ requires 26 iterations.

CHAPTER 8

2. Using $r = \frac{1}{2}$, at $t = 2.0625$ (10 steps):

x:	0.25	0.50	0.75	1.00	1.25
u:	17.34	32.04	41.86	45.31	41.86
Analytical:	17.70	32.71	42.74	46.26	42.74
Error:	0.36	0.67	0.88	0.95	0.88

6. Values at $t = 268.8$ sec:

	Relative effort	$u(4)$	Error	$u(12)$	Error
$\Delta x = 4, r = \frac{1}{2}$:	1	1.25	-0.142	4.75	-0.479
$\Delta x = 4, r = \frac{1}{4}$:	2	1.115	-0.007	4.372	-0.101
$\Delta x = 4, r = \frac{1}{8}$:	4	1.132	-0.024	4.317	-0.046
$\Delta x = 2, r = \frac{1}{2}$:	4	1.115	-0.007	4.372	-0.101
$\Delta x = 1, r = \frac{1}{2}$:	16	1.109	-0.001	4.297	-0.026

    One can conclude that decrease of r is more effective than decrease of Δx in increasing accuracy. (It can be shown that $r = \frac{1}{6}$ is particularly favorable, however.)

9.

x:	0.2	0.4	0.6	0.8
At $x = 0.0468$:	-0.00390	-0.00762	-0.01056	-0.01064
At $x = 0.0936$:	-0.00738	-0.01392	-0.01782	-0.01510
At $x = 0.2340$:	-0.01326	-0.02357	-0.02757	-0.02126

12. Values at $t = 15.12$:

x:	0	1	2	3	4
u:	620.1	778.3	833.6	778.3	620.1

14. Computing with $\Delta x = 2''$, $r = \frac{1}{4}$ in the explicit method, the midpoint reaches 200°F in 876.8 sec.

15. After 4, 8, and 20 time steps, values are

t_4:	0	-0.224	0.504	-0.216	0
t_8:	0	-0.336	0.509	0.336	0
t_{20}:	0	-0.615	0.870	-0.615	0

    The errors are growing, so the method is unstable.

19. a) $\lambda_{max} = 0.9595$; the formula gives the same value.

    b) $\lambda_{max} = -1.3509$; the formula gives -1.3514. The power method has great difficulties with this matrix.

22. The equations are of the form

$$u_{i,j,k}^{n+1} = r\left\{1 \begin{matrix} & 1 & 1 & \\ & & & 1 \\ & 1 & 1 & \end{matrix}\right\}u^n + (1 - 6r)u_{i,j,k}^n$$

where $r = k\,\Delta t/c\rho(\Delta x)^2$ and must be less than $\frac{1}{6}$. There are 125 equations to be solved at each time step (assuming no symmetry), and if $r = \frac{1}{8}$, 16 time steps are needed to reach $t = 15.12$.

24. There are three sets of 125 equations, but each set can be decomposed into 25 subsets of 5 equations, each subset being tridiagonal, so at most 1125 coefficients need to be stored; r can be chosen conveniently at unity, so only two time steps will suffice.

CHAPTER 9 1. The calculated frequency is 409.9 sec^{-1}. If $\Delta x = 70/4$, $\Delta t = 3.050 \times 10^{-4}$ sec. Eight time steps are required for a full cycle. This corresponds to a frequency of 409.84 sec^{-1}.

3. Using $\Delta x = 0.1$, $\Delta t = 0.1$ (for $r = 1$), representative values are

x:	**0.1**	**0.3**	**0.5**
$t = 0.1$:	0.2939	0.7694	0.9511
$t = 0.7$:	−0.1816	−0.4755	−0.5878
$t = 1.6$:	0.0955	0.2500	0.3090

These are the same as the analytical results.

6. Representative values (compare these to Exercise 2(c)):

x:	**L/8**	**L/4**	**3L/8**	**L/2**
$t = \Delta t$:	−0.00651	−0.01140	−0.01434	−0.01531
$t = 2\Delta t$:	−0.01140	−0.02085	−0.02670	−0.02868
$t = 11\Delta t$:	0.01434	0.02670	0.03518	0.03810

Compare these to Exercise 2(d):

$t = \Delta t$:	0.02344	0.4688	0.7031	0.6927
$t = 2\Delta t$:	0.2188	0.4375	0.4115	0.4062
$t = 11\Delta t$:	−0.2031	−0.1615	−0.1406	−0.1302

9. The errors expand (or contract) exactly as in the example in Section 9.3 or in Exercise 8. Whenever the boundary conditions are specified, the error is zero at that boundary.

12. Some representative values:

(0.25, 0.1021): 0.0186; (0.50, 0.0884): 0.0216;
(0.75, 0.0791): 0.0144;
(0.25, 0.2042): 0.0333; (0.50, 0.1768): 0.0408;
(0.75, 0.1581): 0.0269;
(0.25, 0.3062): 0.0402; (0.50, 0.2652): 0.0546;
(0.75, 0.2372): 0.0360.

Comparison is awkward and would require two-way interpolation, but it is clear that the values do not agree. The values by finite differences are 20 to 50% larger.

14. Beginning with points at $i = 0$ where $x = 0.2, 0.4, 0.6,$ and 0.8, intersections of characteristics are

(0.2993, 0.1006), (0.4989, 0.1067), (0.6982, 0.1005),
(0.3915, 0.1946), (0.5950, 0.2003).

Computing along the characteristics gives $u = 0.5986, 0.9978, 1.3964, 0.7830$, and 1.1900. These are identical to the analytical solution, $u = 2x$.

17. Representative values at points (x, y), with $\Delta t = 0.0018$:

Point:	(0.5, 0.5)	(0.5, 1.0)	(1.5, 3.5)
$t = 0.0036$	0.0625	0.1250	0.1458
$t = 0.0072$	−0.0729	−0.1458	−0.4062
$t = 0.0144$	−0.0911	−0.2135	−0.0078

CHAPTER 10

2. From the normal equations:
$$a = \frac{N \sum xy - \sum x \sum y}{N \sum x^2 - \left(\sum x\right)^2}, \qquad b = \frac{\sum x^2 \sum y - \sum x \sum xy}{N \sum x^2 - \left(\sum x\right)^2}.$$

If we substitute $y = \sum Y/N$ and $x = \sum x/N$ into $ax + b = y$, we find that
$$a\frac{\sum x}{N} + b = \frac{\sum Y}{N}.$$

3. $y = 1.908x + 0.025$.

7. $\ln S = 0.18395 + 9.6027 \times 10^{-3}T$, $S = 1.2019e^{9.6027 \times 10^{-3\exp T}}$.

10. $y = 0.104x^2 + 1.183x + 0.992$.

12. $f = -6.366 + 24.086x - 3.611x^2 + 0.122x^3$.

When the degree is increased to 4, β does not decrease:

Degree	1	2	3	4
β:	511	532	85	87

16. For (0, 1): integral $= -\sqrt{1 - x^2}]^1_{-1} = 0$.

For (1, 1): integral $= \left(-\dfrac{x}{2}\sqrt{1 - x^2} + \dfrac{1}{2}\sin^{-1} x \right)\bigg|^1_{-1} = \dfrac{\pi}{2}$.

For (0, 0): integral $= \sin^{-1} x]^1_{-1} = \pi$.

For (1, 2): (Use integration by parts)
$$\text{integral} = -2x^2\sqrt{1 - x^2} + 4\int x\sqrt{1 - x^2}\, dx$$
$$= (-2x^2\sqrt{1 - x^2} - \tfrac{4}{3}\sqrt{(1 - x^2)^3})]^1_{-1} = 0.$$

20. The error of the ninth-degree Maclaurin series is very small near the origin but increases rapidly to 0.04952 at $x = \pm 1$.

The error of P_3 has four maxima (counting the endpoints) and has a maximum error (near $x = \pm 0.4$) of about 0.01975. This is much smaller than the maximum error of the ninth-degree expression.

24. c) $R_{3,3} = \dfrac{1 + (x/2) + (x^2/10) + (x^3/120)}{1 - (x/2) + (x^2/10) - (x^3/120)}$.

Maximum error is -0.00003 at $x = 1$.

29. $R_{4,2} = \dfrac{0.851632 - 0.14202T_2 + 1.1620 \times 10^{-3}T_4}{1 + 5.18619 \times 10^{-3}T_2}$.

Index